Inorganic Chemistry: Reactions and Mechanisms

Inorganic Chemistry: Reactions and Mechanisms

Edited by
Bernard Wilde

Willford Press
www.willfordpress.com

Published by Willford Press,
118-35 Queens Blvd., Suite 400,
Forest Hills, NY 11375, USA

ISBN: 978-1-68285-369-6

Cataloging-in-Publication Data

Inorganic chemistry : reactions and mechanisms / edited by Bernard Wilde.
p. cm.
Includes bibliographical references and index.
ISBN 978-1-68285-369-6
1. Chemistry, Inorganic. 2. Reaction mechanisms (Chemistry). 3. Physical inorganic chemistry.
4. Chemistry, Physical and theoretical. I. Wilde, Bernard.
QD151.3 .I56 2017
546--dc23

For information on all Willford Press publications
visit our website at www.willfordpress.com

WILLFORD PRESS

Printed in the United States of America.

Contents

Preface

The world is advancing at a fast pace like never before. Therefore, the need is to keep up with the latest developments. This book was an idea that came to fruition when the specialists in the area realized the need to coordinate together and document essential themes in the subject. That's when I was requested to be the editor. Editing this book has been an honour as it brings together diverse authors researching on different streams of the field. The book collates essential materials contributed by veterans in the area which can be utilized by students and researchers alike.

Inorganic chemistry is a scientific field that plays a significant role in various industries ranging from paints to medicine. It deals with the behavior and properties of organometallic compounds and other inorganic compounds. This book provides an extensive analysis of the types of reactions and mechanisms in inorganic chemistry, helping the reader to understand how inorganic elements or compounds would behave in a chemical reaction. This book is a valuable compilation of topics, ranging from the basic to the most complex advancements in this field. It covers in detail some existent theories and innovative concepts revolving around this area of study. For all those who are interested in inorganic chemistry, this book can prove to be an essential guide.

Each chapter is a sole-standing publication that reflects each author´s interpretation. Thus, the book displays a multi-facetted picture of our current understanding of applications and diverse aspects of the field. I would like to thank the contributors of this book and my family for their endless support.

Editor

1

A highly efficient polysulfide mediator for lithium-sulfur batteries

Xiao Liang[1], Connor Hart[1], Quan Pang[1], Arnd Garsuch[2], Thomas Weiss[2] & Linda F. Nazar[1]

The lithium-sulfur battery is receiving intense interest because its theoretical energy density exceeds that of lithium-ion batteries at much lower cost, but practical applications are still hindered by capacity decay caused by the polysulfide shuttle. Here we report a strategy to entrap polysulfides in the cathode that relies on a chemical process, whereby a host—manganese dioxide nanosheets serve as the prototype—reacts with initially formed lithium polysulfides to form surface-bound intermediates. These function as a redox shuttle to catenate and bind 'higher' polysulfides, and convert them on reduction to insoluble lithium sulfide via disproportionation. The sulfur/manganese dioxide nanosheet composite with 75 wt% sulfur exhibits a reversible capacity of 1,300 mA h g^{-1} at moderate rates and a fade rate over 2,000 cycles of 0.036%/cycle, among the best reported to date. We furthermore show that this mechanism extends to graphene oxide and suggest it can be employed more widely.

[1] Department of Chemistry, University of Waterloo, 200 University Avenue West, Waterloo, Ontario, Canada N2L 3G1. [2] BASF SE, 67056 Ludwigshafen, Germany. Correspondence and requests for materials should be addressed to L.F.N. (email: lfnazar@uwaterloo.ca).

As the development of portable electronics devices, electric vehicles and large-scale energy storage increases, so do the demands for energy storage batteries with high energy density and long service life. At the same time, it is recognized that traditional lithium-ion batteries are approaching their theoretical energy density limits[1–4]. Lithium–sulfur batteries are one of the most promising candidates to satisfy emerging market demands[5–7], as they possess a theoretical capacity and energy density of 1,675 mA h g^{-1} and 2,500 kW kg^{-1}, respectively, superior to current lithium-ion batteries[8,9]. In addition, they present an inherently low competitive cost due to the high natural abundance of sulfur. These advantages suggest that the lithium–sulfur battery should be capable of energy storage several times greater than conventional lithium-ion batteries, with reduced cost. However, practical applications are currently hindered by several obstacles. These predominantly relate to the insulating nature of sulfur and lithium sulfides, which require addition of conductive additives (hence lowering the active sulfur mass fraction), pronounced capacity fading on cycling and an internal redox shuttle that lowers Coulombic efficiency[10]. Over decades, much effort has been expended to try to solve these problems by trapping the polysulfides within the cathode structure. The most popular approaches are to encapsulate sulfur in the pores of carbon materials or a conductive polymer matrix. Meso/microporous carbons[11–14], carbon spheres[15–17], carbon nanotubes/fibres[18–20], polyaniline[21], polypyrrole[22,23] and poly(3,4-ethylenedioxythiophene)[24] sulfur-host cathodes have all had a positive effect on increasing the cycle life of lithium–sulfur cells. However, such architectures and polymer coatings can only partially retain polysulfides and are limited in their function over time, owing to structural changes that arise from the 80% volume change of the sulfur cathode on discharge. Two-dimensional materials such as graphene oxides (GOs) have also been very successfully employed as cathode hosts to increase the performance of the lithium–sulfur battery, where interaction between the oxygen/nitrogen functional groups and sulfur/polysulfides has been suggested[25–28]. Similarly, binding polysulfides onto hydrophilic metal oxide hosts was shown to significantly aid in maintaining high-capacity retention[29–31]. More recently, high-surface-area polar metallic oxides have been used as a two-in-one approach to provide both a 'sulfiphilic' surface and supply electron transport to effect surface-enhanced redox chemistry[32]; or to spatially locate Li_2S deposition and enhance redox[33]. Organometallic redox mediators are shown to provide better utilization of Li_2S (ref. 34). Additives such as P_2S_5 also effectively control Li_2S deposition and aid in elimination of the polysulfide shuttle[35].

Herein, we present a quite different chemical approach to polysulfide retention in the sulfur cathode, which relies on mediating polysulfide redox through insoluble thiosulfate species in a two-step process. The thiosulfate groups are first created *in situ* by oxidation of initially formed soluble lithium polysulfide (LiPS) species on the surface of ultra-thin MnO_2 nanosheets. As reduction proceeds, the surface thiosulfate groups are proposed to anchor newly formed soluble 'higher' polysulfides by catenating them to form polythionates and converting them to insoluble 'lower' polysulfides. The polythionate complex formed on the surface is thus best described as a transfer mediator. This process curtails active mass loss during the discharge/charge process and supresses the polysulfide shuttle to result in high-performance cathodes with high sulfur loading, capable of high-capacity retention of 92% after 200 cycles at a C/5 rate and cycling over 2,000 cycles at 2C with a capacity decay of 0.036% per cycle.

Results

Synthesis and characterization. The MnO_2 nanosheets were prepared by reducing GO with $KMnO_4$ (ref. 36). Transmission electron microscopy (TEM) and scanning electron microscopy (SEM) images (Fig. 1a,b) reveal their two-dimensional lamellar structure. The selected area electron diffraction pattern of the MnO_2 nanosheets shows weak diffraction rings characteristic of δ-MnO_2, and the high-resolution TEM image in Fig. 1c exhibits a 0.25-nm lattice spacing corresponding to the (100) planes. X-ray diffraction (Fig. 1d) analysis confirms the MnO_2 nanosheets are monoclinic birnessite, δ-MnO_2 (JCPDS-01-080-1098). Birnessite is a phyllomanganate, a manganese oxide containing predominantly Mn^{4+} cations assembled in layers of edge-sharing octahedra. A layer charge deficit arises from the presence of Mn^{3+} cations and/or vacant-layer octahedra and is compensated by the presence of interlayer cations, which are hydrolysable K^+ in this case. Water molecules are also present between the

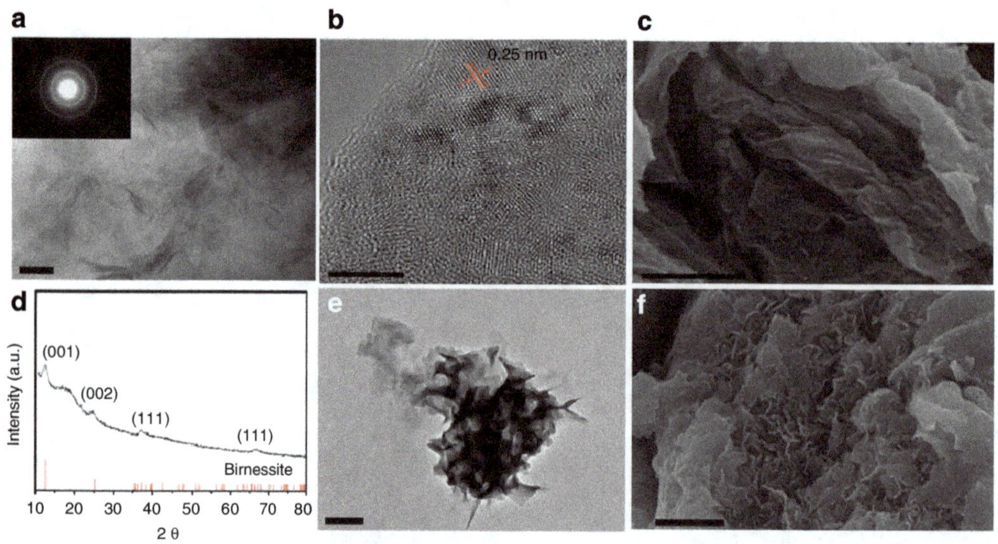

Figure 1 | Morphology and characteristics of the MnO_2 nanosheets and S/MnO_2 composite. (**a**) TEM image of MnO_2 nanosheets and its corresponding selected area electron diffraction pattern, (**b**) high-resolution TEM image of MnO_2 nanosheets, (**c**) SEM image of the MnO_2 nanosheets, (**d**) X-ray diffraction pattern of MnO_2 nanosheets, (**e**) TEM and (**f**) SEM images of the S/MnO_2 nanosheets composite. Scale bars, 50 nm (**a**); 100 nm (**b**,**c**,**f**); 200 nm (**e**).

sheets of MnO_6 octahedra, coordinated both to the Mn and the K, with some fraction of the former being present as hydroxyl groups[37]. Thermogravimetric analysis (TGA, Supplementary Fig. 1a) reveals that the as-prepared MnO_2 nanosheets possess 8 wt% surface-absorbed water and 3 wt% water in the interlayers. Melt diffusion at 155 °C was used to uniformly disperse sulfur onto the surface of the MnO_2 nanosheets[28–30]. To maximize contact, the nanosheets were first pre-dispersed by sonication before being mixed with sulfur nanoparticles (see Methods) in a 25:75 weight ratio (Supplementary Fig. 2). The TEM and SEM images in Fig. 1e,f show a homogeneous sulfur coating on the nanosheet surface after melt diffusion, as confirmed by energy dispersive X-ray spectroscopy mapping (Supplementary Fig. 3). The targeted high sulfur loading of 75 wt% was achieved (TGA analysis of 75S/MnO_2, Supplementary Fig. 1b).

Electrochemical properties. Electrochemical behaviour evaluated at different current densities for the 75S/MnO_2 nanocomposite, showed characteristic Li-S voltage profiles (Fig. 2a)[12,14]. The cells exhibit a capacity of ~1,300 mA h g^{-1} at C/20 (discharge/charge of full theoretical capacity in 20 h). At a 20-fold higher current density (a C rate), only a modest drop in capacity to 950 mA h g^{-1} was observed, suggestive of highly efficient kinetics. Most notable is the lack of overpotential in the initial charging process even on the first cycle, indicative of a good Li_2S interface with the MnO_2 nanosheets. Figure 2b shows the cycling stability of the cells at different rates. At C/5, the initial discharge capacity was 1,120 mA h g^{-1}, with 1,030 mA h g^{-1} being sustained after more than 200 cycles and representing excellent capacity retention of 0.04% per cycle. At higher current densities (1C), the 75S/MnO_2 composite still delivered exceptional reversible capacity of 800 mA h g^{-1} after 200 cycles. This indicates that processes such as loss of electrical contact of Li_2S or sulfur with the conductive host caused by volume changes or the 'shuttle reaction' caused by the solubilized polysulfides are minimized.

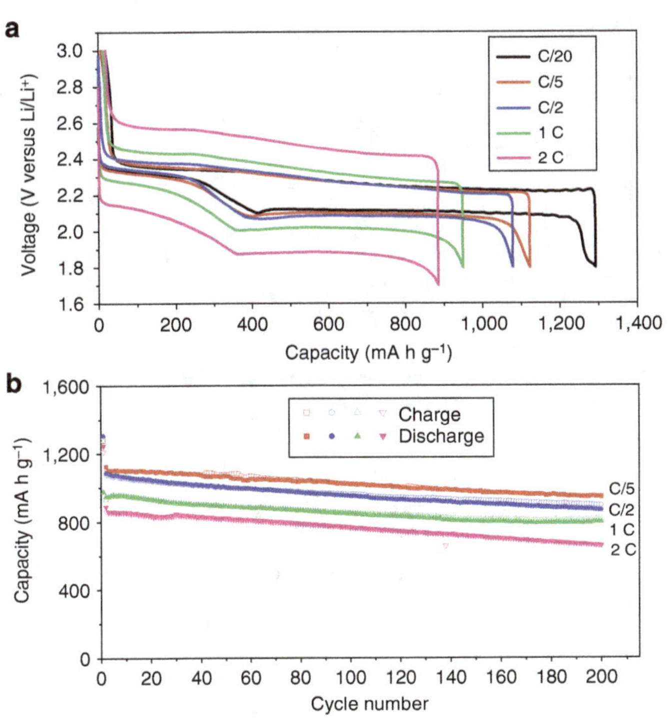

Figure 2 | Electrochemical performance of S/MnO_2. (**a**) Voltage profiles of S/MnO_2 nanosheets at current densities ranging from C/20 to 2C. The highlighted circle shows the lack of overpotential. (**b**) Cycling performance of S/MnO_2 nanosheets at C/5, C/2, 1C and 2C rates. The cells were subjected to an initial slow conditioning cycle at C/20 to allow complete access of the electrolyte to the active material.

***In-situ* visual-electrochemical study of 75S/MnO_2.** To further illustrate the excellent properties discussed above, we examined electrochemical behaviour of 75S/MnO_2 in an optically transparent Li-S cell, exploiting the very strong colour of LiPSs to probe their interaction with the cathode surface. Recent reports have visually demonstrated the binding of LiPSs with metallic metal oxide sulfur hosts such as Ti_4O_7 (ref. 32) and spatially localized deposition of Li_2S on conductive indium-tin oxide[33]. With δ-MnO_2—a poor semiconductor—the interaction is based on a very different principle, as detailed in the next section. The cell using 75S/MnO_2 as the cathode clearly demonstrates the ability of the nanosheet host to control LiPS dissolution. For comparison, results are shown for a S/Ketjen Black (75S/KB) cathode prepared with the same 75% sulfur loading (Fig. 3). Figure 3a reveals that the electrolyte in the 75S/KB cell changed from colourless to bright yellow–green on partial discharge of the cell over 4 h. This point corresponds to the 'knee' in the discharge curve at a capacity of 400 m Ah g^{-1} (a potential of 2.1 V, see Fig. 2a), where the LiPS concentration is expected to be at a maximum. The characteristic colour of medium chain LiPSs (i.e, ~, S_4^{2-}) indicates that the polysulfides diffuse out of the cathode and are solubilized in the electrolyte.

At the end of discharge (12 h), the electrolyte is still yellow, showing LiPSs remain in solution. In contrast, in the 75S/MnO_2 cell the electrolyte exhibits only a faint yellow colour at 4 h. This provides striking visual evidence of very low LiPS content in the electrolyte and demonstrates effective trapping by the MnO_2. On full discharge, the electrolyte is rendered completely colourless, indicating effective conversion to insoluble reduced species, Li_2S_2 and Li_2S. A test of the polysulfide adsorption by the MnO_2 nanosheets was conducted by electrochemical titration to measure the degree of residual LiPS in solution after contact, confirming the strong binding (Supplementary Fig. 4)[38].

XPS study of the interaction of lithium (poly)sulfides and MnO_2 nanosheets. The nature of the interaction of the MnO_2

Figure 3 | Visual confirmation of polysulfide entrapment at specific discharge depths. (**a**) 75S/KB and (**b**) 75S/MnO_2 cells. These were both discharged at C/20 between open circuit potential to 1.8 V under identical conditions.

nanosheets with LiPSs was determined from Mn $3p_{3/2}$ and S2*p* X-ray photoelectron spectroscopy (XPS) analysis. Only the peak position of the lower binding energy component of the $S2p_{3/2}/2p_{1/2}$ spin orbit doublet is described in the subsequent discussion following convention. The results are summarized in Figs 4 and 5.

Li_2S_4 was employed as the representative polysulfide species to probe the system at a partial state of discharge. Its XPS spectrum (Fig. 4a) shows a 1:1 ratio of the two $S2p_{3/2}$ contributions at 161.7 and 163.1 eV, ascribed to the terminal (S_T^{-1}) and bridging sulfur (S_B^0) atoms, respectively. This assignment is in accord with the chain structure of Li_2S_4, where the sulfur atoms at both ends have a formal charge of (-1), while those in the middle bear a formal charge of (0)[32,39]. Figure 4e displays the XPS Mn $2p_{3/2}$ spectra of δ-MnO_2, which is mainly composed of Mn^{4+}, with its near-surface enriched in Mn^{3+} relative to the bulk[40]. The MnO_2 nanosheet spectrum is typical of birnessite[41], showing a strong Mn $2p_{3/2}$ Mn^{4+} contribution that appears as a characteristic multiplet with a maximum at 642.5 eV, and three satellite peaks at higher binding energies of 643.7, 644.8 and 645.7 eV. The ratio of the multiplet peaks is in agreement with calculations[41]. The Mn $2p_{3/2}$ contribution at 641.4 eV corresponds to Mn^{3+}, which makes up ~4.6% of the MnO_2 nanosheets.

To examine the LiPS–Mn interaction, a suspension of vacuum-dried δ-MnO_2 in ether was added to a solution of yellow-green Li_2S_4 in dimethoxyethane, resulting in immediate discolouration. XPS data were collected on the recovered solid ('MnO_2–Li_2S_4'). Its S2*p* spectrum (Fig. 4b) reveals four sulfur environments. Two are the same terminal and bridging S environments as in Li_2S_4. The new significant contribution between 171 and 165 eV can be fit with two sulfur environments. The $S2p_{3/2}$ peak at 167.2 eV is in precise accord with the binding energy of thiosulfate[42], which must arise from a surface redox reaction between Li_2S_4 and δ-MnO_2. This peak is from the 'central' or S=O sulfur in thiosulfate ($[SSO_3]^{2-}$), whereas its 'peripheral' sulfur lies at 161.5 eV, contributing to the S_T^{-1} from residual Li_2S_4 (thus accounting for the change in ratio of the (S_T^{-1}:S_B^0) $2p_{3/2}$ peaks from 1:1 to 2.4:1). The S2*p* XPS spectrum of sodium thiosulfate is provided for reference (Supplementary Fig. 5a). The other $S2p_{3/2}$ peak at 168.2 eV (ascribed to a polythionate complex in the MnO_2–Li_2S_4 material) will be discussed below. Oxidation of Li_2S_4 to thiosulfate is accompanied by a decrease in the Mn oxidation state, as evident in Fig. 4f. The contribution from Mn^{3+} at 641.4 eV in the Mn $2p_{3/2}$ XPS spectrum significantly increases (from ~5% to 35%) and two additional Mn $2p_{3/2}$ peaks arise at lower energy (640.4 and 639.4 eV), readily attributable to Mn^{2+}. The overall redox process is summarized in Fig. 4g. The presence of thiosulfate was also proven by Fourier transform infrared spectroscopy (FTIR). The FTIR spectrum of Li_2S_4 (Supplementary Fig. 6) shows an S–S band at 482 cm^{-1} (ref. 31), whereas MnO_2 has a strong antisymmetric MnO_6 stretching vibration at 519 cm^{-1} (ref. 43). The FTIR spectrum of MnO_2–Li_2S_4 exhibits a new peak at 670 cm^{-1}, characteristic of $S_2O_3^{2-}$ (ref. 44), and one very broad peak at 503 cm^{-1} midway between the contributions expected from S-S and MnO_2. We attribute this significant shift of both species to a strong interaction at the interface.

To further support the thiosulfate formation mechanism, we investigated the interaction of LiPS with other sulfur hosts—GO (ACS Material, USA) and graphene (ACS Material, single layer). The XPS S2*p* spectrum of GO-Li_2S_4 shows the same components as the MnO_2–Li_2S_4 (Fig. 4c), which suggests thiosulfate/polythionate surface species are formed here as well. The characteristic peaks are reduced in intensity compared with

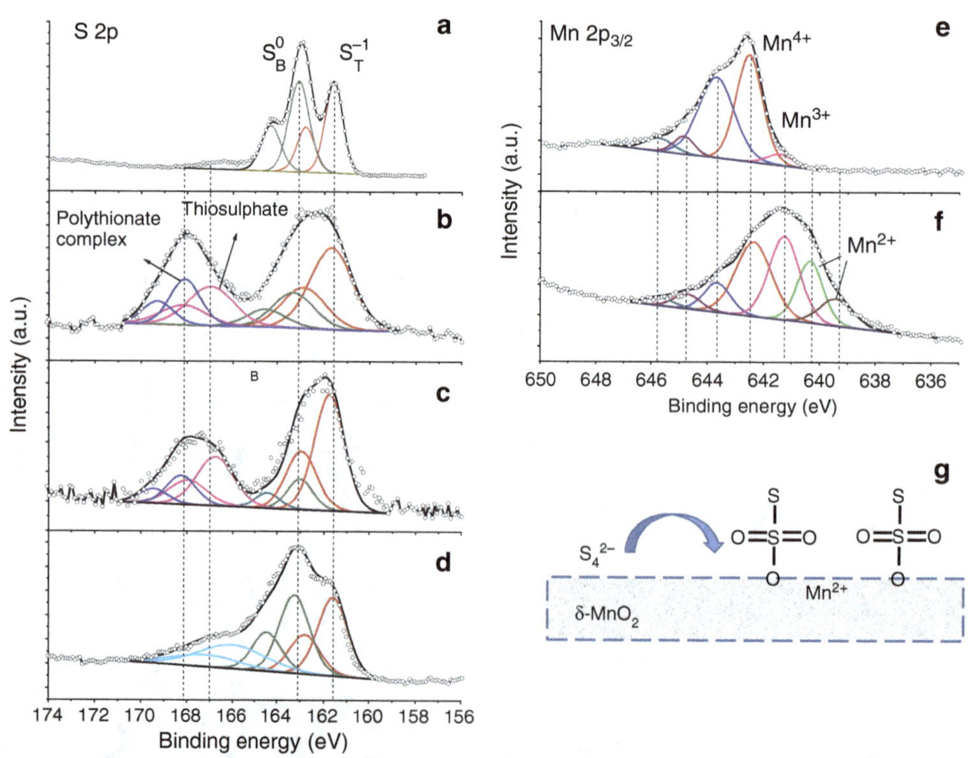

Figure 4 | XPS study of the interaction between sulfur species and MnO_2 nanosheets. S2*p* core spectra of (**a**) Li_2S_4 showing the terminal and bridging sulfur atoms in the expected 1:1 ratio; (**b**) MnO_2-Li_2S_4 and (**c**) GO-Li_2S_4 exhibiting the thiosulfate/polythionate active groups and (**d**) 'graphene'-Li_2S_4 showing their absence; the contribution in turquoise is due to sulfite (from the Li_2S_4)[32]. Mn $2p_{3/2}$ spectra of (**e**) MnO_2 nanosheets and (**f**) MnO_2-Li_2S_4. (**g**) Schematic showing the oxidation of initially formed polysulphide by δ-MnO_2 to form thiosulfate on the surface, concomitant with the reduction of Mn^{4+} to Mn^{2+}.

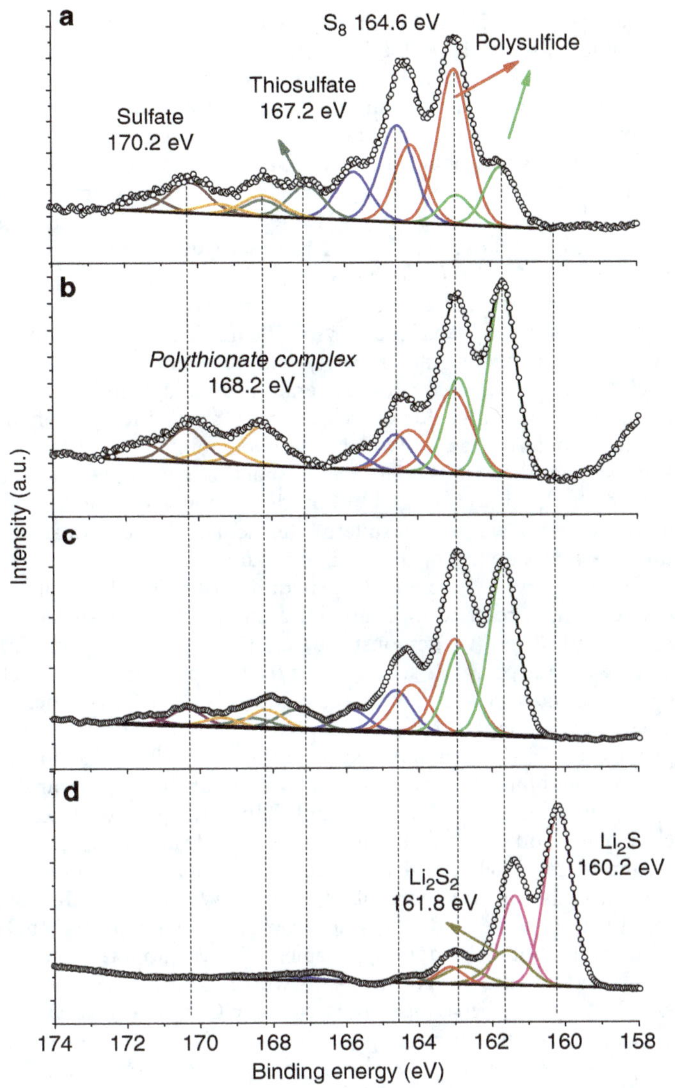

Figure 5 | *Ex-situ* XPS of S/MnO_2 nanosheets electrodes after discharge to specific states. (**a**) Discharged to 2.15 V, (**b**) discharged to 2.15 V and then aged in the cell for 20 h, (**c**) discharged to 800 mA h g^{-1} (the middle of the second plateau) and (**d**) discharged to 1.8 V. All cells were galvanostatically discharged at C/20 with 1 M $LiClO_4$ in DOL/DME (1:1 vol) as the electrolyte. The trace of sulfate at 170.2 eV is due to momentary air exposure during the XPS sample loading process.

δ-MnO_2, as expected. The XPS C 1*s* comparison of the GO and GO–Li_2S_4 confirms the chemical interaction of LiPS in GO–Li_2S_4, showing that 20% of the C-O(H) bonds were reduced to C–C (see Supplementary Fig. 7). Regarding the S2*p* spectrum of graphene–Li_2S_4, it shows two contributions at 161.7 and 163.1 eV, identical to that of Li_2S_4 itself (Fig. 4d). This suggests almost no interaction between graphene and polysulfides. The absence of thiosulfate in graphene–Li_2S_4 implies that the thiosulfate oxygen originates from surface oxy-groups on GO; for MnO_2, these arise from the OH groups that replace some of the O^{2-} in the birnessite structure for charge balance. This additional XPS analysis and insight into thiosulfate formation can explain the origin of the excellent cyclability of the MnO_2 nanosheets and that of GO materials[25,26], as discussed in detail below.

XPS analysis was further conducted on 75S/MnO_2 nanosheet electrodes extracted from cells at different discharge states (Fig. 5a–d). When the cell was discharged to 2.15 V, the presence of long-chain polysulfides is evidenced by strong S2$p_{3/2}$ contributions at 161.7 (S_T^{-1}) and 162.9 eV (S_B^{-2}) (Fig. 5a)[32,39], with their 1:2 ratio suggesting a chain length approximately equal to Li_2S_6 owing to the 2:4 ratio of terminal/bridging sulfur in this LiPS. Elemental sulfur was also detected at this and other intermediate states of discharge, consistent with incomplete reaction and/or the disproportionation of metastable polysulfides such as Li_2S_8 to form Li_2S_6 and sulfur as previously reported[45]. S2*p* peaks between 166 and 170 eV correspond to two sulfur environments: one at 167.2 eV (thiosulfate) and another at 168.2 eV, which is assigned to a polythionate complex as discussed above. These are the same as observed in the S2*p* spectrum of MnO_2–Li_2S_4 in Fig. 4b. The thiosulfate peak vanishes when the electrode is aged in the coin cell for 20 h after discharge to 2.15 V; only the peak due to the polythionate complex remains in this region. Thus, transformation of thiosulfate to the polythionate complex is accompanied by the conversion of longer-chain polysulfides to shorter-chain polysulfides ($\sim Li_2S_3$ based on the 2:1 ratio of the terminal:bridging sulfur; see Supplementary Fig. 8, for concentrations of each specific component). Exactly the same reaction is also observed between sodium thiosulfate and polysulfides (Li_2S_4), as shown in Supplementary Fig. 5b.

We conclude that S–S from longer-chain polysulfide species reacts with thiosulfate to yield the polythionate complex and shorter polysulfides. Prolonged contact time results in more complete conversion. XPS analysis of the electrode collected at the middle of the second plateau (Fig. 5c; 800 mA h g^{-1}) shows that the sulfur environments are the same as the electrode discharged to 2.15 V. Their different relative intensities represent the further consumption of longer-chain polysulfides accompanied by the formation of shorter-chain polysulfides. After the electrode was fully discharged to 1.8 V, this process is essentially complete (Fig. 5d). The contribution from thiosulfate and the polythionate complex is small, whereas a strong contribution from Li_2S (S2$p_{3/2}$ at 160.2 eV) along with $\sim$20% Li_2S_2 (161.7 eV) indicates almost full reduction of sulfur in the electrode. The minority contribution from Li_2S_2 was assigned based on its very similar binding energy as that of Na_2S_2 (161.8 eV[32], and see Supplementary Fig. 9). On charge, the XPS spectrum (Supplementary Fig. 10) is similar to Fig. 5a, revealing the re-appearance of the polythionate complex along with polysulfides and sulfur, and indicating that the process is reversible.

Discussion

We propose that the insoluble $S_2O_3^{2-}$ that forms on the surface serves as an internal mediator to anchor long-chain polysulfides from solution and trigger conversion to lower polysulfides via the reaction shown below (equation (1)):

$$[O-\underset{\underset{O}{\|}}{\overset{\overset{O}{\|}}{S}}-S]^{2-}+Li_2S_x \rightleftarrows [O-\underset{\underset{O}{\|}}{\overset{\overset{O}{\|}}{S}}-(S)_{x-y}-\underset{\underset{O}{\|}}{\overset{\overset{O}{\|}}{S}}-O]^{2-}(I)+Li_2S_y(x\geq 4; y<3) \quad (1)$$

In this mechanism, the polysulfide catenates to the thiosulfate by insertion in the S–S single bond to create the polythionate complex (I) and a lower polysulfide (that is, Li_2S_2 or Li_2S) by an internal disproportionation reaction. This can be explained by the attack of the highly nucleophilic thiosulfate on the bridging S(0) in the polysulfides followed by associative nucleophilic substitution. Sulfur is well known to undergo catenation reactions, such as recently proven for the reaction of sulfur with Li_3PS_4 to form polysulfidophosphates, Li_3PS_{4+n} (ref. 46). Elemental sulfur is also susceptible to nucleophilic attack by HS^- or SO_3^{2-} in aqueous solution, resulting in the formation of chains of polysulfides or polythionate complexes, respectively. This is known as the 'Wackenroder reaction'[47]. The polythionate complex could also be formed with only one $-SO_3$ group (equation (2)):

$$[O-\overset{\overset{\displaystyle O}{\|}}{\underset{\underset{\displaystyle O}{\|}}{S}}-S]^{2-} + Li_2S_x \rightleftarrows [O-\overset{\overset{\displaystyle O}{\|}}{\underset{\underset{\displaystyle O}{\|}}{S}}-(S)_{x-y}-S]^{2-}(II) + Li_2S_y(x \geq 4; y<3) \quad (2)$$

However, the bi-polythionate complex (I) is preferred due to its known stability (albeit in aqueous media)[47]. Based on the above model, the electrochemistry of the Li–S system can be explained as follows: (1) initial formation of Li_2S_n ($4 \leq n \leq 8$) at the start of discharge gives rise to reaction with MnO_2, producing thiosulfate at the surface; (2) further reduction produces more polysulfides ($4<n<8$) that are immediately catenated to form an intermediate polythionate complex and shorter-chain polysulfides ($n<3$). The polythionate complex is expected to be poorly soluble, curtailing the polysulfide shuttle. The process occurs progressively on discharge until full conversion to Li_2S_2/Li_2S is achieved. Deposition of lithium sulfide presumably overlays the signature of polythionate complex and thiosulfate, as XPS only probes <10 nm of the surface.

To evaluate the efficiency of the thiosulfate mediator, long-term cycling was performed at C/5, C/2 and 2C rates. The two former current densities are suitable for practical applications, while the latter is a benchmark for power utilization. For the C/5 cell (Fig. 6a), galvanostatic cycling for over half a year still results in a capacity of 380 mA h g^{-1} after 1,200 cycles. Similarly good performance was achieved at C/2 (Fig. 6b) over 1,500 cycles. After 2,000 cycles at 2C, 245 mA h g^{-1} useable capacity was still available with good Coulombic efficiency (>98.5%) and a very low decay rate of 0.036% per cycle. The rate was periodically changed to investigate the specific capacity that can be delivered at lower C-rates. After 1,000 cycles at 2C a discharge capacity of 715 mA h g^{-1} was obtained at C/20, and even after 2,000 cycles at 2C a reversible discharge capacity of 460 mA h g^{-1} was achieved at C/20 (Fig. 6c). The electrochemical robustness extends to cycling at very high C-rates. On switching the C-rate from 3C to C/5, the discharge capacity (985 mA h g^{-1}) was 96.3% of the original capacity (Supplementary Fig. 11). At the high rate of 4C, a stable capacity of 700 mA h g^{-1} was achieved and most of the original capacity was recovered on switching back to a C-rate. Moreover, the overall morphology and structure of the electrode after 1,000 cycles is very similar to that of the original material, as shown in Supplementary Fig. 12. This suggests that the MnO_2 nanosheet host acts to spatially locate and control both Li_2S/Li_2S_2 and sulfur deposition by providing an active interface via the thiosulfate intermediate and demonstrates the excellent stability of this cathode material.

In summary, by the *in-situ* generation of thiosulfate surface species, which react to form an active complex on the surface of MnO_2 nanosheets, we demonstrate that it is possible to cycle a Li–S cell with a capacity decay as low as 0.036% per cycle over 2,000 cycles, using a classic low viscosity dioxolane (DOL)–dimethyl ether (DME) electrolyte and polyvinylidene fluoride binder. We propose that a new, unique mechanism is responsible, whereby an active polythionate complex serves as an anchor and transfer mediator to inhibit active mass (polysulfide) dissolution into the electrolyte and control the deposition of Li_2S_2 or Li_2S. The resultant cells deliver high rechargeable capacity at practical current densities and high sulfur loading. We furthermore note that the thiosulfate mediator is not restricted to support on MnO_2 nanosheets, but is broadly applicable to other support materials and may be also responsible—in whole or in part—for the excellent cycling behaviour observed for GO composites. Unlike previous strategies to trap polysulfides by physical barriers or simple surface interactions, this chemistry is quite efficient. The discovery and understanding of a transfer mediator, which binds polysulfides and promotes stable redox activity, addresses one of the important challenges that face this chemistry. Along with future anticipated improvements in electrolytes and the lithium-negative electrode, this brings the Li–S battery a step closer to practical realization.

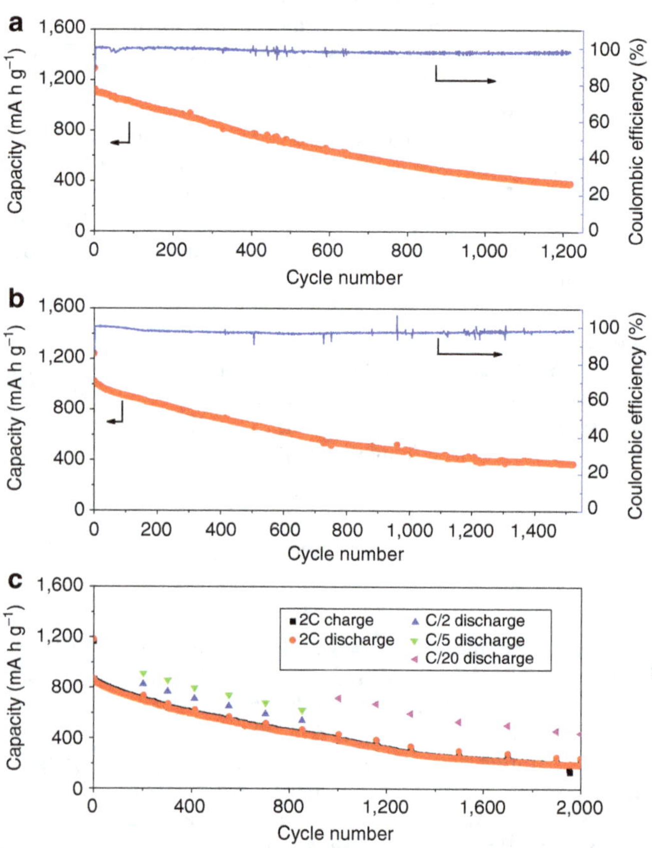

Figure 6 | Long-term cycling of 75S/MnO$_2$ at variable current densities. Capacity and the Coulombic efficiency at (**a**) C/5 over 1,200 cycles, (**b**) C/2 over 1,500 cycles, (**c**) 2C with periodic slow rate for over 2,000 cycles. The capacities of the C/2 and C/5 cells are quite similar to those shown in Fig. 2, demonstrating the reproducibility of the electrochemistry.

Methods

Preparation of the MnO$_2$ nanosheets. The layered MnO_2 nanosheets were synthesized by a one-step facile method using GO as template. Briefly, 20 mg of single-layer GO (ACS Materials) were dispersed in 100 ml deionized (DI) water by sonication. A well-mixed solution of 10 ml DI water and 160 mg of $KMnO_4$ (Sigma-Aldrich) was added into the GO suspension and stirred at room temperature for 30 min. The mixture was transferred into a thermostatic oven at 80 °C for 24 h. The resulting material was washed with DI water.

Preparation of the 75S/MnO$_2$ composite. For sulfur loading onto the MnO_2 nanosheets, the nano-sized sulfur was first synthesized by reacting 255 mg $Na_2S_2O_3$ (Sigma-Aldrich) with 278 μl concentrated HCl and 17 mg polyvinylpyrrolidone (Sigma-Aldrich) in 85 ml DI water. The MnO_2 nanosheets and nano-sized sulfur were dispersed in 40 ml DI water separately by sonication before being mixed to obtain a homogenous suspension. The suspension was filtered and then dried at 60 °C. The 75S/MnO_2 nanosheet composite was obtained by heating the mixture at 155 °C overnight.

Synthesis of Li$_2$S$_4$. Sulfur (Sigma-Aldrich) and Super-Hydride Solution (Aldrich, 1.0 M lithium triethylborohydride in tetrahydrofuran) were mixed together in a 2.75:1 molar ratio until the sulfur was fully dissolved. The resulting solution was dried under vacuum, resulting in precipitation of a yellow powder. A final wash with toluene was conducted, followed by centrifugation to separate the powder from the supernatant to isolate the Li_2S_4 powder.

Preparation of MnO_2-Li_2S_4. The MnO_2 nanosheets (87 mg; 1 mmol) were dried at 90 °C under vacuum overnight and dispersed in 10 ml DME solution by sonication in a sealed vial. The mixture was transferred to a glovebox and combined with 10 ml of 0.1 M Li_2S_4 in DME. The mixture was stirred for 2 h and the product was centrifuged and dried under vacuum overnight. GO-Li_2S_4 and graphene-Li2S4 samples were prepared by the same procedure.

Preparation of MnO_2-Li_2S. 75S/MnO_2 was mixed with DMF and then cast as a slurry onto a carbon paper current collector. A 2325 coin cell was discharged to 1.8 V at a C/20 rate, using 1 M $LiClO_4$ in a 1:1 v/v ratio of DOL:DME as the electrolyte with lithium foil as the anode. The cathode was then removed from the coin cell in a glovebox and rinsed with acetonitrile three times before being dried.

Electrochemical measurements. Cathodes were prepared by casting the DMF slurry containing 75S/MnO_2 nanosheets, 15 wt% Super P carbon and 10 wt % polyvinylidene fluoride onto a carbon paper current collector. Electrochemical studies were carried out using an Arbin cycler in galvanostatic mode, employing electrodes with an average sulfur loading between 0.7 and 1.0 mg cm^{-2} in 2325 coin cells with lithium foil as the anode and Celgard 3501 separator sheets. Cells were operated in a voltage window of 1.8–3.0 V, except for high-rate studies that were conducted between 1.7 and 3.0 V (2C and 3C rates), and between 1.6 and 3.0 V (4C rate), in an electrolyte comprising 1 M LiTFSI in a 1:1 volume of DME:DOL and 2 wt% $LiNO_3$. Specific capacity values were calculated with respect to the mass of sulfur. Electrodes for XPS analysis were prepared by discharging coin cells at C/20 in 1 M $LiClO_4$ (to avoid sulfur contributions from LiFTSI) in a 1:1 v/v ratio of DOL and DME.

Characterization. SEM studies were carried out on a Zeiss Ultra field emission SEM instrument and TEM was performed on a Jeol 2010F TEM/STEM operating at 200 KeV. TGA was used to determine the sulfur content of the material, on a TA Instruments SDT Q600 employing a heating rate of 10 °C/min from room temperature to 500 °C under an air flow. FTIR analysis was performed on a Bruker Tensor 37 spectrometer. For XPS, the samples were sealed in a vial before being quickly transferred to the chamber of a ultra-high vacuum Imaging XPS Microprobe system for analysis (Thermo VG Scientific ESCALab 250). All spectra were fitted with Gaussian–Lorentzian functions and a Shirley-type background using CasaXPS software. S2*p* peaks were fit using two equal full-width half maximum S2*p* doublets with 2:1 area ratios and splittings of 1.2 eV. The binding energy values were all calibrated using the C 1s peak at 285.0 eV.

References

1. Choi, N. S. *et al.* Challenges facing lithium batteries and electrical double-layer capacitors. *Angew. Chem. Int. Ed.* **51,** 9994–10024 (2012).
2. Yoo, H. D., Markevich, E., Salitra, G., Sharon, D. & Aurbach, D. On the challenge of developing advanced technologies for electrochemical energy storage. *Mater. Today* **17,** 110–121 (2014).
3. Manthiram, A. Materials challenges and opportunities of lithium ion batteries. *J. Phys. Chem. Lett.* **2,** 176–184 (2011).
4. Ellis, B., Lee, K. T. & Nazar, L. F. Positive electrode materials for Li-ion and Li-batteries. *Chem. Mater.* **22,** 691–714 (2010).
5. Bruce, P. G., Freunberger, S. A., Hardwick, L. J. & Tarascon, J. M. Li–O_2 and Li–S batteries with high energy storage. *Nat. Mater.* **11,** 19–29 (2012).
6. Yin, Y. X., Xin, S., Guo, Y. G. & Wan, L. J. Lithium–sulphur batteries: electrochemistry, materials, and prospects. *Angew. Chem. Int. Ed.* **52,** 2–18 (2013).
7. Evers, S. & Nazar, L. F. New approaches for high energy density lithium-sulfur batteries. *Acc. Chem. Res.* **46,** 1135–1143 (2013).
8. Peled, E., Gorenshtein, A., Segal, M. S. & Sternberg, Y. Rechargeable lithium sulfur battery. *J. Power Sources* **26,** 269–271 (1989).
9. Yang, Y., Zheng, G. Y. & Cui, Y. Nanostructured sulphur cathodes. *Chem. Soc. Rev.* **42,** 3018–3032 (2013).
10. Mikhaylik, Y. V. & Akridge, J. R. Polysulfide shuttle study in the Li/S battery system. *J. Electrochem. Soc.* **151,** A1969–A1976 (2004).
11. Ji, X. L., Lee, K. T. & Nazar, L. F. A highly ordered nanostructured carbon-sulphur cathode for lithium-sulphur batteries. *Nat. Mater.* **8,** 500–506 (2009).
12. Schuster, J. *et al.* Spherical ordered mesoporous carbon nanoparticles with high porosity for lithium-sulphur batteries. *Angew. Chem. Int. Ed.* **51,** 3591–3595 (2012).
13. He, G., Ji, X. L. & Nazar, F. L. High 'C' rate Li-S cathodes: sulphur imbibed bimodal porous carbons. *Energy Environ. Sci.* **4,** 2878–2883 (2011).
14. Liang, C. D., Dudney, N. J. & Howe, J. Y. Hierarchically structured sulphur/carbon nanocomposite material for high-energy lithium battery. *Chem. Mater.* **21,** 4724–4730 (2009).
15. He, G. *et al.* Tailoring porosity in carbon nanospheres for lithium sulphur battery cathodes. *ACS Nano* **7,** 10920–10930 (2013).
16. Jayaprakash, N., Shen, J., Moganty, S. S., Corona, A. & Archer, L. A. Porous hollow carbon@sulphur composites for high-power lithium-sulphur batteries. *Angew. Chem. Int. Ed.* **50,** 5904–5908 (2011).
17. Zhang, B., Qin, X., Li, G. R. & Gao, X. P. Enhancement of long stability of sulphur cathode by encapsulating sulphur into micropores of carbon spheres. *Energy Environ. Sci.* **3,** 1531–1537 (2010).
18. Elazari, R., Salitra, G., Garsuch, A., Panchenko, A. & Aurbach, D. Sulphur impregnated activated carbon fiber cloth as a binder-free cathode for rechargeable Li-S batteries. *Adv. Mater.* **23,** 5641–5644 (2011).
19. Zheng, G., Yang, Y., Cha, J. J., Hong, S. S. & Cui, Y. Hollow carbon nanofiber encapsulated sulphur cathodes for high specific capacity rechargeable lithium batteries. *Nano Lett.* **11,** 4462–4467 (2011).
20. Guo, J., Xu, Y. & Wang, C. Sulphur-impregnated disordered carbon nanotubes cathode for lithium-sulphur batteries. *Nano Lett.* **11,** 4288–4294 (2011).
21. Zhou, W. D., Yu, Y. C., Chen, H., DiSalvo, F. J. & Abruña, H. D. Yolk-shell structure of polyaniline coated sulphur for lithium-sulphur batteries. *J. Am. Chem. Soc.* **135,** 16736–16743 (2013).
22. Fu, Y. Z. & Manthiram, A. Orthorhombic bipyramidal sulphur coated with polypyrrole nanolayers as a cathode material for lithium-sulphur batteries. *J. Phys. Chem. C* **116,** 8910–8915 (2012).
23. Liang, X. *et al.* A nano-structured and highly ordered polypyrrole-sulphur cathode for lithium-sulphur batteries. *J. Power Sources* **196,** 6951–6955 (2011).
24. Yang, Y. *et al.* Improving the performance of lithium–sulphur batteries by conductive polymer coating. *ACS Nano* **5,** 9187–9193 (2011).
25. Ji, L. *et al.* Graphene oxide as a sulphur immobilizer in high performance lithium/sulphur cells. *J. Am. Chem. Soc.* **133,** 18522–18525 (2011).
26. Song, M. K., Zhang, Y. G. & Cairns, E. J. A long-life, high-rate lithium/sulphur cell: a multifaceted approach to enhancing cell performance. *Nano Lett.* **13,** 5891–5899 (2013).
27. Zhou, G. M. *et al.* Fibrous hybrid of graphene and sulphur nanocrystals for high-performance lithium–sulphur batteries. *ACS Nano* **7,** 5367–5375 (2013).
28. Qiu, Y. C. *et al.* High-rate, ultralong cycle-life lithium/sulfur batteries enabled by nitrogen-doped graphene. *Nano Lett.* **14,** 4821–4827 (2014).
29. Seh, Z. W. *et al.* Sulphur TiO_2 yolk-shell nanoarchitecture with internal void space for long-cycle lithium-sulphur batteries. *Nat. Commun.* **4,** 1331 (2013).
30. Ji, X., Evers, S., Black, R. & Nazar, L. F. Stabilizing lithium-sulphur cathodes using polysulphide reservoirs. *Nat. Commun.* **2,** 325 (2011).
31. Evers, S., Yim, T. & Nazar, L. F. Understanding the nature of absorption/adsorption in nanoporous polysulfide sorbents for the Li-S battery. *J. Phys. Chem. C.* **116,** 19653–19658 (2012).
32. Pang, Q., Kundu, D., Cuisinier, M. & Nazar, L. Surface-enhanced redox chemistry of polysulphides on a metallic and polar host for lithium-sulphur batteries. *Nat. Commun.* **5,** 4759 (2014).
33. Yao, H. B. *et al.* Improving lithium–sulphur batteries through spatial control of sulphur species deposition on a hybrid electrode surface. *Nat. Commun.* **5,** 3943 (2014).
34. Meini, S., Elazari, R., Rosenman, A. & Aurbach, D. The use of redox mediators for enhancing utilization of Li_2S cathodes for advanced Li-S battery systems. *J. Phys. Chem. Lett.* **5,** 915–918 (2014).
35. Lin, Z., Liu, Z. C., Fu, W. J., Dudney, N. J. & Liang, C. D. Phosphorous pentasulfide as a novel additive for high-performance lithium-sulfur batteries. *Adv. Funct. Mater.* **23,** 1064–1068 (2013).
36. Zhao, G. X. *et al.* synthesizing MnO_2 nanosheets from graphene oxide templates for high performance pseudosupercapacitors. *Chem. Sci.* **3,** 433–437 (2012).
37. Ma, R. Z., Bando, Y., Zhang, L. Q. & Sasaki, T. Layered MnO_2 nanobelts: hydrothermal synthesis and electrochemical measurements. *Adv. Mater.* **16,** 918–922 (2004).
38. Nazar, L. *et al.* Rational design of sulphur host materials for Li-S batteries: correlating lithium polysulphide adsorptivity and self-discharge capacity loss. *Chem. Commun.* doi: 10.1039/C4CC08980D (2014).
39. Kartio, I. J., Basilio, C. I. & Yoon, R. H. An XPS study of sphalerite activation by copper. *Langmuir* **14,** 5274–5278 (1998).
40. Najdoski, M., Koleva, V., Demiri, S. & Stojkovikj, S. A simple chemical method for deposition of electrochromic potassium manganese oxide hydrate thin films. *Mater. Res. Bull.* **47,** 2239–2244 (2012).
41. Nesbitt, H. W. & Banerjee, D. Interpretation of XPS Mn(2p) spectra of Mn oxyhydroxides and constraints on the mechanism of MnO_2 precipitation. *Amer. Miner.* **83,** 305–315 (1998).
42. Lindberg, B. J. *et al.* Molecular spectroscopy by means of ESCA II. Sulphur compounds. Correlation of electron binding energy with structure. *Phys. Scr.* **1,** 286–298 (1970).
43. Julien, C. M., Massot, M. & Poinsignon, C. Lattice vibrations of manganese oxides Part I. Periodic structures. *Spectrochim. Acta A* **60,** 689–700 (2004).
44. Miller, F. A. & Wilkins, C. H. Infrared spectra and characteristic frequencies of inorganic ions. *Anal. Chem.* **24,** 1253–1294 (1952).
45. Cuisinier, M. *et al.* Sulphur speciation in Li-S batteries determined by operando X-ray absorption spectroscopy. *J. Phys. Chem. Lett.* **4,** 3227–3232 (2013).

46. Lin, Z., Liu, Z. C., Fu, W. J., Dudney, N. J. & Liang, C. D. Lithium polysulfidophosphates: a family of lithium-conducting sulphur-rich compounds for lithium-sulphur batteries. *Angew. Chem. Int. Ed.* **125,** 1–5 (2013).
47. Holleman-Wiberg: *Inorganic Chemistry* (eds Wiberg, N.) 514 (Academic Press, 2001).

Acknowledgements

The research was supported by the BASF International Scientific Network for Electrochemistry and Batteries. We thank Dr Carmen Andrei, MacMaster University, and the Canadian Centre for Electron Microscopy for help with acquisition of the TEM images, and NSERC for platform funding through a Discovery Grant to LFN.

Author contributions

X.L. and L.F.N. designed this study. X.L. prepared materials and carried out the electrochemical experiments. X.L., C.H. and Q.P. carried out of the XPS investigation. X.L. carried out data processing and prepared figures. X.L. and L.F.N. wrote the manuscript. T.W. first suggested the proposed mechanism and all of the authors contributed to the scientific discussion.

Additional information

Competing financial interests: The authors declare no competing financial interests.

2

Evidence for an oxygen evolving iron–oxo–cerium intermediate in iron-catalysed water oxidation

Zoel Codolà[1], Laura Gómez[2], Scott T. Kleespies[3], Lawrence Que Jr[3], Miquel Costas[1] & Julio Lloret-Fillol[1]

The non-haem iron complex α-[$Fe^{II}(CF_3SO_3)_2$(mcp)] (mcp = (*N,N*′-dimethyl-*N,N*′-bis (2-pyridylmethyl)-1,2-*cis*-diaminocyclohexane) reacts with Ce^{IV} to oxidize water to O_2, representing an iron-based functional model for the oxygen evolving complex of photosystem II. Here we trap an intermediate, characterized by cryospray ionization high resolution mass spectrometry and resonance Raman spectroscopy, and formulated as [(mcp)Fe^{IV}(O)(μ-O)$Ce^{IV}(NO_3)_3]^+$, the first example of a well-characterized inner-sphere complex to be formed in cerium(IV)-mediated water oxidation. The identification of this reactive Fe^{IV}-O-Ce^{IV} adduct may open new pathways to validate mechanistic notions of an analogous Mn^{V}-O-Ca^{II} unit in the oxygen evolving complex that is responsible for carrying out the key O-O bond forming step.

[1] Institut de Química Computacional i Catàlisi (IQCC) and Departament de Química, Universitat de Girona, Campus de Montilivi, 17071 Girona, Spain.
[2] Department of Chemistry and Center for Metals in Biocatalysis, University of Minnesota, 207 Pleasant St S.E., Minneapolis, Minnesota 55455, USA.
[3] Serveis Tècnics de Recerca (STR), Parc Cientific i Tecnològic, Universitat de Girona, 17003 Girona, Spain. Correspondence and requests for materials should be addressed to L.Q. (email: larryque@umn.edu) or to M.C. (email: miquel.costas@udg.edu) or to J.L.-F. (email: julio.lloret@udg.edu).

High-valent iron-oxo species are implicated in important biological and synthetic oxidative transformations[1–3]. Some of these species have also shown a remarkable competence towards water oxidation to form dioxygen, a reaction that involves O-O bond formation[4–7]. The reaction is particularly appealing because it has been recognized as one of the bottleneck challenges for the success of artificial photosynthesis, and the identification of iron compounds as efficient catalysts opens novel directions in the quest for earth-abundant systems that could be used on a large scale[8–11]. In this context, the study of the water oxidation (WO) reaction with chemical oxidants catalysed by well-defined coordination complexes can clarify fundamental principles underlying the mechanisms of O-O bond formation and cleavage[12,13], the latter being at the heart of chemical processes by which aerobic organisms respire[14]. Most remarkably, given the strong parallels between the oxidation chemistry of iron and manganese compounds[15], it can be envisioned that fundamental chemical insights extracted from studies of O-O bond formation by iron coordination compounds can shed light on intermediates and mechanisms operating at the Oxygen Evolving Complex (OEC).

We and others have recently shown that non-haem iron complexes containing tetradentate nitrogen-based ligands are competent catalysts for chemically driven WO[4,5,16,17]. In our systematic studies, we determined that the presence of two *cis* labile sites is required for WO activity and that electronic effects modulate the catalytic efficiency, providing basic principles for catalyst design[4,5]. Herein, we present evidence for a catalytically relevant reaction intermediate that precedes the oxidation of the water molecule in the case of α-[$Fe^{II}(CF_3SO_3)_2$(mcp)], **1-α** (Table 1), which we have identified as a particularly active catalyst. This transient species is characterized by high resolution mass spectrometry (CSI-HRMS) and resonance Raman spectroscopy (rR) as a heterodimetallic oxo-bridged iron-cerium complex formed by the reaction of the oxoiron(IV) complex α-[$Fe^{IV}(O)(H_2O)$(mcp)]$^{2+}$ and Ce^{IV}. The Fe(O)(μ-O)Ce$(OH_2)_n$ core of this intermediate bears a strong structural resemblance to the Mn(O)(μ-O)Ca$(OH_2)_2$ unit found in the natural oxygen evolving centre. Most remarkably, the Fe(O)(μ-O)Ce$(OH_2)_n$ unit is catalytically competent in water oxidation.

Results

Catalytic activity of 1-α and 1-β in water oxidation. [$Fe^{II}(CF_3SO_3)_2$(mcp)] is known to form two topological isomers, **1-α** and **1-β** (Fig. 1)[18–21], which have previously been shown to behave differently as catalysts for C-H and C=C bond oxidations[19,21]. Given this sensitivity, the abilities of both complexes to act as water oxidation catalysts were explored. In a typical reaction, an aqueous solution of **1-α** or **1-β** (0.5 ml, 12.5 μM, final concentration) was added to an aqueous solution of CAN (9.5 ml, 125 mM, pH = 0.8, final concentration and pH). Gases evolved from these reactions were monitored by manometry and quantified by gas chromatography with a thermal conductivity detector (GC-TCD). Under these conditions (Table 1), **1-α** was highly active and yielded 160 ± 10 and 380 ± 20 TON of O_2, (mol of O_2/mol of catalyst) when using 100 and 12.5 μM catalyst concentrations, respectively. In contrast, very small amounts of O_2 were detected when the same experiment was carried out with its topological isomer **1-β** (4 ± 1 and 5 ± 2 TON, respectively). Therefore, despite the fact that the two complexes differed only in their ligand topology, their activities as water oxidation catalysts were found to be drastically different. Besides O_2, only small amounts of CO_2 were present at the end of the reaction (< 1 TON), indicating that no major ligand oxidation occurred during catalytic O_2 evolution.

Spectroscopic characterization. Reactions were then analysed spectroscopically to gain insight into the origin of the marked differences in reactivity (Fig. 2). The reaction of **1-α** or **1-β** with CAN (3 eq.) in H_2O (final pH = 1) resulted in the formation of **α/β**-[$Fe^{IV}(O)(H_2O)$(mcp)]$^{2+}$, **2-α** or **2-β**, respectively. Both exhibited near-IR bands characteristic of Fe^{IV}(O) complexes, namely $\lambda_{max} = 769$ nm ($\varepsilon = 270\ M^{-1}cm^{-1}$) for **2-α** and $\lambda_{max} = 778$ nm ($\varepsilon = 280\ M^{-1}cm^{-1}$) for **2-β** (Fig. 2a,b and Supplementary Fig. 1-2). The use of 2.5 equivalents of Ce^{IV} to fully transform **1-α,β** to **2-α,β** may be an indication of the high redox potential necessary to oxidize Fe^{III}-OH_2 to Fe^{IV}(O) (Fig. 3). Taking into consideration the Nernst equation, the redox potential for the Fe^{III}-OH_2/Fe^{IV}(O) couple under the low pH reaction conditions is estimated at ~1.4 V versus NHE, which matches that obtained in recent DFT calculations[22].

Formation of **2-α** was further confirmed and monitored by CSI-HRMS (Fig. 4a and Supplementary Figs 3–7) and ^{1}H-NMR spectroscopy (see Supplementary Figs 8 and 9). The excellent agreement between speciation observed by UV–vis and ESI-MS shows a clean transformation between **1-α** and **2-α** (Fig. 3). Importantly, the CSI-HRMS spectra of **2-α** showed an intense peak at $m/z = 545.110 \pm 0.003$ that can be assigned to {[Fe^{IV}(O)(mcp)](CF_3SO_3)}$^+$ on the basis of its m/z value and isotope distribution pattern. Consistently, the peak moved to

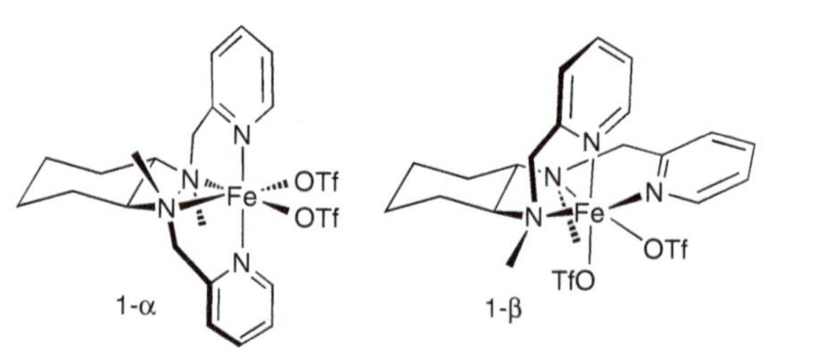

Figure 1 | Iron water oxidation catalysts. Schematic structure of the iron complexes used in this study.

Table 1 | Water oxidation catalytic activities for 1-α and 1-β.

Catalyst	[cat] (μM)	TON O_2*	TOF (h^{-1})†	TON O_2‡	TON CO_2‡
1-α	100	110 ± 10	580 ± 30	160 ± 10	0.3 ± 0.2
	12.5	360 ± 20	1,020 ± 90	380 ± 20	0.4 ± 0.2
1-β	100	4 ± 1	9 ± 1	4 ± 1	0.2 ± 0.1
	12.5	8 ± 2	25 ± 11	5 ± 2	1.1 ± 0.1

Reactions performed in Milli-Q water at 25 °C. The catalyst (0.5 ml) was injected through a septum into the reaction vessel containing CAN aqueous solution, (9.5 ml, 131 mM; final concentration 125 mM). The initial pH value was 0.8.
*TON (turnover number) = (n (O_2) produced / n(catalyst)) obtained by manometry.
†TOF (TON · h^{-1}) measured 5 min after the addition of the catalyst.
‡Values measured by GC-TCD after 3 h of reaction.

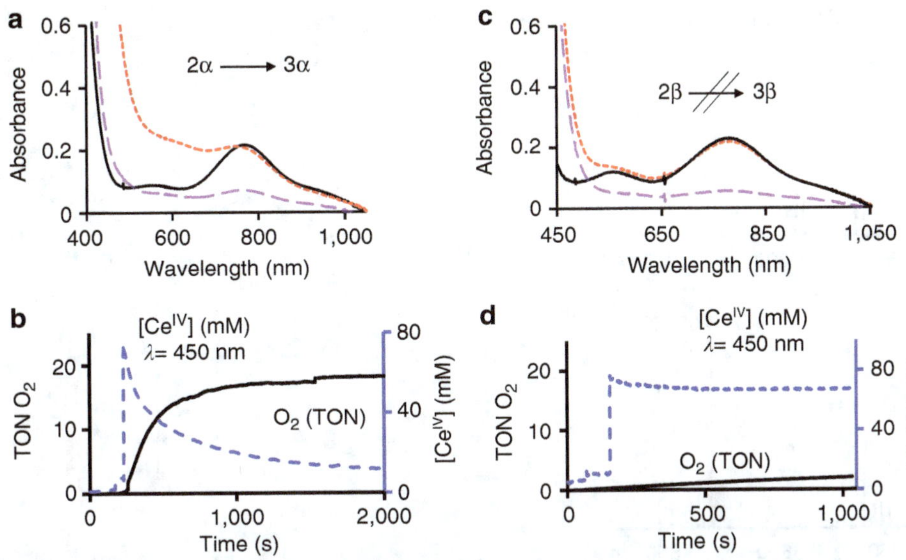

Figure 2 | Real-time manometry/UV-vis monitoring of oxygen evolution and cerium(IV) consumption. UV-vis spectra of (**a**) **2-α** and (**c**) **2-β** (solid black line) formed by the respective reactions of **1-α** and **1-β** (1 mM) in Milli-Q H_2O at 25 °C with 3 equiv. of CAN, on addition of 75 equiv. of CAN (dotted red line) and after CAN consumption (dashed purple line). Time courses for the reaction of (**b**) **1-α** and (**d**) **1-β** with 75 equiv. CAN monitoring, [Ce^{IV}] (right axis, dotted blue line), and O_2 evolution (left axis, solid black line). Kinetic traces for [Ce^{IV}] were monitored at 450 nm. $\Delta[O_2]$ (TON) was monitored by a pressure transducer and quantified by GC-TCD.

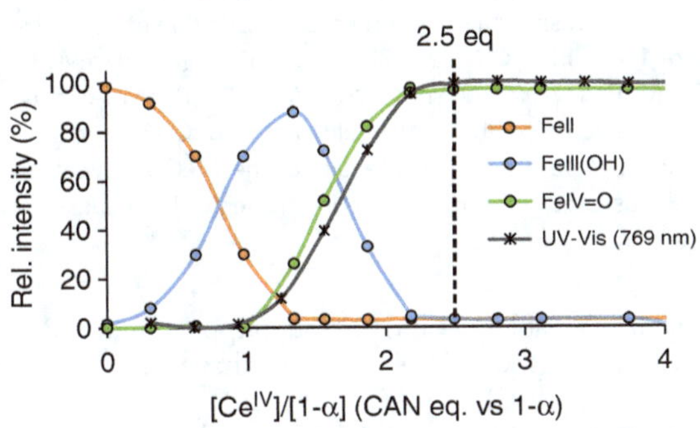

Figure 3 | $Fe^{II}(OH_2)$ to $Fe^{IV}(O)$ titration with Ce^{IV} monitored by UV-vis and HRMS. Distribution of **Fe^{II}**, **Fe^{III}** and **Fe^{IV}** species according to relative ion peak intensities obtained by HRMS in the titration of **2-α** with Ce^{IV}. The UV-vis titration (formation of $Fe^{IV}=O$ monitored at 769 nm) was also superimposed in the plot (black crosses).

547.117 ± 0.003 when **2-α** was generated in $H_2{}^{18}O$, due to the incorporation of ^{18}O into the oxo ligand. In addition, a second strong signal at $m/z = 413.162 \pm 0.003$ was observed, which shifts to 417.172 ± 0.003 when $H_2{}^{18}O$ was used. Therefore, the peak was assigned to the $[Fe^{IV}(O)(OH)(mcp)]^+$ ion (Supplementary Figs 3–7). These ions may correspond to the respective loss of H_2O and CF_3SO_3H from the parent $\{[Fe^{IV}(O)(OH_2)(mcp)](CF_3SO_3)\}^+$ ion, thus identifying **2-α** as $[Fe^{IV}(O)(OH_2)(mcp)]^{2+}$. **2-α** could be also prepared by reacting $[Fe^{II}(Cl)_2(mcp)]$ with CAN (3 eq.), providing further evidence for water and not a triflate ion as the sixth ligand. In addition, support for a terminal oxo ligand in **2-α** was obtained by rR spectroscopy ($\lambda_{exc} = 413.1$ nm, -8 °C, in 1:1 H_2O:CH_3CN), where a resonance enhanced feature was observed at 822 cm^{-1}. This feature downshifted by 40 cm^{-1} with the use of $H_2{}^{18}O$ (Fig. 5a), consistent with its assignment to the Fe=O stretch of **2-α**[23,24].

Decay of **2-α** ($\tau_{1/2} = 20$ min at 25 °C) did not result in O_2 evolution, showing that it cannot be the WO species. In fact, further reaction of **2-α** with CAN was necessary to generate the WO species. To investigate the nature of the latter, the reaction of **2-α** with an excess of CAN was monitored simultaneously by UV-vis spectroscopy, manometry and CSI-HRMS (Fig. 6, Supplementary Fig. 10). UV-vis spectroscopy and CSI-HRMS were used to monitor the time-dependent evolution of the oxoiron(IV) species, as well as the consumption of Ce^{IV}, while O_2 evolution within the same time period was determined by manometry and GC-TCD.

Reaction of CAN (75 equiv.) with **2-α** resulted in O_2 evolution concomitant with Ce^{IV} consumption. After all Ce^{IV} was consumed, O_2 evolution stopped and 18.4 TON of O_2 (Fig. 2c, Supplementary Fig. 10) were formed. Therefore, 98% of the added CAN was converted productively to O_2, (maximum TON = 18.75 based on the initial Ce^{IV} with $n(Ce^{IV})/n(O_2) = 4$). At this point, 30% of the initial amount of **2-α** remained in solution, as judged by its absorbance at $\lambda = 769$ nm (Supplementary Fig. 10d). The CSI-HRMS spectrum of the solution showed intense peaks at $m/z = 414.1696$ and 739.0270, values that are within experimental error of the exact masses for $[Fe^{III}(mcp)(OH)_2]^+$ and $[(mcp)Fe(O)(OH)Ce(NO_3)_3]^+$ species, respectively; associated isotope distribution patterns are also consistent with these formulations. The addition of a second aliquot of CAN (75 equiv.) restarted O_2 evolution, producing an additional 11 TON of O_2 until Ce^{IV} depletion ceased and complete disappearance of the 769 nm band (**2-α**) was observed (Supplementary Figs11,12).

The reaction of **2-α** with CAN (75 equiv.) elicited the immediate appearance of a broad absorption band in the 500-650 nm range and a hypsochromic shift of the near-IR feature of **2-α** (Fig. 2a, Supplementary Fig. 10b). These changes were not observed on addition of triflic acid (TfOH), $Ce^{III}(NO_3)_3$ or $Sc(OTf)_3$ to **2-α**. Interestingly, the new chromophore gradually diminished in intensity concomitant with Ce^{IV} consumption, and the near-IR band of **2-α** re-emerged (albeit with a reduced intensity because of partial decomposition) (Fig. 2a and Supplementary Fig. 10). These observations provide evidence for the accumulation of a novel reaction intermediate, **3-α**.

The formation of **3-α** was further corroborated by a kinetic analysis of the reaction of **2-α** with CAN at 25 °C and pH = 1.

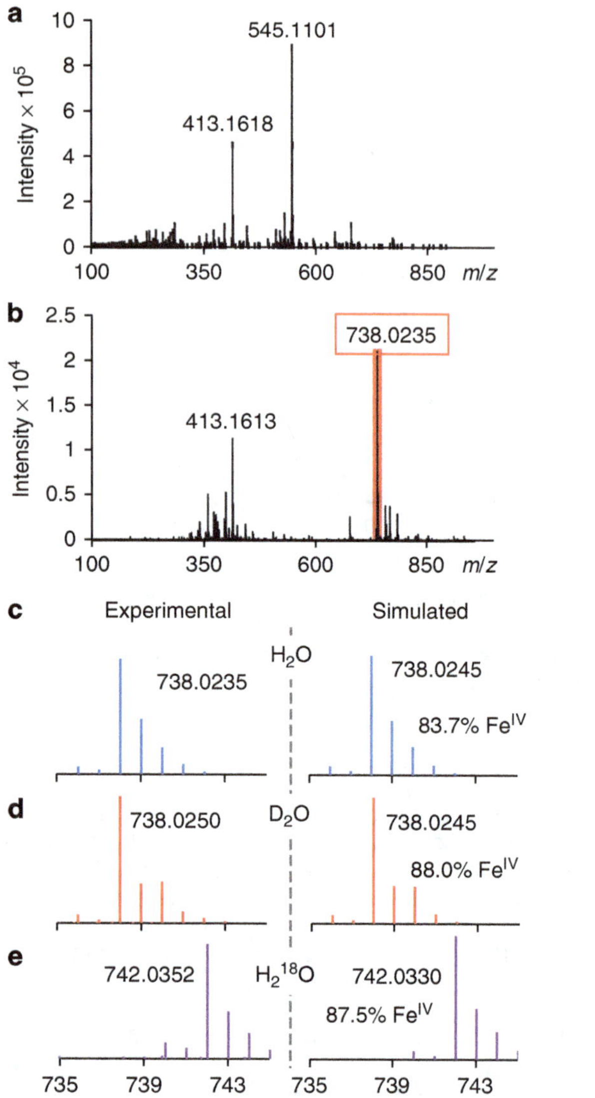

Figure 4 | Characterization of 3-α by high resolution mass spectrometry (**a**) CSI-HRMS spectrum of **2-α**, recorded immediately after the oxidation of **1-α** (1 mM) by CAN (3 mM). (**b**) CSI-HRMS spectrum obtained during the WO reaction, on the addition of 75 equiv. of CAN to **2-α**. (**c-e**) CSI-HRMS features associated with **3-α** obtained in H_2O, D_2O and $H_2{}^{18}O$ as solvent. Spectra were recorded by setting the nebulizer and dry gas temperature of the cryospray instrument to 25 °C.

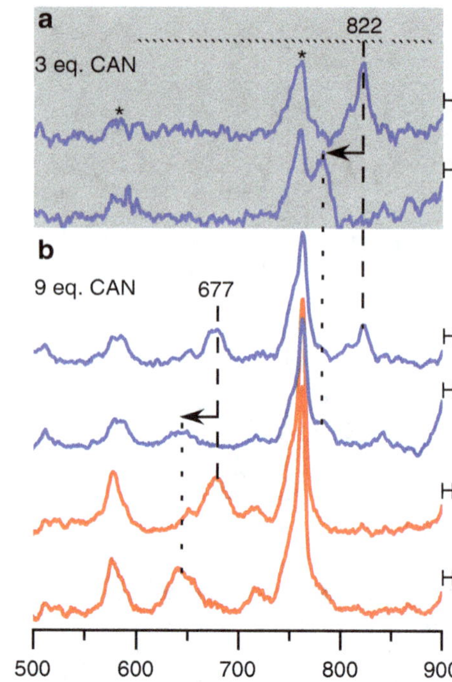

Figure 5 | Characterization of reaction intermediates by Resonance Raman. (**a**) Highlighted in grey, rR spectra of **2-α** (λ_{ex} 413.1 nm, 100 mW) prepared on addition of 3 equiv. CAN to a solution of **1-α** (5 mM) in 1:1 H_2O:MeCN. (**b**) rR spectra of **3-α** prepared on addition of 9 equiv. CAN to a solution of **1-α** in 1:1 H_2O:MeCN. Blue: $\lambda_{ex} = 413.1$ nm, 100 mW, [Fe] = 5 mM. Red: $\lambda_{ex} = 514.5$ nm, 100 mW, [Fe] = 8 mM. All spectra were collected from liquid solutions maintained at −8 °C. Asterisks denote features arising from CAN. The intensity of the MeCN solvent peak at 922 cm^{-1} was used to normalize the intensities of the peaks among the various spectra.

The initial rates of Ce^{IV} consumption and O_2 evolution by **1-α** (0.1 mM) showed a saturation behaviour with respect to [Ce^{IV}]. In addition, the reaction rate exhibited a linear dependence on **2-α** (0.01–1 mM), and in [Ce^{IV}] at [Ce^{IV}] in the 1.25–12.5 mM range, with a reaction order close to one for both reagents at large Ce/Fe ratios (under pseudo-first order conditions). At [Ce^{IV}] > 12.5 mM, reaction rates exhibited saturation, becoming independent of [Ce^{IV}] (see Supplementary Figs 13–19). This kinetic behaviour indicates a pre-equilibrium binding of Ce^{IV} to **2-α** leading to the accumulation of an intermediate, namely **3-α**, the evolution of which represents the rate determining step of the reaction. We assessed the viability of this notion by analysing the reaction between Ce^{IV} and **2-α** to form **3-α** in 1:1 v-v CH_3CN:H_2O solution at −8 °C, a temperature at which O_2 evolution and Ce^{IV} consumption basically ceased. We found that the formation of **3-α** is very fast and reversible, with an equilibrium constant K_{eq} of 120(±25) that could be estimated from the titration data (See Supplementary Figs 20–24, Supplementary Tables 1 and 2 and Supplementary Methods). Notably, analogous kinetic observations have been made in WO mediated by [$Fe^{II}(CF_3SO_3)_2(^{Me,X}Pytacn)$] (^{Me,X}Pytacn = 1-(2′-pyridylmethyl)-4,7-dimethyl-1,4,7-triazacyclononane, X = H, Cl, CO_2Et, NO_2), suggesting the possibility of a common pathway for WO by these two non-haem iron WO catalysts[5].

Like **1-α**, the topological isomer **1-β** reacted with CAN (3 eq.) to form the $S = 1$ β-[$Fe^{IV}(O)(OH_2)(mcp)]^{2+}$ (**2-β**, $\lambda_{max} = 753$ nm, $\varepsilon = 280\ M^{-1}\ cm^{-1}$) as judged by UV–vis, CSI-HRMS and ^{1}H-NMR (See Supplementary Fig. 1, right, Supplementary Figs 3, 8 and 9). However, in contrast to **2-α**, the reaction of **2-β** with CAN (75 equiv.) did not result in a significant decrease in the concentration of CAN nor was O_2 evolution observed (Fig. 2d). Moreover, the formation of a putative **3-β** was not observed. Given the structural similarity between **2-α** and **2-β**, the failure of the latter to elicit water oxidation, together with the kinetic similarities between **2-α** and the catalytically active [$Fe^{II}(CF_3SO_3)_2(^{Me,X}Pytacn)$] complexes[5], strongly suggests that the formation of **3-α** is required for WO to proceed.

Characterization of intermediate 3-α. CSI-HRMS experiments shed some light on the nature of **3-α** that was formed in the reaction of **2-α** with Ce^{IV} under catalytic conditions ([Fe] = 1 mM; [CAN] = 75 mM, Fig. 4, Supplementary Figs 25 and 26). The addition of CAN caused the disappearance of the peak at $m/z = 545.110$ corresponding to **2-α** (Fig. 4b) and the appearance of a new peak at m/z 738.0235 and designated as $\mathbf{M_3}$, which was not observed when **1-β** was treated with excess of CAN (Supplementary Fig. 27).

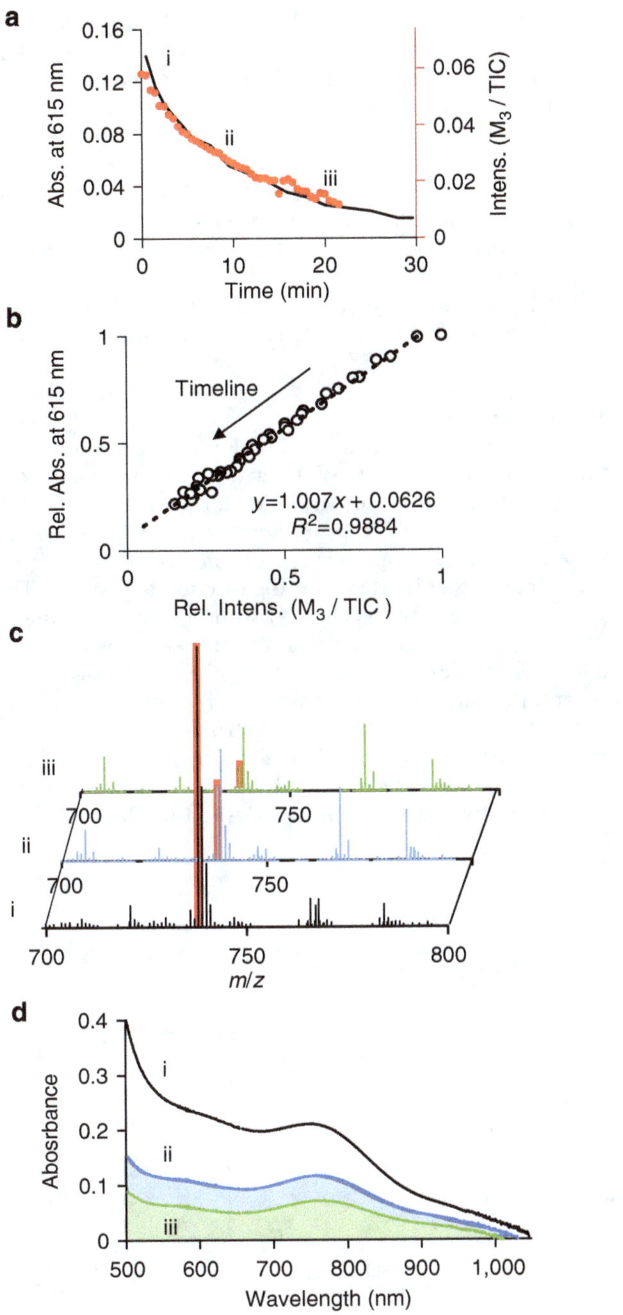

Figure 6 | Decay of 3-α monitored by UV-vis and HRMS. (**a**) Time course for the decay of the 615 nm chromophore (line, left axis) and the **M_3** MS peak intensity (dots, right axis) during the reaction of **1-α** (1 mM) with CAN (75 mM) at 25 °C. (**b**) Correlation between the relative absorbance (Abs^t/Abs^0) at 615 nm and ion count [$(\mathbf{M_3}/TIC)^t/(\mathbf{M_3}/TIC)^0$] of **$M_3$**. ($t$ = time of measurement;. TIC = total ion count). (**c**,**d**) Mass and visible spectra observed at 10 s (i), 10 min (ii) and 20 min (iii) after the addition of CAN (75 equiv.) to **2-α** (1 mM in Milli-Q water, 25 °C).

The composition of **M_3** was deduced with the aid of isotopic labelling. When $H_2{}^{18}O$ was used as a solvent, **M_3** upshifted by four units (Fig. 4e). As the oxygen atoms of the nitrates do not undergo fast exchange with $H_2{}^{18}O$ (see Supplementary Fig. 28), this upshift suggests that **M_3** must contain two oxygen atoms capable of facile exchange with water. In contrast, experiments with D_2O showed that **M_3** does not have exchangeable protons (Fig. 4d). Furthermore, the isotopic distribution pattern of **M_3** could only be correctly simulated by taking into account the natural abundance isotopes of one Fe and one Ce atom. Specifically, the ^{54}Fe isotope (5.8% relative to ^{56}Fe) is responsible for the **M_3**–2 peak, while the ^{142}Ce isotope (11.1% relative to ^{140}Ce) contributes significantly to the **M_3** + 2 peak (Supplementary Fig. 29). Taken together, these results allow **M_3** to be formulated as $\{[Fe^{IV}(O)_2(mcp)Ce^{IV}(NO_3)_3]\}^+$ (calc. m/z 738.0245, found m/z 738.0235) (For further details see Supplementary Methods). This formulation requires the replacement of one hydrogen atom from the $[Fe^{IV}(O)(OH)(mcp)]^+$ ion (m/z 413.1616) by the $[Ce^{IV}(NO_3)_3]^+$ moiety to give rise to **M_3**.

Collision induced dissociation (CID) MS/MS experiments at different collision energies (CE) were performed for **M_3** (m/z = 738.0). At relative low energies (12 eV), the CID fragmentation of **M_3** + H led to the loss of a neutral HNO_3 (m/z = 676.0). Only at CE higher than 24 eV did the parent peak at 738.0 disappear, but the m/z 676.0 ion (corresponding to neutral loss of HNO_3 from **M_3** + H) persisted at even higher collision energies (up to 39 eV). These results indicate the robustness of the $\{Fe^{IV}$-O-$Ce^{IV}\}$ moiety in the gas phase (Supplementary Fig. 30).

Decisively, Fig. 6 shows that the decay of **M_3** (monitored by CSI-HRMS) parallels the decrease in the intensity of the 615 nm chromophore attributed to **3-α**. Such a correlation indicates that **M_3** in fact corresponds to **3-α**. Taken together, the spectroscopic results and the kinetic model that arises from analysing Ce^{IV} consumption and O_2 evolution establishes **3-α** as the last detectable and catalytically competent intermediate in water oxidation.

The molecular structure derived from the MS analysis indicates that **3-α** corresponds to **M_3**, but the ion may also contain additional water ligands bound at Ce(IV) in the solution state that would be readily lost in the gas phase. This proposal is supported by the fact that Ce(IV) forms kinetically labile and dynamic complexes with a coordination number of 8 to 9. Furthermore, recent Ce(IV) speciation and quantum dynamic studies have shown a fast ligand exchange at the coordination sphere of Ce(IV)[25–27].

More detailed insight into the nature of the iron-cerium adduct was provided by rR spectroscopy of liquid samples maintained at − 8 °C in a 1:1 H_2O:MeCN solvent mixture. UV–vis and mass spectrometry characterization of **2-α** and **3-α** in this solvent mixture indicates that their identities are the same as in pure water (Supplementary Table 3). This temperature was employed to stabilize **3-α** sufficiently so as to be characterized by rR. Furthermore, under these conditions, O_2 production also correlates with decay of **3-α** (Supplementary Fig. 31). As shown in Fig. 5, the rR spectrum (λ_{ex} = 413.1 nm, Fig. 5a) of **2-α** prepared with 3 equiv. of Ce^{IV} exhibits a feature at 822 cm^{-1} that downshifts to 782 cm^{-1} on replacing $H_2{}^{16}O$ by $H_2{}^{18}O$, consistent with its assignment to the $\nu_{Fe=O}$ of **2-α**. Near-UV excitation was required to observe this feature as **2-α** has insignificant absorbance in the visible region, as found for other non-haem oxoiron(IV) complexes[23,24].

Interestingly, the rR spectra of **3-α** obtained with the same excitation wavelength (λ_{ex} = 413.1 nm, Fig. 5b) also show the 822 cm^{-1} vibration and its downshift to 782 cm^{-1} upon $H_2{}^{18}O$ substitution. This feature may arise from either **2-α** or **3-α** or both species. To clarify, we compared the intensities of the 822 cm^{-1} peaks for the **2-α** (3-CAN) and **3-α** (9-CAN) samples relative to the adjacent 762-cm^{-1} nitrate feature as an internal standard, after taking into account that the 9-CAN sample has three times the amount of nitrate as the 3-CAN sample. Indeed, the two samples are found to have 822 cm^{-1} peaks of comparable intensity. Given an association constant of 120(± 25) M^{-1} (in agreement with a K_{eq} of 90(± 10) M^{-1} under catalytic conditions, at 25 °C in H_2O, obtained by the kinetic analysis of O_2 evolution, Supplementary Fig. 19) for the interaction of **2-α** with CAN from UV–vis titration experiments, it is expected that **3-α** would account for ~80% of the iron in the Raman samples. Therefore, the 822 cm^{-1} $\nu_{Fe=O}$ vibration observed in the 9-CAN samples must arise mainly from **3-α**.

Notably, a second feature of similar intensity appears at 677 cm^{-1} with 413.1 nm excitation in the sample of **3-α** (Fig. 5b). This lower-frequency feature persists when **3-α** is probed with 514.5 nm excitation, a wavelength at which only **3-α** has an absorption feature, but the 822 cm^{-1} vibration is not observed, consistent with the absorption features associated with the $Fe^{IV}=O$ unit[26,27]. With either excitation wavelength, the 677 cm^{-1} peak downshifts to 643 cm^{-1} on replacing $H_2{}^{16}O$ by $H_2{}^{18}O$, but is not affected when the experiment was carried out in D_2O (Supplementary Figs 32–33). The 34 cm^{-1} downshift is consistent with a diatomic Fe–O vibration, but the lower frequency indicates a weaker Fe–O bond that is also resonance-enhanced with 514.5 nm excitation on Ce-adduct formation. Fe–O modes in this frequency range have previously been reported for complexes with $Fe^{III}(\mu\text{-}O)_2Fe^{IV}$ (666 cm^{-1})[28,29] and $Fe^{IV}{}_2(\mu\text{-}O)_2$ (674 cm^{-1})[30] cores, respectively, suggesting the possibility of forming an analogous $Fe^{IV}(\mu\text{-}O)_2Ce^{IV}$ core, which would be consistent with the elemental composition of **3-α** determined by CSI-HRMS. To test whether this feature could arise from an $Fe^{IV}(\mu\text{-}O)_2Ce^{IV}$ diamond core, rR experiments on **3-α** were performed in a 1:1 $H_2{}^{16}O$:$H_2{}^{18}O$ solvent mixture. Under these conditions, the rR spectrum exclusively shows peaks at 677 and 643 cm^{-1}, with no peak found at an intermediate frequency that could be associated with a mixed-labelled diamond core (See Supplementary Fig. 32)[31]. The $H_2{}^{16}O$:$H_2{}^{18}O$ experiment thus excludes formation of an $Fe^{IV}(\mu\text{-}O)_2Ce^{IV}$ core. We have also considered the possibility that the 677 cm^{-1} vibration may correspond to the Fe–OH–Ce mode of an $O=Fe^{IV}\text{–}OH\text{–}Ce^{IV}$ core, as earlier postulated[5]. However, two points argue against this possibility: (a) the lack of a D_2O effect on the 677 cm^{-1} vibration[32] and (b) more importantly, the fact that Fe–OH–Fe vibrations are typically found in the range of 400–500 cm^{-1} (refs 32,33) much lower than observed for **3-α**. Consideration of the above points together with its mass spectral formulation leads us to favour an $O=Fe^{IV}\text{–}O\text{–}Ce^{IV}$ core for **3-α**. An analogous heterodimetallic $Fe^{IV}\text{–}O\text{–}Sc^{III}$ unit has been characterized crystallographically by Fukuzumi and Nam[34], but unfortunately there is no Raman data reported for comparison.

Discussion

Identification of **3-α** with an $O=Fe^{IV}\text{–}O\text{–}Ce^{IV}$ core as the key reaction intermediate in WO reactions introduces unconsidered mechanistic scenarios for the oxidation of water by Ce^{IV} (Fig. 7). Most obviously, Ce^{IV} appears not to behave simply as an outer-sphere oxidant, but instead forms an inner-sphere $Fe^{IV}(\mu\text{-}O)Ce^{IV}$ intermediate that is crucial for the reaction to proceed. Therefore, Ce^{IV}-driven Fe-based WO activity not only requires the presence of two *cis*-labile sites but also a structure of the iron complex that allows formation of the $Fe^{IV}(\mu\text{-}O)Ce^{IV}$ intermediate. In this regard, the contrasting reactivities of **1-α** and **1-β** may arise from differences in their steric or electronic properties. A possible rationale comes from the differences in the nature of the donor *trans* to the *cis*-labile sites of the α and α isomers. While the α isomer has two tertiary amines *trans* to the labile positions, the *β*

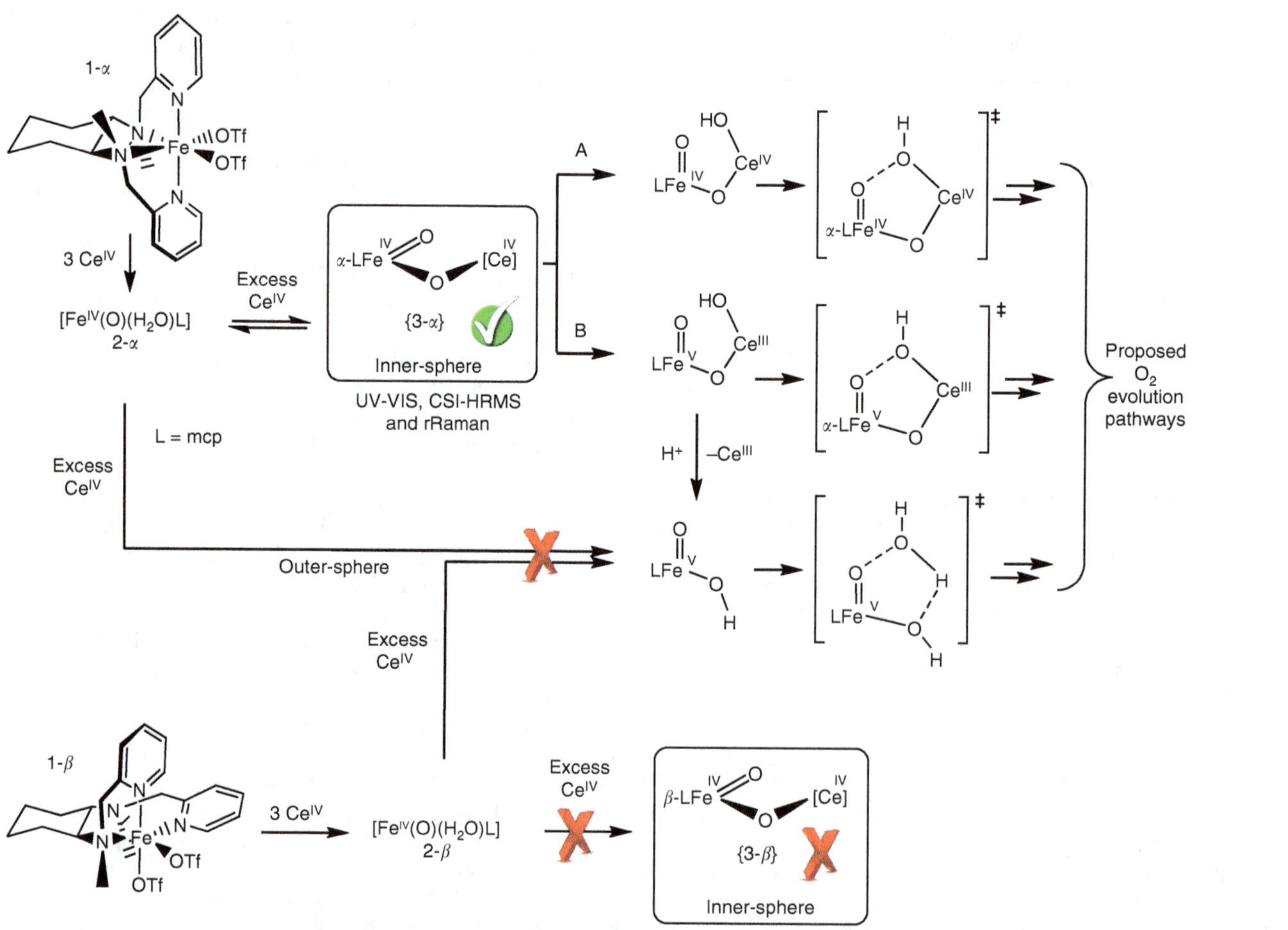

Figure 7 | Reactivity differences between topological isomers α-β. The mechanism is based on the formation of **3-α** as key intermediate for the O_2 evolution. Mechanism **a** describes the oxo/oxyl radical coupling between $Fe^{IV}(O)$ and $Ce^{IV}(OH)$. Path (**b**) describes the formation of $Fe^{V}(O)$ and the subsequent nucleophilic attack of a water molecule.

isomer has one tertiary amine and one pyridine. From our earlier survey of non-haem iron WO catalysts, the more active complexes are similar to **1-α** with two tertiary amines *trans* to the two labile sites, whereas **1-β**, like the $[Fe^{II}(CF_3SO_3)_2(tpa)]$ complex (tpa = tris(2-pyridylmethyl)amine), has lower WO activity[4]. In contrast, close inspection of the crystallographic data for **1-β** (ref. 19) shows that the N-Me groups point towards the *cis*-labile sites in such a way that might inhibit for formation of oxo-bridged dinuclear species (Supplementary Fig. 34). Although examination of simple β-$[(mcp)Fe^{IV}(\mu\text{-}O)Ce^{IV}(NO_3)_3]^+$ models seems to discard this possibility, a survey of the literature reveals that **1-α** can form a number of oxo-bridged diferric complexes that have been crystallographically characterized, but dimeric species derived from **1-β** are at present not known[35].

How intermediate **3-α** evolves further to produce O_2 remains unclear[36]. In Fig. 7 are postulated two possible mechanism an oxo/oxyl radical coupling mechanism to form the O–O bond (Path **a**, Fig. 7)[12] and the alternative hypothesis entailing electron transfer from Fe^{IV} to Ce^{IV} to generate a highly electrophilic $\{Fe^V(O)(\mu\text{-}O)Ce^{III}\}$ oxidant (Fig. 7, path **b**). Water binding to the Ce^{III} would position it well for electrophilic attack by the $Fe^V=O$ moiety to form the O–O bond. This pathway is analogous to that proposed for **1**-catalysed C-H and C = C oxidations with Ce^{IV} as an oxidant[37]. Also supporting the notion of a highly electrophilic oxidant are the observed effects of substituents on the pyridine of the supporting ligand of a related catalyst on WOC efficacy, where more electron withdrawing substituents on the pyridine ligand enhance both TON and TOF for water oxidation[5]. Nevertheless, at this point it is not possible to discard the other scenarios unequivocally.

The $[Fe^{IV}(O)(\mu\text{-}O)Ce^{IV}]$ intermediate we have characterized resembles the corresponding $Mn^V(O)(\mu\text{-}O)Ca^{II}(OH_2)$ unit proposed in some mechanistic models for the OEC (Fig. 8),[38–40] making **3-α** an iron-based functional OEC model. In fact, the resemblance becomes stronger with electron transfer from Fe^{IV} to Ce^{IV} to generate an $[Fe^V(O)(\mu\text{-}O)Ce^{III}]$ oxidant. The strong similarities between iron and manganese chemistry in biologically relevant oxidation reactions suggest that this analogy should not be unexpected. Also relevant is the fact that Ce^{III} can replace Ca^{II} in biomolecules or act as an antagonist in pharmacology[41–43]. For example, it has been found that lanthanides, Ce in particular, are essential for the growth of *Methylacidiphilum fumariolicum* SolV[41], while Ca^{II} can be replaced by Ce^{III} in the pyrroloquinoline quinone (PQQ)-dependent methanol dehydrogenanse (MDH) enzymes to elicit superior catalytic properties[41,42].

The mechanisms for O–O bond formation currently being considered for the OEC can be grouped in two main paths: (a) a nucleophilic attack of a water molecule to a $Mn^V{\equiv}O$ (or a Mn^{IV}-oxyl, $Mn^{IV}=O\cdot$) unit, and (b) an oxo/oxyl radical coupling mechanism[40,44,45]. Remarkable mechanistic insight into the O–O bond formation reaction has been obtained from studies of Mn (refs 38,39,46) and Co (refs 47–50) tetrametallic cubane model complexes. Up to now these cubane OEC models generally support the oxo/oxyl radical coupling mechanism to form the O–O bond. However, there are no catalytically active WO models that include in their structure a Lewis acid to mimic the role of Ca^{II} in PSII (refs 38,39,46,51,52), despite the fact that a Lewis acid (Ca^{II}, but also Sr^{II})[53,54] has been proven to be absolutely required for the operation of the OEC. Given their obvious interest, the absence of heterodimetallic complexes that are active in water oxidation may reflect the extraordinary difficulty in accessing well-defined models where both a first row transition metal centre and a Lewis acid work together in water solution[50]. The success of our effort to trap and characterize **3-α** as having an $[O=Fe^{IV}\text{-}O\text{-}Ce^{IV}]$ core that resembles the fundamental $Mn^V(O)(\mu\text{-}O)Ca(OH_2)_2$ structural motif of the OEC unit is therefore remarkable from an inorganic synthesis perspective. More importantly, the identification of an $[O=Fe^{IV}\text{-}O\text{-}Ce^{IV}]$ core that is active in water oxidation provides a model compound to study the formation of the O–O bond in detail.

In conclusion, this work provides experimental evidence and characterization of an Fe^{IV}–O–Ce^{IV} species as the last detectable intermediate in Fe-catalysed WO reactions. Although Ru–O–Ce intermediates have been proposed in water oxidation reactions by different research groups[55–58], to the best of our knowledge this is the first direct experimental characterization of a heterodimetallic core in a synthetic WO catalyst. Furthermore, the Fe^{IV}–O–Ce^{IV} active species we have described in this work can be construed as the closest structural and functional model for the essential heteodimetallic Mn^V–O–Ca^{II} centre involved in the water oxidation event in PSII.

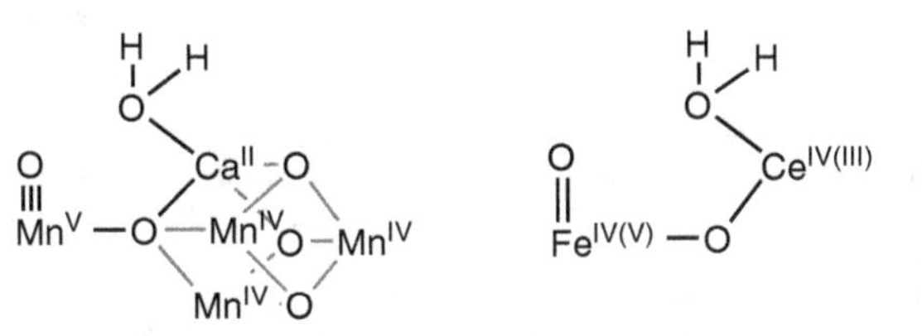

Figure 8 | Structure of the active species and analogy to the proposed intermediate of the OEC. Analogy between a proposed oxygen evolving complex intermediate and the iron-cerium adduct (**3-α**), both active species for the water oxidation, containing a metal in a high oxidation state and a Lewis acid, which may orientate the water molecule for the nucleophilic attack.

Methods

Materials. Reagents were purchased from commercial sources and used as received, without any further purification. Cerium(IV) ammonium nitrate (CAN) (≥99.99%) and trifluoromethanesulfonic acid (ReagentPlus grade ≥99%) were purchased from Sigma-Aldrich. Solvents were purchased from SDS and Scharlab, purified and dried by passing through an activated alumina purification system (MBraun SPS-800) and stored in an anaerobic glovebox under N_2. Complexes **1-α** and **1-β** were prepared as previously described. Water (18.2 MΩ cm) was purified with a Milli-Q Millipore Gradient AIS system.

Physical methods. UV–vis-NIR spectra were recorded on an Agilent 8453 diode array spectrophotometer (190–1,100 nm range) in 1-cm quartz cells. A cryostat from Unisoku Scientific Instruments was used for temperature control. The amount of gas generated was measured with a differential pressure transducer sensor (Honeywell-ASCX15DN, ± 15 psi). Each reaction had its own reference reaction, which was connected to the other port of the sensor. Further details of the equipment are available elsewhere[4,5].

High resolution mass spectra. High resolution mass spectra were recorded on a Bruker MicrOTOF-Q II instrument with ESI or Cryospray ionization sources at Serveis Tècnics of the University of Girona. Samples were introduced into the mass spectrometer ion source by direct infusion using a syringe pump and were externally calibrated using sodium formate. The instrument was operated in the positive ion mode.

Resonance raman spectroscopy. Resonance Raman (rR) spectra were collected via excitation with Kr^+ and Ar^+ lasers (Spectra-Physics BeamLok 2060-RM) and an Acton AM-506M3 monochromator equipped with a Princeton Instruments ACTON PyLON LN/CCD-1340 × 400 detector. Low-temperature spectra in H_2O:CH_3CN solution mixtures were obtained at 265 K using a 90° backscattering geometry. Raman frequencies were calibrated to indene before data collection. Rayleigh scattering was attenuated using a holographic notch filter (Kaiser Optical Systems) for each excitation wavelength. The monochromator slit width was set for a band pass of 4 cm^{-1} for all spectra. All spectra were collected using a laser excitation power of 100 mW. The plotted spectra are averages of 96 scans with collection times of 20 s. All spectra were intensity-corrected to the 922 cm^{-1} peak of the CH_3CN solvent.

Kinetic studies. The required amount of iron complex was dissolved with 2.7 ml of Milli-Q water in a quartz UV–vis cell at 25 °C. Then, the $Fe^{IV}(O)(H_2O)$ intermediate was generated by the addition of 3 equiv. of CAN dissolved in 50 μl of diluted HOTf (3:10, HOTf:H_2O). Once fully formed, Milli-Q water (0.25 ml) was added to the solution. A second addition of CAN in 0.3 ml of HOTf:H_2O (at pH ~ 0) was added. This second addition leads to the desired final concentration of Ce^{IV} (ranged from 1.25 to 18.75 mM) and iron complex (0.1 mM) for the kinetic studies at pH = 1. Cerium(IV) consumption was monitored at $\lambda = 420$ nm. Oxygen evolution was monitored in parallel by a pressure transducer. The initial rate method was applied to obtain rate constants in the kinetic studies to avoid interference of the consumption of the catalysts during the catalytic reactions. All the kinetic data were treated following the same procedure. Only data obtained within the first 10 s of reaction was used, which corresponds to the decay of 20% of the starting Ce absorbance. From these data, the self-decay of a Ce^{IV} blank experiment (same concentration, pH = 1) was subtracted. For the transformation of the absorbance to concentration, blank samples for every Ce^{IV} concentration were linearized and the equation was used as a calibration curve.

References

1. Gunay, A. & Theopold, K. H. C—H bond activations by metal oxo compounds. *Chem. Rev.* **110,** 1060–1081 (2010).
2. Shaik, S., Hirao, H. & Kumar, D. Reactivity of high-valent iron–oxo species in enzymes and synthetic reagents: a tale of many states. *Acc. Chem. Res.* **40,** 532–542 (2007).
3. Liu, X. & Wang, F. Transition metal complexes that catalyze oxygen formation from water: 1979–2010. *Coord. Chem. Rev.* **256,** 1115–1136 (2012).
4. Lloret-Fillol, J. *et al.* Efficient water oxidation catalysts based on readily available iron coordination complexes. *Nat. Chem.* **3,** 807–813 (2011).
5. Codolà, Z. *et al.* Electronic effects on single-site iron catalysts for water oxidation. *Chem. Eur. J.* **19,** 8042–8047 (2013).
6. Ellis, W. C., McDaniel, N. D., Bernhard, S. & Collins, T. J. Fast water oxidation using iron. *J. Am. Chem. Soc.* **132,** 10990–10991 (2010).
7. Hoffert, W. A., Mock, M. T., Appel, A. M. & Yang, J. Y. Incorporation of hydrogen-bonding functionalities into the second coordination sphere of iron-based water-oxidation catalysts. *Eur. J. Inorg. Chem.* 3846–3857 (2013).
8. Wiechen, M., Berends, H.-M. & Kurz, P. Water oxidation catalysed by manganese compounds: from complexes to 'biomimetic rocks'. *Dalton Trans.* **41,** 21–31 (2012).
9. Cao, R., Lai, W. & Du, P. Catalytic water oxidation at single metal sites. *Energy Environ. Sci.* **5,** 8134–8157 (2012).
10. Yagi, M. & Kaneko, M. Molecular catalysts for water oxidation. *Chem. Rev.* **101,** 21–36 (2000).
11. Nocera, D. G. The Artificial Leaf. *Acc. Chem. Res.* **45,** 767–776 (2012).
12. Kundu, S. *et al.* O–O bond formation mediated by a hexanuclear iron complex supported on a stannoxane core. *Chem. Eur. J.* **18,** 2787–2791 (2012).
13. Kodera, M. *et al.* Reversible O–O bond scission of peroxodiiron(III) to high-spin oxodiiron(IV) in dioxygen activation of a diiron center with a bis-tpa dinucleating ligand as a soluble methane monooxygenase model. *J. Am. Chem. Soc.* **134,** 13236–13239 (2012).
14. Ferguson-Miller, S. & Babcock, G. T. Heme/copper terminal oxidases. *Chem. Rev.* **96,** 2889–2908 (1996).
15. Company, A., Lloret-Fillol, J. & Costas, M. in *Comprehensive Inorganic Chemistry II* (eds Reedijk, J. & Poeppelmeier, K.) 2nd edn (Elsevier 487–564, 2013).
16. Hong, D. *et al.* Water oxidation catalysis with nonheme iron complexes under acidic and basic conditions: homogeneous or heterogeneous? *Inorg. Chem.* **52,** 9522–9531 (2013).
17. Chen, G., Chen, L., Ng, S.-M., Man, W.-L. & Lau, T.-C. Chemical and visible-light-driven water oxidation by iron complexes at pH 7–9: evidence for dual-active intermediates in iron-catalyzed water oxidation. *Angew. Chem. Int. Ed.* **52,** 1789–1791 (2013).
18. Knof, U. & von Zelewsky, A. Predetermined chirality at metal centers. *Angew. Chem. Int. Ed.* **38,** 302–322 (1999).
19. Costas, M. & Que, Jr L. Ligand topology tuning of iron-catalyzed hydrocarbon oxidations. *Angew. Chem. Int. Ed.* **41,** 2179–2181 (2002).
20. Costas, M., Tipton, A. K., Chen, K., Jo, D.-H. & Que, Jr. L. Modeling rieske dioxygenases: the first example of iron-catalyzed asymmetric cis-dihydroxylation of olefins. *J. Am. Chem. Soc.* **123,** 6722–6723 (2001).
21. Hong, S. *et al.* Ligand topology effect on the reactivity of a mononuclear nonheme iron(IV)-oxo complex in oxygenation reactions. *J. Am. Chem. Soc.* **133,** 11876–11879 (2011).
22. Acuña-Parés, F., Codolà, Z., Costas, M., Luis, M. & Lloret-Fillol, J. Unraveling the mechanism of water oxidation catalyzed by nonheme iron complexes. *Chem. Eur. J.* **20,** 5696–5707 (2014).
23. Jackson, T. A. *et al.* Axial ligand effects on the geometric and electronic structures of nonheme oxoiron(IV) complexes. *J. Am. Chem. Soc.* **130,** 12394–12407 (2008).
24. Wang, D. *et al.* Nonheme oxoiron(IV) complexes of pentadentate N5 ligands: spectroscopy, electrochemistry, and oxidative reactivity. *Chem. Sci.* **4,** 282–291 (2013).
25. Ikeda-Ohno, A., Hennig, C., Weiss, S., Yaita, T. & Bernhard, G. Hydrolysis of tetravalent cerium for a simple route to nanocrystalline cerium dioxide: an in situ spectroscopic study of nanocrystal evolution. *Chem. Eur. J.* **19,** 7348–7360 (2013).
26. Lutz, O. M. D., Hofer, T. S., Randolf, B. R., Weiss, A. K. H. & Rode, B. M. A QMCF-MD investigation of the structure and dynamics of Ce4 + in aqueous solution. *Inorg. Chem.* **51,** 6746–6752 (2012).
27. Piro, N. A., Robinson, J. R., Walsh, P. J. & Schelter, E. J. The electrochemical behavior of cerium(III/IV) complexes: Thermodynamics, kinetics and applications in synthesis. *Coord. Chem. Rev.* **260,** 21–36 (2014).
28. Wilkinson, E. C. *et al.* Raman signature of the Fe_2O_2 "Diamond" core. *J. Am. Chem. Soc.* **120,** 955–962 (1998).
29. Skulan, A. J., Hanson, M. A., Hsu, H.-f., Que, Jr. L. & Solomon, E. I. Spectroscopic Study of $[Fe_2O_2(5\text{-}Et_3\text{-}TPA)_2]^{3+}$: nature of the Fe_2O_2 diamond core and its possible relevance to high-valent binuclear non-heme enzyme intermediates. *J. Am. Chem. Soc.* **125,** 7344–7356 (2003).
30. Xue, G., Fiedler, A. T., Martinho, M., Münck, E. & Que, Jr. L. Insights into the P-to-Q conversion in the catalytic cycle of methane monooxygenase from a synthetic model system. *Proc. Natl Acad. Sci. USA* **105,** 20615–20620 (2008).
31. Que, Jr L. & Tolman, W. B. Bis(μ-oxo)dimetal "Diamond" cores in copper and iron complexes relevant to biocatalysis. *Angew. Chem. Int. Ed.* **41,** 1114–1137 (2002).
32. Zheng, H., Zang, Y., Dong, Y., Young, V. G. & Que, Jr. L. Complexes with $Fe^{III}_2(\mu\text{-}O)(\mu\text{-}OH)$, $Fe^{III}_2(\mu\text{-}O)_2$, and $[Fe^{III}_3(\mu_2\text{-}O)_3]$ cores: structures, spectroscopy, and core interconversions. *J. Am. Chem. Soc.* **121,** 2226–2235 (1999).
33. Cranswick, M. A. *et al.* Protonation of a peroxodiiron(III) complex and conversion to a diiron(III/IV) intermediate: implications for proton-assisted O-O bond cleavage in nonheme diiron enzymes. *Inorg. Chem.* **51,** 10417–10426 (2012).
34. Fukuzumi, S. *et al.* Crystal structure of a metal ion-bound oxoiron(IV) complex and implications for biological electron transfer. *Nat. Chem.* **2,** 756–759 (2010).
35. Stubna, A. *et al.* A structural and Mössbauer study of complexes with $Fe_2(\mu\text{-}O(H))_2$ cores: stepwise oxidation from $Fe^{II}(\mu\text{-}OH)_2Fe^{II}$ through $Fe^{II}(\mu\text{-}OH)_2Fe^{III}$ to $Fe^{III}(\mu\text{-}O)(\mu\text{-}OH)Fe^{III}$. *Inorg. Chem.* **43,** 3067–3079 (2004).
36. Halfen, J. A. *et al.* Reversible cleavage and formation of the dioxygen O-O bond within a dicopper complex. *Science* **271,** 1397–1400 (1996).
37. Garcia-Bosch, I. *et al.* Iron-catalyzed C-H hydroxylation and olefin cis-dihydroxylation using a single-electron oxidant and water as the oxygen-atom source. *Chem. Eur. J.* **18,** 13269–13273 (2012).
38. Kanady, J. S., Tsui, E. Y., Day, M. W. & Agapie, T. A synthetic model of the Mn3Ca subsite of the oxygen-evolving complex in photosystem II. *Science* **333,** 733–736 (2011).
39. Tsui, E. Y., Tran, R., Yano, J. & Agapie, T. Redox-inactive metals modulate the reduction potential in heterometallic manganese-oxido clusters. *Nat. Chem.* **5,** 293–299 (2013).
40. Cox, N., Pantazis, D. A., Neese, F. & Lubitz, W. Biological water oxidation. *Acc. Chem. Res.* **46,** 1588–1596 (2013).
41. Pol, A. *et al.* Rare earth metals are essential for methanotrophic life in volcanic mudpots. *Environ. Microbiol.* **16,** 255–264 (2014).
42. Jakupec, M. A., Unfried, P. & Keppler, B. K. Pharmacological properties of cerium compounds. *Reviews of Physiology, Biochemistry and Pharmacology*, vol. 153 (Springer 101–111, 2005).
43. Nakagawa, T. *et al.* A catalytic role of XoxF1 as La3 + -dependent methanol dehydrogenase in Methylobacterium extorquens AM1. *PLoS ONE* **7,** e50480 (2012).
44. Rapatskiy, L. *et al.* Detection of the water-binding sites of the oxygen-evolving complex of photosystem II using W-band ^{17}O electron–electron double resonance-detected nmr spectroscopy. *J. Am. Chem. Soc.* **134,** 16619–16634 (2012).
45. Siegbahn, P. E. M. Water oxidation mechanism in photosystem II, including oxidations, proton release pathways, O—O bond formation and O_2 release. *Biochim. Biophys. Acta* **1827,** 1003–1019 (2013).
46. Dismukes, G. C. *et al.* Development of bioinspired Mn_4O_4 – cubane water oxidation catalysts: lessons from photosynthesis. *Acc. Chem. Res.* **42,** 1935–1943 (2009).
47. Symes, M. D., Surendranath, Y., Lutterman, D. A. & Nocera, D. G. Bidirectional and unidirectional PCET in a molecular model of a cobalt-based oxygen-evolving catalyst. *J. Am. Chem. Soc.* **133,** 5174–5177 (2011).
48. McCool, N. S., Robinson, D. M., Sheats, J. E. & Dismukes, G. C. A Co_4O_4 "Cubane" water oxidation catalyst inspired by photosynthesis. *J. Am. Chem. Soc.* **133,** 11446–11449 (2011).
49. Berardi, S. *et al.* Photocatalytic water oxidation: tuning light-induced electron transfer by molecular Co_4O_4 cores. *J. Am. Chem. Soc.* **134,** 11104–11107 (2012).

50. Evangelisti, F., Güttinger, R., Moré, R., Luber, S. & Patzke, G. R. Closer to photosystem II: A Co_4O_4 cubane catalyst with flexible ligand architecture. *J. Am. Chem. Soc.* **135,** 18734–18737 (2013).
51. Leeladee, P. *et al.* Valence tautomerism in a high-valent manganese–oxo porphyrinoid complex induced by a lewis acid. *J. Am. Chem. Soc.* **134,** 10397–10400 (2012).
52. Lacy, D. C., Park, Y. J., Ziller, J. W., Yano, J. & Borovik, A. S. Assembly and properties of heterobimetallic $Co^{II/III}$-Ca^{II} complexes with aquo and hydroxo ligands. *J. Am. Chem. Soc.* **134,** 17526–17535 (2012).
53. Ono, T.-a. & Inoue, Y. Discrete extraction of the Ca atom functional for O_2 evolution in higher plant photosystem II by a simple low pH treatment. *FEBS Lett.* **227,** 147–152 (1988).
54. Boussac, A. & Rutherford, A. W. Nature of the inhibition of the oxygen-evolving enzyme of photosystem II induced by sodium chloride washing and reversed by the addition of calcium(2 +) or strontium(2 +). *Biochemistry* **27,** 3476–3483 (1988).
55. Murakami, M. *et al.* Catalytic mechanism of water oxidation with single-site ruthenium–heteropolytungstate complexes. *J. Am. Chem. Soc.* **133,** 11605–11613 (2011).
56. Yoshida, M., Masaoka, S., Abe, J. & Sakai, K. Catalysis of mononuclear aquaruthenium complexes in oxygen evolution from water: a new radical coupling path using hydroxocerium(IV) species. *Chem. Asian J.* **5,** 2369–2378 (2010).
57. Kimoto, A., Yamauchi, K., Yoshida, M., Masaoka, S. & Sakai, K. Kinetics and DFT studies on water oxidation by Ce^{4+} catalyzed by $[Ru(terpy)(bpy)(OH_2)]^{2+}$. *Chem. Commun.* **48,** 239–241 (2012).
58. Wasylenko, Derek J. *et al.* Berlinguette electronic modification of the $[Ru^{II}(tpy)(bpy)(OH_2)]^{2+}$ scaffold: effects on catalytic water oxidation. *J. Am. Chem. Soc.* **45,** 16094–16106 (2010).

Acknowledgements

We thank Dr. Xavi Ribas and Dr. Anna Company for reading this work and for helpful comments. We also thank the European Research Foundation for project FP7-PEOPLE-2010-ERG-268445 (J.L.-F), MICINN for project CTQ2009-08464 (M.C.) and for a Ramon y Cajal contract (J.L.-F.), Generalitat de Catalunya for an ICREA Academia Award and the European Research Foundation for Project ERC-2009-StG-239910 (M.C.). The work carried out at the University of Minnesota was supported by US National Institutes of Health grant GM38767 to L.Q. We appreciate financial support from INNPLANTA project INP-2011-0059-PCT-420000-ACT1 to Dr. X. Ribas.

Author contributions

J.L.-F. and M.C. devised the initial concept for the work and designed the experiments. Z.C. carried out the WO experiments and trapped the intermediate; L.G. performed the CSI-HRMS studies and analysed the data. S.T.K. obtained the resonance Raman spectra and analysed the data together with L.Q. Z.C., L.Q., M.C. and J.L.-F. wrote the manuscript.

Additional information

Competing financial interests: The authors declare no competing financial interests.

3

Methane hydrate formation in confined nanospace can surpass nature

Mirian E. Casco[1], Joaquín Silvestre-Albero[1], Anibal J. Ramírez-Cuesta[2], Fernando Rey[3], Jose L. Jordá[3], Atul Bansode[4], Atsushi Urakawa[4], Inma Peral[5], Manuel Martínez-Escandell[1], Katsumi Kaneko[6] & Francisco Rodríguez-Reinoso[1]

Natural methane hydrates are believed to be the largest source of hydrocarbons on Earth. These structures are formed in specific locations such as deep-sea sediments and the permafrost based on demanding conditions of high pressure and low temperature. Here we report that, by taking advantage of the confinement effects on nanopore space, synthetic methane hydrates grow under mild conditions (3.5 MPa and 2 °C), with faster kinetics (within minutes) than nature, fully reversibly and with a nominal stoichiometry that mimics nature. The formation of the hydrate structures in nanospace and their similarity to natural hydrates is confirmed using inelastic neutron scattering experiments and synchrotron X-ray powder diffraction. These findings may be a step towards the application of a smart synthesis of methane hydrates in energy-demanding applications (for example, transportation).

[1] Laboratorio de Materiales Avanzados, Departamento de Química Inorgánica-Instituto Universitario de Materiales, Universidad de Alicante, Ctra. San Vicente-Alicante s/n, E-03690 San Vicente del Raspeig, Spain. [2] Oak Ridge National Laboratory, Chemical and Engineering Materials Division, 1 Bethel Valley Road, Oak Ridge, Tennessee 37831, USA. [3] Instituto de Tecnología Química, Universidad Politécnica de Valencia-CSIC, Avda. de los Naranjos s/n, 46022 Valencia, Spain. [4] Institute of Chemical Research of Catalonia (ICIQ), 43007 Tarragona, Spain. [5] ALBA Light Source, 08290 Cerdanyola del Vallés, Barcelona, Spain. [6] Research Center for Exotic Nanocarbons, Shinshu University, 4-17-1 Wakasato, Nagano 380-8553, Japan. Correspondence and requests for materials should be addressed to J.S.-A. (email: joaquin.silvestre@ua.es).

Natural methane hydrates are crystalline solids that form when methane and water comes into contact under thermodynamically favourable conditions, that is, high pressure (typically more than 6 MPa) and relatively low temperature (slightly below room temperature), giving rise to an ice-like hydrogen-bonded structure[1]. Under these conditions, natural hydrates crystallize in a cubic structure known as type sI. This structure is constituted by cages of two different types: (i) six large cages having 12 pentagonal and 2 hexagonal faces (denoted by $5^{12}6^{2}$) formed by 24 water molecules and (ii) two small cages having 12 pentagonal faces (denoted by 5^{12}) formed by 20 water molecules[2]. These cages have an average cavity radius of 0.433 and 0.395 nm, respectively, which is enough to host one methane molecule per cage, resulting in a nominal stoichiometry $1CH_4 \cdot 5.75\ H_2O$.

Actual prospections have estimated vast deposits of methane hydrates in deep-sea sediments and the permafrost (that is, the estimated amount of methane contained in the form of hydrates in the southern US Atlantic margin is as high as 30 billion metric tons (Gt)[3,4]. Actual depletion of fossil fuel reserves all around the world has addressed the attention of governments and scientists towards natural methane hydrates reservoirs, based on the assumption that they could be a major energy resource in the near future (if 100% of the cavities are filled with methane, one volume of hydrate would dissociate into a maximum of 180 volumes (STP) of methane). The amount of energy in the form of hydrate may be twice that of all other fossil fuels combined[4].

Methane, the major component of natural gas (NG), has high hydrogen to carbon ratio and it involves both a greater energy density per unit mass and less carbon dioxide emissions (around 40% less) during combustion than gasoline or diesel. These properties led to the nomination of NG as premium fuel in a low-carbon energy future. Actual technology for NG storage is based either on compressed natural gas (CNG), in automobile industry, or on liquefied natural gas (LNG), for large distance transportation. Despite the large energy density of these technologies, mainly liquefied NG, the use of high pressure (25 MPa and 25 °C) or low temperature (− 162 °C and 0.1 MPa) becomes undesirable from the safety and energy saving view. An alternative solution concerns the use of porous materials to store NG (adsorbed natural gas (ANG)) at a similar high density, but substantially lower pressure. Unfortunately, actual storage on porous solids, mainly metal-organic frameworks and carbon materials, exhibits certain limitations, at least in a gravimetric basis, to fulfil the new methane program from the US Department of Energy (see DOE MOVE program at https://arpa-e-foa.energy.gov)[5–10].

As described above, methane hydrates are constituted by methane molecules entrapped (typically one molecule per cavity) within hydrogen-bonded water cavities. When a small molecule like methane is encaged in hydrates, the guest methane molecules are separated *ca.* 0.5 nm by the water cages, that is, the energy density in hydrates is approximately the same as that of the compressed gas (that is, methane gas at 0 °C and 18 MPa). Taking into account these premises, NG storage and transportation in the form of gas hydrates could be envisaged as a promising alternative, provided that they can be synthesized under mild conditions (low pressures and mild temperatures) and within reasonable time scale.

With this in mind, here we report the nucleation and growth of methane hydrates in the confined nanospace of activated carbons. High-pressure methane adsorption measurements show that these nanostructures can be synthesized under mild conditions, with ultrafast kinetics and with a stoichiometry that mimics nature. The formation of methane hydrate nanocrystals with sI structure is confirmed for the first time by using inelastic neutron scattering (INS) experiments and synchrotron X-ray powder diffraction (XRPD). These results provide some bright light into the formation mechanism and into the prospective of this promising field.

Results

Methane hydrate formation in activated carbon. Methane adsorption experiments either in wet or dry conditions were carried out using a petroleum pitch-derived activated carbon, labelled PP-AC. It is worth mentioning some of the characteristics of this sample to better understand the adsorption behaviour[11]. The pore walls of the PP-AC consist of wrinkled graphenes that infers efficient slit-shaped pores, as it can be observed in Supplementary Fig. 1. The well-developed graphene layers being quite different from conventional activated carbon can be clearly observed[12,13].

The PP-AC sample exhibits a large surface area, around 3,670 $m^2\,g^{-1}$, and a total pore volume (V_t) of 2.44 $cm^3\,g^{-1}$. Micropore (V_0) and mesopore (V_{meso}) volume in the PP-AC sample deduced from the nitrogen adsorption isotherm at − 196 °C are 1.20 and 1.24 $cm^3\,g^{-1}$, respectively. Pore size distribution after application of the quenched solid density functional theory method to the nitrogen adsorption data reflects that sample PP-AC contains a well-defined micro- (25%) and mesoporous (75%) structure with a narrow peak at 1 nm and a wide shoulder with peak maxima at 2 and 3 nm (see Supplementary Fig. 1). In addition, mercury porosimetry shows a wide macropore size distribution, ranging 0.1–100 μm, with a maximum at 10 μm. The presence of highly developed porous structure combining micro-, meso- and macropores would be of paramount importance to elucidate the hydrate formation mechanism on confined nanospaces.

Water adsorption isotherm of PP-AC at 25 °C is reported in Supplementary Fig. 2. At low relative pressures, the amount adsorbed is nil, as corresponds to a hydrophobic material. Condensation takes place at high relative pressures ($p/p_0 > 0.8$) with a saturation capacity of 1.8 $g\,g^{-1}$ of dry carbon, corresponding to a 93% filling of the total pore volume (under the assumption that density of the adsorbed water is 1.0 $g\,cm^{-3}$). The water adsorption isotherm has a remarkable hysteresis loop that comes from a metastable assembly structure of water in the adsorption branch[14,15]. Our initial hypothesis would be that the small thickness of the metastable water layer (that provides a very high-interfacial liquid water–methane contact area), together with the low interaction of water molecules with the hydrophobic carbon surface could speed up the methane hydrate formation process[16,17]. Consequently, the methane hydrate formation on sample PP-AC was examined for different amounts of adsorbed water. The available nanoporosity of water-adsorbed PP-AC sample was measured by N_2 adsorption at − 196 °C after freezing water adsorbed in nanopores (Supplementary Fig. 3). Here we introduce the weight ratio of adsorbed water to carbon, R_w; $R_w = 1.8$ corresponds to fully saturated pores with water. The R_w was changed up to $R_w = 4.1$, that is, from unsaturated samples (R_w of 0.6 and 1.1) to oversaturated samples (R_w of 2.9 and 4.1). Supplementary Figure 3 shows that the larger the R_w the lower the residual porosity is. For R_w of 0.6 and 1.1, the pores are partially filled with water to 66 and 80% total pore filling (total pore volume measured by nitrogen at − 196 °C), whereas for R_w superior to 1.8, the pores (micropores and mesopores; see Supplementary Fig. 1b) are almost saturated with water (even macropores must be filled with a metastable water film), in agreement with the water adsorption isotherm.

Methane adsorption capacity under the water-supplying conditions by varying R_w was evaluated using a volumetric apparatus as described in the experimental section. Figure 1a shows the adsorption isotherms of PP-AC in a gravimetric basis

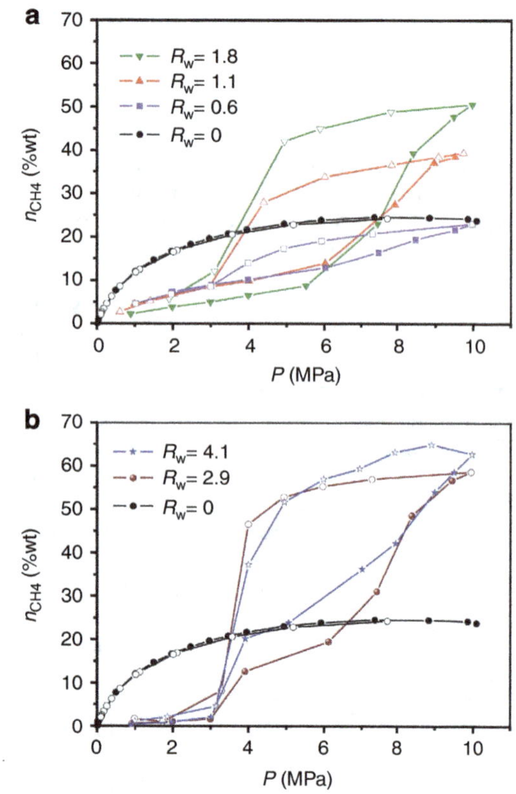

Figure 1 | Methane adsorption in the presence of water. Methane adsorption/desorption isotherms at 2 °C for sample PP-AC dry and wet (**a**) $R_w = 0.6$, 1.1, 1.8; and (**b**) $R_w = 2.9$, 4.1, in a gravimetric basis (g_{CH4} $g^{-1}_{dry\ carbon}$ × 100). Closed symbols represent adsorption branch and open symbols represent desorption branch.

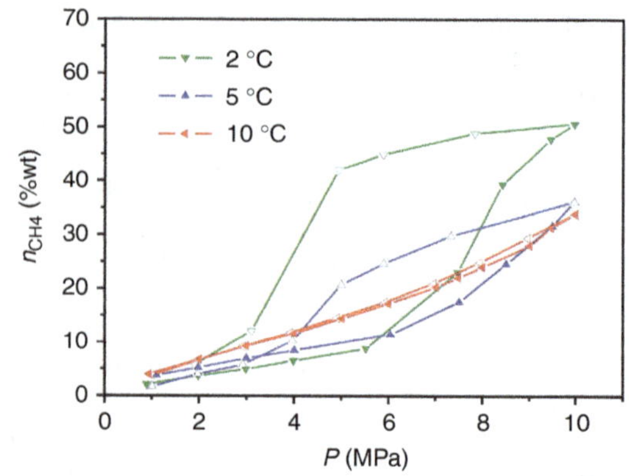

Figure 2 | Effect of temperature on methane hydrates formation. Methane adsorption/desorption isotherms for sample PP-AC with $R_w = 1.8$ at different adsorption temperatures. Closed symbols represent adsorption branch and open symbols represent desorption branch.

(wt.% methane per weight of dry carbon) for R_w 0.6, 1.1 and 1.8 (dry sample was included for the sake of comparison). At this point, it is interesting to highlight that the methane adsorption capacity for the dry carbon at 10 MPa (24 wt.% or 0.24 g g^{-1}) is completely reversible and it is among the highest values ever reported in the literature, even if compared with carbons of similar textural properties or metal-organic frameworks[6–10,18]. Concerning the wet samples and independently of R_w, the methane adsorption capacity is low compared with the dry sample at least at low pressures (below 3 MPa), most probably due to pore blocking by preadsorbed water. However, for pressures higher than 3 MPa, the scenario changes completely. In all cases a sigmoidal behaviour can be clearly appreciated, the higher the R_w value, the more remarkable the adsorption increment is. All adsorption isotherms exhibit a large hysteresis loop that increases with R_w, thus indicating a different adsorption and desorption mechanisms for methane in these samples. In any case, methane is almost completely released below 4 MPa.

A closer look to the effect of the coexistent water amount in the adsorption behaviour reflects the presence of a critical R_w value above which the adsorption capacity of the wet samples surpasses that for the dry sample, thus suggesting an enhancement role of preadsorbed water in methane adsorption. Previous studies reported in the literature have anticipated the beneficial effect of water on methane adsorption, although a deep evaluation of the adsorption process (adsorption kinetics at different steps, phase identification, chemical stoichiometry and so on) and the adsorption mechanism responsible for this behaviour has not been clearly elucidated[19–21].

Figure 1b shows the methane adsorption isotherms for heavily water-supplied samples with higher $R_w = 2.9$ and 4.1 (oversaturated samples). Interestingly, on these samples, in addition to the inflexion point at high pressures (above 6–8 MPa) observed in Fig. 1a, there is an additional adsorption step at relatively lower pressures (around 3 MPa). Despite the scarce number of studies dealing with methane adsorption on wet-activated carbons, to our knowledge, this is the first time that a two-step adsorption process can be discerned for activated carbons. Previous studies described in the literature have only described the sigmoidal shape at 4–6 MPa[19–21]. In addition to the two-step adsorption process, both samples exhibit a large hysteresis loop, the desorption branch being quite stable down to *ca.* 4 MPa, indicating the formation of the stable phase of the methane and water mixture. The transition of a metastable island state to the stable solid-like state on the adsorption branch with the methane pressure should occur. In fact, the pressure associated with the first jump in the adsorption isotherm perfectly fits with the formation pressure for natural hydrates in bulk water at 2 °C ($P_w = 3.14$ MPa)[1]. Last but not the least, it is important to highlight that the saturated adsorption capacity in both heavily water-supplied samples is extraordinarily large, 59 wt.% (0.59 g g^{-1}) and 63 wt.% (0.63 g g^{-1}) for R_w 2.9 and 4.1, respectively, these values being over twofold the methane adsorption capacity of the dry sample. As described above, these values (on wet and dry samples) constitute the best adsorption capacity values for methane ever reported in the literature[8–10].

Despite the beneficial effect of preadsorbed water on methane adsorption, Supplementary Fig. 4 clearly reflects that the adsorption capacity achieves a plateau with R_w, that is, above a certain amount of water supplied, the system does not experience any improvement at all since it behaves like a bulk system. At this point, it is noteworthy to mention that the methane adsorption isotherm on bulk water (in the absence of confinement effects) gives nil adsorption over the whole-pressure range evaluated due to the following: (i) the small gas–liquid interphase and (ii) the slow kinetics of the nucleation and growth of synthetic hydrates, at least in the absence of confined effect as it will be discussed later.

Another important factor in the nucleation and growth of synthetic hydrates is the formation temperature. As described above, natural hydrates are formed in nature on demanding conditions of high pressure and low temperature. To evaluate the effect of temperature in the adsorption behaviour reported for water preadsorbed in confined nanopores, sample PP-AC saturated under $R_w = 1.8$ was exposed to methane up to 10 MPa at different adsorption temperatures. As it can be observed in

Fig. 2, the amount adsorbed decreases and the hysteresis loop becomes progressively less marked on a slight increase in the adsorption temperature up to 10 °C, where the hysteresis loop completely vanishes. This observation suggests that the complex methane adsorption process on confined nanospaces, that is, nucleation and growth of methane hydrates, is highly sensitive to the adsorption temperature, ~2 °C being the upper limit for nucleation.

The major problem associated with storing methane in the form of hydrates, that is, synthetic methane hydrates, concerns its extremely slow growth kinetics. Indeed, the blank experiment in the absence of confinement effects, that is, in the absence of activated carbon, gives no methane adsorption over the whole-pressure range evaluated (up to 10 MPa). Only under special conditions, for example, water-in-crude oil emulsion, methane hydrate can be formed under realistic time scale (within hours) due to the increase gas–liquid contact area, although using extremely high pressures (10 MPa)[15,16,22].

Interestingly, these kinetic limitations can be overcome by taking advantage of the confinement effects on carbon nanospaces[23]. It is well known that confinement of materials in nanometre-scale spaces induces unique phenomena in respect to the bulk phase due to the enhanced interaction potential of molecules and ions within the nanospace walls. These phenomena called quasi-high-pressure effect has been used by some of the authors to induce high-pressure gas-phase reactions, that is, production of metallic sulphur one-dimensional chain and KI nanocrystals, at ambient pressure[23,24]. To evaluate the effect of the confinement in the hydrate formation kinetics at the two steps of the adsorption process described above, Fig. 3 reports the pressure–time change profile until equilibrium for three different points of the methane adsorption/desorption isotherm on sample PP-AC with $R_w = 4.1$, two points corresponding to the adsorption branch (4 and 8 MPa) and one point corresponding to the desorption branch (3 MPa).

Interestingly, the first adsorption step at relatively 'low pressures' (4 MPa) is surprisingly fast, pressure reaching equilibrium in <1 h. This observation suggests that the complex adsorption process taking place in this pressure range (possible hydrate formation in large pores) is relatively fast. Longer equilibrium times (around 14 h) are required for the high-pressure points of the isotherms, thus suggesting the presence of certain kinetic restrictions at the methane adsorption stage. Most probably, the first jump in the adsorption isotherm, mainly observed in oversaturated samples, must correspond to methane hydrate formation in large pores (macropores and large mesopores where a metastable water layer is present), this process being relatively fast. Once these structures are formed, excess methane must diffuse throughout the pore network filled by hydrate crystallites to reach the inner pores (small mesopores and micropores), where hydrate nucleation and growth is slightly hindered, thus explaining the different growing kinetics.

In summary, these observations suggest that confinement effects on carbon nanopores not only enhance the methane adsorption capacity but rather infer an intensive improvement in the hydration formation kinetics.

***In situ* Ineslatic Neutron Scattering**. As described along the manuscript, previous studies have anticipated the beneficial effect of water in the methane adsorption capacity of activated carbons[19–21]. Although it has been speculated no evidence of methane hydrate formation has been described in the literature up to date. To dissipate any doubt about the possibility of methane hydrate formation in nanoconfined spaces, *in situ* INS measurements have been performed at the Rutherford Appleton facilities at the United Kingdom. There is an intrinsic advantage in using INS with hydrogenous materials due to the incoherent scattering cross-section of the proton, which is almost two orders of magnitude greater than in any other nucleus, thus allowing for a relatively simple access to the self-dynamics of molecular hydrogen. INS experiments were performed using D_2O in spite of H_2O to reduce parasitic scattering from the water framework. Before any adsorption experiment, a background spectrum was collected to subtract any hydrogen interference from the carbon surface. After the background, deuterated sample was contacted with methane (CH_4) at two different pressures (4 and 6 MPa) for 1 h at 2 °C. Shortly after that, the reactor was properly cooled

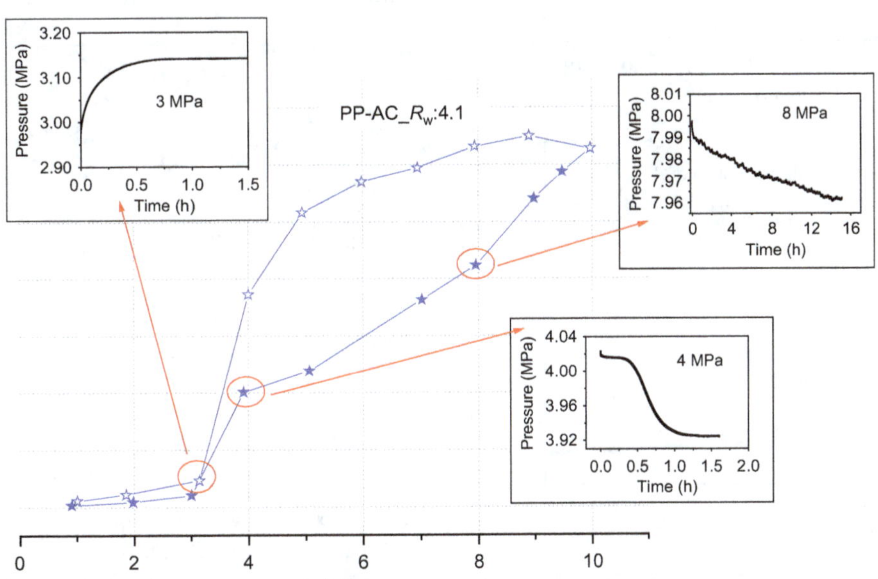

Figure 3 | Methane hydrates formation kinetics. Pressure change profile for methane until equilibrium at two different points of the adsorption isotherm (4 and 8 MPa) and at one point of the desorption branch (3 MPa) for sample PP-AC $R_w = 4.1$ at 2 °C (closed symbols represent adsorption and open symbols represent desorption).

down to −268 °C for the neutron scattering measurements using the TOSCA spectrometer at the Rutherford Appleton facilities. The methane spectrum on dry carbon was also measured for the sake of comparison.

Figure 4a shows the quantum rotational spectra for activated carbon PP-AC dry and wet (deuterated) at 4 and 6 MPa (in the dry sample only 6 MPa was evaluated). As it can be observed in Fig. 4a, methane adsorbed on the dry-activated carbon gives rise only to the elastic peak at an energy transfer of 0 meV. The spectrum of methane in the dry-activated carbon is dominated by its molecular recoil and all spectroscopic information is washed out. This is due to the light mass of methane and the presence of weak intermolecular interactions. However, the situation changes drastically in the deuterated (wet) samples. In this case, the rotational spectra show, in addition to the elastic band described above, additional contributions on both sites due to inelastic scattering of neutrons. When encaged in deuterated ice network, methane molecules behave almost like free rotors even at −268 °C. These inelastic rotational lines appear at energy transfers of $E = \pm$ 0.99 meV (7.98 cm^{-1}), $E_1 = 2.3$ meV (18.54 cm^{-1}) and $E_2 = 3.3$ meV (26.61 cm^{-1}), independently of the methane pressure applied. The rotational constant of methane as a free rotor is $B_{CH4} = h/2\theta = 0.655$ meV. The first peak corresponding to the transition from the rotational ground state ($J = 0$) to the first excitation state ($J = 1$) should appear at 1.31 meV (10.56 cm^{-1}). The shift observed in this peak in our experiment (by 20%) is in accordance with the observations by Tse *et al.*[25] for synthetic hydrates on non-confined spaces. Other two peaks are assigned to the rotational transitions $J = 1 \rightarrow J = 2$ and $J = 0 \rightarrow 2$, respectively. Interestingly, these values are coincident with those found in natural methane hydrates from the Pacific see-floor, in which the methane molecule was considered like an almost free rotor[26]. The almost free rotation of methane is a clear indication that guest molecules occupy positions in the centre of the cages of the hydrate lattice. At off-centre positions, the effect of a non-spherical symmetry potential due to the dipole–octopole interactions would be observed. These peaks above 5 meV are attributed to translational motion of the methane molecule in the two kind of water cages, the small ones denoted as 5^{12} and the large ones denoted $5^{12}6^2$ (ref. 25). Consequently, the observation of these inelastic quantum rotational bands is the first direct evidence of methane hydrate formation in the confinement space of nanoporous carbons, even at 4 MPa. The relatively 'low' pressure and the relatively fast adsorption kinetics (<1 h) constitute a step stone towards the application of synthetic methane hydrates for methane storage in on-board applications or for overseas transportation.

Figure 4b shows the effect of the temperature for the INS measurement on the relative intensity of the rotational transitions. An increase in temperature from −268 to −233 °C gives rise to a line broadening, suggesting the transformation from quantum rotations to translation-coupled rotation, that is, classical rotational diffusion.

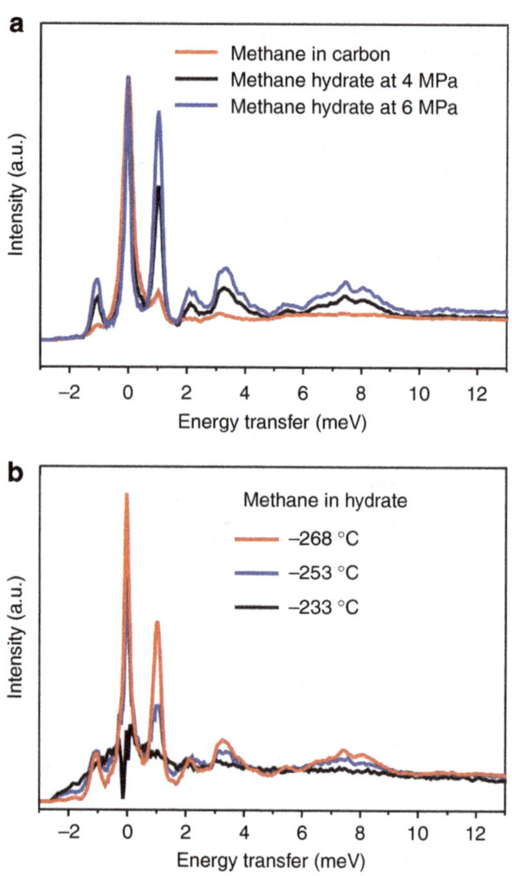

Figure 4 | Inelastic neutron scattering of methane hydrates. Rotational spectra of hydrogen coming from the methane molecule in wet (deuterated)-activated carbon PP-AC (**a**) at different pressures and (**b**) at different temperatures. The spectrum of dry-activated carbon containing 6 MPa methane is included for the sake of comparison.

***In situ* synchrotron X-ray diffraction.** To further confirm the methane hydrate formation and to give some insight into the crystal structure of the synthesized nanocrystals, synchrotron XRPD experiments were performed. The XRPD patterns were collected in the high-pressure end station of the MSPD beamline at synchrotron ALBA in Spain using the PP-AC sample over-saturated with D_2O ($R_w = 4.1$). The spectra of the 'wet' sample in the absence of methane and at room temperature exhibit the expected pattern for an amorphous carbon with no well-defined peaks over the whole range evaluated (see Fig. 5a). Incorporation of 3 MPa of methane at 2 °C, just below the threshold pressure for hydrate formation, gives rise to the characteristic peaks of the hexagonal structure of deuterated ice (denoted by I in the spectra), these peaks being more prominent after a subsequent cooling down to −3 °C. Interestingly, no diffraction peaks attributed to

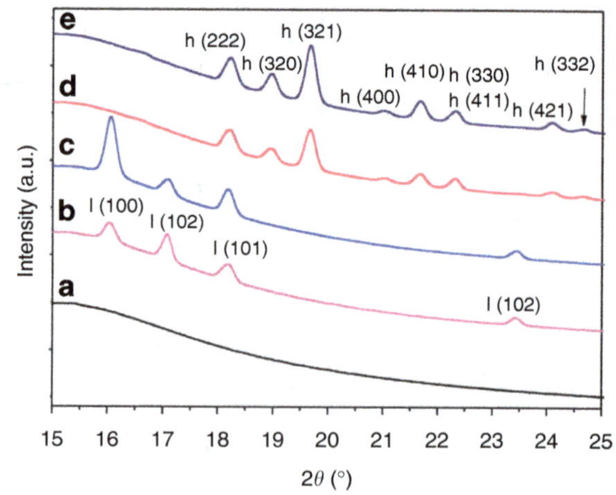

Figure 5 | X-ray powder diffraction of methane hydrates. Synchrotron XRPD patterns of (**a**) wet-deuterated activated carbon PP-AC at 25 °C (in the absence of methane), (**b,c**) wet-deuterated carbon at 3 MPa of methane and 2 and −3 °C, respectively, and (**d,e**) wet-deuterated carbon at 5 MPa of methane and 2 and −3 °C, respectively. Reflections corresponding to ice and hydrate are market as I and h, respectively, followed by their Miller indices.

Table 1 | Methane hydrate stoichiometry.

R_w	n_w(moles)*	N_m(moles)†	$xCH_4 \cdot yH_2O$
Unsaturated samples			
0.6	3.33×10^{-2}	8.08×10^{-3}	$1\ CH_4 \cdot 4.1\ H_2O$
1.1	6.11×10^{-2}	1.70×10^{-2}	$1\ CH_4 \cdot 3.6\ H_2O$
Saturated samples			
1.8	1.01×10^{-1}	2.76×10^{-2}	$1\ CH_4 \cdot 3.7\ H_2O$
Oversaturated samples			
2.9	6.11×10^{-2} (1.1 g of water)	1.16×10^{-2}	$1\ CH_4 \cdot 5.3\ H_2O$
	1.01×10^{-1} (1.8 g of water)	2.45×10^{-2}	$1\ CH_4 \cdot 4.1\ H_2O$
4.1	1.28×10^{-1} (2.3 g of water)	2.21×10^{-2}	$1\ CH_4 \cdot 5.8\ H_2O$
	1.01×10^{-1} (1.8 g of water)	2.43×10^{-2}	$1\ CH_4 \cdot 4.2\ H_2O$

Estimated from the methane adsorption isotherms on sample PP-AC as a function of R_w.
*Moles of water estimated from the corresponding weight increase.
†Moles of methane adsorbed estimated from the corresponding isotherm at 10 MPa (for oversaturated samples an intermediate value has been estimated corresponding to the first jump in the isotherm).

methane hydrate formation were observed under these conditions, in close agreement with the methane adsorption isotherms. The crystallite size of the ice formed was calculated from the peak width after application of the Scherrer equation using a Si standard (NIST material 640c) as a reference. The estimated average crystallite size was 15 nm in both cases, independently of the temperature and the time.

Subsequently, the sample was heated up to 5 °C until complete melting of the ice (no peaks in the spectra could be observed), and then the capillary was pressurized with methane up to 5 MPa (well above the first pressure threshold) and the temperature was lowered to 2 °C. Under these conditions, new XRPD peaks suddenly appeared (in <10 min), corresponding to the sI structure of the methane hydrate[27–29]. The formation process was completed in 30 min, and peak intensities did not change when continuing the process for another 30 min. Interestingly, a subsequent decrease in the capillary temperature down to -3 °C did not produce any change in the resulting pattern, and no additional ice formation could be observed. This observation suggests that most of the water incorporated into the carbon was involved in the methane hydrate phase. The analysis of the peak widths shows an average crystallite size of 16 nm at both temperatures, very close to the ice crystallite size previously formed.

In summary, XRPD measurements constitute the first experimental evidence that methane hydrate sI nanocrystals can be grown in the cavities of activated carbons within minutes by taking advantage of the nanoconfinement effects. The nanocrystals observed under these conditions must correspond to methane hydrates grown in the metastable water layer present on large pores (macropores) with an average crystallite size of 16 nm, which correspond to several unit cells (unit cell of the sI methane structure is 1.2 nm). Gas pressures above 5–6 MPa favours the methane hydrate nucleation and growth in inner cavities (mesopores and wide micropores), with slower kinetics and a marked hysteresis loop, as described above, although these additional nanostructures may not be detected by X-ray measurements due to their limited crystal size in small cavities.

Stoichiometry of confined methane hydrates. Last but not least, the stoichiometry of the methane hydrates synthesized in the confined nanospace has been estimated from the amount of methane adsorbed at 10 MPa for each R_w, taking into account the amount of water adsorbed under each condition. In the case of oversaturated conditions, it is assumed that 1.8 g of water per gram of sample are adsorbed in micro- and mesopores (according to the water vapour adsorption isotherm; see Supplementary Fig. 2), whereas the remaining water is placed in larger pores (macropores). In addition, in the oversaturated samples the amount of methane adsorbed has also been split into the two steps of the adsorption isotherm, so that the stoichiometry of the nanostructures can be estimated in the low and in the high-pressure regime. As it can be observed in Table 1, samples under less saturated conditions exhibit an empirical formula of $1CH_4 \cdot (3.6–4.1)H_2O$. Taking into account that natural hydrates have a stoichiometry of $1CH_4 \cdot 5.75H_2O$, the under-stoichiometric behaviour at these conditions suggests (i) either the formation of non-stoichiometric hydrates in small pores, due to steric hindrance[30] or (ii) the presence of additional free-adsorbed methane in narrow cavities (below 1 nm), non-forming hydrates. Interestingly, the scenario changes completely for oversaturated samples. In these samples, the empirical formula has been estimated for the two steps in the adsorption isotherm (3–4 MPa range and 6–8 MPa range), that is, the amount of methane adsorbed in each step has been considered independently for the calculations as described above. As observed in Table 1, the stoichiometry of the structure formed in the first jump is extremely close to natural hydrates $1CH_4 \cdot (5.3–5.8)H_2O$, corresponding to those hydrates formed in large pores. Concerning the second step in the methane adsorption isotherm (6–8 MPa), a similar under-stoichiometric behaviour is observed, $1CH_4 \cdot (4.1–4.2)H_2O$. Taking into account the slow kinetics for this process together with the non-stoichiometric behaviour, this observation provides more insight into the adsorption mechanism: methane hydrates in large macropores are formed within minutes above 3 MPa, in close agreement with the stoichiometry and formation pressure of natural hydrates, whereas larger pressures (above 6 MPa) and slower sorption kinetics, within hours, identify the methane hydrate nucleation and growth in smaller pores (mainly mesopores and wide micropores). In any case, the presence of free-adsorbed methane in narrow pores cannot be excluded to explain the non-stoichiometric behaviour.

Discussion

We have experimentally identified for the first time the growth of synthetic methane hydrates in the nanospace of carbon materials by using *in situ* INS measurements and synchrotron XRPD. Quantum rotational spectra (elastic and ineslastic scattering of neutrons) perfectly fit those found in natural methane hydrates from the Pacific see-floor, while XRPD measurements clearly identify the sI structure of methane hydrate. Taking advantage of the confinement effect on carbon nanospaces synthetic hydrates can be prepared under less demanding conditions, thus surpassing nature. Confinement effects allow faster growth kinetics (only a few minutes are required to growth these structures) and lower nucleation pressures (below 4 MPa), while keeping the stoichiometry of natural hydrates ($1CH_4 \cdot 5.75\ H_2O$). In addition, synthetic hydrates are fully reversible over the whole-pressure range evaluated, with a large working capacity in a 'narrow' pressure window, that is, 3–10 MPa. The total methane adsorption capacity in these oversaturated systems is extraordinarily large with value around 59 wt.% (0.59 g g^{-1}) and 63 wt.% (0.63 g g^{-1}) for samples with $R_w = 2.9$ and 4.1, respectively, these values being over twofold the methane adsorption capacity of the dry sample. The large adsorption capacity for methane, the fast kinetics, the complete reversibility of the nucleation and growth process and the large working capacity of these systems under a 'narrow' pressure window make methane hydrates grown in confined nanospaces a promising alternative for NG storage and transportation.

Methods

Sample preparation. The nanoporous carbon sample used in the present study was developed using a mesophase pitch prepared by the pyrolysis of a vacuum residue (VR). The pyrolysis conditions (10 bar of N_2 at 460 °C for 1.5 h) were defined to achieve a pitch with 93% mesophase content, labelled PP-AC. After the pyrolysis, the pitch was grounded in a ball mill to obtain a particle size <500 µm. The powder was physically mixed with anhydrous KOH (activating agent) in a 6:1 ratio (wt.%/wt.%) and immediately heat treated under a nitrogen flow (100 ml min^{-1}) at 800 °C for 2 h (heating rate 5 °C min^{-1}). The resulting material was washed, first with a HCl solution (10 wt.%) and finally with distilled water. The final sample, labelled PP-AC, was dried overnight at 100 °C.

To make the hydrate structures, the nanoporous carbon sample was humidified under water-supplying conditions denoted by R_w (R_w, which represents the mass of water to dry carbon). The lower R_w values (0.6, 1.1 and 1.8) were achieved by placing the dry carbon in a close container with 90% relative humidity (relative humidity was obtained using a water solution of 34 wt.% glycerine). The larger R_w values (2.9 and 4.1) were reached by adding drops of water directly to the sample.

Sample characterization. Textural properties of the synthesized samples were evaluated by gas physisorption of nitrogen at −196 °C. Gas adsorption measurements were performed in a homemade fully automated manometric equipment designed and constructed by the Advanced Materials Group (LMA), now commercialized as N_2GSorb-6 (Gas to Materials Technologies; www.g2mtech.com). The sample was previously degassed for 4 h at 250 °C under vacuum (10^{-3} Pa). Nitrogen adsorption data were used to determine the following: (i) the total pore volume (V_t) at a relative pressure of 0.95, (ii) the BET-specific surface area (S_{BET}) and (iii) the micropore volume (V_{N2}) by application of Dubinin–Radushkevich equation. The difference between V_t and V_{N2} is considered to be the mesopores volume (V_{meso}). The pore size distribution was evaluated by applying the quenched solid density functional theory method (slit-shape pore-equilibrium model) to the nitrogen adsorption data. The macroporosity of the samples was tested by Hg porosimetry measurements using an Autopore IV 9500 v1.07 porosimeter (Micromeritics, USA). Water adsorption isotherms were measured at 2 °C in a homemade fully automated equipment designed and constructed by the LMA group, now commercialized as Vstar by Quantachrome Instruments. High-pressure analyses were performed in a homemade fully automated manometric equipment designed and constructed by the LMA group, now commercialized as iSorbHP by Quantachrome Instruments. CH_4 adsorption capacity was measured at 2 °C and up to 10 MPa. Dry samples were outgassed at 250 °C for 4 h, while the wet samples were frozen at −20 °C before the outgassing treatment to avoid water loss. Transmission electron microscopyanalyses were performed on a JEOL JEM-2100 F microscope.

INS experiment. INS experiments were performed at TOSCA spectrometer at the pulse neutron source of ISIS at Rutherford Appleton Laboratory in United Kingdom. Before the experiment, 2 g of dry PP-AC were humidified with deuterated water up to a carbon/water ratio, $R_w = 4.1$. The wet sample (*ca.* 10.2 g) was wrapped in an Al-foil and loaded into the high-pressure reactor supplied by ISIS. The reactor and the stainless steel pipelines were surrounded by a resistance wire that allows a good temperature control. The sample holder was placed in a 2 m reactor where neutron beam converges. Before the analysis, the sample was kept in contact with methane gas at 2 °C for 1 h. Shortly after that the reactor was properly cooled down to −268 °C with a cryostat. Finally, the reactor was impacted by the neutron beam (150 µA) at −268, −253 and −233 °C.

Synchrotron XRPD experiments. XRPD experiments were collected at the high-pressure end station of the MSPD beamline at synchrotron ALBA in Spain, using a Rayonix SX165CCD camera and a wavelength of 0.4243 Å. The experiments were performed on an *ad hoc* capillary reaction cell (fused silica capillary, inner diameter 247 µm, outer diameter 662 µm). Before the experiment, the D_2O-containing carbon sample was placed inside of the capillary connected to the methane gas cylinder (purity 3.5) via a pressure regulator. An Oxford Cryostream 700 was used to control the temperature of the sample. *In situ* XRPD measurements were performed at 0, 3 and 5 MPa of methane and two different temperatures, −3 and 2 °C.

References

1. Sloan Jr., E. D. & Koh, C. A. *Clathrate Hydrates of Natural Gases,* 3rd edn (CRC Press, 2007).
2. Gutt, C. *et al.* The structure of deuterated methane-hydrate. *J. Chem. Phys.* **113,** 4713–4721 (2000).
3. Holbrook, W. S., Hoskins, H., Wood, W. T., Stephen, R. A. & Lizarralde, D. Methane hydrate and free gas on the Blake Ridge from vertical seismic profiling. *Science* **273,** 1840–1843 (1996).
4. Sloan Jr., E. D. Fundamental principles and applications of natural gas hydrates. *Nature* **426,** 353–363 (2003).
5. Rodríguez-Reinoso, F., Almansa, C. & Molina-Sabio, M. Contribution to the evaluation of density of methane adsorbed on activated carbon. *J. Phys. Chem. B* **109,** 20227–20231 (2005).
6. Kockrick, E. *et al.* Ordered mesoporous carbide derived carbons for high pressure gas storage. *Carbon* **48,** 1707–1717 (2010).
7. Klein, N. *et al.* A mesoporous metal-organic framework. *Angew. Chem. Int. Ed.* **48,** 9954–9957 (2009).
8. Makal, T. A., Li, J.-R., Lu, W. & Zhou, H.-C. Methane storage in advanced porous materials. *Chem. Soc. Rev.* **41,** 7761–7779 (2012).
9. Peng, Y. *et al.* Methane storage in metal-organic frameworks: Current records, surprise findings, and challenges. *J. Am.Chem. Soc.* **135,** 11887–11894 (2013).
10. Casco, M. E. *et al.* High-pressure methane storage in porous materials: are carbon materials in the pole position? *Chem. Mater* **27,** 959–964 (2015).
11. Ramos-Fernández, J. M., Martínez-Escandell, M. & Rodríguez-Reinoso, F. Production of binderless activated carbon monoliths by KOH activation of carbon mesophase materials. *Carbon* **46,** 384–386 (2008).
12. Marsh, H. & Rodríguez-Reinoso, F. *Activated Carbon* (Elsevier, 2006).
13. Kubo, T. *et al.* Diffusion-barrier-free porous carbon monoliths as a new form of activated carbon. *ChemSusChem* **5,** 2271–2277 (2012).
14. Kaneko, K., Itoh, T. & Fujimori, T. Collective interactions of molecules with an interfacial solid. *Chem. Lett.* **41,** 466–475 (2012).
15. Nakamura, M., Ohba, T., Branton, P., Kanoh, H. & Kaneko, K. Equilibrium-time and pore-width dependent hysteresis of water adsorption isotherm on hydrophobic microporous carbons. *Carbon* **48,** 305–308 (2010).
16. Vysniauskas, A. & Bishnoi, P. R. A kinetic study of methane hydrate formation. *Chem. Eng. Sci.* **38,** 1061–1072 (1983).
17. Junhong, Q. & Tianmin, G. Kinetics of methane hydrate formation in pure water and inhibitor containing systems. *Chin. J. Chem. Eng* **10,** 316–322 (2002).
18. Liu, J., Zhou, Y., Sun, Y., Su, W. & Zhou, L. Methane storage in wet carbon of tailored pore sizes. *Carbon* **49,** 3731–3736 (2011).
19. Perrin, A., Celzard, A., Marêché, J. F. & Furdin, G. Methane storage within dry and wet activated carbons: a comparative study. *Energy Fuels* **17,** 1283–1291 (2003).
20. Zhou, L., Liu, L., Su, W., Sun, Y. & Zhou, Y. Progress in studies of natural gas storage with wet adsorbents. *Energy Fuels* **24,** 3789–3795 (2010).
21. Celzard, A. & Marêché, J. F. Optimal wetting of activated carbons for methane hydrate formation. *Fuel* **85,** 957–966 (2006).
22. Webb, E. B. *et al.* High pressure rheology of hydrate slurries formed from water-in-oil emulsions. *Energy Fuels* **26,** 3504–3509 (2012).
23. Urita, K. *et al.* Confinement in carbon nanospace-induced production of KI nanocrystals of high-pressure phase. *J. Am. Chem. Soc.* **133,** 10344–10347 (2011).
24. Fujimori, T. *et al.* Conducting linear chains of sulphur inside carbon nanotubes. *Nat. Commun.* **4,** 2162 (2013).
25. Tse, J. S., Ratcliffe, C. L., Powell, B. M., Sears, V. F. & Handa, Y. P. Rotational and translational motions of trapped methane. Incoherent inelastic neutron scattering of methane hydrate. *J. Phys. Chem. A* **101,** 4491–4495 (1997).
26. Gutt, C. *et al.* Quantum rotations in natural methane-clathrates from the Pacific sea-floor. *Europhys. Lett.* **48,** 269–275 (1999).
27. Stern, L. A., Kirby, S. H. & Durham, W. B. Peculiarities of methane clathrate hydrate formation and solid-state deformation, including possible superheating of water ice. *Science* **273,** 1843–1848 (1996).
28. Gutt, C. *et al.* The structure of deuterated methane hydrate. *J. Chem. Phys.* **113,** 4713–4721 (2000).

29. Everett, S. M. *et al.* Kinetics of methane hydrate decomposition studies via in situ low temperature X-ray powder diffraction. *J. Phys. Chem. A* **117,** 3593–3598 (2013).
30. Miyawaki, J. *et al.* Macroscopic evidence of enhanced formation of methane nanohydrates in hydrophobic nanospaces. *J. Phys. Chem. B* **102,** 2187–2192 (1998).

Acknowledgements

We acknowledge UK Science and Technlology Facilities Council for the provision of beam time on the TOSCA spectrometer (Projects RB1410624 and RB122099) and financial support from the European Commission under the 7th Framework Programme through the 'Research Infrastructures' action of the 'Capacities' Programme (NMI3-II Grant number 283883). J.S.-A. and F.R. acknowledges the financial support from MINECO: Strategic Japanese-Spanish Cooperation Program (PLE2009-0052), Concert Project-NASEMS (PCIN-2013-057) and Generalitat Valenciana (PROMETEO/2009/002). F.R. and J.L.J. thank the financial support from MINECO (MAT2012-38567-C02-01, Consolider Ingenio 2010-Multicat CSD-2009-00050 and SEV-2012-0267). K.K. thanks Grant-in-Aid for Scientific Research (A) (2424-1038), Japan. A.B. and A.U. thank the financial support from MINECO (SEV-2013-0319). J.L.J. and I.P. thank synchrotron ALBA for beamtime availability.

Author contributions

M.E.C. prepared the sample and performed the experiments and analysis. J.S.-A., M.M.-E., K.K. and F.R.-R. conducted the research. M.E.C., J.S.-A., A.J.R.-C. F.R. designed and performed the experiment at ISIS. Among these authors A.J.R.-C. was in-charge of TOSCA instrument. J.L.J. and I.P. were responsible for the XRPD measurements in ALBA. A.B. and A.U. designed the high-pressure cell for XRPD. M.E.C. and J.S.-A wrote the paper. All the authors critically revised and approved the manuscript.

Additional information

Competing financial interests: The authors declare no competing financial interests.

4

Porous coordination polymers with ubiquitous and biocompatible metals and a neutral bridging ligand

Shin-ichiro Noro[1,2,3,4], Junya Mizutani[2], Yuh Hijikata[5], Ryotaro Matsuda[3,6], Hiroshi Sato[6], Susumu Kitagawa[6], Kunihisa Sugimoto[7], Yasutaka Inubushi[8,†], Kazuya Kubo[1,2] & Takayoshi Nakamura[1,2]

The design of inexpensive and less toxic porous coordination polymers (PCPs) that show selective adsorption or high adsorption capacity is a critical issue in research on applicable porous materials. Although use of Group II magnesium(II) and calcium(II) ions as building blocks could provide cheaper materials and lead to enhanced biocompatibility, examples of magnesium(II) and calcium(II) PCPs are extremely limited compared with commonly used transition metal ones, because neutral bridging ligands have not been available for magnesium(II) and calcium(II) ions. Here we report a rationally designed neutral and charge-polarized bridging ligand as a new partner for magnesium(II) and calcium(II) ions. The three-dimensional magnesium(II) and calcium(II) PCPs synthesized using such a neutral ligand are stable and show selective adsorption and separation of carbon dioxide over methane at ambient temperature. This synthetic approach allows the structural diversification of Group II magnesium(II) and calcium(II) PCPs.

[1] Research Institute for Electronic Science, Hokkaido University, Sapporo 001-0020, Japan. [2] Graduate School of Environmental Earth Science, Hokkaido University, Sapporo 060-0810, Japan. [3] PRESTO, Japan Science and Technology Agency (JST), 4-1-8 Honcho, Kawaguchi 332-0012, Japan. [4] Creative Research Institute (CRIS), Hokkaido University, Sapporo 001-0021, Japan. [5] Institute of Transformative Bio-Molecules (WPI-ITbM), Nagoya University, Chikusa-ku, Nagoya 464-8602, Japan. [6] Institute for Integrated Cell-Material Science (WPI-iCeMS), Kyoto University, Kyoto 615-8510, Japan. [7] Research and Utilization Division, Japan Synchrotron Radiation Research Institute, Hyogo 679-5198, Japan. [8] Synthesis Research Laboratory, Kurashiki Research Center, Kuraray Co. Ltd., Sakazu, Kurashiki, Okayama 710-0801, Japan. † Present address: EVAL Research and Development Department, EVAL Division, Vinyl Acetate Company, Kuraray Co. Ltd., 7471, Tamashimaotoshima, Kurashiki, Okayama 713-8550, Japan. Correspondence and requests for materials should be addressed to S.-i.N. (email: noro@es.hokudai.ac.jp) or to T.N. (email: tnaka@es.hokudai.ac.jp).

Porous coordination polymers (PCPs) or metal-organic frameworks (MOFs) constructed from metal ions and organic bridging ligands have attracted much attention as intriguing porous materials because of their high structural regularity and diversity, easy modification of frameworks, high porosity and structural flexibility[1–13]. One of the main trends in work on these materials is the development and improvement of their porous functions, such as storage[14,15], separation[16–18] and catalysis[19–21]. Towards industrial and biological applications, however, we must consider costs and safety of PCPs in addition to their porous performances. Generally, first-transition-row divalent metal ions such as Co(II), Ni(II), Cu(II) and Zn(II) have been utilized as metal sources of PCPs. On the other hand, Group II metal ions, Mg(II) and Ca(II), belonging to the light metal ions are in abundant supply on the Earth and biofriendly[22,23]. If Group II Mg(II) or Ca(II) ions are embedded in PCP frameworks instead of first-transition-row metal ions, the resulting PCPs could become cheaper and have higher biocompatibility.

Nevertheless, the examples of Mg(II) and Ca(II) PCPs are extremely limited to date compared with transition metal PCPs[23–29], because neutral bridging ligands have not been available for the construction of Group II Mg(II) and Ca(II) PCPs. Transition metal ions are classed as medium acids according to the hard and soft acids and bases theory. According to this theory, a variety of organic bridging ligands containing anionic carboxylate or imidazolate, and neutral pyridine or imidazole units, tend to be coordinated to them, which makes the PCP frameworks diversified. Furthermore, the combination of anionic and neutral organic bridging ligands is much more effective in diversifying transition metal PCP frameworks. By contrast, Mg(II) and Ca(II) ions are hard acids and, therefore, organic ligands with hard basicity, that is, anionic multicarboxylates have been their sole partners. This is one of the reasons why Mg(II) and Ca(II) PCPs have not been thoroughly investigated.

To overcome this problem, we introduce a new partner, 4,4′-bipyridine-*N*,*N*′-dioxide (bpdo), which is a neutral and charge-polarized bridging ligand with pyridine-*N*-oxide (pyo) moieties, to Group II Mg(II) and Ca(II) ions. Although the structure of the neutral bpdo ligand is similar to that of the neutral ligand 4,4′-bipyridine (4,4′-bpy), we expect that it can coordinate to hard Group II Mg(II) and Ca(II) ions through negatively charged oxygen atoms with hard basicity, despite a totally neutral ligand (Fig. 1a). In this study, using the neutral and charge-polarized bridging ligand bpdo, we demonstrate rational design and syntheses of Mg(II) and Ca(II) PCPs connected by neutral organic ligands. The resulting Mg(II) and Ca(II) PCPs, which are the uncommon examples including neutral bridging ligands, are stable after the removal of guest molecules and show selective adsorption and separation of CO_2 over CH_4 at ambient temperature. Our finding contributes to the structural diversification of PCP frameworks with ubiquitous and biocompatible metals.

Results

Charge distribution. First, we evaluated the hard basicity of the bpdo ligand. Generally, the bond between a hard acid and a hard base is dominated by an ionic interaction. The bpdo ligand has two coordinated oxygen atoms, which are better coordination sites for hard acids than nitrogen atoms because of their larger negative charge. We calculated the charge distribution of a bpdo ligand with other typical and similar organic ligands, pyo, 4,4′-bpy, benzoate (bza^-) and 4,4′-biphenyldicarboxylate (4,4′-$bpdc^{2-}$) to compare the strengths of ionic interactions. It is well known that pyo and their derivatives show a variety of resonance structures. We investigated which resonance structures mainly contribute to the actual structure. Figure 1b shows five resonance structures of a pyo derivative and Fig. 1c shows electron occupation numbers for each bond and natural bond orbital (NBO) charges for each atom in bpdo, which were obtained by the density functional theory. The negative charges are localized on the oxygen atoms, and little π-bond character is observed between the nitrogen and oxygen atoms, supporting the main contribution of the resonance structures (I), (II) and (III). Figure 1d and Supplementary Fig. 1 show NBO charges. The oxygen atoms of the neutral bpdo and pyo ligands have negative charges of -0.543 e and -0.568 e, respectively, which are smaller than those for anionic carboxylate-type bza^- and 4,4′-$bpdc^{2-}$ ligands (-0.795 e and -0.805 e, respectively) but larger than the negative charges of the nitrogen atoms of the neutral 4,4′-bpy ligand (-0.446 e). All calculation results conclude that the neutral ligands with pyo moieties should act as stronger hard bases than 4,4′-bpy and be good partners for hard Group II Mg(II) and Ca(II) ions.

Structural characterization. To experimentally prove this new strategy, we next attempted to synthesize Group II Mg(II) and Ca(II) coordination polymers using the neutral bpdo ligand. The reaction of $Mg(AcO)_2\bullet 4H_2O$ with bpdo yielded the two-dimensional coordination polymer $[Mg_2(AcO)_4(bpdo)]_n$ (**1**). Further, the reaction of $MgCl_2\bullet 6H_2O$ or $Ca(NO_3)_2\bullet 4H_2O$ with 1,4-benzenedicarboxylic acid (1,4-H_2bdc) and bpdo in *N*,*N*-dimethylformamide (DMF) afforded the three-dimensional PCPs, $\{[Mg_2(1,4\text{-bdc})_2(bpdo)]\bullet 2DMF\}_n$ (**2** ⊃ 2DMF) and $\{[Ca(1,4\text{-bdc})(bpdo)]\bullet 0.5DMF\}_n$ (**3** ⊃ 0.5DMF). As expected, **1**, **2** ⊃ 2DMF and **3** ⊃ 0.5DMF included the neutral bpdo ligands that bridge the metal centres. The crystal structures of **1**, **2** ⊃ 2DMF and **3** ⊃ 0.5DMF were determined using single-crystal X-ray crystallography at 173 K. **1** has an octahedral Mg(II) environment with four AcO^- oxygen atoms and two bpdo oxygen atoms. The Mg–O(AcO^-) bond distances (2.017(6)–2.085(6) Å) are slightly shorter than the Mg–O(bpdo) bond distances (2.154(4) and 2.161(4) Å). The Mg(II) centres are bridged by two carboxylate parts and one pyo part to form one-dimensional Mg(II) chains with the corner-shared MgO_6 octahedral units running along the *b*-axis. These one-dimensional chains are further linked by bpdo ligands, resulting in the two-dimensional coordination network (Fig. 2a).

In **2** ⊃ 2DMF, the Mg(II) centre has an octahedral environment with four 1,4-bdc^{2-} oxygen atoms and two bpdo oxygen atoms (Fig. 2b). The corner-shared MgO_6 octahedral units interconnected by both organic moieties form infinite chains (Supplementary Fig. 2). These chains are further bridged by both bpdo and 1,4-bdc^{2-} ligands, affording a three-dimensional porous coordination framework with one-dimensional channels (Fig. 2c, channel size: 4.5×4.1 Å^2) occupied by guest DMF molecules. The accessible volume of the fully desolvated **2** is ca 35.6%, which was calculated using the PLATON programme (probe radius: 1.2 Å)[30].

In **3** ⊃ 0.5DMF, the Ca(II) centre has an octahedral environment with four 1,4-bdc^{2-} oxygen atoms and two bpdo oxygen atoms arranged in a *cis* conformation (Fig. 2d). The CaO_6 octahedral units are interconnected by two carboxylate parts to form one-dimensional Ca(II) chains (Supplementary Fig. 3). These chains are bridged by 1,4-bdc^{2-} ligands, affording two-dimensional layers parallel to the *ac* plane. These layers are further connected by bpdo ligands, forming a three-dimensional porous coordination framework with one-dimensional channels (Fig. 2e, channel size: 3.4×3.2 Å^2) that contain guest DMF

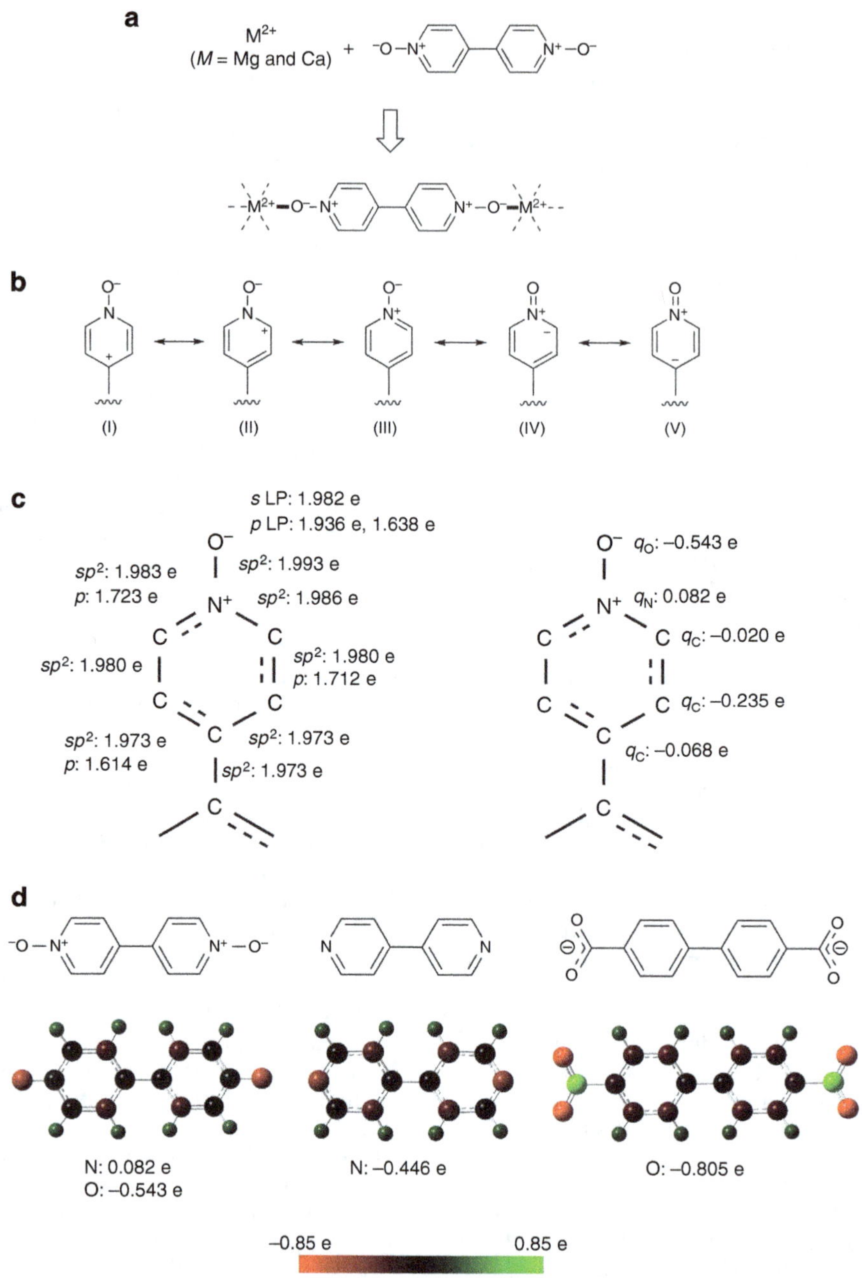

Figure 1 | Designing organic ligands for magnesium(II) and calcium(II) porous coordination polymer. (**a**) New neutral partner, 4,4′-bipyridine-*N*,*N*′-dioxide (bpdo), for hard Group II Mg(II) and Ca(II) ions. (**b**) Five resonance structures in a pyridine-*N*-oxide derivative. (**c**) Electron occupation numbers for each bond and natural bond orbital (NBO) charges for each atom in bpdo. (**d**) NBO charges for selected atoms in organic bpdo, 4,4′-bipyridine and 4,4′-biphenyldicarboxylate ligands.

molecules. From the calculation using the PLATON programme, the accessible volume of the fully desolvated **3** was found to be ca 21.1%, smaller than that in **2** (ref. 30).

It was confirmed from theoretical and experimental results that the neutral organic bpdo ligand has hard basicity and is a good partner to hard Group II Mg(II) and Ca(II) ions. It should be noted that the NBO charge of oxygen atoms in the bpdo ligand (-0.543 e) is considerably smaller than -1 e, which is the maximum value when only the resonance structures of (I), (II) and (III) contribute to the electronic structure of bpdo. This means that the negative charge on the oxygen atoms of the bpdo ligand may further increase by ionic interactions with cationic Group II metal ions. To clarify this, infrared spectra were measured (Supplementary Figs 4 and 5); these showed that the N–O stretching band of bpdo in **2** $\supset$ 2DMF (1,225 cm^{-1}) shifts to a lower wavenumber than that of free bpdo ligand (1,242 cm^{-1}), suggesting an enhanced single bond character and negative charge accumulation by coordination to Mg(II) ions. Furthermore, the NBO charge of the model structure of **2**, $[Mg_4(bza)_4(OH)_4(bpdo)(H_2O)_8]$ (Supplementary Fig. 6), was calculated. As shown in Supplementary Fig. 7, the coordinated bpdo ligand has oxygen atoms with an NBO charge of -0.660 e, 21.5% higher than that of coordination-free bpdo, whereas the NBO charge of the oxygen atoms in the carboxylate ligand, bza^{-}, is only changed by -1.1 to 2.1% before and after the coordination, indicating that the hard basicity of the bpdo ligand indeed increases by coordination to cationic Group II Mg(II) ions. The N–O bond of bpdo in the model structure increases in length by further negative charge accumulation on bpdo oxygen atoms (Supplementary Table 1). On the other hand, the

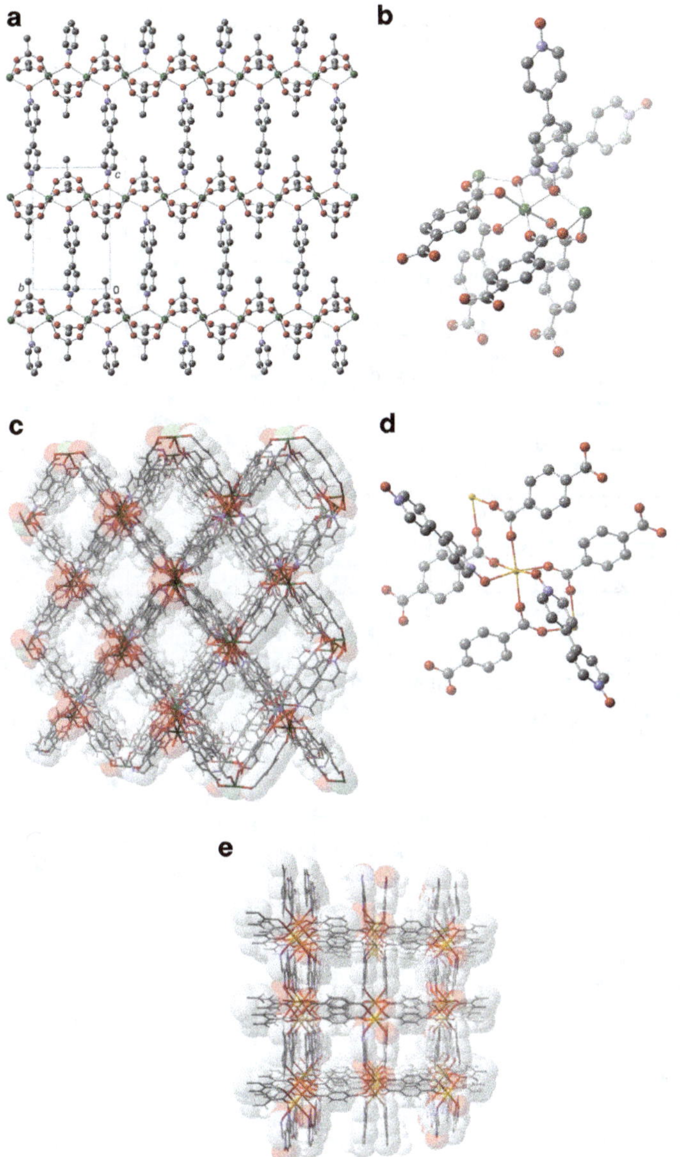

Figure 2 | Crystal structures. (**a**) Two-dimensional structure of **1**. (**b**) Coordination environment around the Mg(II) centre in **2** ⊃ 2DMF (DMF = *N,N*-dimethylformamide). (**c**) Porous structure in **2** ⊃ 2DMF (the guest DMF molecules are omitted). (**d**) Coordination environment around the Ca(II) centre in **3** ⊃ 0.5DMF. (**e**) Porous structure in **3** ⊃ 0.5DMF (the guest DMF molecules are omitted). In **a**,**b** and **d**, the hydrogen atoms are omitted for clarity. Green represents magnesium; orange, calcium; red, oxygen; blue, nitrogen; grey, carbon and white, hydrogen.

corresponding N–O stretching band position in **3** ⊃ 0.5DMF (1,240 cm^{-1}) is almost the same as that of the free bpdo ligand, because a Ca(II) ion (effective ion radius: 1.00 Å (ref. 31)) with lower charge density than a Mg(II) ion (effective ion radius: 0.720 Å (ref. 31)) cannot induce further charge accumulation. These results indicate that **2** has more polar channels than **3**, which affects the adsorption properties (vide infra).

Framework stability. To evaluate the framework stability, thermogravimetry-differential thermal analyses (TG-DTA) and powder X-ray diffraction (XRD) analyses of the desolvated **2** and **3** were performed. **2** was prepared by heating the EtOH-exchanged sample at 373 K under vacuum, whereas **3** was obtained by the direct removal of DMF molecules from **3** ⊃ 0.5DMF at 423 K under vacuum. TG-DTA analysis suggests that **2** and **3** are stable up to 573 and 473 K, respectively (Supplementary Figs 8 and 9). The powder XRD patterns of the desolvated **2** and **3** (Supplementary Figs 10 and 11) show that the original frameworks are stable after the removal of guest molecules.

Porosity. We first measured the adsorption/desorption isotherms for N_2 and CO_2 at low temperature to investigate the fundamental porous properties. **1** shows no CO_2 adsorption at 195 K (Supplementary Fig. 12), consistent with the fact that **1** has no permanent pores in the X-ray crystal structure. In the PCPs **2** and **3**, CO_2 and N_2 gases are adsorbed (Supplementary Figs 13 and 14). The N_2 adsorption isotherms show type-I curves, indicative of permanent micropores. The Dubinin–Radushkevich plots give the micropore volumes of 0.19 and 0.085 $cm^3 g^{-1}$ for **2** and **3**, respectively. The CO_2 adsorption/desorption isotherms in **3** also show type-I curves, whereas **2** shows a stepwise adsorption curve (Supplementary Fig. 14). Coincident infrared/adsorption measurements were performed for various CO_2 pressures at 195 K to confirm the structural change. If the structural change occurs, an increase in infrared intensity at one point should be observed, with a decrease in infrared intensity at another point. In **3**, the intensities of bands derived from the framework decrease with increasing adsorption (Supplementary Fig. 15), whereas the intensities of some bands increase at first and second adsorption events in **2** (Supplementary Fig. 16), suggesting the structural change of **2** associated with first and second CO_2 adsorption. To investigate the structural changes of **2** in more detail, coincident XRD/adsorption measurements were conducted. Upon CO_2 adsorption and desorption, pronounced changes in the XRD patterns were observed (Fig. 3 and Supplementary Fig. 17). The XRD pattern does not change dramatically in the initial CO_2 adsorption (ads1 to ads5). After ads5, the XRD pattern changes and the structural change finishes at ads10. The structure again begins to change after ads12, and the peak positions in the XRD pattern at ads19 where the second adsorption event is completed are similar to those of the desolvated **2**. Further gradual CO_2 adsorption after ads19 leads to the slight structural change. In the desorption process, the reverse change in the XRD patterns was observed. Because the crystal structure and the change in the XRD patterns upon CO_2 adsorption and desorption in **2** are similar to those observed in $[Cr(OH)(1,4\text{-bdc})]_n$ (MIL-53)[32], we speculate that the pore contraction and re-expansion occur during CO_2 adsorption/desorption in **2** (Supplementary Fig. 18).

Adsorption selectivity. The low-pressure CO_2, CH_4, N_2, O_2, Ar and H_2 adsorption isotherms for **2** at 298 K are compared in Fig. 4a. **2** adsorbs very small amounts of N_2, O_2, Ar and H_2, and a small amount of CH_4 at 298 K. On the other hand, the uptake of CO_2 at 97 kPa is ~12 times higher than the uptake of N_2. The effective interaction with polar CO_2 gas causes a high selectivity for the adsorption of CO_2 over other gases, which is a prerequisite for the application of a separation material. The adsorption of gas mixtures in porous materials can be reliably estimated from the ideal adsorbed solution theory (IAST), which is a precise method used to describe gas-mixture adsorption in representative zeolites and PCPs from the experimental single-component gas adsorption isotherms[33]. The predicted adsorption selectivities for CO_2/CH_4 and CO_2/N_2 for a bulk gas composition of $CO_2:CH_4 = 40:60$ (typical composition of biogas) and $CO_2:N_2 = 10:90$ (typical composition of flue gas) are 21–16 and 87–67, respectively (Fig. 4b), high enough (> 8) for the potential feasibility of the practical procedure[34]. The IAST selectivity for other compositions (including typical compositions in the air or

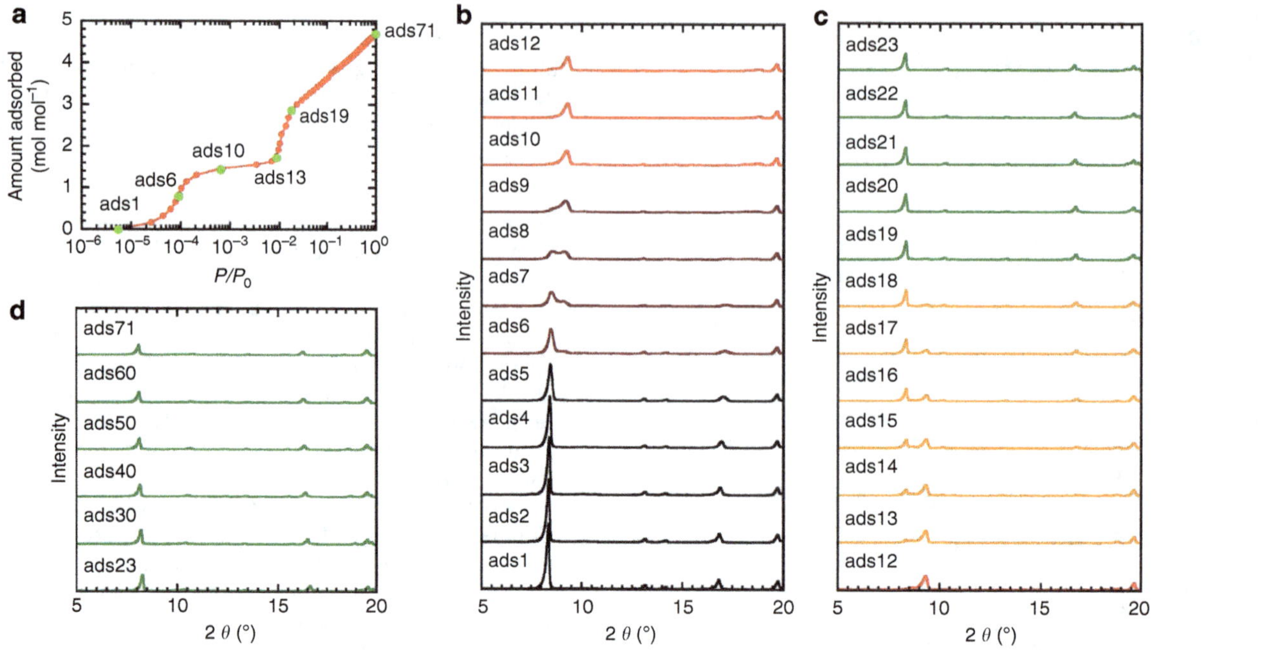

Figure 3 | Coincident X-ray diffraction (XRD)/adsorption measurements of 2. (**a**) CO_2 adsorption isotherm at 195 K. (**b–d**) XRD patterns measured at each point (ads1–ads23, ads30, ads40, ads50, ads60 and ads71).

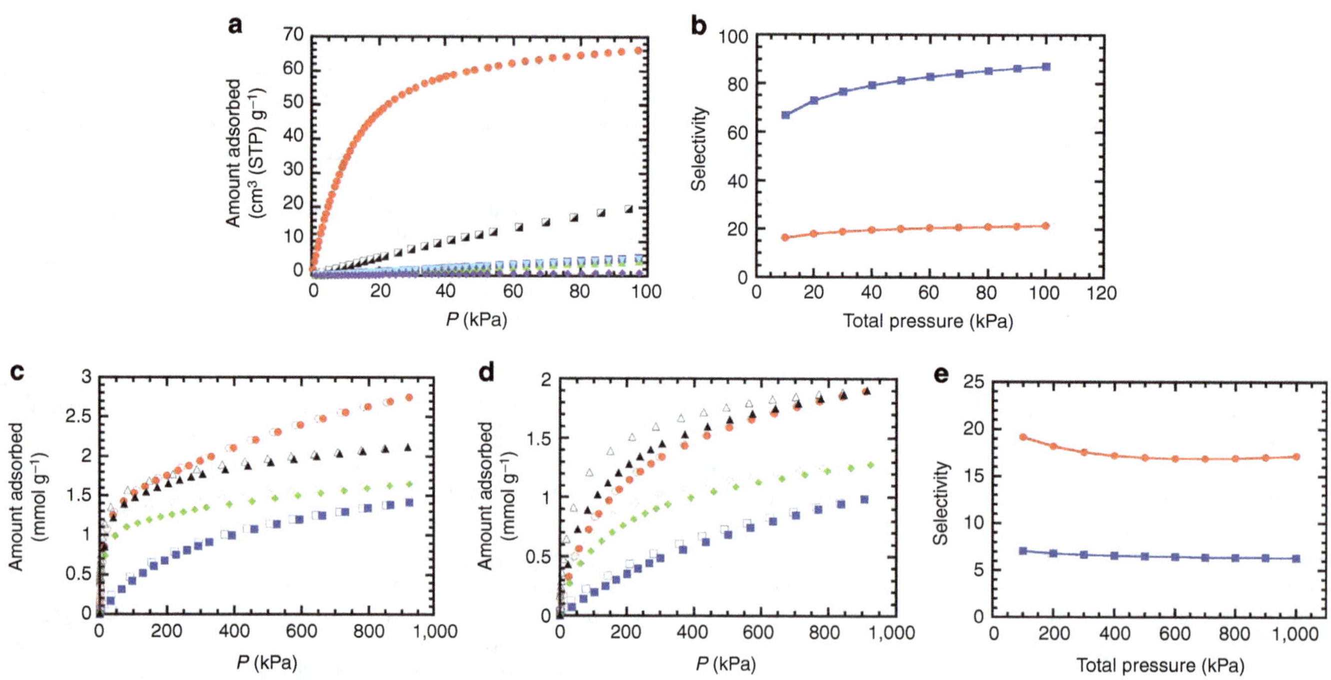

Figure 4 | Gas adsorption properties. (**a**) Low-pressure adsorption isotherms for CO_2 (red circle), CH_4 (black half-filled square), N_2 (blue square), O_2 (green triangle), Ar (sky-blue inverted triangle) and H_2 (purple rhombus) in **2** at 298 K. Because the amounts adsorbed for N_2, O_2 and Ar are almost the same, the symbols overlap. (**b**) Ideal adsorbed solution theory (IAST)-predicted selectivity for CO_2/CH_4 (red) and CO_2/N_2 (blue) in **2** based on the data measured at 298 K for a bulk gas composition of CO_2:CH_4 = 40:60 (typical composition of biogas) and CO_2:N_2 = 10:90 (typical composition of flue gas). (**c,d**) High-pressure total adsorption/desorption isotherms for CO_2 (red circle), CH_4 (blue square), C_2H_6 (green rhombus), and C_2H_4 (black triangle) in **2** (**c**) and **3** (**d**) at 298 K. Adsorption, closed symbols; desorption, open symbols. (**e**) IAST-predicted selectivity for CO_2/CH_4 in **2** (red) and **3** (blue) based on the data measured at 298 K for a bulk gas composition of CO_2:CH_4 = 40:60 (typical composition of biogas).

ternary mixture) is also high (Supplementary Figs 19–24). In addition, high-pressure adsorption measurements were performed at 298 K (Fig. 4c,d), which confirmed the favourable adsorption of CO_2 and C2 hydrocarbons (C_2H_6 and C_2H_4) relative to CH_4 (see Supplementary Figs 25 and 26). The IAST selectivity of CO_2 relative to CH_4 is still larger than 8 for **2** and slightly smaller than 8 for **3** (Fig. 4e), indicating the practical implementability for biogas purification (vide infra).

Recently, Snurr *et al.* have used five adsorbent evaluation criteria from the chemical engineering literature for the potential use of PCPs/MOFs in CO_2 separation processes[35]. We focus on flue gas separation, which is currently an important research area. The typical composition of flue gas, that is, the CO_2/N_2 ratio, is assumed to be 10:90 and the adsorption/desorption pressures are set to 100/10 kPa, respectively. Based on this condition, **2** was evaluated and compared not only with other PCPs/MOFs but also with commercially available inorganic and organic adsorbents (see Supplementary Table 2). As shown in Supplementary Table 2, **2** has high working capacity (or regenerability) and selectivity under these conditions, resulting in a high sorbent selection parameter (*S*) that combines the working capacity and the adsorption selectivity. The *S* value obtained (314) is considerably higher than those of zeolites (163 for zeolite 5A and 128 for zeolite 13X) and one of the best among the PCPs/MOFs.

The regenerability of the CO_2 adsorption process in **2** was measured at 298 K (Supplementary Fig. 27). Significantly, the CO_2 adsorption ability of **2** is maintained over repeated cycling, and the material can be regenerated by short vacuum processing without additional heating, in contrast to zeolites, which require high temperatures for complete regeneration (Supplementary Figs 28 and 29). The isosteric heat of CO_2 adsorption, Q_{st}, shows values 32–36 kJ mol^{-1} (zero-coverage $Q_{st} = 34.9$ kJ mol^{-1}, Supplementary Fig. 30), which are smaller than those of amine-functionalized PCPs and zeolites[36,37]. Such a moderate Q_{st} value is a strong advantage for the implementation of low-energy regeneration for CO_2 separation.

To evaluate the gas separation ability of adsorbents, it is important to study not only the single-gas adsorption properties and IAST simulation under equilibrium conditions, but also mixed gases under flowing (kinetic) conditions. Thus, we conducted a breakthrough experiment, which is a typical way of evaluating the gas separation ability of adsorbents under flowing gas conditions that are related to the pressure swing adsorption (PSA) process. We measured the gas separation properties of **2** and **3** for the CO_2:CH_4 = 40:60 mixture at 298 K and under a total pressure of 0.8 MPa. The separation of CO_2 from CO_2/CH_4 mixtures in biogas has become increasingly important in recent times. The PSA process is a well-known technology and the selective binding of the target gas and gas adsorption kinetics are important factors in the design of adsorbent materials. PSA typically runs at ambient temperature and under a total pressure of 0.4–0.8 MPa, and the cycle time is several minutes. As shown in Fig. 5, the gas detected first by gas chromatography was CH_4 only, with no detection of CO_2, indicating high selectivity for CO_2 over CH_4 under flowing conditions. After a few minutes, the process reached the breakpoint, and then returned to the original gas ratio. It therefore seems that **2** and **3** have favourable characteristics for CO_2 capture, yielding good separation behaviour.

Although high water stability and hydrophobicity are also one of important factors in using gas separation materials under practical conditions, syntheses of PCPs/MOFs showing high CO_2 selectivity and no degradation of frameworks under humid environments have been just started[38–40]. Here we tested the stability of frameworks after exposure of **2** and **3** to water vapour. **2** adsorbs a large amount of H_2O (Supplementary Fig. 31), and the adsorbed amount of CO_2 in **2** after the H_2O adsorption shows a decrease of 65% compared with the as-synthesized **2** (Supplementary Fig. 32), suggesting that **2** is not stable in humid conditions. On the other hand, a few amount of H_2O is adsorbed to **3** at 37% relative humidity, and the adsorbed amount of CO_2 is unchanged after the H_2O adsorption experiment (Supplementary Fig. 33), indicating higher water stability of **3** than **2**. However, it is necessary to further improve the stability to water and evaluate gas separation properties under humid conditions.

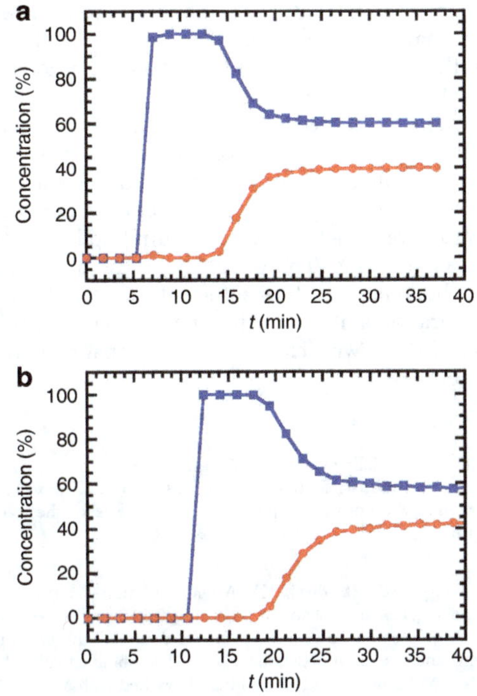

Figure 5 | Gas separation performance. (**a**,**b**) Breakthrough curves for a CO_2/CH_4 mixture (40:60 (vol)) for **2** (**a**) and **3** (**b**). The red circle is CO_2 and the blue square is CH_4. These were measured at 298 K, the total pressure was 0.8 MPa and the space velocity was 3 min^{-1}.

Discussion

To our knowledge, almost all PCPs with Group II Mg(II) and Ca(II) ions contain only anionic organic ligands; there are few examples including bridging neutral organic ligands. **2** and **3** are the uncommon examples of Group II Mg(II) and Ca(II) PCPs including bridging neutral organic ligands. Note that we can use a variety of neutral organic ligands with pyo moieties, as shown in Supplementary Fig. 34a. In contrast to a monodentate neutral pyridine ligand, even the most simple pyo ligand is capable of bridging metal cations because the coordinated oxygen has two lone-pair electrons. Furthermore, pyo ligands with carboxylate substituents (for example, pyridine-4-carboxylate-*N*-oxide; see Supplementary Fig. 34b) are also useful building blocks for the construction of Group II Mg(II) and Ca(II) PCPs. These considerations strongly suggest that our proposed strategy to use pyo-type ligands will be quite effective for the diversification of Group II Mg(II) and Ca(II) PCP frameworks. We can expect the pyo-type ligands to also be good partners to monovalent Group I alkali Na(I) and K(I) ions. Although successful studies on syntheses of coordination polymers with Group I alkali metal ions have recently been performed using the neutral cyclodextrin and polyether ligands[22,41–43], there are still few PCPs containing Group I alkali Na(I) and K(I) ions. Their hard acidity impedes the coordination of neutral organic ligands, and the monovalence restricts the number of anionic multicarboxylate ligands (such as 1,4-bdc^{2-} and 1,3,5-benzenetricarboxylate) coordinating to one Group I alkali Na(I) and K(I) ion, which causes difficulty in the formation of porous frameworks. Because of their neutrality and hard basicity, the pyo-type ligands are good candidates to resolve this problem. Therefore, we conclude that the structural diversification of

PCPs with Group I Na(I) and K(I) ions and Group II Mg(II) and Ca(II) ions that are more abundant on the Earth and more biofriendly than commonly used transition metal ions would be realized by utilizing pyo-type ligands.

In conclusion, we succeeded in finding a new partner, charge-polarized neutral 'bpdo' ligand, for Group II Mg(II) and Ca(II) ions in order to realize the structural diversification of Mg(II) and Ca(II) PCPs. The bpdo ligand has hard basicity, derived from its polarized structure, which causes the neutral bpdo to bridge hard Mg(II) and Ca(II) ions. The obtained Mg(II) and Ca(II) PCPs with neutral bpdo ligands show sufficient framework stability and good gas separation ability. Our finding opens up the science and engineering of PCPs with cheap and biocompatible Group I and Group II metal ions.

Methods

Materials. All commercially available starting materials were purchased from Wako Pure Chemical Industries, Ltd., and used as received. The solvents for the syntheses were used without further purification. bpdo was synthesized according to the literature[44].

Synthesis of [$Mg_2(AcO)_4$(bpdo)]$_n$ (1). A DMF solution (5 ml) of $Mg(AcO)_2$•$6H_2O$ (107 mg, 0.50 mmol) and bpdo (94 mg, 0.50 mmol) was heated at 423 K for 6 h. The obtained colourless microcrystals were filtered, dispersed in a DMF solution (20 ml) and heated at 373 K for 10 min. The obtained colourless microcrystals of [$Mg_2(AcO)_4$(bpdo)]$_n$ (**1**) were filtered, washed with MeOH and dried at 298 K under vacuum for 2 h. Yield: 97 mg, 82%. Elemental analysis: Calcd for **1** ($C_{18}H_{20}Mg_2N_2O_{10}$): C = 45.71; H = 4.26; N = 5.92. Found: C = 44.10; H = 4.32; N = 5.44%.

Synthesis of {[Mg_2(1,4-bdc)$_2$(bpdo)]•2DMF}$_n$ (2 ⊃ 2DMF). A DMF solution (5 ml) of $MgCl_2$•$6H_2O$ (203 mg, 1.00 mmol), 1,4-H_2bdc (166 mg, 1.0 mmol) and bpdo (94 mg, 0.50 mmol) was heated at 423 K for 24 h. The obtained colourless microcrystals of {[Mg_2(1,4-bdc)$_2$(bpdo)]•2DMF}$_n$ (**2** ⊃ 2DMF) were filtered, washed with DMF and dried at 298 K under vacuum for 2 h. Yield: 292 mg, 41%. Elemental analysis: Calcd for **2** ⊃ 1.8DMF•0.6H_2O ($C_{31.4}H_{29.8}Mg_2N_{3.8}O_{12.4}$): C = 53.31; H = 4.25; N = 7.52. Found: C = 52.69; H = 4.43; N = 7.63%.

Synthesis of {[Ca(1,4-bdc)(bpdo)]•0.5DMF}$_n$ (3 ⊃ 0.5DMF). A DMF solution (40 ml) of $Ca(NO_3)_2$•$4H_2O$ (708 mg, 3.00 mmol), 1,4-H_2bdc (498 mg, 3.00 mmol) and bpdo (1.13 g, 6.00 mmol) was heated at 423 K for 48 h. The obtained colourless microcrystals of {[Ca(1,4-bdc)(bpdo)]•0.5DMF}$_n$ (**3** ⊃ 0.5DMF) were filtered, washed with DMF and dried at 298 K under vacuum for 2 h. Yield: 1.20 g, 97%. Elemental analysis: Calcd for **3** ⊃ 0.3DMF ($C_{18.9}H_{4.1}CaN_{2.3}O_{6.3}$): C = 54.79; H = 3.43; N = 7.78. Found: C = 53.44; H = 3.68; N = 7.73%.

Physical measurements. Elemental analyses (C, H and N) were performed using a Yanaco CHN corder MT-6. The infrared spectra were recorded using KBr disks on a Thermo Nicolet 6700 FT-IR spectrometer with a resolution of 4 cm^{-1}. TG-DTA analysis was performed using a Rigaku ThermoPlus2/TG-DTA8129 over the temperature range r.t.–500 °C under a N_2 flow at a heating rate of 10 °C min^{-1}. Powder XRD data of microcrystals were collected using a Rigaku RINT-Ultima III diffractometer employing Cu Kα radiation. The adsorption and desorption isotherms for CO_2 (195, 288 and 298 K), CH_4 (298 K), N_2 (77 and 298 K), O_2 (298 K), Ar (298 K) and H_2 (298 K) were recorded on a BELSORP-max volumetric adsorption instrument (BEL Japan, Inc., Supplementary Table 3). The adsorption isotherm measurements for H_2O at 298 K were performed using a BELSORP-aqua volumetric adsorption instrument (BEL Japan, Inc.). The high-pressure adsorption and desorption isotherms for CO_2, CH_4, C_2H_4 and C_2H_6 (298 K) were measured with a BELSORP-HP volumetric adsorption equipment (BEL Japan, Inc., Supplementary Tables 4 and 5). We estimated the total adsorbed amount using the simple equation (1)[45]:

$$N_{total} = N_{ex} + \rho_{bulk} V_{pore} \quad (1)$$

where N_{total} is the total adsorbed amount, N_{ex} is the surface excess amount, ρ_{bulk} is the bulk density of the gases and V_{pore} is the pore volume of the PCPs. Total adsorption was calculated using NIST Thermochemical Properties of Fluid Systems: CO_2, CH_4, C_2H_4 and C_2H_6 densities between 0 and 2,000 kPa were fitted using a sixth-order polynomial, then multiplied by the pore volume of each material[46]. The IAST of Prausnitz and Myers was used to estimate the composition of the adsorbed phase from pure component isotherm data[33]. Experimental isotherm data were fitted to the single-site or dual-site Langmuir–Freundlich model (Supplementary Table 6). Selectivities were calculated using the following expression (2):

$$S = \frac{x_i y_j}{x_j y_i} \quad (2)$$

where x_i is the mole fraction of component i in the adsorbed phase and y_i is the mole fraction of component i in the bulk. The coverage-dependent isosteric heat of adsorption was evaluated by first fitting the temperature-dependent isotherm data (288 and 298 K, Supplementary Figs 35 and 36) to a virial-type expression[47], which can be written as follows:

$$\ln P = \ln N + \frac{1}{T}\sum_{i=0}^{m} a_i N^i + \sum_{i=0}^{n} b_i N^i \quad (3)$$

where P is the pressure, N is the quantity of CO_2 adsorbed, T is the temperature, a_i and b_i are virial coefficients, and m and n are the number of virial coefficients required for adequate fitting of the isotherms. Then, the isosteric heat of adsorption was evaluated using the following expression (4):

$$Q_{st} = -R\sum_{i=0}^{m} a_i N^i \quad (4)$$

where R is the universal gas constant. The zero-coverage isosteric heat of adsorption is given by:

$$Q_{st} = -Ra_0 \quad (5)$$

Repeated adsorption measurements were performed at 298 K using the BELSORP-max. The samples after the adsorption measurement were degassed at 298 K under vacuum for 1 min and used for the next measurement. Molecular Sieves 13X was used as a reference sample. Coincident infrared/adsorption measurements were carried out at 195 K using a JASCO model VIR-200 Fourier transform infrared spectrometer connected to a BELSORP-18PLUS volumetric adsorption instrument (BEL Japan, Inc.) with neat samples. Those apparatuses were synchronized with each other and each infrared spectrum was obtained at each equilibrium point of the adsorption/desorption isotherms. The initial infrared spectrum before CO_2 dosage was used as the reference for obtaining the background-subtracted spectra. Coincident XRD/adsorption measurements were performed at 195 K using a Rigaku UltimaIV with Cu Kα radiation connected to a BELSORP-18PLUS. Those apparatuses were synchronized with each other and each powder XRD pattern was obtained at each equilibrium point of the adsorption/desorption isotherms. Breakthrough curve measurements were performed using a hand-made gas flowing system. The sample cell was filled with sample powders, and the temperature of the cell was controlled by a refrigerant circulating system. The gas ratio was CO_2:CH_4 = 40:60 (vol) and the measurements were executed at 0.80 MPa of total pressure at 298 K with a space velocity of 3 min^{-1}. The pressure of CO_2 was 0.32 MPa.

Computational details. Geometry optimizations of organic ligands and a model of **2** were carried out using density functional theory with the B3LYP functional[48–51]. We constructed a finite model based on the X-ray crystal structure of **2**, where four oxygen atoms of bpdo and eight oxygen atoms of bridging carboxylate ligands were replaced by hydroxide ions and water molecules to retain the coordinative environments of Mg, as shown Supplementary Fig. 6. In an optimization process, we fixed the four Mg atoms to the relative positions observed in the crystal structure, whereas all geometries of organic ligands were optimized. The 6-31G(d, p) basis sets were employed for all atoms, where one set of diffuse functions was added to the O atoms in carboxylate ligands, bpdo ligands and hydroxide ions. All geometry optimizations, NBO analysis and evaluation of NBO atomic charges were performed using the Gaussian 09 package[52].

Crystal structure determination. Single-crystal XRD measurements of **1**, **2** ⊃ 2DMF and **3** ⊃ 0.5DMF were performed using a Rigaku RAXIS–RAPID imaging plate diffractometer with graphite-monochromated Mo Kα radiation (λ = 0.71075 Å). The data were corrected for Lorentz and polarization effects. The structures were solved using direct methods (SIR2004 (**1**) and SHELXS-97 (**2** ⊃ 2DMF and **3** ⊃ 0.5DMF)) and expanded using Fourier techniques[53,54]. All nonhydrogen atoms were refined anisotropically. All hydrogen atoms were refined using the riding model. The refinements were carried out using full-matrix least-squares techniques on F^2 using SHELXL-97 (**1** and **2** ⊃ 2DMF) and SHELXL-2014 (**3** ⊃ 0.5DMF)[54] (Supplementary Table 7 and Supplementary Data 1–3). For **3** ⊃ 0.5DMF, the thermal displacement parameters for one of the disordered bpdo molecules were restrained using the *DELU* and *SIMU* commands, and the disordered DMF solvent was refined using the *DFIX, DANG, FLAT* and *EADP* commands. For **1** and **2** ⊃ 2DMF, all calculations were performed using the CrystalStructure software package[55]. For **3** ⊃ 0.5DMF, all calculations were performed using the WinGX software package[56].

References

1. Yaghi, O. M. *et al.* Reticular synthesis and the design of new materials. *Nature* **423,** 705–714 (2003).
2. Kitagawa, S., Kitaura, R. & Noro, S. Functional porous coordination polymers. *Angew. Chem., Int. Ed.* **43,** 2334–2375 (2004).

3. Bradshaw, D., Claridge, J. B., Cussen, E. J., Prior, T. J. & Rosseinsky, M. J. Design, chirality, and flexibility in nanoporous molecule-based materials. *Acc. Chem. Res.* **38,** 273–282 (2005).
4. Férey, G. Hybrid porous solid: past, present, future. *Chem. Soc. Rev.* **37,** 191–214 (2008).
5. Morris, R. E. & Wheatley, P. S. Gas storage in nanoporous materials. *Angew. Chem. Int. Ed.* **47,** 4966–4981 (2008).
6. Zacher, D., Schmid, R., Wöll, C. & Fischer, R. A. Surface chemistry of metal–organic frameworks at the liquid–solid interface. *Angew. Chem. Int. Ed.* **50,** 176–199 (2011).
7. Lu, G. *et al.* Imparting functionality to a metal–organic framework material by controlled nanoparticle encapsulation. *Nat. Chem.* **4,** 310–316 (2012).
8. Cohen, S. M. Postsynthetic Methods for the Functionalization of Metal–Organic Frameworks. *Chem. Rev.* **112,** 970–1000 (2012).
9. Haldar, R. & Maji, T. K. Metal–organic frameworks (MOFs) based on mixed linker systems: structural diversities towards functional materials. *CrystEngComm* **15,** 9276–9295 (2013).
10. Brozek, C. K. & Dinca, M. Ti^{3+}-, $V^{2+/3+}$-, $Cr^{2+/3+}$-, Mn^{2+}-, and Fe^{2+}-Substituted MOF-5 and Redox Reactivity in Cr- and Fe-MOF-5. *J. Am. Chem. Soc.* **135,** 12886–12891 (2013).
11. Gao, W.-Y. *et al.* Crystal engineering of an nbo topology metal–organic framework for chemical fixation of CO_2 under ambient conditions. *Angew. Chem. Int. Ed.* **53,** 2615–2619 (2014).
12. Takashima, Y., Long, D.-L. & Cronin, L. Towards imaging electron density inside metal-organic framework structures. *Chem. Commun.* **50,** 2271–2274 (2014).
13. Yamada, T., Otsubo, K., Makiura, R. & Kitagawa, H. Designer co-ordination polymers: dimensional crossover architectures and proton conduction. *Chem. Soc. Rev.* **42,** 6655–6669 (2013).
14. Makal, T. A., Li, J. R., Lu, W. G. & Zhou, H. C. Methane storage in advanced porous materials. *Chem. Soc. Rev.* **41,** 7761–7779 (2012).
15. Torad, N. L., Hu, M., Imura, M., Naito, M. & Yamauchi, Y. Large Cs adsorption capability of nanostructured Prussian Blue particles with high accessible surface areas. *J. Mater. Chem.* **22,** 18261–18267 (2012).
16. Sumida, K. *et al.* Carbon dioxide capture in metal–organic frameworks. *Chem. Rev.* **112,** 724–781 (2012).
17. Gascon, J. *et al.* Practical approach to zeolitic membranes and coatings: state of the art, opportunities, barriers, and future perspectives. *Chem. Mater.* **24,** 2829–2844 (2012).
18. Sato, H. *et al.* Self-accelerating CO Sorption in a soft nanoporous crystal. *Science* **343,** 167–170 (2014).
19. Lee, J. *et al.* Metal–organic framework materials as catalysts. *Chem. Soc. Rev.* **38,** 1450–1459 (2009).
20. Meng, L. *et al.* Crystal engineering of a microporous, catalytically active fcu topology MOF using a custom-designed metalloporphyrin linker. *Angew. Chem. Int. Ed.* **51,** 10082–10085 (2012).
21. Gu, X., Lu, Z.-H., Jiang, H.-L., Akita, T. & Xu, Q. Synergistic catalysis of metal–organic framework-immobilized Au–Pd nanoparticles in dehydrogenation of formic acid for chemical hydrogen storage. *J. Am. Chem. Soc.* **133,** 11822–11825 (2011).
22. Fromm, K. M. Coordination polymer networks with s-block metal ions. *Coord. Chem. Rev.* **252,** 856–885 (2008).
23. Miller, S. R. *et al.* A rare example of a porous Ca-MOF for the controlled release of biologically active NO. *Chem. Commun.* **49,** 7773–7775 (2013).
24. McDonald, T. M. *et al.* Capture of carbon dioxide from air and flue gas in the alkylamine-appended metal–organic framework mmen-Mg_2(dobpdc). *J. Am. Chem. Soc.* **134,** 7056–7065 (2012).
25. Caskey, S. R., Wong-Foy, A. G. & Matzger, A. J. Dramatic Tuning of carbon dioxide uptake via metal substitution in a coordination polymer with cylindrical pores. *J. Am. Chem. Soc.* **130,** 10870–10871 (2008).
26. Senkovska, I. & Kaskel, S. Solvent-Induced pore-size adjustment in the metal-organic framework $[Mg_3(ndc)_3(dmf)_4]$ (ndc = naphthalenedicarboxylate). *Eur. J. Inorg. Chem.* **2006,** 4564–4569 (2006).
27. Dincă, M. & Long, J. R. Strong H_2 binding and selective gas adsorption within the microporous coordination solid $Mg_3(O_2C\text{-}C_{10}H_6\text{-}CO_2)_3$. *J. Am. Chem. Soc.* **127,** 9376–9377 (2005).
28. Banerjee, D., Zhang, Z., Plonka, A. M., Li, J. & Parise, J. B. A calcium coordination framework having permanent porosity and high CO_2/N_2 selectivity. *Cryst. Growth Des.* **12,** 2162–2165 (2012).
29. Mallick, A. *et al.* Fine-tuning the balance between crystallization and gelation and enhancement of CO_2 uptake on functionalized calcium based MOFs and metallogels. *J. Mater. Chem.* **22,** 14951–14963 (2012).
30. Spek, A. L. Structure validation in chemical crystallography. *Acta Crystallogr., Sect. D* **65,** 148–155 (2009).
31. Shannon, R. D. Revised effective ionic radii and systematic studies of interatomic distances in halides and chalcogenides. *Acta Crystallogr. Sect. A* **32,** 751–767 (1976).
32. Serre, C. *et al.* An explanation for the very large breathing effect of a metal–organic framework during CO_2 adsorption. *Adv. Mater.* **19,** 2246–2251 (2007).
33. Myers, A. L. & Prausnitz, J. M. Thermodynamics of mixed-gas adsorption. *AIChE. J.* **11,** 121–127 (1965).
34. Hamon, L., Jolimaître, E. & Pirngruber, G. D. CO_2 and CH_4 separation by adsorption using Cu-BTC metal–organic framework. *Ind. Eng. Chem. Res.* **49,** 7497–7503 (2010).
35. Bae, Y.-S. & Snurr, R. Q. Development and evaluation of porous materials for carbon dioxide separation and capture. *Angew. Chem. Int. Ed.* **50,** 11586–11596 (2011).
36. McDonald, T. M., D'Alessandro, D. M., Krishna, R. & Long, J. R. Enhanced carbon dioxide capture upon incorporation of *N,N′*-dimethylethylenediamine in the metal–organic framework CuBTTri. *Chem. Sci* **2,** 2022–2028 (2011).
37. Dunne, J. A., Rao, M., Sircar, S., Gorte, R. J. & Myers, A. L. Calorimetric heats of adsorption and adsorption isotherms. 2. O_2, N_2, Ar, CO_2, CH_4, C_2H_6, and SF_6 on NaX, H-ZSM-5, and Na-ZSM-5 zeolites. *Langmuir* **12,** 5896–5904 (1996).
38. Nugent, P. *et al.* Porous materials with optimal adsorption thermodynamics and kinetics for CO_2 separation. *Nature* **495,** 80–84 (2013).
39. Fracaroli, A. M. *et al.* Metal–organic frameworks with precisely designed interior for carbon dioxide capture in the presence of water. *J. Am. Chem. Soc.* **136,** 8863–8866 (2014).
40. Shekhah, O. *et al.* Made-to-order metal-organic frameworks for trace carbon dioxide removal and air capture. *Nat. Commun.* **5,** 4228 (2014).
41. Smaldone, R. A. *et al.* Metal-organic frameworks from edible natural products. *Anegw. Chem. Int. Ed.* **49,** 8630–8634 (2010).
42. Gassensmith, J. J. *et al.* Strong and reversible binding of carbon dioxide in a green metal–organic framework. *J. Am. Chem. Soc.* **133,** 15312–15315 (2011).
43. Forgan, R. S. *et al.* Nanoporous carbohydrate metal–organic frameworks. *J. Am. Chem. Soc.* **134,** 406–417 (2012).
44. Chamayou, A.-C. & Janiak, C. A > 22 Å long metal-complex anon and a 1D cationic coordination polymer, both based on 4,4′-bipyridine-*N,N′*-dioxide, yielding a 3D and 2-fold interpenetrated fsc net of the hydrogen-bonded metal-organic 1D polymer. *Inorg. Chim. Acta* **363,** 2193–2200 (2010).
45. Neimark, A. V. & Ravikovitch, P. I. Calibration of pore volume in adsorption experiments and theoretical models. *Langmuir* **13,** 5148–5160 (1997).
46. Lemmon, E. W., McLinden, E. W. & Friend, D. G. in *NIST Chemistry WebBook, NIST Standard Reference Database Number 69* (eds Linstrom, P. J. & Mallard, W. G.) 20899 (National Institute of Standards and Technology, 2003).
47. Czepirski, L. & Jagiello, J. Virial-type thermal equation of gas–solid adsorption. *Chem. Eng. Sci.* **44,** 797–801 (1989).
48. Becke, A. D. Density-functional thermochemistry. III. The role of exact exchange. *J. Chem. Phys.* **98,** 5648–5652 (1993).
49. Lee, C., Yang, W. & Parr, R. G. Development of the Colle-Salvetti correlation-energy formula into a functional of the electron density. *Phys. Rev. B* **37,** 785–789 (1988).
50. Becke, A. D. Density-functional exchange-energy approximation with correct asymptotic behavior. *Phys. Rev. A* **38,** 3098–3100 (1988).
51. Miehlich, B., Savin, A., Stoll, H. & Preuss, H. Results obtained with the correlation energy density functionals of becke and Lee, Yang and Parr. *Chem. Phys. Lett.* **157,** 200–206 (1989).
52. Frisch, M. J. *et al. Gaussian 09, Revision D.01* (Gaussian, Inc., 2009).
53. Burla, M. C. *et al. SIR2004*: an improved tool for crystal structure determination and refinement. *J. Appl. Crystallogr* **38,** 381–388 (2005).
54. Sheldrick, G. M. A short history of *SHELX*. *Acta Crystallogr. Sect. A* **64,** 112–122 (2008).
55. *CrystalStructure 4.0.1: Crystal Structure Analysis Package*; Rigaku Corp.: Tokyo, Japan (2000-2010).
56. Farrugia, L. J. *WinGX* suite for small-molecule single-crystal crystallography. *J. Appl. Crystallogr.* **32,** 837–838 (1999).

Acknowledgements

This work was supported by PRESTO-JST and a Grant-in-Aid for Scientific Research (No. 24550068) from the Ministry of Education, Culture, Sports, Science and Technology, Japan, and the Shitagau Noguchi Foundation.

Author contributions

S.-i.N. designed this study, interpreted the results and wrote the paper. S.-i.N. and J.M. performed the experiments. Y.H. carried out the calculations. R.M., H.S. and S.K. performed the coincident infrared/adsorption and XRD/adsorption measurements. K.S. carried out the structural analysis. Y.I. performed the high-pressure adsorption and breakthrough curve measurements. All authors discussed the results and commented on the manuscript.

Additional information

Accession codes: The X-ray crystallographic coordinates for structures reported in this Article have been deposited at the Cambridge Crystallographic Data Centre (CCDC), under deposition numbers 997086 to 997088. These data can be obtained free of charge from the Cambridge Crystallographic Data Centre via www.ccdc.cam.ac.uk/data_request/cif.

Competing financial interests: The authors declare no competing financial interests.

5

A monotopic aluminum telluride with an Al = Te double bond stabilized by *N*-heterocyclic carbenes

Daniel Franz[1], Tibor Szilvási[2], Elisabeth Irran[1] & Shigeyoshi Inoue[1]

Aluminum chalcogenides are mostly encountered in the form of bulk aluminum oxides that are structurally diverse but typically consist of networks with high lattice energy in which the chalcogen atoms bridge the metal centres. This makes their molecular congeners difficult to synthesize because of a pronounced tendency for oligomerization. Here we describe the isolation of the monotopic aluminum chalcogenide $(L^{Dip}N)AlTe(L^{Et})_2$ (L^{Dip} = 1,3-(2,6-diisopropylphenyl)-imidazolin-2-imine, L^{Et} = 1,3-diethyl-4,5-dimethyl-imidazolin-2-ylidene). Unique features of $(L^{Dip}N)AlTe(L^{Et})_2$ are the terminal position of the tellurium atom, the shortest aluminum–tellurium distance hitherto reported for a molecular complex and the highest bond order reported for an interaction between these elements, to the best of our knowledge. At elevated temperature $(L^{Dip}N)AlTe(L^{Et})_2$ equilibrates with dimeric $\{(L^{Dip}N)AlTe(L^{Et})\}_2$ in which the chalcogen atoms assume their common role as bridges between the metal centres. These findings demonstrate that $(L^{Dip}N)AlTe(L^{Et})_2$ comprises the elusive Al = Te double bond in the form of an *N*-heterocyclic carbene-stabilized species.

[1] Institut für Chemie, Anorganische und Analytische Chemie, Technische Universität Berlin, Straße des 17 Juni 135, Sekr. C2, 10623 Berlin, Germany.
[2] Department of Inorganic and Analytical Chemistry, Budapest University of Technology and Economics, Szent Gellért tér 4, 1111 Budapest, Hungary.
Correspondence and requests for materials should be addressed to S.I. (email: shigeyoshi.inoue@tu-berlin.de).

Aluminum is the third most abundant element in the earth's crust surpassed in quantity only by oxygen and silicon and it is most often found among the elements of the group 13 family. The metal is easily oxidized and mostly encountered in the form of aluminum oxides, which are ionic compounds with high lattice energies. The structural diversity of aluminum oxides has led them to be used in a variety of applications in science and technology[1–3]. Among the heavier aluminum chalcogenides, tellurium compounds are explored the least, which is mainly due to this element's low abundance on our planet. Applications of aluminum tellurides in chemical synthesis have barely been reported[4]; however, attention to the heaviest non-radioactive chalcogen has intensified lately because of its use for electronic devices[5,6]. Considering the general importance of inorganic aluminum chalcogenides to applied research, it is of prime interest to understand the aggregation processes that create the bulk materials from single atoms and to clarify the nature of the chemical bond between the elements. Theoretical calculations, as well as low-temperature matrix isolation techniques have been used to get an initial insight into the ultrafast bond formations involved in those processes, but these methods have considerable limitations[7–9]. One of the best methods to overcome these limitations is to study well-defined molecular complexes that are stable in the condensed phase at ambient temperatures. Astonishingly, complexes of the general formula R–Al = E (**I**, Fig. 1, R = monoanionic ligand, E = chalcogen) with a double bond between the elements are scarcely found in the literature[10] and oligomeric compounds $(RAlE)_n$ ($n = 2, 3, 4, \ldots$) dominate the field[11–15].

Interestingly, the double-bonded species could be regarded as a molecular mimic of an elusive intermediate in the processes of aggregation and polymerization that may lead to bulk inorganic aluminum chalcogenides reminiscent of the formation of macromolecular polyolefins from monomolecular vinylic precursors. An apparent explanation for the high reactivity of **I** is the electropositive nature of the aluminum atom, which renders the Al = E moiety a highly polarized Lewis acid–base pair. In consequence, the combination of an electrophilic site at the Al centre and a nucleophilic lone pair located at the chalcogen atom results in self-oligomerization. By the use of uncharged electron pair donor ligands, one can decrease the Lewis acidity of the metal centre to thermodynamically stabilize the Al = E functionality, a technique often complemented by the addition of a Lewis acid that binds to the chalcogen atom. This results in donor–acceptor-stabilized complexes of type **II** (Fig. 1). Further stabilization can be achieved by introduction of an additional Lewis base (**III**, Fig. 1).

This technique has led to the isolation of a monoalumoxane complex of the highly unstable R–Al = O functionality, for which two Lewis bases and one Lewis acid were used to stabilize the Al and O atom, respectively[16]. Though donor–acceptor stabilization is an elaborate and useful synthetic tool to generate complexes with considerable multiple bond character, the steric and electronic stabilization at the terminal chalcogen atom of R–Al = E species significantly limits the investigation of their reactivity. In this work, we describe the monotopic aluminum telluride $(L^{Dip}N)AlTe(L^{Et})_2$ (**3**, L^{Dip} = 1,3-(2,6-diisopropylphenyl)-imidazolin-2-imine, L^{Et} = 1,3-diethyl-4,5-dimethyl-imidazolin-2-ylidene), an electron-precise aluminum chalcogenide that bears a terminal atom of the group 16 family. The unique bonding situation of **3** results in the shortest aluminum–tellurium distance and the highest bond order reported for this type of molecular complex. Furthermore, **3** dimerizes to $\{(L^{Dip}N)AlTe(L^{Et})\}_2$ (**5**) at elevated temperature. We interpret these results in terms of an Al = Te double bond to be present in **3**.

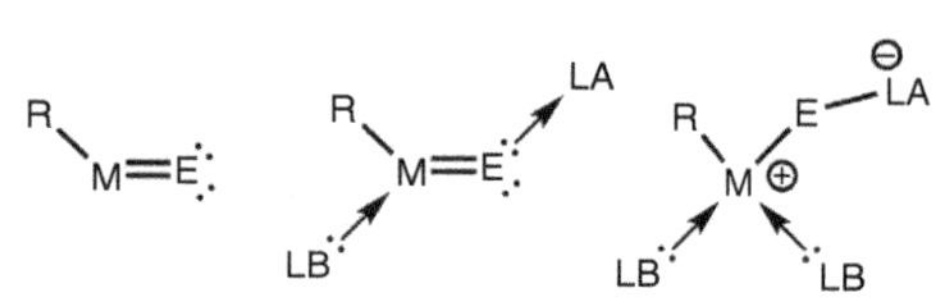

Figure 1 | General types of molecular group 13 metal chalcogenides with double-bond character. The parent complex (**I**), as well as the donor-acceptor-stabilized species **II** and **III** (R = anionic ligand, LA = Lewis acid, LB = Lewis base).

Results

Synthesis of the monotopic aluminum telluride 3. The synthesis of the aluminum telluride **3** starts out with the previously reported[17,18] dimeric aluminum dihydride **1** (Fig. 2). The introduction of tellurium to the system was accomplished by addition of two equivalents of the convenient tellurium transfer reagent $^nBu_3PTe$ ($^nBu = n$-butyl) to produce the aluminum ditelluride **2** via a dehydrogenative three-electron redox process in which the chalcogen is reduced from the oxidation state of + 2 to − 1. An X-ray study on **2** revealed that the four-membered Al_2N_2 ring is decorated with a bridging ditelluride group. Though we could not obtain crystals of sufficient quality to allow for the discussion of structural parameters, the connectivity as in the structural formulation **2** was verified (Supplementary Fig. 16).

The Al–Te–Te–Al structural motif is without precedence in the literature. Notably, it resembles the Al–Te–Al moiety reported for a pyrazolato aluminum telluride, wherein one tellurium atom bridges the aluminum centres[19].

Bicyclic systems, such as the $Al_2N_2Te_2$ moiety in **2**, obviously possess high ring strain because the geometry imposes internal bond angles significantly $< 109.5°$, which aggravates typical tetrahedral coordination. Thus, it seemed reasonable that the cycle might be prone to the cleavage by nucleophilic attack at the metal centres to furnish a monotopic aluminum–tellurium complex. We chose *N*-heterocyclic carbene (NHC) as a nucleophile because NHCs are strong σ-electron pair donors that have proven viable for the taming of numerous elusive main group element compounds on a molecular level[20–24]. Indeed, the reaction of the ditelluride **2** with two equivalents of L^{Et} yielded the monotopic telluride **3** (Fig. 2). During this dehydrogenative redox process, the tellurium atom is further reduced to the − 2 oxidation state with concomitant formation of dihydrogenated NHC ($L^{Et}(H_2)$, hydrogenated at the formerly carbenoid atom). The abstraction of hydrogen from iminoboranes using NHC to produce $NHC(H_2)$ had been previously investigated[25].

The single-crystal X-ray analysis unambiguously shows that the aluminum centre in **3** is bonded to one imino group, two NHC ligands and a terminal tellurium atom (Fig. 3a). The aluminum centre is coordinated in a distorted tetrahedral manner with the chalcogen atom tilted away from the bulkier diisopropylphenyl (Dip) moieties of the imino group towards one of the sterically less hindered carbene ligands. The Al–Te bond length was determined to be 2.5130(14) Å, which is decreased in comparison with the typical range found for molecular aluminum tellurides with four-coordinate aluminum atoms (2.562–2.750 Å)[11,15,26–28]. Remarkably, the Al–Te distance in **3** is even shorter than the 2.549(1) Å found for the ditopic $Te\{Al(CH(SiMe_3)_2)_2\}_2$ with three-coordinate aluminum centres that marked the hitherto shortest AlTe contact reported for a molecular complex[29]. The terminal position of the tellurium atom in **3** is a unique finding because chalcogens commonly function as bridging ligands in aluminum compounds or are coordinated to another

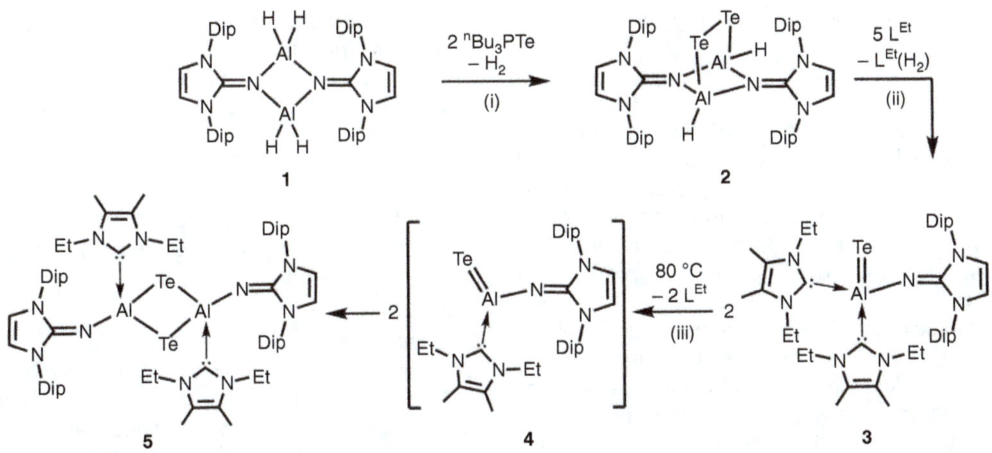

Figure 2 | Synthetic overview for this work. Conversion of dimeric aluminum dihydride **1** via ditelluride **2** to monotopic aluminum telluride **3** and its dimerization via intermediate **4** to tellurium-bridged **5** (Dip = 2,6-diisopropylphenyl, L^{Et} = 1,3-diethyl-4,5-dimethyl-imidazolin-2-ylidene). (i) Toluene, rt, 48 h. (ii) Benzene, rt, 48 h. (iii) Benzene, 80 °C, 22 h; rt = ambient temperature.

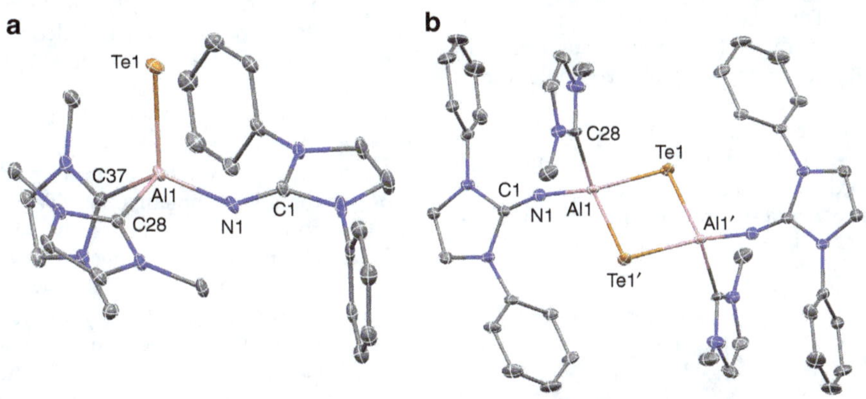

Figure 3 | Ellipsoid plots derived from single-crystal X-ray diffraction data. Molecular structures of **3** (**a**) and **5** (**b**) in the solid state. Hydrogen atoms, isopropyl groups and non-N-bonded methyl groups have been omitted for reasons of clarity. Thermal ellipsoids are at the 30% probability level. Selected bond lengths (Å) and bond angles (°): **3**: Al1-Te1 = 2.5130(14), Al1-N1 = 1.804(4), Al1-C28 = 2.084(5), Al1-C37 = 2.077(5), N1-C1 = 1.259(6); Te1-Al1-N1 = 118.34(14), Te1-Al1-C28 = 102.65(13), Te1-Al1-C37 = 119.08(14), Al1-N1-C1 = 139.8(4). **5**: Al1-Te1 = 2.6211(15), Al1-Te1′ = 2.6143(14), Al1-N1 = 1.765(5), Al1-C28 = 2.086(5), N1-C1 = 1.253(7); Al1-Te1-Al1′ = 80.86(5), Te1-Al1-Te1′ = 99.14(5), N1-Al1-Te1 = 115.83(15), N1-Al1-Te1′ = 121.13(16), Al1-N1-C1 = 164.3(4).

Lewis acid. The closest distance between an aluminum and a tellurium centre in **3** of molecules adjacent in the crystal lattice amounts to 7.31 Å.

Theoretical study of monotopic aluminum telluride 3. We calculated a Wiberg Bond Index (WBI) of 1.20 for the AlTe bond in **3**, which indicates significant double-bond character, particularly, if the strong polarization along the AlTe vector is taken into account (selected natural population analysis (NPA) charges for **3**: Al = + 1.24, Te = − 0.95). Notably, it is commonly acknowledged that WBI values have a tendency to underestimate bond orders for heavier main group element bonds. For comparison the WBI_{AlTe} for the ditopic $Te\{Al(CH(SiMe_3)_2)_2\}_2$ (*vide supra*) was calculated using the same theoretical method as for **3** and found to be smaller than one (0.91). This suggests that a trigonal-planar coordination environment of the aluminum centre is not a prerequisite for relevant AlTe double-bond character. Moreover, the WBI_{AlTe} of 1.20 for **3** is in line with that reported for the highly polarized Ge = O double bond in a germanone (WBI_{GeO} = 1.252) and the Si = O double bond in a sila-urea (WBI_{SiO} = 1.29), respectively[30,31]. Remarkably, the two latter compounds bear trigonal-planar-coordinate metal centres. The molecular orbital analysis of **3** reveals the highest occupied molecular orbital (HOMO) to comprise a lone pair with π-symmetry that is primarily located at the tellurium atom (Fig. 4). The HOMO − 1 shows a π-symmetric orbital lobe expanding between the aluminum and the tellurium centre that hints towards an interaction of higher bond order between these atoms. An expected AlTe σ-bond is observed in the HOMO − 3 (Supplementary Fig. 21). Antibonding π-interaction among the phenyl rings of the Dip groups is found for the lowest unoccupied molecular orbital (LUMO), which suggests that electron donation to **3** would preferably start out at the ligand system rather than the metal site.

Natural Resonance Theory (NRT) investigation of **3** reveals a major contribution (77%) of resonance structures that feature an Al = Te double bond (that is, **3**) and only a minor share (23%) of zwitterionic species with an Al–Te single bond as **3A** or **3B** (Fig. 5). This result is in accordance with a recent NRT study on NHC-stabilized sila-acylium ions that are isoelectronic to the aluminum telluride **3** (ref. 32). Interestingly, though solid batches of **3** can be stored in an inert atmosphere for weeks the compound decomposes in deuterated tetrahydrofurane solution within few days to yield the protonated iminato ligand ($L^{Dip}NH$, Supplementary Fig. 6).

Thermal dimerization of 3 to tellurium-bridged 5. When we heated a solution of **3** in benzene dimerization to **5** occurred (Fig. 2) in which the four-membered Al_2Te_2 cycle shows the chalcogen-bridged connectivity pattern common for complexes of aluminum with group 16 ligand atoms[14,15,26–28]. Monitoring the process with NMR spectroscopy in C_6D_6 solution revealed that the system equilibrates at 80 °C in about 24 h with the dimer (**5**) and the monomer (**3**) coexisting in an approximate ratio of 1:1 (Supplementary Figs 11 and 12). After **5** has been isolated and stripped from solvent, it decomposes at ambient temperature within hours to a considerable degree (Supplementary Figs 13 and 14). Presumably, the stability of **5** depends on the presence of **3** and free NHC in solution. In comparison with monotopic **3**, the Al–Te distances in **5** are expanded and amount to 2.6143(14) and 2.6211(15) Å (Fig. 3b). This increase in bond length corresponds to the significantly lower calculated bond index of the AlTe interaction in **5** ($WBI_{AlTe} = 0.75$) and clearly implies a polarized aluminum tellurium single bond (selected NPA charges for **5**: Al = + 1.21, Te = − 0.79). Thus, the immense impact of the chalcogen atom's coordination mode (bridging versus terminal) on the type of bonding to the aluminum atom (single versus double bond) is underlined.

DFT analysis of the possible reaction mechanism for the transformation of **3** into **5** revealed intermediate complex **4** (Fig. 2) with a three-coordinate aluminum centre. The formation of **4** from **3** is a barrierless process that is usual for NHC-coordinated main group element structures[32] and underlines the donor–acceptor-bonding mode[33] in NHC–Al bonds. The dissociated pair of complex **4** and free L^{Et} is only 15.0 kcal mol^{-1} higher in Gibbs free energy than its precursor **3**, while a comparably low energy gain of 2.9 kcal mol^{-1} is obtained in the overall process of forming the dimer **5** and free L^{Et} from **3** (Supplementary Fig. 22). This small difference in the thermodynamic stabilities of **3** and **5** allows for assuming dynamic equilibrium between the two compounds that explains that **3** does not fully convert into **5** on heating.

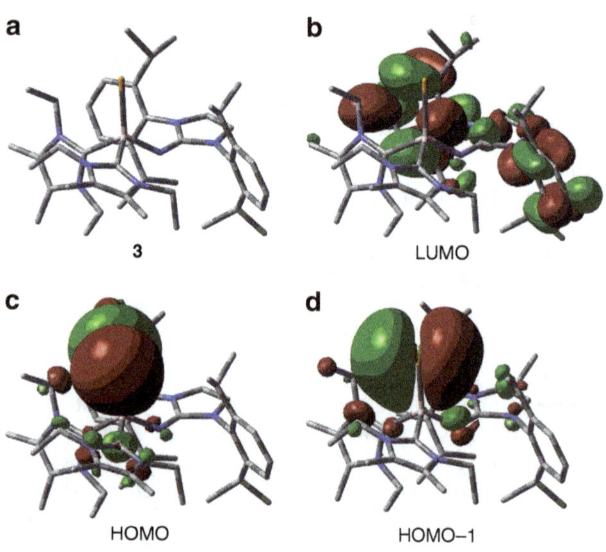

Figure 4 | Molecular orbital analysis. Optimized structure and selected molecular orbitals of **3**, the LUMO, HOMO and HOMO-1 ((**a-d**) stick models). Te, yellow; Al, pink; N, blue; C, grey. Hydrogen atoms have been omitted for clarity.

Discussion

As a consequence of the terminal position of the tellurium atom in **3**, the AlTe bond order is markedly increased with respect to complexes in which the chalcogen atom assumes a bridging position. Backed by the results of our theoretical study (WBI, NPA and NRT), we presume that **3** represents the elusive R–Al = Te species (*vide supra*) in the form of an NHC-stabilized congener. Though the aluminum centre in **3** is four-coordinate, the WBI_{AlTe} is comparable to those of polarized metal chalcogen double bonds. Upon thermal dimerization of **3** into **5**, the WBI_{AlTe} of the dimer is vastly decreased, which again underlines the importance of a terminal chalcogen atom for a bond order larger than one. Consequently, an Al = Te double bond is unlikely to be found for complexes following the approach of donor–acceptor stabilization (type **II**, **III**; Fig. 1). The calculated intermediate in the process of dimerization, that is, **4**, which bears a three-coordinate aluminum centre exhibits a higher WBI_{AlTe} than **3** (Supplementary Table 1). There is no experimental proof for its existence, but it is conceivable that the isolation of a more stable congener to **4** may be accomplished by the use of bulkier NHC ligands. However, our conversions of the ditelluride **2** with imidazolin-2-ylidenes more sterically congested than L^{Et} did not yield respective analogues to neither **3** nor **4** or **5**.

The isolation of the monotopic aluminum telluride **3** markedly contributes to the field of aluminum element multiple bonds that are surprisingly unexplored to date. The thermal transformation of **3** into dimeric **5** may serve as a simplified model on the molecular level that could contribute to the elucidation of the complex process of how isolated atoms aggregate to three-dimensional networks in bulk aluminum chalcogenides.

Methods

Experimental. All experiments and manipulations were carried out under dry oxygen-free nitrogen using a Schlenk Line or in an MBraun Glove Box Workstation containing an atmosphere of purified nitrogen. Glass junctions were coated with the polytetrafluoroethylene (PTFE)-based grease Merkel Triboflon III. Solvents were dried by standard methods and freshly distilled before the use. NMR spectra were recorded on a Bruker Avance 400 or an Avance III 500 spectrometer. Chemical shift values (δ) are given in parts per million and are referenced to (residual) solvent signals (^1H), and $^{13}C\{^1H\}$ NMR. s = singlet, br = broadened, d = doublet, t = triplet, sept = septet, n.o. = not observed, n.r. = not resolved. $\{L^{Dip}NAlH_2\}_2$ (**1**), $^nBu_3PTe$ and L^{Et} were synthesized according to the published procedures[17,34,35]. Reagents purchased from commercial sources were used as received if not stated otherwise. High-resolution mass spectrometry (HRMS) was performed on a Thermo Fisher Scientific LTQ Orbitrap XL machine using an atmospheric-pressure chemical ionization (APCI) or an electrospray ionization (ESI)

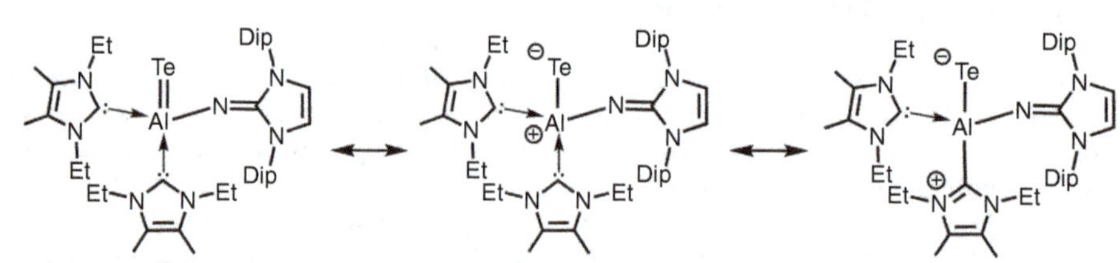

3 3A 3B

Figure 5 | NRT analysis. Selected resonance structures for the aluminum telluride **3**.

ion source and positive detection mode. Elemental analyses (via combustion) were carried out by the microanalytical laboratory of the Institut für Chemie, Technische Universität Berlin. For depiction of selected NMR spectra see Supplementary Figs 1–14. For NMR spectroscopic characterization of $L^{Et}(H_2)$ see Supplementary Fig. 15 and Supplementary Methods. For details to X-ray analyses see Supplementary Tables 6–8 and Supplementary Methods.

Computational. Geometry optimizations were carried out at B97-D/6-31G(d)[Al: cc-pVTZ, Te: cc-pVTZ-PP] level of theory in all cases[36–39]. All stationary points were verified via frequency calculations. All analyses (WBI, NPA, NRT and orbital) occurred at the same level. In case of the reaction mechanism calculation, accurate single-point energy calculations at ωB97X-D/cc-pVTZ[Te: cc-pVTZ-PP] level were performed[40]. Thermodynamic corrections were calculated at 353 K to model reaction conditions. Benzene solvent was considered via the polarizable continuum model, in the framework of the solvation model based on density method[41]. All calculations were carried out with the GAUSSIAN 09 program package[42]. For geometry optimized structures of **2**–**5** see Supplementary Figs 17–20 (depiction) and Supplementary Tables 2–5 (cartesian geometry).

Synthesis and isolation of {L^{DiP}NAl(H)Te}$_2$ (2). In a Glove Box Workstation, a 25-ml Schlenk flask equipped with a PTFE-coated magnetic stirrer bar was charged with {L^{DiP}NAlH$_2$}$_2$ (**1**, 898 mg, 1.04 mmol) and ^{n}Bu$_3$PTe (733 mg, 2.22 mmol). At the Schlenk Line, toluene (9 ml) was added via syringe to the stirring mixture of solids and the reaction was continued for 48 h (note: gradual evolution of hydrogen gas). A blue suspension was obtained that was diluted with hexane (13 ml). The solid was collected on a frit and the filter cake washed thoroughly with hexane (12 ml) before it was compressed using a pressure gradient. The product was dried in vacuum for 3 h after which 906 mg of **2** in the form of a blue powder was obtained from the frit (0.81 mmol, 78%).

Crystals for X-ray diffraction analysis formed after storing a dilute solution of **2** in toluene/hexane (that is, the combined filtrates from the work-up procedure) at 0 °C for 2 weeks or by gas-phase diffusion exchange of solvent between a solution of **2** in fluorobenzene and a reservoir of hexane for a period of 1 week. Crystals retrieved from either batch analysed with insufficient data quality for the discussion of structural parameters see Supplementary Data 1.

^{1}H NMR (500.1 MHz, C$_6$D$_6$): δ = 7.23 (t, J_{HH} = 8 Hz, 4 H, C$_6$*H*$_3$-4), 7.08 (d, J_{HH} = 8 Hz, 8 H, C$_6$*H*$_3$-3,5), 5.88 (s, 4 H, NC*H*), 3.14 (n.r., 8 H, C*H*(CH$_3$)$_2$), 2.30 (br, 2 H, Al*H*), 1.47 (d, J_{HH} = 7 Hz, 24 H, CH(C*H*$_3$)$_2$), 1.06 (d, J_{HH} = 7 Hz, 24 H, CH(C*H*$_3$)$_2$); ^{13}C{^{1}H} NMR (125.8 MHz, C$_6$D$_6$): δ = 148.8 (NCN), 148.4 (C$_{Aryl}$), 133.4 (C$_{Aryl}$), 130.6 (C$_6$H$_3$-3,5), 124.6 (C$_6$H$_3$-4), 117.0 (NCH), 29.2 (CH(CH$_3$)$_2$), 25.9 (CH(CH$_3$)$_2$) and 23.7 (CH(CH$_3$)$_2$).

HRMS (APCI), found (%), {calculated} and assigned: *m/z* = 1118.3701 (17), {1118.3707}, [M]$^+$; the analyte was provided as a solution in C$_6$D$_6$/toluene.

Elemental analysis calculated (%) for C$_{54}$H$_{74}$Al$_2$N$_6$Te$_2$ [1116.39]: C 58.10, H 6.68, N 7.53; found: C 57.47, H 6.70, N 7.47; the analyte was provided wrapped in silver foil.

Synthesis and isolation of (L^{DiP}N)AlTe(L^{Et})$_2$ (3). In a Glove Box Workstation a solution of L^{Et} (148 mg, 0.97 mmol) in 2.0 ml benzene was prepared in a 4 ml SUPELCO Screw Top Vial equipped with a PTFE-coated magnetic stirrer bar. Under vigorous stirring, powderous {L^{DiP}NAl(H)Te}$_2$ (**2**, 185 mg, 0.17 mmol) was gradually added from a glass tube. The inner walls of the tube were washed with benzene (0.5 ml) to complete the transfer. Stirring was continued for 15 min and a yellow solution had formed. Agitation of the reaction mixture was stopped and the open vial was placed in a glass vessel that contained a reservoir of hexane (22 ml). The vessel was closed to separate the system from the Glove Box Workstation atmosphere for gas-phase diffusion exchange of solvent. After a period of 48 h, the pale yellow supernatant was decanted from the yellow crystals that had formed (the volume of the liquid phase was determined to 3.5 ml). The residual solid was washed by adding hexane (3.5 ml), briefly stirring the suspension and removing the supernatant via syringe along with smaller particles that did not sediment within half a minute. A small amount of residual solvent was evaporated under reduced pressure and the crystal fraction was homogenized with a spatulum before it was dried in vacuum for 1 h. From the vial 217 mg (74%) of **3** in the form of a yellow solid was collected.

The crystal used for the reported X-ray diffraction analysis was obtained by gas-phase diffusion exchange of solvent between a solution of **3** in fluorobenzene and a reservoir of Et$_2$O for a period of 1 week, Supplementary Data 2. A crystal that had formed from a saturated solution of **3** in C$_6$D$_6$ analysed to (**3**)$_2$ · (C$_6$D$_6$)$_7$ (data not reported). Compound **3** slowly decomposes in THF-*d*$_8$ solution at ambient temperature (Supplementary Figs 4 and 6) but it is significantly more stable in C$_6$D$_6$ solution (Supplementary Figs 7 and 8). Solid batches of **3** were stored in a Glove Box Workstation at ambient temperature for weeks without notable change in the NMR spectroscopic analysis; however, it was observed in some instances that the substance was covered with a thin layer of brown colour. Thus, storage of **3** in an inert atmosphere at low temperature is recommended.

^{1}H NMR (400.1 MHz, THF-*d*$_8$): δ = 7.23 (t, J_{HH} = 8 Hz, 2 H, C$_6$*H*$_3$-4), 7.11 (d, J_{HH} = 8 Hz, 4 H, C$_6$*H*$_3$-3,5), 6.14 (s, 2 H, NC*H*), 4.67 (m, 4 H, C*H*$_2$CH$_3$), 3.68 (m, 4 H, C*H*$_2$CH$_3$), 3.51 (sept, J_{HH} = 7 Hz, 4 H, C*H*(CH$_3$)$_2$), 2.01 (s, 12 H, CC*H*$_3$), 1.29 (d, J_{HH} = 7 Hz, 12 H, CH(C*H*$_3$)$_2$), 1.13 (d, J_{HH} = 7 Hz, 12 H, CH(C*H*$_3$)$_2$), 0.74 (t, J_{HH} = 7 Hz, 12 H, CH$_2$C*H*$_3$); ^{1}H NMR (400.1 MHz, C$_6$D$_6$): δ = 7.23–7.13 (m, 6 H, C$_6$*H*$_3$-3,4,5)^a, 6.11 (s, 2 H, NC*H*), 4.81 (m, 4 H, C*H*$_2$CH$_3$), 3.87–3.73 (m, 8 H, C*H*$_2$CH$_3$ and C*H*(CH$_3$)$_2$), 1.54 (d, J_{HH} = 7 Hz, 12 H, CH(C*H*$_3$)$_2$), 1.51 (s, 12 H, CC*H*$_3$), 1.31 (d, J_{HH} = 7 Hz, 12 H, CH(C*H*$_3$)$_2$), 0.73 (t, J_{HH} = 7 Hz, 12 H, CH$_2$C*H*$_3$); ^{13}C{^{1}H} NMR (125.8 MHz, THF-*d*$_8$): δ = 149.6 (C$_{Aryl}$), 142.5 (NCN$_{imino}$), 138.4 (C$_{Aryl}$), 128.2 (C$_6$H$_3$-4), 125.0 (C$_{Aryl}$), 124.0 (C$_6$H$_3$-3,5), 114.9 (NCH), 42.4 (CH$_2$CH$_3$)^b, 29.0 (CH(CH$_3$)$_2$), 25.3 (CH(CH$_3$)$_2$)^c, 24.4 (CH(CH$_3$)$_2$), 15.2 (CH$_2$CH$_3$), 8.3 (CCH$_3$), n.o. (NCN$_{carbene}$)^d. a = the resonance superimposes with the residual solvent signal which explains its high intensity in Supplementary Fig. 7; b = the ^{1}H^{13}C-HSQC NMR experiment verifies that the two signals of the C*H*$_2$CH$_3$ protons in the ^{1}H NMR spectrum correlate to one resonance in the ^{13}C{^{1}H} NMR analysis produced by the CH$_2$CH$_3$ carbon atoms; c = the resonance superimposes with the signal produced by the NMR solvent but is identified via ^{1}H^{13}C-HSQC NMR analysis; d = observation of the aluminum-bonded ^{13}C atoms is hampered by the quadrupolar momentum of the ^{27}Al nucleus.

HRMS (ESI), found (%), {calculated} and assigned: *m/z* = 864.4519 (100), {864.4487}, [M + H]$^+$; 1016.5832 (99), {1016.5800}, [M + L^{Et} + H]$^+$; the analyte was provided as a solution in fluorobenzene.

Elemental Analysis calculated (%) for C$_{45}$H$_{68}$AlN$_7$Te [861.67]: C 62.73, H 7.95, N 11.38; found: C 62.54, H 8.14, N 11.41; the analyte was provided wrapped in silver foil.

Synthesis and isolation of {L^{DiP}NAl(L^{Et})Te}$_2$ (5). In a Glove Box Workstation, an NMR sample tube was charged with (L^{DiP}N)AlTe(L^{Et})$_2$ (**3**, 72 mg, 0.08 mmol) and benzene (0.9 ml). At the Schlenk Line, the mixture was frozen in liquid nitrogen and the tube was flame-sealed under vacuum. After thawing in a water bath at ambient temperature the reaction vessel was wrapped in aluminum foil (as a safety consideration) and stored in an oven at 80 °C for 22 h. In a Glove Box Workstation, the tube was unsealed and the yellow solution was withdrawn via syringe. The inner glass walls and residual solid were washed with benzene (0.6 ml) and the liquid phases were combined in a 9-ml Snap Cap Vial equipped with a PTFE-coated magnetic stirrer bar. Piercing the polyethylene cap with a steel cannula hexane (4.0 ml) was added dropwise via syringe to the vigorously stirred solution over a period of 8 min resulting in the formation of a pale yellow precipitate. The solid was allowed to sediment for 12 h, and then the supernatant (5.0 ml) was withdrawn via syringe and filtered through a 0.45-μm PTFE filter put between the outlet and the steel cannula. The vacuum-dried precipitate (30 mg) was identified as spectroscopically clean **3** via NMR analysis (THF-*d*$_8$). The filtrate was stored in a Snap Cap Vial at − 33 °C for 4 days. A crystalline solid had formed (*vide infra*) and the liquid phase (4.8 ml) was removed via syringe. The crystal fraction was dried in vacuum for 30 min to yield an off-white powder. From the glass vessel 24 mg (38%) of product were transferred and verified as spectroscopically clean **5** via NMR analysis (C$_6$D$_6$).

The crystal used for the X-ray diffraction study of **5** was obtained by slow diffusion of hexane into the filtered (0.45 μm PTFE filter) reaction mixture via the gas phase for a period of 3 days, Supplementary Data 3. Afterwards, the batch (crystals + mother liquor) was stored for an additional 2 days in an enclosed glass vessel (screw-cap vial) during which it recrystallized from smaller to larger entities before a crystal was picked. A crystal retrieved from the work-up process before phase separation (*vide supra*) analysed to **5**(C$_6$H$_6$)$_4$ (data not reported). The procedure was reproduced several times; however, in one case removal of the cold solution from the freezer and brief storage at ambient temperature was necessary to initialize the final crystallization process at low temperature.

After isolation as a vaccum-dried solid, **5** slowly decomposes at ambient temperature. A batch stored in the Glove Box Workstation for 20 h showed significant signs of decomposition upon NMR spectroscopic reanalysis (Supplementary Fig. 13). Suitable elemental analysis was found for a batch that had been briefly washed with cold hexane (2 ml) and stored at − 30 °C for 12 days. After that period it was briefly dried in vacuum and submitted for combustion analysis without further delay. NMR spectroscopic analysis (C$_6$D$_6$) verified that the isolated product did not decompose during the prolonged low-temperature storage episode. In C$_6$D$_6$ solution **5** is sufficiently stable for the recording of NMR spectra but shows signs of decomposition after few hours at ambient temperature (Supplementary Fig. 14).

^{1}H NMR (500.1 MHz, C$_6$D$_6$): δ = 7.23 (t, J_{HH} = 8 Hz, 4 H, C$_6$*H*$_3$-4), 7.11 (d, J_{HH} = 8 Hz, 8 H, C$_6$*H*$_3$-3,5), 5.93 (s, 4 H, NC*H*), 4.07 (m, 8 H, C*H*$_2$CH$_3$), 3.42 (sept, J_{HH} = 7 Hz, 8 H, C*H*(CH$_3$)$_2$), 1.67 (s, 12 H, CC*H*$_3$), 1.27–1.20 (m, 60 H, CH(C*H*$_3$)$_2$ and CH$_2$C*H*$_3$); ^{13}C{^{1}H} NMR (125.8 MHz, C$_6$D$_6$): δ = 149.0 (C$_6$H$_3$-2,6), 137.9 (C$_6$H$_3$-1), 137.2 (NCN$_{imino}$), 128.4 (C$_6$H$_3$-4)^a, 123.5 (C$_6$H$_3$-3,5), 123.2 (CCH$_3$), 113.7 (NCH), 41.7 (CH$_2$CH$_3$), 28.7 (CH(CH$_3$)$_2$), 24.6 (CH(CH$_3$)$_2$), 24.3 (CH(CH$_3$)$_2$), 17.0 (CH$_2$CH$_3$), 8.0 (CCH$_3$), n.o. (NCN$_{carbene}$)^b. a = the resonance partially superimposes with the signal produced by the solvent but is assigned via ^{1}H^{13}C-HSQC NMR experiment; b = observation of the aluminum-bonded ^{13}C atoms is hampered by the quadrupolar momentum of the ^{27}Al nucleus.

Elemental analysis calculated (%) for $C_{72}H_{104}Al_2N_{10}Te_2$ [1418.86]: C 60.95, H 7.39, N 9.87; found: C 60.86, H 7.55, N 9.81; the analyte was provided wrapped in silver foil.

References

1. Levin, I. & Brandon, D. Metastable alumina polymorphs: crystal structures and transition sequences. *J. Am. Ceram. Soc.* **81,** 1995–2012 (1998).
2. Busca, G. The surface of transitional aluminas: a critical review. *Catal. Today* **226,** 2–13 (2014).
3. Lee, W. & Park, S.-J. Porous anodic aluminum oxide: anodization and templated synthesis of functional nanostructures. *Chem. Rev.* **114,** 7487–7556 (2014).
4. Kambe, N., Kondo, K. & Sonoda, N. Reduction of aromatic nitro, nitroso, hydroxylamino, azo, and azoxy compounds with hydrogen telluride from aluminum telluride and water. *Angew. Chem. Int. Ed. Engl.* **19,** 1009–1010 (1980).
5. Chivers, T. & Laitinen, R. S. Tellurium: a maverick among the chalcogens. *Chem. Soc. Rev.* **44,** 1725–1739 (2015).
6. Xia, M. *et al.* Aluminum-centered tetrahedron-octahedron transition in advancing Al-Sb-Te phase change properties. *Sci. Rep.* **5,** 8548 (2015).
7. Timoshkin, A. Y. & Schaefer, H. F. Theoretical studies of $[MYR_2]_n$ isomers ($M = B$, Al, Ga; $Y = N$, P, As; $R = H$, CH_3): structures and energetics of monomeric and dimeric compounds ($n = 1, 2$). *J. Phys. Chem. A* **112,** 13180–13196 (2008).
8. Himmel, H.-J., Downs, A. J. & Greene, T. M. Reactions of ground state and electronically excited atoms of main group elements: a matrix perspective. *Chem. Rev.* **102,** 4191–4241 (2002).
9. Young, N. A. Main group coordination chemistry at low temperatures: a review of matrix isolated group 12 to group 18 complexes. *Coord. Chem. Rev.* **257,** 956–1010 (2013).
10. Fischer, R. C. & Power, P. P. π-bonding and the lone pair effect in multiple bonds involving heavier main group elements: developments in the new millennium. *Chem. Rev.* **110,** 3877–3923 (2010).
11. Schulz, S. *et al.* A simple synthesis of $[(Cp^*Al)_4]$ and its conversion to the heterocubanes $[(Cp^*AlSe)_4]$ and $[(Cp^*AlTe)_4]$ ($Cp^* = \eta^5\text{-}C_5(CH_3)_5$). *Angew. Chem. Int. Ed. Engl.* **32,** 1729–1731 (1993).
12. Wehmschulte, R. J. & Power, P. P. A new synthetic route to organoalumoxanes $(RAlO)_n$: Synthesis of $(Mes^*AlO)_4$ ($Mes^* = -C_6H_2\text{-}2,4,6\text{-}t\text{-}Bu_3$) and its reactions with AlR_3 ($R = Me$ or Et). *J. Am. Chem. Soc.* **119,** 8387–8388 (1997).
13. Wehmschulte, R. J. & Power, P. P. Low-temperature synthesis of aluminum sulfide as the solvate $Al_4S_6(NMe_3)_4$ in hydrocarbon solution. *J. Am. Chem. Soc.* **119,** 9566–9567 (1997).
14. Zhu, H. *et al.* The selective preparation of an aluminum oxide and its isomeric C–H-activated hydroxide. *J. Am. Chem. Soc.* **127,** 10170–10171 (2005).
15. González-Gallardo, S., Cruz-Zavala, A. S., Jancik, V., Cortés-Guzmán, F. & Moya-Cabrera, M. Preparation of telluro- and selenoalumoxanes under mild conditions. *Inorg. Chem.* **52,** 2793–2795 (2013).
16. Neculai, D. *et al.* Formation and characterization of the first monoalumoxane, $LAlO \cdot B(C_6F_5)_3$. *Angew. Chem. Int. Ed.* **41,** 4294–4296 (2002).
17. Franz, D., Irran, E. & Inoue, S. Synthesis, characterization and reactivity of an imidazolin-2-iminato aluminium dihydride. *Dalton Trans.* **43,** 4451–4461 (2014).
18. Franz, D. & Inoue, S. Activation of elemental sulfur by aluminum dihydride: isolation of mono- and bis(hydrogensulfide) complexes of aluminum. *Chem. Eur. J.* **20,** 10645–10649 (2014).
19. Zheng, W. *et al.* Alumoxane hydride and aluminum chalcogenide hydride compounds with pyrazolato ligands. *Angew. Chem. Int. Ed.* **39,** 4276–4279 (2000).
20. De Frémont, P., Marion, N. & Nolan, S. P. Carbenes: synthesis, properties, and organometallic chemistry. *Coord. Chem. Rev.* **253,** 862–892 (2009).
21. Wang, Y. & Robinson, G. H. Unique homonuclear multiple bonding in main group compounds. *Chem. Commun.* 5201–5213 (2009).
22. Wang, Y. & Robinson, G. H. Carbene stabilization of highly reactive main-group molecules. *Inorg. Chem.* **50,** 12326–12337 (2011).
23. Hopkinson, M. N., Richter, C., Schedler, M. & Glorius, F. An overview of N-heterocyclic carbenes. *Nature* **510,** 485–496 (2014).
24. Wang, Y. *et al.* Stabilization of elusive silicon oxides. *Nat. Chem.* **7,** 509–513 (2015).
25. Franz, D. & Inoue, S. Systematic investigation of the ring-expansion reaction of N-heterocyclic carbenes with an iminoborane dihydride. *Chem. Asian. J.* **9,** 2083–2087 (2014).
26. Gardiner, M. G., Raston, C. L. & Tolhurst, V.-A. Alane reduction of selenium and tellurium: tertiary amine stabilised dimeric chalcogenides, *trans*-$[\{Me_3N(H)Al(\mu\text{-}E)\}_2]$ ($E = Se$, Te). *J. Chem. Soc., Chem. Commun.* 2501–2502 (1995).
27. Cui, C., Roesky, H. W., Noltemeyer, M. & Schmidt, H.-G. Synthesis of organoaluminum chalcogenides $[RAl(\mu\text{-}E)]_2$ ($R = N(SiMe_3)C(Ph)C(SiMe_3)_2$, $E = Se$, Te) from aluminum dihydride $[RAlH(\mu\text{-}H)]_2$. *Organometallics* **18,** 5120–5123 (1999).
28. Jancik, V. *et al.* Phosphane-catalyzed reactions of $LAlH_2$ with elemental chalcogens; preparation of $[LAl(\mu\text{-}E)_2AlL]$ [$E = S$, Se, Te, $L = HC\{C(Me)N(Ar)\}_2$, $Ar = 2,6\text{-}iPr_2C_6H_3$]. *Eur. J. Inorg. Chem.* 3508–3512 (2004).
29. Uhl, W. & Schütz, U. Monomeric $[(Me_3Si)_2CH]_2Al\text{–}Te\text{–}Al[CH(SiMe_3)_2]_2$ with a bent Al–Te–Al group. *Z. Naturforsch.* **49b,** 931–934 (1994).
30. Li, L. *et al.* A stable germanone as the first isolated heavy ketone with a terminal oxygen atom. *Nat. Chem.* **4,** 361–365 (2012).
31. Xiong, Y., Yao, S., Müller, R., Kaupp, M. & Driess, M. From silicon(II)-based dioxygen activation to adducts of elusive dioxasiliranes and sila-ureas stable at room temperature. *Nat. Chem.* **2,** 577–580 (2010).
32. Ahmad, S. U., Szilvási, T., Irran, E. & Inoue, S. An NHC-stabilized silicon analogue of acylium ion: synthesis, structure, reactivity, and theoretical studies. *J. Am. Chem. Soc.* **137,** 5828–5836 (2015).
33. Frenking, G. *et al.* New bonding modes of carbon and heavier group 14 atoms Si–Pb. *Chem. Soc. Rev.* **43,** 5106–5139 (2014).
34. Zingaro, R. A., Steeves, B. H. & Irgolic, K. Phosphine tellurides. *J. Organomet. Chem.* **4,** 320–323 (1965).
35. Kuhn, N. & Kratz, T. Synthesis of imidazol-2-ylidenes by reduction of imidazol-2(3*H*)-thiones. *Synthesis* 561–562 (1993).
36. Grimme, S. Semiempirical GGA-type density functional constructed with a long-range dispersion correction. *J. Comput. Chem.* **27,** 1787–1799 (2006).
37. Frisch, M. J., Pople, J. A. & Binkley, J. S. Self-consistent molecular orbital methods 25. Supplementary functions for Gaussian basis sets. *J. Chem. Phys.* **80,** 3265–3269 (1984).
38. Dunning, T. H. Gaussian basis sets for use in correlated molecular calculations. I. The atoms boron through neon and hydrogen. *J. Chem. Phys.* **90,** 1007–1023 (1989).
39. Schuchardt, K. L. *et al.* Basis Set Exchange: A community database for computational sciences. *J. Chem. Inf. Model* **47,** 1045–1052 (2007).
40. Chai, J.-D. & Head-Gordon, M. Long-range corrected hybrid density functionals with damped atom–atom dispersion corrections. *Phys. Chem. Chem. Phys.* **10,** 6615–6620 (2008).
41. Marenich, A. V., Cramer, C. J. & Truhlar, D. G. Universal solvation model based on solute electron density and a continuum model of the solvent defined by the bulk dielectric constant and atomic surface tensions. *J. Phys. Chem. B* **113,** 6378–6396 (2009).
42. Frisch, M. J. *et al. Gaussian 09, Revision B.01* (Gaussian, Inc., 2009).

Acknowledgements

Financial support of the Alexander von Humboldt foundation (Sofja Kovalevskaja Program) and computational time under the grant of the New Széchenyi Plan TAMOP-4.2.2/B-10/1-2010-0009 are gratefully acknowledged. We are indebted to Carsten Eisenhut for a generous supply of L^{Et}.

Author contributions

D.F. conceived and performed the synthetic experiments. T.S. designed and performed the theoretical analyses. E.I. collected and solved the X-ray diffraction data. S.I. conceived and supervised the study. D.F., T.S. and S.I. discussed the results and co-wrote the manuscript.

Additional information

Accesion Codes: The X-ray crystallographic coordinates for structures reported in this Article have been deposited at the Cambridge Crystallographic Data Centre (CCDC), under deposition number CCDC 1415006-1415008. These data can be obtained free of charge from The Cambridge Crystallographic Data Centre via www.ccdc.cam.ac.uk/data_request/cif.

Competing financial interests: The authors declare no competing financial interests.

6

Coordination templated [2 + 2 + 2] cyclotrimerization in a porous coordination framework

Yong-Sheng Wei[1], Mei Zhang[1], Pei-Qin Liao[1], Rui-Biao Lin[1], Tai-Yang Li[1], Guang Shao[1], Jie-Peng Zhang[1] & Xiao-Ming Chen[1]

Controlling chemical reactions by the supramolecular confinement effects of nanopores has attracted great attention. Here we show that open metal sites in porous coordination frameworks can constitute more powerful and strict templates for precision syntheses. A Fe(III) dicarboxylate framework functionalized with triangularly arranged metal sites is used to accomplish [2+2+2] cyclotrimerization reactions for organonitrile, alkyne and alkene monomers bearing a geometrically suitable pyridyl group. *In situ* single-crystal X-ray diffraction facilitates the direct observation of such a coordination templated reaction, before cylcotrimerization, the monomer coordinates at the Fe(III) centre by its pyridyl donor, which forces three unsaturated groups to gather around a position very similar with that of the desired covalent cyclic trimer. After the reaction, the trimers serve as tripodal ligands to perfectly fix the Fe(III) ions and the whole crystal to generate an exceptionally rigid and porous material with large surface area coupled with guest-proof zero thermal expansion.

[1] MOE Key Laboratory of Bioinorganic and Synthetic Chemistry, School of Chemistry & Chemical Engineering, Sun Yat-Sen University, Guangzhou 510275, China. Correspondence and requests for materials should be addressed to J.-P.Z. (email: zhangjp7@mail.sysu.edu.cn).

Besides many useful applications such as storage, separation, catalysis, and so on, porous materials may also serve as nano-reactors to control the molecular structures and morphologies of the products by restricting the space around the guest reactants/products, and even form host-guest nanocomposites with interesting properties[1–3]. Compared with the conventional bulk reactions, the structrues, molecular weights and stereoregularity of polymer products obtained in porous hosts[4], especially periodic/crystalline ones with microporous channels[5–7], can be regulated to much higher levels. When the reactant and products are both small molecules, organic reactions such as Diels-Alder addition[8], isomerization[9] and photodimerization[10] can be more rationally controlled[11], because the guest molecules can fit the periodicity of the pore environment to enable good host-guest matching[12]. More importantly, some reaction stages may be directly visualized by single-crystal X-ray diffraction (SCXRD)[11,13,14], although keeping the sample single-crystallinity after physical/chemical changes is always a great challenge[15–23].

Because the pore and the guest recognize each other by hydrogen bonding, π–π stacking, and more commonly other weaker supramolecular interactions, their sizes and shapes must be commensurate so that the guest reactants can be fixed at the right positions to achieve good control over the chemical reaction. In other words, the host always loses porosity after carrying out such precision chemical syntheses. On the other hand, by virtue of the strong and highly directional coordination bonds, metal ions and organic bridging ligands can self-assemble to build crystalline open frameworks (well known as porous coordination polymers or metal-organic frameworks) with overwhelmingly high porosity and robustness compared with other molecular materials[24–29]. In many cases, the coordination sites of the metal ions are not fully used by the bridging ligands. The unoccupied or open metal sites can bind guest molecules strongly at specific positions/directions, which should be ideal for development of strictly controlled reactors/reactions. Herein we prove this concept by using a prototypical material $[Fe_3(\mu_3\text{-}O)(bdc)_3X(H_2O)_2]$ (MIL-88B (refs 30–32); $H_2bdc = 1,4$-benzenedicarboxylic acid, $X = OH^-/Cl^-/F^-$) as the host coordination template. A series of facile and strictly selective [2 + 2 + 2] coordination templated cyclotrimerization (CTC) reactions have been realized. Unlike other molecular reactors, the reaction can drastically improve the gas adsorption performance of the host crystal.

Results

Synthesis and structure. MIL-88B is well known for its highly flexible behaviours towards various organic molecules[30–32]. An interesting structural feature of MIL-88B is that every three adjacent Fe(III) ions locating at the same channel (from three $Fe_3(\mu_3\text{-}O)$ clusters) defines a regular triangle with converging Fe-H_2O/X vectors. The highly directional coordination bonds of the octahedral Fe(III) ions with such an alignment may fix three molecules in an ideal configuration to facilitate the [2 + 2 + 2] CTC reactions (Fig. 1). Considering the geometry of such a Fe_3 coordination template, we selected 4-cyanopyridine (4-pyCN), 4-ethynylpyridine (4-pyC_2H) and 4-vinylpyridine (4-pyC_2H_3), each bearing a coordination donor at one end and an unsaturated group at the opposite end, as the reactants or monomers.

The CTC reactions were accomplished by loading the monomers in microcrystalline MIL-88B, removing excess monomers outside the microcrystal host, and then heating at 140 °C within 12 h, or simply heating a mixture of MIL-88B and excess monomers at the same condition. Powder X-ray diffraction (PXRD) showed that, the unit-cell volumes of the resultant crystalline products are even larger than that of the most-open form of MIL-88B (Supplementary Figs 2 and 3)[31,32]. The desired trisubstituted *s*-triazine, benzene, and cyclohexane derivatives, namely 2,4,6-tri(pyridin-4-yl)-1,3,5-triazine (tpt), 1,3,5-tri(pyridin-4-yl)benzene (tpb) and 1,3,5-tri(pyridin-4-yl)cyclohexane (tpc), were first detected by infrared (IR) spectroscopy and electrospray ionization mass spectrometry (ESI-MS) analyses of the resultant crystalline products (Supplementary Figs 4 and 5). The formation of these cyclic trimers inside the Fe_3 coordination templates was further confirmed by SCXRD. It should be noted that, MIL-88B has been always synthesized as microcrystalline powders[30,33], but we now succeeded to grow large single crystals by using hydrofluoric acid to replace the conventionally used NaOH reactant. Finally, the organic products were recovered from the acid-digested crystalline products and fully characterized by IR, ESI-MS, ^1H and ^{13}C nuclear magnetic resonance (NMR) spectroscopy (Supplementary Figs 6-11).

On the basis of the SCXRD data, the chemical formulae of the product crystals obtained by the CTC reactions are determined as $[Fe_3(\mu_3\text{-}O)(bdc)_3(tpt)]X$ (MIL-88B-tpt), $[Fe_3(\mu_3\text{-}O)(bdc)_3(tpb)]X$ (MIL-88B-tpb) and $[Fe_3(\mu_3\text{-}O)(bdc)_3(tpc)]X$ (MIL-88B-tpc), respectively, (Fig. 2 and Supplementary Table 1). Besides the original 6-connected **acs** networks of $[Fe_3(\mu_3\text{-}O)(bdc)_3]$, each Fe_3 coordination template embeds a tpt/tpb/tpc ligand, which coordinates with the three Fe(III) ions by its three pyridyl ends (Fig. 2 and Supplementary Fig. 1). The coordination networks of the product crystals possess a rare binodal (3,9)-connected **nia-d** topology[34], regarding the tpt/tpb/tpc ligands and the Fe_3 clusters as 3- and 9-connected nodes, respectively. Interestingly, the unit-cell volumes of the product crystals are ca. 145, 18 and 9% larger than those of the dry, as-synthesized, and most-open forms of MIL-88B, respectively[31], because the large tpt/tpb/tpc ligands force the host framework to expand to an extent not possible for lattice guest molecules.

Reaction mechanism. To verify the CTC reaction mechanism, control experiments were performed. Using 3-cyanopyridine (3-pyCN) and benzonitrile (PhCN) as monomers (Fig. 3d,e), the trimeric product of 3-pyCN/PhCN was not detected by either IR or ESI-MS of the reaction mixtures (Supplementary Figs 12 and 13). When a mixture of 4-pyCN and PhCN was used to perform the same reaction, only tpt was detected among four possible trimeric products (Fig. 3f and Supplementary Fig. 14). These results confirmed that MIL-88B is not able to catalyse the trimerization reactions involving monomers with unsuitable coordination sites and molecular geometry. More importantly, when excess 4-pyCN monomers were used for the CTC reaction, tpt cannot be detected out of the Fe_3 coordination templates (Supplementary Fig. 15), meaning that in MIL-88B-tpt, the trimeric product tpt is solely generated at the Fe_3 coordination templates. Note that the CTC reactions can be furnished in very high yields (>90%) without using excess monomers (Supplementary Figs 3–5), confirming the rule of the Fe_3 coordination template and exclude the possibilities of cyclotrimerization at other sites such as a single metal centre locating at defects and/or outer surface of the crystals.

To reveal more details for the CTC reaction mechanism, we carried out an *in situ* SCXRD study. The unit-cell volume of the crystal stabilized at a value of 16% larger than that of as-synthesized MIL-88B but 1.0% smaller than MIL-88B-tpt, after heated to 110 °C for 24 h. Structure analysis revealed a stable composition of $[Fe_3(\mu_3\text{-}O)(bdc)_3(4\text{-}pyCN)X(H_2O)]$ (denoted as 4-pyCN@MIL-88B, Supplementary Table 1), indicating that one third of terminal H_2O/X ligands are substituted by 4-pyCN molecules. From the crystal structure (Fig. 2b), it can be clearly seen that 4-pyCN coordinates at the Fe(III) ion using its pyridyl

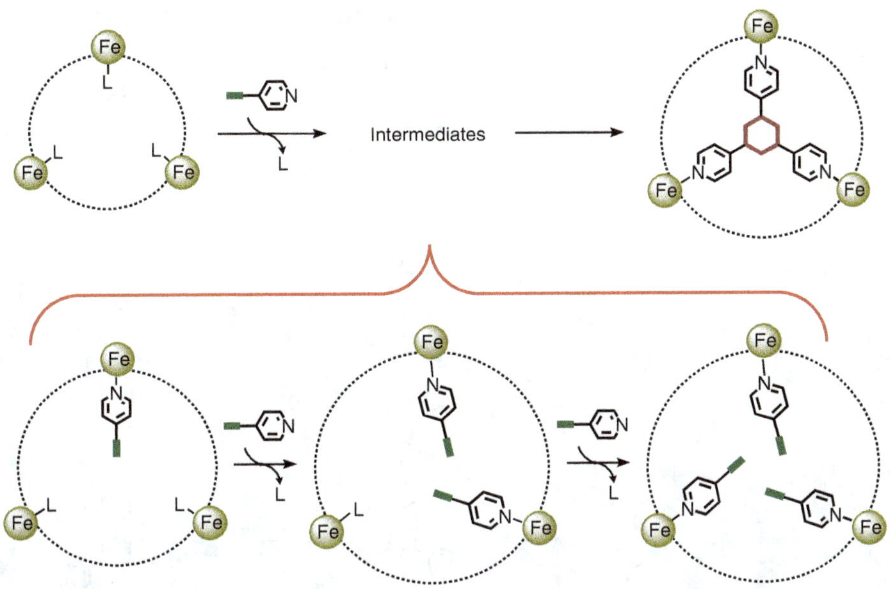

Figure 1 | The proposed CTC reaction mechanism. The monomer molecule bearing an unsaturated functional group (thick green stick) and an anchor (the pyridyl group) replaces the terminal ligand (L) originally coordinated at the Fe(III) site. The convergent and triangular arrangement of the Fe(III) sites forces every three unsaturated groups of the monomer ligands to gather in a close configuration suitable for the cyclotrimerization reaction. The ligand replacement occurs at a step-by-step manner.

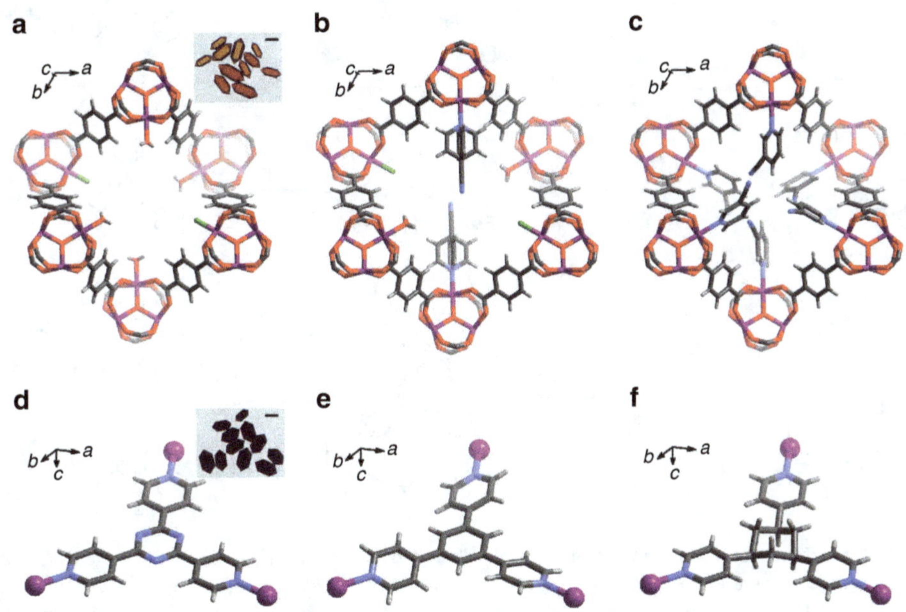

Figure 2 | X-ray single-crystal structures. (**a**) MIL-88B. (**b**) 4-pyCN@MIL-88B. (**c**) 3-pyCN@MIL-88B. (**d-f**) The trimers tpt, tpb and tpc embedded in MIL-88B-tpt, MIL-88B-tpb and MIL-88B-tpc, respectively. Color codes: dark grey, C; light grey, H; blue, N; red, O; purple, Fe; green, F. The pyridyl ring of 4-pyCN is two-fold disordered at conformations parallel and perpendicular with the Fe_3 plane. Some other crystal structures are symmetrically disordered, only one possible structure is shown for clarity. Inset: photographs of corresponding compounds (Scale bars, 0.10 mm).

N donor, so that its cyano group is pointed at the centre of the Fe_3 coordination template. The pyridyl ring of 4-pyCN in 4-pyCN@MIL-88B showed obvious conformational disorder, because the pyridyl C-H moieties has no specific interaction with other parts. In contrast, the pyridyl and triazine rings of tpt in MIL-88B-tpt are strictly coplanar (Fig. 2d). Further increasing the heating time and/or temperature resulted in monotonic and small increase of the unit-cell volume until a stable value virtually identical with that of MIL-88B-tpt, which was not changed after heated at 180 °C for 24 h. As shown in Supplementary Fig. 16, similar phenomena can be also observed by PXRD. Structure analysis confirmed that the crystal at this stage is identical with MIL-88B-tpt (Supplementary Table 1), meaning that the cyclotrimerization reaction was completed.

For comparison, 3-pyCN was also used for the *in situ* SCXRD study. Structure analysis of the crystal above the melting point of 3-pyCN revealed a chemical formula of $[Fe_3(\mu_3\text{-}O)(bdc)_3(3\text{-pyCN})_3]X$ (denoted as 3-pyCN@MIL-88B, Supplementary Table 1), meaning that each Fe(III) ion is coordinated by one 3-pyCN, or all terminal H_2O/X ligands are completely substituted. Because the cyano group locates at the meta-position of the pyridyl ring, there is no steric hindrance among three 3-pyCN ligands in the same Fe_3 coordination template (Fig. 2c). Unlike 4-pyCN, the pyridyl ring of 3-pyCN is almost perpendicular with

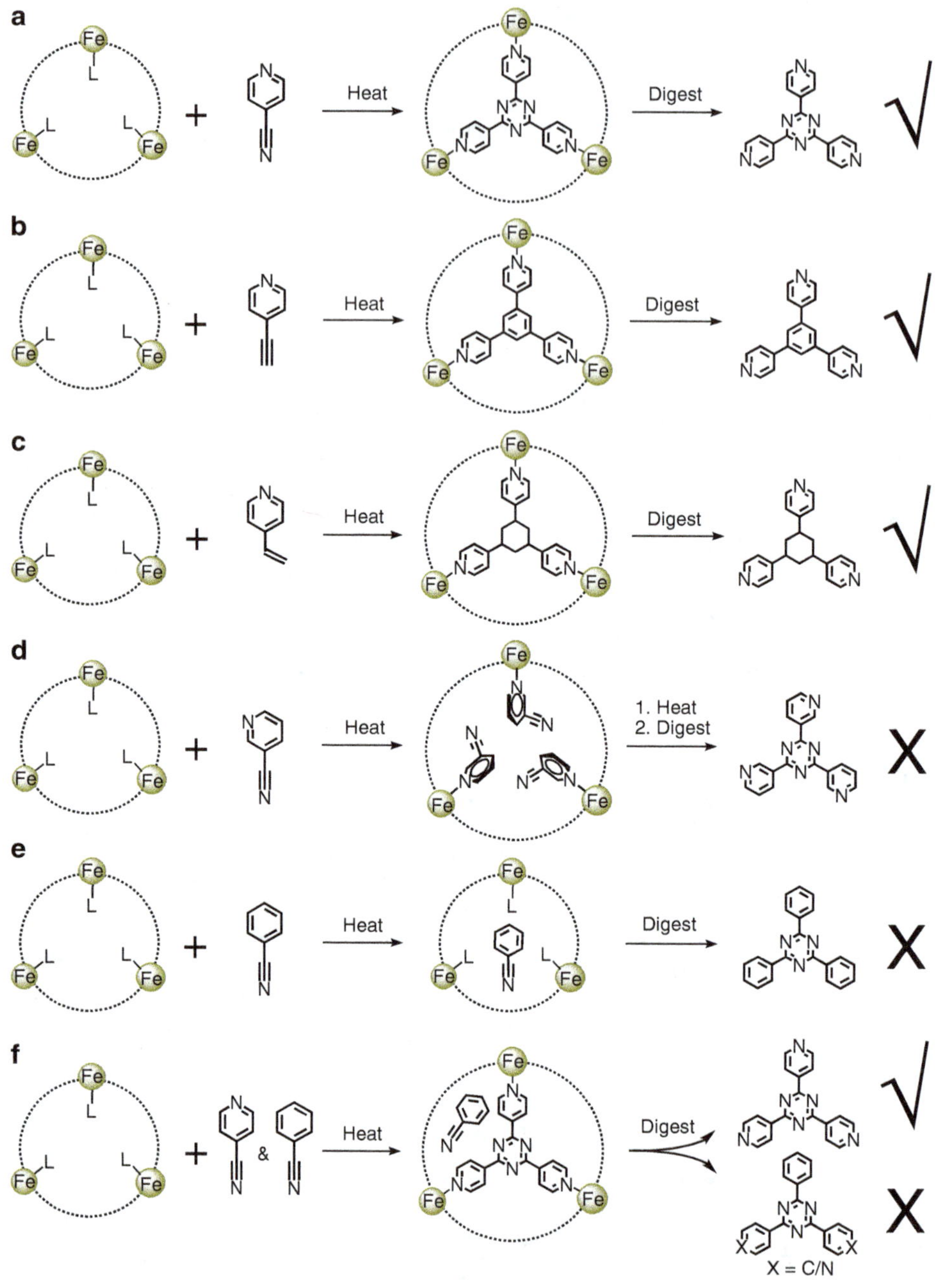

Figure 3 | Summary of the CTC reactions for various monomers. (**a–c**) When 4-pyCN, 4-pyC_2H, and 4-pyC_2H_3 were used as monomers, the corresponding trimeric products were observed by ESI-MS, IR, ^1H NMR, ^{13}C NMR and SCXRD. (**d**) When 3-pyCN was used as the monomer, the coordination of 3-pyCN molecules on the Fe(III) centre was observed by SCXRD, but the expected trimeric product could not be detected by ESI-MS, IR or SCXRD. (**e**) When PhCN was used as the monomer, the expected trimeric product was not detected by ESI-MS, IR or SCXRD. (**f**) When a mixture of 4-pyCN and PhCN was used as monomers, only tpt was detected by ESI-MS among four possible trimeric products.

the Fe_3 plane, because with this conformation the cyano group can form weak hydrogen bonds (N···C 3.81(4) Å) with the pyridyl ring of a neighbour locating at another Fe_3 coordination template (Supplementary Fig. 17). In other words, the cyano groups of 3-pyCN monomers are evenly distributed in 3-pyCN@MIL-88B and fixed at positions distinctly unfit for the trimerization reaction. Consequently, the corresponding triazine product of 3-pyCN cannot be obtained by further increasing the heating time/temperature.

Modification of flexibility and porosity. Due to the additional tripodal ligands, the product crystals obtained by the CTC reactions display very different framework flexibility and adsorption property compared with MIL-88B. Thermogravimetry analyses and variable-temperature PXRD showed that MIL-88B-tpt is a rigid framework stable up to 400 °C (Supplementary Figs 18 and 19). Variable-temperature SCXRD at 100–455 K showed that guest and temperature can barely affect the unit-cell parameters of MIL-88B-tpt, giving zero thermal expansion across the *ab*-plane and very small negative thermal expansion along the *c*-axis (Fig. 4a and Supplementary Figs 19b and 20), which is exceptional for such a porous structure[35–37]. Further, guest-free MIL-88B-tpt showed a type-I N_2 isotherm, corresponding with a Brunauer–Emmett–Teller surface area of 1,113 $m^2 g^{-1}$ and a Langmuir surface area of 1,331 $m^2 g^{-1}$, as well as a pore volume of 0.56 $cm^3 g^{-1}$. In contrast, guest-free MIL-88B showed a type-II N_2 isotherm with very small uptake, surface area, and pore

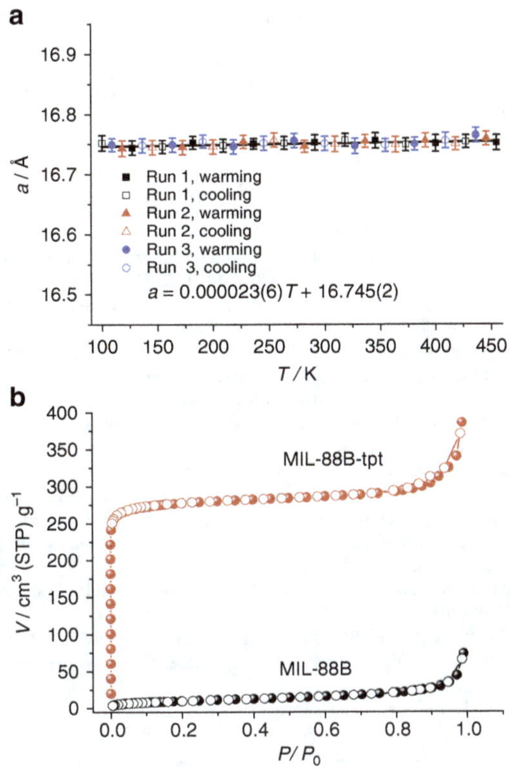

Figure 4 | Framework flexibility and gas adsorption property of the host-guest composite. (**a**) Temperature- and guest-dependences of the *a*-axis length of MIL-88B-tpt obtained in several heating/cooling cycles. The as-synthesized crystal of MIL-88B-tpt was used to start the measurement. Obviously, the *a*-axis length is nearly unaffected by guest and temperature, giving zero thermal expansion across the *ab*-plane. (**b**) N_2 adsorption (filled) and desorption (open) isotherm of MIL-88B and MIL-88B-tpt.

volume (Fig. 4b), because the shrunk coordination framework of MIL-88B can be hardly opened by N_2 molecules[38], although it can undergo significant solvent-induced breathing.

Discussion

In conventional approaches, cyano, alkyne and alkene groups are quite different in reactivity. Cyclotrimerization of organonitriles to form 1,3,5-triazine derivatives is stereo selective, but both catalyst and high reaction temperature (>200 °C) are required[39,40]. The synthesis of benzene derivatives through catalysed cyclotrimerization of alkynes, which usually forms 1,2,4- and 1,3,5-trisubstituted isomers as a mixture, concomitant with polymers as byproducts, has been studied extensively[41]. Without catalyst, alkenes can easily polymerize to form long chains. On the other hand, when two alkene molecules are properly aligned (usually in a crystal) with separation less than 4.1 Å, photo/thermochemical [2+2] cycloaddition can occur to generate cyclobutane with high stereo selectivity[9,15]. Nevertheless, by using weak supramolecular interactions and pore confinement effects, three alkene molecules can be hardly fixed in a configuration suitable for [2+2+2] cyclotrimerization, so that cyclohexanes are generally obtained by hydrogenation of benzenes. Obviously, MIL-88B greatly facilitates [2+2+2] cyclotrimerization reactions of organonitrile, alkyne, and alkene monomers bearing a geometrically suitable pyridyl group, highlighting the coordination templating effects of its open metal sites.

The results of *in situ* SCXRD showed that the intermediate state with a formula of $[Fe_3(\mu_3\text{-}O)(bdc)_3(4\text{-}pyCN)_3]X$, that is, three 4-pyCN monomers simultaneously coordinate at the same Fe_3 coordination template, cannot be observed by diffraction techniques that are suitable for characterizing thermodynamic equilibrium structures[15–19]. Obviously, the supramolecular trimer $(4\text{-}pyCN)_3$ has a larger size compared with the covalent trimer tpt, meaning that the unit-cell volume of $[Fe_3(\mu_3\text{-}O)(bdc)_3(4\text{-}pyCN)_3]X$ needs to be larger than that of MIL-88B-tpt. However, the unit-cell volume of the most-open form of MIL-88B, obtained by inclusion of large organic molecules[31,32], is still smaller than that of MIL-88B-tpt, indicating that $[Fe_3(\mu_3\text{-}O)(bdc)_3(4\text{-}pyCN)_3]X$ with an even larger unit-cell volume should be energetically unfavourable, at least at the thermodynamic equilibrium state. In other words, such a theoretical intermediate state may be too reactive, which occurs transiently and transforms to the trimeric product immediately. Therefore, the CTC reaction might be mainly controlled by the ligand substitution reaction rather than the organic reaction. Indeed, it has been demonstrated that other organic molecules, including pyridine (with a stronger coordination ability compared with 4-pyCN), always serve as pure lattice guests, and cannot substitute the H_2O/X ligand in all MIL-88 series at room temperature[31,32]. On the other hand, while there is no conventional catalytic site in MIL-88B, the 4-pyCN to tpt transformation occurs at a relatively low temperature, which implies that the flexible coordination network and the Fe_3 coordination template, having a large tendency to shrink at the transient intermediate state $[Fe_3(\mu_3\text{-}O)(bdc)_3(4\text{-}pyCN)_3]X$, might compress the supramolecular trimer $(4\text{-}pyCN)_3$ and force it to form the covalent trimer tpt.

In summary, we demonstrated that with a judicious geometric consideration of the host-guest structure, the open metal sites in a porous coordination framework can be rationally used to fix guest molecules to facilitate chemical reaction difficult or even impossible for conventional methods and other nano-reactors. Importantly, the template crystal can retain single-crystallinity throughout the supramolecular and molecular reaction processes, allowing the direct observation of several key reaction stages. Also interestingly, the precisely controlled reaction can drastically improve the framework rigidity and gas adsorption performance of the crystalline host. These results should be instructive for further exploration of new chemical reactions and functions using porous crystals as a platform.

Methods

Materials and physical measurements. Reagents were commercially available and used without further purification except otherwise stated. The monomer $4\text{-}pyC_2H_3$ was distilled to remove polymerization inhibitor hydroquinone before use. ESI-MS was obtained using a Shimadzu LCMS-2010A mass spectrometer, for which samples were digested by sonication in a mixture of hydrochloric acid and MeOH after removal of the residual monomers by sublimation under vacuum. Elemental analyses (EA) were conducted using an Elementar Vario EL analyzer using guest-free samples (with a little water adsorbed from air). IR spectra were recorded on a Bruker TENSOR 27 FT-IR spectrometer in the 400–4,000 cm^{-1} region with KBr pellets. ^1H NMR and ^{13}C NMR spectra were recorded on a Bruker AVANCE III 400 MHz NMR spectrometer. Chemical shifts were quoted in p.p.m. referenced to the appropriate solvent peak or 0.0 p.p.m. for tetramethylsilicane. PXRD patterns were collected on a Bruker D8-Advance diffractometer with Cu Kα radiation and a LynxEye detector. Thermogravimetry analyses were performed on a TA Q50 system under N_2 at a heating rate of 5 °C min. N_2 sorption isotherms were measured by a Micromeritics ASAP 2020M physisorption analyzer.

Synthesis of MIL-88B. Microcrystals[30,33]: A mixture of $FeCl_3 \cdot 6H_2O$ (4.32 g, 16.0 mmol) and H_2bdc (2.66 g, 16.0 mmol) were dispersed in *N,N*-dimethylformamide (DMF; 80.0 ml) with 2 M NaOH (6.4 ml). The mixture was placed in a 100-ml vial and sealed with a screw cap, which was heated at 100 °C for 12 h. Then, the orange solid was filtered, washed with DMF, and dried in air overnight to give $[Fe_3(\mu_3\text{-}O)Cl(H_2O)_2(bdc)_3] \cdot$ guest (yield 60%). The content of chloride ion was determined by potentiometric titration with the acid-digested samples. Large single crystals: a mixture of $FeCl_3 \cdot 6H_2O$ (0.081 g, 0.3 mmol), H_2bdc (0.050 g, 0.3 mmol), hydrofluoric acid (0.1 ml, 49 wt% in H_2O), and DMF (3.0 ml) was stirred for 30 min in air, then transferred and sealed in a 20-ml vial with a screw cap, which was heated at 90 °C for 3 days and then cooled to room temperature, giving orange block crystals, $[Fe_3(\mu_3\text{-}O)F_{0.74}(OH)_{0.26}(H_2O)_2(bdc)_3] \cdot$ guest (yield 50%). The content of fluoride ion was determined by the fluoride selective electrode with the acid-digested samples.

Coordination templated cyclotrimerization reactions. The as-synthesized microcrystals of MIL-88B were exchanged with hot MeOH using a Soxhlet extractor to remove the unreacted $FeCl_3$ and H_2bdc in the channels. The reaction method without using excess monomers: the monomer was loaded into MIL-88B (0.100 g) by mixing it with 4-pyCN (1.00 g) and heated at 110 °C for 24 h, immersing it in an anhydrous toluene solution (20.0 mL) of 4-pyC_2H (0.100 g) and heated at 50 °C for 6 h under N_2, or by immersing it in 4-pyC_2H_3 (1.0 ml) at room temperature for 3 days; After removal of the excess monomers by sublimation for 4-pyCN or washing with CH_2Cl_2 for 4-pyC_2H and 4-pyC_2H_3, the monomer-loaded MIL-88B was transferred and sealed in a 1-ml glass tube, heated in an oven at 140 °C for 12 h and then cooled to room temperature, giving black products (yields: 92%, 94% and 91%, respectively). The reaction method using excess monomers: a mixture of MIL-88B (0.100 g) and the monomer, that is, 4-pyCN (3.00 g) or DMF (3.0 ml) solution of 4-pyC_2H (0.140 g) or 4-pyC_2H_3 (3.0 ml) sealed in a 25-ml teflon-lined stainless container, and the products were finally washed by DMF and dried in air (yield > 90%). MIL-88B-tpt: EA calc. (%) for $[Fe_3(\mu_3\text{-}O)(bdc)_3(tpt)]Cl \cdot 9H_2O$ ($C_{42}H_{42}ClFe_3N_6O_{22}$): C, 43.14; H, 3.62; N, 7.19; found: C, 44.09; H, 3.92; N, 7.03. MIL-88B-tpb: EA calc. (%) for $[Fe_3(\mu_3\text{-}O)(bdc)_3(tpb)]Cl \cdot 14H_2O$ ($C_{45}H_{55}ClFe_3N_3O_{27}$): C, 42.46; H, 4.36; N, 3.30; found: C, 42.41; H, 4.51; N, 3.30. MIL-88B-tpc: EA calc. (%) for $[Fe_3(\mu_3\text{-}O)(bdc)_3(tpc)]Cl \cdot 7H_2O$ ($C_{45}H_{47}ClFe_3N_3O_{20}$): C, 46.88; H, 4.11; N, 3.64; found: C, 47.41; H, 4.24; N, 3.57. For samples used for SCXRD studies, the as-synthesized single crystals of MIL-88B were directly used, and the reactions were carried out at 180 °C for 24 h.

Separation of covalent trimers. The host-guest hybrid crystals were firstly dissolved in concentrated hydrochloric acid with stirring for 20 min. After removing the undissolved white solid (H_2bdc) by filtration, pH of the solution was adjusted to ~11 with aqueous NaOH (1 M). After that, the aqueous solution was extracted with CH_2Cl_2 for three times. The combined organic phase was dried over anhydrous Mg_2SO_4 and concentrated to remove solvent under reduced pressure, finally giving white products.

Single-crystal X-ray diffraction analyses. Single-crystal X-ray diffraction data were collected on an Oxford Gemini-S Ultra CCD-detector or a Rigaku XtaLAB P300DS-detector diffractometer (Cu Kα). The measurement temperatures were controlled by a dry nitrogen flow using a Rigaku Gas Flow GN2 apparatus. For the *in situ* SCXRD study, a single crystal of MIL-88B and excess amount of monomer were sealed in a glass capillary. Nevertheless, identical single-crystal structure can be also obtained without using excess monomer. The diffraction data were collected just above the melting point of the monomer to avoid the interference of diffraction of crystalline monomer. All structures were solved by the direct method and refined with the full-matrix least-squares technique on F^2 by the *SHELXTL* software package. The terminal coordinated molecules/anions, namely, H_2O, OH^- and F^- in MIL-88B, H_2O, OH^-, F^- and 4-pyCN in 4-pyCN@MIL-88B, are statistically and symmetrically disordered at the three terminal sites (crystallography three-fold symmetry) of one $Fe_3(\mu_3\text{-}O)(bdc)_3(L^T)_3$ (L^T = terminal ligand). Since the donor atoms occupy essentially the same positions and cannot be distinguished by X-ray diffraction, they were refined using the EADP and EXYZ constraints. In 4-pyCN@MIL-88B and MIL-88B-tpc, the pyridyl rings of 4-pyCN and tpc are statistically/dynamically two-fold disordered at conformations parallel and perpendicular with the Fe_3 plane. In MIL-88B-tpb, the pyridyl rings of tpb are statistically/dynamically/symmetrically two-fold disordered with molecular planes about 18° with the Fe_3 plane. In 3-pyCN@MIL-88B, the coordinated 3-pyCN molecule is symmetrically four-fold disordered because the bent molecule is locates at a two-fold axis and a mirror plane. Moreover, in MIL-88B-tpc, the non-planar cyclohexane ring is in the *e,e,e*-conformation and symmetrically two-fold disordered at the crystallography mirror plane. The SHELXTL restraint instructions DFIX, SIMU, ISOR and FLAT were applied to these disordered groups to keep their geometries and atomic displacement parameters reasonable. All host-framework non-hydrogen atoms were refined anisotropically. Hydrogen atoms were placed geometrically. The PLATON SQUEEZE treatment was applied because all lattice guests are extremely disordered and cannot be modelled. Detailed structure determination parameters and crystallographic data are given in Supplementary Table 1.

References

1. Farina, M., Silvestro, G. D. & Sozzani, P. in *Comprehensive Supramolecular Chemistry*. (ed. Reinhoudt, D.) 371 (Pergamon: Oxford, 1996).
2. Sivaguru, J. *et al.* Asymmetric photoreactions within zeolites: role of confinement and alkali metal ions. *Acc. Chem. Res.* **36,** 509–521 (2003).
3. Tajima, K. & Aida, T. Controlled polymerizations with constrained geometries. *Chem. Commun. (Camb)* 2399–2412 (2000).
4. Soler-Illia, G. J. A. A. & Azzaroni, O. Multifunctional hybrids by combining ordered mesoporous materials and macromolecular building blocks. *Chem. Soc. Rev.* **40,** 1107–1150 (2011).
5. Distefano, G. *et al.* Highly ordered alignment of a vinyl polymer by host–guest cross-polymerization. *Nat. Chem.* **5,** 335–341 (2013).
6. Uemura, T., Uchida, N., Higuchi, M. & Kitagawa, S. Effects of unsaturated metal sites on radical vinyl polymerization in coordination nanochannels. *Macromolecules* **44,** 2693–2697 (2011).
7. Uemura, T., Yanai, N. & Kitagawa, S. Polymerization reactions in porous coordination polymers. *Chem. Soc. Rev.* **38,** 1228–1236 (2009).
8. Ikemoto, K., Inokuma, Y. & Fujita, M. Diels-Alder via molecular recognition in a crystalline molecular flask. *J. Am. Chem. Soc.* **133,** 16806–16808 (2011).
9. Yang, S.-Y. *et al.* Crystallographic snapshots of the interplay between reactive guest and host molecules in a porous coordination polymer: stereochemical coupling and feedback mechanism of three photoactive centers triggered by UV-induced isomerization, dimerization, and polymerization reactions. *J. Am. Chem. Soc.* **136,** 558–561 (2013).
10. Zheng, S.-L., Pham, O., Vande Velde, C. M. L., Gembicky, M. & Coppens, P. Competitive isomerization and dimerization in co-crystals of 1,1,6,6-tetraphenyl-2,4-hexadiyne-1,6-diol and sorbic acid: a new look at stereochemical requirements for [2 + 2] dimerization. *Chem. Commun. (Camb)* 2538–2540 (2008).
11. Inokuma, Y., Kawano, M. & Fujita, M. Crystalline molecular flasks. *Nat. Chem.* **3,** 349–358 (2011).
12. MacGillivray, L. R. *et al.* Supramolecular control of reactivity in the solid state: from templates to ladderanes to metal-organic frameworks. *Acc. Chem. Res.* **41,** 280–291 (2008).
13. Bloch, W. M. *et al.* Capturing snapshots of post-synthetic metallation chemistry in metal-organic frameworks. *Nat. Chem.* **6,** 906–912 (2014).
14. Ikemoto, K., Inokuma, Y., Rissanen, K. & Fujita, M. X-ray snapshot observation of palladium-mediated aromatic bromination in a porous complex. *J. Am. Chem. Soc.* **136,** 6892–6895 (2014).
15. Kole, G. K. & Vittal, J. J. Solid-state reactivity and structural transformations involving coordination polymers. *Chem. Soc. Rev.* **42,** 1755–1775 (2013).
16. Zhang, J.-P., Liao, P.-Q., Zhou, H.-L., Lin, R.-B. & Chen, X.-M. Single-crystal X-ray diffraction studies on structural transformations of porous coordination polymers. *Chem. Soc. Rev.* **43,** 5789–5814 (2014).
17. Kawano, M. & Fujita, M. Direct observation of crystalline-state guest exchange in coordination networks. *Coord. Chem. Rev.* **251,** 2592–2605 (2007).
18. Suh, M. P. & Cheon, Y. E. Recent advances in the dynamics of single crystal to single crystal transformations in metal-organic open frameworks. *Aust. J. Chem.* **59,** 605–612 (2006).
19. Halder, G. J. & Kepert, C. J. Single crystal to single crystal structural transformations in molecular framework materials. *Aust. J. Chem.* **59,** 597–604 (2006).
20. Yang, S. *et al.* Selectivity and direct visualization of carbon dioxide and sulfur dioxide in a decorated porous host. *Nat. Chem.* **4,** 887–894 (2012).
21. Liu, T.-F. *et al.* Stepwise synthesis of robust metal-organic frameworks via postsynthetic metathesis and oxidation of metal nodes in a single-crystal to single-crystal transformation. *J. Am. Chem. Soc.* **136,** 7813–7816 (2014).
22. Xiao, B. *et al.* Chemically blockable transformation and ultraselective low-pressure gas adsorption in a non-porous metal organic framework. *Nat. Chem.* **1,** 289–294 (2009).
23. Martí Gastaldo, C. *et al.* Side-chain control of porosity closure in single- and multiple-peptide-based porous materials by cooperative folding. *Nat. Chem.* **6,** 343–351 (2014).
24. Furukawa, H., Cordova, K. E., O'Keeffe, M. & Yaghi, O. M. The chemistry and applications of metal-organic frameworks. *Science* **341,** 1230444 (2013).
25. Batten, S. R. *et al.* Terminology of metal-organic frameworks and coordination polymers (IUPAC Recommendations 2013). *Pure Appl. Chem.* **85,** 1715–1724 (2013).
26. Zhou, H.-C. & Kitagawa, S. Metal-organic frameworks (MOFs). *Chem. Soc. Rev.* **43,** 5415–5418 (2014).
27. Devic, T. & Serre, C. High valence 3p and transition metal based MOFs. *Chem. Soc. Rev.* **43,** 6097–6115 (2014).
28. Kubota, R., Tashiro, S., Shiro, M. & Shionoya, M. In situ X-ray snapshot analysis of transient molecular adsorption in a crystalline channel. *Nat. Chem.* **6,** 913–918 (2014).
29. Cohen, S. M. Modifying MOFs: new chemistry, new materials. *Chem. Sci.* **1,** 32–36 (2010).
30. Surble, S., Serre, C., Mellot-Draznieks, C., Millange, F. & Ferey, G. A new isoreticular class of metal-organic-frameworks with the MIL-88 topology. *Chem. Commun. (Camb)* 284–286 (2006).
31. Serre, C. *et al.* Role of solvent-host interactions that lead to very large swelling of hybrid frameworks. *Science* **315,** 1828–1831 (2007).

32. Horcajada, P. *et al.* how linker's modification controls swelling properties of highly flexible iron(III) dicarboxylates MIL-88. *J. Am. Chem. Soc.* **133,** 17839–17847 (2011).
33. McKinlay, A. C. *et al.* Nitric oxide adsorption and delivery in flexible MIL-88(Fe) metal–organic frameworks. *Chem. Mater.* **25,** 1592–1599 (2013).
34. Zheng, S.-T. *et al.* Entrapment of metal clusters in metal–organic framework channels by extended hooks anchored at open metal sites. *J. Am. Chem. Soc.* **135,** 10270–10273 (2013).
35. Zhou, H.-L. *et al.* Direct visualization of a guest-triggered crystal deformation based on a flexible ultramicroporous framework. *Nat. Commun.* **4,** 2534 (2013).
36. Phillips, A. E., Goodwin, A. L., Halder, G. J., Southon, P. D. & Kepert, C. J. Nanoporosity and exceptional negative thermal expansion in single-network cadmium cyanide. *Angew. Chem. Int. Ed.* **47,** 1396–1399 (2008).
37. Goodwin, A. L., Chapman, K. W. & Kepert, C. J. Guest-dependent negative thermal expansion in nanoporous prussian blue analogues $M^{II}Pt^{IV}(CN)_6 \cdot x\{H_2O\}$ ($0 \leq x \leq 2$; M = Zn, Cd). *J. Am. Chem. Soc.* **127,** 17980–17981 (2005).
38. Vuong, G.-T., Pham, M.-H. & Do, T.-O. Direct synthesis and mechanism of the formation of mixed metal Fe_2Ni-MIL-88B. *CrystEngComm* **15,** 9694–9703 (2013).
39. Zhu, X. *et al.* A superacid-catalyzed synthesis of porous membranes based on triazine frameworks for CO_2 separation. *J. Am. Chem. Soc.* **134,** 10478–10484 (2012).
40. Li, M.-X., Miao, Z.-X., Shao, M., Liang, S.-W. & Zhu, S.-R. Metal-Organic frameworks constructed from 2,4,6-Tris(4-pyridyl)-1,3,5-triazine. *Inorg. Chem.* **47,** 4481–4489 (2008).
41. Agenet, N. *et al.* Cotrimerizations of Acetylenic Compounds. in *Organic Reactions* **68**, Ch 1: 1–302 (John Wiley & Sons, Inc., 2004).

Acknowledgements

This research was supported by the '973 Project' (2014CB845602 and 2012CB821706) and NSFC (21225105, 21290173, and 21473260).

Author contributions

J-.P.Z. conceived and designed the experiments. Y-.S.W., M.Z., and T-.Y.L. conducted synthetic experiments. Y-.S.W., M.Z., R-.B.L. and J-.P.Z. performed the physical measurements and crystal-structure analyses. P-.Q.L. performed sorption measurements. Y-.S.W., G.S., J-.P.Z., and X-.M.C. co-wrote the manuscript. All authors discussed the results and commented on the manuscript.

Additional information

Accession codes. The X-ray crystallographic coordinates for structures reported in this study have been deposited at the Cambridge Crystallographic Data Centre (CCDC), under deposition numbers 1415803-1415810. These data can be obtained free of charge from The Cambridge Crystallographic Data Centre via www.ccdc.cam.ac.uk/data_request/cif.

Competing financial interests: The authors declare no competing financial interests.

7

Dynamic formation of single-atom catalytic active sites on ceria-supported gold nanoparticles

Yang-Gang Wang[1,2], Donghai Mei[1], Vassiliki-Alexandra Glezakou[1], Jun Li[2,3] & Roger Rousseau[1]

Catalysis by gold supported on reducible oxides has been extensively studied, yet issues such as the nature of the catalytic site and the role of the reducible support remain fiercely debated topics. Here we present *ab initio* molecular dynamics simulations of an unprecedented dynamic single-atom catalytic mechanism for the oxidation of carbon monoxide by ceria-supported gold clusters. The reported dynamic single-atom catalytic mechanism results from the ability of the gold cation to strongly couple with the redox properties of the ceria in a synergistic manner, thereby lowering the energy of redox reactions. The gold cation can break away from the gold nanoparticle to catalyse carbon monoxide oxidation, adjacent to the metal/oxide interface and subsequently reintegrate back into the nanoparticle after the reaction is completed. Our study highlights the importance of the dynamic creation of active sites under reaction conditions and their essential role in catalysis.

[1] Institute for Interfacial Catalysis, Pacific Northwest National Laboratory, Richland, Washington 99352, USA. [2] Department of Chemistry, Tsinghua University, Beijing 100084, China. [3] Environmental Molecular Sciences Laboratory, Pacific Northwest National Laboratory, Richland, Washington 99352, USA. Correspondence and requests for materials should be addressed to R.R. (email: roger.rousseau@pnnl.gov) or to J.L. (email: junli@tsinghua.edu.cn).

Ever since Haruta, Hutchings and others[1–3] found that nanosized gold particles, supported by metal oxides, were highly reactive for a variety of important catalytic reactions, gold has been widely used as a catalyst. Extensive studies have been devoted to understanding the chemical nature of high reactivity of oxide supported gold (Au) particles[4–10]. Though it is generally recognized that the Au–oxide interfacial perimeter plays an important role, the chemical nature of the active sites and the details of the reaction mechanism are still hotly debated. This is partly because of the complexity of the systems, whose reactivity is affected by many factors, such as the structural and electronic properties of Au particles, the electronic interaction between the support and Au particles, the reactants and products, and finite temperature effects[4]. Nearly all of the theoretical studies on the Au/oxide system were based on carefully selected model clusters of surfaces, where the tendency of Au nanoparticles and surface to exhibit reconstruction, large amplitude motions and disorder at finite temperatures were not fully accounted for. As a prominent example, ceria (CeO_2) has been found to be a particularly effective support for gold particles in catalysing reactions such as CO oxidation and water–gas shift reactions[9,10]. Generally, the high reactivity is attributed to the excellent redox properties of ceria, such as easy formation and diffusion of oxygen defects, high capacity of oxygen storage and release, and facile acceptance/release and transport of excess electrons in a polaronic state derived from a narrow 4f-band[9,10]. Various studies[11–14] have identified the interface sites between ceria support and gold particles as vital in the catalytic reactions. Recent evidence[15–17] has led to the claim that the active sites are, in actuality, positively charged single gold atoms bound to the support by O linkages. Furthermore, environmental transmission electron microscopy analyses[18] also showed that gold nanoparticles on a CeO_2 support change their shapes at the interfacial area in response to alternating oxidizing and reducing atmospheres. These observations highlight the necessity of understanding the nature of reactive sites on ceria-supported Au nanocatalysts, at actual reaction conditions, that is, at elevated temperature and in the presence of dynamic reacting species.

In our recent report[19], we showed that a stable, high-symmetry, tetrahedral Au_{20} cluster on a rutile TiO_2 support exhibited liquid-like motion on charge transfer and CO adsorption. The large adsorbate-induced structural reconstruction also allowed for facile transfer of adsorbed CO to the metal/oxide interface, where the CO_2 formation reaction occurs. Connected with this phenomenon was the finding that charge transfer between the nanoparticle and support enabled low-energy redox reactions during the catalytic cycle. These ideas motivated us to consider how generalizable these findings are to other reducible oxides and whether other related phenomena could be involved.

In this work, we present results from density functional theory (DFT)-based *ab initio* molecular dynamics (AIMD) simulations to explore the reactivity and dynamics of ceria-supported Au catalysts for the CO oxidation reaction. Our simulations take into account both the dynamics of atoms in the Au metal particle, ceria support, reactants and products, as well as the excess charge carriers within the partially reduced ceria surface. We find that, in the presence of CO, a cationic Au^+-CO species is created. This species migrates directly onto the ceria support and, via a strong coupling with the redox properties of the ceria, it enables low-energy barriers for redox steps in the catalytic cycle. After the catalytic cycle is complete, the Au single atom is reintegrated into the Au nanoparticle, indicating that the actual catalytic species only exists as a transient under operating conditions.

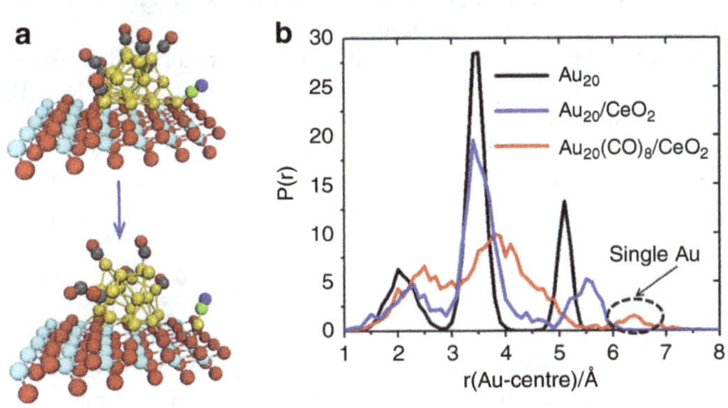

Figure 1 | MD simulations. **(a)** Initial and final configuration of the AIMD simulation for a $CeO_2(111)$ supported Au_{20} cluster with eight CO adsorption. **(b)** The probability distribution functions $P(r_{cm})$ of the Au atoms relative to the centre-of-mass of the Au_{20} based on ~20 ps MD simulations at 700 K. (Yellow sphere: Au; cyan sphere: Ce; red sphere or blue sphere: O; green sphere or grey sphere: C)

Results

The behaviour of Au particles on a ceria support. As is known, the geometric and electronic properties of Au particles on the oxide support strongly affect the catalytic activity of Au nano-catalyst[4,20–23]. Thus, we first performed AIMD simulations for ceria-supported Au_{20} cluster to illustrate how the finite temperature and the nature of the adsorbed species influence the morphology of gold nanoparticle, before examining the catalytic mechanism and the effect of particle size. Considering the extensive existence of oxygen vacancies on CeO_2, we have removed one surface oxygen atom to model a partially reduced CeO_2 support, on which we deposited the Au_{20} cluster, and allowed it to bind to the surface during a 20-ps AIMD trajectory at $T = 700$ K. In a second step, we added an additional eight CO molecules on the cluster and allowed the system to further equilibrate for another 20 ps. This approach guaranteed a well-thermalized and unbiased structural model of CO adsorbed on a CeO_2-supported Au_{20} cluster.

The distribution of the distance of Au atoms relative to the centre-of-mass of the Au_{20} cluster, $P(r_{cm})$ shown in Fig. 1a, can be used to characterize the structural fluctuation within each system. For the isolated Au_{20} cluster in the gas phase, its tetrahedral configuration is retained at the simulation temperature of 700 K, and the $P(r_{cm})$ exhibits three well-defined peaks clearly separated by zero probability density between them, corresponding to three types of Au atoms. However, when Au_{20} is supported on a $CeO_2(111)$ surface, Fig. 1b, it exhibits a distinct structural reconstruction and loses its tetrahedral configuration, which is manifested in $P(r_{cm})$ as broader peaks, compared with the isolated Au_{20} cluster. The particle–surface interaction are strong due to a gold-to-oxide charge transfer, resulting in partially oxidized Au atoms at the interface as evidenced by the appearance of unpaired electrons localized in the upper layer of Ce atoms when the nanoparticle is adsorbed.

The gold nanoparticle, however, shows appreciably enhanced structural flexibility when CO is co-adsorbed. It is found that Au_{20} exhibits a strong adsorption-induced surface reconstruction and extrudes many low-coordinated sites to bind CO. As a result, the function $P(r_{cm})$ for the Au nanoparticle exhibits significantly broader peaks with non-zero density between them. We further observed that CO diffusion proceeds in the same way as reported previously for Au/TiO_2 (ref. 19); CO does not move freely from one Au site to another, but rather diffuses as an adsorbed Au–CO species (see Fig. 2 and Supplementary Fig. 1). Since CO oxidation is generally reported to happen at the gold-oxide interface, this finding is vital for understanding how the adsorbed CO species is

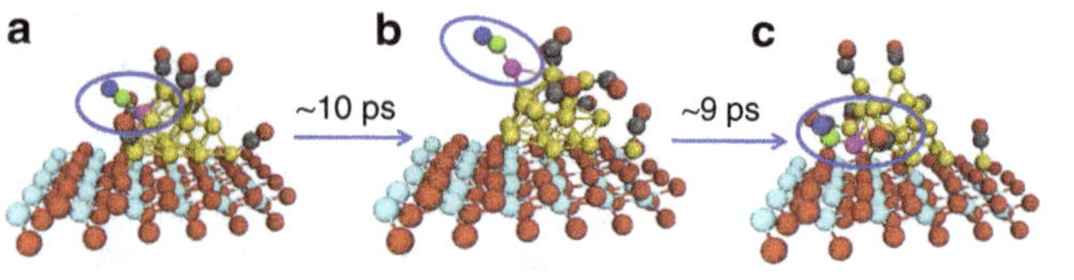

Figure 2 | Diffusion of Au-CO unit. (**a**–**c**) Selected snapshots of the MD trajectory for Au_{20}/CeO_2 with a circled Au-CO unit to show the diffusion process. (yellow or pink sphere: Au; cyan sphere: Ce; red sphere or blue sphere: O; green sphere or grey sphere: C)

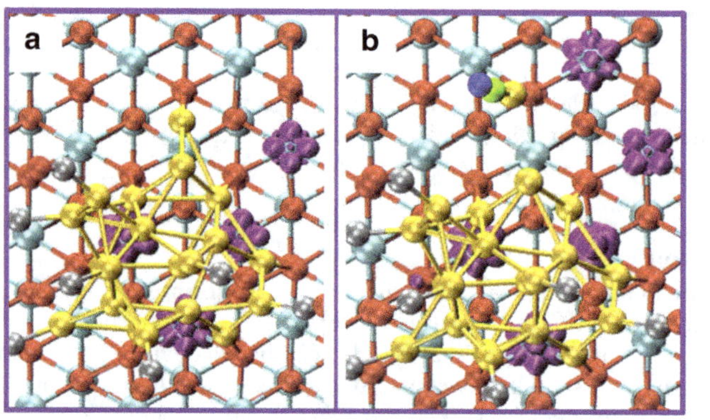

Figure 3 | Three-dimensional isosurface of the spin densities. (**a**) Spin densities (purple areas)for the Au/CeO_2 without an isolated Au-CO unit at the interfacial area. (**b**) Spin densities with an isolated Au-CO unit at the interfacial area. Note, the configuration in **a** is geometry-optimized configuration by removing one CO from the configuration in **b**.

transported to the perimeter site (Fig. 2a–c). These results indicate that the Au nanoparticle is rather flexible in geometry and can change its morphology on CO adsorption, consistent with previous theoretical studies[19,24–26]. However, the ceria-supported system exhibits one unique feature. As shown in Fig. 1, an Au–CO species migrates away from the gold nanoparticle directly onto the ceria support adjacent to the metal/oxide interfacial area. In the function $P(r_{cm})$ for Au_{20}, the isolated Au atom could also be identified by a peak at around 6.4 Å that is well separated from the rest of the probability function.

To understand the nature of this specific single-Au atom species, we consider its electronic properties with and without CO adsorption. When CO is removed from the isolated Au–CO, the Au atom is readily reintegrated into the Au nanoparticle (see Fig. 3a), even in the course of a simple local geometry optimization. By monitoring the spin density, we observed four excess electrons in the CeO_2 lattice, two more than the number that an oxygen defect can create, indicating that the Au nanoparticle transfers additional charges to the surface[27]. Once the CO is adsorbed at the Au atom, and the Au–CO moiety becomes isolated from the rest of the Au nanoparticle, an additional excess electron becomes localized on an adjacent Ce site in the topmost layer of the support. Concurrently, the charge of this Au atom increases from $+0.10$ $|e|$ to $+0.42$ $|e|$ This indicates that by moving to the oxygen site, the single Au atom transfers one electron to the support, resulting in a species that can be formally described as an adsorbed Au^+–CO. Our results are consistent with previous findings[28,29] that the single Au atom on CeO_2 surface can transfer its 6s electron to the unoccupied 4f orbital of Ce. We note that the observed excess electrons are mobile on the AIMD timescale, showing hops between Ce sites in the topmost surface layer on the order of 2 ps, in accord with our previous work on partially reduced ceria[30].

owing to the nature of ligand-to-metal electron donation of the Au–CO bond, CO is expected to energetically favour binding at the positively charge Au site. To understand the formation of single Au atom, we further considered the formation energies of the single Au atom with or without CO adsorption:

$$E_1 = E(Au_{n-1} + Au) - E(Au_n) \tag{1}$$

$$E_2 = E(Au_{n-1} + Au{-}CO) - E(Au_n{-}CO) \tag{2}$$

where E_1 is the energy change by taking one Au atom from the Au_n nanoparticle to the next-nearest oxygen site on ceria surface and E_2 is the energy associated with taking one Au-CO unit from the Au_n nanoparticle to the next-nearest oxygen site. Since in our AIMD simulations CO adsorption can only cause a low-coordinated Au atom to be isolated from the Au nanoparticle at the interface, we only consider a single four-coordinated Au site at the interface. For the Au_{20} cluster on ceria surface, the calculated E_1 is 1.21 eV, indicating that for the bare Au nanoparticle it is energetically unfavourable for the Au atom to break away from the Au nanoparticle. While the calculated E_2 is -0.81 eV, it indicates that forming an isolated Au^+–CO could be thermodynamically favourable. Our work is consistent with the recent findings[31] that CO adsorption on a CeO_2-supported Au_4 cluster leads to the formation of Au–CO species. Considering the CO oxidation on CeO_2-supported Au nanoparticles is reported to proceed at room temperature[4,32], we further considered the possible reaction paths for Au–CO dissociation into the adjacent oxygen site on Au_{20}/CeO_2 using climbing image nudged-elastic-band method (see Supplementary Fig. 2). The barrier is estimated to be ~ 0.20–0.62 eV, which indicates that the isolated Au–CO can easily be form at room temperature. Combined with the fact that Au–CO species on the Au cluster are mobile (Fig. 2), we also conclude that even if CO does not initially adsorb directly at the interface site, the isolated Au^+–CO will eventually form.

Inasmuch as in our previous simulation[19] we did not observe any isolated Au–CO on TiO_2 support, we checked the single-atom formation energies on Au_{20}/TiO_2 system. The isolated Au atom without adsorbed CO is not stable at the oxygen site and slips to the adjacent five-coordinated Ti site spontaneously during the geometry optimization, in accord with Metiu's recent report[33,34]. However, its formation energies are positive ($E_2 = 1.25$ eV at the oxygen site; $E_1 = 0.96$ eV and $E_2 = 0.90$ eV at the Ti site), implying that stabilizing the oxidized Au^+–CO on a TiO_2 support is unfavourable. This is traced back to the fact that rutile TiO_2 is less capable of accepting an electron from an Au–CO to stabilize a cationic species and highlights the point that the CeO_2 support, as a stronger oxidizer, can.

Generally, the reactivity for an oxidation reaction is strongly affected by the electronic chemical potential (work function, W) of the catalyst. We found that creation of the isolated Au^+–CO increased W from 4.8 to 5.1 eV, which indicates that the Ceria is more likely to take electrons by becoming a stronger oxidizing agent, and it is itself more easily reduced. This implies that the single Au atom at the interfacial area creates an activation centre for oxidation processes and, thus, is expected to be more reactive, in accord with recent reports on highly reactive single-atom catalysts on metal oxide[35,36]. In the next section, we further consider the possible reaction mechanism for CO oxidation at the isolated Au species on CeO_2 surface and directly compare with the process at the cluster/oxide interface

Reaction mechanism for CO oxidation. Recent theoretical studies have proposed[9,14,37] three CO oxidation mechanisms on CeO_2-supported Au particles: (1) CO oxidation by co-adsorbed O_2 on Au particles, (2) CO oxidation by adsorbed O_2 at the interfacial site and (3) CO oxidation by a lattice oxygen ion from CeO_2 substrate. In the present study, we examined a large number of adsorption configurations of O_2 on Au_{20} cluster or interfacial sites, but we found that the O_2 hardly binds at these sites ($E_{ads} < -0.2$ eV). These results suggest that CO oxidation with preadsorbed O_2 at the Au nanoparticle or the interface sites are less likely to be operative for larger Au particles, and lower reduction levels of the support, as is expected to be the case for a reaction that is run under oxidizing conditions. Therefore, we consider the reaction of CO with the lattice oxygen ion, in accord with extensive experimental studies[18,38–40] that strongly indicate that CO oxidation proceeds via a Mars-van Krevelen mechanism (that is, a redox) mechanism.

We first considered the oxidation of CO that preadsorbs on the gold nanoparticle, comparable with previous theoretical studies[14]. The detailed pathways and energetics are shown in Supplementary Fig. 3. We found that CO easily reacts with the lattice oxygen ion, and forms a bent CO_2^- intermediate. This is a highly exothermic process ($\Delta E = -0.94$ eV) with a low-energy barrier ($\Delta E^\dagger = 0.25$ eV). However, the subsequent CO_2 desorption is quite unfavourable, with a desorption energy of 1.29 eV. As the lattice oxygen ion is being extracted, there is a concurrent reduction of the surface that leaves two excess electrons localized on the Ce 4f states. After the formation of an oxygen defect, O_2 adsorption from the gas phase at this site is highly favourable in energy and the subsequent oxidization of the second CO to CO_2 proceeds with a low-energy barrier ($\Delta E^\dagger = 0.45$ eV). Overall, the rate determining step for the oxidation of CO preadsorbing on the gold particle involves desorption of the first CO_2 with a high-enough energy barrier such that the reaction would not be anticipated to proceed at room temperature. However, the result is in direct contradiction with the experimental findings[32,41] that show the reaction of CO oxidation with lattice oxygen ion in the interfacial area at the room temperature. In short, for the current Au_{20} system on a mildly reduced ceria support, none of the three previously proposed routes corroborate this experimental observation.

As the adsorption of CO at the perimeter sites could lead to the formation of an Au^+–CO near the interfacial area, we further consider the reaction mechanism for CO oxidation at this specific isolated site. The reaction pathway and associated energetics are shown in Fig. 4. For convenience of discussion, we also track the Bader charges of the isolated Au and the number of Ce^{3+} throughout the catalytic process for each stationary point on the potential energy surface, shown in Table 1. In the proposed mechanism, we start the reaction process from the Au nanoparticle without an isolated Au site (Configuration i in Fig. 4). First, the adsorption of CO at the low-coordinate site leads to the formation of an isolated Au^+–CO unit (Configuration ii). By overcoming a barrier of 0.77 eV, the CO can react with the adjacent oxygen ion, forming a bent CO_2^- intermediate (Configuration iii). Further, this bent intermediate only requires 0.27 eV of energy to release the CO_2 molecule into the gas phase (Configuration iv). In this process, the single Au atom migrates towards and occupies the oxygen defect after CO_2 desorption. The Bader charge of the Au atom significantly decreases from +0.40 to −0.56, while the number of Ce^{3+} ions remains unchanged. These results indicate that in tandem with the removal of the lattice oxygen ion, the two resulting excess electrons are transferred to the single Au atom, leading to a conversion of the oxidation state from Au^+ to a Au^- that fills the oxygen vacancy. In other words, this single atom stabilizes the reduction of the ceria surface leading to more facile CO_2 desorption. This observation is consistent with our previous analysis of work function, which suggests in the presence of a charged Au^+ + CO, Ceria is locally a stronger oxidizing agent. However, the resulting Au^- species, which fills the oxygen vacancy, is in direct contrast with the CO oxidation on the Au nanoparticle where the two resulting excess electrons become localized at two Ce^{3+} ions.

To complete the catalytic cycle, we consider the second half of the CO oxidation reaction. Once the single Au atom has moved into the oxygen defect, it is harder to bind CO ($E_{ads} = -0.16$ eV) or O_2 ($E_{ads} = -0.10$ eV) molecules. At first glance, this would indicate that the catalyst is deactivated for further CO oxidation, which is consistent with a recent study[17] showing that Au at the oxygen defect has no further reactivity for CO oxidation. However, further CO oxidation can be achieved by extracting the Au atom out of the vacancy back to the Au cluster, so that it could bind the second CO while the vacancy becomes available for O_2 adsorption. We noted that our AIMD simulations for Au_{20}/CeO_2 with one oxygen vacancy did not show any Au atom permanently occupying the oxygen vacancy, which implies that it may also be possible that the single Au atom may be able to reintegrate with the Au cluster (Configuration vi). With this in mind, we further performed CI-NEB calculations, as well as, constrained AIMD computation of a potential of mean force for the single Au reintegrating into the Au cluster without the presence of CO. We found that the single Au atom, at the oxygen vacancy, can move up and reintegrate within into Au cluster with a barrier of 0.89 eV on the potential energy surface and 0.81 eV on the free energy surface. This process is slightly exothermic with Au preferring energetically to be reintegrating into the Au cluster by −0.10 eV. Note these relatively low barriers are consistent with room temperature activity. Once the Au is back to the Au cluster, the second CO can form another stable Au^+–CO species near the interfacial area (Configuration vii), by exothermic adsorption energy of −1.12 eV. During these processes, the Bader charge of the single Au atom increases from −0.54 to 0.40 e^-. The surface accumulates two more Ce^{3+} ions, indicating that by moving out of the oxygen defect, the Au ion returns two excess electrons to the lattice. This result implies that the driving force for the Au moving out of the oxygen defect comes from the attraction of the Au cluster, and it also explains the recent experimental findings[42,43] that the clustering of Au from an isolated Au precursor results in an increase of the CO oxidation activity. Recent theoretical studies[17,31] also showed that the small Au_n cluster (Au_2 or Au_3) are catalytically reactive for CO oxidation without filling one Au into the oxygen defect site. Since the oxygen vacancy is recreated, the O_2 molecule in the gas phase could easily fill into the oxygen defect with a high binding energy of −1.51 eV, eliminating the two excess electrons and forming a peroxo O_2^{2-} species (Configuration viii). As a result, the CO at the single Au site only needs to overcome a small barrier of 0.65 eV to react with the peroxo O_2^{2-} species, releasing the second CO_2 (Configuration ix). During this process, the Au atom gradually loses its adsorbed CO and returns to the Au nanoparticle, completing the catalytic cycle. Overall, the proposed mechanism suggests that the rate-limiting step for CO oxidation is the removal of the single Au out of the oxygen vacancy. The barrier of the rate-limiting step at the single atom site is 0.89 eV, which is far lower than that at the gold particles (1.29 eV), suggesting that the transient monatomic species are the active sites in CeO_2-supported Au catalyst.

For a direct comparison with the experimental observation for the room temperature CO oxidation[32,41], we made a simple estimation of the reaction rate in the form of an attempt frequency multiplied by a Boltzmann factor, that is, $v = v_0\exp(-E_a/k_BT)$ and assuming $v_0 = 10^{13}\,s^{-1}$, a reasonable value for most oxides as suggested by Marrochelli and Yildiz[44]. From this formula, the range of temperatures at which the reaction process could occur, within a minimum range of rate

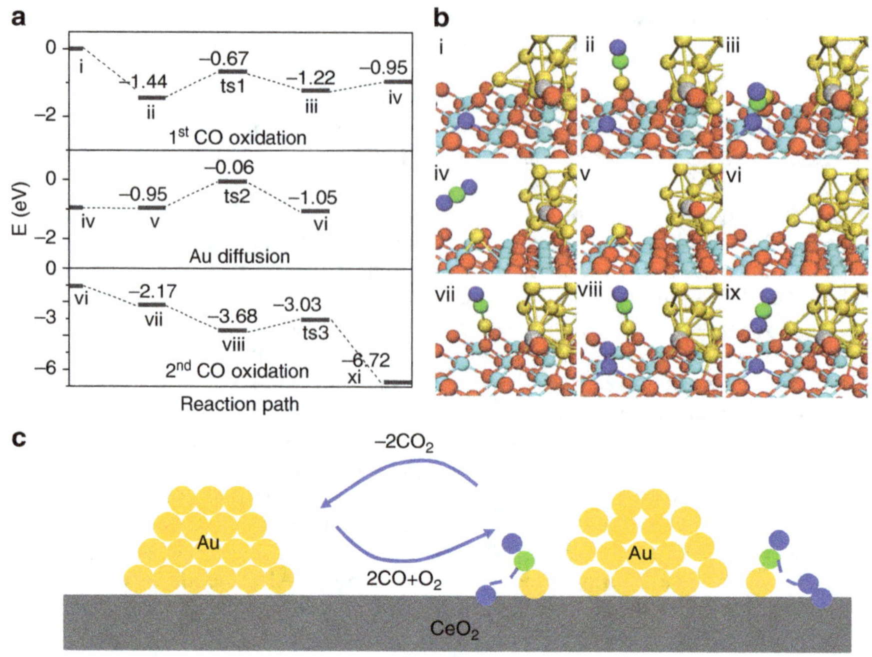

Figure 4 | Proposed reaction mechanism for CO oxidation at single Au site. (**a**) Reaction pathways, (**b**) configurations of reactants, intermediates and products, and (**c**) schematic representation of the dynamic behavior at the interface. The catalytic processes proceed as the following steps: (1) i-ii, CO adsorption leads to the formation of isolated Au-CO; (2) ii-ts1-iii, CO preadsorbing at the single Au site reacts with the lattice oxygen ion, forming a bent intermediate; (3) iii-iv-v, CO_2 desorbs into the gas phase and at the same time the Au atom fills into the oxygen defect; (4) v-ts2-vi, the Au atom at the oxygen defect moves back to the Au cluster; (5) vi-vii, the second CO adsorption leads to the formation of isolated Au-CO; (6) vii-viii, the O_2 adsorption at the oxygen defect. (7) viii-ts3-xi, the second CO reacts with the adsorbed O_2 species, forming a CO_2 molecule. At the same time, the Au returns the Au nanoparticle, completing the catalytic cycle. (The configurations for transition states are shown in Supplementary Fig. 4. Yellow sphere: Au; cyan sphere: Ce; red sphere or blue sphere: O; green sphere: C.

Table 1 | The Bader charges of the single Au atom and the number of Ce^{3+} ions in the system.

	First CO oxidation				Second CO oxidation				
Intermediate	**i**	**ii**	**iii**	**iv**	**v**	**vi**	**vii**	**viii**	**ix**
q(Au)/e –	0.10	0.42	0.04	– 0.56	– 0.54	0.11	0.40	0.46	0.10
N(Ce^{3+})	4	5	5	5	5	6	7	5	4

constant $0.01 \sim 1\,s^{-1}$, are estimated to be 300–340 K, consistent with room temperature activity. We also performed constrained MD simulations to obtain the free energy barrier at 300 K for the migration of the Au atom towards the Au cluster (Supplementary Fig. 5). This simulation allows for the sampling of collective motions, such as the restructuring of the entire Au particle on reintegration of the Au ad-atom, and thus would be expected to better account for both the energy and entropic components of the process. The free energy barrier is estimated to be 0.81 eV and based on the transition state theory formalism $v = k_BT/h \exp(-\Delta G^{\dagger}/k_BT)$ the rate constant at 300 K is estimated to be $0.1\,s^{-1}$, indicating that the process could occur several times per minute at room temperature.

Above, we proposed that the Au cluster could drive the Au atom out of the vacancy. We also noted that an isolated Au atom at the oxygen defect could also be directly pulled out of the oxygen defect by an adsorbed CO, with a barrier of 0.85 eV, and a highly exothermic reaction energy of − 1.06 eV. Since the initially CO adsorption is very weak (only − 0.16 eV), this process will resemble an Eley–Rideal mechanism (the energy barrier is 0.85–0.16 = 0.69 eV) with a relatively low rate at reaction conditions where CO oxidation is typically performed. However, it may still happen under CO-rich conditions, where the high partial pressure of CO would significantly increase the sticking possibility of CO at the single Au site, and would enhance the reaction probability for the second CO oxidation step without changing the subsequent steps in reaction mechanism. In addition, recent theoretical studies by Song *et al.*[45,46] also proposed a relevant mechanism for CO oxidation and propylene oxidation at an isolated Au site, which are compatible with our findings.

Since the single Au atom is induced by CO adsorption at the interfacial area, it is its mobility that ultimately determines where the CO oxidation proceeds. In Supplementary Fig. 6, we show the reaction paths and energetics for Au diffusion to another oxygen site away from the Au nanoparticle. With the adsorption of CO, the single Au atom has to overcome a high barrier of 1.14 eV to diffuse from the interface to the next oxygen site, indicating that the diffusion of Au–CO unit on the ceria surface is improbable at ambient conditions. As the barrier for CO reacting with the lattice oxygen is smaller (0.77 eV) than the diffusion barrier, the Au^{+}–CO is expected to oxidize appreciably faster than it can diffuse. We further consider a single Au atom away from the Au nanoparticle without CO adsorption. Similarly, the barrier for the diffusion away from the interface is much higher (1.59 eV). Especially, when considering the inverse diffusion, the Au atom at the oxygen site only needs to overcome a very low barrier of 0.37 eV to diffuse back towards the Au nanoparticle. The process

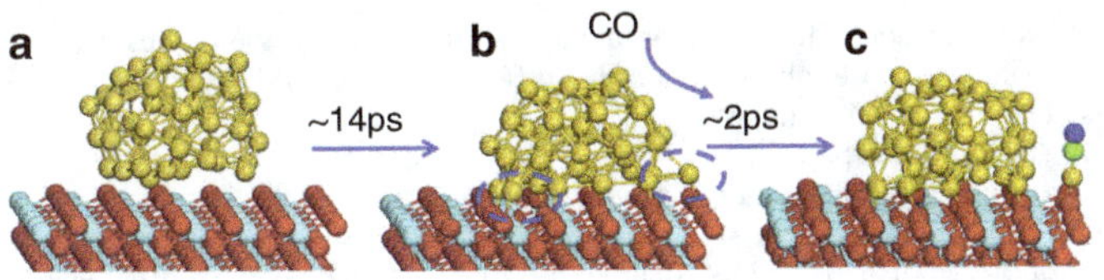

Figure 5 | Snapshots of MD simulation for CeO_2-supported Au_{50} cluster. (**a**) Initial configuration of Au_{50}/CeO_2. (**b**) the snapshot of Au_{50}/CeO_2 at ~14 ps. (**c**) the snapshot for the MD simulation with one CO initially adsorbed at the low-coordinated site. The low coordinate sites in (**b**) are circled in dotted line. See Supplementary Information for more snapshots to demonstrate the formation of low-coordinated sites frequently during the MD simulation.

is highly exothermic with a reaction energy of − 1.22 eV and is the result of the above-mentioned propensity for the Au atoms to reintegrate with the nanoparticle when CO is not present. Taken together, these results demonstrate that single Au atom catalysis for CO oxidation would occur predominantly at the interfacial area.

Nanoparticle size effects. Since our AIMD simulations for Au_{20}/CeO_2 only show the formation of the isolated Au^+-CO species at the low-coordinated sites, one may ask if the formation of the isolated Au^+-CO species is determined by the specific model cluster and how applicable are these findings to large Au particles. Therefore, we further performed AIMD simulations for a bigger Au_{50} cluster (~ 1.5 nm) on CeO_2 support, which initially had no low-coordinate site at the interface area, shown in Fig. 5. We observed that the low-coordinate sites formed at the interface area, revealing that for even 1 nm Au particles, or larger, the low-coordination sites should always exist at the interface. We further considered the formation energy of the isolated Au-CO species on the CeO_2 surface. In Table 2, we list the single-Au formation energies for Au_{10}, Au_{20} and Au_{50}. Without CO adsorption, the formation energies are endothermic for all three Au clusters, indicating that the creation of an isolated Au ad-atom is energetically unfavourable in all cases. Conversely, in the presence of CO, the adsorption energies at isolated Au^+ sites for all clusters become exothermic, illustrating how formation of this species near the metal/oxide interface becomes thermodynamically favourable. We find that the above-mentioned formation energies are all within the same energy range with a variance of 0.3 eV and no discernible trend indicating that this general result is not particularly influenced by the cluster size. We further performed a short (~5 ps) AIMD simulation for CeO_2-supported Au_{10} and Au_{50}, where CO molecules are initial located at a low-coordination interfacial Au site, to compare with our Au_{20} results. In agreement with our analysis above, we observe spontaneous formation of an Au^+ + CO atom at the interface for both cases within only a few ps, see Fig. 5c and Supplementary Fig. 7.

In summary, we find that, for both smaller and larger Au nanoparticles, the surface-adsorbed Au^{+-}-CO can be formed readily by CO adsorption. We speculate that under realistic conditions, small-size Au particles, with potentially a higher ratio of low coordinate sites at the interface than the large particles, should produce more single Au atom on CO adsorption. This conclusion is also consistent with our AIMD simulations on Au_{10}/CeO_2 with four adsorbed CO molecules, where we observed two single-Au atoms at the interface (Supplementary Fig. 7).

Discussion

In the present study, we performed static DFT calculations and *ab initio* molecular dynamic simulations to gain insight on the reaction mechanism of CO oxidation at the interfacial area.

Table 2 | The formation energy of single Au atom at interfacial area.

	Au_{10}	Au_{20}	Au_{50}
E_1 (eV)	1.55	1.21	1.50
E_2 (eV)	− 0.50	− 0.81	− 0.73

A new, dynamic, single-atom catalytic mechanism is established for CO oxidation on Au/CeO_2 catalysts, arising from a transient Au^+-CO species. During the reaction process, a single Au^+ atom acts as charge acceptor for enhanced CO adsorption and transport. This species considerably lowers the barriers for CeO_2 reduction, effectively promoting the CO oxidation reaction. However, the active species, an Au^+-CO ion, emerges only in the presence of CO, and otherwise reintegrates with the Au nanoparticle. It is this monoatomic species that is able to couple its redox processes effectively with those of the support to enhance the redox processes, whilst the Au nanoparticle shows little evidence of coupling of its charge state with that of the oxide during the catalytic cycle.

Our study touches on a very important concept in the heterogeneous catalysis, namely that the actual catalytic active centre may be created only under reaction conditions and may not necessarily be easy to identify *ex situ*. The current study adds a layering to this concept by suggesting that, in the present case, the formation of the active site is a dynamic process occurring at the support oxide/metal particle interface. Our results promote the concept that the dynamic behaviour induced by the reactants at the interfacial area may be an overlooked aspect on the catalysis on reducible supports.

Methods

DFT parameters. All DFT calculations were performed using the CP2K package[47]. The exchange-correlation energy was described by the generalized-gradient approximation with spin-polarized Perdew-Burke-Ernzerh functional[48]. The wavefunctions were expanded in an optimized double-ζ Gaussian basis sets[30,49] with an auxiliary plane wave basis set with a cutoff energy of 500 Rydberg[50]. Core electrons have been modelled by scalar relativistic norm-conserving pseudo potentials[51] with 12, 6, 1 and 4 valence electrons for Ce, O, H and C, respectively. Brillouin zone integration is performed with a reciprocal space mesh consisting of only the gamma point. The DFT + U method, based on the Mullikan 4f state population analysis, was used to describe the Ce 4f electrons. A *U* value of 7.0 eV was used to reproduce the correct band gap of 3.2 eV, gap state location of 1.3 eV above the valence band, and work function of 4.6 eV for CeO_2, which ensure that the redox chemistry is reproduced correctly. The detail tests and discussion can be found in our recent study[30] and in the Supplementary methods. We observed many localized 4f electrons on CeO_2-supported Au clsuter, which are mobile and can be localized at many different Ce sites in the simulations. This may lead to 0.1–0.3 eV difference for the energetics, as is discussed in our previous study[30]. The location and energy of the transition state was performed using the climbing image nudged-elastic-band method (CI-NEB)[52,53] including 9 replicas. The convergence criterion for the maximum force is set as 2×10^{-3} atomic units. Vibrational analysis was further used to confirm the transition states corresponded to true saddle points on the potential energy surface with only one imaginary frequency.

Computational models. The $CeO_2(111)$-p(5 × 6) surface with one oxygen defect was used to model the CeO_2 substrate, consisting of 4 O-Ce-O tri-layers (12 atomic layers), and only the bottom Ce atomic-layer were frozen while the remaining layers were allowed to relax. The slab was repeated periodically with a vacuum layer of 20 Å in the direction of the surface normal. A tetrahedral Au_{20} cluster, which has been well characterized in previous studies, was chosen to model Au particles. The benchmark test of current approach with respect to a family of Au_{20} isomers can be found in our recent study[19]. To consider the nanoparticle size effect, Au_{10} (planar) and Au_{50} cluster were chose to model the different size of gold nanoparticles (the coordinates are listed in Supplementary Data 1).

AIMD simulations. All *ab initio* MD simulations were performed by sampling the canonical ensemble employing Nose-Hoover thermostats[54,55] with a time step of 0.5 fs at a finite temperature of 300 or 700 K. On the basis of simple geometry optimization, CeO_2-supported Au_{20} cluster is estimated to be able to bind up to 10

CO molecules and the average binding energy is 0.87 eV per CO. This result is consistent with our recent study[19] on TiO_2-supported Au_{20} cluster where we also found the system can at most bind 10 CO. Therefore, in our MD simulations, we considered both low coverage and high coverage of CO on the Au cluster to explore the behaviour of CO-induced reconstruction. Since the relatively short timescales of AIMD limit sampling to only very fast, low-energy barrier events and preclude the observation of slow processes, the high temperature simulations could more rapidly explore a large volume of phase space. On the basis of a simple estimation according to $10^{13} \cdot \exp(-E_a/RT)$, the rate constant for a reaction with an energy barrier of 0.5 eV at 700 K is around 0.0025 ps^{-1}, this means that the process can happen only one time every 400 ps. Therefore, only the reaction process with the energy barrier lower than 0.5 eV may be observed in our MD simulations at 700 K due to the limitation of the timescale (10–30 ps).

References

1. Haruta, M., Kobayashi, T., Sano, H. & Yamada, N. Novel gold catalysts for the oxidation of carbon monoxide at a temperature far below 0 °C. *Chem. Lett.* **2,** 405–408 (1987).
2. Haruta, M. Size-and support-dependency in the catalysis of gold. *Catal. Today* **36,** 153–166 (1997).
3. Hutchings, G. J. Gold catalysis in chemical processing. *Catal. Today* **72,** 11–17 (2002).
4. Hvolbæk, B. *et al.* Catalytic activity of Au nanoparticles. *Nano Today* **2,** 14–18 (2007).
5. Vilhelmsen, L. B. & Hammer, B. Identification of the catalytic site at the interface perimeter of au clusters on rutile TiO_2(110). *ACS Catal.* **4,** 1626–1631 (2014).
6. Green, I. X., Tang, W., Neurock, M. & Yates, Jr J. T. Insights into catalytic oxidation at the Au/TiO_2 dual perimeter sites. *Acc. Chem. Res.* **47,** 805–815 (2013).
7. Wang, J. & Hammer, B. Oxidation state of oxide supported nanometric gold. *Top. Catal.* **44,** 49–56 (2007).
8. Farnesi Camellone, M. & Marx, D. Nature and role of activated molecular oxygen species at the gold/titania interface in the selective oxidation of alcohols. *J. Phys. Chem. C* **118,** 20989–21000 (2014).
9. Zhang, C., Michaelides, A. & Jenkins, S. J. Theory of gold on ceria. *Phys. Chem. Chem. Phys.* **13,** 22–33 (2010).
10. Paier, J., Penschke, C. & Sauer, J. Oxygen defects and surface chemistry of ceria: quantum chemical studies compared to experiment. *Chem. Rev.* **113,** 3949–3985 (2013).
11. Rodriguez, J. A. Gold-based catalysts for the water–gas shift reaction: active sites and reaction mechanism. *Catal. Today* **160,** 3–10 (2011).
12. Zhou, Z., Kooi, S., Flytzani-Stephanopoulos, M. & Saltsburg, H. The role of the interface in CO Oxidation on Au/CeO_2 multilayer nanotowers. *Adv. Funct. Mater.* **18,** 2801–2807 (2008).
13. Longo, A. *et al.* Structure of the metal–support interface and oxidation state of gold nanoparticles supported on ceria. *J. Phys. Chem. C* **116,** 2960–2966 (2011).
14. Kim, H. Y., Lee, H. M. & Henkelman, G. CO oxidation mechanism on CeO_2-Supported Au nanoparticles. *J. Am. Chem. Soc.* **134,** 1560–1570 (2012).
15. Guan, Y. & Hensen, E. Cyanide leaching of Au/CeO_2: highly active gold clusters for 1, 3-butadiene hydrogenation. *Phys. Chem. Chem. Phys.* **11,** 9578–9582 (2009).
16. Flytzani-Stephanopoulos, M. Gold atoms stabilized on various supports catalyze the water–gas shift reaction. *Acc. Chem. Res.* **47,** 783–792 (2013).
17. Camellone, M. F. & Fabris, S. Reaction mechanisms for the CO oxidation on Au/CeO_2 catalysts: Activity of substitutional Au^{3+}/Au^{+} cations and deactivation of supported Au+ adatoms. *J. Am. Chem. Soc.* **131,** 10473–10483 (2009).
18. Ta, N. *et al.* Stabilized gold nanoparticles on ceria nanorods by strong interfacial anchoring. *J. Am. Chem. Soc.* **134,** 20585–20588 (2012).
19. Wang, Y.-G., Yoon, Y., Glezakou, V.-A., Li, J. & Rousseau, R. The role of reducible oxide–metal cluster charge transfer in catalytic processes: new insights on the catalytic mechanism of CO oxidation on Au/TiO_2 from ab initio molecular dynamics. *J. Am. Chem. Soc.* **135,** 10673–10683 (2013).
20. Lopez, N. & Nørskov, J. K. Catalytic CO oxidation by a gold nanoparticle: a density functional study. *J. Am. Chem. Soc.* **124,** 11262–11263 (2002).
21. Yoon, B. *et al.* Charging effects on bonding and catalyzed oxidation of CO on Au_8 clusters on MgO. *Science* **307,** 403–407 (2005).
22. Remediakis, I. N., Lopez, N. & Nørskov, J. K. CO oxidation on rutile-supported Au nanoparticles. *Angew. Chem. In. Ed.* **117,** 1858–1860 (2005).
23. Gates, B. C. Supported gold catalysts: new properties offered by nanometer and sub-nanometer structures. *Chem. Commun.* **49,** 7876–7877 (2013).
24. Pal, R. *et al.* Chemisorption-Induced 2D–3D–2D structural transitions in gold heptamer:$(CO)_nAu_7^-$(n= 1–4). *J. Phys. Chem. Lett.* **2,** 2288–2293 (2011).
25. Li, W.-K., Chu, L.-N., Gong, X.-Q. & Lu, G. A comparative DFT study of adsorption and catalytic performance of Au nanoparticles at anatase and brookite TiO_2 surfaces. *Surf. Sci.* **605,** 1369–1380 (2011).
26. Zhai, H. J. *et al.* Chemisorption-induced structural changes and transition from chemisorption to physisorption in $Au_6(CO)_n^-$ (n= 4– 9). *J. Phys. Chem. C* **112,** 11920–11928 (2008).
27. Zhang, C., Michaelides, A., King, D. A. & Jenkins, S. J. Positive charge states and possible polymorphism of gold nanoclusters on reduced ceria. *J. Am. Chem. Soc.* **132,** 2175–2182 (2010).
28. Liu, Z. P., Jenkins, S. J. & King, D. A. Origin and activity of oxidized gold in water-gas-shift catalysis. *Phys. Rev. Lett.* **94,** 196102 (2005).
29. Zhang, C., Michaelides, A., King, D. A. & Jenkins, S. J. Structure of gold atoms on stoichiometric and defective ceria surfaces. *J. Chem. Phys.* **129,** 194708 (2008).
30. Wang, Y.-G., Mei, D., Li, J. & Rousseau, R. DFT+U study on the localized electronic states and their potential role during H_2O dissociation and CO oxidation processes on CeO_2(111) surface. *J. Phys. Chem. Lett.* **4,** 2256–2363 (2013).
31. Ghosh, P., Camellone, M. & Fabris, S. Fluxionaality of Au clusters at ceria surfaces during CO oxidation: relationships among reactivity, size, cohesion, and surface defects from DFT simulations. *J. Phys. Chem. C* **117,** 23082–23089 (2013).
32. Zhang, S. *et al.* CO oxidation activity at room temperature over Au/CeO_2 catalysts: disclosure of induction period and humidity effect. *ACS Catal.* **4,** 3481–3489 (2014).
33. Chrétien, S. & Metiu, H. Density functional study of the interaction between small Au clusters, Aun (n= 1–7) and the rutile TiO_2 surface. II. Adsorption on a partially reduced surface. *J. Chem. Phys.* **127,** 244708 (2007).
34. Chrétien, S. & Metiu, H. Density functional study of the charge on Aun clusters (n=1–7) supported on a partially reduced rutile TiO2 (110): are all clusters negatively charged? *J. Chem. Phys.* **126,** 104701 (2007).
35. Qiao, B. *et al.* Single-atom catalysis of CO oxidation using Pt1/FeOx. *Nat. Chem.* **3,** 634–641 (2011).
36. Yang, X.-F. *et al.* Single-Atom catalysts: a new frontier in heterogeneous catalysis. *Acc. Chem. Res.* **46,** 1740–1748 (2013).
37. Song, W. & Hensen, E. J. M. A computational DFT study of CO oxidation on a Au nanorod supported on CeO_2(110): on the role of the support termination. *Catal. Sci. Technol.* **3,** 3020–3029 (2013).
38. Shapovalov, V. & Metiu, H. Catalysis by doped oxides: CO oxidation by $Au_xCe_{1-x}O_2$. *J. Catal.* **245,** 205–214 (2007).
39. Widmann, D., Leppelt, R. & Behm, R. Activation of a Au/CeO_2catalyst for the CO oxidation reaction by surface oxygen removal/oxygen vacancy formation. *J. Catal.* **251,** 437–442 (2007).
40. Qian, K. *et al.* Influences of CeO_2 microstructures on the structure and activity of Au/CeO_2/SiO_2 catalysts in CO oxidation. *J. Mol. Catal. A-Chem.* **306,** 40–47 (2009).
41. Deng, W., Carpenter, C., Yi, N. & Flytzani-Stephanopoulos, M. Comparison of the activity of Au/CeO_2 and Au/Fe_2O_3 catalysts for the CO oxidation and the water-gas shift reactions. *Top. Catal.* **44,** 199–208 (2007).
42. Aguilar-Guerreroa, V. & Gates, B. C. Genesis of a highly active cerium oxide-supported gold catalyst for CO oxidation. *Chem. Commun.* **30,** 3210–3212 (2007).
43. Aguilar-Guerrero, V., Lobo-Lapidus, R. J. & Gates, B. C. Genesis of a cerium oxide supported gold catalyst for CO oxidation: transformation of mononuclear gold complexes into clusters as characterized by X-ray absorption spectroscopy. *J. Phys. Chem. C* **113,** 3259–3269 (2009).
44. Marrocchelli, D. & Yildiz, B. First-principles assessment of H_2S and H_2O reaction mechanisms and the subsequent hydrogen absorption on the CeO2(111) surface. *J. Phys. Chem. C* **116,** 2411–2424 (2012).
45. Song, W. & Hensen, E. J. M. Structure sensitivity in CO oxidation by a single Au atom supported on ceria. *J. Phys. Chem. C* **117,** 7721–7726 (2013).
46. Song, W. *et al.* Selective propylene oxidation to acrolein by gold dispersed on $MgCuCr_2O_4$ spinel. *ACS. Catal.* 5, 1100–1111, 2015).
47. VandeVondele, J. *et al.* Quickstep: fast and accurate density functional calculations using a mixed Gaussian and plane waves approach. *Comp. Phys. Commun.* **167,** 103–128 (2005).
48. Perdew, J. P., Burke, K. & Ernzerhof, M. Generalized gradient approximation made simple. *Phys. Rev. Lett.* **77,** 3865–3868 (1996).
49. VandeVondele, J. & Hutter, J. Gaussian basis sets for accurate calculations on molecular systems in gas and condensed phases. *J. Chem. Phys.* **127,** 114105 (2007).
50. Lippert, B. G., Hutter, J. & Parrinello, M. A hybrid Gaussian and plane wave density functional scheme. *Mol. Phys.* **92,** 477–488 (1997).
51. Goedecker, S., Teter, M. & Hutter, J. Separable dual-space Gaussian pseudopotentials. *Phys. Rev. B* **54,** 1703–1710 (1996).
52. Mills, G., Jónsson, H. & Schenter, G. K. Reversible work transition state theory: application to dissociative adsorption of hydrogen. *Surf. Sci.* **324,** 305–337 (1995).
53. Henkelman, G., Uberuaga, B. P. & Jónsson, H. A climbing image nudged elastic band method for finding saddle points and minimum energy paths. *J. Chem. Phys.* **113,** 9901 (2000).

54. Nosé, S. A unified formulation of the constant temperature molecular dynamics methods. *J. Chem. Phys.* **81,** 511 (1984).
55. Hoover, W. G. Canonical dynamics: equilibrium phase-space distributions. *Phys. Rev. A* **31,** 1695–1697 (1985).

Acknowledgements

This work was supported by the US Department of Energy, Office of Science, Office of Basic Energy Sciences, Division of Chemical Sciences, Geosciences & Biosciences and performed at Pacific Northwest National Laboratory (PNNL). PNNL is a multi-programme national laboratory operated by Battelle for the US Department of Energy. J.L. and Y.-G.W. were also financially supported by NKBRSF (2011CB932400) and NSFC (91026003, 21101098) of China. Computational resources were provided at W.R. Wiley Environmental Molecular Sciences Laboratory (EMSL), a national scientific user facility located at PNNL and sponsored by the US Department of Energy, Office of Science, Office of Biological and Environmental Research and at the National Energy Research Scientific Computing Center (NERSC) located at Lawrence Berkeley National Laboratory.

Author contributions

Y.-G.W. performed the calculations. D.M. and V.-A.G. contributed to the MD simulations. R.R., J.L. and V.-A.G. conceived the work and supervised the project. Y.-G.W., V.-A.G., R.R. and J.L. co-wrote the manuscript. All authors discussed the results and commented on the manuscript.

Additional information

Competing financial interests: There are no competing financial interests.

8

Chiral recognition and selection during the self-assembly process of protein-mimic macroanions

Panchao Yin[1,2], Zhi-Ming Zhang[3], Hongjin Lv[4], Tao Li[5], Fadi Haso[1,2], Lang Hu[1,2], Baofang Zhang[1,2], John Bacsa[4], Yongge Wei[6], Yanqing Gao[3], Yu Hou[4], Yang-Guang Li[3], Craig L. Hill[4], En-Bo Wang[3] & Tianbo Liu[1,2]

The research on chiral recognition and chiral selection is not only fundamental in resolving the puzzle of homochirality, but also instructive in chiral separation and stereoselective catalysis. Here we report the chiral recognition and chiral selection during the self-assembly process of two enantiomeric wheel-shaped macroanions, $[Fe_{28}(\mu_3\text{-}O)_8(Tart)_{16}(HCOO)_{24}]^{20-}$ (Tart = D- or L-tartaric acid tetra-anion). The enantiomers are observed to remain self-sorted and self-assemble into their individual assemblies in their racemic mixture solution. The addition of chiral co-anions can selectively suppress the self-assembly process of the enantiomeric macroanions, which is further used to separate the two enantiomers from their mixtures on the basis of the size difference between the monomers and the assemblies. We believe that delicate long-range electrostatic interactions could be responsible for such high-level chiral recognition and selection.

[1] Department of Polymer Science, The University of Akron, Akron, Ohio 44325-3909, USA. [2] Department of Chemistry, Lehigh University, Bethlehem, Pennsylvania 18015, USA. [3] Key Laboratory of Polyoxometalate Science of Ministry of Education, Northeast Normal University, Changchun, Jilin 130024, China. [4] Department of Chemistry, Emory University, Atlanta, Georgia, 30322, USA. [5] X-ray Science Division, Advanced Photon Source, Argonne National Laboratory, Argonne, Illinois 60439, USA. [6] Department of Chemistry, Tsinghua University, Beijing 100084, China. Correspondence and requests for materials should be addressed to E.W. (email: wangeb889@nenu.edu.cn) or to T.L. (email: tliu@uakron.edu).

The homochirality of basic biological molecules such as amino acids and sugars, an intriguing puzzle in modern chemistry, is still not clearly understood[1-5]. Current research on the mechanism of homochirality mainly focuses on chiral recognition and chiral selection through studies of chirality transfer and amplification induced by chemical reactions, force fields[3,6-12], physical properties of solid samples (including crystallization)[13-15], amino-acid clustering in gas phase[16], molecular interaction on surfaces[17-20] and supramolecular assemblies of chiral molecules in bulk and solutions[21-30]. A few sketchy theories have been developed to explain how homochirality originated, was amplified in macromolecules and supramolecular assemblies and was transmitted to other biomolecules/assemblies during the evolution of life[2,31]. This process certainly involves recognition and competition between the enantiomers during their organization into supramolecular structures[32,33]. The limitation of stereoselective synthetic technology and the rare examples of the natural biomacromolecule enantiomers make it difficult to study chiral recognition/selection. Through manipulating intermolecular hydrogen bonding and aromatic stacking interactions, Aida and Meijer *et al.* observed the self-sorted behaviour in the solution of organic enantiomers during their formation of one-dimensional supramolecular structure[27-29]. However, the investigation of physical interactions (for example, electrostatic interaction) and spherical supramolecular assemblies (good analogues of virus capsid) in chiral recognition and selection is still rare.

Macroions are soluble ions that are a few nanometres in size; examples include various polyoxometalate molecular clusters, metal–organic nanocages, dendrimers and some biomacromolecules. They exhibit solution behaviour distinct from that associated with simple ions and large colloids[34-36]. When bearing moderate charges, these macroions can strongly attract each other *via* counterion-mediated interactions and sometimes also hydrogen bonding, forming stable, robust supramolecular assemblies in polar solvents, that is, single-layered, hollow, spherical 'blackberry' structures[36]. This self-assembly allows relatively simple building blocks to organize themselves into larger supramolecular structures, a phenomenon that has parallels with the formation of viral capsids (both in morphology and kinetics)[37]. Furthermore, two almost identical macroions remain self-sorted and assemble into homogeneous 'blackberry' structures in solutions containing both, demonstrating a high level of self-recognition mimicking biosystems[38,39].

Here two inorganic chiral macroions, $[Fe_{28}(\mu_3\text{-}O)_8(Tart)_{16}(HCOO)_{24}]^{20-}$ (Tart = D-(S,S)-(-)-tartaric acid tetra-anion, **D-Fe$_{28}$**; Tart = L-(R,R)-(+)-tartaric acid tetra-anion, **L-Fe$_{28}$**), are used as simple models, for the first time, to study the role of one type of universal physical interaction, electrostatic interaction, in chiral recognition and selection behaviour of biomolecules in solution. The kinetics study of the self-assembly indicates that the two enantiomers recognize each other and form their individual 'blackberry' structures. The addition of chiral co-anions is critical to selectively suppress the self-assembly process of one of the enantiomers. This study provides a new way to experimentally achieve chiral separation.

Results

Molecular structures of the chiral macroanions. The nano-scaled anionic ferric wheel $[Fe_{28}(\mu_3\text{-}O)_8(Tart)_{16}(HCOO)_{24}]^{20-}$ consists of four $\{Fe_7\}$ subunits connected by four Tart linkers (Fig. 1 and Supplementary Figs 1–12 and Supplementary Tables 1–5)[40]. If the building units, the Tart ligands, are treated as 'amino acids', the **Fe$_{28}$** cluster can be considered as a protein-mimic molecule involving 'polymerization' of 16 such 'amino acids' via the connections to Fe^{3+}. Therefore, it is straightforward to understand chiral recognition and chiral selection among biomacromolecules from the corresponding studies of **Fe$_{28}$** chiral clusters. Small-angle X-ray scattering, zeta-potential measurements and structural analysis confirm that **Fe$_{28}$** clusters preserve their molecular structures in aqueous solution and stay as macroanions with associated counterions around their surface without involving any hydrophobic or intermolecular chemical interactions (Supplementary Figs 13–17).

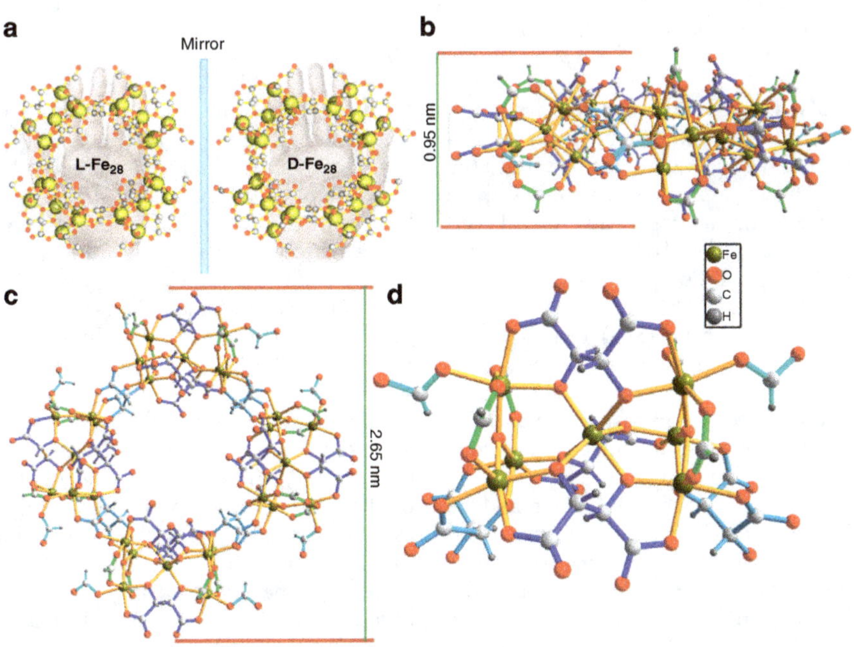

Figure 1 | Molecular structures of enantiomers of Fe$_{28}$ clusters. (**a**) Ball-stick representation of molecular structures of enantiomers of Fe$_{28}$. (**b**,**c**) Side and top view of the molecular structures of **L-Fe$_{28}$** (molecular units highlighted in blue are tartaric acid fragments). (**d**) Ball-stick representation of $\{Fe_7\}$ subunit of the cluster. Colour code of the fragment: blue and teal, two coordination modes of Tart; bright green and turquoise, two coordination modes of methanoate group.

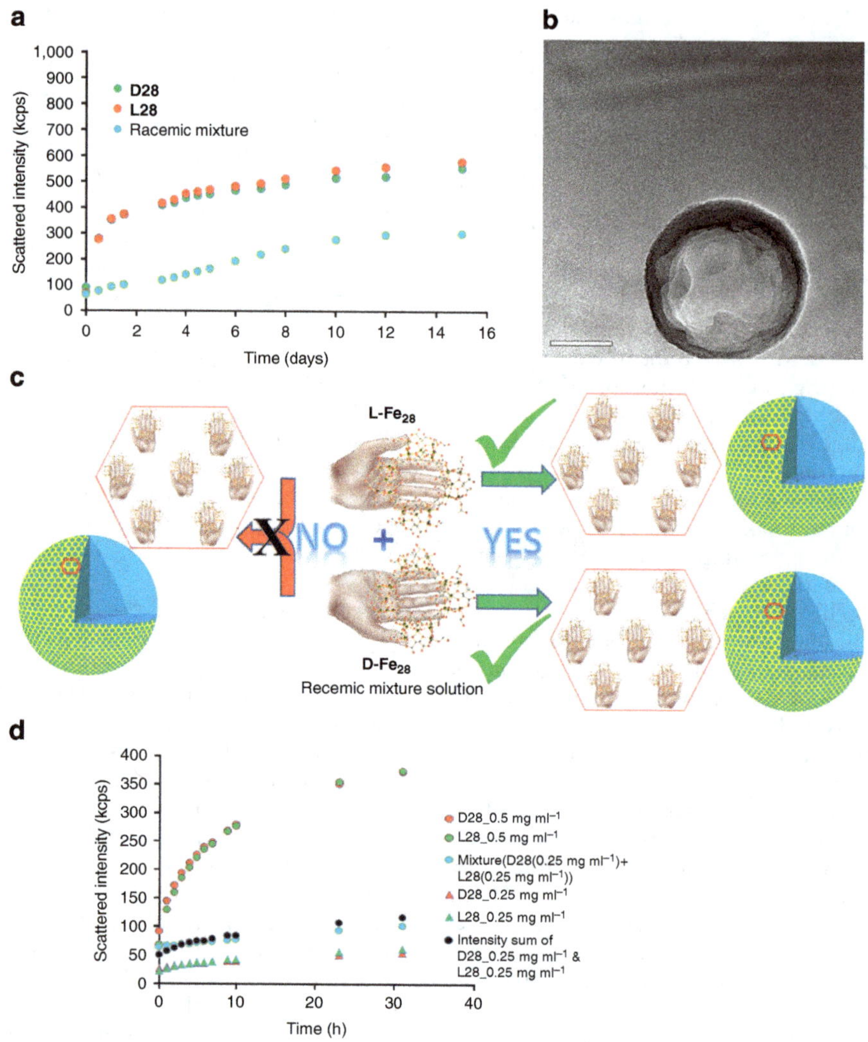

Figure 2 | Self-recognition behaviour among the enantiomers of Fe_{28} clusters. (**a**) SLS results of **Ba-D-Fe_{28}**, **Ba-L-Fe_{28}** and their racemic mixture solutions. (**b**) TEM image of the assembly in the aged solution of **Ba-D-Fe_{28}** (0.5 mg ml^{-1}). Scale bar, 50 nm. (**c**) Graphical representation of the chiral recognition behaviour. (**d**) SLS monitoring of different solutions for 31 h.

Chiral recognition between macroanionic enantiomers. With the aim of using simple macroions to understand some fundamental processes and features in biological systems, we study the racemic mixture solutions of the **Fe_{28}** enantiomers to determine whether they form homogeneous or heterogeneous 'blackberry' structures without any disturbance. With Ba^{2+} being counter-ions, **Ba-D-Fe_{28}** (0.5 mg ml^{-1}), **Ba-L-Fe_{28}** (0.5 mg ml^{-1}) and their racemic mixture (D/L = 1:1, 0.25 mg ml^{-1} for each) in water, as monitored by the static light scattering (SLS) technique, all show a slow self-assembly process and reach equilibrium in ~10 days (Fig. 2a). The two enantiomers self-assemble at the same speed, as confirmed by time-resolved SLS measurements. Since the self-assembly rate is proportional to the macroionic concentration (Supplementary Fig. 18 and Supplementary Note 1), the time-resolved SLS curve for the mixed solution should be similar to the two individual enantiomeric solutions if no recognition exists between the two enantiomers. However, the rate is much slower in the solution of racemic mixtures and the equilibrium state achieved is at a much lower scattered intensity. This observation implies that the two enantiomers can recognize each other and form their individual assemblies (Fig. 2c)[38]. Laser light scattering (LLS)[41] and transmission electron microscopy (TEM) studies of the above three solutions indicate that hollow spherical assemblies of hydrodynamic radius (R_h) as ca 65 ± 4 nm form during the self-assembly process (Fig. 2b and Supplementary Figs 19 and 20).

The two individual enantiomer solutions (0.5 mg ml^{-1}) and their mixtures (D/L = 1:1; 0.25 mg ml^{-1} for each) were monitored by SLS for the first 31 h after preparation to check whether the lag phase period, known as the dimer/oligomer formation stage, is responsible for the self-recognition behaviour[37]. The first two solutions self-assemble at almost the same rate and much faster than that of the mixed solution, confirming that the self-recognition starts at the early stage of the assembly (Fig. 2d). The scattered intensity from the D/L mixed solution (0.25 mg ml^{-1} each, blue dots in Fig. 2d) also grows slower than the sum of the scattered intensities of two individual enantiomer solutions at 0.25 mg ml^{-1} (black dots in Fig. 2d), indicating that the presence of the other type of monomer seems to negatively impact the assembly process, possibly due to the ineffective collisions among the macroions.

The rigid nature of the monomer structure and the high energy barrier for dimer/oligomer formation are expected to be important for the self-recognition. Olson *et al.*[42] studied the chiral recognition theoretically and suggested that only homochiral oligomers were observed at the beginning and homogeneous assemblies were obtained at the end, which is consistent with our observations. SLS studies on the solutions of

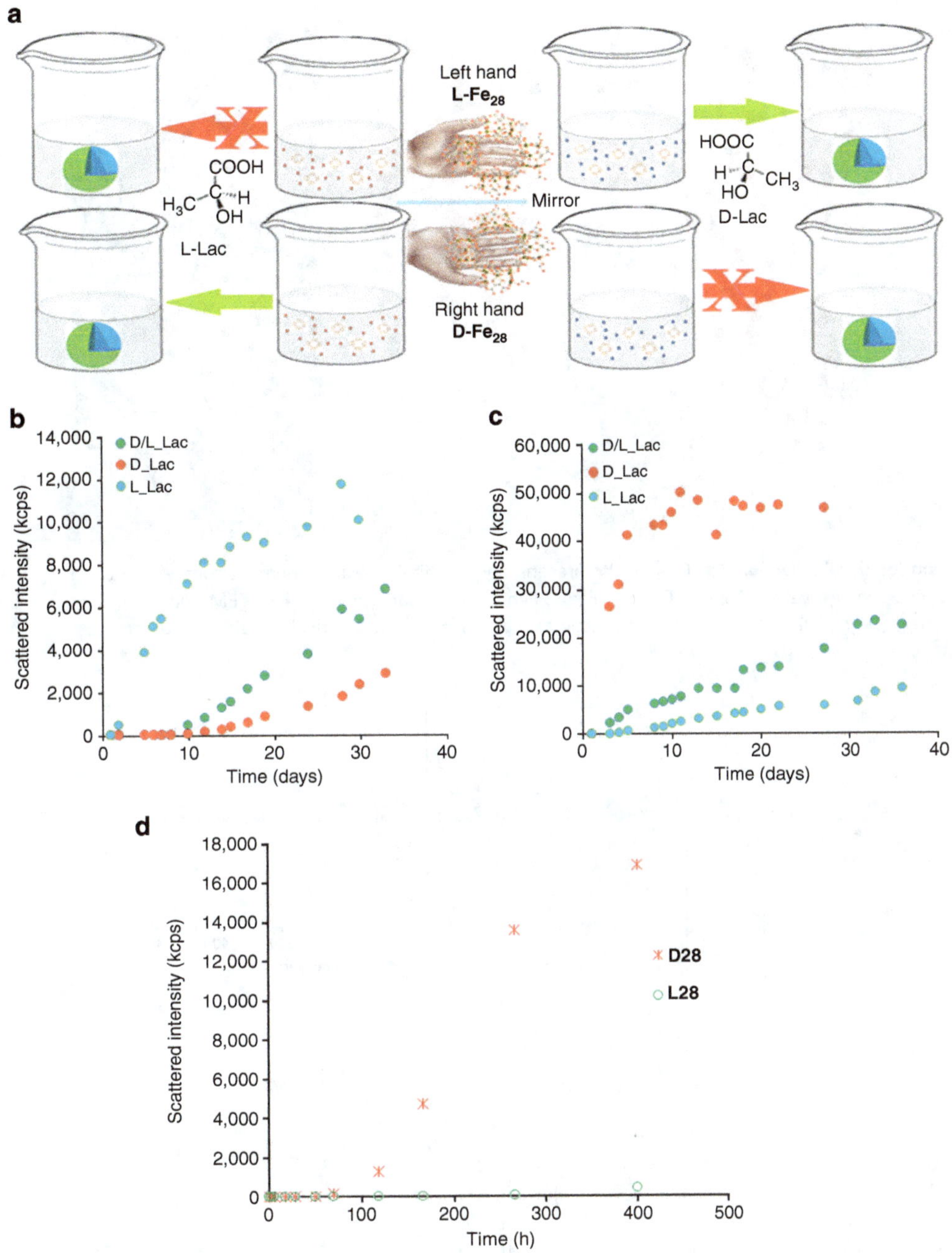

Figure 3 | Chiral selection during the self-assembly process of $\mathbf{Fe_{28}}$. (**a**) Graphical representation of the chiral selection experiments by the introduction of lactic acid molecules. (**b**,**c**) SLS results of **Ba-D-Fe$_{28}$** (0.5 mg ml^{-1}) and **Ba-L-Fe$_{28}$** (0.5 mg ml^{-1}) in three different buffers, respectively. (**d**) SLS monitoring of Ba-D-Fe$_{28}$ and Ba-L-Fe$_{28}$ (0.5 mg ml^{-1}) in the buffer of L-lactic acid (33 μg ml^{-1}) over a period of 400 h.

another completely different pair of inorganic macroanionic enantiomers $\{[\alpha\text{-}P_2W_{15}O_{55}(H_2O)]Zr_3(\mu_3\text{-}O)(H_2O)(\text{L-tartH})[\alpha\text{-}P_2W_{16}O_{59}]\}^{15-}$ (L-CH) and $\{[\alpha\text{-}P_2W_{15}O_{55}(H_2O)]Zr_3(\mu_3\text{-}O)(H_2O)(\text{D-tartH})[\alpha\text{-}P_2W_{16}O_{59}]\}^{15-}$ (D-CH) with dimension $2.4 \times 1.0 \times 0.9\,\text{nm}^3$ (ref. 43) and their racemic mixture, respectively, confirm the chiral recognition during the self-assembly process of chiral macroions (Supplementary Figs 21 and 22, and Supplementary Note 2). Dimers/oligomers are considered as the intermediate state in the self-assembly process[37]. Owing to the morphological compatibility, the formation energy for homo-oligomers might be slightly lower than the hybrid oligomers. This tiny difference will be continuously amplified during the multiple steps of association and it leads to the dominance of homogeneous blackberries over heterogeneous ones.

Chiral selection via the introduction of chiral co-anions. Small chiral organic molecules (for example, lactic acid) are found to selectively suppress the self-assembly process of the enantiomers of **Ba-Fe$_{28}$**. As weak acids, lactic acids partially exist as anions in aqueous solution[44]. For **Ba-D-Fe$_{28}$** and **Ba-L-Fe$_{28}$** (0.5 mg ml^{-1}) in three aqueous solutions containing D-lactic acid (33 μg ml^{-1}), L-lactic acid (33 μg ml^{-1}) and D/L-lactic acid (16.5 μg ml^{-1} for each) as chiral buffers, respectively, the **D-Fe$_{28}$** macroanions are observed to self-assemble rapidly in the solution of L-lactic acid with a short lag phase period (<1 day). However, in the solution of D-lactic acid, the assembly process is slower with a longer lag phase of ~10 days (Fig. 3a,b and Supplementary Fig. 23). The assembly speed of the **D-Fe$_{28}$** in the buffer of mixed D/L-lactic acids stays in the middle of the above two solutions with a lag

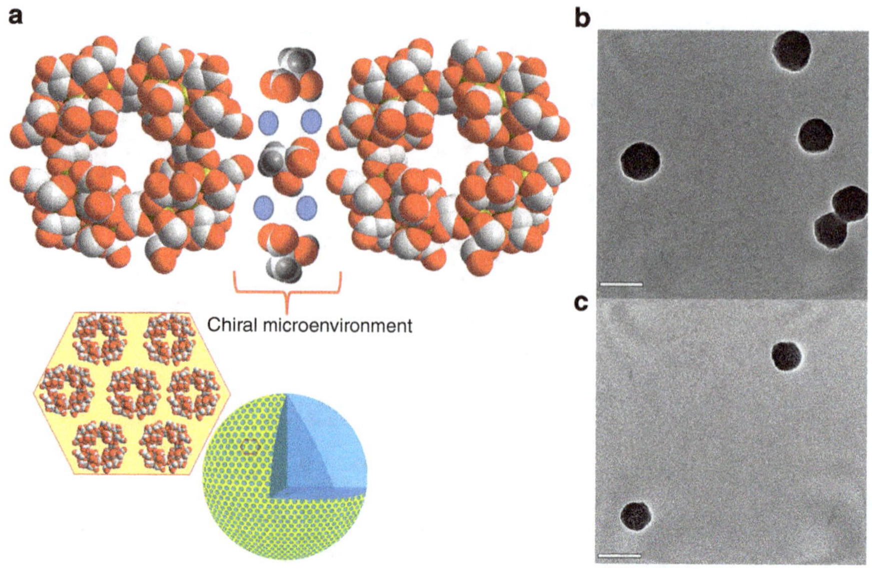

Figure 4 | The mechanism for chiral selection. (**a**) Graphical representation of chiral microenvironment confined by two Fe_{28} clusters in their dimeric structure. Colour code of the spheres: yellow, Fe; red, O; blue, Ba^{2+}; light grey, C; dark grey, H. (**b,c**) TEM images of the assemblies in Ba-D-Fe_{28} (0.5 mg ml^{-1}) and Ba-L-Fe_{28} (0.5 mg ml^{-1}) in the buffer of L-lactic acid (33 μg ml^{-1}), respectively. Scale bar, 100 nm.

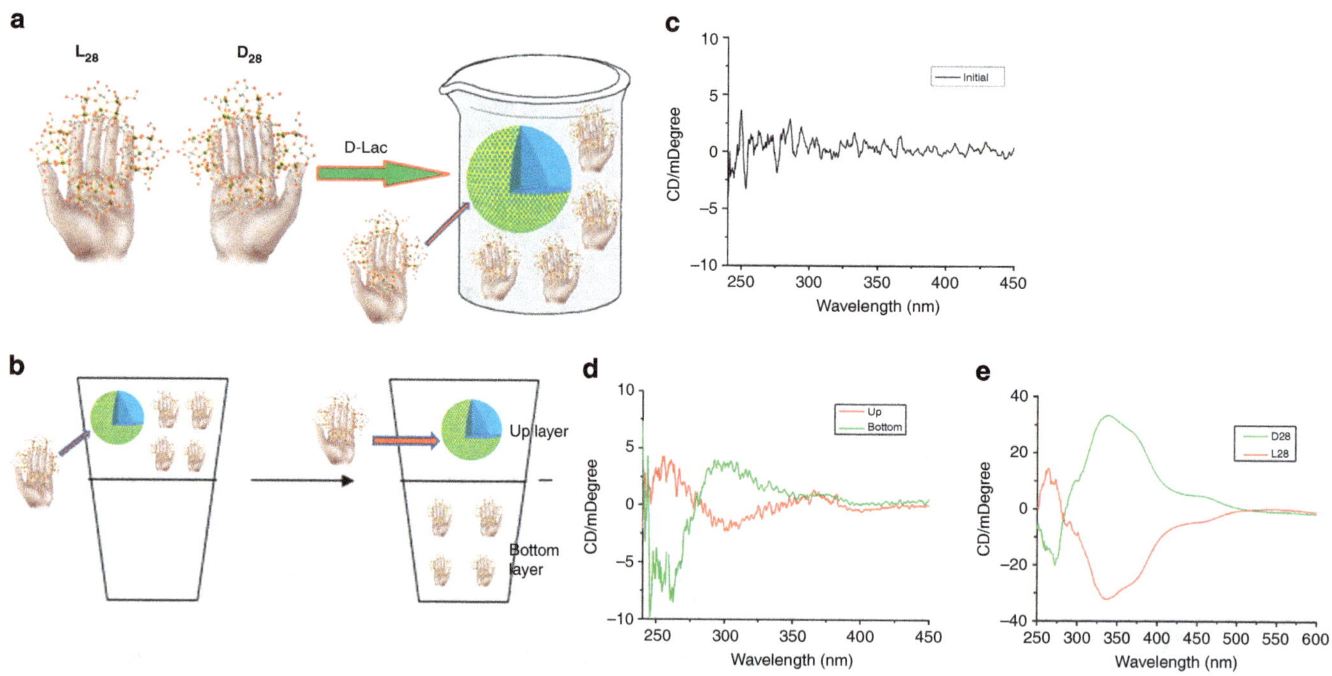

Figure 5 | Chiral separation from the synergistic effect of chiral recognition and selection. (**a**) Graphical representation of the self-assembly of the two enantiomers in the D-lactic buffer. (**b**) Graphical representation of ultrafiltration experiment. (**c**) CD spectrum of the solution before ultrafiltration. (**d**) CD spectra of upper layer (red) and bottom layer (green) solutions after ultrafiltration. (**e**) CD spectra of aqueous solutions of pure Ba-D-Fe_{28} (green) and pure Ba-L-Fe_{28} (red).

phase of ~6 days (Fig. 3b). Identically, the assembly speed of **L-Fe_{28}** is the fastest and with the shortest lag phase time (<1 day) in the solution of D-lactic acid, followed by the D/L-lactic acid mixed solution and the L-lactic acid solution (Fig. 3a,c and Supplementary Fig. 23). The explanation for these observations is that D-lactic acid selectively suppresses the self-assembly of **D-Fe_{28}** while L-lactic acid suppresses that of **L-Fe_{28}**. To observe the counter-cations playing a critical role on controlling the assembly of **Fe_{28}** macroanions is expected; however, the current results show that the co-anions (here lactic acids) are also involved in the assembly process. A small preference from the environment can lead to a complete chiral selection during assembly. Similar to the chiral recognition study, the chiral selection was proved to start during the lag phase period of the self-assembly process. The lag phase periods for the solutions of **D-Fe_{28}** and **L-Fe_{28}** (0.5 mg ml^{-1}) in the buffer of L-lactic acid (33 μg ml^{-1}) were determined as ~50 and ~270 h, respectively (Fig. 3d). In the buffer with L-lactic acid, the formation of **D-Fe_{28}** dimers/oligomers at the beginning of the self-assembly is much faster than that of **L-Fe_{28}**.

Further studies were conducted using the buffers containing other chiral molecules, neutral arabinose and anionic tartaric acid, to determine whether the anionic feature of added chiral organic molecules is critical for the chiral selection. **Ba-D-Fe_{28}**

and **Ba-L-Fe$_{28}$** self-assemble at a similar rate in both D-arabinose and L-arabinose (78 μg ml^{-1}) buffers (Supplementary Fig. 24), confirming that no chiral selection can be observed in the buffers of arabinose. However, time-resolved SLS studies on **D-Fe$_{28}$** and **L-Fe$_{28}$** in D-(S,S)-(−)- and L-(R,R)-(+)-tartaric acid buffers, respectively, indicate that the self-assembly process of **D-Fe$_{28}$** can be suppressed by L-tartaric acid in the buffer (31 μg ml^{-1}) and that of **L-Fe$_{28}$** can be suppressed by D-tartaric acid in the buffer (31 μg ml^{-1}) (Supplementary Figs 25 and 26). The study with buffers containing three different chiral organic molecules demonstrates the significance of co-anions in chiral selection during the self-assembly process of chiral macroanions (Supplementary Table 6).

The chiral selection during the self-assembly process could result from the compatibility between chiral co-anions (lactic acid or tartaric acid) and the chiral microenvironment confined by the macroions. In the lag phase period, discrete macroanionic clusters, surrounded by counterions and loosely by some co-anions, collide with each other to slowly form dimer (or oligomer) structures. It is well known that the counterions mediate the attraction in confined spaces between the clusters, the so-called chiral microenvironment (Fig. 4a)[45]. The new results indicate that the loosely associated co-anions are also involved in dimer formation. The compatibility of lactic acid/tartaric acid and the chiral microenvironment determines the rate of self-assembly very likely by affecting both the pre-exponential frequency (reactant encounter probability) as was the activation energy for formation of dimers/oligomers. Experimental results confirm that the D-lactic acid is compatible with the chiral microenvironments of **L-Fe$_{28}$**, and *vice versa*. These preferred pairs (for example, D-lactic acid (L-tartaric acid)/**L-Fe$_{28}$**) very likely have slightly lower activation energies for dimer/oligomer formation, leading to shorter lag phase periods, and consequently more dominant blackberry formation. LLS[41] and TEM measurements indicate that the preferred pairs form larger blackberry structures ($R_h \sim 63 \pm 4$ nm) than the unpreferred pairs (for example, D-lactic acid/**D-Fe$_{28}$**) do ($R_h \sim 40 \pm 2$ nm) (Fig. 4b,c and Supplementary Figs 27 and 28). The observation indicates that the macroions with preferred partners interact more strongly with each other in the assembly structures, resulting in a shorter inter-macroionic distance and consequently smaller curvature of the spherical shell, that is, larger blackberry sizes.

Chiral separation from the enantiomer mixture. The individual enantiomers can be separated from their racemic mixtures by the introduction of the chiral anionic molecules due to the synergistic impact of both chiral recognition and chiral selection (Supplementary Note 3). For a racemic mixture of **Ba-Fe$_{28}$** (0.25 mg ml^{-1} for each) in the buffer of D-lactic acid (66 μg ml^{-1}) after blackberry formation, the enantiomers were easily separated by filter membranes (molecular weight cutoff, 30 kDa). **L-Fe$_{28}$** macroions self-assemble dominantly, while most **D-Fe$_{28}$** stay as single clusters (Fig. 5a,b). The upper layer solution, expected to contain only large assemblies, shows the signal of **L-Fe$_{28}$**, while the bottom layer solution, expected to be dominated by monomers, shows the signal of **D-Fe$_{28}$** (Fig. 5c–e). This indicates that the two enantiomers were successfully separated.

Discussion

In summary, chiral macroanions demonstrate chiral recognition behaviour by forming homogeneous blackberry structure via long-range electrostatic interactions between the individual enantiomers in their racemic mixture solutions. Adding chiral co-anions suppresses the self-assembly of one enantiomer while maintaining the assembly of the other one. This leads to a natural chiral selection and chiral amplification process, indicating that some environmental preferences can lead to a complete chiral selection. The fact that the relatively simple inorganic macroions exhibit chiral recognition and selection during their assembly process indicates that the related features of biomacromolecules might be due to their macroionic nature via long-range electrostatic interactions.

Methods

Buffer preparation. Solution **A** was prepared by adding 100 ml of formic acid to 50 g of $Ba(OH)_2 \cdot 8H_2O$ dissolved in 500 ml water. Solution **B** was prepared by adding 90 ml of propanoic acid to a solution of 15 g $Ca(OH)_2$ dissolved in 500 ml water.

Synthesis of Ba-L-Fe$_{28}$. To 30 ml of solution **A**, 8 ml of 1 M $Fe(NO_3)_3$ aqueous solution and 0.25 g of L-tartaric acid were added. Another 4.0 g of $Ba(OH)_2 \cdot 8H_2O$ was added to the mixture solution. The obtained solution was heated to 50 °C with stirring for 3 h. The resulting solution was filtered hot to remove any precipitate and left in a refrigerator at 4 °C for crystallization. A mixture of dark red crystals and chunky crystalline powder started to form after 3 days, and the high-quality dark red crystals were collected by filtration after 10 days (yield 0.68 g dark red crystals; ca 32.5%, based on Fe). One single crystal was submitted to structural analysis by X-ray crystallography and the bulk sample was analysed by elemental analysis. Ba-L-Fe$_{28}$: $Ba_{10}[Fe_{28}(\mu_3\text{-}O)_8(Tart)_{16}(HCOO)_{24}] \cdot 47H_2O$. Anal. Calcd (%): Ba, 18.73; Fe, 21.33; Found: Ba, 17.95; Fe, 20.60.

Synthesis of Ba-D-Fe$_{28}$. To 30 ml of solution **A**, 8 ml of 1 M $Fe(NO_3)_3$aqueous solution and 0.25 g of D-tartaric acid were added. Another 4.0 g of $Ba(OH)_2 \cdot 8H_2O$ was added to the mixture solution. The obtained solution was heated to 50 °C with stirring for 3 h. The resulting solution was filtered hot to remove any precipitate and left in a refrigerator at 4 °C for crystallization. A mixture of dark red crystals and chunky crystalline powder started to form after 3 days, and the high-quality dark red crystals were collected by filtration after 10 days (yield 0.54 g dark red crystals; ca 25.2%, based on Fe). One single crystal was submitted to structural analysis by X-ray crystallography and the bulk sample was analysed by elemental analysis. **Ba-D-Fe$_{28}$**: $Ba_{10}[Fe_{28}(\mu_3\text{-}O)_8(Tart)_{16}(HCOO)_{24}] \cdot 57H_2O$. Anal. Calcd (%): Ba, 18.29; Fe, 20.82; Found: Ba, 18.70; Fe, 20.81.

Synthesis of Ca-L-Fe$_{28}$. To 30 ml of solution **B**, 6 ml of 1 M $Fe(NO_3)_3$ aqueous solution and 0.30 g of L-tartaric acid were added. Another 0.45 g of $Ca(OH)_2$ was added to this solution. The resulting solution was heated to 90 °C with stirring for 2 h then cooled down to room temperature and filtered. Black crystals were obtained after 15 days. Yield, 31% (based on Fe). Ca-L-Fe$_{28}$: $Ca_6H_6[Fe_{28}(\mu_3\text{-}O)_8(C_4H_2O_6)_{16}(CH_3CH_2COO)_{22}(H_2O)_2] \cdot 76H_2O$. Anal. Calcd (%): Ca, 3.30; Fe, 21.46; Found: Ca, 3.15; Fe, 20.47.

Synthesis of Ca-D-Fe$_{28}$. To 30 ml of solution **B**, 6 ml of 1 M $Fe(NO_3)_3$ aqueous solution and 0.30 g of D-tartaric acid were added. Another 0.45 g of $Ca(OH)_2$ was added to this solution. The resulting solution was heated to 90 °C with stirring for 2 h then cooled down to room temperature and filtered. Black crystals were obtained after 3 days. Yield, 28% (based on Fe). Ca-D-Fe$_{28}$: $Ca_6H_6[Fe_{28}(\mu_3\text{-}O)_8(C_4H_2O_6)_{16}(CH_3CH_2COO)_{22}(H_2O)_2] \cdot 76H_2O$. Anal. Calcd (%): Ca, 3.30; Fe, 21.46; Found: Ca, 3.06; Fe, 20.27.

Static light scattering. The solutions for light scattering testing were prepared by dissolving the solid samples in D.I. water with sonication and dialysed against pure D.I. water to remove the extra counterion salts with a 1K dialysis membrane overnight. The resulting solution was transferred in a syringe and pushed to pass through 450 nm filters and into 20 ml glass vials. A commercial Brookhaven Instrument LLS spectrometer equipped with a solid-state laser operating at 532 nm was used for measurement of both SLS and DLS. SLS experiments were performed at scattering angles (θ) between 20° and 100°, at 2° intervals. However, due to the large fluctuations in scattered intensities at low scattering angles, data from 20° to 40° in the final analysis were removed. Derived from Rayleigh–Gans–Debye equation[46], partial Zimm plot was used to analyse the SLS data to obtain the radius of gyration (R_g). The partial Zimm plot stems from the following approximation: $1/I = C(1 + R_g^2/q^2/3)$. Here R_g is determined from the slope and the intercept of a plot of $1/I$ versus q^2.

Dynamic light scattering. DLS measures the intensity–intensity time correlation function by means of a BI-9000AT multi-channel digital correlator. The field correlation function $|g^{(1)}(\tau)|$ was analysed by the constrained regularized CONTIN method[41] to yield information on the distribution of the characteristic linewidth Γ from $|g^{(1)}(\tau)| = \int G(\Gamma)e^{-\Gamma\tau}d\Gamma$. The normalized distribution function of the characteristic linewidth, $G(\Gamma)$, so obtained, can be used to determine an average apparent translational diffusion coefficient, $D_{app} = \Gamma/q^2$. The hydrodynamic radius R_h is related to D via the Stokes–Einstein equation: $R_h = kT/(6\pi\eta D)$ where k is the Boltzmann constant and η the viscosity of the solvent at temperature T. From DLS measurements, we can obtain the particle size distribution in solution from a plot of $\Gamma G(\Gamma)$ versus R_h. The R_h of the particles is obtained by extrapolating $R_{h,app}$ to zero scattering angle.

CD spectrum. The solution samples were put in a 1-cm cuvette and the solutions were diluted by D.I. water until the HT in the whole wavelength range is <800 mV. The solutions with appropriate concentrations were scanned five times from 600 to 220 nm and the resulting data were averaged after subtracting the baseline of pure solvent (water).

Ultrafiltration. Ultrafiltration was performed using Macrosep Advance Centrifugal Device (30 K MWCO) to separate blackberry structures and discrete molecular clusters. A sample solution was poured into the device and the device is rotated at a speed of 4,000*g* for 50 min. The upper layer and bottom layer were collected for CD spectrum testing. The racemic mixture solution with D-lactic acid was filtered 20 days after its preparation. Several drops of acetic acid were used to help dissolve the red solid on the filter membrane; this redissolved material was combined with the upper layer solution.

References

1. Blackmond, D. G. The origin of biological homochirality. *Cold Spring Harb Perspect Biol* **2,** a002147 (2010).
2. Lough, W. J. & Wainer, I. *Chirality in Natural and Applied Science* (Blackwell Science–CRC, 2002).
3. Rikken, G. L. & Raupach, E. Enantioselective magnetochiral photochemistry. *Nature* **405,** 932–935 (2000).
4. Cronin, J. R. & Pizzarello, S. Enantiomeric excesses in meteoritic amino acids. *Science* **275,** 951–955 (1997).
5. Mason, S. F. Origins of biomolecular handedness. *Nature* **311,** 19–23 (1984).
6. Breslow, R. A likely possible origin of homochirality in amino acids and sugars on prebiotic earth. *Tetrahedron Lett.* **52,** 2028–2032 (2011).
7. Bailey, J. *et al.* Circular polarization in star-formation regions: implications for biomolecular homochirality. *Science* **281,** 672–674 (1998).
8. Ribo, J. M., Crusats, J., Sagues, F., Claret, J. & Rubires, R. Chiral sign induction by vortices during the formation of mesophases in stirred solutions. *Science* **292,** 2063–2066 (2001).
9. Toyoda, K., Miyamoto, K., Aoki, N., Morita, R. & Omatsu, T. Using optical vortex to control the chirality of twisted metal nanostructures. *Nano Lett.* **12,** 3645–3649 (2012).
10. Yeom, J. *et al.* Chiral templating of self-assembling nanostructures by circularly polarized light. *Nat. Mater.* **14,** 66–72 (2014).
11. Puchot, C. *et al.* Nonlinear effects in asymmetric synthesis. Examples in asymmetric oxidations and aldolization reactions. *J. Am. Chem. Soc.* **108,** 2353–2357 (1986).
12. Soai, K., Shibata, T., Morioka, H. & Choji, K. Asymmetric autocatalysis and amplification of enantiomeric excess of a chiral molecule. *Nature* **378,** 767–768 (1995).
13. Addadi, L. & Weiner, S. Crystal, asymmetry and life. *Nature* **411,** 753–754 (2001).
14. Viedma, C. Chiral symmetry breaking during crystallization: complete chiral purity induced by nonlinear autocatalysis and recycling. *Phys. Rev. Lett.* **94,** 065504 (2005).
15. Viedma, C., McBride, J. M., Kahr, B. & Cintas, P. Enantiomer-specific oriented attachment: formation of macroscopic homochiral crystal aggregates from a racemic system. *Angew. Chem. Int. Ed.* **52,** 10545–10548 (2013).
16. Cooks, R. G., Zhang, D., Koch, K. J., Gozzo, F. C. & Eberlin, M. N. Chiroselective self-directed octamerization of serine: implications for homochirogenesis. *Anal. Chem.* **73,** 3646–3655 (2001).
17. Kuhnle, A., Linderoth, T. R., Hammer, B. & Besenbacher, F. Chiral recognition in dimerization of adsorbed cysteine observed by scanning tunnelling microscopy. *Nature* **415,** 891–893 (2002).
18. Meyer, V. R. & Rais, M. A vivid model of chiral recognition. *Chirality* **1,** 167–169 (1989).
19. Berthod, A. (ed.) *Chiral Recognition in Separation Methods: Mechanisms and Applications* 1st edn (Springer, 2010).
20. De Cat, I. *et al.* Induction of chirality in an achiral monolayer at the liquid/solid interface by a supramolecular chiral auxiliary. *J. Am. Chem. Soc.* **134,** 3171–3177 (2012).
21. Crego-Calama, M. & Reinhoudt, D. N. (eds). *Supramolecular Chirality* 1st edn (Springer, 2006).
22. Azeroual, S. *et al.* Mirror symmetry breaking and chiral amplification in foldamer-based supramolecular helical aggregates. *Chem. Commun.* **48,** 2292–2294 (2012).
23. Lawrence, D. S., Jiang, T. & Levett, M. Self-assembling supramolecular complexes. *Chem. Rev.* **95,** 2229–2260 (1995).
24. Zhang, L., Qin, L., Wang, X., Cao, H. & Liu, M. Supramolecular chirality in self-assembled soft materials: regulation of chiral nanostructures and chiral functions. *Adv. Mater.* **26,** 6959–6964 (2014).
25. George, S. J. *et al.* Asymmetric noncovalent synthesis of self-assembled one-dimensional stacks by a chiral supramolecular auxiliary approach. *J. Am. Chem. Soc.* **134,** 17789–17796 (2012).
26. Helmich, F., Smulders, M. M. J., Lee, C. C., Schenning, A.P.H.J. & Meijer, E. W. Effect of stereogenic centers on the self-sorting, depolymerization, and atropisomerization kinetics of porphyrin-based aggregates. *J. Am. Chem. Soc.* **133,** 12238–12246 (2011).
27. Sato, K., Itoh, Y. & Aida, T. Homochiral supramolecular polymerization of bowl-shaped chiral macrocycles in solution. *Chem. Sci.* **5,** 136–140 (2014).
28. Wolffs, M. *et al.* Influence of the solvent and the enantiomeric purity on the transition between different supramolecular polymers. *Chem. Eur. J.* **18,** 15057–15064 (2012).
29. Ishida, Y. & Aida, T. Homochiral supramolecular polymerization of an "S"-shaped chiral monomer: translation of optical purity into molecular weight distribution. *J. Am. Chem. Soc.* **124,** 14017–14019 (2002).
30. Micali, N. *et al.* Selection of supramolecular chirality by application of rotational and magnetic forces. *Nat. Chem.* **4,** 201–207 (2012).
31. Guijarro, A. & Yus, M. *The Origin of Chirality in the Molecules of Life: A Revision from Awareness to the Current Theories and Perspectives of this Unsolved Problem* (RSC Publishing, 2008).
32. Alkorta, I., Picazo, O. & Elguero, J. Theoretical studies on chiral discrimination. *Curr. Org. Chem.* **10,** 695–714 (2006).
33. Alkorta, I. & Elguero, J. in *Practical Aspects of Computational Chemistry*. (eds Leszczynski, J. & Shukla, M. K.) 37–86 (Springer, 2010).
34. Liu, T., Diemann, E., Li, H., Dress, A. W. M. & Muller, A. Self-assembly in aqueous solution of wheel-shaped Mo154 oxide clusters into vesicles. *Nature* **426,** 59–62 (2003).
35. Yin, P., Li, D. & Liu, T. Solution behaviors and self-assembly of polyoxometalates as models of macroions and amphiphilic polyoxometalate-organic hybrids as novel surfactants. *Chem. Soc. Rev.* **41,** 7368–7383 (2012).
36. Liu, T. B. Hydrophilic macroionic solutions: what happens when soluble ions reach the size of nanometer scale? *Langmuir* **26,** 9202–9213 (2010).
37. Zhang, J., Li, D., Liu, G., Glover, K. J. & Liu, T. B. Lag periods during the self-assembly of {Mo72Fe30} macroions: connection to the virus capsid formation process. *J. Am. Chem. Soc.* **131,** 15152–15159 (2009).
38. Liu, T. *et al.* Self-recognition among different polyprotic macroions during assembly processes in dilute solution. *Science* **331,** 1590–1592 (2011).
39. Yin, P. *et al.* Self-recognition of structurally identical, rod-shaped macroions with different central metal atoms during their assembly process. *J. Am. Chem. Soc.* **135,** 4529–4536 (2013).
40. Zhang, Z.-M. *et al.* Enantiomerically pure chiral {Fe28} wheels. *Angew. Chem. Int. Ed.* **48,** 1581–1584 (2009).
41. Provencher, S. W. CONTIN: a general purpose constrained regularization program for inverting noisy linear algebraic and integral equations. *Comput. Phys. Commun.* **27,** 229–242 (1982).
42. Olson, A. J., Hu, Y. H. E. & Keinan, E. Chemical mimicry of viral capsid self-assembly. *Proc. Natl Acad. Sci. USA* **104,** 20731–20736 (2007).
43. Fang, X., Anderson, T. M. & Hill, C. L. Enantiomerically pure polytungstates: chirality transfer through zirconium coordination centers to nanosized inorganic clusters. *Angew. Chem. Int. Ed.* **44,** 3540–3544 (2005).
44. Clayden, J., Greeves, N. & Warren, S. *Organic Chemistry* Edn. 2nd (Oxford Univ. Press, 2012).
45. Eeckhaut, A. V. & Michotte, Y. (eds.) *Chiral Separations by Capillary Electrophoresis* (CRC, 2009).
46. Hiemenz, P. C. & Rajagopalan, R. *Principles of Colloid and Surface Chemistry* (Marcel Dekker, 1997).

Acknowledgements

T.L. acknowledges the support from the National Science Foundation (CHE1305756), Lehigh University and the University of Akron. E.W. is grateful for support from the National Natural Science Foundation of China (21101022). C.L.H. thanks the National Science Foundation (CHE1124862) for support. We are thankful for the use of the Advanced Photon Source, an Office of Science User Facility operated for the US Department of Energy (DOE) Office of Science by Argonne National Laboratory; the use of this facility was supported by the US DOE under Contract No. DE-AC02-06CH11357. We acknowledge Dr Xiaobing Zuo for insightful discussion.

Author contributions

P.Y., E.-B.W. and T.Liu designed the project. Z.-M.Z., E.-B.W., H.L., Y.H., Y.G. and C.L.H. performed the synthetic experiments and related structural analysis. J.B., Y.W., and Y.-G.L. provided detailed refinements on the crystals structures. P.Y. prepared the solution samples, completed most of the light scattering studies, and performed TEM and

CD measurements. P.Y. and T.Li carried out small-angle X-ray scattering experiments and analysed the data. F.H. and B.Z. helped P.Y. with part of the light scattering studies. L.H. performed zeta-potential analysis. P.Y. and T.Liu analysed all the data and wrote the manuscript.

Additional information

The X-ray crystallographic coordinates for structures reported in this study have been deposited at the Cambridge Crystallographic Data Centre (CCDC) under deposition numbers 1044500–1044503. These data can be obtained free of charge from The CCDC via www.ccdc.cam.ac.uk/data_request/cif.

Competing financial interests: The authors declare no competing financial interests.

9

Conformation and dynamics of the ligand shell of a water-soluble Au_{102} nanoparticle

Kirsi Salorinne[1], Sami Malola[2], O. Andrea Wong[3], Christopher D. Rithner[3], Xi Chen[1], Christopher J. Ackerson[3] & Hannu Häkkinen[1,2]

Inorganic nanoparticles, stabilized by a passivating layer of organic molecules, form a versatile class of nanostructured materials with potential applications in material chemistry, nanoscale physics, nanomedicine and structural biology. While the structure of the nanoparticle core is often known to atomic precision, gaining precise structural and dynamical information on the organic layer poses a major challenge. Here we report a full assignment of ^{1}H and ^{13}C NMR shifts to all ligands of a water-soluble, atomically precise, 102-atom gold nanoparticle stabilized by 44 para-mercaptobenzoic acid ligands in solution, by using a combination of multidimensional NMR methods, density functional theory calculations and molecular dynamics simulations. Molecular dynamics simulations augment the data by giving information about the ligand disorder and visualization of possible distinct ligand conformations of the most dynamic ligands. The method demonstrated here opens a way to controllable strategies for functionalization of ligated nanoparticles for applications.

[1] Department of Chemistry, Nanoscience Center, University of Jyväskylä, FI-40014 Jyväskylä, Finland. [2] Department of Physics, Nanoscience Center, University of Jyväskylä, FI-40014 Jyväskylä, Finland. [3] Department of Chemistry, Colorado State University, Fort Collins, Colorado 80523, USA. Correspondence and requests for materials should be addressed to H.H. (email: hannu.hakkinen@phys.jyu.fi).

Inorganic nanoparticles, including metals, semiconductors and metal oxides, comprise a common set of structures exhibiting an inorganic core 'passivated' by an organic shell[1,2]. Ligated inorganic nanoparticles currently provoke widespread fundamental interest in their structural, optical and magnetic properties, which differ fundamentally from bulk counterparts. These nanomaterials already find application in biology[3], medicine, solar energy[4] and display panels[5]. Despite their widespread study, the structure of inorganic nanoparticles, especially of their surfaces, is often still a matter of conjecture[6]. While the structure of the inorganic fraction of nanoparticles is straightforwardly determined by methods such as transmission electron microscopy and/or powder diffraction[7,8], the nature of the ligand/metal interface is often obscure.

In special cases where synthesis is successfully refined to produce 'atomically precise' nanoparticles that can form well-ordered crystals, single-crystal X-ray methods can result in total structure determination[9–11]. Single-crystal X-ray methods reveal that the surfaces of nanoparticles can be markedly rearranged by ligand interactions, often into surprisingly low-symmetry structures[12]. These single-crystal X-ray structures highlight the shortcomings of conventional transmission electron microscopy methods for determining nanoparticle structure.

Multidimensional nuclear magnetic resonance (NMR) spectroscopy can reveal a three-dimensional structure with fidelity comparable to single-crystal X-ray crystallography. Indeed, in determining the complex structures of proteins, the method is ideal for addressing the structures and dynamical interactions of proteins or protein fragments that are recalcitrant to crystallization[13]. Importantly, NMR provides a sensitive method to probe the structure of the ligand shell of ligand-passivated inorganic nanoparticles in solution phase, and it has indeed been recently applied to a few cases of atomically precise gold nanoclusters stabilized by thiolates[14–17].

Single-crystal X-ray diffraction studies[9] yielded the total structure of $Au_{102}(pMBA)_{44}$ in 2007, revealing a complex ligand layer with 22 symmetry-unique ligand environments in C_2 overall symmetry (pMBA stands for para-mercapto benzoic acid). Since then, its electronic structure[18–20], spectroscopic[21] properties and chiral[22] properties as well as the propensity for thiolate ligand exchange[12] have been studied, and this nanoparticle has been functionalized for site-specific labelling of enteroviruses by utilizing covalent[23] and weak[24] interactions with viral capsid proteins. NMR spectroscopy has previously revealed the protein-like complexity of the $Au_{102}(pMBA)_{44}$ ligand shell[15], yielding a broad one-dimensional (1D) ^{1}H spectrum comprising several overlapping signals. Here data from multidimensional NMR spectroscopy (nuclear Overhauser effect spectroscopy (NOESY), rotating frame nuclear Overhauser effect spectroscopy (ROESY), total correlated spectroscopy (TOCSY) ^{1}H NMR and heteronuclear single-quantum correlation spectroscopy (HSQC) ^{1}H–^{13}C correlations), combined with theoretical calculations for ^{1}H and ^{13}C NMR shifts based on the known crystal structure of $Au_{102}(pMBA)_{44}$, augmented by molecular dynamics (MD) simulations of the ligand dynamics in water, allowed for a full assignment of the 1D NMR signals to the 22 symmetry-unique ligands. The successful strategy disclosed here opens a new avenue for structural and dynamical studies of ligated inorganic nanoparticles in the solution phase, and is generalizable to both organic and aqueous media. Resolving the structure and dynamics of the ligand layer with molecular precision can lead to highly controllable and accurate methods for functionalization of the ligated nanoparticles for various applications where the essential chemistry between the nanoparticle and environment takes place in solution phase.

Results

Solid-state structure. The atomic structure of $Au_{102}(pMBA)_{44}$ determined from the single-crystal X-ray experiment[9] is shown in Fig. 1a. Within the C_2 overall symmetry, the cluster contains 22 symmetry-unique pMBA ligands. The X-ray structure reveals that the pMBA ligands interact with the inorganic core through two types of gold–thiolate motifs—short RS-Au-SR and long RS-Au-(SR)-Au-SR units[9]. Although crystal packing has an effect on the ligand orientation due to the intermolecular interactions between neighbouring clusters in the solid state, the same ligand symmetry equivalence was previously suggested in the ^{1}H NMR spectrum of $Au_{102}(pMBA)_{44}$ cluster in solution[15]. Therefore, an intrinsic ordering of the ligands on the cluster surface in solution may be similar to that observed in the solid state. The assignment of ligands to their symmetry-equivalent site on the cluster surface cannot be done solely based on the 1D proton spectrum due to high complexity of broad and overlapping signals. Thus, two-dimensional (2D) NMR techniques together with density functional theory (DFT) calculations and MD simulations were undertaken for full assignment of the symmetry-equivalent pMBA ligands in the 1D proton spectrum of $Au_{102}(pMBA)_{44}$ cluster.

2D NMR experiments and chemical shift assignment. NMR spectroscopy is sensitive to the magnetic environment of atomic nuclei of the markers (such as ^{1}H and ^{13}C). The magnetic environment is modified by shielding, which is influenced by the electronegativity of neighbouring atoms, the interaction between

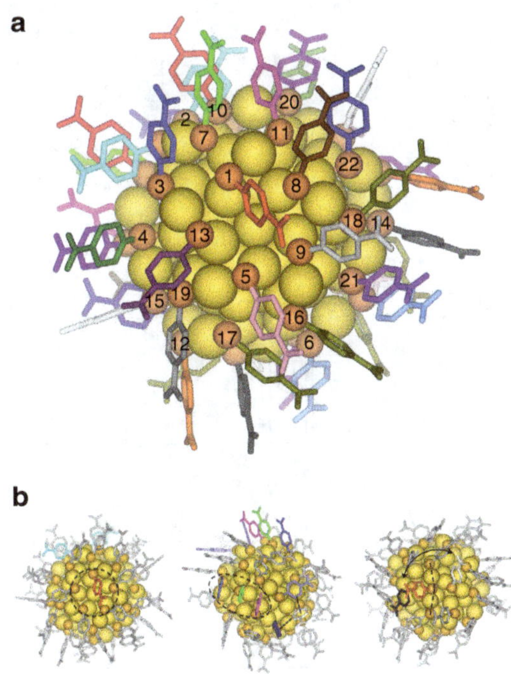

Figure 1 | Crystal structure and selected special ligand environments of $Au_{102}(pMBA)_{44}$. **(a)** Total structure, with numbering and unique colouring of all the 22 symmetry-unique ligands. The same numbering scheme is used throughout this paper. Gold is yellow and sulfur is orange. **(b)** Selected special ligand environments on cluster surface: ligand-to-gold core (pMBA-1 and -2; red and cyan, respectively; left panel), face-to-face (pMBA-11, -7 and -3; magenta, green and blue, respectively; middle panel) and edge-to-face (pMBA-15, -22 and-20; white, navy and purple, respectively; middle panel) and dynamic ligands in the long gold-thiolate unit (pMBA-12, -15 and-14; orange, white and dark grey, respectively; right panel). For clarity, only the highlighted ligands have been coloured.

markers and nearby metals, as well as orientation and distance of neighbouring functional groups. Owing to the restricted mobility of ligands bound at the cluster surface, the local magnetic environment becomes more pronounced as each of the symmetry-equivalent ligand will experience unique and distinguishable shielding based on their different local chemical environment on the cluster surface. Therefore, a set of 2D NMR experiments were recorded to resolve the individual chemical shifts of the 22 pMBA ligands in the 1D proton spectrum (Fig. 2a) and to elucidate neighbouring ligands and ligand order on the cluster surface.

pMBA has two chemically different protons (H_a closer to sulfur and H_b closer to COOH group; see inset to Fig. 2a), which give rise to two doublets in the proton spectrum. As seen in Fig. 2a, the 1D proton spectrum is extremely broad and peaked around the chemical shifts of free pMBA at 7.40 and 7.23 p.p.m. in the reference solvent. Figure 2b shows the theoretically calculated chemical shifts for H_a and H_b based on the structure of the ligand layer in the X-ray crystal data.

The 22 symmetry-unique pMBA ligands were distinguished by using the TOCSY technique (see Fig. 3 and a full-page version in Supplementary Fig. 1). This collection method allowed for the determination of through-bond correlations and, therefore, enabled the separation of chemical shifts belonging to each pMBA ligand. HSQC gave correlations to which carbon the particular proton is attached, thus, identifying protons H_a and H_b. In addition, from the HSQC spectra, multiple correlations for C_a and C_b carbons (a1, a2, b1 and b2) for rigid (as opposed to dynamic) ligands could be identified (Fig. 4 and Supplementary Fig. 2).

Combined, TOCSY and HSQC gave self-consistent information resolving the chemical shift shielding arising from the local chemical environment of each symmetry-equivalent pMBA ligand (Supplementary Table 1). Assignment of the chemical shift of each proton was based initially on the TOCSY spectrum. This enabled numbering of the ligands sequentially from downfield to upfield. The ligands are numbered thus pMBA-1, pMBA-2, ..., where pMBA-1 is the most downfield ligand (Figs 1a and 2a)

Through-space correlations were obtained with NOESY and ROESY experiments, which were complementary and afforded information about connectivity of neighbouring ligands within 5 Å proximity (Fig. 5 and Supplementary Figs 3 and 4).

Assignment of ligand arrangement on the cluster surface. Initial assignment of each of the 22 symmetry independent pMBA ligands to their respective proton chemical shifts was accomplished by comparison of the experimental TOCSY and

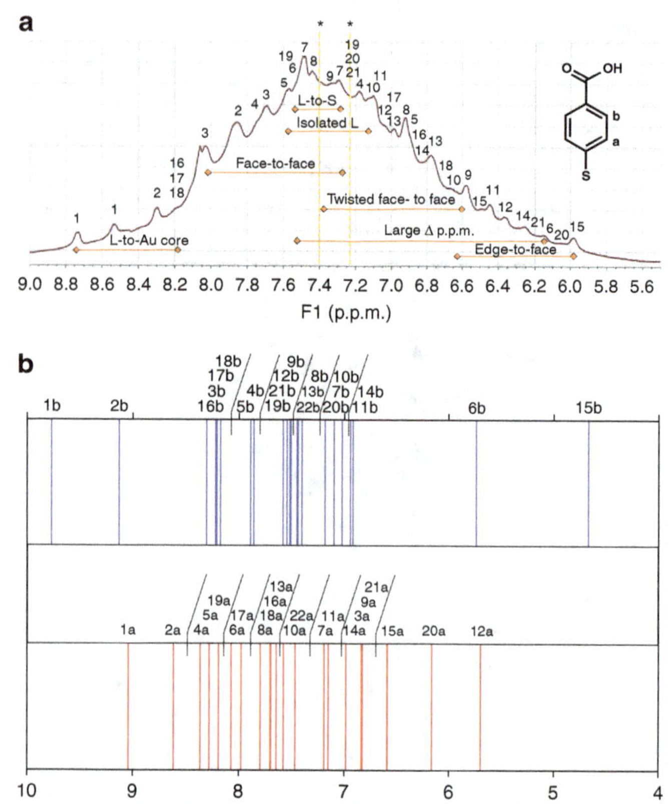

Figure 2 | Ligand assignments in the proton NMR spectrum. (**a**) ^{1}H NMR spectrum of $Au_{102}(pMBA)_{44}$ cluster in 0.3 M D_2O-NaOH with full assignment of the 22 symmetry-equivalent pMBA ligands based on TOCSY spectrum. Strong correlation peaks have been marked with black and weaker signals in light blue (see TOCSY spectrum in Fig. 3). Special ligand environments affecting the chemical shift shielding have been identified in the spectrum (orange double arrows). Asterisk (*) denotes chemical shifts of free pMBA at 7.40 and 7.23 p.p.m. in 0.3 M D_2O-NaOH (ref. 15). (**b**) DFT-calculated proton chemical shifts for H_a (red) and H_b (blue).

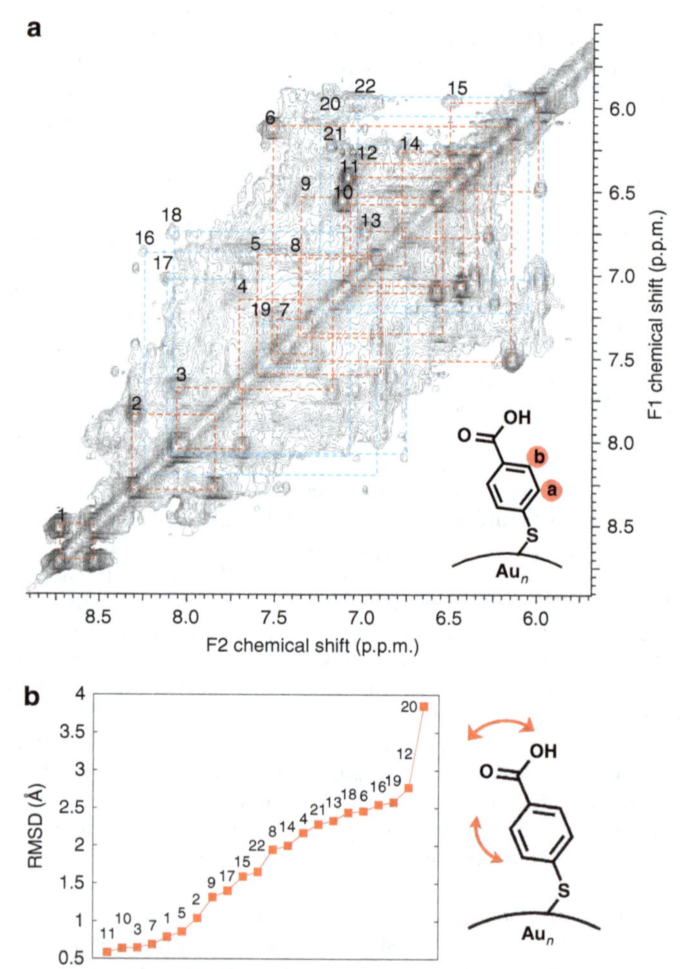

Figure 3 | TOCSY data and dynamics of ligands. (**a**) TOCSY spectrum of $Au_{102}(pMBA)_{44}$ cluster in 0.3 M D_2O-NaOH with full assignment of the 22 symmetry-equivalent pMBA ligands. Strong correlation peaks have been marked with red and weaker correlation peaks are marked with blue. (**b**) Root mean squared deviations (r.m.s.d.) of all the symmetry-unique 22 ligands in MD simulations of $Au_{102}(pMBA)_{44}$ in water at room temperature.

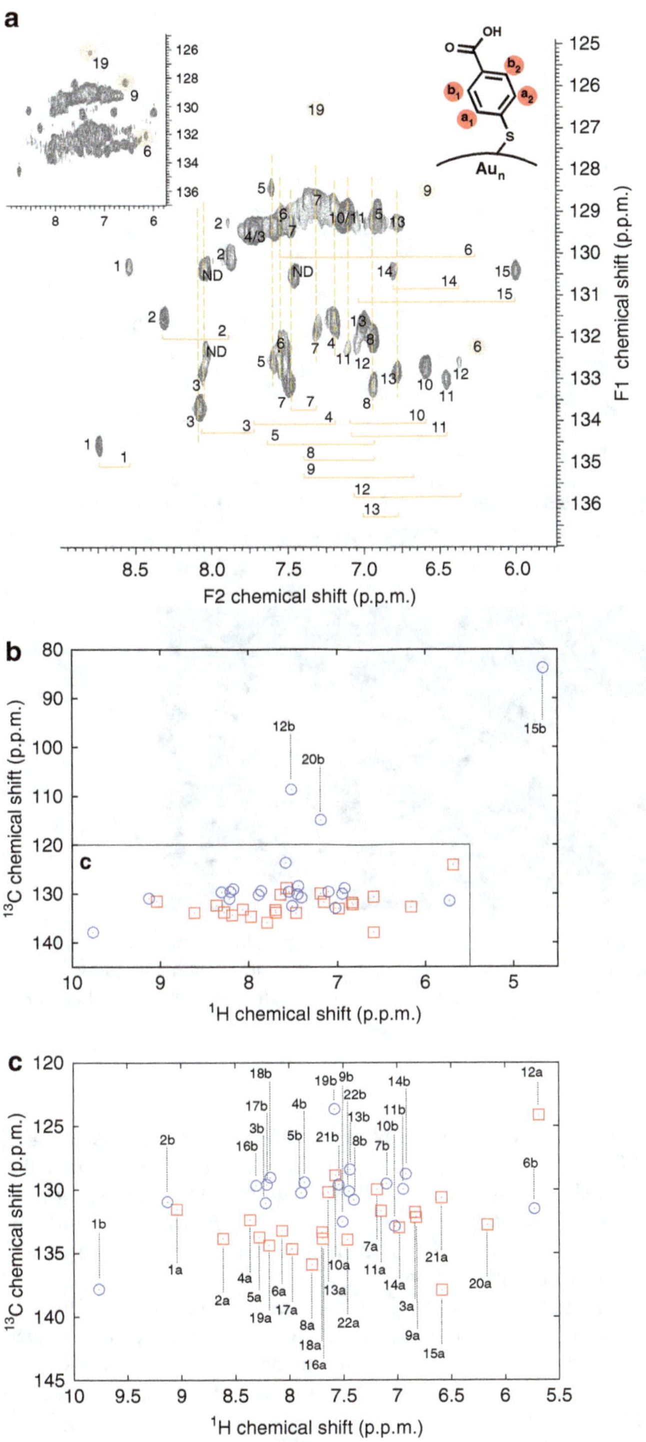

Figure 4 | HSQC data. (**a**) HSQC spectrum of $Au_{102}(pMBA)_{44}$ cluster in 0.3 M D_2O-NaOH showing assignment of ^1H-^{13}C correlation peaks. For some of the less-mobile ligands, correlation peaks for both a_1 and a_2 and/or b_1 and b_2 carbons could be identified. Inset shows the entire spectrum with higher signal-to-noise level and ND denotes 'not determined'. (**b,c**) Calculated ^1H-^{13}C correlation chemical shifts. (**c**) The congested area in **b** in a larger scale.

HSQC correlation peaks to DFT-calculated proton and carbon chemical shifts for each of the symmetry-equivalent pMBA ligand based on the crystal structure of $Au_{102}(pMBA)_{44}$ as a model (Fig. 2b and Supplementary Table 1). The dynamics of ligands in solution was taken into consideration by MD simulation studies, which revealed that the dynamical behaviour of the ligands was affected by interactions with the neighbouring ligands, the available space around the ligands and constraints set by the parent gold–thiolate unit. The final assignment of the 22 pMBA ligands was made based on special ligand environments (see below), neighbouring ligand interactions and ligand dynamics.

Special ligand environments. From the calculated spectra, a set of five special ligand environments could be identified arising from shielding effects imposed by the local chemical environment (marked by horizontal bars in Fig. 2a). These five environments are, in order of deshielding: (1) ligand-to-gold interaction; (2) ligand-to-sulfur interaction; (3) isolated ligand; (4) face-to-face aromatic interaction; and (5) edge-to-face aromatic interaction. Ligand-to-gold interactions were the most deshielded and gave the highest downfield shifts (8.8–8.2 p.p.m.), whereas isolated ligand and ligand-to-sulfur interactions fell in the same region (7.6–7.1 p.p.m.) as free pMBA (7.40 and 7.23 p.p.m.). Face-to-face aromatic interactions were likewise more deshielded and shifted downfield (8.0–7.3 p.p.m.) in comparison with free pMBA, while edge-to-face aromatic interactions shifted the most upfield (6.6–5.9 p.p.m.). Large difference (Δ) between H_a and H_b shifts was observed for ligands that experienced both shielding and deshielding ligand environments. For example, combination of edge-to-face and face-to-face interactions yielded this large difference, marked by 'large Δ p.p.m.' in Fig. 2a.

Correlating the experimental TOCSY assignments with the calculated chemical shifts revealed two pMBA ligands (pMBA-1 and -2) having the most downfield shifts, which were assigned as ligands in the model having distinct ligand-to-gold interactions (Fig. 1b, left). pMBA-1 is located at the five-fold symmetry axis of the gold core. A second special ligand environment was traced to face-to-face aromatic interactions of three ligands (pMBA-3, -7 and -11) assembling into a 'stack of three' that gave downfield shifts both in the experimental and calculated data (Fig. 1b, middle). The different chemical environment imposed by the static π-stacking interactions was clearly seen as multiple strong HSQC correlations for C_a and C_b carbons. Owing to face-to-face aromatic interactions from both sides, ligand pMBA-7 in the middle of the stack showed two HSQC correlations for both C_a and C_b carbons (129.4/132.5 and 129.1/131.9 p.p.m.), whereas only the C_a carbons of ligands pMBA-3 (132.8 and 133.6 p.p.m.) and -11 (129.0 and 132.3 p.p.m.) at the top and bottom of the stack were affected (Fig. 4). The most upfield shift, on the other hand, was identified as an edge-to-face aromatic interaction (pMBA-15), which also correlated well with the calculated model (Fig. 1b, middle). The last special ligand environment was observed with a pMBA ligand (pMBA-6) that had both symmetry-equivalent ligands attached to the same RS-Au-SR unit having the largest Δ value (1.39 p.p.m.) between protons H_a and H_b that resulted from both edge-to-face and face-to-face aromatic interactions with the neighbouring ligands.

Through-space interactions. Whereas chemical shift shielding afforded information about the nature of the ligand interactions, through-space correlations gave information to which neighbouring ligands these interactions are formed. NOESY and ROESY spectra therefore afforded a connectivity map of neighbouring ligands, which could be used together with the previous assignment based on the local ligand environment (Fig. 5). The strongest NOESY and ROESY correlations were associated with ligands pMBA-1 and -15, which also suggested lesser mobility of these ligands and the neighbouring ligands they were interacting with. Thus, pMBA-1 showed NOESY/ROESY

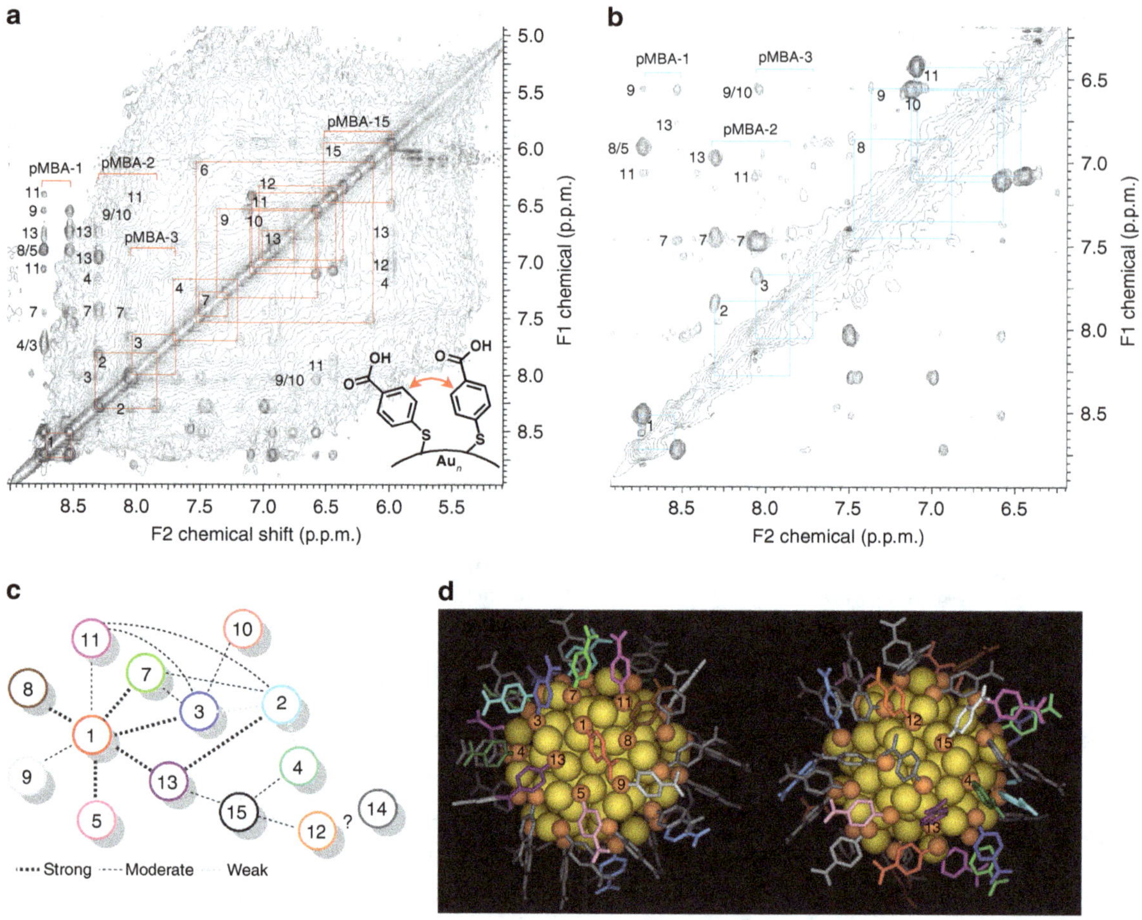

Figure 5 | Through-space correlation data. (**a**) NOESY and (**b**) ROESY spectra of $Au_{102}(pMBA)_{44}$ cluster in 0.3 M D_2O-NaOH showing the assignment (red and blue squares) of the through-space correlation peaks between neighbouring pMBA ligands. (**c**) Connectivity map of the neighbouring ligands based on the NOESY and ROESY correlations with strong, moderate and weak correlations highlighted accordingly. (**d**) The respective ligand positions in two areas of the surface, centred around ligands pMBA-1 and pMBA-15.

correlations to ligands pMBA-4/-3, -7, -8/-5, -9, -11 and -13 (Fig. 4). Of these the symmetry-equivalent site was already established for pMBA-3, -7 and -11 corroborating the earlier assignment. pMBA-7 also showed ROESY correlations to pMBA-1, -2 and -3 confirming their proximity and agreement with the model. pMBA-11 likewise gave correlations to pMBA-1 and -2 that were in close ranges. It should be noted that pMBA-1 and -11 are attached to the same RS-Au-SR unit and are therefore likely to interact. The same holds for ligands pMBA-2 and -3.

The assignment of the remaining ligands surrounding pMBA-1 at the five-fold symmetry axis was based on the best possible agreement of the experimental and calculated chemical shifts. pMBA-13 was assigned first as having a distinct upfield shift in the experimental spectrum (6.98–6.76 p.p.m., $\Delta = 0.20$ p.p.m.), which was most likely due to an edge-to-face aromatic interaction to a ligand that was assigned as pMBA-4 showing up more downfield (7.70–7.18 p.p.m.). Both ligands pMBA-4 and -13 had a NOESY correlation to pMBA-2 and -15, which were in close distance and agreed well with the model. The order of ligands pMBA-5, -8 and -9 was made based on the subtle changes in their chemical shift shielding, since they all appeared at the 'middle range' (7.62–6.56 p.p.m., $\Delta = 0.5$–0.8 p.p.m.) and their symmetry-equivalent site was determined based on the best agreement with the experimental and calculated data (Supplementary Table 1).

Ligand pMBA-3 gave NOESY/ROESY correlations to ligands pMBA-2, -7 and -11, whose proximity was already confirmed and established, but also to pMBA-9/-10. This interaction was assigned to pMBA-10 due to its closer distance and more probable interaction to pMBA-3 based on the model in comparison with pMBA-9.

Finally, the last through-space interactions were established for ligand pMBA-15 that showed NOESY correlations to ligands pMBA-4, -12 and -13. Of these, the assignment of pMBA-4 and -13 was already made. Therefore, the symmetry-equivalent site for pMBA-12 was assigned as the centre position of the long RS-Au-(SR)-Au-SR unit, also agreeing well with the experimental and calculated data (Fig. 1b, right). MD simulation studies showed this ligand to be one of the most dynamic ones, due to the flexibility of the long gold–thiolate unit (Figs 3b and 6 and Supplementary Table 2).

Ligand dynamics. MD simulations were conducted to assess the different degrees of mobility and orientation dynamics of the ligands. The dynamics is visualized in Supplementary Movies 1 and 2 that visualize the dynamics of the particle over 30 ns in water. A more detailed analysis of the dynamics of three selected ligands are given in Fig. 6, Supplementary Table 2 and Supplementary Data 1. We observed that the extent of the chemical shift in the NMR spectra had a clear correlation to

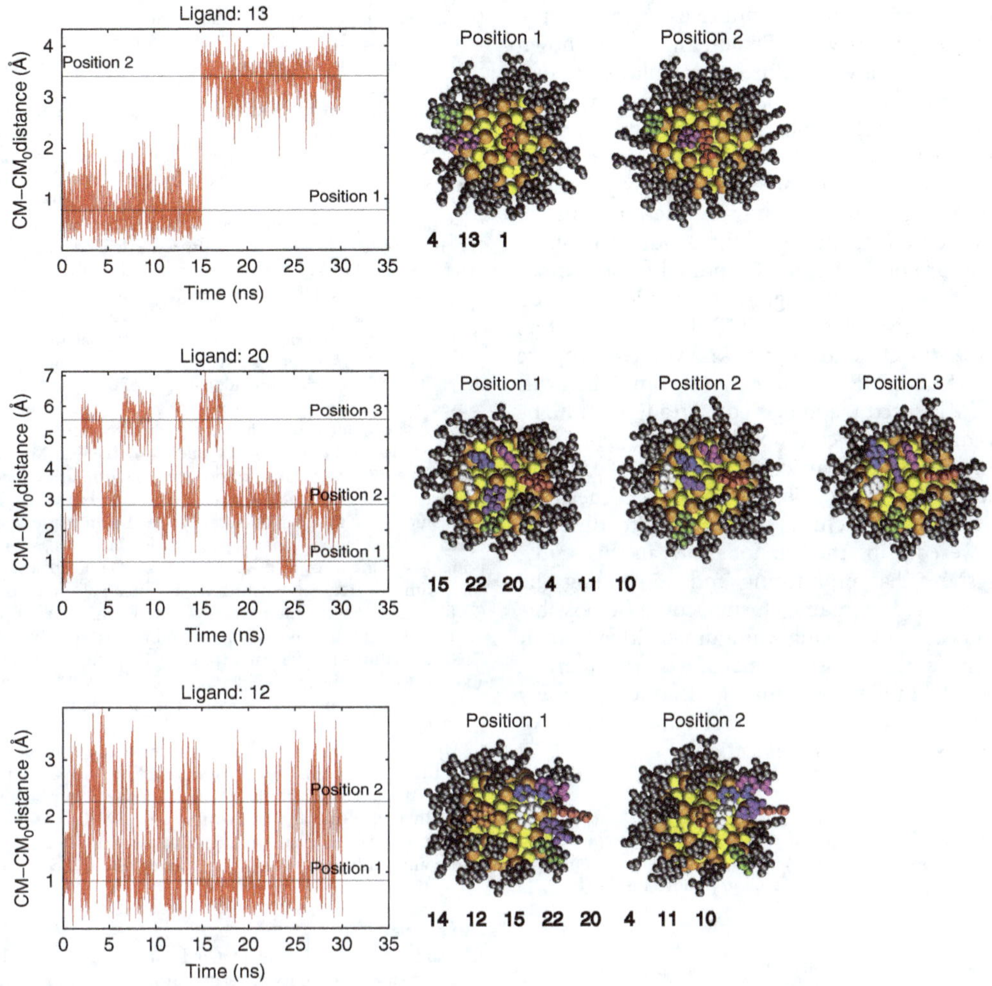

Figure 6 | Dynamics of the centre of mass (CM) of selected ligands during the MD simulation. Distances are plotted with respect to the initial position of the CM. Structures visualize the differences in the orientation of the ligands at each position 1, 2 and 3. The colours of the ligands are as in Fig. 1.

ligand mobility (Fig. 3). More static ligands gave stronger correlation peaks, whereas ligands that showed higher mobility in the MD simulation gave weaker correlation peaks (Supplementary Table 1). Strong TOCSY correlations were found for 15 pMBA ligands, of which 14 were assigned above. The remaining strong correlation (pMBA-14) in the upfield region of the proton spectrum (6.79–6.28 p.p.m.) was assigned based on its good agreement with the calculated data. The symmetry-equivalent site for pMBA-14 is a close neighbour to pMBA-12, which was established as one of the most mobile ligands being attached on the long gold–thiolate unit, and therefore does not show through-space correlation to pMBA-14, although correlation to the other closest neighbour pMBA-15 was observed (Fig. 1b, right; and Fig. 3).

The remaining weak TOCSY correlations were assigned by comparing the experimental chemical shifts with the calculated ones. These ligands also showed high mobility in the MD simulation studies. Ligands pMBA-16, -17 and -18 all gave similar TOCSY correlations in the downfield region with high Δ value (8.25–6.76 p.p.m., $\Delta = 1.10$–1.39 p.p.m.) indicating more than one type of interaction (Fig. 2). The observed downfield shifts can be explained by the interaction with the gold core that is very similar for all the three ligands. Ligand pMBA-19 appeared at the 'middle range' (7.57–7.29 p.p.m., $\Delta = 0.28$ p.p.m.), whereas good agreement with the calculated data was found for ligands pMBA-20 and -21 in the upfield region of the spectrum (7.24–6.06 p.p.m., $\Delta = 0.95$–1.18 p.p.m.). This is reasonable since pMBA-19 is nearly isolated on the cluster surface while pMBA-20 and pMBA-21 both have edge-to-face and face-to-face characteristics. The last remaining ligand pMBA-22 was assigned to the symmetry-equivalent site at the other end of the long gold–thiolate unit.

Discussion

We were able to interpret fully the very complicated ^{1}H and ^{13}C NMR spectra of the inorganic nanoparticle $Au_{102}(pMBA)_{44}$ in solution by using a combination of multidimensional NMR methods, DFT calculations and MD simulations. With help of DFT calculations and MD simulations we showed that the ligands have their own intrinsic orientation in close connection to the arrangement seen in the crystal structure. We also showed that different ligands can have considerably different dynamical behaviour, which is confirmed by both MD simulations and the observed signal strengths of the measured NMR shifts. We mapped the ligand–ligand interactions and ligand environments to the proton spectrum, which can be generalized to also other ligand-protected inorganic clusters and nanoparticles. Strong ligand–ligand interactions were seen to reduce dynamics of the ligands and cause splitting of the signal in HSQC, which meant that four signals could be measured for the carbon atoms of the phenyl ring that are bound to the hydrogen.

Our observations are important for future research of the ligated inorganic clusters in many ways. First, many of the known ligand-protected clusters have a well-defined molecular composition, including nanoparticles with perhaps 500 metal atoms and a well-developed plasmon, but lack empirical structure information regarding the ligand shell and ligand–metal interactions. NMR spectra for many of these clusters are reported, with severe congestion in the aromatic region of the spectra when the clusters are ligated by phenyl-containing ligands[14,16,25]. Our work gives 'fingerprints' of typical proton NMR shifts expected for a number of different classes of weak ligand–ligand interactions, which should prove useful in future analyses of NMR data from clusters where the solid-state X-ray structure is not known. We propose that the multidimensional NMR measurements combined with DFT calculation of NMR spectra, augmented by traditional high-resolution electron microscopy, can now allow for empirical structural models of entire nanoparticles (including the ligand shell and ligand–metal interaction). Second, unsolved questions like reason for the different structural symmetry with different ligands could be answered using the methods used in this study. Finally, the results show that monitoring and controlling the ligand exchange of protected inorganic clusters could be possible with a single ligand accuracy using multidimensional NMR methods, allowing more rapid assessment of regio-isomerism than is presently possible. In the future this can lead to extremely highly controllable and accurate methods for functionalization of the ligated nanoparticles for various applications.

Methods

Sample preparation. $Au_{102}(pMBA)_{44}$ was synthesized as previously reported[15]. For NMR acquisition, a 1.0-ml saturated solution of $Au_{102}(pMBA)_{44}$was prepared in 0.3 M NaOH in D_2O. The solution was centrifuged to remove insoluble $Au_{102}(pMBA)_{44}$.

NMR spectroscopy. NMR data were acquired at 25 °C on a Varian Inova 500 MHz spectrometer equipped with three analytical channels and a 5-mm proton detect, triple resonance probe (HCN probe). Varian and Agilent softwares, VnmrJ2.2 and VnmrJ4.2, respectively, were used for spectrometer control and data processing. All pulse sequences were used without modifications made from the software release. The ^{1}H-detected dimension was acquired as a downsampled 8-kHz free induction decay (FID) in quadrature mode with hardware digital signal processing from a 500-kHz oversampled data stream. The ^{1}H $\pi/2$ pulse width was 7.5 μs. 1D ^{1}H data were acquired with a $\pi/4$ pulse, 2.0 s acquisition time, 16 K complex points and 7.0 s recycle time. In the case of the 2D data acquisitions, the ^{1}H directly detected dimension had a 0.25 s acquisition time, 2 K complex points and a 2.0 s recycle time. Some spectra were acquired with the residual water signal pre-saturated by a 1.5-s, 25-Hz continuous wave pulse centred on the residual water signal. The first two acquired data points were linear back-predicted. The 2D, indirectly detected dimension was modulated in phase-sensitive mode by using States-TPPI scheduling. In the case of the ^{1}H–^{1}H homonuclear NMR experiments, the indirectly acquired dimension was 8 kHz in 200 complex points. In the case of the ^{1}H–^{13}C{^{13}C} HSQC, the ^{13}C indirectly acquired dimension was 12.5 kHz in 128 complex points. In all 2D data sets, the interferogram was extended by a factor of two through linear forward prediction, weighted with a time-shifted Gaussian function for resolution enhancement and then the full, 2D data sets were zero-filled and Fourier-transformed. The homonuclear experiments had a final matrix size of 2 × 2 K complex points and 4 Hz per point, and the HSQC was 2 K × 512 complex points and 4 and 24 Hz per point, respectively. The HSQC data were acquired by taking 256 signal averaging transients per increment. TOCSY data were acquired by using 0.024 and 0.0640-s 8-kHz DIPSI-2 isotropic mixing sequences and taking 32 signal averaging transients per increment. The TOCSY with 0.064-s mixing was used in our results. NOESY data were acquired by providing 0.080, 0.150 and 0.200 s dipolar contact time and taking 64 signal averaging transients per increment. The nuclear Overhauser effect build-up was examined qualitatively to maximize cross-peak intensity without interference from spin diffusion. The 0.200-s mixing period provided the best cross-peak intensity for simple distance constraints in our results.

Computational methods. The NMR shifts were calculated for a theoretically relaxed structure obtained from the solved crystal structure of $Au_{102}(pMBA)_{44}$ (ref. 9). The structure was relaxed by using DFT as implemented in code-package GPAW[26] that uses a real-space grid with projector-augmented waves (PAW). The electron–electron interactions were treated by using the Perdew–Burke–Ernzerhof (PBE)[27] exchange-correlation functional, and the wave functions and electron density were described in a real-space grid of 0.2 Å grid spacing. The relaxation of the atomic structure was stopped when the residual forces on atoms were below 0.1 eV Å^{-1}.

NMR shifts were calculated using Gauge-Including Projector-Augmented Wave (GIPAW) formalism as implemented in pseudopotential- and planewave-based DFT code-package Quantum Espresso[28]. We used ultrasoft PBE-PAW pseudopotentials (http://www.quantum-espresso.org/pseudopotentials/), a kinetic energy cutoff of 950 eV for wave functions and 9,520 eV for the charge density and potential. The energy convergence criterion for the self-consistency was set to $1.4 \cdot 10^{-11}$ eV. Total chemical shifts were defined as $\sigma = \sigma(\text{ref}) - \sigma(\text{calc})$, where σ(ref) was fixed to 31.099 p.p.m. for ^{1}H shifts and for 166.049 p.p.m. for ^{13}C shifts. Both of the selected reference values give a total average of all the calculated ^{1}H and ^{13}C shifts that is comparable to the experimental results. From the calculated NMR shifts we mainly concentrate on the symmetrically averaged values of 22 unique pMBA ligands as seen also in the experiment. A couple of clear overestimations in the shifts for the specific individual atoms of the ligands were found, most probably caused by the unique ligand arrangement and ligand–ligand interactions that are static in NMR calculations but not in the experiment.

Dynamics of the ligands was studied by classical MD simulations in a water environment for 30 ns time at room temperature using the GROMACS[29] package. To describe the $Au_{102}(pMBA)_{44}$ cluster, we used an AMBER-compatible molecular mechanics force field for thiolate-protected gold clusters that has been recently developed and tested in our group (Pohjolainen *et al*, submitted manuscript). The principles of the force field were as follows: (1) no constraints included in the gold core, with the atoms interacting via a simple pair potential (Lennard–Jones type); (2) for transferability, the charges of gold atoms were set to zero; (3) for the Au-S interface, equilibrium bond lengths and angles were averaged from our previous DFT calculations for the $Au_{102}(pMBA)_{44}$ structure, and force constants were adopted from a previous work[30]; and (4) for pMBA parameters, atom types were chosen from existing AMBER atom types, and partial charges were calculated from RESP fit.

Dynamical behaviour of the 22 symmetrically unique ligands were estimated from the MD trajectory both visually and by calculating an averaged root mean squared deviations for all the ligands separately, in which we concentrated on the atoms that were relevant for the measured NMR shifts. The reference structure in root mean squared deviation calculations were taken after initial thermalization but before any conformational changes were seen in the ligand layer.

References

1. Daniel, M.-C. & Astruc, D. Gold nanoparticles: assembly, supramolecular chemistry, quantum-size-related properties, and applications toward biology, catalysis, and nanotechnology. *Chem. Rev.* **104,** 293–346 (2004).
2. Alivisatos, A. P. Perspectives on the physical chemistry of semiconductor nanocrystals. *J. Phys. Chem.* **100,** 13226–13239 (1996).
3. Howes, P. D., Chandrawati, R. & Stevens, M. M. Bionanotechnology. Colloidal nanoparticles as advanced biological sensors. *Science* **346,** 1247390 (2014).
4. Kamat, P. V. Quantum dot solar cells. Semiconductor nanocrystals as light harvesters. *J. Phys. Chem. C* **112,** 18737–18753 (2008).
5. Shirasaki, Y., Supran, G. J., Bawendi, M. G. & Bulović, V. Emergence of colloidal quantum-dot light-emitting technologies. *Nat. Photon.* **7,** 13–23 (2013).
6. Zherebetskyy, D. *et al.* Hydroxylation of the surface of PbS nanocrystals passivated with oleic acid. *Science* **344,** 1380–1384 (2014).
7. Juhás, P., Cherba, D. M., Duxbury, P. M., Punch, W. F. & Billinge, S. J. L. Ab initio determination of solid-state nanostructure. *Nature* **440,** 655–658 (2006).
8. Van Aert, S., Batenburg, K. J., Rossell, M. D., Erni, R. & Van Tendeloo, G. Three-dimensional atomic imaging of crystalline nanoparticles. *Nature* **470,** 374–377 (2011).
9. Jadzinsky, P. D., Calero, G., Ackerson, C. J., Bushnell, D. A. & Kornberg, R. D. Structure of a thiol monolayer-protected gold nanoparticle at 1.1 A resolution. *Science* **318,** 430–433 (2007).
10. Liu, J. *et al.* Ligand-stabilized and atomically precise gold nanocluster catalysis: a case study for correlating fundamental electronic properties with catalysis. *Chemistry* **19,** 10201–10208 (2013).
11. Tsukuda, T. & Häkkinen, H. (eds). *Protected Metal Clusters: From Fundamentals to Applications* (Elsevier, 2015).
12. Heinecke, C. L. *et al.* Structural and theoretical basis for ligand exchange on thiolate monolayer protected gold nanoclusters. *J. Am. Chem. Soc.* **134,** 13316–13322 (2012).
13. Cavanagh, J., Fairbrother, W. J., Palmer, III A. G. & Skelton, N. I. (eds). *Protein NMR Spectroscopy: Principles and Practice* (Elsevier, 1995).
14. Qian, H., Zhu, M., Gayathri, C., Gil, R. R. & Jin, R. Chirality in gold nanoclusters probed by NMR spectroscopy. *ACS Nano* **5,** 8935–8942 (2011).
15. Wong, O. A., Heinecke, C. L., Simone, A. R., Whetten, R. L. & Ackerson, C. J. Ligand symmetry-equivalence on thiolate protected gold nanoclusters determined by NMR spectroscopy. *Nanoscale* **4,** 4099–4102 (2012).
16. Chen, Y. *et al.* Crystal structure of barrel-shaped chiral Au130(pMBT)50 nanocluster. *J. Am. Chem. Soc.* **137,** 10076–10079 (2015).

17. Marbella, L. E. & Millstone, J. E. NMR techniques for noble metal nanoparticles. *Chem. Mater.* **27,** 2721–2739 (2015).
18. Walter, M. *et al.* A unified view of ligand-protected gold clusters as superatom complexes. *Proc. Natl Acad. Sci. USA* **105,** 9157–9162 (2008).
19. Li, Y., Galli, G. & Gygi, F. Electronic structure of thiolate-covered gold nanoparticles: Au102(pMBA)44. *ACS Nano* **2,** 1896–1902 (2008).
20. Gao, Y., Shao, N. & Zeng, X. C. Ab initio study of thiolate-protected Au102 nanocluster. *ACS Nano* **2,** 1497–1503 (2008).
21. Hulkko, E. *et al.* Electronic and vibrational signatures of the $Au_{102}(pMBA)_{44}$ cluster. *J. Am. Chem. Soc.* **133,** 3752–3755 (2011).
22. Knoppe, S. *et al.* Chiral phase transfer and enantioenrichment of thiolate-protected Au102 clusters. *J. Am. Chem. Soc.* **136,** 4129–4132 (2014).
23. Marjomäki, V. *et al.* Site-specific targeting of enterovirus capsid by functionalized monodisperse gold nanoclusters. *Proc. Natl Acad. Sci. USA* **111,** 1277–1281 (2014).
24. Martikainen, M. *et al.* Hydrophobic pocket targeting probe for enteroviruses. *Nanoscale* **7,** 17457–17467 (2015).
25. Salorinne, K. *et al.* Non-destructive size determination of thiol-stabilized gold nanoclusters in solution by DOSY NMR spectrosocopy. *Anal. Chem.* **85,** 3489–3492 (2013).
26. Enkovaara, J. *et al.* Electronic structure calculations using GPAW: a real-space implementation of the projector-augmented wave method. *J. Phys. Condens. Matter* **22,** 253202 (2010).
27. Perdew, J. P., Burke, K. & Ernzerhof, M. Generalized gradient approximation made simple. *Phys. Rev. Lett.* **77,** 3865–3868 (1996).
28. Giannozzi, P. *et al.* QUANTUM ESPRESSO: a modular and open-source software project for quantum simulations of materials. *J. Phys. Condens. Matter* **21,** 395502 (2009).
29. Berendsen, H. J. C., van der Spoel, D. & van Drunen, R. GROMACS: a message-passing parallel molecular dynamics implementation. *Comp. Phys. Commun.* **91,** 43–56 (1995).
30. Heikkilä, E. *et al.* Atomistic simulations of functional Au144(SR)60 gold nanoparticles in aqueous environment. *J. Phys. Chem. C* **116,** 9805–9815 (2012).

Acknowledgements

This work was supported by NSF CAREER 1455099 (C.J.A.), NIH R21 EB014520 (C.J.A.) and the Finnish Academy grant 266492 (H.H.). The computational resources were provided by CSC—the Finnish IT Center for Science. We thank E. Pohjolainen and G. Groenhof for their contribution in parameterization and testing the GROMACS force field for $Au_{102}(pMBA)_{44}$.

Author contributions

O.A.W. prepared the $Au_{102}(pMBA)_{44}$ samples; C.D.R. measured the NMR data; K.S. analysed the experimental NMR spectra, made the ligand assignments and wrote the first version of the manuscript; S.M. calculated the theoretical NMR shifts and contributed to ligand assignments; X.C. ran the GROMACS MD calculations for $Au_{102}(pMBA)_{44}$ in water; C.J.A. and H.H. supervised the study and finalized the manuscript.

Additional information

Competing financial interests: The authors declare no competing financial interests.

10

Efficient purification of ethene by an ethane-trapping metal-organic framework

Pei-Qin Liao[1], Wei-Xiong Zhang[1], Jie-Peng Zhang[1] & Xiao-Ming Chen[1]

Separating ethene (C_2H_4) from ethane (C_2H_6) is of paramount importance and difficulty. Here we show that C_2H_4 can be efficiently purified by trapping the inert C_2H_6 in a judiciously designed metal-organic framework. Under ambient conditions, passing a typical cracked gas mixture (15:1 C_2H_4/C_2H_6) through 1 litre of this C_2H_6 selective adsorbent directly produces 56 litres of C_2H_4 with 99.95% + purity (required by the C_2H_4 polymerization reactor) at the outlet, with a single breakthrough operation, while other C_2H_6 selective materials can only produce *ca.* $\leqslant$1 litre, and conventional C_2H_4 selective adsorbents require at least four adsorption-desorption cycles to achieve the same C_2H_4 purity. Single-crystal X-ray diffraction and computational simulation studies showed that the exceptional C_2H_6 selectivity arises from the proper positioning of multiple electronegative and electropositive functional groups on the ultramicroporous pore surface, which form multiple C-H···N hydrogen bonds with C_2H_6 instead of the more polar competitor C_2H_4.

[1] MOE Key Laboratory of Bioinorganic and Synthetic Chemistry, School of Chemistry and Chemical Engineering, Sun Yat-Sen University, Guangzhou 510275, P.R. China. Correspondence and requests for materials should be addressed to J.-P.Z. (email: zhangjp7@mail.sysu.edu.cn).

As the most important chemical product, ethene (C_2H_4) is generally obtained through steam cracking and thermal decomposition of naphtha or ethane (C_2H_6) (ref. 1). Besides being obtained as a byproduct of petroleum refining, C_2H_6 is also isolated on an industrial scale from natural gas (CH_4 70 ~ 96%, C_2H_6 1 ~ 14% and CO_2 0 ~ 8%) (ref. 2). As a result of their similar physical properties, it is difficult to separate C_2H_6, C_2H_4 and CO_2 (refs 3–5). In industry, C_2H_6 and C_2H_4 are separated by cryogenic high-pressure distillation, typically at 7–28 bar and 183–258 K using very high towers consisting of over 150 trays, which is very energy consuming ($7\,GJ\,t^{-1}$) and constitutes a notable portion of the ethylene cost[6,7]. To save energy, separation methods effective at the ambient temperature and pressure are highly demanded[8–11]. Passing the gas mixture through a fixed-bed adsorber can be a very simple and promising approach to afford low energy consumption and high product purity.

Because unsaturated hydrocarbons like to coordinate with metal ions, C_2H_4 can be selectively bound and separated from its saturated counterpart C_2H_6 (refs 12–16). Compared with other types of porous materials, porous metal-organic frameworks (MOFs) are unique for their diversified/designable framework structures and pore surfaces, including the ease of introducing open metal sites (OMSs), which have shown great potentials for C_2H_4/C_2H_6 separation[17–22]. In the fix-bed separation process, the un-adsorbed C_2H_6 first breakthrough, and C_2H_4 enriched in the stationary phase is later obtained by heating and/or inert-gas blowing. Because the un-adsorbed C_2H_6 residing in the mobile phase contaminates the desired product C_2H_4 during the desorption stage, the highest C_2H_4 purity produced by a full adsorption–desorption cycle can just reach 99%+ (refs 13,17,23,24), and at least four such cycles are necessary to achieve 99.95%+ (ref. 25), the lower limit required by the C_2H_4 polymerization reactor[26–28]. Obviously, this problem can be solved by using a C_2H_6 selective adsorbent, which not only improves the C_2H_4 purity but also reduces energy consumption. The simple separation operation and device (just one adsorption process in a single breakthrough operation) are also necessary for onsite supply of purified C_2H_4. However, such an unusual adsorption behaviour has been only reported for a few low-polarity or hydrophobic MOFs[29–36], and their C_2H_4/C_2H_6 separation performances (that is, C_2H_6/C_2H_4 selectivities) are poor, because the polarities of C_2H_4 and C_2H_6 are very similar and can be hardly distinguished by hydrophobic adsorbents.

As C_2H_6 possesses the lowest polarity (quadrupole moment) compared with similar molecules such as C_2H_4 and CO_2 (Supplementary Table 1)[37]; polar adsorbents are generally selective for the latter gases. However, considering that the electropositive and electronegative portions locate quite differently among these gas molecules, we speculated that by rational utilization of polar functional groups, it is still possible to design a MOF with optimized pore size/shape and surface electrostatic distribution that can bind C_2H_6 much stronger than for C_2H_4. Herein, we report the design, structure and gas adsorption/separation properties of such a C_2H_6-trapping MOF, which is useful for not only direct producing highly pure C_2H_4 from C_2H_4/C_2H_6 mixtures, but also efficient separation of four-component $CH_4/C_2H_4/C_2H_6/CO_2$ mixtures and extraction of C_2H_6 from natural gas.

Results

Synthesis, structure and stability. Bis(5-amino-1*H*-1,2,4-triazol-3-yl)methane (H_2batz) with two 3-amino-1,2,4-triazole rings bridged by a methylene group was designed as a new ligand combining multiple nitrogen atoms as hydrogen-bonding acceptors and methylene groups as dipole repulsion groups, as well as short bridging lengths for construction of an ultra-microporous framework. Reaction of H_2batz and $Zn(OH)_2$ in dilute aqueous ammonia produced a porous metal-azolate framework $[Zn(batz)]\cdot 0.5H_2O$ (MAF-49 · H_2O). Single-crystal X-ray diffraction (SCXRD) analysis of MAF-49 · H_2O (Supplementary Table 2 and Supplementary Data 1) showed that each Zn(II) is tetrahedrally coordinated by four triazolate nitrogen atoms from three $batz^{2-}$ ligands (Supplementary Fig. 1), and each $batz^{2-}$ ligand coordinates to three Zn(II) ions in a bisimidazolate mode, giving a three-dimensional (3D) coordination framework with narrow 1D zigzag channels (Fig. 1a). Since only four of the eight nitrogen donors of $batz^{2-}$ are utilized according to the coordination requirement of Zn(II), the pore surface of MAF-49 is rich with electronegative nitrogen atoms, although some of them form intra-framework N–H · · · N hydrogen bonds to reduce their abilities as hydrogen-bonding acceptors. Notably, the narrowest section of the 1D channel ($3.3 \times 3.0\,Å^2$) is approximately a folded four-membered ring defined by a pair of free amino groups (with their lone electron pairs) and a pair of methylene groups with a *cis*-configuration, which is occupied by a guest H_2O molecule with two O–H · · · N and two C–H · · · O hydrogen bonds (Fig. 1b).

Thermogravimetry and powder XRD showed that MAF-49 · H_2O can be readily activated and is stable to 450 °C in nitrogen (Supplementary Fig. 2), in boiling water for at least 1 month and in aqueous acid/base ($4 \leq pH \leq 12$) at room temperature for at least 1 week (Supplementary Fig. 3), which is extraordinary among MOFs and can be partly explained by the strong metal-azolate coordination bonds[38]. SCXRD showed that complete dehydration leads to a slight framework expansion (0.17% in volume, Supplementary Table 2 and Supplementary Data 2).

Gas adsorption property and mechanism. Single-component adsorption isotherms for CH_4, C_2H_6, C_2H_4 and CO_2 were measured for guest-free MAF-49 at 298 K, 307 K and 316 K (Fig. 2a and Supplementary Fig. 4). According to their different isotherm shapes, it can be judged that the host–guest binding follows $C_2H_6 > C_2H_4 > CO_2 > CH_4$. The gas adsorption enthalpies were calculated quantitatively by Virial analyses (Fig. 2b and Supplementary Fig. 5), which are $60\,kJ\,mol^{-1}$, $48\,kJ\,mol^{-1}$, $30\,kJ\,mol^{-1}$ and $25\,kJ\,mol^{-1}$ for C_2H_6, C_2H_4, CO_2, and CH_4, respectively, at zero-coverage. The mixed gas adsorption isotherms for equimolar C_2H_6/C_2H_4, C_2H_6/CO_2 and C_2H_6/CH_4 mixtures were simulated by the ideal adsorbed solution theory[39], in which the single-component adsorption isotherms were fitted by the Langmuir – Freundlich model (Supplementary Fig. 6). At total pressure of 100 kPa and a temperature of 316 K, the C_2H_6/C_2H_4, C_2H_6/CO_2 and C_2H_6/CH_4 selectivities of these mixtures were calculated as *ca.* 9, 40 and 170, respectively

a
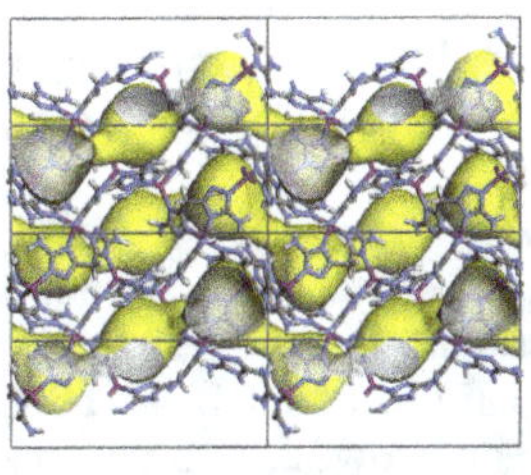
b
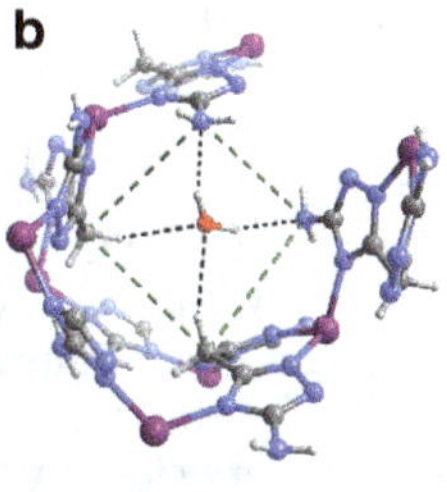

Figure 1 | X-ray crystal structure of MAF-49 · H_2O. (**a**) Framework (Zn purple, C dark grey, H light grey, N blue) and pore surface (yellow/grey curved surface) structures. Guest molecules are omitted for clarity. (**b**) Local environment and hydrogen-bonding interactions of the narrowest channel neck (highlighted by green dashed lines).

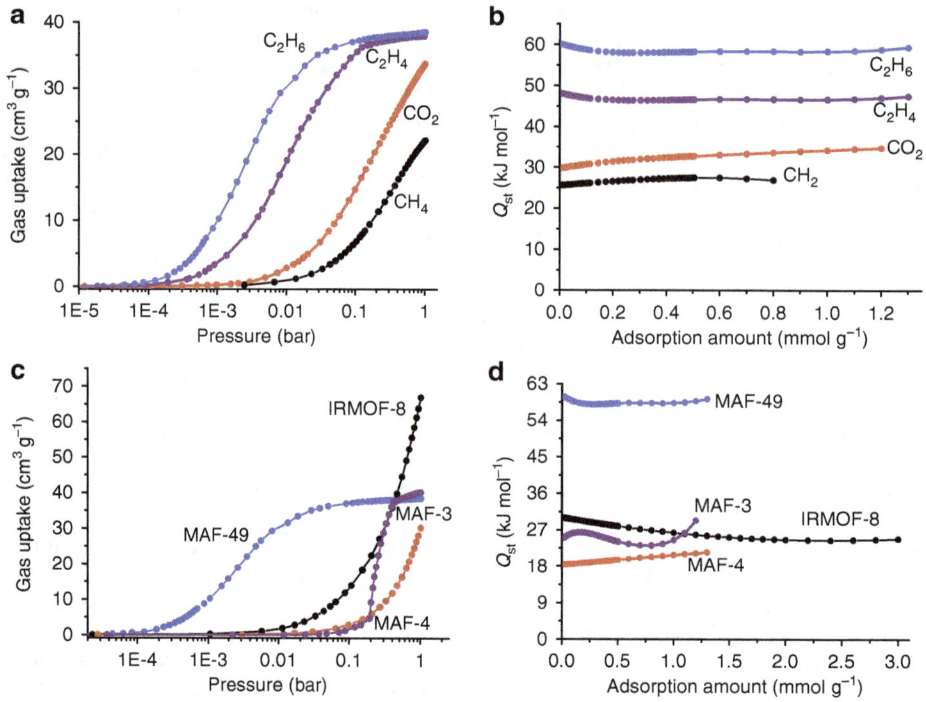

Figure 2 | Single-component gas adsorption properties. (**a**) Gas adsorption isotherms for C_2H_6, C_2H_4, CO_2 and CH_4 in MAF-49 at 316 K. (**b**) The coverage-dependent C_2H_6, C_2H_4, CO_2 and CH_4 adsorption enthalpy obtained by the Virial method. (**c**) C_2H_6 adsorption isotherms of MAF-49, MAF-3, MAF-4 and IRMOF-8 measured at 316 K. (**d**) Coverage-dependent C_2H_6 adsorption enthalpy of MAF-49, MAF-3, MAF-4 and IRMOF-8.

(Supplementary Fig. 7). Notably, the C_2H_6/C_2H_4 selectivity of MAF-49 is much higher than the highest value in the literature (2.4 for IRMOF-8 at 318 K) (ref. 31). Except for CH_4 with obviously lower molecular weight and boiling point, which interacts weakly with all adsorbents, the binding strength order of MAF-49 for other three heavier gases is unusual. Among a variety of physical properties of the four gases, only the polarizability trend is consistent with the binding trend (Supplementary Table 1). Nevertheless, the small differences of their polarizabilities are not enough to explain the large variation of their adsorption enthalpies, especially for C_2H_6 and C_2H_4. Notably, the C_2H_6 adsorption enthalpy is significantly higher than reported values, while the C_2H_4 one is moderate[18].

To elucidate the very different C_2H_6, C_2H_4 and CO_2 affinities of MAF-49, their preferential host-guest structures and energy changes were calculated by grand canonical Monte Carlo simulation and further periodic density functional theory optimization. The obtained binding energies of the final host–guest structures are -56.7, -45.5 and -41.3 kJ mol^{-1} for C_2H_6, C_2H_4 and CO_2, respectively. However, to adsorb these gas molecules, the host framework undergoes different structural distortions from the guest-free form and consumes energies of $+0.2$ kJ mol^{-1}, $+0.3$ kJ mol^{-1} and $+5.6$ kJ mol^{-1}, respectively. Taking both the host-guest binding and host-framework distortion into consideration, the total energies or adsorption enthalpies can be calculated as -56.5 kJ mol^{-1}, -45.2 kJ mol^{-1} and -35.7 kJ mol^{-1} for C_2H_6, C_2H_4, and CO_2, respectively, which are consistent with the experimental values (Supplementary Table 3). In the density functional theory optimized host–guest structures, it can be seen that C_2H_6, C_2H_4 and CO_2 are all adsorbed in or very close to the narrowest channel neck, but they interact very differently with the pore surface.

C_2H_6 forms three strong C–H···N hydrogen bonds and three weak C–H···N electrostatic interactions with MAF-49 (Fig. 3a,d, Supplementary Table 4). Specifically, one methyl group interacts with two amino groups and an coordinated triazolate nitrogen atom of the narrowest channel neck, forming one very short and directional (C6-H61···N8) and one unsymmetrical-bifurcated/three-centred (C6-H62···N8A/N3A) hydrogen bonds, in which the H···N separations (2.15 Å) are much shorter than the sum of van der Waals radii of nitrogen (1.55 Å) and hydrogen (1.20 Å) atoms. The third strong hydrogen bond involves the hydrogen atom (H71) of another methyl group and a coordinated triazolate nitrogen atom (C8-H71···N1A), which is approximately centro-symmetric with the strongest one (C6-H61···N8) about the molecular centre and fits well with the most stable stagger conformation of C_2H_6. Besides, the less polar part of the pore surface, that is, two methylene groups of the batz^{2-} ligand (C3), fits well with the guest C_2H_6 molecule in the context of both molecular shape and electrostatic potential.

For C_2H_4, two less strong C–H···N hydrogen bonds and two very weak C–H···N electrostatic interactions were observed (Fig. 3b,e and Supplementary Table 4). The strongest one involves one methylene group and one amino group at the narrowest channel neck (C6-H61···N8), while the secondary one involves another methylene group and an uncoordinated triazolate nitrogen atom (C7-H71···N2A), which are also approximately centro-symmetric about the molecular centre. These two C–H···N hydrogen bonds are similar in geometry with the first and third strongest ones for C_2H_6. However, their H···N separations (2.54–2.65 Å) are obviously longer, albeit C_2H_4 is more polar (Supplementary Table 4). The *cis*-configuration of the two electronegative amino groups and two electropositive methylene groups of the narrow channel neck is crucial for the very different host–guest interactions. Obviously, the molecular geometry of C_2H_4 prevents the two hydrogen atoms of a methylene group to form two strong hydrogen bonds with the narrow channel neck like H_2O and C_2H_6. Furthermore, there is significant steric hindrance and electrostatic repulsion between the two C–H moieties of the two methylene groups from the host channel neck and the guest C_2H_4 (C3···C6 = 3.88 Å,

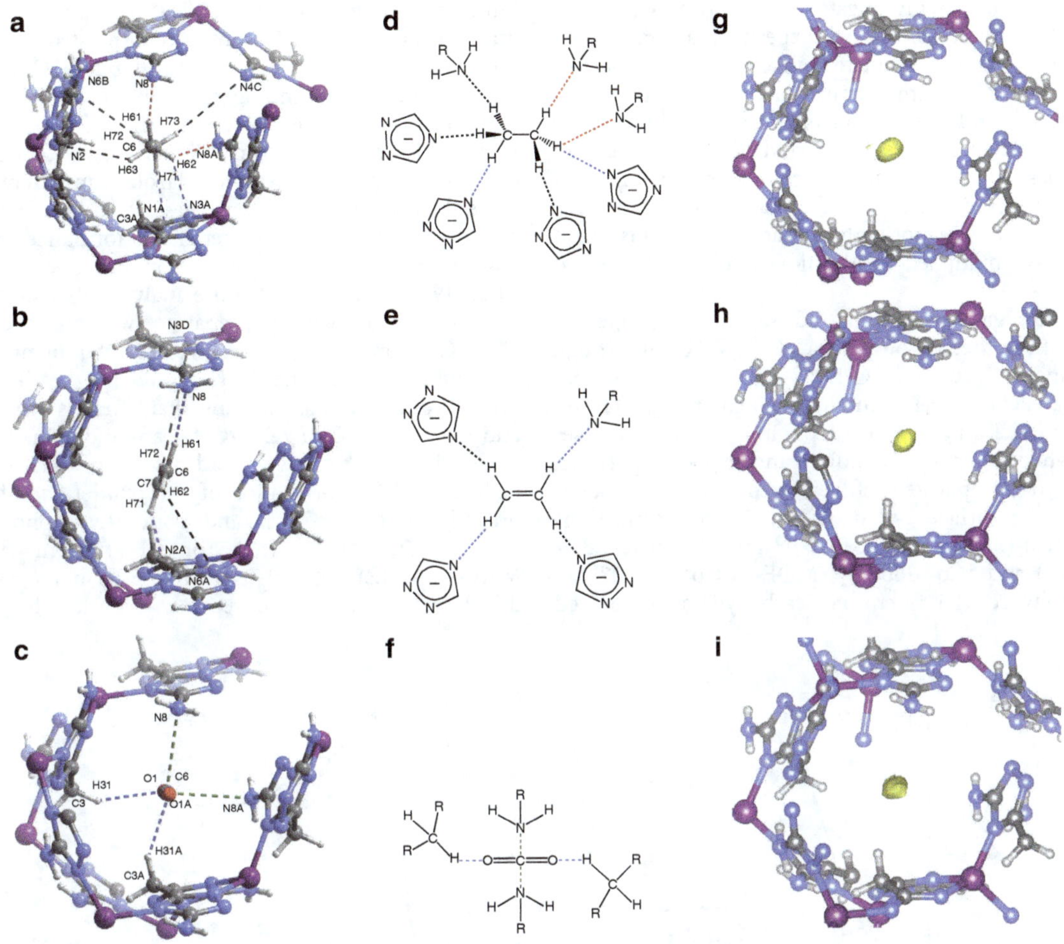

Figure 3 | Host-guest fittings and interactions. Preferential adsorption sites for (**a**) C_2H_6, (**b**) C_2H_4 and (**c**) CO_2 in MAF-49 revealed by computational simulations (Zn purple, C dark grey, H light grey, N blue). Schematic representation of the corresponding host-guest interactions for (**d**) C_2H_6, (**e**) C_2H_4 and (**f**) CO_2. Strong (H···N/O<2.3 Å), weak (2.3 Å<H···N/O<2.8 Å) and almost negligible (H···N/O>2.8 Å) C-H···N interactions are displayed as red, blue and black dashed lines, respectively. 3D electron density maps (F_o-F_c contoured at 0.80 e Å^{-3} in yellow) of MAF-49 loaded with trace amounts of (**g**) C_2H_6, (**h**) C_2H_4 and (**i**) CO_2.

Supplementary Fig. 8), which pushes the guest away from the best position for forming a strong C–H···N hydrogen bond with the *p*-position amino group. Conversely, the methylene group of the host channel neck fits well with the threefold symmetric methyl group of C_2H_6 (Fig. 3a). For the less strong C–H···N hydrogen bonds and other weak electrostatic attractive interactions, C_2H_6 also fits much better with the locations of the electronegative nitrogen atoms, as compared with those for C_2H_4 (Fig. 3 and Supplementary Table 4). These observations indicate that the proper locations of both the electronegative nitrogen atoms and the electropositive methylene groups play critical roles in distinguishing C_2H_6 and C_2H_4 with large adsorption enthalpy difference.

In the simulated host–guest structure for CO_2, the guest carbon atom locates exactly at the centre of host channel neck, forming short contacts with two free amino groups simultaneously (N···C = 2.91 Å), while two oxygen atoms of CO_2 interact with two methylene groups, respectively, through weak C–H···O hydrogen bonds (C···O = 3.33, H···N = 2.45 Å, ∠C–H···N = 135°) (Fig. 3c,f, Supplementary Table 4). Although these host–guest interactions seem relatively strong, the channel neck diameter (measured by the separation of the *p*-position amino and methylene groups, N8···C3 3.60 Å) significantly expanded from the guest-free state (3.13 Å), while it changes little after loading C_2H_6 (3.18 Å) and C_2H_4 (3.31 Å), indicating that there is significant steric hindrance and repulsive effect between the CO_2 molecule and the host framework, and the very short C···N separation is actually the result enforced by the contraction action of the channel neck (Supplementary Fig. 9). It should be noted that all carbon atoms of C_2H_6 and C_2H_4 reside on one side of the quadrangular channel neck, resulting in much smaller steric hindrance effects compared with CO_2 (Supplementary Fig. 10).

To confirm the simulation results and directly visualize the host–guest interactions, we carried out SCXRD analyses for MAF-49 loaded with trace amounts of C_2H_6, C_2H_4 and CO_2 (denoted as MAF-49·C_2H_6, MAF-49·C_2H_4 and MAF-49·CO_2, respectively, see Supplementary Table 2 and Supplementary Data 3–5). Compared with the unit-cell volume of guest-free MAF-49, those of MAF-49·C_2H_6 and MAF-49·C_2H_4 showed minor shrinkage (<0.2%), while that of MAF-49·CO_2 showed relatively large expansion (1.4%). Further, the N8···C3 separation order of MAF-49, MAF-49·C_2H_6, MAF-49·C_2H_4 and MAF-49·CO_2 is consistent with that predicted by computational simulations (Supplementary Fig. 10). In all host–guest crystal structures, the residue electron density peaks can be unambiguously found inside the narrow host channel neck (Fig. 3g–i). Furthermore, in the final crystal structures, all guest molecules locate very similar or identical with those predicted by computational simulations (Supplementary Fig. 10).

Mixed gas separation. To investigate the practical separation performance of MAF-49, breakthrough experiments were carried out at 313 K and 1 bar. To evaluate and compare the performances of the materials unambiguously, identical column and flow rate were used, and the parameters of each column were optimized (all columns have similar voidage, Supplementary Table 5). Besides, we used the specific injection amount ($mmol\ g^{-1}$) of the mixed gas as the abscissa, meaning that the breakthrough time (s) was not only divided by the adsorbent weight (g) but also multiplied by the flow rate of the injected mixed gas ($mmol\ s^{-1}$)[40].

To compare the gas adsorption and separation properties of MAF-49 with other protopytical MOFs, breakthrough experiments using an equimolar $C_2H_6/C_2H_4/CO_2/CH_4$ mixture injection were carried out (Fig. 4 and Supplementary Figs 11 and 12). For MAF-49, a clean and sharp separation of all four gases was observed, while other MOFs showed much poor separation performances and complicated effluent sequences dependent on their pore surface structures. With transition-metal OMSs, $[Cu_3(btc)_2]$ (HKUST-1, H_3btc = benzene-1,3,5-tricarboxylic acid) and $[Co_2(dobdc)]$ (MOF-74-Co/CPO-27-Co, H_4dobdc = 2,5-dihydroxyl-1,4-benzenedicarboxylic acid) showed binding strength orders $C_2H_4 > C_2H_6 > CO_2$. Because the main-group-metal OMS tends to form strong interaction with the oxygen atom of CO_2, $[Mg_2(dobdc)]$ (MOF-74-Mg/CPO-27-Mg) showed a binding strength order $CO_2 > C_2H_4 > C_2H_6$. Without pore surface active site, $[Zr_6O_4(OH)_4(bdc)_{12}]$ (UiO-66, H_2bdc = 1,4-benzenediarboxylic acid) and $[Zn(mim)_2]$ (MAF-4 or ZIF-8, Hmim = 2-methylimidazole) can barely distinguish the three heavier gases. Nevertheless, the low-polarity adsorbent MAF-4 exhibits slightly better performance compared with UiO-66, and exhibits a separation order similar with that of MAF-49. As expected from the analyses of adsorption isotherms, MAF-49 can also clearly separate two-component C_2H_4/C_2H_6, C_2H_6/CO_2 and C_2H_6/CH_4 mixtures (Supplementary Fig. 13). It should be noted that C_2H_6 could not be detected before its breakthrough points, meaning that C_2H_6 is efficiently extracted and high-purity $C_2H_4/CO_2/CH_4$ can be obtained directly.

Considering that selective adsorption of C_2H_6 over C_2H_4 could be beneficial for purification of C_2H_4 under fixed-bed adsorption/breakthrough processes, and some hydrophobic MOFs[29–36], such as $[Zn(bim)_2]$ (MAF-3 or ZIF-7, Hbim = benzimidazole), MAF-4 and $[Zn_4O(ndc)_3]$ (IRMOF-8, H_2ndc = naphthalene-2,6-dicarboxylic acid), were recently reported to exhibit such a

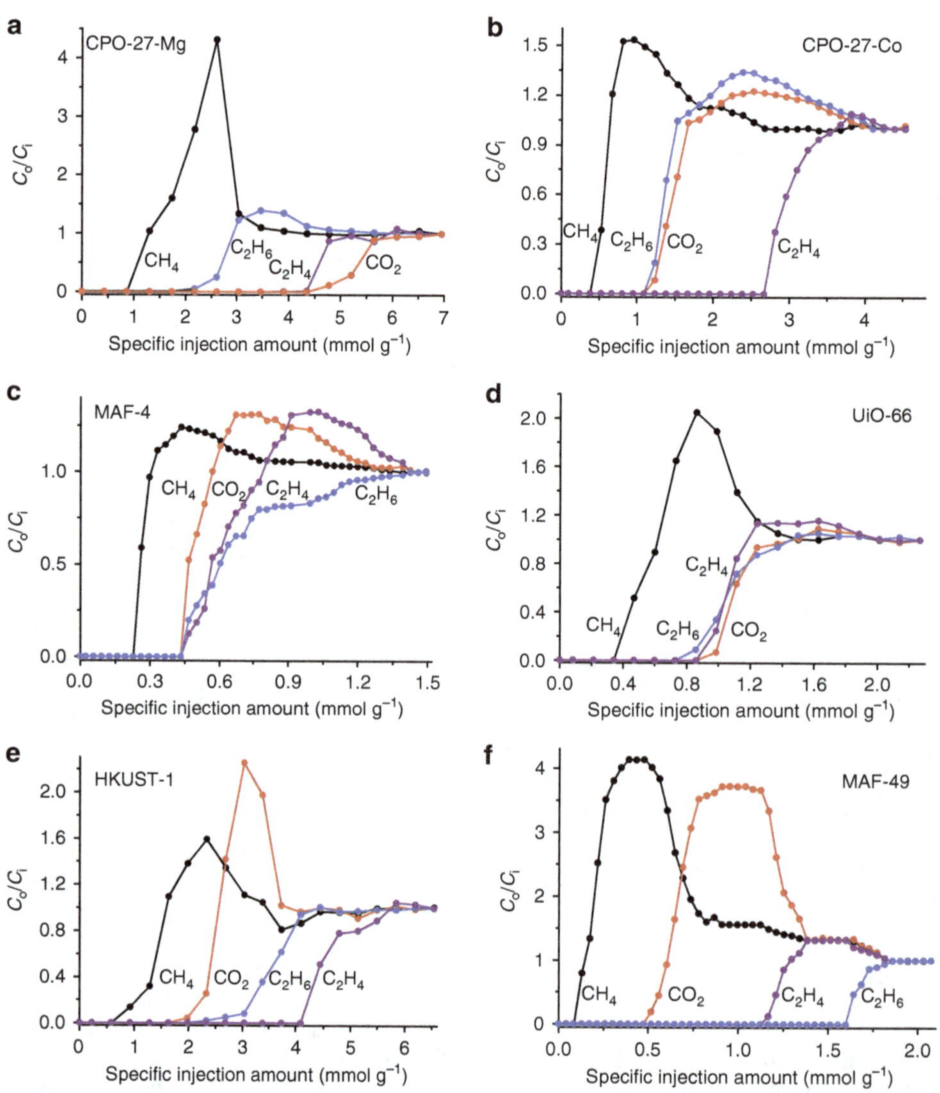

Figure 4 | Four-component gas mixture separation. Breakthrough curves of $CH_4/CO_2/C_2H_4/C_2H_6$ mixture (1:1:1:1 (vol)) for (**a**) CPO-27-Mg, (**b**) CPO-27-Co, (**c**) MAF-4, (**d**) UiO-66, (**e**) HKUST-1 and (**f**) MAF-49 measured at 313 K and 1 bar. Lines are drawn to guide eyes. C_i and C_o are the concentrations of each gas at the inlet and outlet, respectively.

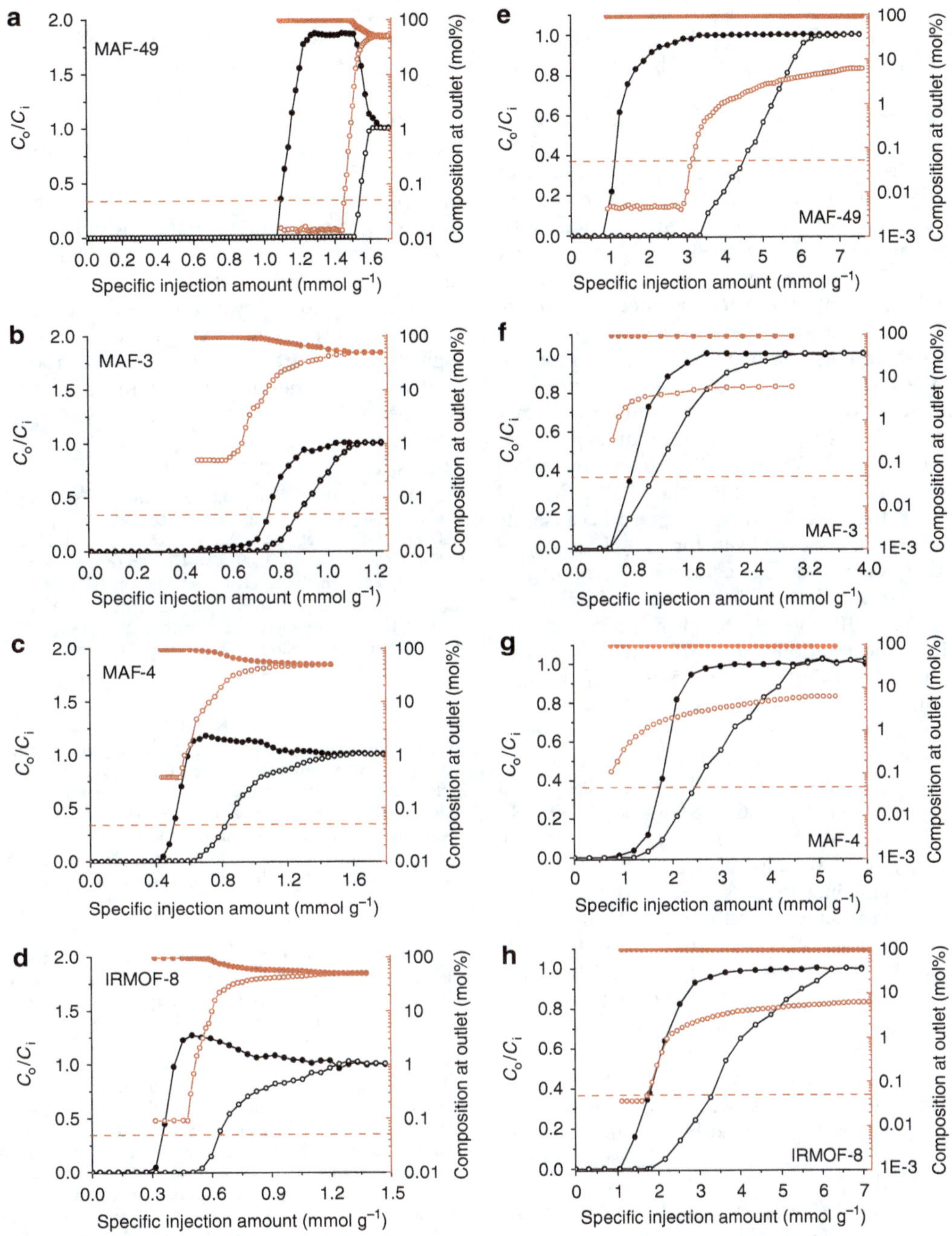

Figure 5 | C_2H_4/C_2H_6 separation performances. C_2H_4/C_2H_6 (1:1) mixture breakthrough curves of **(a)** MAF-49, **(b)** MAF-3, **(c)** MAF-4 and **(d)** IRMOF-8, and C_2H_4/C_2H_6 (15:1) mixture breakthrough curves of **(e)** MAF-49, **(f)** MAF-3, **(g)** MAF-4 and **(h)** IRMOF-8 measured at 313 K and 1 bar. Solid symbols: C_2H_4, Open symbols: C_2H_6. Lines are drawn to guide eyes. C_i and C_o are the concentrations of each gas at the inlet and outlet, respectively. Horizontal red dashed lines highlight C_2H_6 composition at outlet of 0.05%, that is, C_2H_4 purity of 99.95%.

property, we compared the C_2H_4/C_2H_6 adsorption and separation properties of these MOFs with MAF-49 in detail. Single-component C_2H_6 adsorption isotherms were measured for MAF-3, MAF-4 and IRMOF-8, which show adsorption enthalpies of 25 kJ mol^{-1}, 18 kJ mol^{-1} and 30 kJ mol^{-1}, respectively (Fig. 2c and Supplementary Fig. 14), at zero loading, being much lower than that of MAF-49. Although the C_2H_6 uptake at 1 bar for MAF-49 (38 cm^3 g^{-1}) is lower than that of the more porous adsorbents IRMOF-8 (91 cm^3 g^{-1}), MAF-4 (48 cm^3 g^{-1}) and MAF-3 (41 cm^3 g^{-1}), its C_2H_6 uptake at 0.06 bar (36 cm^3 g^{-1}) is *ca.* 4 times that of IRMOF-8 (9 cm^3 g^{-1}), 19 times that of MAF-4 (1.9 cm^3 g^{-1}) and 45 times that of MAF-3 (0.8 cm^3 g^{-1}) (Fig. 2d). Considering that a purity of 100% is impossible and the C_2H_6 concentration before its breakthrough point is lower than the detection limit of the conventionally used thermal conductivity detector, the gas stream at the column outlet was analysed with a mass spectrometer (MS). For a 1:1 C_2H_4/C_2H_6 mixture injection (Fig. 5a and Supplementary Fig. 15a), the breakthrough points of C_2H_4 and C_2H_6 for MAF-49 were observed by thermal conductivity detector at 1.09 and 1.44 mmol g^{-1}, respectively, during which the C_2H_6 concentration was determined as 0.014–0.016% by MS, corresponding to a C_2H_4 purity of 99.986–99.984% (Fig. 5a and Supplementary Figs 15a and 16). Under identical conditions, the highest C_2H_4 purities achieved by MAF-3, MAF-4 and IRMOF-8 are only 99.5%, 99.6% and 99.9% (C_2H_6 concentrations of 0.5%, 0.4% and

0.1%), respectively, reflecting their much lower C_2H_6/C_2H_4 selectivity compared with MAF-49 (Fig. 5b–d, Supplementary Table 6 and Supplementary Figs 15–19). Nevertheless, such C_2H_4 purities are obviously higher than those reported for C_2H_4 selective adsorbent materials (99% +)[13,17,23,24], which exemplify the feasibility of using C_2H_6 selective adsorbents for purifying C_2H_4, because the desired gas can be continuously purified by passing through the column and directly obtained from the first effluents. Indeed, desorbing the MAF-49 column saturated by 1:1 C_2H_4/C_2H_6 mixture can give C_2H_6 with 99%+ purity with a peak value of only 99.7% (Supplementary Fig. 20). A realistic comparison for the C_2H_4 purification performance of different adsorbents, of relevance to industrial operations, can be obtained by comparing the breakthrough amount of C_2H_4 (denoted as productivity) with the desired purity in a single breakthrough operation (for the calculation method see Supplementary Methods). For the MAF-49 column, 0.28 mmol g^{-1} or 0.44 mol l^{-1} of C_2H_4 with 99.95%+ purity can be recovered from a 1:1 C_2H_4/C_2H_6 mixture injection. For the MAF-3, MAF-4 and IRMOF-8 columns, their productivities are zero because the C_2H_4 effluents are not pure enough. Even for a C_2H_4 purity of 99% +, the productivity of the MAF-49 column (0.32 mmol g^{-1} or 0.47 mol l^{-1}) is still much higher than the others (the largest value is 0.11 mmol g^{-1} or 0.10 mol l^{-1}) (Supplementary Table 6).

Since the C_2H_6 concentration in C_2H_4/C_2H_6 mixtures produced by hydrocarbon cracking is just *ca.* 5–9% (refs 41–43), breakthrough experiments using a 15:1 C_2H_4/C_2H_6 mixture injection were carried out. The lowest C_2H_6 impurity or highest C_2H_4 purities achieved by the MAF-49, MAF-3, MAF-4 and IRMOF-8 columns are decreased to 0.005%, 0.4%, 0.1% and 0.04% or improved to 99.995%, 99.6%, 99.9% and 99.96%, respectively (Fig. 5e–h and Supplementary Figs 21–24). Obviously, using C_2H_6 selective adsorbents, the C_2H_4 purity can be increased by lengthening the adsorbent bed (increasing adsorbent amount), which is simpler and more convenient than the C_2H_4 selective adsorbents[13,17,25]. For the 15:1 C_2H_4/C_2H_6 mixture injection and the C_2H_4 output purity of 99.95% +, the MAF-49 column gave a C_2H_4 productivity of 1.68 mmol g^{-1} or 2.48 mol l^{-1}, which is about 30 or 50 times that of IRMOF-8 (0.06 mmol g^{-1} or 0.05 mol l^{-1}), in the gravimetric or volumetric point-of-view, respectively (Supplementary Table 7). Note that for C_2H_4 purification, the adsorbent volume is more practical than its weight because the fixed-bed equipment does not need to move during operation. For lower C_2H_4 purities such as 99.5% + and 99% +, the C_2H_4 productivities of MAF-49 and IRMOF-8 were also increased (Supplementary Tables 6 and 7), because the adsorber needs more time to reach adsorption saturation for the mixture gas containing low-concentration C_2H_6. Nevertheless, the C_2H_4 productivity of MAF-49 improved more significantly than for IRMOF-8 at all purity standards (Supplementary Tables 6 and 7), because the former material exhibits much higher C_2H_6 uptakes at the low pressure region (Fig. 2c). In contrast, the MAF-3 and MAF-4 columns only showed slightly increased C_2H_4 purities (did not reach 99.95% +) at a C_2H_4/C_2H_6 feeding ratio of 15:1 (Fig. 2c), because lengthening the adsorber is not so effective to improve the effluent purity by using adsorbents with weak impurity affinity. For C_2H_4 purities of 99.5%+ and 99%+, the C_2H_4 productivities of the MAF-3 and MAF-4 columns obtained by using a 15:1 C_2H_4/C_2H_6 input were unexpectedly lower than for the 1:1 C_2H_4/C_2H_6 mixture (Supplementary Tables 6 and 7), which can be attributed to the extremely low C_2H_6 adsorption ability of the adsorbents at the low pressure region. Also, the partial pressures of C_2H_4 and C_2H_6 in the 15:1 C_2H_4/C_2H_6 mixture are not beneficial for utilizing the differential gate-opening effect of MAF-3 (ref. 44).

Discussion

In summary, we reported a unique adsorbent material showing selective adsorption of C_2H_6 over more polar analogous molecules such as C_2H_4 and CO_2, which can be useful for extraction of C_2H_6 from natural gases and particularly valuable for direct production of high-purity C_2H_4 from C_2H_4/C_2H_6 mixtures. The key to this C_2H_6 selectivity is a combination of multiple hydrogen-bonding acceptors and dipole repulsion groups locating at appropriate positions on the pore surface of a very narrow channel, which not only allows multiple attractive interactions for C_2H_6 but also enforces C_2H_4 to adopt a position that can only form fewer and weaker attractive interactions. In short, this work provides not only a new MOF with exceptionally high C_2H_4 separation/purification performances, but also a new molecular design strategy for developing next-generation adsorbents.

Methods

Materials and general methods. Reagents and solvents were commercially available and were used without further purification, H_2batz (ref. 45), MAF-3 (ref. 30), MAF-4 (ref. 46), IRMOF-8 (ref. 47), HKUST-1 (ref. 48), CPO-27-Mg (ref. 49), CPO-27-Co (ref. 49) and UiO-66 (ref. 50) were synthesized according to the literature methods. Elemental analyses (C, H, N) were performed with a Vario EL elemental analyzer. Thermogravimetry analysis was performed under N_2 with temperature increased with 5 °C min^{-1} using a TA-Q50 system. Powder XRD patterns were collected on a Bruker D8 Advance diffractometer (Cu Kα) at room temperature.

Synthesis of [Zn(batz)]·0.5H$_2$O (MAF-49·H$_2$O). A mixture of $Zn(OH)_2$ (0.100 g, 1.0 mmol), H_2batz (0.180 g, 1.0 mmol), aqueous ammonia (25%, 4 ml) and water (4 ml) was stirred for 15 min in air, then transferred and sealed in a 15 ml Teflon reactor, which was heated in an oven at 160 °C for 72 h. The oven was cooled to room temperature at a rate of 5 °C h^{-1}. The resulting colourless block crystals were filtered, washed and dried in air (yield *ca.* 86%). Anal. Calcd (%) for $C_5H_7N_8O_{0.5}Zn$: C, 23.77; H, 2.79; N, 44.36. Found: C, 23.97; H, 2.82; N, 44.13. Guest-free MAF-49 was obtained by heating the as-synthesized sample under high vacuum at 150 °C for 12 h.

Single-crystal X-ray crystallography. Diffraction intensities were collected on a Pilatus XtaLAB P300DS diffractometer with graphite-monochromated Mo Kα radiation. Absorption corrections were applied by using the multi-scan programme REQAB. The structures were solved by the direct method and refined with the full-matrix least-squares technique using the SHELXTL programme package. It should be noted that, because the molecular centres of the very short C_2H_6 and C_2H_4 molecules do not locate at the centre of the two-fold symmetric host channel neck as predicted by computational simulations, their molecular geometries have to be restricted during refinement of the crystal structures. Also, the positions of their hydrogen atoms were added according to the computational simulation result. Because of disorder and low occupancies of the gas molecules, anisotropic thermal parameters were only applied to all non-hydrogen atoms of the host framework. Crystal data for the compounds were summarized in Supplementary Table 2. Electron density maps were generated using the output from standard SHELXL refinements in a number of ways using WinGX and VESTA 3.0.8.

Gas sorption measurement. The sorption isotherms were measured with an automatic volumetric adsorption apparatus (BELSORP-max). The as-synthesized sample (about 200 – 300 mg) was placed in the sample tube and dried for 12 h at 320 °C to remove the remnant solvent molecules prior to measurement. CO_2 (99.999%), C_2H_4 (99.95%), CH_4 (99.999%) and C_2H_6 (99.99%) were used for all adsorption isotherm and breakthrough experiments (Supplementary Fig. 25). The temperatures were controlled by a water bath (298, 307 and 316 K).

References

1. Matar, S. & Hatch, L. F. *Chemistry of Petrochemical Processes* 2nd edn (Gulf Publishing Company, 2000).
2. Baker, R. W. Future directions of membrane gas separation technology. *Ind. Eng. Chem. Res.* **41,** 1393–1411 (2002).
3. Horike, S. *et al.* Dense coordination network capable of selective CO_2 capture from C1 and C2 hydrocarbons. *J. Am. Chem. Soc.* **134,** 9852–9855 (2012).
4. Zhou, J. *et al.* Removal of C_2H_4 from a CO_2 stream by adsorption: a study in combination of ab initio calculation and experimental approach. *Energy Fuels* **20,** 778–782 (2006).
5. Liao, P.-Q., Zhu, A.-X., Zhang, W.-X., Zhang, J.-P. & Chen, X.-M. Self-catalysed aerobic oxidization of organic linker in porous crystal for

on-demand regulation of sorption behaviours. *Nat. Commun.* **6,** 6350 (2015).
6. Worrell, E., Phylipsen, D., Einstein, D. & Martin, N. Energy Use and Energy Intensity of the U.S. Chemical Industry. Report No. LBNL-44314 (Energy Analysis Department, Environmental Energy Technologies Division, Ernest Orlando Lawrence Berkeley National Laboratory, University of California, Berkeley, California 9472, USA, 2000).
7. Ren, T., Patel, M. & Blok, K. Olefins from conventional and heavy feedstocks: energy use in steam cracking and alternative processes. *Energy* **31,** 425–451 (2006).
8. Safarik, D. J. & Eldridge, R. B. Olefin/paraffin separations by reactive absorption: a review. *Ind. Eng. Chem. Res.* **37,** 2571–2581 (1998).
9. Yang, R. T. *Adsorbents: Fundamentals and Applications* 191–230 (John Wiley & Sons, Inc., 2003).
10. Tanaka, K., Taguchi, A., Hao, J., Kita, H. & Okamoto, K. Permeation and separation properties of polyimide membranes to olefins and paraffins. *J. Membr. Sci.* **121,** 197–207 (1996).
11. Bux, H., Chmelik, C., Krishna, R. & Caro, J. Ethene/ethane separation by the MOF membrane ZIF-8: molecular correlation of permeation, adsorption, diffusion. *J. Membr. Sci.* **369,** 284–289 (2011).
12. Uchida, S. *et al.* Selective sorption of olefins by halogen-substituted macrocation-polyoxometalate porous ionic crystals. *Chem. Mater.* **24,** 325–330 (2012).
13. Li, B. *et al.* Introduction of π-complexation into porous aromatic framework for highly selective adsorption of ethylene over ethane. *J. Am. Chem. Soc.* **136,** 8654–8660 (2014).
14. Uchida, S., Kawamoto, R., Tagami, H., Nakagawa, Y. & Mizuno, N. Highly selective sorption of small unsaturated hydrocarbons by nonporous flexible framework with silver ion. *J. Am. Chem. Soc.* **130,** 12370–12376 (2008).
15. Denysenko, D., Grzywa, M., Jelic, J., Reuter, K. & Volkmer, D. Scorpionate-type coordination in MFU-4l metal–organic frameworks: small-molecule binding and activation upon the thermally activated formation of open metal sites. *Angew. Chem. Int. Ed.* **53,** 5832–5836 (2014).
16. Aguado, S., Bergeret, G., Daniel, C. & Farrusseng, D. Absolute molecular sieve separation of ethylene/ethane mixtures with silver zeolite A. *J. Am. Chem. Soc.* **134,** 14635–14637 (2012).
17. Bloch, E. D. *et al.* Hydrocarbon separations in a metal-organic framework with open iron(ii) coordination sites. *Science* **335,** 1606–1610 (2012).
18. Herm, Z. R., Bloch, E. D. & Long, J. R. Hydrocarbon separations in metal–organic frameworks. *Chem. Mater.* **26,** 323–338 (2014).
19. Geier, S. J. *et al.* Selective adsorption of ethylene over ethane and propylene over propane in the metal-organic frameworks M_2(dobdc) (M = Mg, Mn, Fe, Co, Ni, Zn). *Chem. Sci.* **4,** 2054–2061 (2013).
20. Zhang, Y. *et al.* Highly selective adsorption of ethylene over ethane in a MOF featuring the combination of open metal site and π-complexation. *Chem. Commun.* **51,** 2714–2717 (2015).
21. Bao, Z. *et al.* Adsorption of ethane, ethylene, propane, and propylene on a magnesium-based metal-organic framework. *Langmuir* **27,** 13554–13562 (2011).
22. Yang, S. *et al.* Supramolecular binding and separation of hydrocarbons within a functionalized porous metal–organic framework. *Nat. Chem.* **7,** 121–129 (2015).
23. He, Y., Krishna, R. & Chen, B. Metal-organic frameworks with potential for energy-efficient adsorptive separation of light hydrocarbons. *Energy Environ. Sci.* **5,** 9107–9120 (2012).
24. Rege, S. U., Padin, J. & Yang, R. T. Olefin/paraffin separations by adsorption: π-Complexation *vs.* kinetic separation. *AIChE J.* **44,** 799–809 (1998).
25. Silva, F. A. D. & Rodrigues, A. E. Propylene/propane separation by vacuum swing adsorption using 13X zeolite. *AIChE J.* **47,** 341–357 (2001).
26. Brookhart, M., Findlater, M., Guironnet, D. & Lyons, T. W. Synthesis of para-xylene and toluene. US Patent 13/875,610 (2013).
27. Vogler, D. E. & Sigrist, M. W. Near-infrared laser based cavity ringdown spectroscopy for applications in petrochemical industry. *Appl. Phys. B* **85,** 349–354 (2006).
28. Meyers, R. & Meyers, R. A. *Handbook of Petrochemicals Production Processes* (McGraw-Hill Prof Med/Tech, 2005).
29. Böhme, U. *et al.* Ethene/ethane and propene/propane separation via the olefin and paraffin selective metal–organic framework adsorbents CPO-27 and ZIF-8. *Langmuir* **29,** 8592–8600 (2013).
30. Gücüyener, C., van den Bergh, J., Gascon, J. & Kapteijn, F. Ethane/ethene separation turned on its head: selective ethane adsorption on the metal – organic Framework ZIF-7 through a gate-opening mechanism. *J. Am. Chem. Soc.* **132,** 17704–17706 (2010).
31. Pires, J., Pinto, M. L. & Saini, V. K. Ethane selective IRMOF-8 and its significance in ethane–ethylene separation by adsorption. *ACS Appl. Mater. Interfaces* **6,** 12093–12099 (2014).
32. Cai, J. *et al.* A doubly interpenetrated metal–organic framework with open metal sites and suitable pore sizes for highly selective separation of small hydrocarbons at room temperature. *Cryst. Growth Des.* **13,** 2094–2097 (2013).
33. He, Y. *et al.* A microporous metal–organic framework for highly selective separation of acetylene, ethylene, and ethane from methane at room temperature. *Chem. Eur. J.* **18,** 613–619 (2012).
34. He, Y. *et al.* A robust doubly interpenetrated metal-organic framework constructed from a novel aromatic tricarboxylate for highly selective separation of small hydrocarbons. *Chem. Commun.* **48,** 6493–6495 (2012).
35. Das, M. C. *et al.* A Zn_4O-containing doubly interpenetrated porous metal-organic framework for photocatalytic decomposition of methyl orange. *Chem. Commun.* **47,** 11715–11717 (2011).
36. He, Y. *et al.* A microporous lanthanide-tricarboxylate framework with the potential for purification of natural gas. *Chem. Commun.* **48,** 10856–10858 (2012).
37. Li, J.-R., Sculley, J. & Zhou, H.-C. Metal–organic frameworks for separations. *Chem. Rev.* **112,** 869–932 (2011).
38. Zhang, J.-P., Zhang, Y.-B., Lin, J.-B. & Chen, X.-M. Metal azolate frameworks: from crystal engineering to functional materials. *Chem. Rev.* **112,** 1001–1033 (2012).
39. Myers, A. L. & Prausnitz, J. M. Thermodynamics of mixed-gas adsorption. *AIChE. J.* **11,** 121–126 (1965).
40. Liao, P.-Q. *et al.* Monodentate hydroxide as a super strong yet reversible active site for CO_2 capture from high-humidity flue gas. *Energy Environ. Sci.* **8,** 1011–1016 (2015).
41. Sedighi, M., Keyvanloo, K. & Darian, T. J. Olefin production from heavy liquid hydrocarbon thermal cracking: kinetics and product distribution. *Iran. J. Chem. Chem. Eng.* **29,** 135–147 (2010).
42. Marquevich, M., Coll, R. & Montané, D. Steam reforming of sunflower oil for hydrogen production. *Ind. Eng. Chem. Res.* **39,** 2140–2147 (2000).
43. Guo, J., Wu, X., Jing, S., Zhang, Q. & Zheng, D. Vapor-liquid equilibrium of ethylene + mesitylene system and process simulation for ethylene recovery. *Chin. J. Chem. Eng.* **19,** 543–548 (2011).
44. van den Bergh, J. *et al.* Understanding the anomalous alkane selectivity of ZIF-7 in the separation of light alkane/alkene mixtures. *Chem. Eur. J.* **17,** 8832–8840 (2011).
45. Dippold, A. A., Feller, M. & Klapoetke, T. M. 5,5'-dinitrimino-3,3'-methylene-1H-1,2,4-bistriazole - a metal free primary explosive combining excellent thermal stability and high performance. *Cent. Eur. J. Energ. Mater.* **8,** 261–278 (2011).
46. Huang, X.-C., Lin, Y.-Y., Zhang, J.-P. & Chen, X.-M. Ligand-directed strategy for zeolite-type metal–organic frameworks: Zinc(II) imidazolates with unusual zeolitic topologies. *Angew. Chem. Int. Ed.* **45,** 1557–1559 (2006).
47. Feldblyum, J. I., Wong-Foy, A. G. & Matzger, A. J. Non-interpenetrated IRMOF-8: synthesis, activation, and gas sorption. *Chem. Commun.* **48,** 9828–9830 (2012).
48. Chui, S.S.-Y., Lo, S.M.-F., Charmant, J. P. H., Orpen, A. G. & Williams, I. D. A chemically functionalizable nanoporous material $[Cu_3(TMA)_2(H_2O)_3]_n$. *Science* **283,** 1148–1150 (1999).
49. Caskey, S. R., Wong-Foy, A. G. & Matzger, A. J. Dramatic tuning of carbon dioxide uptake via metal substitution in a coordination polymer with cylindrical pores. *J. Am. Chem. Soc.* **130,** 10870–10871 (2008).
50. Cavka, J. H. *et al.* A new zirconium inorganic building brick forming metal organic frameworks with exceptional stability. *J. Am. Chem. Soc.* **130,** 13850–13851 (2008).

Acknowledgements

This work was supported by the '973 Project' (2014CB845602 and 2012CB821706) and NSFC (21225105, 21290173 and 21473260).

Author contributions

J.-P.Z. designed the research. P.-Q.L. performed the syntheses and measurements. P.-Q.L. and W.-X.Z. performed the SCXRD analyses. P.-Q.L., J.-P.Z. and X.-M.C. wrote the manuscript.

Additional information

Accession codes: The X-ray crystallographic coordinates for structures reported in this Article have been deposited at the Cambridge Crystallographic Data Centre (CCDC), under deposition number CCDC 1421354-1421358. These data can be obtained free of charge from the Cambridge Crystallographic Data Centre via www.ccdc.cam.ac.uk/data_request/cif.

Competing financial interests: The authors declare no competing financial interests.

11

Fabrication of three-dimensionally interconnected nanoparticle superlattices and their lithium-ion storage properties

Yucong Jiao[1,2], Dandan Han[2], Yi Ding[1], Xianfeng Zhang[1,2], Guannan Guo[1], Jianhua Hu[1], Dong Yang[1] & Angang Dong[2]

Three-dimensional superlattices consisting of nanoparticles represent a new class of condensed materials with collective properties arising from coupling interactions between close-packed nanoparticles. Despite recent advances in self-assembly of nanoparticle superlattices, the constituent materials have been limited to those that are attainable as monodisperse nanoparticles. In addition, self-assembled nanoparticle superlattices are generally weakly coupled due to the surface-coating ligands. Here we report the fabrication of three-dimensionally interconnected nanoparticle superlattices with face-centered cubic symmetry without the presynthesis of the constituent nanoparticles. We show that meso-porous carbon frameworks derived from self-assembled supercrystals can be used as a robust matrix for the growth of nanoparticle superlattices with diverse compositions. The resulting interconnected nanoparticle superlattices embedded in a carbon matrix are particularly suitable for energy storage applications. We demonstrate this by incorporating tin oxide nanoparticle superlattices as anode materials for lithium-ion batteries, and the resulting electrochemical performance is attributable to their unique architectures.

[1] State Key Laboratory of Molecular Engineering of Polymers and Department of Macromolecular Science, Fudan University, Shanghai 200433, China.
[2] Department of Chemistry and Collaborative Innovation Center of Chemistry for Energy Materials, Fudan University, Shanghai 200433, China.
Correspondence and requests for materials should be addressed to D.Y. (email: yangdong@fudan.edu.cn) or to A.D. (email: agdong@fudan.edu.cn).

Three-dimensional (3D) superlattices consisting of nanoparticles (NPs) are emerging as a new and important class of nanostructured materials[1], the properties of which can be rationally tuned by manipulating the size, shape and composition of the constituent NPs. In particular, interparticle interactions in NP superlattices can lead to collective properties, which are significantly different from the isolated NPs[1–3]. Existing methods of growing NP superlattices rely on the self-assembly of monodisperse colloidal NPs induced by solvent evaporation or antisolvent destabilization[1]. Recent progress on the self-assembly of NPs has been witnessed by the growth of a rich array of single- and multicomponent NP superlattices[4–10], which have found wide applications in electronic and optoelectronic devices, catalysis and energy storage.

The self-assembly of colloidal NPs is a complicated process, involving various driving forces and interactions (for example, van der Waals, Coulombic, dipolar)[1]. As a consequence, successful assembly of long-range ordered NP superlattices requires exquisite control over many parameters, such as NP size distribution, surface-coating ligands and solvent evaporation kineticsand so on. In addition, self-assembled NP superlattices typically suffer from poor electrical conductivity due to the large interparticle spacing maintained by the capping ligands[1,11]. Therefore, post-surface treatment such as ligand exchange is required to enhance electronic coupling, which unfortunately can lead to severe structural defects such as cracks[12,13]. Another major bottleneck hampering prospects of NP superlattices is the limited choice of monodisperse NP building blocks, despite recent progress in colloidal synthesis[1,14].

In this work we report an approach that can overcome the aforementioned limitations associated with self-assembly methods, enabling the growth of 3D interconnected, strongly coupled NP superlattices without the presynthesis of the constituent NPs. The resulting connected NP superlattices embedded in 3D continuous carbon frameworks represent a new class of superlattice materials, which are demonstrated to be remarkably promising for energy storage applications. SnO_2 NP superlattices are chosen as a model system to study Li-ion storage properties, which exhibit exceptional cycling stability and rate capability when used as Li-ion battery (LIB) anodes.

Results

Fabrication procedure. Figure 1a schematically illustrates the overall fabrication procedure. Briefly, starting with self-assembled Fe_3O_4 NP supercrystals, we obtain 3D continuous, highly ordered mesoporous carbon frameworks on the carbonization of the surface-coating oleic acid (OA) ligands followed by the removal of Fe_3O_4 NPs. The subsequent impregnation of carbon frameworks with desired precursors leads to 3D NP superlattices on hydrolysis and/or thermal annealing. The constituent NP composition can be modulated by selecting appropriate precursors, while the NP size can be tailored by controlling that of the original Fe_3O_4 NPs. The reason to select Fe_3O_4 NPs as the starting material is that the scalable production of Fe_3O_4 NP superlattices can be routinely achieved by self-assembly[14,15]. In addition, Fe_3O_4 is an inexpensive and thermally stable material, which can be readily removed by acid etching.

It is worth to mention that although the structure of our mesoporous carbon frameworks appears to be similar to that of previous mesoporous carbons prepared from hard template (that is, silica opals or mesoporous silica)[16–22], we should emphasize that this is the first preparation of ordered mesoporous carbons from inorganic nanocrystals. There are several unique features associated with this preparation strategy, leading to mesoporous carbons with distinctive structural and textural properties. First, the surface-coating OA ligands, which are a necessary component for the growth and self-assembly of Fe_3O_4 NPs, concurrently serve as the carbon source, yielding mesoporous carbons having ultrathin pore walls ($\sim$2 nm) and the same topological structure as the Fe_3O_4 NPs supercrystal template. This is in sharp contrast with previous hard-templating approaches where additional carbon precursors (that is, polymers or prepolymers) have to be applied to fill the voids of the template and the resulting mesoporous carbons usually possess an inverse structure of the template[16–22]. Furthermore, unlike previous approaches where silica opals or mesoporous silica only act as template[16–22], the Fe_3O_4 NPs used here also serve as a graphitization catalyst (to be discussed below), allowing for partially graphitic frameworks at low temperatures (500 °C). More importantly, the unique structure of our carbon frameworks enables interconnected, strongly coupled NP superlattices, which could not be readily accessible by conventional self-assembly methods.

Mesoporous carbon frameworks. Monodisperse Fe_3O_4 NPs ($\sim$11 nm) stabilized by OA are synthesized by a literature method[14] (Supplementary Fig. 1a), and Fe_3O_4 NP superlattices are grown by a conventional drying-mediated self-assembly process (Supplementary Fig. 1b)[4]. The content of the surface-coating OA ligands determined from thermogravimetric analysis (TGA) is $\sim$15.9 wt% (Supplementary Fig. 1c), corresponding to the area ligand coverage of $\sim$3.83 nm^{-2} (that is, $\sim$1455 OA molecules per NP, Supplementary Note 1). The prior addition of squalane ($\sim$1 wt%) into the NP solution is found to favour the gram-scale growth of micrometer-sized NP supercrystals with well-developed facets[15], although squalane is not critical for the subsequent formation of mesoporous carbon frameworks (Supplementary Fig. 1d). The as-assembled NP supercrystals are then heated at 500 °C in an argon atmosphere, producing grey powders arising from the carbonization of OA (Supplementary Fig. 2a). The carbon species constitute $\sim$11 wt% of the carbonized NP supercrystals, determined from TGA (Supplementary Fig. 2b). Scanning electron microscopy (SEM) and high-resolution SEM (HRSEM) establish that both the faceted morphology and the long-range NP ordering are well retained in carbonized Fe_3O_4 NP supercrystals (Fig. 1b,c and Supplementary Fig. 2c–f). These results strongly suggest that cracks and other structural defects, which have been observed in many thermally treated NP superlattices reported previously[23,24], are largely prevented in our experiments, probably due to the 3D crystal-like morphology as well as the long-range structural ordering of our Fe_3O_4 NP supercrystals. The small-angle X-ray scattering (SAXS) pattern of carbonized Fe_3O_4 NP supercrystals exhibits at least four scattering peaks, which can be assigned to the 111, 220, 311, 420 reflections of a well-crystallized face-centered cubic (fcc) structure (Fig. 1d, black curve), consistent with the highly ordered superlattice structure observed in HRSEM (Fig. 1c and Supplementary Fig. 2f). The unit cell parameter calculated from the SAXS data is 22.8 nm.

HNO_3 or HCl etching is employed to treat carbonized NP supercrystals. The complete removal of Fe_3O_4 NPs yields black carbon powders after washing and drying (Supplementary Fig. 3a), which display a highly ordered porous structure as revealed by SEM (Supplementary Fig. 3b) and transmission electron microscopy (TEM, Fig. 1e,f and Supplementary Fig. 3c,d). SAXS (Fig. 1d, red curve) and SEM (Fig. 1e, inset) indicate that the porous carbon frameworks possess the same fcc structure and faceted morphology inherited from Fe_3O_4 NP supercrystals, while the pore size ($\sim$10 nm) is slightly smaller than the diameter of Fe_3O_4 NPs, probably caused by the framework shrinkage during etching and/or post drying

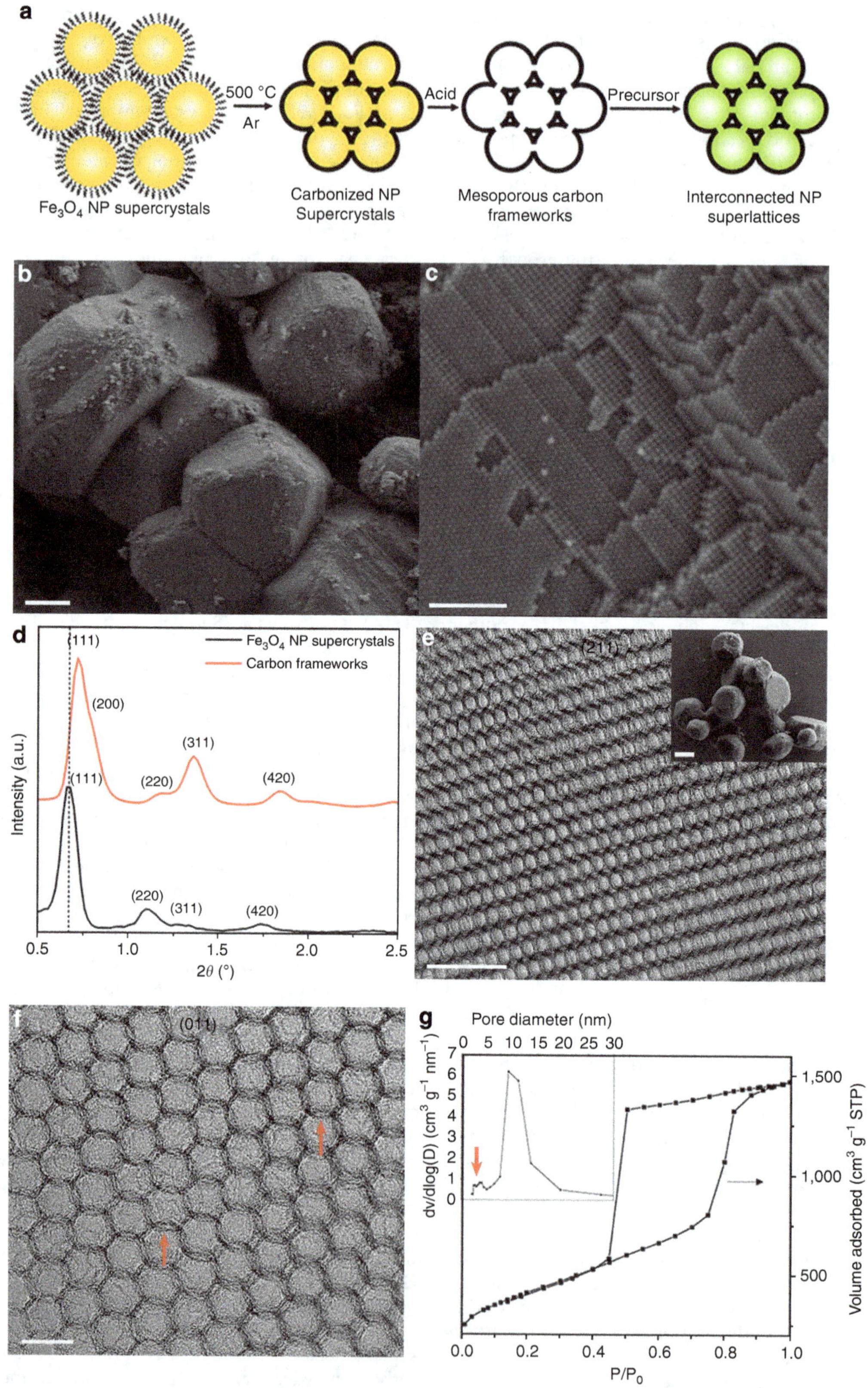

Figure 1 | Fabrication of three-dimensionally interconnected NP superlattices from mesoporous carbon frameworks. (**a**) Schematic illustration of the fabrication procedure (cross-sectional view). (**b**,**c**) SEM and HRSEM images of carbonized Fe_3O_4 NP supercrystals, respectively. Scale bars, 1 μm and 200 nm, respectively. (**d**) SAXS patterns of carbonized Fe_3O_4 NP supercrystals and mesoporous carbon frameworks, respectively. (**e**,**f**) TEM images of mesoporous carbon frameworks with different lattice projections. Scale bars, 50 and 20 nm, respectively. The inset in (**e**) is a low-magnification SEM image of mesoporous carbon frameworks. Scale bar, 1 μm. The red arrows in (**f**) indicate the interconnected windows. (**g**) N_2 adsorption-desorption isotherms and the corresponding pore size distribution (inset) of mesoporous carbon frameworks. The red arrow indicates the small pores corresponding to the interconnected windows observed in (**f**).

processes. Interestingly, high-resolution TEM (HRTEM) and Raman spectroscopy suggest that the pore walls of carbon frameworks are partially graphitic (Supplementary Fig. 4), which is remarkable considering the low carbonization temperature (500 °C). We attribute the formation of partially graphitic frameworks at such a low temperature to the use of Fe_3O_4 NPs, as transition metals such as Fe and Co and the corresponding metal oxides have been widely used as graphitization catalysts[25–27]. We surmise that the carbonaceous species arising from the thermal decomposition of OA molecules leads to the partial reduction of Fe_3O_4 NPs (most likely surface atoms) into metallic iron, which concurrently promotes the graphitization of the surrounding carbonaceous layers into partially graphitic pore walls. Despite the thin wall thickness (~2 nm, Supplementary Fig. 4b), the graphitization degree of carbon frameworks can be further increased by heat treatment at 1,200 °C in argon, converting the pore walls into few-layer graphene while retaining structural ordering (Supplementary Fig. 4c). More interestingly, tilted experiments along the [011] zone axis in TEM clearly shows that the adjacent pores are connected through small windows with dimensions of ~3 nm (Fig. 1f, indicated by red arrows), leading to a 3D continuous porous structure. Presumably, the formation of the interconnected windows is attributed to the slight sintering of neighbouring Fe_3O_4 NPs occurring during the carbonization process.

The porous structure of carbon frameworks is further characterized by N_2 adsorption–desorption isotherms, which show a type-IV curve with a sharp capillary condensation step occurring in the relative pressure (P/P_0) range of 0.8–0.85 (Fig. 1g), a typical feature of mesoporous materials[28]. The Brunauer–Emmett–Teller surface area and the pore volume are determined to be ~1,500 $m^2 g^{-1}$ and ~2.5 $cm^3 g^{-1}$, respectively. The pore size distribution curve determined by using the Barrett–Joyner–Halenda model suggests a bimodal porous structure (Fig. 1g, inset). The large pores at ~10 nm correspond to the removed Fe_3O_4 NPs, while the small pores in the range of 2–4 nm are ascribed to the interconnected windows observed in TEM.

Interconnected NP superlattices. The 3D continuous porosity and high surface area of carbon frameworks are expected to facilitate the diffusion of precursors, which is crucial for the subsequent growth of NP superlattices. As a proof-of-concept demonstration, tetraethoxysilane is chosen as the precursor for SiO_2 NP superlattices, which are obtained by repeated precursor infiltration followed by hydrolysis. Remarkably, detailed structural features such as facets and surface terraces originated from Fe_3O_4 NP supercrystals are also observed in the product (Fig. 2a,b and Supplementary Fig. 5), indicating a high degree of 3D NP ordering. The flexibility offered by the approach and the robustness of carbon frameworks allow the ready tuning of NP compositions by simply selecting appropriate precursors. For example, pure carbon NP superlattices can be realized by impregnating carbon frameworks with an aqueous solution of sucrose followed by drying and calcination (Fig. 2c,d). Energy-dispersive X-ray spectroscopy (EDS) and elemental mapping confirm that the resulting superlattices are composed of carbon NPs (Supplementary Fig. 6). To our knowledge, this is the first preparation of 3D carbon NP superlattices, as monodisperse, ligand-capped carbon NPs are not yet accessible for self-assembly[29].

To further illustrate the generality of the approach, we fabricate superlattices of crystalline metal oxide NPs such as SnO_2 by using $SnCl_2$ as the precursor. Thermal annealing at 350 °C in argon is carried out to crystallize the embedded NPs. TEM suggests that the crystallization process does not disrupt the ordered structure of SnO_2 NP superlattices (Fig. 3a and Supplementary Fig. 7), while SAXS confirms the fcc geometry with a lattice constant of 21.2 nm (Fig. 3b). Powder X-ray diffraction (XRD, Fig. 3c) and HRTEM (Fig. 3d) establish that the constituent SnO_2 NPs possess a high crystallinity with a tetragonal crystal structure. In addition to SnO_2, other types of metal oxide nanocrystal superlattices can be prepared in a similar way using metal alkoxides or anhydrous inorganic salts as precursors. For instance, 3D superlattices consisting of anatase TiO_2 nanocrystals are formed by impregnating carbon frameworks with titanium isopropoxide (TIP) followed by hydrolysis and thermal annealing (Supplementary Fig. 8). Moreover, NP superlattices of multicomponent phases, such as mixed oxides and metal phosphates, are also accessible by using a pre-mixed precursor (Supplementary Fig. 9). Apparently, the NP composition can be tailored by changing the ratio of the two precursors. For example, $Ti_{0.3}Sn_{0.7}O_2$ NP superlattices are obtained when a mixture of $SnCl_2$ and TIP is employed as the precursor, and the uniform distribution of Ti and Sn as revealed by elemental mapping indicates that the homogenous filling of $Ti_{0.3}Sn_{0.7}O_2$ NPs within the carbon framework (Supplementary Fig. 9c–e). The ability to fabricate such multiphase NP superlattices is particularly important, as they are generally hard to be prepared by self-assembly due to the challenge to obtain the corresponding monodisperse NPs.

Li-ion storage properties. One unique structural feature of NP superlattices produced by our approach is that the neighbouring NPs are connected with each other through interconnected windows within carbon frameworks, as shown in Fig. 3d where SnO_2 NP connections can be clearly observed (indicated by red arrows). Such connection configuration could substantially enhance electronic coupling between close-packed NPs[30,31], which is highly desirable for applications requiring high electrical conductivity. In comparison to self-assembled NP superlattices, another important feature of our NP superlattices is that the constituent NPs are embedded in a continuous and partially graphitic carbon matrix. Such carbon coating could further facilitate electron transport and accommodate volume variations of NPs, which is particularly beneficial for energy storage devices[32–34]. To demonstrate this, SnO_2 NP superlattices are chosen as anode materials for LIBs, which exhibit excellent electrochemical performance in terms of cycling stability and rate capability. The reason to select SnO_2 NP superlattices is that SnO_2 has attracted increasingly greater attention for Li-ion storage due to its high theoretical capacity (780 $mAh\,g^{-1}$) and environmental benignity[35–44].

To investigate their electrochemical performance, LIB anodes based on SnO_2 NP superlattices are cycled on the basis of the half-cell configuration. The carbon content of SnO_2 NP superlattices is ~28 wt%, determined from TGA (Supplementary Fig. 10a). Electron microscopies indicate that SnO_2 NP superlattices are uniformly distributed in the electrode with well-retained structural ordering before cycling (Supplementary Fig. 10b–d). Fig. 4a shows the cyclic voltammograms of SnO_2 NP superlattices in the potential range from 3.0 to 0.005 V (versus Li/Li^+) at a scan rate of 0.5 $mV\,s^{-1}$. The irreversible peak at 0.75 V in the first lithiation process is attributed to the reduction of SnO_2 into Sn as well as the formation of a solid electrolyte interphase (SEI)[39]. The two peaks at 0.7 and 1.3 V in the first delithiation process correspond to the de-alloying of Li_xSn and the partially reversible reaction of SnO_2 with Li^+, respectively, consistent with previous results[37,38]. The cycling stability of SnO_2 NP superlattices is studied by galvanostatic charge/discharge. The first discharge process leads to a capacity of 1,570 $mAh\,g^{-1}$ at a relatively high current density of 600 $mA\,g^{-1}$, whereas the

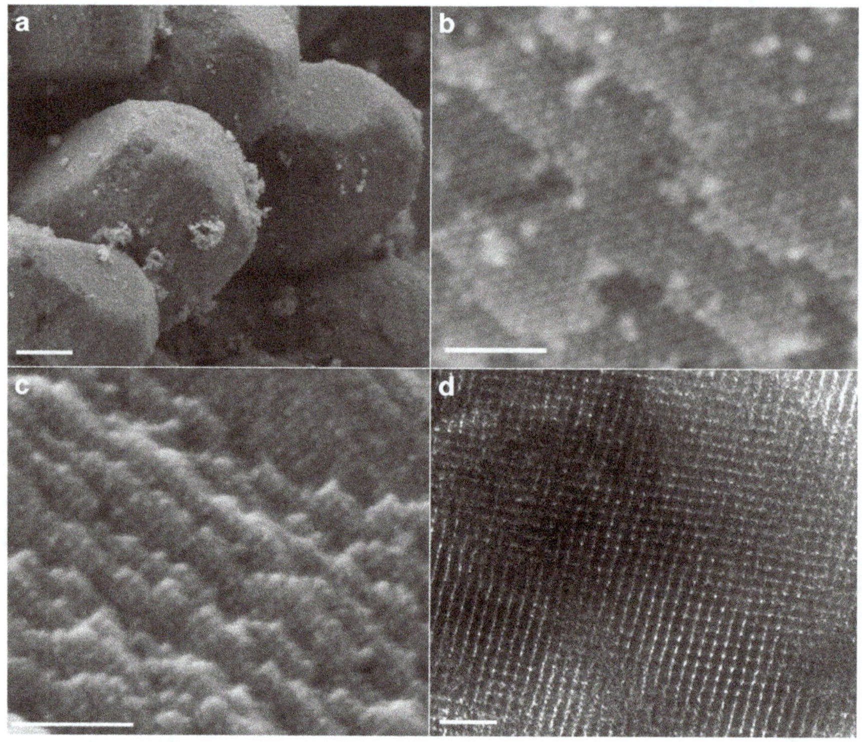

Figure 2 | Representative electron microscopy images of various NP superlattices. (**a**) SEM image of SiO_2 NP superlattices, showing the faceted morphology. Scale bar, 500 nm. (**b**) HRSEM image of SiO_2 NP superlattices, showing the surface terraces and long-range NP ordering. Scale bar, 100 nm. (**c,d**) SEM and TEM images of 3D carbon NP superlattices, respectively. Scale bars, 100 and 50 nm, respectively.

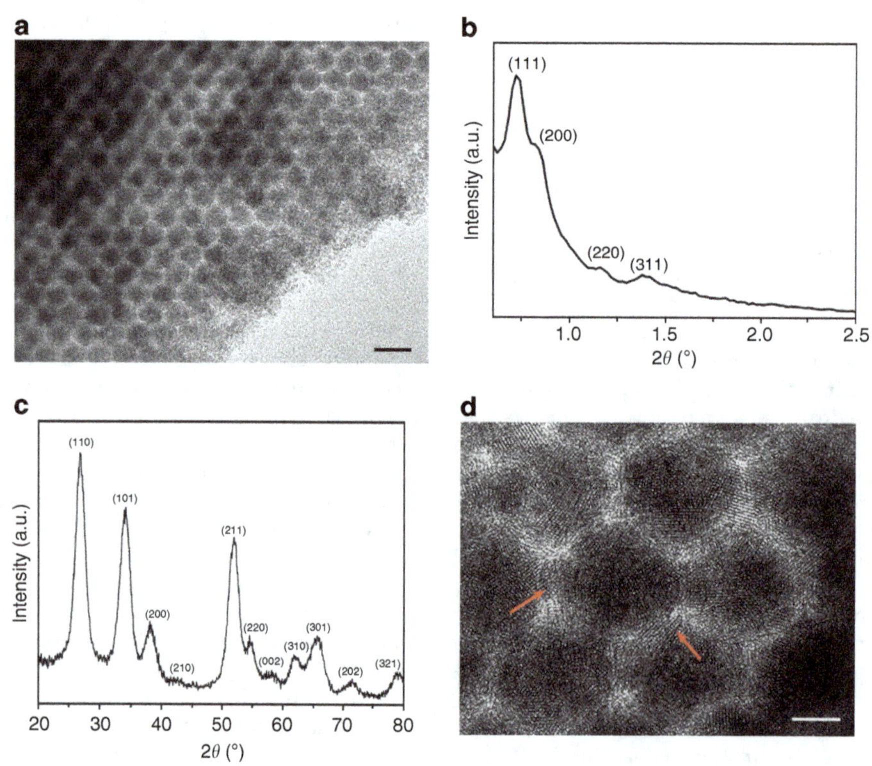

Figure 3 | Structural characterization of SnO_2 NP superlattices. (**a**) TEM image of SnO_2 NP superlattices. Scale bar, 20 nm. (**b,c**) SAXS pattern and XRD pattern of SnO_2 NP superlattices, respectively. (**d**) HRTEM image of SnO_2 NP superlattices, showing the high crystallinity of the embedded SnO_2 NPs. Scale bar, 5 nm. The red arrows indicate NP interconnections.

subsequent charge process delivers a capacity of 676 mAh g^{-1} (Supplementary Fig. 11). This irreversibility is probably caused by the SEI formation as well as the lithium insertion into carbon frameworks[36]. Despite the first-cycle capacity loss commonly observed for SnO_2 anodes[36,37], our SnO_2 NP superlattices exhibit excellent cycling stability, as manifested by the retention of a

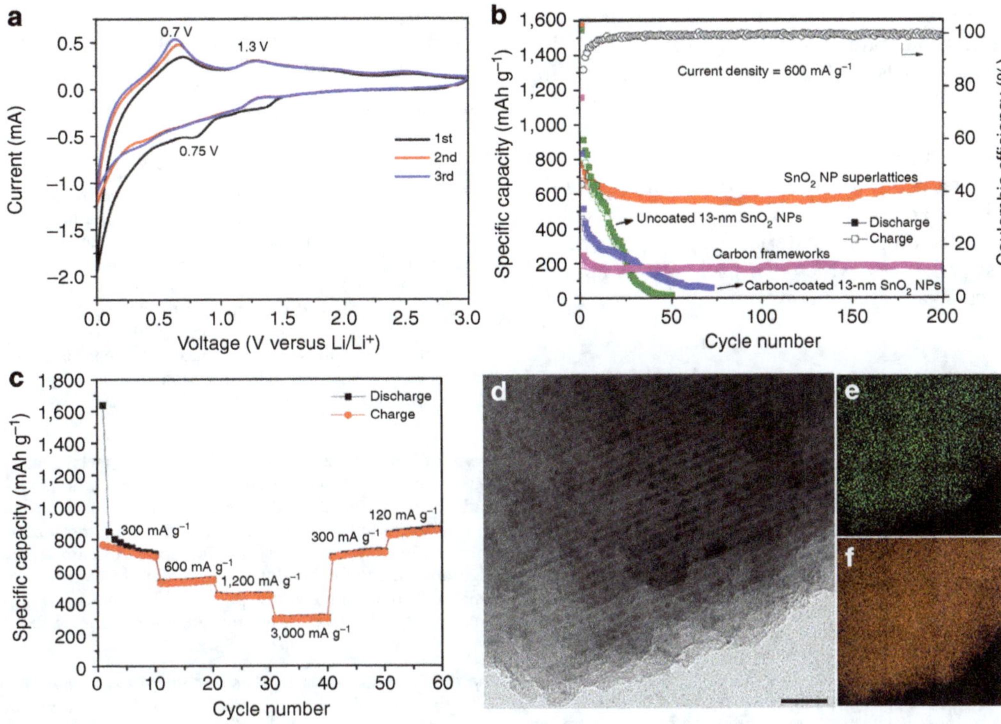

Figure 4 | Electrochemical characterization of SnO_2 NP superlattices. (**a**) Representative cyclic voltammograms at a scan rate of $0.5\,mV\,s^{-1}$. (**b**) Cycling performance at a current density of $600\,mA\,g^{-1}$ and the corresponding Coulombic efficiency. The cycling performance of 13-nm SnO_2 NPs with and without carbon coating as well as carbon frameworks tested under the same conditions are also included for comparison. (**c**) Rate-capability test at current densities ranging from 120 to $3{,}000\,mA\,g^{-1}$. (**d**) TEM image and (**e,f**) the corresponding EDS elemental mapping of SnO_2 NP superlattices after 200 cycles, showing the preservation of ordered structure without NP aggregation. Scale bar, 50 nm.

specific capacity of $640\,mAh\,g^{-1}$ after 200 cycles and a stabilized Coulombic efficiency over 98% from the 10th cycle (Fig. 4b). LIB anodes based on mesoporous carbon frameworks only are also tested under the same conditions, which exhibit a stable capacity at $185\,mAh\,g^{-1}$ after 200 cycles (Fig. 4b, pink curve), indicating that the high capacity of our SnO_2 NP superlattices is primarily attributed to the embedded SnO_2 NPs. To better evaluate the charge/discharge performance of our SnO_2 NP superlattices, we synthesize colloidal SnO_2 NPs with a similar diameter (~ 13 nm, Supplementary Fig. 12)[45], which are cycled under the same conditions in control experiments. As expected, the uncoated SnO_2 NPs exhibit fast capacity decay within 50 cycles (Fig. 4b, olive curve), presumably caused by the pulverization and/or aggregation of SnO_2 NPs in the absence of carbon protection. In comparison, the carbon-coated SnO_2 NPs (carbon content: ~26 wt%) exhibit better cycling stability, but their capacity rapidly fades below $100\,mAh\,g^{-1}$ after 50 cycles (Fig. 4b, blue curve). The cycling performance of our SnO_2 NP superlattices is further evaluated at different current densities. As shown in Fig. 4c, the electrode is able to deliver a capacity of $300\,mAh\,g^{-1}$ even at a current density of $3{,}000\,mA\,g^{-1}$, and a capacity above $850\,mAh\,g^{-1}$ is recovered when the current density is switched back to $120\,mA\,g^{-1}$. Such cycling stability and rate capability outperform those of most SnO_2 NP anodes reported previously[38–44].

Discussion

The superior battery performance of SnO_2 NP superlattices is attributed to their unique structural characteristics. First, our NP superlattices exist as micrometer-sized secondary particles inherited from the original Fe_3O_4 NP supercrystals, which are believed to be the ideal architectures for LIBs due to the reduced interfacial areas between active materials and electrolyte[46,47]. Second, and perhaps most importantly, the 3D graphitic carbon frameworks combined with interconnected NPs provide a continuous electron pathway, facilitating electronic connectivity within the components of the electrode. Moreover, the inherent flexibility of carbon frameworks can buffer large volume expansions of the embedded SnO_2 NPs during cycling[35,36], reducing the strain of the entire electrode.

To investigate the structural evolution of our SnO_2 NP superlattices during cycling, ex situ XRD, SEM, TEM, and EDS elemental mapping studies are performed after 200 cycles. In accordance with previous results for SnO_2-based anodes[35,43], XRD confirms the conversion of SnO_2 NPs into Sn NPs after cycling (Supplementary Fig. 13a), while the broad diffraction peaks imply that the resulting Sn NPs are small in size without agglomeration. SEM reveals that the secondary particle morphology is largely preserved after cycling (Supplementary Fig. 13b,c), while TEM confirms that the Sn-based NPs are homogeneously confined within carbon frameworks with good structural ordering without aggregation (Fig. 4d–f and Supplementary Fig. 13d). These results strongly suggest that our carbon frameworks effectively suppress the agglomeration and pulverization of Sn-based NPs, which could explain the superior cycling stability of our SnO_2 NP superlattices. It should also be noted that the LIB performance could be further improved by enhancing the electrical conductivity of carbon frameworks before precursor infiltration, which can be realized by increasing the graphitization degree via heat treatment as mentioned above.

In summary, we have developed a new approach to fabricate three-dimensionally interconnected, strongly coupled NP superlattices with fcc packing symmetry. The generality of the approach is illustrated by the growth of NP superlattices with various compositions including oxides, carbon, mixed oxides, and phosphates. We also demonstrate the successful use of SnO_2 NP superlattices as anode materials for LIBs, and the excellent cycling stability and rate capability are attributed to their unique architectures (that is, 3D continuous and graphitic carbon coating, NP interconnections, and supercrystal morphology). Given the fact that the fabrication procedure does not require the presynthesis of monodisperse NP building blocks, we anticipate that in the future a wider range of materials can be prepared as interconnected NP superlattices, which can find various applications in electronics, catalysis, and energy storage.

Methods

Materials. Oleic acid (OA, 90%), squalane, and 1-octadecene (ODE, 90%) were purchased from Aldrich. Sodium oleate was obtained from TCI. Iron chloride hexahydrate ($FeCl_3.6H_2O$), titanium isopropoxide (TIP), anhydrous tin chloride ($SnCl_2$), anhydrous tin tetrachloride ($SnCl_4$) tetraethoxysilane (TEOS), anhydrous zirconium tetrachloride ($ZrCl_4$), Cetyl trimethyl ammonium bromide (CTAB), and triethyl phosphate were purchased from Aladdin. All chemicals were used as received without further purification.

Synthesis and self-assembly of Fe_3O_4 NPs. Monodisperse, OA-stabilized Fe_3O_4 NPs with a diameter of ~ 11 nm were synthesized according to the literature method[14]. In a typical synthesis, 72 g of iron oleate and 11.4 g of OA were dissolved in 400 g of ODE in a three-neck flask, and the resulting solution was heated at 320 °C under a N_2 atmosphere for 1 h. After cooling down to room temperature, ethanol and isopropanol were added to precipitate Fe_3O_4 NPs, and the precipitates were dispersed in hexane to form a stable colloidal solution with a concentration of ~ 10 mg ml^{-1}. To self-assemble Fe_3O_4 NP supercrystals, squalane (~ 1 wt%) was added to the NP solution and the solvent was then allowed to evaporate under ambient conditions. The complete solvent evaporation yielded faceted Fe_3O_4 NP supercrystals with dimensions on the micrometer scale.

Fabrication of three-dimensionally interconnected NP superlattices. The as-assembled Fe_3O_4 NP supercrystals were heated in a quartz tube furnace at 500 °C for 2 h in argon, converting the surface-coating ligands into carbon. The carbonized NP supercrystals were then refluxed in a HNO_3 or HCl solution to remove the embedded Fe_3O_4 NPs. The resulting mesoporous carbon frameworks retrieved by centrifugation were washed with deionized water, yielding black powders after drying. To grow interconnected NP superlattices, the dried porous carbon frameworks were infiltrated with desired precursors via wet impregnation followed by hydrolysis and/or thermal annealing.

Preparation of SiO_2 NP superlattices. 5 mg of dried mesoporous carbon powder was dissolved in 1 ml of TEOS. After stirring for 6 h, the precipitated powder collected by centrifugation was washed with ethanol to remove the excess TEOS. The hydrolysis of TEOS was induced by the addition of ammonia hydroxide, leading to the conversion of TEOS into SiO_2. This impregnation/washing cycle was repeated twice in order to completely fill carbon frameworks with SiO_2 NPs.

Preparation of carbon NP superlattices. 3D carbon NP superlattices were synthesized using sucrose as the precursor. Briefly, 0.5 g of sucrose and 0.1 g of concentrated H_2SO_4 were first dissolved in 1 ml of H_2O to form a mixture, into which 5 mg of dried mesoporous carbon powder was added with constant stirring. After 2 h, the precipitated powder collected by centrifugation was washed with H_2O to remove the excess sucrose and H_2SO_4. After that, the product was placed in an oven at 100 °C for 2 h and then 160 °C for another 2 h. The black powder was then impregnated in a diluted sucrose solution (0.05 g of sucrose, 0.02 g of H_2SO_4, and 1 ml of H_2O) for 2 h. After washing and drying, the product was heated at 600 °C in argon for 2 h to carbonize sucrose, yielding 3D carbon NP superlattices.

Preparation of SnO_2 NP superlattices. 50 mg of mesoporous carbon powder was mixed with a $SnCl_2$ solution dissolved in ethanol (0.1 g ml^{-1}) and the resulting suspension was stirred for 2 h. After that, the black powder was collected by centrifugation and the precipitated powder was washed with ethanol to remove the excess $SnCl_2$. The precursor hydrolysis was initiated by the addition of ammonium hydroxide. This impregnation/washing cycle was repeated twice in order to completely load carbon frameworks with SnO_2. Thermal annealing at 350 °C in argon for 1 h was carried out to crystallize the embedded SnO_2 NPs.

Preparation of TiO_2 NP superlattices. TIP was used as the precursor for the fabrication of TiO_2 NP superlattices. Briefly, 50 mg of dried mesoporous carbon powder was mixed with a TIP solution dissolved in isopropanol (0.5 g ml^{-1}). After stirring for 6 h, the black powder was collected by centrifugation and the precipitated powder was washed with isopropanol to remove the excess TIP. The subsequent hydrolysis of TIP was initiated by exposure to air, leading to the conversion of TIP into TiO_2. This impregnation/washing cycle was repeated twice in order to completely fill carbon frameworks with TiO_2. After drying, the product was heated at 350 °C in argon for 1 h to crystallize the embedded TiO_2 NPs.

Preparation of $Ti_xSn_{1-x}O_2$ NP superlattices. A homogeneous mixture of TIP and $SnCl_2$ with different molar ratios dissolved in ethanol was used as the precursor for the growth of $Ti_xSn_{1-x}O_2$ NP superlattices. In a typical synthesis of $Ti_{0.3}Sn_{0.7}O_2$ NP superlattices, 0.4 mmol of TIP and 0.6 mmol of $SnCl_2$ were dissolved in 1 ml of ethanol to form a homogeneous solution, into which 5 mg of dried mesoporous carbon powder was added with stirring. After 24 h, the product was centrifuged and washed with ethanol to remove the excess precursor. The precursor hydrolysis was initiated by the addition of ammonium hydroxide. This impregnation/washing cycle was repeated twice in order to completely load carbon frameworks with $Ti_{0.3}Sn_{0.7}O_2$ NPs. After drying, the product was heated at 350 °C in argon for 1 h to crystallize the embedded $Ti_{0.3}Sn_{0.7}O_2$ NPs.

Preparation of zirconium phosphate NP superlattices. Zirconium phosphate (ZrP) NP superlattices were prepared by using a homogeneous mixture of $ZrCl_4$ and triethyl phosphate as the precursor. Briefly, 0.45 mmol of $ZrCl_4$ and 0.45 mmol of triethyl phosphate were mixed in 2 ml of ethanol to form a homogeneous solution, into which 5 mg of dried mesoporous carbon powder was added with vigorous stirring for 2 h. After that, the product was centrifuged and washed with ethanol to remove the excess precursor. The precipitated powder was then put in an oven at 110 °C for 2 h to induce the formation of ZrP NP superlattices.

Synthesis of 13-nm SnO_2 NPs in control experiments. In control experiments, colloidal SnO_2 NPs with a similar size as our SnO_2 NP superlattices were synthesized by a literature method[45]. In a typical synthesis, 10 ml of cationic surfactant (CTAB) solution (0.08 mol l^{-1}) was mixed with 10 ml of $NH_3.H_2O$ (25% aqueous solution) to form a homogeneous solution, into which 4.65 g of $SnCl_4$ was added under vigorous stirring. After 4 h, the product was filtered and washed with distilled water for several times. The product was then heated at 400 °C in air for 2 h to increase the crystallinity, leading to SnO_2 NPs with a mean diameter of 13 nm. Sucrose was used as the carbon source for the formation of carbon-coated SnO_2 nanocomposite. In a typical preparation of SnO_2/C nanocomposite with a carbon content of ~ 26 wt%, 0.35 g of SnO_2 NPs was mixed with a sucrose solution (0.4 g in 5 ml H_2O), and the resulting mixture was pre-heated at 180 °C in an oven for 3 h. The dried powder was then heated in a quartz furnace at 500 °C for 5 h in argon, resulting into carbon-coated SnO_2 nanocomposite.

Instrumentation. TEM images, HRTEM images, scanning TEM images, elemental mapping, and EDS spectra were obtained using a Tecnai G2 20 TWIN microscope operated at 200 kV. SEM images and EDS spectra were recorded using a Zeiss Ultra-55 microscope operated at 5 and 10 kV, respectively. XRD was carried out on a Bruker D4 X-ray diffractometer, while SAXS was performed on a Nanostar U small-angle X-ray scattering system using Cu Kα radiation (40 kV, 35 mA). Nitrogen adsorption–desorption isotherms were recorded on a Tristar 3000 instrument. Before measurements, the samples were degassed at 300 °C for 5 h. Raman spectra were collected at room temperature on an XploRA Raman system. TGA measurements were carried out on a Perkin–Elmer Pyris 1 thermogravimetric analyzer.

Electrochemical measurements. The battery performance was evaluated by galvanostatic cycling of 2025-type coin cells assembled in an argon-filled glove box, with SnO_2 NP superlattices as the working electrode and lithium foil as the counter and reference electrode. The electrolyte was a 1.0 M $LiPF_6$ solution in a mixture of ethylene carbonate, dimethyl carbonate and ethyl methyl carbonate (1:1:1 in volume), and a polypropylene film (Celgard-2300) was used as the separator. The working electrodes were prepared by a slurry-coating procedure. The slurry consisted of SnO_2 NP superlattices, acetylene black (Super P) and polyvinylidene fluoride binder with a mass ratio of 7:2:1 dissolved in *N*-methyl-2-pyrrolidinone. This slurry was spread on a copper foil, which acted as a current collector. The electrodes were dried at 90 °C for 4 h in air, and then at 90 °C in vacuum for another 12 h. Cyclic voltammetry was carried out on an electrochemical workstation (Autolab 204 N), while galvanostatic measurements were performed on a Neware cell test instrument, which was cycled between 0.005 and 3.00 V (versus Li/Li^+) at various current densities.

References

1. Talapin, D. V., Lee, J.-S., Kovalenko, M. V. & Shevchenko, E. V. Prospects of colloidal nanocrystals for electronic and optoelectronic applications. *Chem. Rev.* **110,** 389–458 (2010).
2. Panthani, M. G. & Korgel, B. A. Nanocrystals for electronics. *Annu. Rev. Chem. Biomol. Eng.* **3,** 287–311 (2012).
3. Pileni, M. P. Self-assembly of inorganic nanocrystals: fabrication and collective intrinsic properties. *Acc. Chem. Res.* **40,** 685–693 (2007).

4. Murray, C. B., Kagan, C. R. & Bawendi, M. G. Synthesis and characterization of monodisperse nanocrystals and close-packed nanocrystal assemblies. *Annu. Rev. Mater. Sci.* **30,** 545–610 (2000).
5. Shevchenko, E. V., Talapin, D. V., Kotov, N. A., O'Brien, S. & Murray, C. B. Structural diversity in binary nanoparticle superlattices. *Nature* **439,** 55–59 (2006).
6. Kalsin, A. M. *et al.* Electrostatic self-assembly of binary nanoparticle crystals with a diamond-like lattice. *Science* **312,** 420–424 (2006).
7. Mueggenburg, K. E., Lin, X. M., Goldsmith, R. H. & Jaeger, H. M. Elastic membranes of close-packed nanoparticle arrays. *Nat. Mater.* **6,** 656–660 (2007).
8. Dong, A. G., Chen, J., Patrick, M. V., Kikkawa, J. M. & Murray, C. B. Binary nanocrystal superlattice membranes self-assembled at the liquid-air interface. *Nature* **466,** 474–477 (2010).
9. Wang, T. *et al.* Self-assembled colloidal superparticles from nanorods. *Science* **338,** 358–363 (2012).
10. Singh, G. *et al.* Self-assembly of magnetite nanocubes into helical superstructures. *Science* **345,** 1149–1153 (2014).
11. Dong, A. G., Jiao, Y. C. & Milliron, D. J. Electronically coupled nanocrystal superlattice films by in situ ligand exchange at the liquid-air interface. *ACS Nano* **7,** 10978–10984 (2013).
12. Talapin, D. V. & Murray, C. B. PbSe nanocrystal solids for n- and p-channel thin film field-effect transistors. *Science* **310,** 86–89 (2005).
13. Luther, J. M. *et al.* Structural, optical and electrical properties of self-assembled films of PbSe nanocrystals treated with 1,2-ethanedithiol. *ACS Nano* **2,** 271–280 (2008).
14. Park, J. *et al.* Ultra-large-scale syntheses of monodisperse nanocrystals. *Nat. Mater.* **3,** 891–895 (2004).
15. Nagaoka, Y., Chen, O., Wang, Z. W. & Cao, Y. C. Structural control of nanocrystal superlattices using organic guest molecules. *J. Am. Chem. Soc.* **134,** 2868–2871 (2012).
16. Cychosz, K. A. *et al.* Characterization of the pore structure of three-dimensionally ordered mesoporous carbons using high resolution gas sorption. *Langmuir* **28,** 12647–12654 (2012).
17. Gierszal, K. P. & Jaroniec, M. Carbons with extremely large volume of uniform mesopores synthesized by carbonization of phenolic resin film formed on colloidal silica template. *J. Am. Chem. Soc.* **128,** 10026–10027 (2006).
18. Lee, H. I. *et al.* Spontaneous phase separation mediated synthesis of 3D mesoporous carbon with controllable cage and window size. *Adv. Mater.* **23,** 2357–2361 (2011).
19. Vu, A. *et al.* Three-dimensionally ordered mesoporous (3DOm) carbon materials as electrodes for electrochemical double-layer capacitors with ionic liquid electrolytes. *Chem. Mater.* **25,** 4137–4148 (2013).
20. Fang, B., Kim, M., Kim, J. H., Lim, S. & Yu, J. Ordered multimodal porous carbon with hierarchical nanostructure for high Li storage capacity and good cycling performance. *J. Mater. Chem.* **20,** 10253–10259 (2010).
21. Jun, S. *et al.* Synthesis of new, nanoporous carbon with hexagonally ordered mesostructure. *J. Am. Chem. Soc.* **122,** 10712–10713 (2000).
22. Kim, T. W., Park, I. S. & Ryoo, R. A synthetic route to ordered mesoporous carbon materials with graphitic pore walls. *Angew. Chem. Int. Ed.* **42,** 4375–4379 (2003).
23. Chen, J. *et al.* Bistable magnetoresistance switching in exchange-coupled $CoFe_2O_4$–Fe_3O_4 binary nanocrystal superlattices by self-assembly and thermal annealing. *ACS Nano* **7,** 1478–1486 (2013).
24. Law, M. *et al.* Structural, optical, and electrical properties of PbSe nanocrystal solids treated thermally or with simple amines. *J. Am. Chem. Soc.* **130,** 5974–5985 (2008).
25. Zhang, L. S., Li, W., Cui, Z. M. & Song, W. G. Synthesis of porous and graphitic carbon for electrochemical detection. *J. Phys. Chem. C* **113,** 20594–20598 (2009).
26. Lee, K. T., Ji, X., Rault, M. & Nazar, L. F. Simple synthesis of graphitic ordered mesoporous carbon materials by a solid-state method using metal phthalocyanines. *Angew. Chem. Int. Ed.* **48,** 5661–5665 (2009).
27. Gao, W., Wan, Y., Dou, Y. & Zhao, D. Synthesis of partially graphitic ordered mesoporous carbons with high surface areas. *Adv. Energy Mater.* **1,** 115–123 (2011).
28. Wan, Y. & Zhao, D. Y. On the controllable soft-templating approach to mesoporous silicates. *Chem. Rev.* **107,** 2821–2860 (2007).
29. Sun, Y. P. *et al.* Quantum-sized carbon dots for bright and colorful photoluminescence. *J. Am. Chem. Soc.* **128,** 7756–7757 (2006).
30. Zhang, H. *et al.* Surfactant ligand removal and rational fabrication of inorganically connected quantum dots. *Nano Lett.* **11,** 5356–5361 (2011).
31. Boneschanscher, M. P. *et al.* Long-range orientation and atomic attachment of nanocrystals in 2D honeycomb superlattices. *Science* **344,** 1377–1380 (2014).
32. Liu, N. *et al.* A pomegranate-inspired nanoscale design for large-volume-change lithium battery anodes. *Nat. Nanotech.* **9,** 187–192 (2014).
33. Chockla, A. M. *et al.* Silicon nanowire fabric as a lithium ion battery electrode material. *J. Am. Chem. Soc.* **133,** 20914–20921 (2011).
34. Jang, B. *et al.* Direct synthesis of self-assembled ferrite/carbon hybrid nanosheets for high performance lithium-ion battery anodes. *J. Am. Chem. Soc.* **134,** 15010–15015 (2012).
35. Han, F., Li, W., Li, M. & Lu, A. Fabrication of superior-performance SnO_2@C composites for lithium-ion anodes using tubular mesoporous carbon with thin carbon walls and high pore volume. *J. Mater. Chem.* **22,** 9645–9651 (2012).
36. Jahel, A., Ghimbeu, C. M., Monconduit, L. & Vix-Guterl, C. Confined ultrasmall SnO_2 particles in micro/mesoporous carbon as an extremely long cycle-life anode material for Li-ion batteries. *Adv. Energy Mater.* **4,** 1400025 (2014).
37. Liang, J. *et al.* Bowl-like SnO_2@carbon hollow particles as an advanced anode material for lithium-ion batteries. *Angew. Chem. Int. Ed.* **53,** 1–6 (2014).
38. Lin, J. *et al.* Graphene nanoribbon and nanostructured SnO_2 composite anodes for lithium Ion batteries. *ACS Nano* **7,** 6001–6006 (2013).
39. Ding, S. J. *et al.* Formation of SnO_2 hollow nanospheres inside mesoporous silica nanoreactors. *J. Am. Chem. Soc.* **133,** 21–23 (2011).
40. Qiao, H. *et al.* Sonochemical synthesis of ordered SnO_2/CMK-3 nanocomposites and their lithium storage properties. *ACS Appl. Mater. Interfaces* **3,** 3704–3708 (2011).
41. Lee, K. T., Lytle, J. C., Ergang, N. S., Oh, S. M. & Stein, A. Synthesis and rate performance of monolithic macroporous carbon electrodes for lithium-ion secondary batteries. *Adv. Funct. Mater.* **15,** 547–556 (2005).
42. Kim, C., Noh, M., Choi, M., Cho, J. & Park, B. Critical size of a nano SnO_2 electrode for Li-secondary battery. *Chem. Mater.* **17,** 3297–3301 (2005).
43. Li, X. *et al.* Tin oxide with controlled morphology and crystallinity by atomic layer deposition onto graphene nanosheets for enhanced lithium storage. *Adv. Funct. Mater.* **22,** 1647–1654 (2012).
44. Zhang, X., Jiang, B., Guo, J., Xie, Y. & Tang, L. Large and stable reversible lithium-ion storages from mesoporous SnO_2 nanosheets with ultralong lifespan over 1000 cycles. *J. Power Sources* **268,** 365–371 (2014).
45. Wang, Y. D., Ma, C. L., Sun, X. D. & Li, H. D. Preparation and characterization of SnO_2 nanoparticles with a surfactant-mediated method. *Nanotechnology* **13,** 565–569 (2002).
46. Lee, S. H. *et al.* Self-assembled Fe_3O_4 nanoparticle clusters as high-performance anodes for lithium ion batteries via geometric confinement. *Nano Lett.* **13,** 4249–4256 (2013).
47. Oszajca, M. F., Bodnarchuk, M. I. & Kovalenko, M. V. Precisely engineered colloidal nanoparticles and nanocrystals for Li-ion and Na-ion batteries: model systems or practical solutions? *Chem. Mater.* **26,** 5422–5432 (2014).

Acknowledgements

D.Y. is grateful for financial support from Natural National Science Foundation of China (51103026, 51373035, and 51373040), the Shanghai Scientific and Technological Innovation Project (11JC1400600 and 124119a2400), and International Science and Technology Cooperation Program of China (2014DFE40130). A.D. acknowledges the financial support from Fudan University, the '1000 Youth Talents' Plan, Natural National Science Foundation of China (21373052), and National Basic Research Program of China (973 program: 2014CB845602).

Author contributions

Y.J. and A.D. conceived and designed the experiments. Y.J. and D.H. performed the experiments. Y.D., X.Z., G.G. and J.H. collected and analysed the XRD and TEM data. Y.J., D.Y. and A.D. wrote the paper. All authors discussed the results and commented on the manuscript.

Additional information

Competing financial interests: The authors declare no competing financial interests.

12

A modular design of molecular qubits to implement universal quantum gates

Jesús Ferrando-Soria[1], Eufemio Moreno Pineda[1], Alessandro Chiesa[2], Antonio Fernandez[1], Samantha A. Magee[1], Stefano Carretta[2], Paolo Santini[2], Iñigo J. Vitorica-Yrezabal[1], Floriana Tuna[1], Grigore A. Timco[1], Eric J.L. McInnes[1] & Richard E.P. Winpenny[1]

The physical implementation of quantum information processing relies on individual modules—qubits—and operations that modify such modules either individually or in groups—quantum gates. Two examples of gates that entangle pairs of qubits are the controlled NOT-gate (CNOT) gate, which flips the state of one qubit depending on the state of another, and the $\sqrt{\mathrm{iSWAP}}$ gate that brings a two-qubit product state into a superposition involving partially swapping the qubit states. Here we show that through supramolecular chemistry a single simple module, molecular $\{Cr_7Ni\}$ rings, which act as the qubits, can be assembled into structures suitable for either the CNOT or $\sqrt{\mathrm{iSWAP}}$ gate by choice of linker, and we characterize these structures by electron spin resonance spectroscopy. We introduce two schemes for implementing such gates with these supramolecular assemblies and perform detailed simulations, based on the measured parameters including decoherence, to demonstrate how the gates would operate.

[1] School of Chemistry and Photon Science Institute, The University of Manchester, Oxford Road, Manchester M13 9PL, UK. [2] Dipartimento di Fisica e Scienze della Terra, Università di Parma, viale delle Scienze 7/a, Parma 43123, Italy. Correspondence and requests for materials should be addressed to R.E.P.W. (email: richard.winpenny@manchester.ac.uk).

The inability of conventional computers to solve certain problems efficiently, such as the simulation of quantum systems[1,2], is one of the main driving forces for the implementation of quantum computing and quantum information processing (QIP)[3,4], which exploit the laws of quantum mechanics. At the theoretical level, several algorithms have been shown to outperform classical computers in certain computational tasks such as factoring large numbers into primes[5] and searching of unsorted directories[6]. A variety of quantum systems have shown an excellent performance as the basic units for quantum information (qubits)[7–11]. However, they are difficult to link controllably into useful arrays producing the two-qubit entangling quantum gates (QGs) crucial for any quantum algorithm. Among the most important entangling QGs are the controlled NOT-gate (CNOT) and the $\sqrt{iSWAP}$ gate. The $\sqrt{iSWAP}$ gate brings the two-qubit state $|01\rangle$ to the superposition $(|01\rangle + i|10\rangle)/\sqrt{2}$. The CNOT gate flips the state of the target qubit if, and only if, the control qubit is in the $|1\rangle$ state; this implies that each qubit has to respond inequivalently to an external stimulus. All other quantum gates, however complex, can be constructed from these two-qubit gates and single-qubit gates.

Therefore the major challenge for the physical implementation of QIP is bringing together qubits in an organized, scalable and addressable way to make such QGs[3,4]. Here we demonstrate that supramolecular chemistry could have a major impact in addressing this challenge.

Molecular nanomagnets have been proposed as qubits[12–26] and coherence times of individual molecular qubits have been studied and have improved[24,25]. Supramolecular chemistry[27] allows us to bring together, with great control, complex molecules into arrays, and recently this has been exploited to tune two-qubit interactions built from molecular nanomagnets[28]. Hence supramolecular chemistry is a promising tool to build multi-qubit devices; if this could be made scalable this is a competitive route towards the realization of a quantum computer.

Demonstrations of CNOT gates based on molecular nanomagnets have been published[18,19,23]. The interaction between two dissimilar qubits produces a splitting in the low-lying two-qubit states that allows selective addressing of the transition by means of resonant EPR pulses, while keeping the other components of the wave-function frozen. In this way, the excitation of the target depends on the state of the control qubit and a CNOT gate is implemented. The drawback of this approach is the permanent direct coupling between the qubits, which makes these proposals hard to scale as with a permanent coupling the state of the qubits experiences an unwanted many-body spontaneous evolution in time (as in NMR QIP schemes[29]), whose harmful effects increase with the number of qubits. Conversely, a switchable indirect coupling between the two qubits would make the register much more easily scalable as when the switch is in the off state the unwanted spontaneous evolution is suppressed.

We have proposed using $\{Cr_7Ni\}$ heterometallic rings as qubits[14]. They are two-level systems ($S = 1/2$ ground state) that have sufficiently long phase memory times to allow many gate operations before state degradation occurs[25], and assemblies of $\{Cr_7Ni\}$ rings have been made which show a permanent coupling between the spins[26]. Here we report two-qubit assemblies which include switchable links, that allow us to propose a fully modular supramolecular design strategy[27,28] towards quantum computation schemes based on either the CNOT or $\sqrt{iSWAP}$ gate. The strategy uses supramolecular chemistry to tailor the individual components, spatial configuration and hence the properties of the resulting supramolecules. These supramolecules have been studied by electrochemistry, continuous-wave and pulsed EPR spectroscopy to understand their static and dynamic spin properties. These measured parameters have been used to perform detailed simulations of the performance of both the CNOT and $\sqrt{iSWAP}$ gates including the effects of decoherence. The two gates we propose are based on either global or local control of the qubit–qubit interaction. The first proposal exploits uniform magnetic pulses to manipulate two inequivalent Cr_7Ni qubits in an asymmetric supramolecule, which would implement the CNOT gate. The second exploits the local electric control of a redox-active linker, which might be addressed by a tip to reversibly switch on and off the qubit–qubit interaction, to implement the $\sqrt{iSWAP}$ gate. The scalability of these approaches is discussed in the last part of the paper.

Results

Syntheses and structural characterization. We have earlier reported the selective functionalization of $[^nPr_2NH_2][Cr_7NiF_8(O_2C^tBu)_{16}]$ **1** (ref. 30) with *iso*-nicotinate (O_2C-py) to obtain $[^nPr_2NH_2][Cr_7NiF_8(O_2C^tBu)_{15}(O_2C\text{-py})]$ **2** (ref. 26) (Supplementary Fig. 1a). This synthetic method can be extended to produce $[^nPr_2NH_2][Cr_7NiF_8(O_2C^tBu)_{15}(O_2C\text{-terpy})]$ **3** and $[^nPr_2NH_2][Cr_7NiF_8(O_2C^tBu)_{15}(O_2C\text{-Ph-terpy})]$ **4** (Supplementary Fig. 1b,c), from the controlled reaction of **1** with 4-carboxy-2,2′:6′,2″-terpyridine (O_2C-terpy) and 4′-(4-carboxyphenyl)-2,2′:6′,2″-terpyridine (O_2C-Ph-terpy), respectively (full experimental details are given in the Supplementary Methods)[31]. **2**, **3** and **4** are hereafter abbreviated as $\{Cr_7Ni\text{-}O_2C\text{-py}\}$, $\{Cr_7Ni\text{-}O_2C\text{-terpy}\}$ and $\{Cr_7Ni\text{-}O_2C\text{-Ph-terpy}\}$, respectively. They consist of $Cr^{III}{}_7Ni^{II}$ rings, containing an octagon of metal centres, with the inner rim bridged by fluoride ions and the outer rim by carboxylates: the functionalized O_2C-py **2**, O_2C-terpy **3** and O_2C-Ph-terpy **4** ligands sit on a Cr…Ni edge of the octagon. This provides us with three supramolecular qubits with two different denticities, which allow us to assemble different QGs by appropriate choice of the central node (Fig. 1).

Reaction of equimolar quantities of **2** and **3** with cobalt(II) thiocyanate in a mixture of Et_2O/acetone leads to $[\{Cr_7Ni\text{-}O_2C\text{-py}\} \rightarrow Co(SCN)_2 \leftarrow \{Cr_7Ni\text{-}O_2C\text{-terpy}\}]$ **5** in good yield, which has been characterized by X-ray single crystal diffraction (Fig. 2a; for all crystallographic information see Supplementary Data 1). Compound **5** contains two inequivalent Cr_7Ni qubits coordinated to a central cobalt(II) ion, which has a six-coordinate CoN_6 octahedral environment with a *cis*-arrangement of the two thiocyanate N atoms. The Co-N bond distances are typical of high spin cobalt(II) ions[32], with the bonds to the thiocyanate ligands shorter (average 2.056(12) Å) than those to terpy or pyridine N-donors (2.137(6) to 2.224(4) Å). Hence the d^7 Co^{II} site ($S_{Co} = {}^3/_2$) has a $^4T_{1g}$ ground term (using O_h symmetry labels)[33], which leads to a well-isolated effective $S_{eff} = {}^1/_2$ ground state at low temperature (see Characterization of assemblies via EPR spectroscopy below). The *cis* coordination geometry at the Co^{II} node means the $\{Cr_7Ni\text{-}O_2C\text{-terpy}\}$- and $\{Cr_7Ni\text{-}O_2C\text{-py}\}$-based qubits are arranged in an almost orthogonal orientation. Therefore the two qubits in **5** are symmetry inequivalent, as required for implementing a CNOT gate (Fig. 2b).

Reaction of two equivalents of **2** with a preformed oxo-centred pivalate-bridged triangular cluster with terminal pyridine groups $[Ru^{III}{}_2Co^{II}O(O_2C^tBu)_6(py)_3]$ **6**, hereafter abbreviated as $[Ru^{III}{}_2Co^{II}]$, in acetone gives $[\{Cr_7Ni\text{-}O_2C\text{-py}\} \rightarrow [Ru^{III}{}_2Co^{II}O(^tBuCO_2)_6(py)] \leftarrow \{Cr_7Ni\text{-}O_2C\text{-py}\}]$ **7** (Fig. 3a), where two of the terminal pyridine ligands of **6** were replaced by the *iso*-nicotinate group of **2**. Reaction of two equivalents of **3** or **4** and either cobalt(II) perchlorate or tetrafluoroborate gives $[\{Cr_7Ni\text{-}O_2C\text{-terpy}\} \rightarrow Co \leftarrow \{Cr_7Ni\text{-}O_2C\text{-terpy}\}][X]_2$ $[X = ClO_4^-$ **8a** or BF_4^- **8b**$]$ and $[\{Cr_7Ni\text{-}O_2C\text{-Ph-terpy}\} \rightarrow Co \leftarrow \{Cr_7Ni\text{-}O_2C\text{-Ph-terpy}\}][X]_2$ $[X = ClO_4^-$ **9a** or BF_4^- **9b**$]$ (Fig. 3b,c) (full experimental details are given in the Supplementary Methods). The

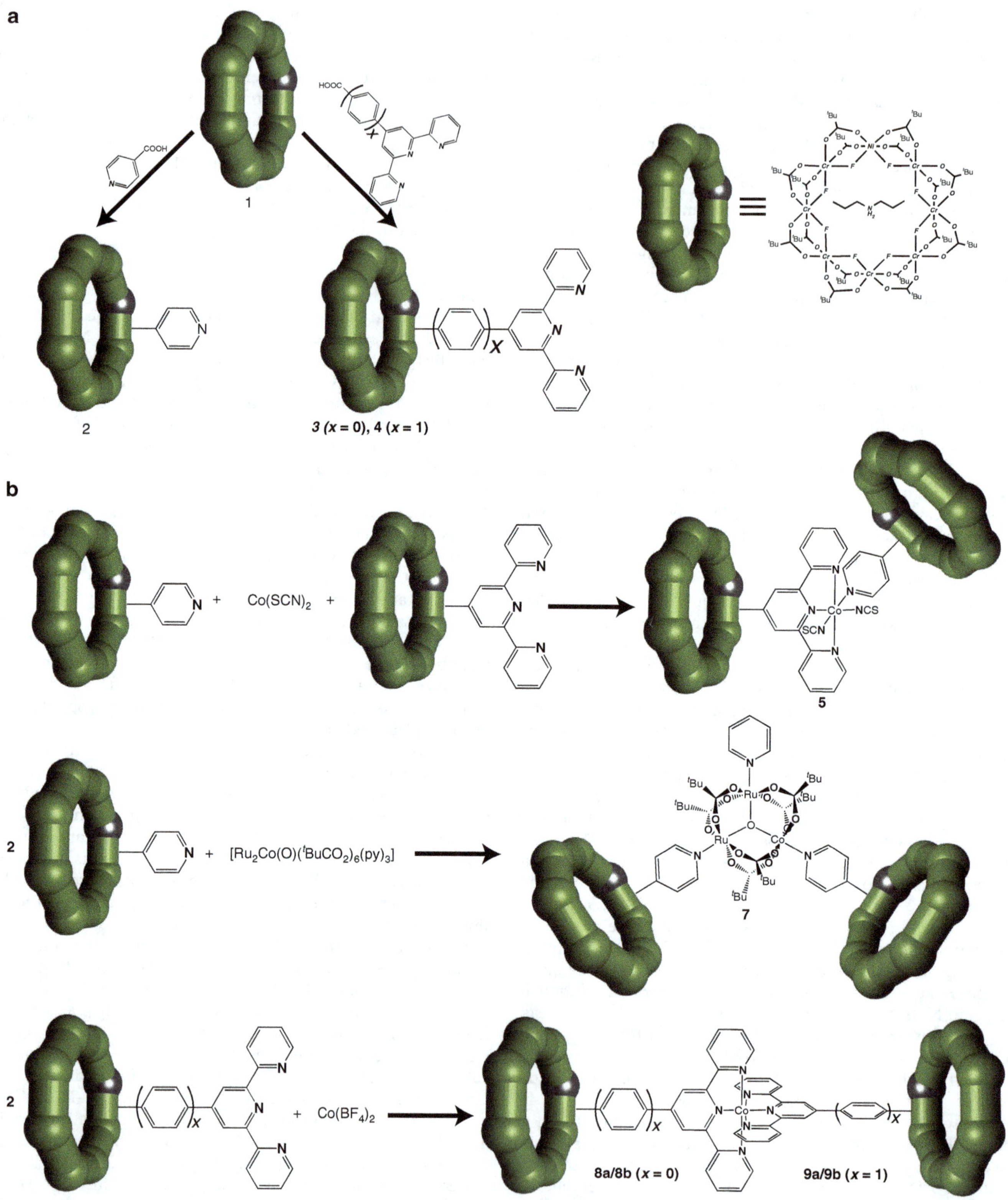

Figure 1 | Supramolecular design strategy for the construction of two-qubits assemblies. (**a**) Synthesis of functionalized {Cr_7Ni} rings (where $x = 0$ and 1). (**b**) Linkage of {Cr_7Ni} coordination cages to different central nodes.

architecture in **7**, **8** and **9** contains two equivalent qubits, separated by a redox-switchable centre[34,35], which makes it suitable for implementation of a $\sqrt{}$iSWAP gate (Fig. 3d).

The crystal structure of **7** consists of two {Cr_7Ni-O_2C-py} rings linked through the *iso*-nicotinate groups to a [$Ru^{III}{}_2Co^{II}O$ ($^tBuCO_2)_6$(py)] triangle (Fig. 3a)[34,36]. The metal ions are statistically disordered over the three sites within the triangular M_3O unit. They have a six-coordinate, octahedral MO_5N environment formed by four carboxylate oxygen atoms from the bridging pivalate ligands in the equatorial plane, with the central oxide and one nitrogen atom from *iso*-nicotinate or pyridine groups occupying the axial positions. The three metal ions are positioned at the corners of an isosceles triangle with two distinct intermetallic distances of 3.296 and 3.315 Å. The central oxide lies within the M_3 plane of the metal atoms, while the six pivalate bridging ligands lie above and below the M_3O plane and the pyridine N-donors rest perpendicular to this plane.

The crystal structure of **8a** and **9a** are made up of cationic [{Cr_7Ni-O_2C-terpy}Co^{II}{Cr_7Ni-O_2C-terpy}] and [{Cr_7Ni-O_2C-Ph-terpy}Co^{II}{Cr_7Ni-O_2C-Ph-terpy}] units together with counter-balancing ClO_4^- anions (Fig. 3b,c). Both compounds have a central cobalt(II) ion with a six-coordinate CoN_6 octahedral environment, with the six Co-N bonds in the range expected for a low spin cobalt(II) ion[37]. The average value of the two N-donors from the central pyridines of the terpy are shorter (1.923(6) **8a** and 1.920(8) **9a** Å) than the other Co-N contacts (2.118(7) **8a** and 1.986(9) **9a** Å). This feature leads to an axially compressed octahedral environment for the low-spin d^7 Co^{II} ions giving a $S_{Co} = {}^1/_2$ ground state. Compounds **8a** and **9a** differ in the shortest Co…M(ring) contact (8.675(2) **8a** and 10.979(5) **9a** Å) and the staggered **8a** or eclipsed **9a** arrangement of rings in the assemblies.

The two Cr_7Ni qubits in each of **7–9** are linked by a redox-switchable central node[34,38]. Cyclic voltammetry on **7–9**, and on the central nodes in isolation, show a one-electron reversible oxidation in each case (measured in CH_2Cl_2, 0.1 M $n$$Bu_4NPF_6$). For **6** and **7** the half-wave potential ($E_{1/2}$) = −0.31 (**6**), −0.35 (**7**) V versus Fc^+/Fc (Fc = ferrocene)[39], and this oxidation is assigned to the [$Ru^{III}_2Co^{II}$] to [$Ru^{III,IV}_2Co^{II}$] couple[34]. For **8b**, **9b** and reference complexes [Co(HO_2C-terpy)$_2$][BF_4]$_2$ **10** and [Co(HO_2C-Ph-terpy)$_2$][BF_4]$_2$ **11**, $E_{1/2}$ = −0.30 (**8b**), −0.27 (**9b**), −0.27 (**10**) and −0.24 (**11**) V, which are assigned to the $Co^{II/III}$ couple. The anodic to cathodic peak separation values are similar to those of ferrocene under the same conditions (see Electrochemistry section in Supplementary Methods, Supplementary Fig. 4 and Supplementary Table 3).

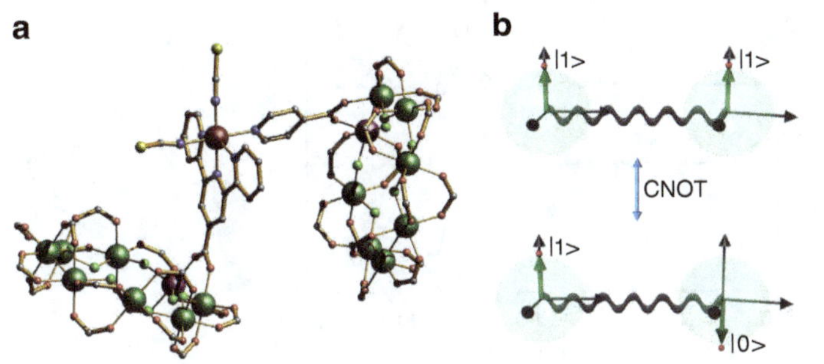

Figure 2 | Molecular structure of the asymmetric two-qubit assembly for proposed implementation of a CNOT gate. (**a**) Molecular structure of [{Cr_7Ni-O_2C-py}→Co(SCN)$_2$←{Cr_7Ni-O_2C-terpy}] **5**. $^nPr_2NH_2^+$ cations are not shown (H atoms and tert-butyl groups are omitted for clarity). Colour code: Co, dark red; Cr, green; Ni, purple; Ru, brown; N, cyan; O, red; S, yellow; C, grey; F, pale green. (**b**) Schematic representation of the effect of the CNOT gate on a pair of qubits, initialized in the computational basis states |11⟩ and |10⟩, respectively. The CNOT flips the target qubit if the control is set to |1⟩.

Characterization of assemblies via EPR spectroscopy. To study the interactions between the molecular components in the supramolecular structures **5**, **7**, **8b** and **9b** we performed magnetometry and multi-frequency continuous wave EPR studies. The measured magnetometry data are essentially the sum of the Cr_7Ni qubits and the linking nodes (Supplementary Figs 5–9), and therefore are uninformative other than confirming that the interactions are very weak. EPR spectroscopy is much more sensitive to such weak interactions.

The EPR spectra of a powder of **5** at 5K (Fig. 4a and Supplementary Fig. 10) have a complex multiplet structure because the $S_{eff} = 1/2$ of the Co(II) node has a very anisotropic effective g-tensor and also gives rise to very anisotropic exchange tensors (J) with the {Cr_7Ni} rings. This structure is best resolved at W-band (94 GHz; Fig. 4a). The spectra can be simulated[40] with the effective three-spin Hamiltonian (1), with two different

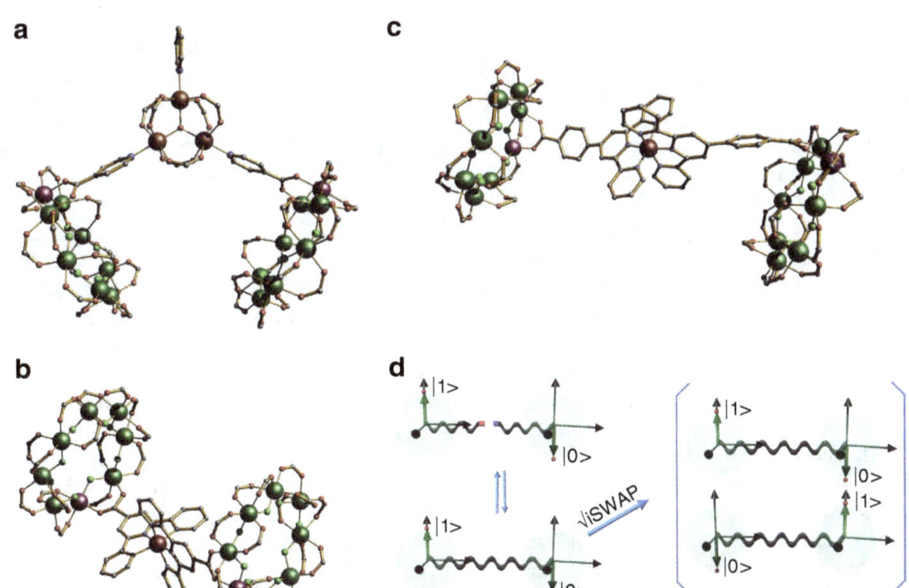

Figure 3 | Molecular structures of the redox-active two-qubit assemblies for proposed implementation of an $\sqrt{i}$SWAP gate. (**a**) [{Cr_7Ni-O_2C-py}→[$Ru^{III}_2Co^{II}O(^tBuCO_2)_6$(py)]←{$Cr_7Ni$-$O_2C$-py}] **7**. (**b**) the cation of [{Cr_7Ni-O_2C-terpy}→Co←{Cr_7Ni-O_2C-terpy}][ClO_4]$_2$ **8a**. (**c**) the cation of [{Cr_7Ni-O_2C-Ph-terpy}→Co←{Cr_7Ni-O_2C-Ph-terpy}][ClO_4]$_2$ **9a**. $^nPr_2NH_2^+$ cations and ClO_4^- anions are not shown (H atoms and tert-butyl groups are omitted for clarity). Colour codes as Fig. 2. (**d**) Schematic representation of the effect of the $\sqrt{i}$SWAP gate on a pair of qubits, initialized in the computational basis state |10⟩. The gate brings |10⟩ to the equal-weight superposition $(|10\rangle + i|01\rangle)/\sqrt{2}$. In the scheme proposed below, it operates as soon as the inter-qubit interaction is turned on (double arrow).

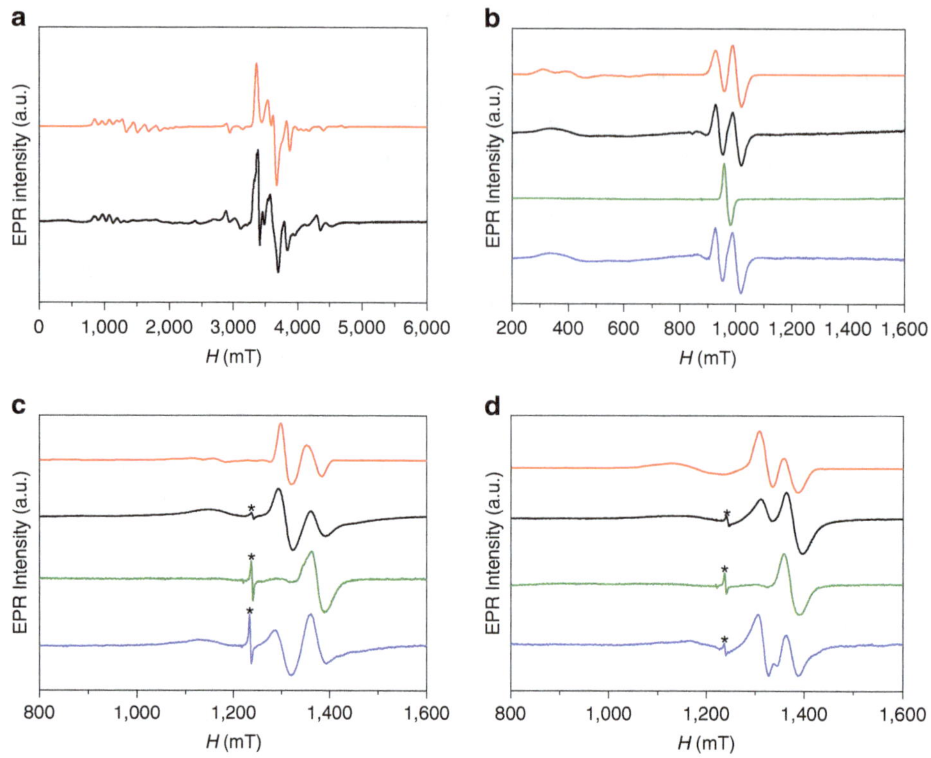

Figure 4 | EPR spectroscopy of QG assemblies. (**a**) Experimental powder W-band (94 GHz) EPR spectrum of **5** at 5K (black) and simulation (red) using Hamiltonian (1) and parameters in Table 1. (**b**) Experimental K-band (24 GHz) EPR spectra of **7** frozen solution (black), and simulation (red) using Hamiltonian (2) and parameters given in the text: experimental spectra after oxidation of **7** to **7b^{ox}** with $[FeCp_2](PF_6)$ (ca. 3 mM, 1:1) (green), and after reducing **7b^{ox}** with cobaltocene (3 mM, 1:1) (blue). (**c**,**d**) Experimental Q-band (34 GHz) EPR spectra of **8b** and **9b**, respectively, in frozen solution (black), and simulations (red) using Hamiltonian (2) and parameters given in the text; experimental spectra after oxidation to **8b^{ox}** and **9b^{ox}** with $AgBF_4$ (ca. 3 mM, 1:1) (green), and after reducing using with cobaltocene (3 mM, 1:1) (blue). The sharp peak marked * is a trace radical impurity.

anisotropic exchange interactions (J_{12} and J_{23}) between the $S_{eff} = ½$ of the Co(II) ion (S_2) and the two $S = ½$ of {Cr$_7$Ni-O$_2$C-terpy} (S_1) and {Cr$_7$Ni-O$_2$C-py} (S_3) rings, respectively (Fig. 4a).

$$\hat{H} = -2\left[\left(\hat{S}_1 \cdot \underline{J}_{12} \cdot \hat{S}_2\right) + \left(\hat{S}_2 \cdot \underline{J}_{23} \cdot \hat{S}_3\right)\right] + \mu_B\left(\hat{S}_1 \cdot \underline{g}_1 + \hat{S}_2 \cdot \underline{g}_2 + \hat{S}_3 \cdot \underline{g}_3\right) \cdot \mathbf{B} \qquad (1)$$

The multiplets centred at ca. 1, 1.5 and 3.5 T (Fig. 4a) mark the effective g values of the Co(II) ion as $g = 6.50$, 4.25 and ca. 2, while the Cr$_7$Ni rings have well defined $g = 1.78$, 1.78 and 1.74 (the unique value being for the orientation perpendicular to the ring)[30]. The large difference (Δg) between the $g = 6.5$ and 4.3 orientations of the Co(II) ion and those of the rings, and the weak exchange interactions ($\mu_B \Delta g B \gg |J|$) is useful as it allows us to treat the problem as an *ABX* spin system. The multiplet structure of the $g = 6.5$ and 4.3 features is then due to weak exchange with two different $S = 1/2$, giving doublets-of-doublets from which we can read the J_y and J_z components of J_{12} and J_{23}. The high field region is more complicated, due to the much smaller difference between the third Co(II) g value and those of the {Cr$_7$Ni} rings. The remaining parameters (the final components of J_{12}, J_{23} and g_2) were obtained by simulation (Fig. 4a). The parameters in Table 1 are labelled to a common reference frame, that is, taking into account the orthogonal orientation of the two {Cr$_7$Ni} rings.

7, **8b** and **9b**, contain redox-active linking nodes and their EPR spectra change with the oxidation state. Hence, EPR was performed on both the reduced and oxidized forms at 5K as frozen solutions, allowing cycling of the redox state (Fig. 4b–d). Spectra were measured on the as prepared samples, then solutions were warmed to room temperature and the oxidized forms **7^{ox}**, **8b^{ox}** and **9b^{ox}** generated *in situ* by addition of $[FeCp_2](PF_6)$ or

Table 1 | Spin-Hamiltonian parameters used for simulation of the W-band EPR spectra of 5.

	g_x	g_y	g_z		J_x (cm^{-1})	J_y (cm^{-1})	J_z (cm^{-1})
{Cr$_7$Ni-O$_2$C-terpy}	1.78	1.78	1.74	J_{12}	−0.14	0.34	0.17
{Cr$_7$Ni-O$_2$C-py}	1.74	1.78	1.78	J_{23}	−0.07	0.17	0.34
Co(II)	1.78	4.25	6.50				

$\underline{J}_{12}$, $\underline{J}_{23}$ and $\underline{g} = \{g_x, g_y, g_z\}$ are diagonal tensors.

$AgBF_4$ followed by freezing and measurement. To complete the cycle, the solutions were thawed again and **7**, **8b** and **9b** regenerated by reduction with cobaltocene. We have also measured spectra of the isolated nodes in the paramagnetic Co(II) form, viz. complex **6**, and $[Co(HO_2C\text{-terpy})_2](BF_4)_2$ **10** and $[Co(HO_2C\text{-Ph-terpy})_2](BF_4)_2$ **11** (Supplementary Figs 2 and 3). Like **5**, **6** also contains a ^4T high-spin Co(II) ion with an $S_{eff} = 1/2$ ground state, giving $g = 5.61$, 4.05, 2.77 (Supplementary Fig. 11), while **10** and **11** have low-spin Co(II) hence $S = 1/2$, giving $g = 2.047$, 2.076, 2.195 (Supplementary Fig. 12) and $g = 2.022$, 2.111 and 2.215 (Supplementary Fig. 13), respectively.

For the supramolecular structures **7**, **8b** and **9b**, the exchange coupling is nicely resolved in K- and Q-band (24 and 34 GHz) EPR spectra (Fig. 4b–d). In each case there is weak coupling with respect to the difference in Zeeman energies ($\mu_B \Delta g B \gg |J|$)[41]. This, and the equivalence of the two Cr$_7$Ni rings, gives AB_2 spin systems. The {Cr$_7$Ni} ring resonances (the '*B*' spins) are split into doublets (Fig. 4b–d), giving a direct measure of J (the exchange splittings of the central node resonances are not resolved). The spectra can be simulated with the simple Hamiltonian (2) using an isotropic exchange interaction between the central nodes (S_2) and the two {Cr_7Ni} rings (S_1 and S_3), even for **7**, which has an effective spin 1/2 centre at the node.

$$\hat{H} = -2J\left(\hat{S}_1 \cdot \hat{S}_2 + \hat{S}_2 \cdot \hat{S}_3\right) + \mu_B\left(\hat{S}_1 \cdot \underline{g}_1 + \hat{S}_2 \cdot \underline{g}_2 + \hat{S}_3 \cdot \underline{g}_3\right) \cdot \mathbf{B} \quad (2)$$

Simulations with the isotropic exchange interaction J as the only variable gives $J = -0.026$, -0.026 and $-0.024\,cm^{-1}$ for **7**, **8b** and **9b**, respectively; the g values were fixed to those measured for individual components. The EPR spectra of $\mathbf{7^{ox}}$, $\mathbf{8b^{ox}}$ and $\mathbf{9b^{ox}}$ are very simple, resembling isolated {Cr_7Ni} rings, because the oxidized forms of the central nodes are diamagnetic. Reduction of $\mathbf{7^{ox}}$, $\mathbf{8b^{ox}}$ and $\mathbf{9b^{ox}}$ back to **7**, **8b** and **9b** regenerates the original spectrum.

Implementing interesting quantum algorithms requires qubits with long phase memory times, such that they can be manipulated many times without errors. Individual {Cr_7Ni} heterometallic rings and simpler paramagnetic centres have shown long enough phase memory times (T_M)[21,22,24,25] to perform coherent electron spin manipulations before state degradation occurs. To check that this key property is preserved when they are incorporated in the supramolecular assemblies, we have performed pulsed EPR measurements at resonances corresponding to both the central nodes and the {Cr_7Ni} rings (Supplementary Figs 14–26). For the latter we find similar values for all compounds, and in both oxidation states where relevant (T_M = 683, 749, 767, 790, 984, 750 and 1,031 ns for **5**, **7**, $\mathbf{7^{ox}}$, **8b**, $\mathbf{8b^{ox}}$, **9b** and $\mathbf{9b^{ox}}$, respectively, measured at Q-band and 3K; Supplementary Table 4), demonstrating that the phase memory times are not strongly influenced by structural or magnetic differences in the linkage of the rings. In addition, values of T_M measured at resonances fields corresponding to the central node (in the paramagnetic state) are also similar for all compounds (500–700 ns at Q-band and 3K; Supplementary Tables 5–12). Hence, the phase memory time of the central node does not represent a limitation (see also the Discussion below) for the applicability of our schemes for implementing a universal set of quantum gates. These results are used in the simulations below, and are extremely promising, as these T_M times are sufficiently long for spin manipulation even when they are integrated in a supramolecular assembly.

CNOT gate with uniform magnetic pulses. In the following, we introduce two quantum computation schemes, based on either local or non-local control of the inter-qubit interaction employing the structures described and the parameters obtained by EPR spectroscopy. First, we show that compound **5** is suitable for a CNOT gate, using uniform magnetic pulses as the only manipulation tool[42]. As we are treating three interacting inequivalent doublets this produces 2^3 energy levels. We define the computational basis within the low-energy subspace where Co^{II} is frozen into its $S_z = -1/2$ state, which corresponds to the four lowest levels shown in red in Fig. 5a. The four levels correspond to arrangements of the spins of the two qubits, S_1 and S_3, having the relative orientations $|\downarrow\downarrow\rangle$, $|\downarrow\uparrow\rangle$, $|\uparrow\downarrow\rangle$ and $|\uparrow\uparrow\rangle$, respectively, which we label as $|00\rangle$, $|01\rangle$, $|10\rangle$ and $|11\rangle$ in Fig. 5a.

In a field of a few Teslas the eigenstates are factorized, that is, the eigenfunctions of the rings and the Co^{II} are not entangled. Hence it is possible to implement high-fidelity single-qubit rotations by EPR pulses resonant with low-energy gaps (see, for example, the shorter arrow in Fig. 5a). The combination of the inequivalent and anisotropic ring-Co exchange interaction and of the perpendicular arrangement of the two rings makes the two qubits significantly inequivalent. In particular, there is a sizeable gap (about 33 μeV in a field of 5 T) between the $|00\rangle \rightarrow |01\rangle$ and $|00\rangle \rightarrow |10\rangle$ transitions, which enables independent single-qubit rotations. We note that a distribution of the spin Hamiltonian parameters arising from a reasonable g or J strain (~1%, see ref. 43) would yield a broadening of the energy levels with s.d. ~5 μeV, thus keeping the two transitions distinguishable in realistic conditions, possibly with the help of narrow and/or composite pulses[44,45].

A controlled phase-shift (C_ϕ) gate is obtained by a pulse resonant with the transition corresponding to the longer arrow in Fig. 5a where the cobalt(II) ion is temporarily in its $S_z = +1/2$ state (blue energy levels), followed by a repetition of the same pulse that would bring the state back with an additional phase ϕ. The value of ϕ is controlled by the phase difference between the first and the second pulse. For a pulse with field along the y direction, the implementation of this gate is fast, owing to the

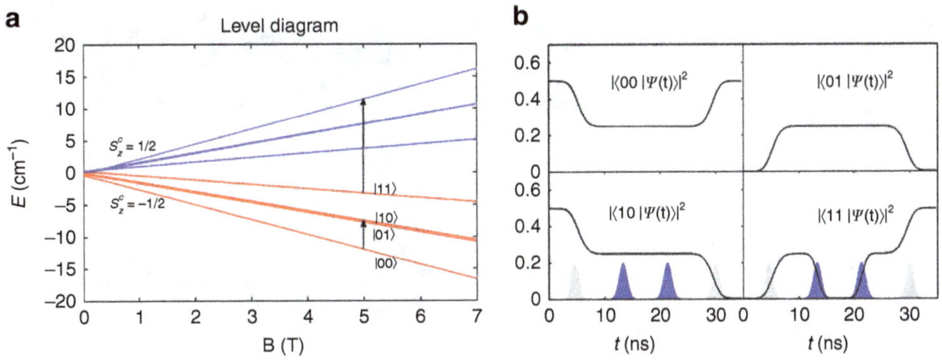

Figure 5 | A CNOT gate based on the structure of 5. (**a**) Field-dependence of the energy levels of **5** resulting from the Hamiltonian (1). The low-energy group of levels (red), where Co is frozen into its $S_z = -1/2$ state, defines the computational basis. The high-energy group of levels, where the Co spin is inverted, is exploited to perform two-qubit gates. (**b**) Simulation of the pulse sequence implementing CNOT as $R_y(\pi/2)C_Z\,R_y(-\pi/2)$, where $R_y(\alpha)$ is a rotation of the target qubit by an angle α around the y axis and C_Z is a controlled-Z gate. We illustrate the gate by starting at time $t = 0$ with a superposition state $|\psi\rangle = (|0,0\rangle + |1,0\rangle)/\sqrt{2}$, which transforms under a CNOT gate (with the left qubit acting as control) into $(|0,0\rangle + |1,1\rangle)/\sqrt{2}$. The latter state is actually obtained by the pulse sequence implementing $R_y(\pi/2)C_ZR_y(-\pi/2)$, with a fidelity of 99.7%. The envelope of the pulses implementing the two R_y rotations and the C_Z are outlined at the bottom. Note that performing the C_Z gate (two central pulses) requires temporarily leaving the computational subspace. The intensity of the oscillating field at the pulse maximum is 50 G, and we assume a static field of 5 T directed along z.

large value of g_y for Co^{II} ion. Moreover, the relatively large exchange interaction in equation (1) allows us to employ large oscillating fields (50 G), since the desired transition is spectroscopically well resolved from all the others. Consequently, the C_ϕ gate can be performed in only about 12 ns, with fidelities close to 99.99 % (see Computational Details section in Supplementary Methods).

The CNOT gate can then be obtained by the sequence of gates $R_y(\pi/2)\ C_Z\ R_y(-\pi/2)$, where $R_y(\alpha)$ is a rotation of the target Cr_7Ni qubit by an angle α around the y axis (a single-qubit operation) and where C_Z is the phase-shift gate described above with $\phi=\pi$. We have numerically solved the time-dependent Schrödinger equation for the Hamiltonian (1) in presence of this pulse sequence. Results are reported in Fig. 5b and show that this two-qubit gate can be obtained with very high fidelity in only 30 ns. As noted above, the two qubits are significantly inequivalent, even if the g tensor of the rings is nearly isotropic. This makes this complex well suited for the quantum simulation of antisymmetric Hamiltonians (Supplementary Figs 28 and 29 for an illustrative example).

$\sqrt{\text{iSWAP}}$ gate with local electric control. The redox properties of the central node in **7**, **8** and **9** can be exploited to perform the universal $\sqrt{\text{iSWAP}}$ gate on the $\{Cr_7Ni\}$ qubits, whose effect on the computational basis is given by:

$$|01> \rightarrow (|01>+i\,|10>)/\sqrt{2};$$
$$|10> \rightarrow (|10>+i\,|01>)/\sqrt{2};$$
$$|00> \rightarrow |00> \text{ and } |11> \rightarrow |11>.$$

When the central node is in the diamagnetic state ($[Ru^{III,IV}{}_2Co^{II}]$ **7** or Co^{III} **8–9**), the two qubits are decoupled and only single-qubit operations can be performed. Conversely, when the paramagnetic oxidation state, $[Ru^{III}{}_2Co^{II}]$ **7** or Co^{II} **8–9**, is present the system behaves as a trimer described by the Hamiltonian (2). Previously, Lehmann *et al.*[16] have proposed that two spins $S=1/2$ connected by a redox-active unit can be exploited for the implementation of the $\sqrt{\text{SWAP}}$ gate by switching the redox unit with a scanning tunnelling microscope (STM) tip at an appropriate potential. In this way one electron can be added or removed from the redox unit very quickly. It was demonstrated that particular sets of parameters of the trimer Hamiltonian lead to a pure $\sqrt{\text{SWAP}}$ evolution of the two qubits after specific time intervals. However, high fidelity for the gate is guaranteed only for fixed ratios between the qubit–qubit exchange (J_{qq}) and the qubit-redox unit exchange (J_{qr}). For the measured parameters of **7**, **8** and **9** ($J_{qr}\equiv J$ and $J_{qq}=0$ in equation (2)) the fidelity of the $\sqrt{\text{SWAP}}$ gate would be low.

Here we propose a more flexible scheme for the $\sqrt{\text{iSWAP}}$ gate based on our measured parameters, which works if the two qubits have the same Zeeman energy but one different to that of the switch in the 'on' state. This is the case here, where the g values of the qubits and switches are very different, $g_{1z}=g_{3z}=1.74$ for $\{Cr_7Ni\}$ rings and $g_{2z}=2.77$ (**7**), 2.195 (**8**) and 2.215 (**9**) for the central node. This difference in the g values means that in magnetic fields of a few T, the ring-central node exchange J is small compared with the difference between the Zeeman energies of the ring and the central node. Hence, the spin state of the central node is nearly frozen in the $S_z=-1/2$ state and has only tiny virtual fluctuations that lead to an effective interaction between the two $\{Cr_7Ni\}$ qubits given by

$$\hat{H}_{qq}=\Gamma\left(\hat{S}_{1x}\hat{S}_{3x}+\hat{S}_{1y}\hat{S}_{3y}\right)+\lambda\left(\hat{S}_{1z}+\hat{S}_{3z}\right) \qquad (3)$$

where the field is along z, S_1 and S_3 refer to the spins of the first and second Cr_7Ni qubits and

$$\Gamma=\frac{2J^2}{B\mu_B(g_{1z}-g_{2z})} \qquad \lambda=J+\frac{J^2}{B\mu_B(g_{1z}-g_{2z})}$$

The feasibility of the scheme relies only on the equivalence of the qubits and on the hierarchy of the interactions $[J \ll \mu_B B(g_{1z}-g_{2z})]$ (see Supplementary Methods for more details).

Therefore when the central node is paramagnetic, the state of the two qubits evolves according to (3), with negligible entanglement with the central node. For specific times, this evolution coincides with the $\sqrt{\text{iSWAP}}$ gate apart from single-qubit rotations along z due to the second term in (3).

This perturbative picture is confirmed by the results of detailed calculations for compound **9** using the full Hamiltonian (2) (Fig. 6). Starting from the $|10>$ logical state, we report in Fig. 6a the time evolution of the trimer wavefunction that would implement the $\sqrt{\text{iSWAP}}$ gate. In a magnetic field of 3 T, after 4 ns the wavefunction has equal-weight contributions from $|10>$ and $|01>$, which is the $\sqrt{\text{iSWAP}}$ gate, while after 8 ns

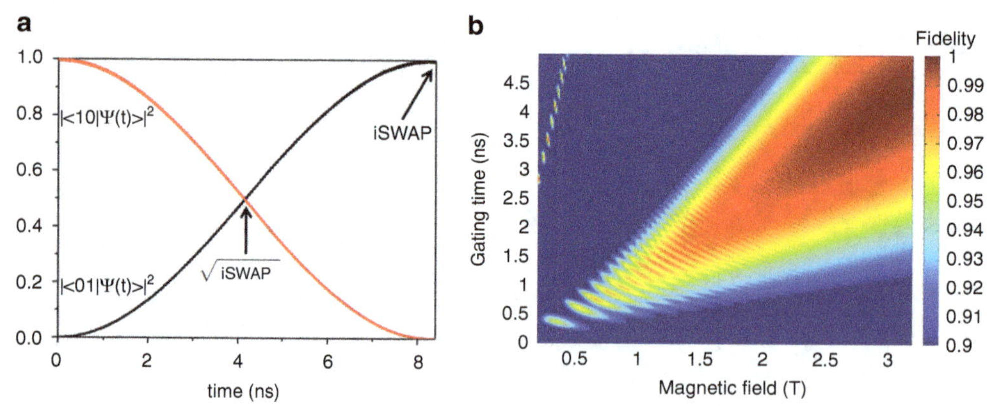

Figure 6 | Simulation of the $\sqrt{\text{iSWAP}}$ gate. (**a**) For **9**, taking $|10>\equiv|1/2,-1/2>\otimes|-1/2>_{Co}$ as the initial state and an applied field of 3 T, we calculate the time-dependence of the oscillation of the trimer wavefunction between $|10>$ and $|01>\equiv|-1/2,1/2>\otimes|-1/2>_{Co}$. Other components are negligible. (**b**) Calculated average fidelity $\bar{F}$ for **9** as a function of the magnetic field B and of the gating time, that is, the time the Co switch is in the on state. The fidelity is defined by $F=|\langle\psi_0|\psi(t_f)\rangle|$, where for a given starting logical state, $|\psi_0>$ is the final state after an ideal gate, whereas $|\psi(t_f)>$ is the actual final state. The average has been made over four random starting states. For each value of the field, the optimal gating time t_f is the one maximizing $\bar{F}$ ca. 4 ns for 3 T as shown in **a**. The oscillations corresponding to the fringes in the picture are associated with fluctuations of the Co spin state. As long as the frequency of these fluctuations is much larger than $1/t_f$ these fluctuations are negligible, that is, the perturbative description of equation (3) is valid.

the states of the two spins are exchanged, that is, we have the $|01>$ state. An extremely good fidelity F (larger than 0.99) for compound **9** is obtained for fields of the order of about 2.5 T or larger, after a suitable gating time t_f of the order of a few nanoseconds (Fig. 6b). For such fields the perturbative picture of equation (3) holds very well, and t_f is proportional to B (Fig. 6b), which is consistent with the form of the effective qubit–qubit coupling $\Gamma \propto 1/B$.

Analogous or better fidelities can be obtained with the parameters derived for compounds **7** and **8**. Indeed, larger values of $|g_{1z}-g_{2z}|$ (as in **7**) increase the validity of the perturbative picture (3), even for smaller values of B. Hence, we can exploit the modular strategy and optimize the performance of the gates by targeted chemical manipulations.

Discussion

Our initial simulations above did not include decoherence. To gain a deeper insight into the performance of the proposed quantum gates, we have performed further simulations that include the effect of both relaxation and pure dephasing, allowing for the finite and measured values of T_1 and T_M, respectively. For the values of T_1 measured by pulsed EPR (>10 μs in all compounds, see the Supplementary Tables 4–12), the effect of relaxation is found to be completely negligible.

Pure dephasing is accounted for by numerically solving the Lindblad master equation for the system density matrix ρ (refs 46,47):

$$\frac{d\rho}{dt} = -i[H,\rho] + \sum_k \frac{1}{2T_M^k}\left(2S_{k,+}S_{k,-}\rho S_{k,+}S_{k,-} - \rho S_{k,+}S_{k,-} - S_{k,+}S_{k,-}\rho\right) \text{ with } \hbar = 1$$

Here the commutator describes the coherent evolution induced by the full system Hamiltonian H, while the second term describes pure dephasing mechanisms. The subscript $k=1$, 2 and 3 labels the spins of the qubits and of the switch, while $S_{k,+} = |1_k\rangle\langle 0_k|$ and $S_{k,-} = |0_k\rangle\langle 1_k|$ are spin 1/2 raising and lowering operators.

Using the values of T_M of the Cr_7Ni qubits measured by pulsed-EPR spectroscopy (between 700 and 1,000 ns, see above) we still find very high fidelities: 99.3% for the implementation of the CNOT gate on compound **5** and 99.6% for the $\sqrt{\text{iSWAP}}$ gate on **9**. As expected, the fidelities are only marginally affected by decoherence because gating times are much shorter than T_M.

It is also worth discussing the role of the decoherence of the switch in the present schemes. As far as the CNOT gate scheme (compound **5**) is concerned, the contribution of the central Co switch to decoherence is similar to that of the rings, because the switch is temporarily excited to the $S_Z = +1/2$ state during the CNOT gates. Conversely, only virtual excitations of the switch are involved in the implementation of the $\sqrt{\text{iSWAP}}$ gate exploiting the redox-active linker. Hence, the fidelity remains very high even for dephasing time of the switch of the order of the gating time (some ns). This difference is shown by the two colour maps reported in Fig. 7. They show the dependence of the fidelity of a $\sqrt{\text{iSWAP}}$ (left) and of a CNOT gate (right) on the dephasing times T_M of the rings and of the switch. It is evident that the fidelity of the $\sqrt{\text{iSWAP}}$ is practically independent on T_M^{Co} whereas its dependence on T_M^{rings} is much more pronounced. Conversely, the implementation of the CNOT on compound **5** leads to a similar dependence of the fidelity on T_M^{Co} and T_M^{rings}.

The chemistry described above can be extended to make one-dimensional (1D) chains incorporating the two-Cr_7Ni supramolecules, linked by either single cobalt sites or oxo-centred metal triangles. Such 1D chains have already been made involving single Cr_7Ni units and [2]- and [3]-rotaxanes and the principles established, especially for 1D chains of rotaxanes[48], should work well. The key steps are to include two functional groups per Cr_7Ni ring, which has already been done for *iso*-nicotinate acid, and by functionalizing the termini of the central organic thread of rotaxanes. Inclusion of two different functionalities is more taxing but entirely feasible (see the detailed schemes in Supplementary Figs 30 and 31). Therefore the chemistry is potentially scalable.

While this challenging chemistry proceeds, it is possible to show that the schemes are theoretically scalable. The extension of a two-qubit QG to a multi-qubit register raises important issues concerning the propagation of errors. In such a register, we can identify two main sources of errors, whose effect increases with the number of qubits: pure dephasing and imperfect operation of the switch arising from a residual inter-qubit interaction still present in the 'off' state.

Errors induced by pure dephasing (decoherence) increase with the overall computational time. However, a finite chain of qubits with interposed switches would allow simultaneous (parallel) manipulation of non-overlapping parts of the register, which drastically reduces the computation time, and hence decoherence-induced errors, with respect to a serial implementation. For instance, a setup based on compound **5** can be manipulated in parallel for interesting classes of quantum simulation algorithms[17]. This requires inclusion of two distinct switches in the structure—Switch_A and Switch_B—with different responses to external stimuli, such that the 1D chain has as a repeat pattern -Qubit-Switch_A-Qubit-Switch_B-.

For compounds **7**, **8** and **9**, the scheme requires a local addressability of individual switches on the chain, which are

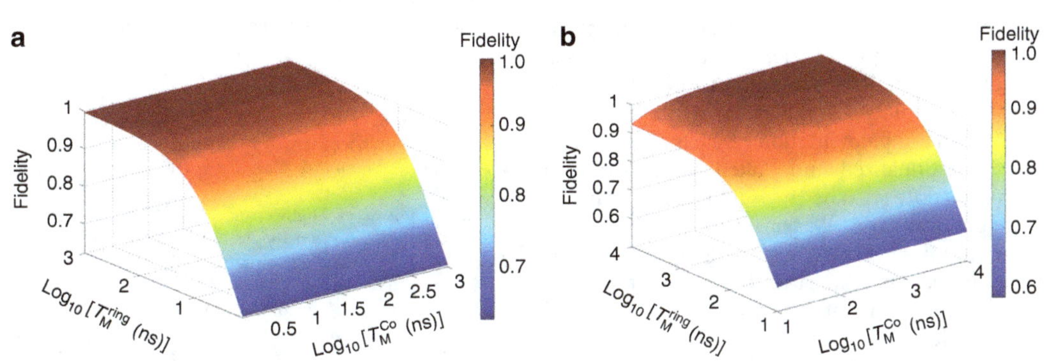

Figure 7 | Effect of decoherence on quantum gates. (**a**) Fidelity of the $\sqrt{\text{iSWAP}}$ as a function of the dephasing times of the rings and of the switch. Remarkably, the fidelity is nearly independent on the T_M^{Co}

separated by about 3 nm. For a first proof-of-principle experiment with a single assembly, an STM tip could be used to provide the best control. A parallel implementation of gates would be possible if different Co switches could be addressed individually at the same time. To achieve a scalable structure, a molecular chain of qubits might be layered onto a surface and addressed by means of a cross-bar architecture similar to that proposed in refs 16,49. It is worth noting that Cr_7Ni rings have already been deposited on surfaces, without significant modification of their magnetic properties[50].

We now examine the scaling of the errors with the number of qubits in the register. As a first step, we consider the effect of pure dephasing on a set of non-interacting qubits subject to Lindblad (Markovian) dynamics. It can be shown (see ref. 51 and Supplementary Information) that the decoherence error $\varepsilon = 1 - F^2$ on N qubits scales at most as $\varepsilon_N = N\varepsilon_1 = \frac{N}{2}(1 - e^{-t_f/2T_M}) \approx N\frac{t_f}{4T_M}$, valid for $t_f \ll T_M$ (where t_f is the optimal gating time). In the parallel quantum-simulator implementation considered in ref. 17, manipulating a chain of N qubits takes the same time as manipulating the shortest chain that contains the two distinct switches (qubit-Switch$_A$-qubit-Switch$_B$-qubit). Therefore t_f is limited by the value it assumes for a chain of three qubits, while in a serial scheme t_f increases linearly with N. By chemical engineering of Cr_7Ni qubits, a T_M of 15 μs has been obtained[25]; this should allow the implementation of around twenty 2-qubit gate operations on a chain consisting of 10 qubits, while keeping the error below 0.1.

Finally, we analyse the consequences of an imperfect operation of the switch and their effect on scalability. For an ideal $\sqrt{iSWAP}$ gate the inter-qubit exchange interaction would be completely turned off when the central node is in its diamagnetic state. Double electron–electron resonance measurements (Supplementary Fig. 27) reveal a very weak, residual interaction between the qubits. The resulting oscillations are on a time scale (ca. 0.2 μs) much longer than our gating times (a few ns) and could be corrected by means of refocusing techniques[29]. In the case of the CNOT scheme (compound **5**), the always 'on' exchange interaction between the rings and the Co ion yields a small residual qubit–qubit coupling (Supplementary Methods). As a consequence, an unwanted slow evolution of the qubits occurs on a timescale T_{UE} of the order of a few hundreds of ns. Although this is longer than the CNOT gate time (about 30 ns), larger values of T_{UE} would be needed for performing sequences of many gates. This can be obtained by modifying the molecule to decrease the Co-ring exchange interaction. For instance, T_{UE} can be increased by a factor of 50 by reducing the Co-ring exchange interaction by a factor of three, which could be achieved chemically by adding an extra phenyl group between the ring and the central node. A similar reduction of the residual qubit–qubit coupling can be obtained by increasing the static magnetic field.

In summary, we have described two different schemes for universal quantum information processing, based on either local or global control of the qubit–qubit interaction. We have demonstrated that the flexibility of molecular $\{Cr_7Ni\}$ qubits makes them suitable for the implementation of each of these two schemes, if properly functionalized and linked by means of a supra-molecular design strategy. The two-qubit units can be controlled either magnetically or electrically, and implement either the CNOT or $\sqrt{iSWAP}$ perfectly entangling gates. Our realistic simulations, based on experimental parameters and including decoherence, show that these gates can be performed with remarkably high fidelity. The modular strategy proposed here offers a degree of control in terms of the magnitude of the coupling between molecular spin qubits, the spatial orientation of the modules and the possibility to have a switchable interaction that represents a significant step forward with respect to previous achievements on the assembly of qubits.

Since future developments of quantum technology are unpredictable, we emphasize the importance of pursuing both these two parallel roads towards the actual realization of a quantum computer.

References

1. Feynman, R. P. Simulating physics with computers. *Int. J. Theor. Phys.* **21,** 467–488 (1982).
2. Lloyd, S. Universal quantum simulators. *Science* **273,** 1073–1078 (1996).
3. Nielsen, M. A. & Chuang, I. L. *Quantum Computation and Quantum Information* (Cambridge University Press, 2000).
4. Ladd, T. D. *et al.* Quantum computers. *Nature* **464,** 45–53 (2010).
5. Shor, P. W. Polynomial-time algorithms for prime factorization and discrete logarithms on a quantum computer. *SIAM J. Comput.* **26,** 1484–1509 (1997).
6. Grover, L. K. Quantum computers can search arbitrarily large databases by a single query. *Phys. Rev. Lett.* **79,** 4709–4712 (1997).
7. Burkard, G., Loss, D. & DiVincenzo, D. P. Coupled quantum dots as quantum gates. *Phys. Rev. B* **59,** 2070–2078 (1999).
8. Hanson, R. & Awschalom, D. D. Coherent manipulation of single spins in semiconductors. *Nature* **453,** 1043–1049 (2008).
9. Balasubramanian, G. *et al.* Ultralong spin coherence time in isotopically engineered diamond. *Nat. Mater.* **8,** 383–387 (2009).
10. Knowles, H. S., Kara, D. M. & Atatüre, M. Observing bulk diamond spin coherence in high-purity nanodiamonds. *Nat. Mater.* **13,** 21–25 (2014).
11. Georgescu, I. M., Ashhab, S. & Nori, F. Quantum simulation. *Rev. Mod. Phys.* **86,** 153 (2014).
12. Leuenberger, M. & Loss, D. Quantum computing in molecular magnets. *Nature* **410,** 789–793 (2001).
13. Meier, F., Levy, J. & Loss, D. Quantum computing with spin cluster qubits. *Phys. Rev. Lett.* **90,** 47901–47904 (2003).
14. Troiani, F. *et al.* Molecular engineering of antiferromagnetic rings for quantum computation. *Phys. Rev. Lett.* **94,** 207208 (2005).
15. Troiani, F., Affronte, M., Carretta, S., Santini, P. & Amoretti, G. Proposal for quantum gates in permanently coupled antiferromagnetic spin rings without need of local fields. *Phys. Rev. Lett.* **94,** 190501 (2005).
16. Lehmann, J., Gaita-Ariño, A., Coronado, E. & Loss, D. Spin qubits with electrically gated polyoxometalate molecules. *Nat. Nanotechnol.* **2,** 312–317 (2007).
17. Santini, P., Carretta, S., Troiani, F. & Amoretti, G. Molecular nanomagnets as quantum simulators. *Phys. Rev. Lett.* **107,** 230502 (2011).
18. Nakazawa, S. *et al.* A synthetic two-spin quantum bit: *g*-engineered exchange-coupled biradical designed for controlled-NOT gate operations. *Angew. Chem. Int. Ed.* **51,** 9860–9864 (2012).
19. Luis, F. *et al.* Molecular prototypes for spin-based CNOT and SWAP quantum gates. *Phys. Rev. Lett.* **107,** 117203 (2011).
20. Aromí, G., Aguilà, D., Gamez, P., Luis, F. & Roubeau, O. Design of magnetic coordination complexes for quantum computing. *Chem. Soc. Rev.* **41,** 537–546 (2012).
21. Warner, M. *et al.* Potential for spin-based information processing in a thin-film molecular semiconductor. *Nature* **503,** 504–509 (2013).
22. Graham, M. J. *et al.* Influence of electronic spin and spin-orbit coupling on decoherence in mononuclear transition metal complexes. *J. Am. Chem. Soc.* **136,** 7623–7626 (2014).
23. Aguilà, D. *et al.* Heterodimetallic [LnLn'] lanthanide complexes: toward a chemical design of two-qubit molecular spin quantum gates. *J. Am. Chem. Soc.* **136,** 14215–14222 (2014).
24. Bader, K. *et al.* Room temperature quantum coherence in a potential molecular qubit. *Nat. Commun.* **5,** 5304 (2014).
25. Wedge, C. J. *et al.* Chemical engineering of molecular qubits. *Phys. Rev. Lett.* **108,** 107204(1)–107204(5) (2012).
26. Timco, G. A. *et al.* Engineering the coupling between molecular spin qubits by coordination chemistry. *Nat. Nanotechnol.* **4,** 173–178 (2008).
27. Lehn, J.-M. *Supramolecular Chemistry: Concepts and Perspectives* (Wiley-VCH, 1995).
28. Ardavan, A. *et al.* Engineering coherent interactions in molecular nanomagnet dimers. *NPJ Quantum Inf.* **1,** 15012 (2015).
29. Jones, J. A. Quantum computing with NMR. *Prog. Nucl. Magn. Reson. Spectrosc.* **59,** 91 (2011).
30. McInnes, E. J. L., Timco, G. A., Whitehead, G. F. S. & Winpenny, R. E. P. Heterometallic rings as a playground for physics and supramolecular building blocks. *Angew. Chem., Int. Ed.* **54,** 14244–14269 (2015).
31. Constable, E. C. *et al.* Expanded ligands: bis(2,2':6',2"-terpyridine carboxylic acid)ruthenium(II) complexes as metallosupramolecular analogues of dicarboxylic acids. *Dalton Trans.* **38,** 4323–4332 (2007).

32. Hayami, S., Komatsu, Y., Shimizu, T., Kamihata, H. & Hoon Lee, Y. Spin-crossover in cobalt(II) compounds containing terpyridine and its derivatives. *Coord. Chem. Rev.* **255,** 1981–1990 (2011).
33. Lloret, F., Julve, M., Cano, J., Ruiz-García, R. & Pardo, E. Magnetic properties of six-coordinated high-spin cobalt(II) complexes: theoretical background and its application. *Inorg. Chim. Acta* **361,** 3432–3445 (2008).
34. Ohto, A., Sasaki, Y. & Ito, T. Mixed-metal trinuclear complexes containing two ruthenium(III) ions and a divalent metal ion, $[Ru_2M(\mu_3\text{-}O)(\mu\text{-}CH_3COO)_6(L)_3]$ (M = Mg, Mn, Co, Ni, Zn; L = H_2O, Pyridine). *Inorg. Chem.* **33,** 1245–1246 (1994).
35. Schubert, U. S., Hofmeier, H. & Newkome, G. R. *Modern Terpyridine Chemistry* (Wiley-VCH, 2006).
36. Cannon, R. D. & White, R. P. Chemical properties of triangular bridged metal complexes. *Prog. Inorg. Chem.* **36,** 195–298 (1988).
37. Murray, K. S. in *Spin-Crossover Materials: Properties and Applications* (ed. Halcrow, M. A.), Ch. 1, 1–54 (John Wiley and Sons, 2013).
38. Aribia, K. B., Moehl, T., Zakeeruddin, S. M. & Grätzel, M. Tridentate cobalt complexes as alternative redox couples for high-efficiency dye-sensitized solar cells. *Chem. Sci.* **4,** 454–459 (2013).
39. Connelly, N. G. & Geiger, W. E. Chemical redox agents for organometallic chemistry. *Chem. Rev.* **96,** 877–910 (1996).
40. Stoll, S. & Schweiger, A. EasySpin, a comprehensive software package for spectral simulation and analysis in EPR. *J. Magn. Reson.* **178,** 42–55 (2006).
41. Fernandez, A. *et al.* g-engineering in hybrid rotaxanes to create *AB* and AB_2 electron spin systems: EPR studies of weak interactions between dissimilar electron spin qubits. *Angew. Chem. Int. Ed.* **54,** 10858–10861 (2015).
42. Chiesa, A. *et al.* Molecular nanomagnets with switchable coupling for quantum simulation. *Sci. Rep.* **4,** 7423 (2014).
43. Park, K., Novotny, M. A., Dalal, N. S., Hill, S. & Rikvold, P. A. Effects of *D*-strain, *g*-strain, and dipolar interactions on EPR linewidths of the molecular magnets Fe_8 and Mn_{12}. *Phys. Rev. B* **65,** 014426 (2001).
44. Wesenberg, J. & Mølmer, K. Robust quantum gates and a bus architecture for quantum computing with rare-earth-ion-doped crystals. *Phys. Rev. A* **68,** 012320 (2003).
45. Cummins, H. K., Llewellyn, G. & Jones, J. A. Tackling systematic errors in quantum logic gates with composite rotations. *Phys. Rev. A* **67,** 042308 (2003).
46. Tempel, D. G. & Aspuru-Guzik, A. Relaxation and dephasing in open quantum systems time-dependent density functional theory: properties of exact functionals from an exactly-solvable model system. *Chem. Phys.* **391,** 130 (2011).
47. Breuer, H. P. & Petruccione, F. *The Theory of Open Quantum Systems* (Oxford University Press, 2002).
48. Whitehead, G. F. S. *et al.* Rings and threads as linkers in metal-organic frameworks and poly-rotaxanes. *Chem. Commun.* **49,** 7195–7197 (2013).
49. Green, J. E. *et al.* A 160-kilobit molecular electronic memory patterned at 10^{11} bits per square centimeter. *Nature* **445,** 414–417 (2007).
50. Corradini, V. *et al.* Magnetic anisotropy of Cr_7Ni spin clusters on surfaces. *Adv. Funct. Mat* **22,** 3706 (2012).
51. Jing, J. & Hu, X. Scaling of decoherence for a system of uncoupled spin qubits. *Sci. Rep.* **5,** 17013 (2015).

Acknowledgements

This work was supported by the EPSRC (UK), the European Commission (Marie Curie Intra-European Fellowship to J.F.-S. (622659) and A.F. (300402)). E.M.P. thanks the Panamanian agency SENACYT-IFARHU for funding. R.E.P.W. thanks the Royal Society for a Wolfson Merit Award. We also thank EPSRC (UK) for funding an X-ray diffractometer (grant number EP/K039547/1) and the National EPR Facility. We thank Diamond Light Source for access to synchrotron X-ray facilities. A.C., S.C. and P.S. acknowledge financial support from the FIRB Project No. RBFR12RPD1 of the Italian Ministry of Education and Research.

Author contributions

J.F.S., E.M.P. and A.F. synthesized the majority of the compounds discussed; S.A.M. synthesized compound **6**. G.T. was involved in design of the synthetic strategies. E.M.P. and I.J.V.-Y. carried out the X-ray single crystal diffraction studies. E.M.P., J.F.-S. and F.T. performed the EPR spectroscopic studies, and J.F.-S. carried out the electrochemical studies reported here. E.J.L.M. helped interpret the EPR spectra. A.C., S.C. and P.S. designed the schemes for quantum gates and A.C. carried out their detailed simulations. R.E.P.W. oversaw the project and wrote the paper with J.F.S., S.C. and P.S., with significant input from all other authors.

Additional information

Accession codes: The X-ray crystallographic coordinates for structures reported in this Article have been deposited at the Cambridge Crystallographic Data Centre (CCDC), under deposition number CCDC 1029608–1029613 and 1415380–1415383. These data can be obtained free of charge from The Cambridge Crystallographic Data Centre via www.ccdc.cam.ac.uk/data_request/cif.

Competing financial interests: The authors declare no competing financial interests.

13

Direct hydrodeoxygenation of raw woody biomass into liquid alkanes

Qineng Xia[1,*], Zongjia Chen[1,*], Yi Shao[1], Xueqing Gong[1], Haifeng Wang[1], Xiaohui Liu[1], Stewart F. Parker[2], Xue Han[3,4], Sihai Yang[4] & Yanqin Wang[1]

Being the only sustainable source of organic carbon, biomass is playing an ever-increasingly important role in our energy landscape. The conversion of renewable lignocellulosic biomass into liquid fuels is particularly attractive but extremely challenging due to the inertness and complexity of lignocellulose. Here we describe the direct hydrodeoxygenation of raw woods into liquid alkanes with mass yields up to 28.1 wt% over a multifunctional $Pt/NbOPO_4$ catalyst in cyclohexane. The superior performance of this catalyst allows simultaneous conversion of cellulose, hemicellulose and, more significantly, lignin fractions in the wood sawdust into hexane, pentane and alkylcyclohexanes, respectively. Investigation on the molecular mechanism reveals that a synergistic effect between Pt, NbO_x species and acidic sites promotes this highly efficient hydrodeoxygenation of bulk lignocellulose. No chemical pretreatment of the raw woody biomass or separation is required for this one-pot process, which opens a general and energy-efficient route for converting raw lignocellulose into valuable alkanes.

[1] Key Laboratory for Advanced Materials, Research Institute of Industrial Catalysis, East China University of Science and Technology, Shanghai 200237, China. [2] ISIS Facility, STFC Rutherford Appleton Laboratory, Chilton, Oxfordshire OX11 0QX, UK. [3] School of Chemistry, University of Nottingham, Nottingham, NG7 2RD, UK. [4] School of Chemistry, University of Manchester, Manchester M13 9PL, UK. * These authors contributed equally to this work. Correspondence and requests for materials should be addressed to H.F.W. (email: hfwang@ecust.edu.cn) or to S.Y. (email: Sihai.Yang@manchester.ac.uk) or to Y.Q.W. (email: wangyanqin@ecust.edu.cn).

Fossil fuel consumption is projected to increase significantly in coming decades, with potentially catastrophic consequences for the environment[1]. Sustainable alternatives to crude oil are imperatively needed to bridge gaps in the supply of chemical fuels and feedstocks[2]. For the production of liquid fuels in particular, the replacement of oil-based routes by renewable biomass has received increasing attention[3–6]. Lignocellulose, as the main component of woody biomass, is composed of cellulose (40–50 wt%; a linear polymer of D-glucopyranose connected by β-1,4-glycosidic linkages), hemicellulose (16–33 wt%; a heteropolymer consisting of many different sugar monomers) and lignin (15–30 wt%; a heavily cross-linked, complex polymer with coumaryl, coniferyl and sinapyl alcohols as monomers)[7,8]. Owing to the complexity of lignocellulosic biomass and its notorious resistance to chemical transformation, energy-efficient and cost-effective production of liquid fuels from lignocellulose remains a mammoth challenge. So far, two strategies have been reported to address this challenge: (i) separation of lignocellulose into isolated sugars and lignin followed by biological or chemical (hydrolysis) processing[9–11]; (ii) thermochemical treatment of lignocellulose to produce upgradeable intermediates, such as bio-oils by pyrolysis or syngas by gasification, coupled with subsequent catalytic upgrading[12,13]. Thermochemical processes offer the total conversion of lignocellulose, but are often non-selective and intractable, and the resultant bio-oils or syngas need to be upgraded for further utilisation. Although hydrolysis-based approaches offer selective production of liquid fuels, they are generally multistep and thus very energy-intensive[14]. Moreover, the lignin by-products generated from the hydrolysis of lignocellulose are usually burned as a low-value fuel[15]. Powerful drivers therefore exist to develop alternative efficient and selective strategies to directly convert raw lignocellulose into liquid fuels.

Direct conversion of raw lignocellulose into alcohols and phenols was realised recently in exceptional cases[16–17]. However, successful direct production of hydrocarbon fuels (that is, total removal of oxygen) is mostly achieved so far from separated components of lignin or cellulose[18–20]. For example, the conversion of lignin into alkanes and methanol has been reported through a two-step process (chemical pretreatment and sequential hydrogenolysis and hydrogenation)[18]. Recently, the one-pot conversion of cellulosic feedstock into liquid alkanes in biphasic reaction systems (organic + water) were also reported over Ir–ReO_x/SiO_2–H–ZSM-5 or tungstosilicic acid–Ru/C catalysts[19,20]. The industrial Shell/GTI hydropyrolysis and Virent Energy System's approaches are also known to directly convert sugars or raw biomass into liquid fuels[21,22]. The former is based on a catalytic thermal–chemical technique, which reacts at very high temperature (350–540 °C) (ref. 21). Virent's approach converts water-soluble oxygenated hydrocarbons into C_{4+} hydrocarbons, alcohols and/or ketones in aqueous phase or vapour phase. This is achieved by aqueous phase reforming of water-soluble oxygenates, followed by condensation and deoxygenation[22]. More recently, a three-catalyst system was reported to convert raw biomass into liquid alkanes and other mono-functional hydrocarbons over layered $LiTaMoO_6$ combined with Ru/C in aqueous phosphoric acid medium[23]. Gaining in-depth understanding on the reaction mechanism is of fundamental importance for the development of improved catalytic systems.

Here we report that, by using a multifunctional Pt/$NbOPO_4$ catalyst, raw woody biomass can be directly converted into liquid alkanes in high yields in a single-phase medium (cyclohexane) with cellulose, hemicellulose and lignin fractions in solid woods being converted into hexane, pentane and alkylcyclohexanes, respectively (Fig. 1), representing direct conversion of raw lignocellulose into liquid alkanes under mild conditions over a single catalyst. Importantly, no chemical pretreatment (for example, hydrolysis and separation) to the raw wood is required for this process, and thus, tremendous energy savings can be potentially gained in comparison with the existing thermochemical- and hydrolysis-based approaches. More significantly, the pathway for this novel catalytic reaction was systematically investigated by control experiments, and the molecular mechanism for the rate-determining step in this conversion studied by *in situ* inelastic neutron scattering and computational studies. These complementary investigations reveal that the NbO_x species promotes the crucial C–O bond cleavage over hydrodeoxygenation of tetrahydrofuran (THF) and phenol (model units of cellulose and lignin, respectively) to hydrocarbons under mild reaction conditions.

Results

Direct hydrodeoxygenation of raw woody biomass. To verify the applicability of this one-pot approach, seven different types of wood sawdusts (<75 μm), including both softwoods and hardwoods, were employed as feedstocks for direct hydrodeoxygenation over the Pt/$NbOPO_4$ catalyst in a cyclohexane medium (Table 1). The reactions were conducted at 190 °C and 5 MPa H_2 for 20 h and over 20 wt% total mass yield of liquid alkanes was achieved for all woods, among which birch wood gave the highest mass yield of 28.1 wt%. Considering that the theoretical mass yield of alkanes from raw woody biomass is limited to ~50 wt% as the removed oxygen accounts for almost half of the mass loss, the yields obtained here are excellent. In addition to C_1–C_6 alkane products, surprisingly appreciable amounts of alkylcyclohexanes (for example, propylcyclohexane and ethylcyclohexane) were also detected (Supplementary Fig. 1), indicating that not only the cellulose and hemicellulose but also the lignin fraction in sawdusts were converted into alkanes. Obviously, the source/texture of lignocellulose had a significant influence on both mass and carbon yields of the alkane products. In general, higher yields of hexanes and pentanes were achieved from softwoods: the carbon yields of hexanes and pentanes on the basis of cellulose and hemicellulose fractions reached 72.8 and 69.3% on average, respectively. These yields are surprisingly high and even comparable to those using isolated cellulose as feedstock[19,20]. Indeed, pure cellulose was tested as model material for carbohydrate fractions in raw woody biomass to confirm the performance of the catalyst. A total of 71.5% yield of hexanes and 8.7% yield of pentanes (by C–C cleavage) were achieved from cellulose conversion with excellent stability (Supplementary Table 1 and Supplementary Fig. 2). On the other hand, the yield of alkylcyclohexanes produced from hardwoods is much higher than that from softwoods, with an average carbon yield of 34.0% from hardwood (here only monomer alkylcyclohexanes were determined). It is worth noting that this yield is very high because there is a large proportion of C–C linkages in lignin structure (30–34% for hardwoods and 43–51% for softwoods on average)[15], which are hardly cleaved under such mild reaction conditions, thus resulting in a maximum theoretical carbon yield of monomer alkylcyclohexanes at 44–49% from hardwoods and 24–32% from softwoods (Supplementary Note 1). This result indicates that the catalyst has excellent performance for the direct hydrogenolysis of C–O–C linkages of lignin and total hydrodeoxygenation of the resultant lignin monomers. To further confirm this, diphenyl ether and phenol, which possess aromatic ether and hydroxyl functionalities, respectively, were tested as model compounds over Pt/$NbOPO_4$. A total of 99.9% yield of cyclohexane was achieved from both substrates, demonstrating the efficient cleavage of ether bond in lignin by this catalyst (Supplementary Table 1).

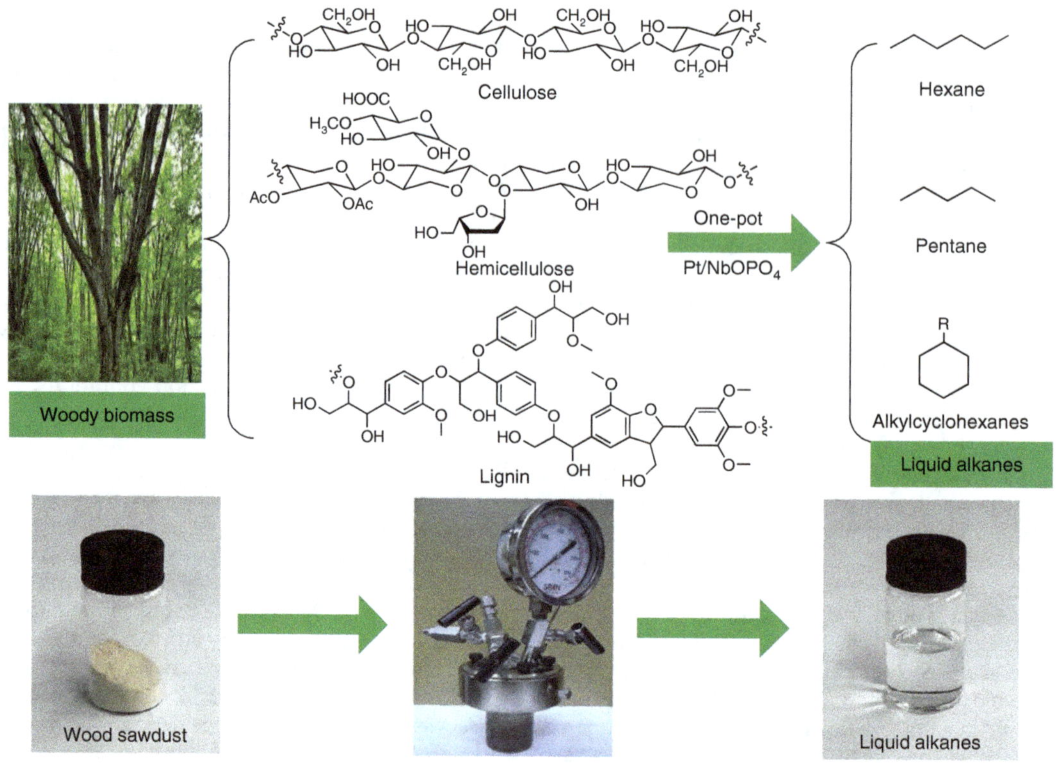

Figure 1 | Schematic representation of direct hydrodeoxygenation of raw woody biomass into liquid alkanes. Raw woody biomass can be directly converted into liquid alkanes over Pt/$NbOPO_4$ catalyst in cyclohexane medium, with cellulose, hemicellulose and lignin fractions in solid woods being converted into hexanes, pentanes and alkylcyclohexanes, respectively.

Table 1 | Summary of direct hydrodeoxygenation of various woody biomass over Pt/$NbOPO_4$.*

Woody biomass	Softwood				Hardwood		
	Pine	White pine	Larch	Fir	Camphor	Birch	Poplar
Lignocellulose content							
Hemicellulose (wt%)	9.3	10.4	13.1	10.3	24.8	25.7	28.2
Cellulose (wt%)	47.2	52.3	49.4	43.5	45.1	48.7	45.3
Lignin (wt%)	32.6	29.5	28.3	33.9	22.6	20.4	22.8
Mass yield/carbon yield (wt%/mol%)†							
Pentanes‡	4.1/80.8	3.6/63.4	4.8/67.2	3.7/65.9	7.3/54.0	10.2/73.1	7.1/46.2
Hexanes§	18.7/74.8	19.3/69.4	20.4/77.7	16.0/69.2	15.2/63.4	13.1/50.7	9.7/40.3
Alkylcyclohexanes‖	2.1/9.6	2.2/11.2	2.6/13.7	2.3/10.1	5.1/33.7	4.8/35.1	5.1/33.3
Total liquid alkanes (wt%)	24.9	25.1	27.8	22.0	27.6	28.1	21.9
Others (wt%)¶	1.7	1.7	2.0	1.4	1.9	2.3	1.6
Residue (wt%)	31.5	27.9	27.1	33.4	17.4	14.8	16.7

*The reactions were conducted at 190 °C and 5 MPa H_2 for 20 h. Feedstock (0.20 g), Pt/$NbOPO_4$ (0.20 g) and cyclohexane (6.46 g) were put into a 50 ml stainless-steel autoclave.
†Mass yields were calculated by the equation: mass yield of pentanes (hexanes, alkylcyclohexanes) = [mass of pentanes (hexanes, alkylcyclohexanes)]/[mass of feedstock input]. Carbon yields were calculated by the equation: carbon yield of pentanes (hexanes, alkylcyclohexanes) = [mass of carbon in pentanes (Hexanes, alkylcyclohexanes)]/[mass of carbon in hemicellulose (cellulose, lignin)].
‡Pentanes include *n*-pentane and iso-pentane.
§Hexanes include *n*-hexane and iso-hexane.
‖Only monomer alkylcyclohexanes were determined here, including methylcyclohexane, ethylcyclohexane, isopropylcyclohexane and propylcyclohexane.
¶Others were mainly CO_2 and C_1–C_4 alkanes.

Clarification of the unique activity of Pt/$NbOPO_4$. A variety of catalysts with different combinations of support and metal (that is, Pt/$NbOPO_4$, Pt/H-ZSM-5, Pt-ReO_x/SiO_2, Pt-ReO_x/C and Pd, Ru, Rh loaded $NbOPO_4$) were tested with birch sawdust as feedstock for this reaction to clarify the unique activity of Pt/$NbOPO_4$ (Table 2). With Pt supported on H-ZSM-5, which posseses similar acidity to $NbOPO_4$ but does not contain transition metal oxide (Supplementary Fig. 3), only 8.7 wt% yields of liquid alkanes was obtained (versus 28.1 wt% for Pt/$NbOPO_4$). This suggests that the NbO_x species of $NbOPO_4$ support has a significant promotion effect in this reaction. Such promotion effect on C–O cleavage was investigated recently in transition metal oxides of NbO_x and ReO_x (refs 24–26). To provide more insight, two other ReO_x support with reduced acidity, ReO_x/SiO_2 and ReO_x/C, were tested for comparison, and alkane yields of 11.4 and 9.8 wt%, respectively, were obtained with a small amount of mono-functional hydrocarbons detected (for example, tetrahydropyran). This result suggests that the promotion effect should be accompanied by sufficient acidity to achieve efficient C–O cleavage in this reaction. The $NbOPO_4$ support fulfills these two requirements (that is, surface NbO_x species and sufficient acidity), and thus posesses the best catalytic activity for this

Table 2 | The results of direct hydrodeoxygenation of birch wood over different catalysts.*

Entry	Catalyst	Mass yield of liquid alkanes (wt%)				
		Pentanes†	Hexanes‡	Alkylcyclohexanes§	Total	Residue
1	$Pt/NbOPO_4$	10.2	13.1	4.8	28.1	14.8
2	Pt/H-ZSM-5	4.9	1.3	2.5	8.7	35.8
3	$Pt\text{-}ReO_x/SiO_2$‖	5.2	3.6	2.6	11.4	26.9
4	$Pt\text{-}ReO_x/C$‖	6.4	1.5	1.9	9.8	ND
5	$Pd/NbOPO_4$	8.1	6.2	3.5	17.8	29.4
6	$Ru/NbOPO_4$	7.3	8.0	3.9	19.2	36.3
7	$Rh/NbOPO_4$	8.2	5.0	4.0	17.2	31.5

*The reactions were conducted at 190 °C and 5 MPa H_2 for 20 h. Feedstock (0.2 g), catalyst (0.2 g), and cyclohexane (6.46 g) were put into a 50 ml stainless-steel autoclave. All metal loading was 5 wt%.
†Pentanes include *n*-pentane and iso-pentane.
‡Hexanes include *n*-hexane and iso-hexane.
§Only monomer alkylcyclohexanes were determined, including methylcyclohexane, ethylcyclohexane, isopropylcyclohexane and propylcyclohexane.
‖The loading amount of Re was 1 by the molar ratio of Re to Pt.

Table 3 | Summary of direct hydrodeoxygenation of birch sawdust and cellulose under various reaction conditions over $Pt/NbOPO_4$.*

Entry	Feedstock	Carbon yield of liquid alkanes (%)		
		Hexanes†	Pentanes‡	alkylcyclohexanes
1§	Birch sawdust	72.4	52.3	34.6
2‖	Birch sawdust	69.1	49.7	35.9
3	Cellulose	71.5	8.7	ND¶
4#	Cellulose	40.3	4.4	ND
5**	Celloluse	65.2	8.2	ND
6††	Cellulose	41.3	4.6	ND
7††,‡‡	Celloluse	37.5	4	ND

*Unless otherwise specified, the reactions were conducted at 190 °C and 5 MPa H_2 for 20 h. Feedstock (0.2 g), catalyst (0.2 g), and cyclohexane (6.46 g) were put into a 50 ml stainless-steel autoclave.
†Hexanes include *n*-hexane and iso-hexane.
‡Pentanes include *n*-pentane and iso-pentane.
§Tridecane was employed as the reaction solvent.
‖0.4 g of birch sawdust, 0.4 g of catalyst, and 6.46 g of cyclohexane were put into a 50 ml stainless-steel autoclave.
¶"ND" is the abbreviation of "not determined".
#Ashes (minerals) obtained by calcination of 1 g birch sawdust at 500 °C for 3 h were added into the reaction system and reaction for 8 h.
**The cellulose was unmilled and used directly.
††The reaction time was 8 h.
‡‡Data shown are for the fourth run of the stability test.

reaction. On the metal side, Pd, Ru and Rh were tested by loading them onto the $NbOPO_4$ support, and moderate yields (17.2–19.2 wt%) of liquid alkanes were achieved from the hydrodeoxygenation of birch wood. The reason that Pt gave the highest yield among the studied metals is due to its supreme activity for H_2 activation and hydrogenation (Supplementary Table 2). From these control experiments, we rationalise the superior performance of $Pt/NbOPO_4$ by a synergistic effect between Pt, NbO_x species and the acidic sites on the support (including both Brönsted acid sites on PO_4 and Lewis acid sites on NbO_x, Supplementary Fig. 4).

Study of the activity and stability of the catalyst. To investigate the applicability and recyclability of the catalyst, $Pt/NbOPO_4$ was tested under various reaction conditions and the results are summerised in Table 3. As the products of this conversion were only alkanes, they act as an additional solvent to drive the sequential reactions without the need of solvent separation or recycling after each run. Cyclohexane was used as a solvent to facilitate the analysis because it does not overlap with any product. Alternatively, other alkane (for example, tridecane) can be used as a solvent for this reaction to afford similar results (Table 3, entry 1). Moreover, the reaction with double solid loadings was attempted, and there was no obvious decline of the alkane yields (Table 3, entry 2), indicating that this process is capable of dealing with higher solid loadings. Mineral poisoning is widely known to reduce the catalyst activity in biomass conversion. To probe more insight, extra ashes obtained by calcination of 1 g birch sawdust at 500 °C for 3 h were added into the system to test the cellulose conversion. The result suggests that a fivefold amount of ash has no influence on the catalytic activity of this catalyst (Table 3, entries 4 and 6). Moreover, the stability of the catalyst at lower alkane yields was tested by shortening the reaction time to 8 h. Small decreases on the yields of hexanes and pentanes were observed after four successive runs (hexanes: from 41.3 to 37.5%; pentanes: from 4.6 to 4.0%; Table 3, entries 6 and 7, Supplementary Fig. 2). Characterisations of the catalyst before and after the reaction showed small reductions on the BET surface area and the Pt dispersion (Supplementary Note 2), consistent with the observed small decreases on the yields of hexanes and pentanes. On the whole, the catalyst showed good and consistent catalytic performance in repeated runs in this one-pot process. This could be due to two reasons: first, the reaction was carried out under mild conditions (190 °C), which retards the significant aggregation of Pt particles (Supplementary Fig. 5); second, the use of non-aqueous single-phase medium (cyclohexane) hinders leaching and structural change of the catalyst (Supplementary Fig. 6). Indeed, the ICP analysis of the reaction solution suggested that the concentration of Pt, P or Nb was all below the detection limit, confirming the absence of catalyst leaching during the reaction.

Studies of the representative reaction pathway. In attempting to reveal the reaction pathway, model compounds were used to simplify the original reaction system for easier detection of intermediates and products. Diphenyl ether and phenol were chosen as the model compounds of the lignin fraction to investigate the cleavage of ether bond (as mentioned above), and cellobiose was used to represent the carbohydrate fraction to elucidate the conversion of cellulose and hemicellulose to alkanes. To better monitor the possible intermediates, the reaction was carried out at a lower temperature of 170 °C, and water was added into the reaction mixture after reaction to extract water-soluable intermediates. After 1 h reaction, small amounts of glucose (**2**), sorbitol (**3**), sorbitan (**4**) and isosorbide (**5**) were detected by HPLC (Supplementary Fig. 7), and a mixture of 1-dehydroxyl-glucose (**6**), 1,6 anhydro glucose (**7**), 2-hydroxymethyl-tetrahydropyran (**8**), 5-methyl-THF-2-methanol (**9**) as well as many other undefined intermediates in the aqueous phase were detected by GC–mass spectrometry (MS) (Supplementary Fig. 8). As no water existed at the beginning of this reaction, this result

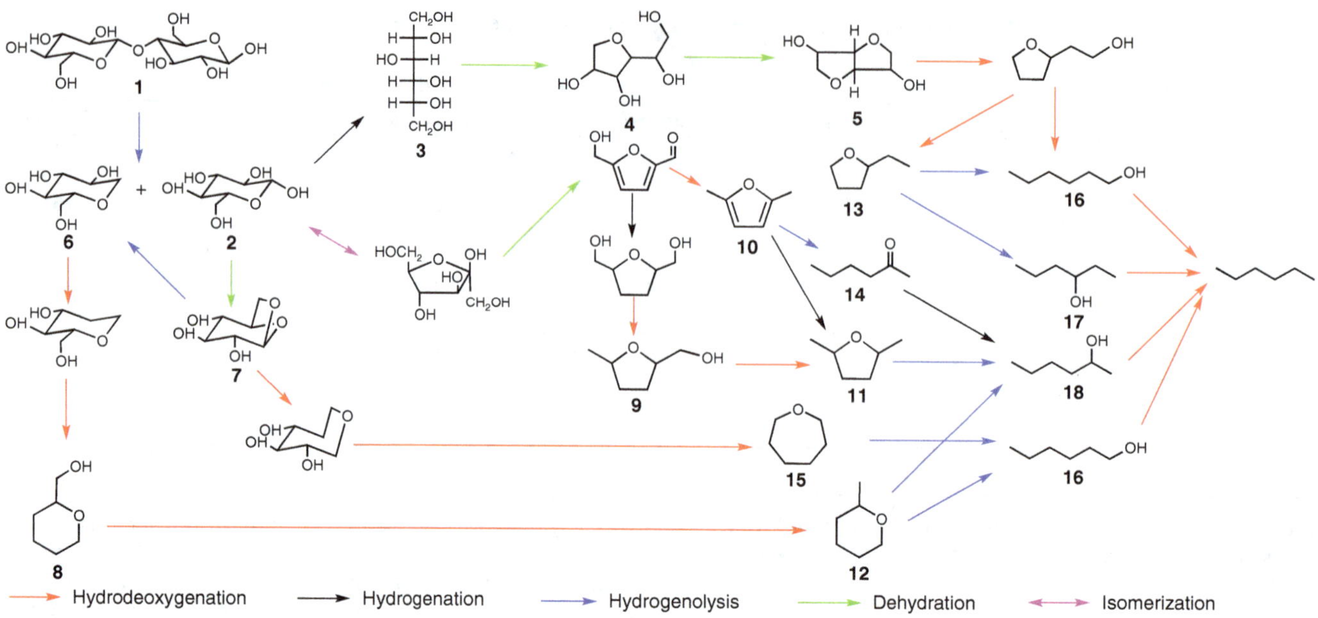

Figure 2 | Representative reaction pathways for the direct hydrodeoxygenation of cellobiose. According to all the intermediates detected, it is possible to deduce the main reaction pathways of the direct conversion of cellobiose to hexane. The reaction occurred by the direct hydrogenolysis of the β-1,4 linkage to D-glucose (**2**) and 1-deoxy-D-glucose (**6**) first, and then **6** was converted to hexane by sequential hydrogenolysis via 2-hydroxymethyl-tetrahydropyran (**8**), 2-methyltetrahydropyran (**12**) and hexanols (**16**) and (**18**), whereas the conversion of **2** has three main reaction pathways: (i) hydrogenated to sorbitol (**3**) and then dehydrated to sorbitan (**4**) and isosorbide (**5**), followed by sequential hydrogenolysis via 2-ethyl-THF (**13**) and hexanols (**16**) and (**17**). (ii) Isomerised to fructose and then dehydrated to 5-hydroxymethylfurfural followed by hydrogenation and sequential hydrogenolysis via 2,5-dimethylfuran (**10**), 2-hexanone (**14**), 5-methyl-THF-2-methanol (**9**), 2,5-dimethyl-THF (**11**) and 2-hexanol (**18**). (iii) Dehydrated to 1,6-anhydroglucose (**7**) followed by sequential hydrogenolysis via oxepane (**15**) and *n*-hexanol (**16**).

indicates that the β-1,4 linkage in cellobiose was cleaved by direct hydrogenolysis rather than hydrolysis, as a result of the excellent performance of Pt/$NbOPO_4$ in hydrogenolysis. When further reacted for 6 h, appreciable amounts of hexane, 2,5-dimethylfuran (**10**), 2,5-dimethyl-THF (**11**), 2-methyltetrahydropyran (**12**), 2-ethyl-THF (**13**), hexanone (**14**), oxepane (**15**) and hexanols (**16–18**) were observed in the organic layer (Supplementary Fig. 9). In the aqueous phase, isosorbide (**5**) and 2-hydroxymethyl-tetrahydropyran (**8**) were the main residuals (Supplementary Fig. 10), indicating that **5** and **8** are the two major intermediates in cellulose conversion. When further reacted for 24 h, no residual was detected in the aqueous phase, and a large amount of **11–13** and **16–18** were observed in the organic phase, suggesting that the ring-opening of THF derivatives and the subsequent hydrodeoxygenation are the rate-determining steps (Supplementary Fig. 11). Therefore, the direct conversion of cellobiose to hexane occurs via the hydrogenolysis of the β-1,4 linkage to **2** and **6** first, which then undergo a combination of hydrogenolysis, dehydration, hydrogenation and isomerisation reactions. The main reaction pathways of cellobiose conversion were proposed in Fig. 2, and was further confirmed by the isolated reactions of the four main intermediates (Supplementary Fig. 12).

Inelastic neutron scattering studies. As discussed above and previously reported[24], the ring-opening and the sequential hydrodeoxygenation of THF derivatives (for example, 5-hydroxymethylfurfural) are problematic and often represent the major barrier for the conversion of bio-derived furans into alkanes. Indeed, it is the rate-determing step for this conversion. Direct visualisation of the interaction between adsorbed THF (as a model compund for various THF derivatives generated from this reaction; the use of THF instead of THF-derivitives gives a clearer interpretation of the results) and the catalyst surface is crucial to understand the molecular details of binding, ring-opening and hydrodeoxygenation of THF into alkanes (butane in this case). INS is a powerful neutron spectroscopy technique to investigate the dynamics of hydrogenous compounds by exploiting the high-neutron scattering cross-section of hydrogen (82.02 barns)[27]. In addition, INS is not subjected to any optical selection rules, and the calculation of INS spectra from DFT calculations is straightforward (Supplementary Note 3). Here, we have successfully combined *in situ* INS and DFT to investigate the vibrational properties of the THF–Pt/Nb_2O_5 system to reveal the mechanism of the challenging hydrodeoxygenation of THF. $NbOPO_4$ has a large amount of surface P–OH groups, the vibrational peaks of which will overlap with signals of adsorbed THF. Therefore, we here used Pt/Nb_2O_5 instead for a clearer interpretation of the experimental observation. It is worth noting that Pt/Nb_2O_5 has a similar catalytic reactivity to Pt/$NbOPO_4$ for this reaction under the same conditions, as evidenced by the direct comparison of the yield and selectivity data for THF conversion in Supplementary Table 3. To the best of our knowledge, this is the first example of using INS/DFT to study the mechanism of catalytic biomass conversion.

The INS spectrum of the bare catalyst gives a clean background with no prominent features (details on the discussion of background spectra are given in Supplementary Note 4 and Supplementary Figs 13 and 14). In comparison, the INS spectrum of the catalyst on THF adsorption at 130 °C shows a significant increase in total intensity, demonstrating the binding of THF to the catalyst surface (Fig. 3a). Comparison of the difference spectrum before and after THF adsorption on the catalyst (that is, signals for adsorbed THF) and that of the solid THF shows a few changes (Fig. 3d). Peaks at low energy (below 200 cm^{-1}), assigned to the translational and rotational modes of THF, shift to lower energy with a continuum profile, suggesting that the

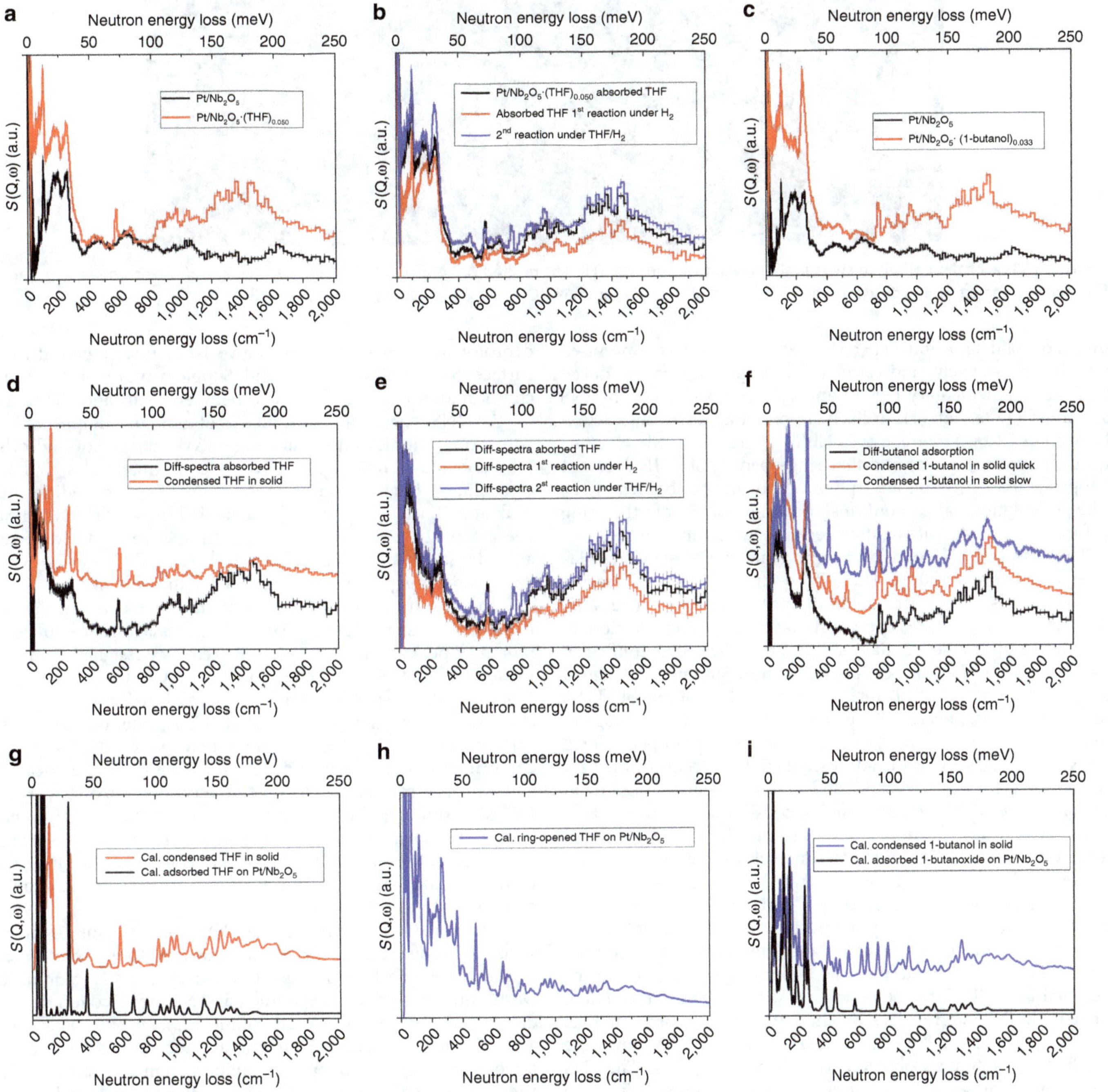

Figure 3 | Inelastic neutron scattering spectra for Pt/Nb_2O_5 during the catalytic hydrodeoxygenation of THF. INS spectrum of the reduced catalyst was used throughout for calculations of the difference spectra. No abscissa scale factor was used throughout this report for INS calculations. $NbOPO_4$ has a large amount of surface -OH groups, which have no specific activity in THF conversion, but will generate additional INS peaks overlapping with signals from substrates. Pt/Nb_2O_5 has a similar catalytic reactivity as $Pt/NbOPO_4$ for the hydrodeoxygenation of THF and was therefore used in INS study instead. Comparison of the experimental INS spectra for bare catalyst and the THF (**a**) and 1-butanol (**c**) adsorbed catalyst. Comparison of the difference plots for experimental INS spectra of bare and THF (**b**) and 1-butanol (**f**) adsorbed catalyst, and the experimental INS spectra of condensed THF (**d**) and 1-butanol (**f**) in the solid state. (**b**) Comparison of the experimental INS spectra for the two hydrodeoxygenation reactions of adsorbed THF on the catalyst, and difference plots (with reduced catalyst as background) are shown in (**e**). The first reaction was carried out at 130 °C in pure H_2 flow for 10 min and the second reaction was carried out at 130 °C in THF/H_2 flow for 5 h, that is fresh THF vapour was fed into the reaction cell continuously with H_2 as carrier gas during the reaction. (**g**) Comparison of the calculated INS spectra for condensed THF in solid and adsorbed THF on the catalyst. (**h**) View of the calculated INS spectrum for ring-opened THF on the catalyst. (**i**) Comparison of the calculated INS spectra for condensed 1-butanol in solid and adsorbed 1-butanoxide on the catalyst.

adsorbed THF molecules are disordered over the catalyst surface, and have restricted translational motion owing to the strong binding to the catalyst. The very strong peak at 251 cm^{-1}, assigned to the torsional mode of the C2–C3 bond of THF ring, reduced significantly in intensity, indicating the loss of this motion on adsorption. The peak at 300 cm^{-1}, likely due to the combination of the lattice mode at ~60 cm^{-1} and the strong mode at 251 cm^{-1}, concurrently reduced in intensity. In addition, the ring deformation mode at 587 cm^{-1} in solid THF shifts to 571 cm^{-1} when adsorbed on the catalyst. A structural

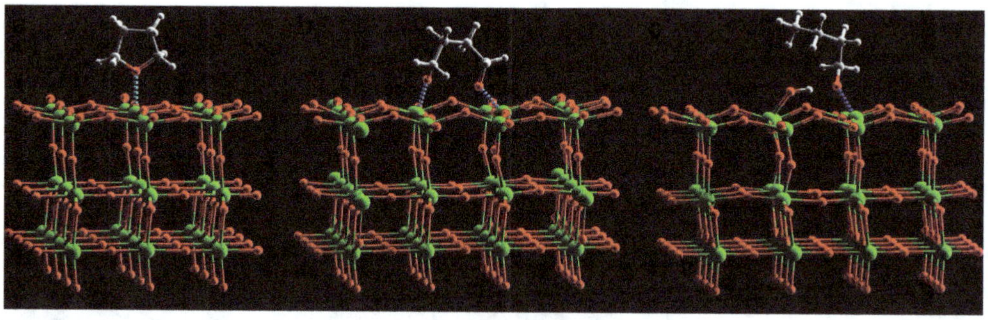

Figure 4 | View of the optimised structural models. (**a**) adsorbed THF, (**b**) reacted (ring-opened) THF and (**c**) 1-butanol after adsorption to generate 1-butanoxide and surface hydroxyl on the (001) plane of the catalyst Pt/Nb_2O_5 (Nb: green; O: red; C; grey; H: white).

model of solid THF and THF adsorbed on Nb_2O_5 were optimised by DFT, respectively, and calculated INS spectra were produced (Fig. 3g; Supplementary Figs 15 and 16). Comparison of the INS spectra suggests that THF is likely adsorbed intact via interaction between its O(δ −) centre to open Nb(δ +) site (O · · · Nb = 2.33 Å) on the surface (Fig. 4a) and that ring-opening of THF does not happen immediately on adsorption and/or in the absence of H_2. The calculation also confirms the downshift of the ring deformation mode but slightly overestimates its magnitude.

The adsorbed THF underwent a first catalytic conversion in H_2 flow for 10 min at 130 °C. The INS spectrum of the first reacted catalyst shows a large decrease in intensity (Fig. 3b,e), suggesting that the adsorbed THF underwent fast catalytic conversion to butane, which was sequentially swept out of the cell as confirmed by MS. In particular, the peaks at 1,244 and 1,308 cm^{-1} (assigned to $-CH_2-$ twisting and internal ring deformation of THF, respectively) disappeared completely, confirming the cleavage of the THF ring. An optimised structural model of ring-opened THF on Nb_2O_5 suggests that the adsorbed THF interacts with the very strong Lewis acid sites (Nb^{5+}), and the ring opens via binding to two adjacent Nb^{5+} centres simultaneously (O · · · Nb = 1.98 Å) (Figs 3h and 4b). Comparison of the calculated INS spectrum for ring-opened THF and experimental difference spectrum, however, does not conclusively suggest the presence of this intermediate bound on the catalyst. This could be due to two possible reasons: (i) the ring-opened intermediate is highly active and was hydrogenated instantly, and thus cannot be captured effectively; (ii) the amount of this intermediate is too low to be detected as no THF feedstock was provided during the reaction. To enrich the intermediate on the catalyst, a second catalytic reaction was conducted in THF/H_2 flow for 5 h, and the production of butane observed continuously by MS. The INS spectrum of the second reacted catalyst indeed shows an increase in intensity even compared with that of THF-adsorbed catalyst, confirming the presence of additional substrates on the catalyst surface (Fig. 3b). The corresponding difference spectra confirm the presence of adsorbed THF, and more importantly, four new peaks at 245, 477, 744 and 805 cm^{-1} were observed (Fig. 3e), indicating the presence of additional hydrogenous species bound on the catalyst.

To reveal the identity of this intermediate/s, the adsorption of 1-butanol on the catalyst was studied because 1-butanoxide bound to Nb^{5+} is predicted to be relatively stable (Fig. 3i). The INS spectrum of 1-butanol adsorbed on the catalyst shows a large increase in intensity, and comparison of the difference spectrum (that is, signals for adsorbed 1-butanol) and that of solid 1-butanol shows significant changes (Fig. 3c,f). The most important one is the disappearance of the peak at 839 cm^{-1} (assigned to the out-of-plane C–O–H bending mode, detailed discussions are given in Supplementary Note 4 and Supplementary Fig. 17), clearly suggesting 1-butanol underwent deprotonation on adsorption to give 1-butanoxide bound to the surface Nb^{5+} sites (Fig. 4c and Supplementary Fig. 18). The conformation-dependent bands between 300 and 600 cm^{-1} in the INS spectrum of solid 1-butanol all disappeared on adsorption, indicating that the alkyl chain of adsorbed 1-butanoxide is not in the all-*trans* conformation as found in crystalline 1-butanol (Supplementary Note 4). In addition, the peak at 257 cm^{-1} (assigned to the methyl torsion) shifts to lower energy at 243 cm^{-1} on binding to the catalyst. Moreover, an additional peak at 1,265 cm^{-1} (assigned to C4 chain deformation) is present for bound 1-butanoxide on Nb^{5+}.

Comparison of the INS spectra of adsorbed 1-butanoxide and the second reacted catalyst gives a clear message: three out of the four new peaks at 245, 744, and 805 cm^{-1} (assigned as methyl torsion, $-CH_2CH_2-$ rocking and $-CH_2CH_2CH_3$ rocking of 1-butanoxide, respectively) of the second reacted catalyst are consistent with the presence of 1-butanoxide bound on the surface (Fig. 3e,f). The remaining new peak at 477 cm^{-1} is conformation-dependent and consistent with the presence of a *gauche* conformer of C4 chain of bound 1-butanoxide[28]. A final INS spectrum for the reactivated catalyst shows no prominent feature (Supplementary Fig. 19), confirming the absence of formation of residual hydrocarbonaceous species and thus demonstrating the high efficiency of catalyst regeneration in cycling experiments.

The calculated and experimental INS spectra for solid THF and 1-butanol show excellent agreement; however, those for the guest-bound-catalysts exhibit a number of discrepancies (Fig. 3d,i). It is worth noting that the experimental data were collected on a disordered system (poorly crystalline mesoporous metal oxide and disordered substrates), wheras the calculations assume fully periodic structures. A poorly crystalline system generates multiple sites with different binding energies, inducing the broadening of INS peaks. The presence of different conformers of the butane chain on the catalyst also induces discrepancy as the calculation was done with the *gauche* conformer with the lowest energy only and the low energy modes are very sensitive to the chain conformation. Indeed, the INS spectra for disordered and crystalline 1-butanol have suggested that the crystallinity of the system can induce significant changes to both peak intensity and positions (Supplementary Fig. 17). Moreover, the calculation used a model with "flat" surface, in which there is no interaction with the surface, other than through the THF/1-butanol oxygen atom. On the real, rough surface, for example, with steps (Supplementary Fig. 5), the hydrogen atoms on C1 and C4 of THF (or C1 and C2 of 1-butanol) will also be enabled to interact with the surface, further enhancing the binding and broadening of the INS bands.

In this study, we concentrate primarily on the experimental observation, which has confirmed that (i) adsorbed THF molecules on the catalyst have, in principle, an intact structure

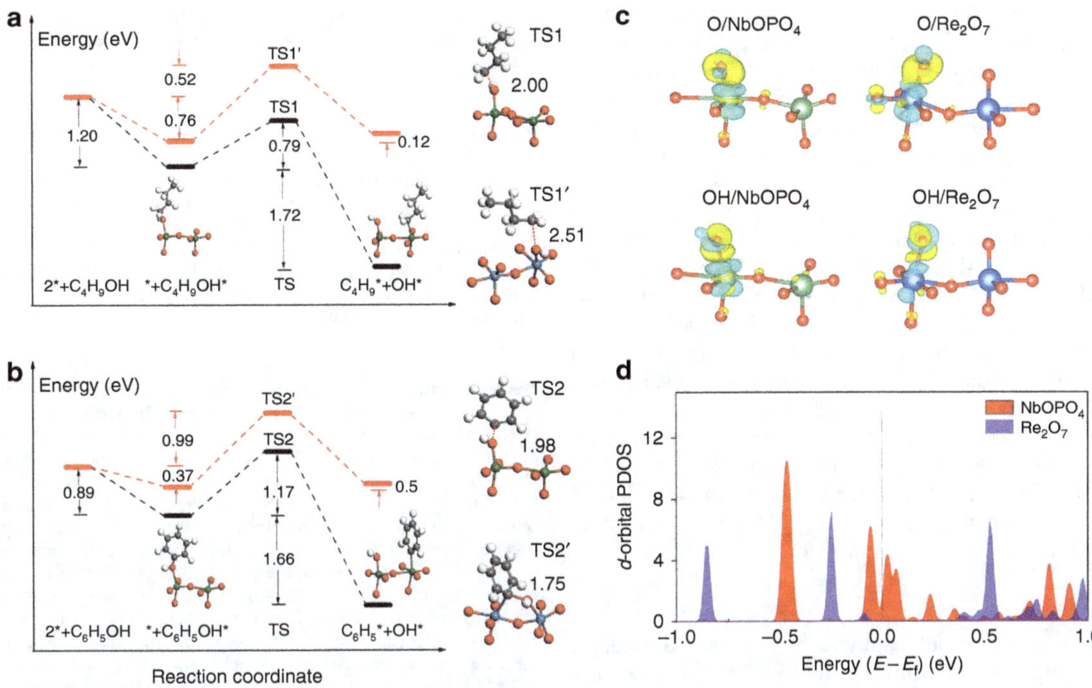

Figure 5 | Calculated results of computational studies. (**a**,**b**) Calculated energy profiles of C-O bond cleavage of C_4H_9OH and C_6H_5OH, respectively. Black and red lines indicate $NbOPO_4(100)$ and $Re_2O_7(010)$, respectively. The structures of the initial and final states on $NbOPO_4(100)$ are shown, whereas the transition states involving C-O bond cleavage of C_4H_9OH and C_6H_5OH are depicted in TS1, TS1',TS2 and TS2', with the elongated C-O bond lengths shown. Emerald balls represent Nb atoms, dark blue for Re, white for H, grey for C and red for O. (**c**) The isosurfaces of charge density difference for O and OH adsorption on $NbOPO_4(100)$ and $Re_2O_7(010)$ surfaces. For clarity, only the local active sites are shown. The regions depicted in yellow indicate charge accumulation and light blue for charge depletion. Emerald balls represent Nb atoms, dark blue for Re, white for H and red for O. (**d**) The *d*-orbital projected density of states for the surface Nb_{5c} and Re_{5c} atoms, demonstrating their relative energies, in which the energy is aligned to the Fermi level (E_F).

with reduced motion (esp. for the C2–C3 torsion and ring deformation modes) owing to the strong binding to the Nb^{5+} sites; (ii) ring-opening of adsorbed THF (cleavage of C–O bond) occurs rapidly in the presence of H_2 as shown by the loss of the internal THF ring deformation mode; (iii) 1-butanoxide bound to the surface Nb^{5+} site is a relatively stable reaction intermediate, consistent with calculations (see below). Therefore, this catalytic conversion of THF follows adsorption, binding, ring-opening, partial hydrogenation and complete hydrodeoxygenation, and the surface Nb^{5+} sites played an important role in this reaction, particularly for the binding and activation of THF substrates. It is worth noting that here the role of Pt is believed to dissociate H_2 and provide [H], and such role is not exclusive in the rate-determining C–O–C bond cleavage step, considering that similar performance can be obtained by substituting Pt with Pd loaded on $NbOPO_4$ for the hydrodeoxygenation of cellulose into alkanes (Supplementary Table 3). However, other possible roles of Pt in the whole reaction such as strengthening acidity through interface interactions[29] may not be fully ruled out, which is beyond the core point of this work and will be studied further in future works.

Computational studies of the catalytic origin of $NbOPO_4$. To probe more insight, first-principle calculations were conducted to examine the crucial C–O bond breaking process of phenol and 1-butanol, which are the model compound of lignin and important intermediate in THF conversion, respectively. Notably, the widely used ReO_x catalyst in biomass conversion was also tested for comparison. The well-ordered flat $NbOPO_4(100)$ and $Re_2O_7(010)$ surfaces, which have exposed five-coordinated Nb_{5c} and Re_{5c} centres as the main binding sites, were chosen as the substrates (Supplementary Methods, Supplementary Figs 20,21 and Supplementary Table 4).

For 1-butanol conversion on $NbOPO_4$, the calculation indicates that 1-butanol efficiently adsorbs on Nb_{5c} with a corresponding adsorption energy of -1.20 eV, which is evidently stronger than that on Re_2O_7 (-0.76 eV), indicating that $NbOPO_4(100)$ possesses a stronger binding ability. Subsequently, with the aid of the surface Nb_{5c}, the C–O bond can break with each OH and butyl occupying a Nb_{5c} site as the product; this process is strongly exothermic by 1.72 eV and gives a barrier of only 0.79 eV (Fig. 5a), implying its feasibility in both thermodynamics and kinetics (Supplementary Note 5 and Supplementary Table 5). In contrast, on Re_2O_7, 1-butanol dissociation is an endothermic process and has to overcome a larger barrier of 1.28 eV (Supplementary Note 5). By comparing the energy profiles (Fig. 5a), it is conclusive that $NbOPO_4$ demonstrates an inherently better performance for 1-butanol deoxygenation than Re_2O_7. Likewise, with respect to C–O bond cleavage of phenol, we performed the same calculation, which yields a similar conclusion that $NbOPO_4$ catalyses phenol dissociation more efficiently with a lower barrier than Re_2O_7 (Fig. 5b and Supplementary Note 5). Therefore, the high catalytic activity of $NbOPO_4$ can be ascribed to better adsorption capability of surface Nb_{5c} and a lower activation barrier in comparison to Re_2O_7. Correlation between the C–O bond dissociation barriers and their corresponding adsorption energies (Supplementary Fig. 22) shows that stronger the bond strength of M_{5c}–O (M = Nb, Re) leads to easier the C–O bond cleavage, which is in line with the principle of Brønsted–Evans–Polanyi relationship[30–32]. In other words, the outstanding binding ability of $NbOPO_4$ is one determining factor in the efficient deoxygenation of cellulose and lignin. We also calculated and compared the adsorption energies of $NbOPO_4$ with two other typical catalysts (Re_2O_7, ZrO_2) toward various intermediate species, such as OH, O, THF, butyl and butoxy (Supplementary

Table 6). Indeed, $NbOPO_4$ demonstrates the strongest binding ability seamlessly, whereas ZrO_2, with the weakest adsorption, is rarely applied for cellulose conversion to alkane in practice.

We are now in a position to elucidate the inherent mechanism of $NbOPO_4$ exhibiting such an exceptional binding ability. An electronic structure analysis was made to understand the bond properties of M_{5c}–O(OH) and the energy level of surface M_{5c}. The isosurface of charge density difference shows electron accumulation between O (or OH) and Nb_{5c} (or Re_{5c}) (Fig. 5c), indicating a typical covalent bond character. Therefore, the binding strength with O(OH) is mainly decided by the size and energy level of the *d* orbital of Nb_{5c} and Re_{5c} cations. The projected density of states on the *d*-orbitals of surface Nb_{5c} and Re_{5c} indicates that the *d* band of Nb_{5c} near the Fermi level (E_F) is more delocalised comparing to Re_{5c}, and more importantly, the energy level of the highest occupied *d* bands were evidently higher for Nb_{5c} with some metallic character across E_F (Fig. 5d). According to the frontier orbital theory, from these *d* bands features one can rationalise such a strong binding ability of $NbOPO_4$ in facilitating deoxygenation reaction.

Discussion

We have presented a one-pot catalytic process for the direct production of liquid alkanes from a wide variety of raw woody biomass over Pt/$NbOPO_4$ catalyst with excellent mass and carbon yields in a cyclohexane medium. This one-pot approach avoids the separation of raw biomass into isolated components and the use of an alkane solvent further simplifies downstream separation, as the alkane products can be used as solvents for the next run. The exceptional activity of the Pt/$NbOPO_4$ catalyst enabled direct conversion of raw woody biomass into liquid alkanes under mild conditions (190 °C) over a single multifunctional catalyst. The superior efficiency of this catalyst for direct hydrodeoxygenation of lignocellulose is found to originate from the synergistic effect between Pt, NbO_x species and acidic sites. This brand new one-pot route requires no chemical pretreatment or separation of the raw woody biomass and thus tremendous energy savings can be potentially gained in comparison to the existing thermochemical and hydrolysis-based approaches for production of liquid fuels and chemical feedstocks from lignocellulose.

Methods

Catalyst preparation. $NbOPO_4$ used here was synthesised by a hydrothermal method at pH = 2 according to literature[33]. M/$NbOPO_4$ (M = Pt, Pd, Rh, Ru) catalysts were prepared by incipient wetness impregnation of $NbOPO_4$ with aqueous solutions of $Pt(NO_3)_2$, $Pd(NO_3)_2$, $Rh(NO_3)_3$ and $RuCl_3$, respectively. After impregnation, the catalysts were dried at 100 °C for 12 h, followed by calcination in air at 500 °C for 3 h. Pt/H-ZSM-5 was prepared by the same procedure. Pt-ReO_x/SiO_2 and Pt-ReO_x/C were prepared by sequential incipient wetness impregnation. Pt was loaded onto the support first, dried and calcined before the process was repeated for Re. The loading of Pt, Pd, Rh, Ru was 5 wt% in all cases, for Pt-ReO_x/SiO_2 and Pt-ReO_x/C, the loading amount of Re was 1 by the molar ratio of Re to Pt. The dispersion of Pt/$NbOPO_4$ catalysts with different Pt loading amount was presented in Supplementary Fig. 23.

Reaction system and product analysis. The direct hydrodeoxygenation of wood sawdusts was conducted in a 50 ml Teflon-lined stainless-steel autoclave. In a typical run, feedstock (0.20 g), catalyst (0.20 g) and cyclohexane (6.46 g) were put into the reactor, which was then sealed, purged three times with H_2 and charged to an initial pressure of 5.0 MPa with H_2. The reactor was then slowly heated to 190 °C under vigorous stirring and held at this temperature for 20 h. After the reaction finished, the reactor was quenched in an ice/water bath. The gas phase was carefully collected in a gas bag and analysed by GC equipped with a packed column, a methaniser (for CO_2 detection) and a flame ionisation detector (FID). The products in the liquid phase were qualitatively analysed by GC–MS and quantitatively analysed by GC–flame ionisation detector equipped with a HP-5 column. The yields of liquid alkanes were determined by adding dodecane as an internal standard after reaction. Mass yields were calculated by the equation: mass yield of pentanes (hexanes, alkylcyclohexanes) = [mass of pentanes (hexanes, alkylcyclohexanes)]/[mass of feedstock input]. Carbon yields were calculated by the equation: carbon yield of pentanes (hexanes, alkylcyclohexanes) = [mass of carbon in pentanes (hexanes, alkylcyclohexanes)]/[mass of carbon in hemicellulose (cellulose, lignin)].

Reaction pathways study. The direct hydrodeoxygenation of cellobiose was carried out in a similar way to that of wood sawdusts. After the reaction was quenched in an ice/water bath, 6.5 g of water was added into the autoclave under vigorous stirring for a few minutes to extract the hydrophilic intermediates and to dissolve cellobiose for total analysis. The aqueous phase was analyzed by HPLC (Agilent 1200 series) equipped with a Shodex SUGAR SC-1011 column and a differential refractive index detector and by GC–MS to monitor the intermediates and unreacted cellobiose. The organic phase was analysed by GC–MS to observe the lipophilic intermediates.

Neutron scattering experiments. INS spectra were recorded on the TOSCA spectrometer at the ISIS Facility at the STFC Rutherford Appleton Laboratory (UK). TOSCA is an indirect geometry crystal analyser instrument that provides a wide dynamic range (16–4,000 cm^{-1}) with resolution optimised in the 50–2,000 cm^{-1} range[34]. In this region TOSCA has a resolution of 1.25% of the energy transfer. The Pt/Nb_2O_5 (34.7 g) catalyst was loaded into an *in situ* catalysis cell with a copper vacuum seal and connected to a gas handling system. The sample was heated at 300 °C (5 °C/min ramping) under He for 3 h to remove any remaining trace water before the experiment. The sample was cooled to room temperature and a weight loss of 0.1 g was noted, assigned to loss of adsorbed water. The samples were cooled to < 15 K during data collection by a closed cycle refrigerator cryostat. The procedure of the *in situ* catalysis experiment with the INS measurements is summarised in Supplementary Fig. 24. INS spectra for condensed THF (2.31 g) and 1-butanol (2.36 g) in the solid state were measured in a flat-plate sample container below 12 K. INS spectra for condensed THF and 1-butanol were used to (i) calculate the amount of adsorbed THF and 1-butanol onto the catalyst in each case by the integration of the peak areas; (ii) identify and compare the vibrational modes for the adsorbed and the free molecules.

Estimation of the amount of adsorbed THF on the catalyst: on the basis of the relative intensities of the INS peak at ~600 cm^{-1} and the known mass (2.31 g) of THF in the condensed sample, there is 0.42 g of adsorbed THF present on the catalyst in the neutron beam.

Estimation of the amount of adsorbed 1-butanol on the catalyst: on the basis of the relative intensities of the INS peak at ~740 cm^{-1} and the known mass (2.36 g) of 1-butanol in the condensed sample, there is 0.30 g of adsorbed 1-butanol present on the catalyst in the neutron beam.

Adsorption of THF was carried out by flowing THF vapour (~200 mbar) in He (1.1 bar, 0.2 l min^{-1}; this flow condition was used throughout the study) over the catalyst at 130 °C for 3 h. The cell was then flushed briefly with pure He flow for 2 min to remove the free and weakly bound THF on the catalyst, sealed and cooled to below 15 K for INS data collection. The adsorbed THF underwent the first catalytic conversion in pure H_2 flow for 10 min at 130 °C, and production of butane was observed instantly by mass spectrometry. The cell was then flushed with He to remove free butane and H_2, sealed and cooled again for INS collection to detect the presence of possible reaction intermediates.

References

1. Armaroli, N. & Balzani, V. The future of energy supply: challenges and opportunities. *Angew. Chem. Int. Ed.* **46,** 52–66 (2007).
2. Murray, J. & King, D. Climate policy: oil's tipping point has passed. *Nature* **481,** 433–435 (2012).
3. Huber, G. W., Iborra, S. & Corma, A. Synthesis of transportation fuels from biomass: chemistry, catalysts, and engineering. *Chem. Rev.* **106,** 4044–4098 (2006).
4. Sutton, A. D. *et al.* The hydrodeoxygenation of bioderived furans into alkanes. *Nat. Chem* **5,** 428–432 (2013).
5. Roman-Leshkov, Y., Barrett, C. J., Liu, Z. Y. & Dumesic, J. A. Production of dimethylfuran for liquid fuels from biomass-derived carbohydrates. *Nature* **447,** 982–985 (2007).
6. Rubin, E. M. Genomics of cellulosic biofuels. *Nature* **454,** 841–845 (2008).
7. McKendry, P. Energy production from biomass (part 1): overview of biomass. *Bioresour. Technol.* **83,** 37–46 (2002).
8. Mäki-Arvela, P., Holmbom, B., Salmi, T. & Murzin, D. Y. Recent progress in synthesis of fine and specialty chemicals from wood and other biomass by heterogeneous catalytic processes. *Catal. Rev. -Sci. Eng.* **49,** 197–340 (2007).
9. Sun, Y. & Cheng, J. Y. Hydrolysis of lignocellulosic materials for ethanol production: a review. *Bioresour. Technol* **83,** 1–11 (2002).
10. Alonso, D. M., Bond, J. Q. & Dumesic, J. A. Catalytic conversion of biomass to biofuels. *Green Chem.* **12,** 1493–1513 (2010).
11. Dugar, D. & Stephanopoulos, G. Relative potential of biosynthetic pathways for biofuels and bio-based products. *Nat. Biotechnol.* **29,** 1074–1078 (2011).
12. Mohan, D., Pittman, C. U. & Steele, P. H. Pyrolysis of wood/biomass for bio-oil: a critical review. *Energy Fuels* **20,** 848–889 (2006).

13. Bridgwater, A. V. Review of fast pyrolysis of biomass and product upgrading. *Biomass Bioenergy* **38,** 68–94 (2012).
14. Huber, G. W., Cortright, R. D. & Dumesic, J. A. Renewable alkanes by aqueous-phase reforming of biomass-derived oxygenates. *Angew. Chem. Int. Ed.* **43,** 1549–1551 (2004).
15. Zakzeski, J., Bruijnincx, P. C. A., Jongerius, A. L. & Weckhuysen, B. M. The catalytic valorization of lignin for the production of renewable chemicals. *Chem. Rev.* **110,** 3552–3599 (2010).
16. Matson, T. D., Barta, K., Iretskii, A. V. & Ford, P. C. One-pot catalytic conversion of cellulose and of woody biomass solids to liquid fuels. *J. Am. Chem. Soc.* **133,** 14090–14097 (2011).
17. Li, C., Zheng, M., Wang, A. & Zhang, T. One-pot catalytic hydrocracking of raw woody biomass into chemicals over supported carbide catalysts: simultaneous conversion of cellulose, hemicellulose and lignin. *Energy Environ. Sci* **5,** 6383–6390 (2012).
18. Yan, N. *et al.* Selective degradation of wood lignin over noble-metal catalysts in a two-step process. *ChemSusChem* **1,** 626–629 (2008).
19. Liu, S., Tamura, M., Nakagawa, Y. & Tomishige, K. One-pot conversion of cellulose into n-hexane over the Ir-ReO_x/SiO_2 catalyst combined with HZSM-5. *ACS Sustain. Chem. Eng.* **2,** 1819–1827 (2014).
20. de Beeck, B. O. *et al.* Direct catalytic conversion of cellulose to liquid straight-chain alkanes. *Energy Environ. Sci.* **8,** 230–240 (2015).
21. Del Paggio, A. A. *et al.* A process for producing hydrocarbons. WO patent WO 2013/074434 A1 (Shell Oil Company, USA 2013).
22. Cortright, R. D. & Blommel, P. G. Synthesis of liquid fuels and chemicals from oxygenated hydrocarbons. US patent US 2008/0300434 A1 (Virent, Inc., USA 2008).
23. Liu, Y. *et al.* One-pot catalytic conversion of raw lignocellulosic biomass into gasoline alkanes and chemicals over $LiTaMoO_6$ and Ru/C in aqueous phosphoric acid. *ACS Sustain. Chem. Eng.* **3,** 1745–1755 (2015).
24. Xia, Q. *et al.* Pd/$NbOPO_4$ multifunctional catalyst for the direct production of liquid alkanes from aldol adducts of furans. *Angew. Chem. Int. Ed.* **53,** 9755–9760 (2014).
25. Burch, R. *et al.* Catalytic hydrogenation of tertiary amides at low temperatures and pressures using bimetallic Pt/Re-based catalysts. *J. Catal.* **283,** 89–97 (2011).
26. Chia, M. *et al.* Selective hydrogenolysis of polyols and cyclic ethers over bifunctional surface sites on rhodium-rhenium catalysts. *J. Am. Chem. Soc.* **133,** 12675–12689 (2011).
27. Mitchell, P. C. H., Parker, S. F., Ramirez-Cuesta, A. J. & Tomkinson, J. *Vibrational Spectroscopy with Neutrons, with Applications in Chemistry, Biology, Materials Science, and Catalysis* (World Scientific, Singapore, 2005).
28. Ohno, K., Yoshida, H., Watanabe, H., Fujita, T. & Matsuura, H. Conformational study of 1-butanol by the combined use of vibrational spectroscopy and ab initio molecular orbitol calculations. *J. Phys. Chem.* **98,** 6924–6930 (1994).
29. Hibbitts, D., Tan, Q. & Neurock, M. Acid strength and bifunctional catalytic behavior of alloys comprised of noble metals and oxophilic metal promoters. *J. Catal.* **315,** 48–58 (2014).
30. Michaelides, A. *et al.* Identification of general linear relationships between activation energies and enthalpy changes for dissociation reactions at surfaces. *J. Am. Chem. Soc.* **125,** 3704–3705 (2003).
31. Wang, S. *et al.* Universal transition state scaling relations for (de)hydrogenation over transition metals. *Phys. Chem. Chem. Phys.* **13,** 20760–20765 (2011).
32. van Santen, R. A. *et al.* Reactivity theory of transition-metal surfaces: a Brønsted – Evans – Polanyi linear activation energy—free-energy analysis. *Chem. Rev.* **110,** 2005–2048 (2010).
33. Zhang, Y. *et al.* Mesoporous niobium phosphate: an excellent solid acid for the dehydration of fructose to 5-hydroxymethylfurfural in water. *Catal. Sci. Technol* **2,** 2485–2491 (2012).
34. Parker, S. F. *et al.* Recent and future developments on TOSCA at ISIS. *J. Phys. Conf. Ser.* **554,** 012003 (2014).

Acknowledgements

This project was supported financially by the National Natural Science Foundation of China (No. 21273071, 21101063, 91545103 and 21073060), the Shanghai Rising-Star Program (14QA1401100), the Fundamental Research Funds for the Central Universities, University of Manchester and the Science and Technology Facilities Council (STFC) in UK. The Rutherford Appleton Laboratory is thanked for access to neutron beam facilities (TOSCA). Computing resources (time on the SCARF cluster for the CASTEP calculations) was provided by STFC's e-Science facility.

Author contributions

Q.N.X., Y.S. and X.H.L. helped in preparation and characterisation of catalysts, and performing the catalytic reactions. Z.J.C., H.F.W. and X.Q.G. used DFT calculations. Q.N.X., S.F.P., X.H. and S.Y. helped in the collection, analysis and DFT modelling of neutron scattering data. Y.Q.W. conceived the overall direction of the project. All the authors discussed the results and co-wrote the manuscript.

Additional information

Competing financial interests: The authors declare no competing financial interests.

14

A facile route to old and new cyclophanes via self-assembly and capture

Mary S. Collins[1], Matthew E. Carnes[1], Bryan P. Nell[1], Lev N. Zakharov[2] & Darren W. Johnson[1]

Cyclophanes are a venerable class of macrocyclic and/or cage compounds that often feature high strain, unusual conformations and quite surprising properties, many of which are legendary in physical organic chemistry. However, the discovery of new, diverse cyclophanes and derivatives has been hindered by syntheses that are traditionally low-yielding, requiring long reaction times, laborious purification steps and often extreme conditions. Herein, we demonstrate a new self-assembly route to a variety of discrete cyclic and caged disulfide structures, which can then be kinetically captured upon sulfur extrusion at room temperature to give a diversity of new thioether (hetera)cyclophanes in high yield. In addition to the synthesis of novel macrocycles (dimers through hexamers), this process provides an improved route to a known macrobicyclic trithiacyclophane. This technique also enables the facile isolation of a tetrahedral macrotricyclic tetrathiacyclophane in two steps at an ambient temperature.

[1] Department of Chemistry and Biochemistry and the Materials Science Institute, University of Oregon, Eugene, Oregon 97403-1253, USA. [2] Center for Advanced Materials Characterization in Oregon (CAMCOR), 1241 University of Oregon, Eugene, Oregon 97403-1241, USA. Correspondence and requests for materials should be addressed to D.W.J. (email: dwj@uoregon.edu).

Since the first synthesis of anti-[2.2]metacyclophane by Pellegrin in 1899 (ref. 1) and [2.2]paracyclophane by Cram *et al.* in 1951 (ref. 2), there has been wide interest in the unusual properties of cyclophanes and related cage molecules. The confined geometries of the often distorted aromatic rings in these classes of molecules have allowed many studies of the fundamental properties of aromaticity itself[2–4]. The well-defined topology and high strain of cyclophanes has found utility in a number of applications including asymmetric catalysis[5], insulating plastics[6], organic electronics[7], metal capture[8,9] and most recognizably in the widely used parylene process in industry. Unfortunately, the field of cyclophane chemistry has been hindered by a lack of high yielding and functional group-tolerant preparative methods for their synthesis. A route to synthesize cyclophanes in a more efficient and selective way would be a substantial advancement to conventional cyclophane syntheses, and may bring forth new fundamental studies on their diverse properties as well as potentially advance applications as single-source precursors for insulating, conducting and dielectric polymers[10]. In this paper, we highlight a new method based on disulfide self-assembly followed by capture using a sulfur extrusion agent that enables high-yield syntheses of old and new macrocyclic cyclophanes, a macrobicyclic cyclophane and two new tetrahedral cyclophanes.

We recently showed self-assembly of a simple *p*-xylyl bridged dithiolate ligand, in the presence of $AsCl_3$, rapidly forms a collection of discrete disulfide macrocycles upon treatment with iodine, rather than the disulfide oligomers/polymers that dominate under standard reaction conditions in the absence of $AsCl_3$ (ref. 11)

While studying this dynamic behaviour in an oxidizing environment, we now reveal the role of the external pnictogen (in this case, $SbCl_3$) to be more powerful than initially surmised. The directing behaviour of the $SbCl_3$ additive exhibits impressive resistance to insoluble disulfide polymer formation, long reactions times and low selectivity for a variety of different ligand systems. Using this self-assembly method, the isolation of previously unattainable disulfide cyclophanes—including trimers, tetramers, pentamers, hexamers and tetrahedra—is now possible in a nearly instantaneous reaction from simple di- and tri-thiol ligands. Furthermore, from the cyclic and polycyclic disulfides, kinetic capture by desulfurization using a phosphoramide reagent provides a direct route to thiacyclophanes, again in exceptionally high yield. Herein we detail applications of our new method to the synthesis of a series of both known and entirely new cyclophanes.

Results

Facile synthesis of macrocyclic thioether cyclophanes. Previous methods to synthesize cyclophanes have relied on intra- and intermolecular macrocyclizations at high dilution and extreme temperatures. In many cases, indiscriminant homocouplings such as Wurtz or McMurry couplings[1,12–15] are used where a mixture of oligomers and polymer is the primary product. In some more selective cases, Wittig reactions[16,17] can be used to make mixed-macrocycles, but these instances still rely on unfavourable ring formation steps. Despite the significant breadth of cyclophane applications, their synthesis is still heavily confined to synthetic methods devised more than 40 years ago. These processes, although often spectacularly successful[18], provide limited selectivity, low yields and require difficult purifications. Retrosynthetically, we imagined a different strategy to make these molecules in which the dynamic covalent chemistry of disulfides could be used to first form discrete disulfide assemblies followed by sulfur extrusion. In this way, the usually low yielding cyclization step is replaced by a self-assembly performed under thermodynamic control in the presence of a pnictogen trichloride additive. The disulfide macrocycles can then be treated with phosphoramides to remove sulfur atoms to contract the large ring resulting in heteracyclophane macrocycles (**1**, **2**) or larger species (*vide infra*).

Seminal work by Otto, Furlan and Sanders established that libraries of disulfide macrocycles can respond to external stimuli

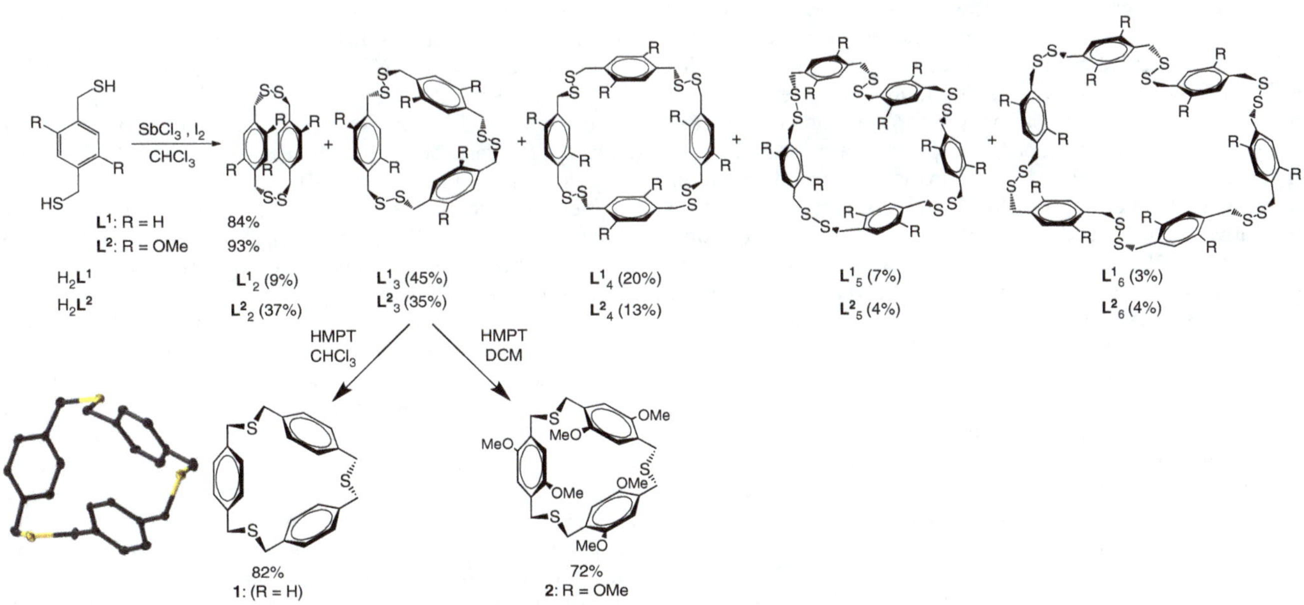

Figure 1 | Synthesis and crystal structure of disulfide and thioether paracyclophanes. Pnictogen-activated iodine oxidation of $H_2\mathbf{L^1}$ and $H_2\mathbf{L^2}$ to form disulfide macrocycles (~1 equiv ligand, 0.5 equiv $SbCl_3$; 2 equiv I_2): $\mathbf{L^{1,2}}_2$ (dimer), $\mathbf{L^{1,2}}_3$ (trimer) and $\mathbf{L^{1,2}}_4$ (tetramer); ($\mathbf{L^{1,2}}_5$ pentamers and $\mathbf{L^{1,2}}_6$ hexamers are not shown, see single crystal structures of $\mathbf{L^1}_5$ and $\mathbf{L^1}_6$ in Supplementary Fig. 1). Overall isolated yields for the self-assembly are 93% and 84%, respectively for $H_2\mathbf{L^1}$ and $H_2\mathbf{L^2}$. The distribution of each disulfide macrocycle formed at ~2 mM reaction conditions are shown. There is no evidence of oligomer/polymer formation by NMR or post GPC purification. Desulfurization of disulfide trimers yields thioether macrocycles **1** and **2** (single crystal X-ray structure representations of **1** shown with thermal ellipsoids at 50%. Hydrogen atoms and solvents of crystallization have been omitted for clarity).

(such as a guest) to amplify specific components of equilibrating 'combinatorial' mixtures[19–22]: the dynamic covalent self-assembly of thiols has emerged as an active area of investigation[23–27]. Our method allows a route to tune the composition of such libraries to form specific, discrete assemblies that can then be isolated using recirculating gel permeation chromatography or crystallization, and ultimately captured by sulfur extrusion to provide a remarkable array of kinetically stable, diverse two-dimensional and three-dimensional (3D) cyclophanes (see experimental details in Supplementary Information). The use of a pnictogen trichloride additive (for example, $SbCl_3$) is critical during the self-assembly step to encourage the formation of discrete disulfide species, perhaps serving an organizational role in bringing together multiple thiolates to form a discrete assembly through a combination of Sb-S and Sb-π interactions[28–34]. This two-step self-assembly and capture process appears to be general and occurs in high yield.

As an example of a facile self-assembly and capture, we turned to constructing thiacyclophane **1** (Fig. 1), which is a challenging target by classical coupling methods. From $[\mathbf{L^1}_3]$, desulfurization with HMPT (hexamethylphosphorous triamide) in chloroform provides **1** in 82% isolated yield after 3 h and quantitative conversion after 6 days by ^{1}H-NMR spectroscopy. The ^{1}H NMR and ^{13}C NMR spectra are simple, exhibiting time-averaged symmetry in solution. The ^{1}H NMR of **1** shows only two signals, corresponding to the aromatic and benzylic protons, while the ^{13}C spectrum shows three peaks arising from the three carbon environments. The desulfurization of unsubstituted **1** would not go to completion in dichloromethane or in the presence of a weaker sulfur extrusion agent, HEPT (hexaethylphosphorous triamide). The crude product, an iridescent white solid, was purified by trituration with water to remove any residual phosphine sulfide and separated using centrifugation. This purification method was found to give the highest isolated yield, serving as an optimal purification technique over separatory funnel extraction. Crystals suitable for X-ray diffraction were grown by vapour diffusion of pentane into chloroform or by slow evaporation of the chloroform solution directly (Fig. 1, bottom). **1** crystallizes in space group $P2_1/c$ without any additional solvent of co-crystallization. In addition, the arene rings in **1** are not coplanar, but rather adopt a face-to-face orientation in the solid state. This is commonly seen in paracyclophanes[35] presumably to relieve strain, in which the arenes orient perpendicular to the macrocycle plane and the methylene protons are located along the macrocyclic equator. The central cavity of **1** has a diameter of 9 Å lengthwise. One of the thioether bond angles is slightly acute (C–S–C angles: 90.69°, 98.58° and 103.55°; typical thiother bond angles are 100–105°; ref. 36) and the average ring distance between the two closest benzene rings is 3.968 Å, too great a distance to suggest any attractive intramolecular aromatic interactions. The unsubstituted trimer **1** is robust in solution and does not appear to decompose with long-term exposure to oxygen, light or silica.

Given the facile conversion of the dithiol ligand $H_2\mathbf{L^1}$ to the thioether trimer **1** in two steps, we attempted a one-pot reaction to directly transform $H_2\mathbf{L^1}$ into trithioether **1** upon treatment with $SbCl_3$, I_2 and phosphoramide. Remarkably, crude ^{1}H-NMR spectra and mass spectrometry data indicate successful one-pot conversion without further optimization of the reaction conditions (Supplementary Figs 12 and 14). It should be emphasized that only in very rare cases are cyclic trimers favoured during traditional cyclophane syntheses and in such cases self-assembly in a single pot is certainly not possible. The ease with which disulfide $[\mathbf{L^1}_3]$ forms during oxidation and then undergoes desulfurization to give trithioether **1** was quite unexpected, with an unoptimized overall two-step yield >50%.

Initial studies also suggest this route exhibits methoxy-functional group tolerance. As a preliminary test, iodine oxidation of 2,5-dimethoxy-1,4-bis(thiomethyl)benzene ($H_2\mathbf{L^2}$) using $SbCl_3$ similarly undergoes selective oxidation to produce $[\mathbf{L^2}_2]$ (dimer), $[\mathbf{L^2}_3]$ (trimer), $[\mathbf{L^2}_4]$ (tetramer), $[\mathbf{L^2}_5]$ (pentamer) and $[\mathbf{L^2}_6]$ (hexamer) in a combined 93% isolated yield (Fig. 1). The ^{1}H NMR spectra of the structures are again quite simple, indicative of high symmetry molecules (Fig. 1 and Supplementary Fig. 2). The aromatic protons exhibit an upfield shift as the macrocycle decreases in size, presumably due to shielding effects by benzene rings that are oriented 'face-to-face'[35]. Unlike the unsubstituted adducts, the formation of higher ordered $[\mathbf{L^2}]$ disulfide-based cyclophanes (that is, tetramer, pentamer, hexamer and so on) require longer reaction times. For example, the intermolecular conversion to larger cyclic disulfides increases after 16–24 h at an ambient temperature. (In addition, this slow conversion to higher ordered cyclophanes, like trimers, tetramers, pentamers and hexamers, can be mediated by the amount of $SbCl_3$, concentration and choice of solvent). Similar to the sulfur extrusion that produces **1**, thiacyclophane **2** forms in 72% yield by desulfurization of $[\mathbf{L^2}_3]$ using HMPT in 4 h in dichloromethane or 2 h in chloroform (Fig. 1).

Macrobicyclic and tetrahedral cyclophanes. The preparation of highly bridged, 3D cyclophanes has suffered from lengthy and tedious syntheses requiring stepwise modification and addition of each bridging component[37]. Larger capsule-like molecules with similar structural features such as cubes and tetrahedra have been synthesized[2,8,38]. Despite numerous exotic and spectacular 3D cyclophanes, limitations in the classical stepwise syntheses have impeded efforts to design and prepare new, large cyclophane structures. Using our self-assembly and capture strategy, we used three-fold symmetric ligands in an attempt to synthesize higher ordered cyclophane structures such as tetrahedra. Pnictogen-activated oxidation of a trisubstituted thiol ligand ($H_3\mathbf{L^3}$) with iodine gives a clean distribution of previously described disulfide dimer $[\mathbf{L^3}_2]$ as well as a distorted tetrahedron $[\mathbf{L^3}_4]$ in a combined 98% isolated yield after purification (Fig. 2). These species were initially identified by diffusion-ordered ^{1}H-NMR spectroscopy where the diffusion coefficients for the dimer and tetrahedron differ (Supplementary Fig. 15). Control reactions using only iodine and $H_3\mathbf{L^3}$ revealed that the pnictogen trichloride is required for fast, selective conversion to discrete $\mathbf{L^3}$ disulfides (Supplementary Figs 23 and 24).

The dynamic covalent behaviour driving the distribution of products during self-assembly appears largely influenced by three components: (1) concentration, (2) solvent effects and (3) reaction time. For example, the presence of $SbCl_3$ influences the $[\mathbf{L^3}_2]$:$[\mathbf{L^3}_4]$ ratio and also encourages immediate formation of both $[\mathbf{L^3}_2]$ and $[\mathbf{L^3}_4]$, which form sluggishly in the absence of the pnictogen as polymer formation competes (Supplementary Figs 23 and 24). In addition, the selectivity at low concentration shows $[\mathbf{L^3}_2]$ forms preferentially, whereas at high concentration, the tetramer $[\mathbf{L^3}_4]$ becomes more favoured (Supplementary Fig. 17). The solvent effects are also observed for the $H_3\mathbf{L^3}$ pnictogen-assisted iodine oxidation. Reactions performed in dichloromethane result in a lowered isolated yield due to the formation of intermediate precipitates, while reactions in chloroform reveal no such precipitation. Last, formation of the tetrahedron is enhanced and can even surpass dimer formation using longer reaction times in a different solvent system, such as benzene (Supplementary Fig. 18). Full details of the scope and limitations of these variables and reaction conditions are currently being investigated and will be reported in a follow-up paper.

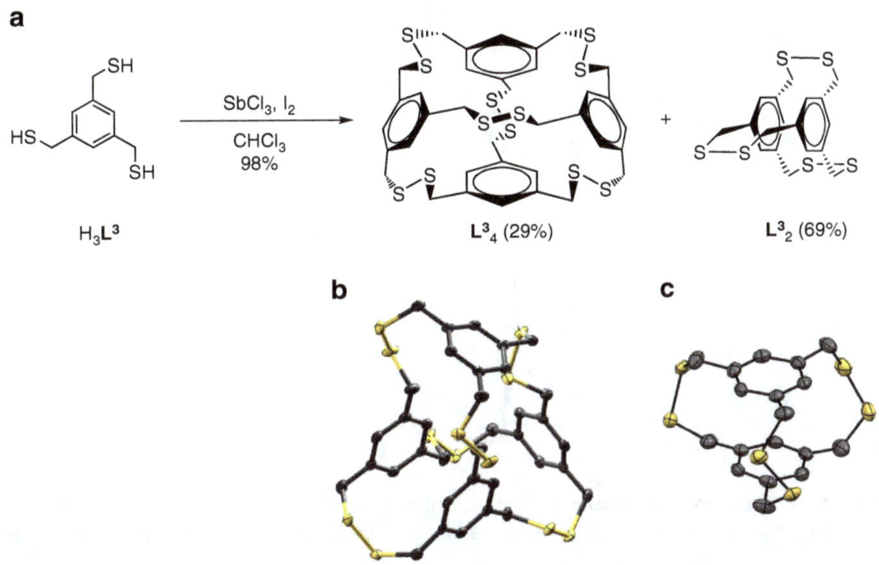

Figure 2 | Synthesis and crystal structure of disulfide dimer and tetrahedron. (**a**) Pnictogen-activated iodine oxidation of $H_3\mathbf{L^3}$ cleanly forms disulfide macrobicycle $\mathbf{L^3}_2$ (dimer, 69%) and tetrahedron $\mathbf{L^3}_4$ (tetramer, 29%) in combined 98% isolated yield at ~2 mM concentration (1 equiv $H_3\mathbf{L^3}$, 2.5 equiv $SbCl_3$; 3 equiv I_2). Single-crystal X-ray structure representations of tetrahedron $\mathbf{L^3}_4$ (**b**) (thermal ellipsoids at 50%) and dimer $\mathbf{L^3}_2$ (**c**) (thermal ellipsoids at 30%). Hydrogen atoms and solvents of crystallization have been omitted for clarity.

Structure of dimer $\mathbf{L^3}_2$. The disulfide dimer, $\mathbf{L^3}_2$, was crystallized by slow evaporation of chloroform. The dimer crystallizes with C_2 symmetry with two symmetrically independent molecules. The molecules are linked in the crystal structure via short intermolecular S···S close contacts. $\mathbf{L^3}_2$ does not co-crystallize with solvent. The C–S–S–C torsion angles in each independent dimer deviate substantially from ideality (90°), adopting highly strained conformations (115.6°, 107.1°, 107.1°) and (120.0°, 120.0°, 116.3°). To better understand the prevalence for similar existing dihedral disulfide bond angles, a systematic survey of the Cambridge Structural Database (CSD) for disulfide bonds at 120° ± 5 was performed (Supplementary Fig. 19). Only 16 structures were identified with torsional angles at nearly 120°; most of them appeared as bridging units in cyclic structures. The rarity of such apparent strain in the disulfide bridges also mirrors the 'stress' exhibited by the ligand seen by the deviation of the methylene carbon from the benzene plane (see experimental details in Supplementary Information). The C6-rings in both molecules are planar within 0.006 and 0.007 Å, but the methylene atoms are significantly 'puckered' out of the C6-ring plane: 0.12(0.1) –0.19(0.1) Å. This is consistent with early work by Houk and Whitesides[39], who suggested that tris-disulfide $\mathbf{L^3}_2$ does not favourably form upon oxidation: 'only polymeric disulfide products were obtained…molecular models suggest that…[$\mathbf{L^3}_2$ is]…seriously strained…[so] it is not surprising that these substances do not form', again indicating the powerful role the pnictogen trichloride plays in the self-assembly process.

To showcase the utility of the sulfur extrusion process, we next sought to prepare the known 2,11,20-trithia[3.3.3](1,3,5)cyclophane[13,40,41] **3** by exposing tris(disulfide) [4.4.4](1,3,5)cyclophane [$\mathbf{L^3}_2$] to a stoichiometric amount of HMPT for 1.5 h at an ambient temperature. The trithiacyclophane **3** was formed in quantitative yield by ^{1}H-NMR spectroscopy and isolated in 95% yield from disulfide [$\mathbf{L^3}_2$] (Fig. 3, Supplementary Fig. 8). Over two steps, this provides an isolated yield of 65%, a significant improvement over the reported 5.3% yield[13] observed by treatment of 1,3,5-tris(bromomethyl)benzene with 1,3,5-tris(mercaptomethyl)benzene $H_3\mathbf{L^3}$ or the 24% optimized yield from 1,3,5-tris(bromomethyl)benzene and Na_2S (ref. 41). Surprisingly, no evidence of ring opened, oligomeric thiother products has been observed using our route (Supplementary Figs 15, 16 and 21). Remarkably, this entire two-step self-assembly and capture method can be performed in one pot: treatment of $H_3\mathbf{L^3}$ with $SbCl_3$, I_2 and HMPT also forms **3** as observed by ^{1}H-NMR spectroscopy (Supplementary Fig. 11). Thioether **3** is known to serve as a starting point to form cyclophane **4** from established procedures[41].

Tetrahedral cages of hexadisulfide $\mathbf{L^3}_4$ and hexathioether 5. Encouraged by the success of sulfur extrusion on the trithioether dimer, we sought to perform sulfur extrusion on the hexadisulfide tetrahedron $\mathbf{L^3}_4$. This would require the extrusion of six sulfur atoms (one on each edge) of the distorted tetrahedron—representing a total of 24 bonds broken/formed in this single step—to proceed with no polymerization and in high yield for each extrusion. We were delighted to discover that sulfur extrusion of all six disulfides to form a tetrahedral hexathioether 'tetrahedrophane' **5**, proceeds by desulfurization with HMPT at an ambient temperature in 4 h in 94% yield (Fig. 4)[42–44]. To the best of our knowledge, this hexathioether assembly is unknown; for related concave spheriphanes, our method reduces an eight-step synthesis to two steps[45]. The formation of hexathioether **5** was confirmed by ^{1}H NMR spectroscopy, exhibiting time-averaged tetrahedral symmetry in $CDCl_3$ (Supplementary Figs 9 and 10), and single-crystal X-ray diffraction (Fig. 4b). The loss of six sulfur atoms decreases the size of the cyclic structure only slightly and the aromatic and methylene singlets display small, yet expected upfield shifts relative to the disulfide analogue in $CDCl_3$.

Diffraction-quality single crystals of each tetrahedron $\mathbf{L^3}_4$ and **5** were grown by vapour diffusion of pentane into chloroform or solvent evaporation. Both $\mathbf{L^3}_4$ and **5** capsules exhibit robust chemical stability and do not show decomposition over time or with exposure to light. $\mathbf{L^3}_4$ crystallizes in the space group $P2_1/c$ and surprisingly does not crystallize with a solvent guest inside its cavity. Instead, $\mathbf{L^3}_4$ 'folds in' on itself to avoid the formation of a large void in the solid state (Fig. 2b). Owing to this conformation, $\mathbf{L^3}_4$ displays a wide range of non-ideal C-S-S-C torsional angles (82.93°, 85.61°, 88.46°, 88.80°, 92.62°, 113.24°). Hexathioether **5** crystallizes in the $P2_1/n$ space group and co-crystallizes with

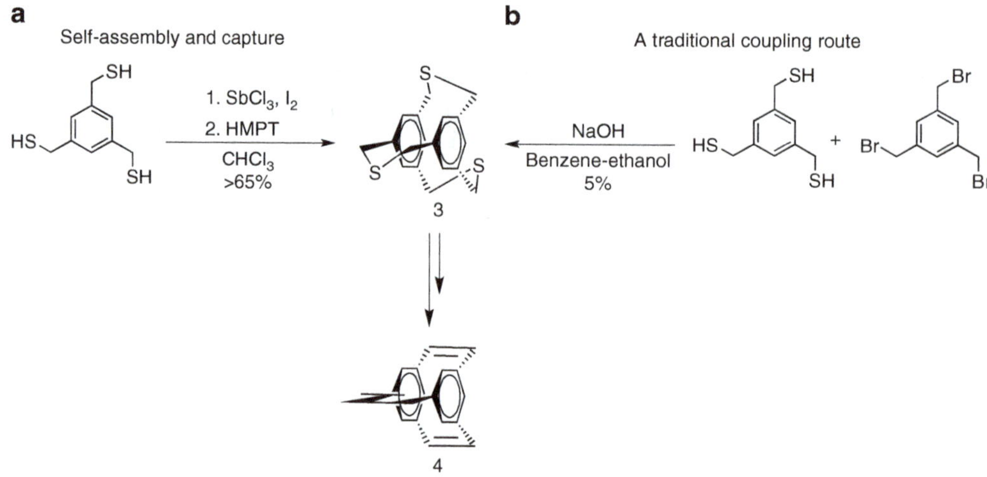

Figure 3 | Self-assembly and capture of a known cyclophane. (**a**) Self-assembly of $H_3\mathbf{L}^3$ (1 mol $SbCl_3$, 1.2 mol I_2; 69%) followed by capture (95%) by sulfur extrusion with HMPT (4 mol) provides known trithiacyclophane **3** in overall 65% yield. (**b**) A traditional stepwise route to **3** proceeds in 5% yield.

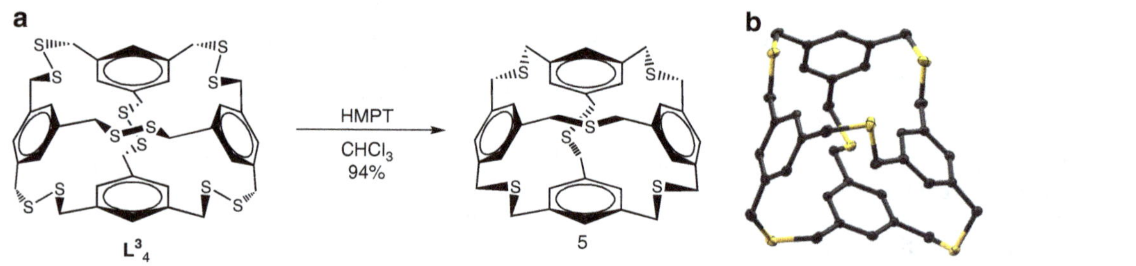

Figure 4 | Synthesis and crystal structure of a thiatetrohedrophane. Desulfurization (**a**) of $\mathbf{L}^3{}_4$ gives tetrathioether [3_4]tetrahedrophane **5** (6 mol HMPT). Single-crystal X-ray structure representation of thiatetrahedrophane **5** (**b**); all thermal ellipsoids at 50%. Hydrogen atoms and solvents of crystallization have been omitted for clarity.

chloroform molecules of solvation, although its small cavity also remains empty. Like $\mathbf{L}^3{}_4$, the thiatetrahedrophane **5** collapses in on itself avoiding unfavourable void space. The C–S–C angles within the cyclophane were in the expected range for cyclic thioethers (98.84°, 99.21°, 100.17°, 102.71°, 102.90°, 102.92°; ref. 36) In addition, the 'folded in' conformation may also enable a slightly favourable aromatic interaction between two adjacent benzene rings, which reveal a centroid-to-centroid distance of 3.7 Å (Fig. 4b).

Discussion

We have shown a selective, high yielding alternative to traditional cyclophane synthesis that uses self-assembly and capture through the use of a pnictogen additive to direct the self-assembly of discrete disulfide (hetera)cyclophanes. The book *Modern Cyclophane Chemistry* cites a need for simpler reaction pathways or procedures for preparing cyclophanes in larger quantities[46]. Using a self-assembly approach eliminates the need for caustic, high-temperature methods and avoids low yields that result from the rapid formation of unfavourable insoluble oligomers. This method complements modern research on dynamic combinatorial covalent libraries of disulfides[19,23] by providing a method to isolate and kinetically capture individual members of such equilibrating systems; in addition, the key pnictogen additive during the self-assembly step avoids the formation of disulfide polymers that plagued early efforts aimed at synthesizing discrete polydisulfides[39,47].

Rather than relying on traditional kinetic syntheses, this route relies on self-assembly to prepare complex cyclophanes as the major product of the reaction. This dynamic covalent method allows the synthesis of formerly inaccessible, higher ordered species such as trimers, tetramers, pentamers, hexamers, macrobicycles and tetrahedra to be performed cleanly and in high yield, and preliminary results indicate that this method is also tolerant to a simple functional group that enhances the solubility in polar solvents. Importantly, the self-assembly reaction is under thermodynamic control. For instance, treatment of ligand $H_2\mathbf{L}^1$ with $SbCl_3$ and I_2 provides the same distribution of macrocyclic disulfides as does reaction of tetrameric disulfide [$\mathbf{L}^1{}_4$] under the same conditions (see stacked NMR plot comparing the two reactions in Supplementary Fig. 25). These disulfide complexes operate as excellent precursors to their thioether analogues by treatment with sulfur extrusion agents, and such thioethers are known to be effective precursors to hydrocarbon cyclophanes, suggesting that this new synthetic method could open the door to the synthesis of myriad new hydrocarbon cyclophanes as well. In conclusion, our supramolecular approach serves as a powerful strategy in which new cyclophanes can be made in cases where strain and complex design are difficult to conceive or do not possess selectivity to be formed via classical coupling techniques. This strategy could be of particular interest to the future synthetic design of macrocycles and cage compounds, as well as toward the myriad applications for which cyclophanes are utilized, such as polymer-based materials (for example, the parylene process)[10], catalysis, strained systems and host–guest binding/separations science[48,49]. We are mindful of the fact that this synthetic method enables the formation of many difficult-to-synthesize cyclophanes that may be of interest as monomers for new polymers or for use directly in the widely used parylene process in industry.

Methods

Single-crystal X-ray diffraction. Full synthetic and characterization details are provided in the supporting information. Single-crystal X-ray diffraction intensities were collected at 100 K (L^3_4, **5** and **1**), 150 K (L^1_6), 173 K (L^1_5), 200 K (L^2_2) and 223(2) K (L^3_2) on a Bruker Apex2 CCD diffractometer using MoKα radiation, $\lambda = 0.71073$ Å, (**5**, **1**, L^2_2) and CuKα radiation, $\lambda = 1.54178$ Å, (L^3_4, L^3_2, L^1_6 and L^1_5). The space groups were determined based on systematic absences. Absorption corrections were applied by SADABS[50]. The structures were solved by direct methods and Fourier techniques and refined on F^2 using full-matrix least-squares procedures. All the non-H atoms were refined with anisotropic thermal parameters. The H atoms in all the structures were refined in calculated positions in a rigid group model. Two S–S bridges in L^3_2 are disordered over two positions in a ratio 0.53/0.47. The -S- bridges in **1** are disordered as well: the three strongest peaks on the residual density map of **1** (values are in the range 2.58–3.69 e Å^{-3}) seem to be related to a second position of the S atom, but its contribution was not taken into account in the final refinement. The Flack parameter for L^3_2 is 0.08(5). The structure of L^2_2 was checked in centro- and non-centrosymmetrical space groups and was determined in non-centrosymmetrical space group $Pna2_1$. The structure of L^2_2 is a racemic twin consisting of left and right molecules in the ratio 0.68/0.32 and having a pseudo-inversion centre. The crystals of L^1_6 and L^1_5 are very thin needles and even using a strong Incoatec Cu IμS source, X-ray diffraction data for these crystals were weak especially at high angles. The data for L^1_6 were collected only up to $2\theta_{max} = 96°$. The data for L^1_5 were collected up to $2\theta_{max} = 135°$, but only reflections with 2θ less than 100° were used in the refinement. As a result, values of R_{int} for both compounds are high. The final X-ray structures for L^1_6 and L^1_5 are not precisely determined, but they clearly confirm the structure and compositions of these compounds and are therefore also included in this paper. All the calculations were performed by the Bruker SHELXTL (v. 6.10; ref. 51) and SHELXL-2013 packages[52].

Crystallographic data for L^3_4. $C_{36}H_{36}S_{12}$, $M = 853.37$, 0.19 × 0.05 × 0.03 mm, $T = 100(2)$ K, monoclinic, space group $P2_1/c$, $a = 18.6720(16)$ Å, $b = 12.2920(10)$ Å, $c = 16.8619(15)$ Å, $\beta = 94.755(6)°$, $V = 3{,}856.8(6)$ Å^3, $Z = 4$, $D_c = 1.470$ Mg m^{-3}, μ(Cu) = 6.524 mm^{-1}, $F(000) = 1{,}776$, $2\theta_{max} = 130.0°$, 22,999 reflections, 6,373 independent reflections($R_{int} = 0.0765$), R1 = 0.0622, wR2 = 0.1623 and GOF = 1.018 for 6,373 reflections (433 parameters) with $I > 2\sigma(I)$, R1 = 0.0762, wR2 = 0.1781 and GOF = 1.018 for all the reflections, max/min residual electron density +0.893/ − 0.607 eÅ^{-3}.

Crystallographic data for 5. $C_{37}H_{37}Cl_3S_6$, $M = 780.38$, 0.19 × 0.09 × 0.07 mm, $T = 100(2)$ K, monoclinic, space group $P2_1/n$, $a = 8.8821(6)$ Å, $b = 16.2160(10)$ Å, $c = 25.2559(15)$ Å, $\beta = 91.188(1)°$, $V = 3{,}636.9(4)$ Å^3, $Z = 4$, $D_c = 1.425$ Mg m^{-3}, μ(Mo) = 0.624 mm^{-1}, $F(000) = 1{,}624$, $2\theta_{max} = 50.0°$, 27,936 reflections, 6,404 independent reflections ($R_{int} = 0.0419$), R1 = 0.0436, wR2 = 0.0975 and GOF = 1.029 for 6,404 reflections (447 parameters) with $I > 2\sigma(I)$, R1 = 0.0610, wR2 = 0.1070 and GOF = 1.029 for all reflections, max/min residual electron density +1.673/ − 0.827 eÅ^{-3}.

Crystallographic data for 1. $C_{24}H_{24}S_3$, $M = 408.61$, 0.24 × 0.16 × 0.09 mm, $T = 100(2)$ K, monoclinic, space group $P2_1/n$, $a = 18.4134(17)$ Å, $b = 5.7476(6)$ Å, $c = 20.3204(18)$ Å, $\beta = 107.774(4)°$, $V = 2047.9(3)$ Å^3, $Z = 4$, $D_c = 1.325$ Mg m^{-3}, μ(Mo) = 0.369 mm^{-1}, $F(000) = 864$, $2\theta_{max} = 50.0°$, 28,649 reflections, 3,596 independent reflections ($R_{int} = 0.0443$), R1 = 0.0982, wR2 = 0.2422 and GOF = 1.134 for 3,596 reflections (244 parameters) with $I > 2\sigma(I)$, R1 = 0.1048, wR2 = 0.2474 and GOF = 1.134 for all the reflections, max/min residual electron density +3.691/ − 0.0915 eÅ^{-3}.

Crystallographic data for L^3_2. $C_{18}H_{18}S_6$, $M = 426.68$, 0.09 × 0.06 × 0.04 mm, $T = 223(2)$ K, monoclinic, space group C2, $a = 7.6361(10)$ Å, $b = 17.165(2)$ Å, $c = 15.419(2)$ Å, $\beta = 103.756(9)°$, $V = 1{,}963.1(5)$ Å^3, $Z = 4$, $D_c = 1.444$ Mg m^{-3}, μ(Cu) = 6.408 mm^{-1}, $F(000) = 888$, $2\theta_{max} = 135.97°$, 8,461 reflections, 3,328 independent reflections ($R_{int} = 0.0689$), R1 = 0.0557, wR2 = 0.1397 and GOF = 1.044 for 3,328 reflections (255 parameters) with $I > 2\sigma(I)$, R1 = 0.0688, wR2 = 0.1486 and GOF = 1.044 for all the reflections, the Flack = 0.08(5), max/min residual electron density +0.255/ − 0.190 eÅ^{-3}.

Crystallographic data for L^1_6. $C_{50}H_{50}Cl_6S_{12}$, $M = 1{,}248.32$, 0.42 × 0.02 × 0.01 mm, $T = 150(2)$ K, monoclinic, space group $P2_1/c$, $a = 17.667(3)$ Å, $b = 5.4025(9)$ Å, $c = 30.355(5)$ Å, $\beta = 102.055(10)°$, $V = 2{,}833.4(8)$ Å^3, $Z = 2$, $D_c = 1.463$ Mg m^{-3}, μ(Cu) = 7.172 mm^{-1}, $F(000) = 1{,}288$, $2\theta_{max} = 95.76°$, 8,328 reflections, 2,619 independent reflections ($R_{int} = 0.1616$), R1 = 0.0798, wR2 = 0.1875 and GOF = 1.023 for 2,619 reflections (307 parameters) with $I > 2\sigma(I)$, R1 = 0.1661, wR2 = 0.2406 and GOF = 1.023 for all the reflections, max/min residual electron density +1.186/ − 0.520 eÅ^{-3}.

Crystallographic data for L^1_5. $C_{42}H_{42}Cl_6S_{10}$, $M = 1{,}080.05$, 0.04 × 0.01 × 0.01 mm, $T = 173(2)$ K, monoclinic, space group $P2_1/c$, $a = 25.778(5)$ Å, $b = 5.4791(14)$ Å, $c = 34.978(7)$ Å, $\beta = 98.066(12)°$, $V = 4{,}891.4(18)$ Å^3, $Z = 4$, $D_c = 1.467$ Mg m^{-3}, μ(Cu) = 7.437 mm^{-1}, $F(000) = 2{,}224$, $2\theta_{max} = 135.36°$, 24,422 reflections, 4,172 independent reflections ($R_{int} = 0.2503$), R1 = 0.1185, wR2 = 0.2866 and GOF = 1.026 for 4,172 reflections (523 parameters) with $I > 2\sigma(I)$, R1 = 0.2530, wR2 = 0.3756 and GOF = 1.026 for all the reflections, max/min residual electron density +0.685/ − 0.589 eÅ^{-3}.

Crystallographic data for L^2_2. $C_{20}H_{24}O_4S_4$, $M = 456.63$, 0.20 × 0.12 × 0.07 mm, $T = 200(2)$ K, orthorhombic, space group $Pna2_1$, $a = 14.437(3)$ Å, $b = 16.405(3)$ Å, $c = 9.0971(17)$ Å, $V = 2{,}154.5(7)$ Å^3, $Z = 4$, $D_c = 1.408$ Mg m^{-3}, μ(Mo) = 0.465 mm^{-1}, $F(000) = 960$, $2\theta_{max} = 50.0°$, 17,105 reflections, 3,781 independent reflections ($R_{int} = 0.0649$), R1 = 0.0490, wR2 = 0.1029 and GOF = 1.038 for 3,781 reflections (254 parameters) with $I > 2\sigma(I)$, R1 = 0.0711, wR2 = 0.1143 and GOF = 1.039 for all the reflections, max/min residual electron density +0.300/ − 0.237 eÅ^{-3}.

References

1. Pellegrin, M. Contribution à l'étude de la réaction de Fittig. *Recl. Trav. Chim. Pays-Bas* **18**, 457–465 (1899).
2. Cram, D. J. & Steinberg, H. Macro Rings. I. Preparation and spectra of the paracyclophanes. *J. Am. Chem. Soc.* **73**, 5691–5704 (1951).
3. Bodwell, G. J. *et al.* Nonplanar aromatic compounds. 8. 1 Synthesis, crystal structures, and aromaticity investigations of the 1,n-dioxa[n](2,7)pyrenophanes. How does bending affect the cyclic π-electron delocalization of the pyrene system? *J. Org. Chem.* **68**, 2089–2098 (2003).
4. Dobrowolski, M. A. *et al.* Interplay of π-electron delocalization and strain in [n](2,7)pyrenophanes. *J. Org. Chem.* **73**, 8001–8009 (2008).
5. Pye, P. J. *et al.* A new planar chiral bisphosphine ligand for asymmetric catalysis: highly enantioselective hydrogenations under mild conditions. *J. Am. Chem. Soc.* **119**, 6207–6208 (1997).
6. Gorham, W. F. A new, general synthetic method for the preparation of linear poly-p-xylylenes. *J. Polym. Sci. A Polym. Chem.* **4**, 3027–3039 (1966).
7. Yu, C.-Y. & Turner, M. L. Soluble poly(p-phenylenevinylene)s through ring-opening metathesis polymerization. *Angew. Chem. Int. Ed.* **45**, 7797–7800 (2006).
8. Gross, J. *et al.* Concave hydrocarbons. *Chem. Eur. J.* **2**, 1585–1595 (1996).
9. Cammann, K. *Das Arbeiten mit ionenselektiven Elektroden* (Springer, 1996).
10. Campbell, G. A. Modern plastics encyclopedia for 1968-69, volume 45: No. 14a, October, 1968, McGraw-Hill Inc., New York, New York 10036. *J. Appl. Polym. Sci.* **13**, 1580–1580 (1969).
11. Collins, M. S. *et al.* Pnictogen-directed synthesis of discrete disulfide macrocycles. *Chem. Commun.* **49**, 6599–6601 (2013).
12. Baker, W., Banks, R., Lyon, D. R. & Mann, F. G. 10. The action of sodium on o-xylylene dibromide. *J. Chem. Soc.* 27–30 (1945).
13. Vögtle, F. Vielfach verbrückte aromatische Verbindungen, I 2.11.20-Trithia[3.3.3](1,3,5)cyclophan. *Justus Liebigs Ann. Chem* **735**, 193–195 (1970).
14. Merner, B. L., Dawe, L. N. & Bodwell, G. J. 1,1,8,8-Tetramethyl[8](2,11)teropyrenophane: Half of an Aromatic Belt and a Segment of an (8,8) Single-Walled Carbon Nanotube. *Angew. Chem. Int. Ed.* **48**, 5487–5491 (2009).
15. Merner, B. L., Unikela, K. S., Dawe, L. N., Thompson, D. W. & Bodwell, G. J. 1,1,n,n-Tetramethyl[n](2,11)teropyrenophanes (n = 7–9): a series of armchair SWCNT segments. *Chem. Commun. (Camb)* **49**, 5930–5932 (2013).
16. Vögtle, F. A new cyclophane synthesis. *Angew. Chem. Int. Ed.* **8**, 274–274 (1969).
17. Vögtle, F. & Effler, A. H. Sterische Wechselwirkungen im Innern cyclischer Verbindungen, X. Heterocyclische [2.2]Metacyclophane. *Chem. Ber* **102**, 3071–3076 (1969).
18. Bodwell, G. J. & Nandaluru, P. R. Olefination reactions in the synthesis of cyclophanes. *Isr. J. Chem.* **52**, 105–138 (2012).
19. Otto, S., Furlan, R. L. E. & Sanders, J. K. M. Selection and amplification of hosts from dynamic combinatorial libraries of macrocyclic disulfides. *Science* **297**, 590–593 (2002).
20. Corbett, P. T. *et al.* Dynamic combinatorial chemistry. *Chem. Rev.* **106**, 3652–3711 (2006).
21. Brisig, B., Sanders, J. K. M. & Otto, S. Selection and amplification of a catalyst from a dynamic combinatorial library. *Angew. Chem. Int. Ed.* **42**, 1270–1273 (2003).
22. Corbett, P. T., Sanders, J. K. M. & Otto, S. Systems chemistry: pattern formation in random dynamic combinatorial libraries. *Angew. Chem. Int. Ed.* **46**, 8858–8861 (2007).
23. Black, S. P., Sanders, J. K. M. & Stefankiewicz, A. R. Disulfide exchange: exposing supramolecular reactivity through dynamic covalent chemistry. *Chem. Soc. Rev.* **43**, 1861–1872 (2014).
24. West, K. R., Bake, K. D. & Otto, S. Dynamic combinatorial libraries of disulfide cages in water. *Org. Lett.* **7**, 2615–2618 (2005).

25. Sarma, R. J., Otto, S. & Nitschke, J. R. Disulfides, imines, and metal coordination within a single system: interplay between three dynamic equilibria. *Chem. Eur. J.* **13,** 9542–9546 (2007).
26. Vial, L., Ludlow, R. F., Leclaire, J., Pérez-Fernández, R. & Otto, S. Controlling the biological effects of spermine using a synthetic receptor. *J. Am. Chem. Soc.* **128,** 10253–10257 (2006).
27. Leclaire, J., Vial, L., Otto, S. & Sanders, J. K. M. Expanding diversity in dynamic combinatorial libraries: simultaneous exchange of disulfide and thioester linkages. *Chem. Commun.* 1959–1961 (2005).
28. Cangelosi, V. M., Zakharov, L. N. & Johnson, D. W. Supramolecular 'transmetalation' leads to an unusual self-assembled P_2L_3 cryptand. *Angew. Chem. Int. Ed.* **49,** 1248–1251 (2010).
29. Vickaryous, W. J., Healey, E. R., Berryman, O. B. & Johnson, D. W. Synthesis and characterization of two isomeric, self-assembled arsenic-thiolate macrocycles. *Inorg. Chem.* **44,** 9247–9252 (2005).
30. Lindquist, N. R., Carter, T. G., Cangelosi, V. M., Zakharov, L. N. & Johnson, D. W. Three's company: co-crystallization of a self-assembled S(4) metallacyclophane with two diastereomeric metallacycle intermediates. *Chem. Commun.* **46,** 3505–3507 (2010).
31. Vickaryous, W. J., Zakharov, L. N. & Johnson, D. W. Self-assembled antimony-thiolate Sb_2L_3 and $Sb_2L_2Cl_2$ complexes. *Main Gr. Chem* **5,** 51–59 (2006).
32. Vickaryous, W. J., Herges, R. & Johnson, D. W. Arsenic-pi interactions stabilize a self-assembled As_2L_3 supramolecular complex. *Angew. Chem. Int. Ed.* **43,** 5831–5833 (2004).
33. Pitt, M. A. *et al.* Multiple weak supramolecular interactions stabilize a surprisingly twisted As_2L_3 assembly. *Chem. Commun.* 3936–3938 (2008).
34. Collins, M. S. *et al.* Self-assembled trinuclear arsenic and antimony macrobicycles. *Chem. Sci.* **6,** 2444–2448 (2015).
35. Tabushi, I., Yamada, H. & Kuroda, Y. Preparations and properties of higher [2n]paracyclophanes, cyclic oligomers of p-xylylene. *J. Org. Chem.* **40,** 1946–1949 (1975).
36. Chatgilialoglu, C. & Asmus, K.-D. *Sulfur-Centered Reactive Intermediates in Chemistry and Biology* (Springer Science & Business Media, 2013).
37. Sekine, Y., Brown, M. & Boekelheide, V. [2.2.2.2.2.2](1,2,3,4,5,6)Cyclophane: superphane. *J. Am. Chem. Soc.* **101,** 3126–3127 (1979).
38. Boekelheide, V. [2n]Cyclophanes: paracyclophane to superphane. *Acc. Chem. Res.* **13,** 65–70 (1980).
39. Houk, J. & Whitesides, G. M. Structure-reactivity relations for thiol-disulfide interchange. *J. Am. Chem. Soc.* **109,** 6825–6836 (1987).
40. Boekelheide, V. & Hollins, R. A. Syntheses of novel tris-bridged cyclophanes. [2.2.2](1,3,5)cyclophane-1,9,17-triene. *J. Am. Chem. Soc.* **92,** 3512–3513 (1970).
41. Boekelheide, V. & Hollins, R. A. [2.2.2](1,3,5)Cyclophane and its derivatives. Extreme example of face-to-face crowding of aromatic rings. *J. Am. Chem. Soc.* **95,** 3201–3208 (1973).
42. Harpp, D. N., Gleason, J. G. & Snyder, J. P. Organic sulfur chemistry. I. The disulfide-phosphine reaction. Desulfurization with tris(diethylamino)phosphine. *J. Am. Chem. Soc.* **90,** 4181–4182 (1968).
43. Harpp, D. N. & Gleason, J. G. Organic sulfur chemistry. X. Selective desulfurization of disulfides. Scope and mechanism. *J. Am. Chem. Soc.* **93,** 2437–2445 (1971).
44. Bodwell, G. J., Bridson, J. N., Chen, S.-L. & Poirier, R. A. Nonplanar aromatic compounds. 5. 1A Strategy for the synthesis of cis-10b,10c-dimethyl-10b, 10c-dihydropyrenes. First crystal structure of a cis-10b,10c-dimethyl-10b, 10c-dihydropyrene. *J. Am. Chem. Soc.* **123,** 4704–4708 (2001).
45. Vögtle, F., Gross, J., Seel, C. & Nieger, M. $C_{36}H_{36}$—Tetrahedral clamping of four benzene rings in a spherical hydrocarbon framework. *Angew. Chem. Int. Ed.* **31,** 1069–1071 (1992).
46. Gleiter, R. & Hopf, H. (eds) *Modern Cyclophane Chemistry* (Weinheim: Wiley-VCH, 2004).
47. Houk, J. & Whitesides, G. M. Characterization and stability of cyclic disulfides and cyclic dimeric bis(disulfides). *Tetrahedron* **45,** 91–102 (1989).
48. Tozawa, T. *et al.* Porous organic cages. *Nat. Mater.* **8,** 973–978 (2009).
49. Jiang, S. *et al.* Porous organic molecular solids by dynamic covalent scrambling. *Nat. Commun.* **2,** 207 (2011).
50. Sheldrick, G. M. SADABS v.2.01, Bruker/Siemens area detector absorption correction program. (Bruker AXS, Madison, WI, USA, 1998).
51. Sheldrick, G. M. SHELXTL V.6.10, (Bruker AXS Inc., Madison, WI, 2000).
52. Sheldrick, G. M. Crystal structure refinement with SHELXL. *Acta Crystallogr. Sect. C* **C71,** 3–8 (2015).

Acknowledgements

We thank the University of Oregon for partial support of this work. D.W.J. is a Scialog Fellow of Research Corporation for Science Advancement (Award ID 20317). We gratefully acknowledge the use of UO CAMCOR facilities, which have been purchased with a combination of federal and state funding. We thank the University of Oregon Professors Ramesh Jasti and Michael M. Haley for helpful discussions, and we thank Holly A. Kuenzi and Eden B. ONeil for performing experiments on reaction scale-up and establishing thermodynamic control for the revised manuscript. We also greatly appreciate the helpful discussions with Professor Graham Bodwell in improving this manuscript.

Author contributions

M.S.C., M.E.C. and D.W.J. conceived and designed the experiments; M.S.C. performed the experiments; L.N.Z. performed the X-ray crystallography studies and wrote the sections of the paper related to single-crystal diffraction; M.S.C., M.E.C., B.P.N. and D.W.J. analysed the data; M.S.C. and D.W.J. co-wrote the paper with editorial assistance from M.E.C.; B.P.N. analysed the data and performed the experiments required during the revisions of this manuscript.

Additional information

Accession code: CCDC 1403626-1403631; CCDC 1421037 contain the supplementary crystallographic data for this paper. These data can be obtained free of charge from The Cambridge Crystallographic Data Centre via www.ccdc.cam.ac.uk/data_request/cif.

Competing financial interests: The authors declare no competing financial interests.

15

A prolific catalyst for dehydrogenation of neat formic acid

Jeff Joseph A. Celaje[1], Zhiyao Lu[1], Elyse A. Kedzie[1], Nicholas J. Terrile[1], Jonathan N. Lo[1] & Travis J. Williams[1]

Formic acid is a promising energy carrier for on-demand hydrogen generation. Because the reverse reaction is also feasible, formic acid is a form of stored hydrogen. Here we present a robust, reusable iridium catalyst that enables hydrogen gas release from neat formic acid. This catalysis works under mild conditions in the presence of air, is highly selective and affords millions of turnovers. While many catalysts exist for both formic acid dehydrogenation and carbon dioxide reduction, solutions to date on hydrogen gas release rely on volatile components that reduce the weight content of stored hydrogen and/or introduce fuel cell poisons. These are avoided here. The catalyst utilizes an interesting chemical mechanism, which is described on the basis of kinetic and synthetic experiments.

[1] Donald P. and Katherine B. Loker Hydrocarbon Institute and Department of Chemistry, University of Southern California, Los Angeles, California 90089-1661, USA. Correspondence and requests for materials should be addressed to T.J.W. (email: travisw@usc.edu).

Many strategies for the conversion of solar energy into chemical bonds involve electrocatalytic (or photocatalytic) cleavage of water to form hydrogen and oxygen. The reducing equivalent, H_2, is thus an energy carrier because it can be re-oxidized, either by combustion to give heat or catalytically in a fuel cell to give electricity. There is a disabling problem with large-scale utilization of hydrogen as a fuel, since it is a gas under ambient conditions, thus limiting its volume-energy density ($0.013\,MJ\,l^{-1}$). As a result, physical methods-based hydrogen storage technologies (compression, cryogenic liquefaction, adsorption) involve low capacity, high costs or safety issues[1]. Therefore, the discovery of highly weight-efficient strategies for on-demand hydrogen release from hydrogen-rich liquids has value. Formic acid (HCO_2H, FA, $7.5\,MJ\,l^{-1}$) is a hydrogen carrier, owing to its ability to release hydrogen under mild conditions with only CO_2 as a by-product[1–6], which can then be recycled, in principle, to give a carbon-neutral fuel cycle[7–9].

To date, many efficient heterogeneous[10–22] and homogeneous[9,23–38] catalysts for FA dehydrogenation have been developed. Heterogeneous catalysts have advantages of separability and reusability[11], while homogeneous catalysts are generally more efficient. The best turnover numbers (TONs) achieved in homogeneous catalysis are (1) >1M, by a catalyst system composed of $[RuCl_2(benzene)]_2$, the ligand diphenylphosphinoethane and a FA/Et_3N adduct as substrate developed by Boddien *et al.*[9] and (2) 983,642, by a system composed of an iron pincer complex and $LiBF_4$ developed by Bielinski *et al.*[35]. The highest turnover frequency achieved is $228{,}000\,h^{-1}$ by an iridium catalyst developed by Hull *et al.*[27]. Supplementary Table 1 has a more complete comparison of homogeneous catalysts for this reaction. In heterogeneous catalysis, the highest TOF achieved is $7{,}256\,h^{-1}$, by palladium nanoparticles immobilized on carbon nanospheres developed by Zhu *et al.*[22]. Also, homogeneous catalysts generally are more selective, producing less carbon monoxide, a common byproduct of FA dehydrogenation. This is essential, because CO is a fuel cell catalyst poison. Still, no known system is stable and reactive through multiple uses, air and water tolerant, selective against CO formation, and functions in neat FA liquid. Each of these is critical to achieving a usable hydrogen generation system based on FA. Herein we report an iridium-based catalytic system that meets all of these criteria.

Results

Reactivity in FA dehydrogenation. Complex **1**, which is easily prepared from known materials (Fig. 1), decomposes FA (500 µl, 12.7 mmol) with NaO_2CH co-catalyst (5 mol%) at 50 p.p.m. loading and 90 °C, resulting in the production of 386 ml of gas (62% conversion; TON = 12,530) after 13 h. The mass balance of FA condenses as a liquid in the reactor out of reach of the catalyst (*vide infra*). The rate of the reaction is constant through ca. 20% of conversion before it accelerates as FA disappears (Supplementary Fig. 1). At the end of the reaction, a pale orange solid (the catalyst system: an iridium complex and sodium formate) remains at the bottom of the reaction vessel. Recharging the reaction flask with FA and reheating to 90 °C results in continued H_2 production without any catalyst regeneration.

The reaction requires base as co-catalyst, but the source of the base is not specific: the reaction rates are similar when 5 mol% NaO_2CH, KO_2CH, KOH, NaOH, LiOH, or nBu_4NOH or 2.5 mol% of Na_2CO_3 or K_2CO_3 is used (Supplementary Table 2). Any of these is converted rapidly to the corresponding formate, which comprises the bulk of the catalytic material and gives it its pale color. Moreover, water does not affect the rate of dehydrogenation significantly—the double logarithmic plot of water concentration versus rate of FA dehydrogenation yields a slope of 0.11(5) (Supplementary Fig. 2).

The catalysts are air stable. Although dehydrogenation is slower when the catalysts are prepared in air, the system remains active, even when the solution is allowed to sit on the bench top for 2 weeks before dehydrogenation rates are measured (Supplementary Table 3). Under these conditions, the catalysts can be re-loaded in an air atmosphere and re-used repeatedly. For example, a reaction flask containing iridium **1** (6.1 mg, 8.9 µmol) and NaO_2CH (185 mg, 2.72 mmol) was charged with FA through 50 cycles (Supplementary Table 4). In this experiment 28.85 l of gas was produced from 25 ml of FA, corresponding to a TON of 66,403 and 89% conversion. Note that these values exclude the FA liquid that condenses in the flask out of reach of the catalyst in each run and accounts for all mechanical leaks or engineering challenges in the laboratory-scale set-up. In a particular, representative single run, we converted FA (2 ml) to gaseous products in 97% conversion with 140 p.p.m. **1** and 280 p.p.m. sodium formate. Over the course of 50 loadings, we measured the initial rates and maximum turnover frequencies during certain runs (Table 1). Interestingly, these increased over the course of 10 cycles before slowing over time.

The iridium catalyst delivers very high TONs at low loading with repeated re-use. For example, we prepared in the drybox a reaction flask containing **1** (90 µg, 0.13 µmol) and NaO_2CH (184 mg, 2.65 mmol) and repeatedly charged it with FA, which was decomposed until a pale yellow solid remained at the bottom of the flask. After 40 cycles over a period of four months, 13.71 l of gas was produced, which corresponds to a TON of 2.16 million (Supplementary Table 5). Although unoptimized, this is the best TON for a FA dehydrogenation catalyst known to date to our knowledge. Under these conditions, the maximum TOF was measured to be $3.7\,s^{-1}$.

The reaction is operationally simple. The catalytic materials are weighed out in a reactor, which is attached to a vent line for the gaseous products. Liquid FA is added, and the reaction is initiated by heating. Upon completion, the catalyst system remains as a pale-coloured precipitate at the bottom of the vessel for re-use.

To be useful in fuel cells, FA decomposition must be selective for H_2 and CO_2 over H_2O and CO, because CO is a poison for polymer electrolyte membrane (PEM) fuel cell catalysts such as

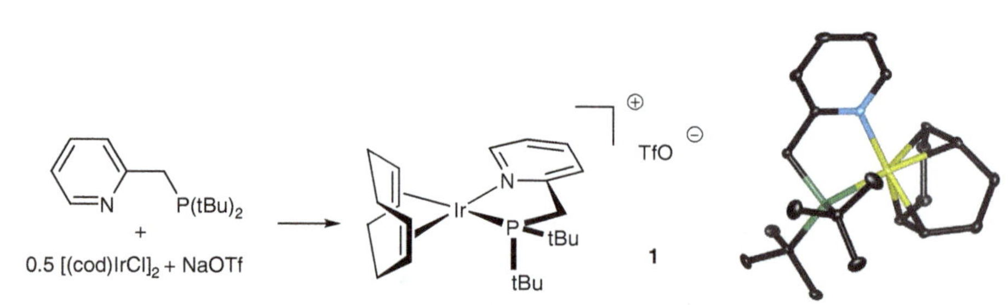

Figure 1 | Synthesis and structure of catalyst precursor cation 1. Elipsoids are drawn at the 50% probability level.

Table 1 | Catalyst performance over iterative uses.

Entry	Loading	Initial rate (μmol s^{-1})	Maximum turnover frequency (h^{-1})
1	1st	0.52	1,378
2	10th	2.35	3,032
3	20th	2.77	2,756
4	30th	2.82	2,618
5	40th	2.51	2,205
6	50th	1.46	1,519

platinum. The composition of gas produced from our conditions was determined by gas chromatography, which showed only H_2 and CO_2 (1:1 ratio) and no detectable CO (<1 part per thousand) (Supplementary Fig. 3). However, further analysis of the product gas by infrared spectroscopy revealed that when the reaction is conducted using neat FA, CO is observed at a concentration near the detection limit of the gas chromatography (Supplementary Fig. 4). It is known that neat FA decomposes in the presence of concentrated acid[39] or at high temperatures to form H_2O and CO[40,41]. We therefore hypothesized that much of the CO produced in our reaction conditions may be formed by thermal, uncatalysed decomposition pathways. Thus, we performed the dehydrogenation in the presence of a portion of water (10 v%), and observed that under these conditions the CO in the bulk gaseous products is <10 p.p.m. by infrared spectroscopy (Supplementary Fig. 5). Moreover, we hypothesized that the thermal decomposition of neat FA might be suppressed by running the reaction using higher sodium formate loading. Indeed, heating 1 ml of neat FA in 26 mg of the iridium precatalyst and 900 mg of sodium formate (50 mol%) to 90 °C for 2 h yields a product mixture with less than 10 p.p.m. CO (Fig. 2). A similarly low level of CO is observed when dehydrogenation is performed at 70 °C for 6 h (Supplementary Fig. 6). While these strategies for CO minimization are known in the FA literature, this collection of demonstrations enables practitioners to select the level of humidity and CO content in the reaction's gas eluent stream simply by adjusting the water and base loading in the FA supply. The optimum of these parameters might be different for any particular fuel cell application, but the reaction affords flexibility to adjust them.

We find that the active catalyst is homogeneous on the basis of physical appearance, clean kinetics, tolerance of liquid mercury and proportional inhibition by phenanthroline[42] (Supplementary Methods). Accordingly, the system exhibits the reactivity and selectivity advantages of homogeneous catalysis. Nonetheless, because the catalytic materials are deposited cleanly in the reactor at the end of the reaction, the system enjoys many of the catalyst separability and reusability benefits of heterogeneous conditions.

Mechanistic studies. Equally remarkable as the reactivity of this new catalytic system is the unique, two-metal mechanism through which it operates. We used three approaches to gain insight into this mechanism: stoichiometric model reactions, reaction kinetics and isotope labelling studies. Figures 3 and 4 present a sketch of a possible mechanism for our system.

Species **1** is a catalyst precursor from which an active catalyst is generated. To determine the nature of this active species, we conducted stoichiometric reactions of **1** (Fig. 3). Species **1** loses its cyclooctadiene ligand as cyclooctene in a solution of either H_2 or buffered FA and dimerizes to form **2**. Complex **2** has analogy to $\{[(P\text{-}N)Ir(CH_2Cl_2)(H)]_2(\mu^2\text{-}H)_2\}^{2+}$ characterized by Pfaltz[43] (P-N = SimplePHOX). In buffered FA conditions, **2** is then converted to a formate-bridged species **3a**. While this species is

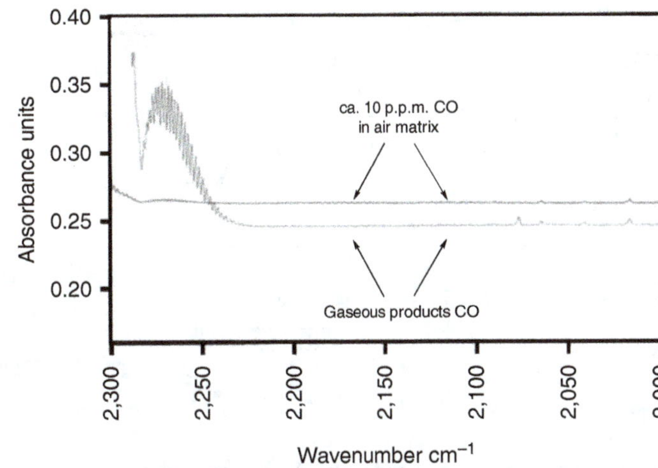

Figure 2 | Gas eluent stream infrared spectrum. The figure shows that compared to a prepared sample with 10 p.p.m. CO in air, the gaseous products from dehydrogenation of neat formic acid saturated in sodium formate contain <10 p.p.m. CO.

observable by NMR, it is not amenable to isolation in our hands. By contrast, its acetate homologue (**3b**) yielded to crystallization, which enabled determination of its structure (Fig. 3). Species **3a** is relevant in catalysis: we observe it by NMR as the minor form of the working catalyst. We see a second, major resting species by NMR, which has a spectrum consistent with structure **4**, featuring three differentiated metal hydride groups. The NMR spectra of intermediates are shown in Supplementary Figs 7–11, which are further analysed in Supplementary Discussion.

Kinetic isotope effect data indicate that both the C–H and O–H groups of FA are involved in (or before) the rate-determining transition state. Table 2 summarizes the reaction rates for four selectively labelled FA isotopologues. The combined isotope effect ($k_{CHOH}/k_{CDOD} = 6.5(2)$) is comparable to the product of the average separate C–H and O–H isotope effects (6.5(4)). This is consistent with a mechanism by which bonds to proton and hydride are transformed in a single kinetically relevant step. Hydrogen loss from **4** involves protonation of an iridium hydride (which comes from FA's C–H group) by a FA group. Further, we observe that in a sample of FA-(O)-d_1, NMR reveals H-D gas as the catalytic product (Supplementary Fig. 12 and Supplementary Discussion). A small portion of H_2 is formed in the process of catalyst initiation, but none is detected during catalysis. This indicates that there is separation of proton and hydride groups throughout the mechanism and refutes the possibility of an iridium dihydride species in the mechanism, because such a species would be likely to enable proton/hydride scrambling via reversible reductive elimination of dihydrogen. This observation also shows that the reaction is irreversible at ambient pressure, so we assign the isotope effects as kinetic.

Eyring analysis reveals activation parameters of $\Delta H^{\ddagger} = +29.0(3)$ kcal mol^{-1} and $\Delta S^{\ddagger} = +16(1)$ eu ($\Delta H^{\ddagger} = 121(12)$ kJ mol^{-1}; $\Delta S^{\ddagger} = +67(4)$ J mol^{-1} K^{-1}) (Supplementary Fig. 13). This strongly favourable entropy of activation is consistent with the release of at least one gaseous product in the rate-determining transition state. We expect that this is H_2 release in the conversion of **4** to **6** because of the strong isotope effects.

The observed rate law for FA dehydrogenation has rate $\sim [Ir]^1[base]^{0.5}[FA]^{-1}$, which is based on the slopes of double logarithmic plots (Supplementary Figs 14–18) recorded both in neat FA and dilute in tetraglyme solution (Table 3). This rate law requires that two sites of the catalyst are activated by a

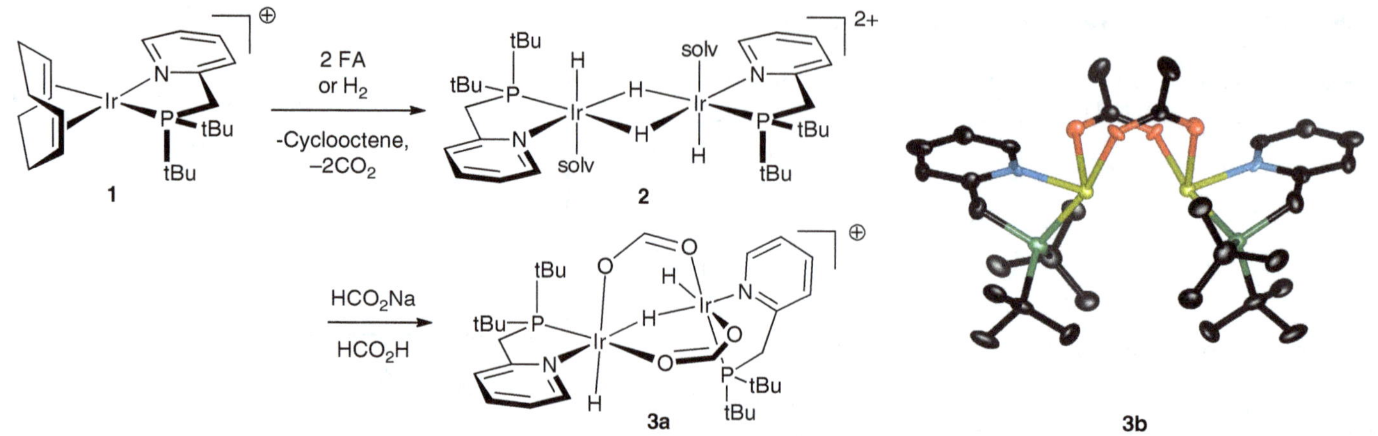

Figure 3 | Catalyst initiation and molecular structure of active catalyst homologue 3b. 3b = $\{[(tBu_2PCH_2(2\text{-}py))Ir(H)]_2(\mu^2\text{-}H)(\mu^2\text{-}\kappa,\kappa'\text{-}O_2CCH_3)_2\}^+$. Hydrogen atoms omitted. Ellipsoids are drawn at the 50% probability level. counterion, trifluoromethanesulfonate; solv, solvent.

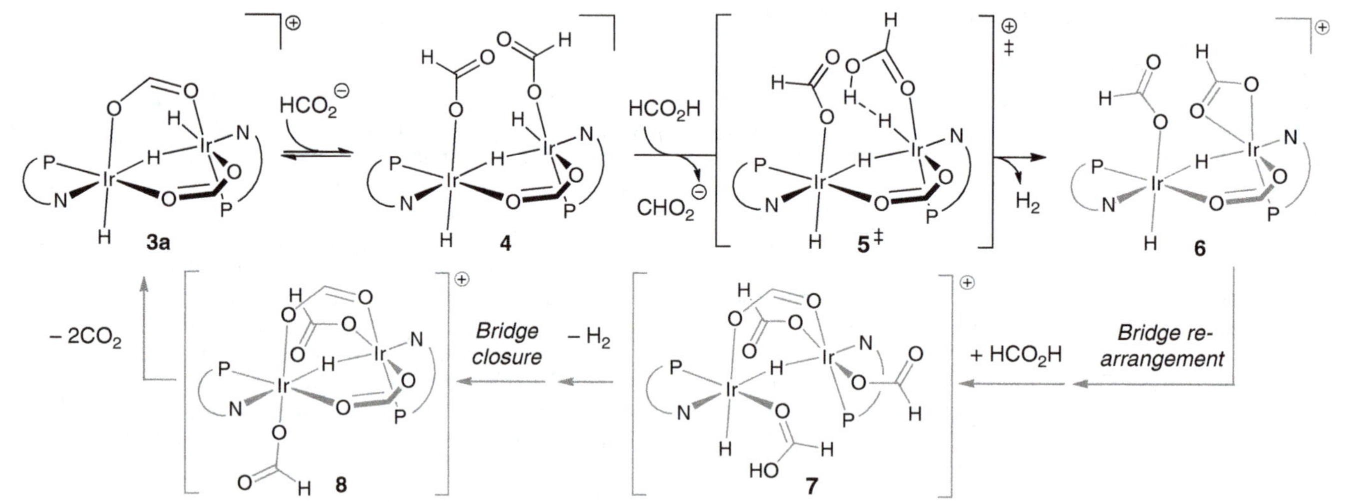

Figure 4 | Proposed mechanism of catalysis. Counterion, trifluoromethanesulfonate. ‡Refers to a proposed transition state. The P-N ligand is 2-((di-tbutylphosphino)methyl)pyridine.

Table 2 | Kinetic isotope effect data.

Compound	k_{rel}	KIE (observed)	
HCO_2H	6.5 (2)	k_{CHOH}/k_{CHOD}	1.8 (3)
HCO_2D	3.6 (2)	k_{CDOH}/k_{CDOD}	1.65 (3)
DCO_2H	1.6 (2)	k_{CHOH}/k_{CDOH}	3.9 (2)
DCO_2D	1.00 (2)	k_{CHOD}/k_{CDOD}	3.6 (2)
		k_{CHOH}/k_{CDOD}	6.5 (2)

Conditions are 50 p.p.m. **1**, 5 mol% base, 86 °C.

single equivalent of formate, thus causing half-order dependence on base. We propose a possible catalytic cycle in Fig. 4. After the first equivalent of H_2 is lost in the conversion of **4** to **6**, a second equivalent forms from the iridium hydride on the complementary metal centre. We propose that the latter is more rapid than the former, and that the single equivalent of formate enables both by opening a formate bridge in dimer **3a**. The rate law also has [Ir] first order, which indicates a dimeric iridium species that does not dissociate once formed. Inverse order in [FA] implies inhibition, but the origins of this inhibition are unclear. Acid is known to favour closure of carboxylate bridges in ruthenium species similar to ours[44], which enables several opportunities for FA inhibition in our mechanism. Moreover, FA has potential roles in the conversion of **4** and as solvent. We are currently studying this complex system of interactions.

Discussion

We show here a new catalytic system for the repeated conversion of FA to CO_2 and hydrogen. This has translation potential because it is the first known homogeneous system, to the best of our knowledge, to operate in neat FA, thus enabling far greater weight content of H_2 release than any other known catalyst for FA dehydrogenation. Moreover, it is the highest turnover system, because, in part, it can be re-used directly with FA substrate that is not rigorously purified or dried. We further propose a mechanism to account for kinetic, thermochemical, stoichiometric and labelling data that we have collected for the catalytic reaction. More detailed mechanistic studies, including computational investigation and ligand variation, are currently under way.

Methods

Synthesis of complex 1. General experimental information and characterization details can be found in the Supplementary Methods. In the drybox under nitrogen, 2-((di-t-butylphosphino)methyl)pyridine[45] (105.3 mg, 0.44 mmol) was dissolved in a dry vial in 5 ml of dry dichloromethane. In another vial containing a Teflon stir bar, chloro(1,5-cyclooctadiene)iridium(I) dimer (149.0 mg, 0.22 mmol) and sodium

Table 3 | Reaction kinetics.

	Neat*	Solution*,†
[Ir]	0.95(3)‡	0.96(4)‖
[base]	0.64(5)¶	0.44(2)#
[FA]	-	– 0.94(9)**

*Data were collected at 86 °C as an average of two runs.
†Tetraglyme was used as solvent. Base was delivered as $(n\text{-Bu})_4NOH$ to generate soluble $(n\text{-Bu})_4N(O_2CH)$.
‡Data were collected using 0.63 M $[NaO_2CH]$ (2.5 mol%) and [Ir] concentrations of 0.63, 1.86, 2.59, 3.25 and 4.41 mM.
‖Data were collected using 13.2 mM $[(n\text{-Bu})_4N(O_2CH)]$ (5 mol%) and [Ir] concentrations of 0.066, 0.13, 0.20, 0.26 and 0.33 mM.
¶Data were collected using 0.66 mM [Ir] and $[NaO_2CH]$ concentrations of 0.26, 0.53, 1.06, 1.59, 2.11 and 2.65 M.
#Data were collected using 0.066 mM [Ir] and $[(n\text{-Bu})_4N(O_2CH)]$ concentrations of 13.2, 26.4, 39.6, 52.8 and 66.0 mM.
**Data were collected using 0.026 [Ir], 13.2 mM $[(n\text{-Bu})_4N(O_2CH)]$ and [FA] concentrations of 265, 331, 398, 530 and 662 mM.

triflate (130 mg, 0.75 mmol) were suspended in 10 ml of dry dichloromethane. The suspension was stirred vigorously and then the phosphinopyridine solution was added slowly dropwise. The phosphinopyridine vial was rinsed with 5 ml of dichloromethane and added to the stirred suspension. After stirring for 1 h, the solution was filtered to remove the sodium chloride byproduct and the excess sodium triflate. The solvent was evaporated under reduced pressure to yield an orange glassy solid. A 5:1 mixture of dry hexanes/ethyl ether (10 ml) was added to the residue and then triturated by sonication. The hexane was decanted and the residue washed with an additional 10 ml of hexanes/ethyl ether. The pure iridium complex was dried under reduced pressure to give an orange solid (235 mg, 77.3%). Recrystallization from dichloromethane and toluene produced crystals suitable for X-ray crystallography (Supplementary Data 1). NMR spectra of complex **1** are shown in Supplementary Figs 19–22.

Dehydrogenation procedures. The dehydrogenation of FA can generally be performed by preparing a stock solution of the catalysts. In the drybox, formate and the iridium precatalyst are dissolved in either FA or tetraglyme solvent. The resulting orange solution slowly turns pale yellow over the course of ca. 1 h. The solution is allowed to sit for several hours or overnight before the catalyst is used for dehydrogenation reactions.

Reactions followed to completion. In the drybox, a 0.5-ml aliquot of a stock solution is transferred into a 5-ml high-pressure reaction flask possessing a side arm and a large-bore plug valve. The flask is then taken out of the drybox and connected to a vent line leading to a gas burette filled with oil (a eudiometer). To follow the reaction to completion, a 1,000-ml gas burette is used. The reaction flask is heated to 90 °C in an oil bath for ca. 15 min before opening the valve. The volume of gas produced over time is recorded. Some portion of the liquid FA vapourizes and re-condenses in the head space of the flask and gas evacuation tube, which prevents complete conversion.

Accurate measurements of initial rates. In the drybox, a 0.5-ml aliquot of a stock solution is transferred into a 5-ml reaction flask possessing a large-bore plug valve and a side arm. This flask is taken out of the drybox and the sidearm is connected to a three-way valve, which is connected to a nitrogen line and a 50-ml gas burette. The tubing and gas burette are purged with nitrogen for ca. 15 min. The reaction flask is then heated in an oil bath to 86 °C. Because the oil bath temperature increases after initial heating, the mixture is heated for ca. 15 min before readings are taken to allow the temperature to equilibrate. The volume of gas formed over time was then recorded. The initial rate of FA decomposition (average of two runs) was obtained from a plot of moles of FA decomposed versus time (20 data points were obtained in each experiment). A sample plot is shown in Supplementary Fig. 23.

Method of initial rates. Because the rate of FA dehydrogenation is constant at the beginning of the reaction (Supplementary Fig. 1), we used the method of initial rates to obtain kinetics data to compare directly the dehydrogenation rates when different bases are utilized, to construct an Eyring plot, to measure kinetic isotope effects, and to determine the effect of water and of different poisons (that is, mercury and phenanthroline). Also, this method was used to study the reaction order in the iridium catalyst, the base and FA to determine the rate law. See the Supplementary Information for full details.

Obtaining the X-ray structure of the catalyst homolog 3b. In the drybox under nitrogen, complex **1** (10 mg, 14.9 µmol) was dissolved in 0.6 ml dichloromethane-d_2 in a J-Young NMR tube. Dry acetic acid (8.6 µl, 149 µmol) was also added to this solution. The tube was then degassed, put under 1 atm head pressure of H_2 gas, and shaken. After ca. 5 min, a ^1H NMR spectrum of the crude reaction mixture was obtained, which confirmed the formation of **2**. The solution was then poured into a dry 1-dram vial. Hexane was carefully layered on top of this dichloromethane solution, and the vial was left in a desiccator for 1 week. A crystal suitable for X-ray diffraction was isolated from the vial (Supplementary Data 2). Although the crystal of **3b** is stable for days, the pure crystal of **3b** re-dissolved in dichloromethane-d_2 appears to be in equilibrium with **2** and potentially any other form of the iridium complex. NMR spectra of complex **3b** are shown in Supplementary Figs 24–27.

References

1. Enthaler, S. Carbon dioxide—the hydrogen-storage material of the future? *ChemSusChem.* **1,** 801–804 (2008).
2. Joo, F. Breakthroughs in hydrogen storage—formic acid as a sustainable storage material for hydrogen. *ChemSusChem.* **1,** 805–808 (2008).
3. Grasemann, M. & Laurenczy, G. Formic acid as a hydrogen source – recent developments and future trends. *Energy Environ. Sci.* **5,** 8171–8181 (2012).
4. Laurenczy, G. & Dyson, P. J. Homogeneous catalytic dehydrogenation of formic acid: progress towards a hydrogen-based economy. *J. Braz. Chem. Soc.* **25,** 2157–2163 (2014).
5. Boddien, A. *et al.* M. Iron-catalyzed hydrogen production from formic acid. *J. Am. Chem. Soc.* **132,** 8924–8934 (2010).
6. Czaun, M. *et al.* Hydrogen generation from formic acid decomposition by ruthenium carbonyl complexes. Tetraruthenium dodecacarbonyl tetrahydride as an active intermediate. *ChemSusChem.* **4,** 1241–1248 (2011).
7. Tanaka, R., Yamashita, M. & Nozaki, K. Catalytic hydrogenation of carbon dioxide using Ir(III)-pincer complexes. *J. Am. Chem. Soc.* **131,** 14168–14169 (2009).
8. Loges, B., Boddien, A., Junge, H. & Beller, M. Controlled generation of hydrogen from formic acid amine adducts at room temperature and application in H_2/O_2 fuel cells. *Angew. Chem. Int. Ed.* **47,** 3962–3965 (2008).
9. Boddien, A., Loges, B., Junge, H. & Beller, M. Hydrogen generation at ambient conditions: application in fuel cells. *ChemSusChem.* **1,** 751–758 (2008).
10. Zhou, X. *et al.* High-quality hydrogen from the catalyzed decomposition of formic acid by Pd–Au/C and Pd–Ag/C. *Chem. Commun.* **44,** 3540–3542 (2008).
11. Huang, Y., Zhou, X., Yin, M., Liu, C. & Xing, W. Novel PdAu@Au/C core-shell catalyst: superior activity and selectivity in formic acid decomposition for hydrogen generation. *Chem. Mater.* **22,** 5122–5128 (2010).
12. Tedsree, K. *et al.* Hydrogen production from formic acid decomposition at room temperature using a Ag-Pd core-shell nanocatalyst. *Nat. Nanotechnol.* **6,** 302–307 (2011).
13. Gu, X., Lu, Z.-H., Jian, H.-L., Akita, T. & Xu, Q. Synergistic catalysis of metal-organic framework-immobilized Au-Pd nanoparticles in dehydrogenation of formic acid for chemical hydrogen storage. *J. Am. Chem. Soc.* **133,** 11822–11825 (2011).
14. Bi, Q.-Y. *et al.* Efficient subnanometric gold-catalyzed hydrogen generation via formic acid decomposition under ambient conditions. *J. Am. Chem. Soc.* **134,** 8926–8933 (2012).
15. Zhang, S., Metin, Ö., Su, D. & Sun, S. Monodisperse AgPd alloy nanoparticles and their superior catalysis for the dehydrogenation of formic acid. *Angew. Chem. Int. Ed.* **52,** 3681–3684 (2013).
16. Wang, Z.-L. *et al.* An efficient CoAuPd/C catalyst for hydrogen generation from formic acid at room temperature. *Angew. Chem. Int. Ed.* **52,** 4406–4409 (2013).
17. Kohsuke, M., Dojo, M. & Yamashita, H. Pd and Pd–Ag nanoparticles within a macroreticular basic resin: an efficient catalyst for hydrogen production from formic acid decomposition. *ACS Catal.* **6,** 1114–1119 (2013).
18. Zhu, Z.-L., Tsumori, N. & Xu, Q. Sodium hydroxide-assisted growth of uniform Pd nanoparticles on nanoporous carbon MSC-30 for efficient and complete dehydrogenation of formic acid under ambient conditions. *Chem. Sci.* **5,** 195–199 (2014).
19. Jian, K., Xu, K., Zou, S. & Cai, W.-B. B-Doped Pd catalyst: boosting room-temperature hydrogen production from formic acid–formate solutions. *J. Am. Chem. Soc.* **136,** 4861–4864 (2014).
20. Chen, Y., Zhu, Q.-L., Tsumori, N. & Xu, Q. Immobilizing highly catalytically active noble metal nanoparticles on reduced graphene oxide: a non-noble metal sacrificial approach. *J. Am. Chem. Soc.* **137,** 106–109 (2015).
21. Zheng, Z., Tachikawa, T. & Majima, T. Plasmon-enhanced formic acid dehydrogenation using anisotropic Pd–Au nanorods studied at the single-particle level. *J. Am. Chem. Soc.* **137,** 948–957 (2015).
22. Zhu, Q.-L., Tsumori, N. & Xu, Q. Immobilizing extremely catalytically active palladium nanoparticles to carbon nanospheres: a weakly-capping growth approach. *J. Am. Chem. Soc.* **137,** 11743–11748 (2015).
23. Gao, Y., Kuncheria, J., Puddephatt, R. J. & Yap, G. P. A. An efficient binuclear catalyst for decomposition of formic acid. *Chem. Commun.* **34,** 2365–2366 (1998).
24. Fellay, C., Dyson, P. J. & Laurenczy, G. A viable hydrogen-storage system based on selective formic acid decomposition with a ruthenium catalyst. *Angew. Chem. Int. Ed.* **47,** 3966–3968 (2008).

25. Himeda, Y. Highly efficient hydrogen evolution by decomposition of formic acid using an iridium catalyst with 4,4'-dihydroxy-2,2'-bipyridine. *Green Chem.* **11,** 2018–2022 (2009).
26. Boddien, A. *et al.* Efficient dehydrogenation of formic acid using an iron catalyst. *Science* **333,** 1733–1736 (2011).
27. Hull, J. G. *et al.* Reversible hydrogen storage using CO_2 and a proton-switchable iridium catalyst in aqueous media under mild temperatures and pressures. *Nat. Chem.* **4,** 383–388 (2012).
28. Barnard, J. H., Wang, C., Berry, N. G. & Xiao, J. Long-range metal-ligand bifunctional catalysis: cyclometallated iridium catalysts for the mild and rapid dehydrogenation of formic acid. *Chem. Sci.* **4,** 1234–1244 (2013).
29. Mellone, I. *et al.* Formic acid dehydrogenation catalyse by ruthenium complexes bearing the tripodal ligands triphos and NP_3. *Dalton Trans.* **42,** 2495–2501 (2013).
30. Oldenhof, S. *et al.* Base-free production of H_2 by dehydrogenation of formic acid using an iridium-bisMETAMORPhos complex. *Chem. Eur. J.* **19,** 11507–11511 (2013).
31. Sponholz, P., Mellmann, D., Junge, H. & Beller, M. Towards a practical setup for hydrogen production from formic acid. *ChemSusChem.* **6,** 1172–1176 (2013).
32. Czaun, M. *et al.* Formic acid as a hydrogen storage medium: ruthenium-catalyzed generation of hydrogen from formic acid in emulsions. *ACS Catal.* **4,** 311–320 (2014).
33. Manaka, Y. *et al.* Efficient H_2 generation from formic acid usine azole complexes in water. *Catal. Sci. Technol.* **4,** 34–37 (2014).
34. Myers, T. & Berben, L. A. Aluminum-ligand cooperation promotes selective dehydrogenation of formic acid to H_2 and CO_2. *Chem. Sci.* **5,** 2771–2777 (2014).
35. Bielinski, E. A. *et al.* Lewis acid-assisted formic acid dehydrogenation using a pincer-supported iron catalyst. *J. Am. Chem. Soc.* **136,** 10234–10237 (2014).
36. Oldenhof, S., Lutz, M., de Bruin, B., van der Vlugt, J. I. & Reek, J. N. H. Dehydrogenation of formic acid by Ir-bisMETAMORPhos complexes: experimental and computational insight into the role of a cooperative ligand. *Chem. Sci.* **6,** 1026–1034 (2015).
37. Thevenon, A., Frost-Pennington, E., Weijia, G., Dalebrook, A. F. & Laurenczy, G. Formic acid dehydrogenation catalyzed by Tris(TPPTS) ruthenium species: mechanism of the initial "fast" cycle. *ChemCatChem.* **6,** 3146–3152 (2015).
38. Bertini, F., Mellone, I., Ienco, A., Peruzzini, M. & Gonsalvi, L. Iron(II) complexes of the linear *rac*-tetraphos-1 ligand as efficient homogeneous catalysts for sodium bicarbonate hydrogenation and formic acid dehydrogenation. *ACS Catal.* **5,** 1254–1265 (2015).
39. DeRight, R. E. The decomposition of formic acid by sulfuric acid. *J. Am. Chem. Soc.* **55,** 4761–4764 (1933).
40. Yu, J. & Savage, P. E. Decomposition of formic acid under hydrothermal conditions. *Ind. Eng. Chem. Res.* **37,** 2–10 (1998).
41. Akiya, N. & Savage, P. E. Role of water in formic acid decomposition. *AIChE J.* **44,** 405–415 (1998).
42. Hagen, C. M., Vieille-Petit, L., Laurenczy, G., Süss-Fink, G. & Finke, R. G. Supramolecular triruthenium cluster-based benzene hydrogenation catalysis: fact or fiction? *Organometallics* **24,** 1819 (2005).
43. Gruber, S., Neuburger, M. & Pfaltz, A. Characterization and reactivity studies of dinuclear iridium hydride complexes prepared from iridium catalysts with N,P and C,N ligands under hydrogenation conditions. *Organometallics* **32,** 4702–4711 (2013).
44. Morris, D. J., Clarkson, G. J. & Wills, M. Insights into hydrogen generation from formic acid using ruthenium complexes. *Organometallics* **28,** 4133–4140 (2009).
45. Beddie, C., Wei, P. & Douglas, S. Titanium pyridyl-phosphinimide complexes: synthesis, structure, and ethylene polymerization catalysis. *Can. J. Chem.* **84,** 755–761 (2006).

Acknowledgements

This work was sponsored by the National Science Foundation (CHE-1054910) and the Hydrocarbon Research Foundation. We thank the NSF (DBI-0821671, CHE-0840366, CHE-1048807) and NIH (1 S10 RR25432) for sponsorship of research instrumentation. Fellowship assistance from the Sonosky Foundation of the USC Wrigley Institute (J.J.A.C., Z.L.) and a USC Provost Fellowship (E.A.K.) is gratefully acknowledged. We thank Dr Ralf Haiges for help with X-ray crystallography and Prof G.K. Surya Prakash, Dr Miklos Czaun and Huong Dang for help with gas chromatography. We thank Prof Michael Richmond (UNT) for insightful discussions.

Author contributions

J.J.A.C., Z.L. and T.J.W. conceived and designed the experiments, analysed the data, and wrote the paper. J.J.A.C., Z.L., E.A.K. and N.J.T. executed the experiments.

Additional information

Accession codes: The X-ray crystallographic coordinates for the structures reported in this Article have been deposited in the Cambridge Crystallographic Data Centre (CCDC) under deposition numbers CDCC #1415049 (**1**) and #1415050 (**3b**).

Competing financial interests: The authors declare no competing financial interests.

16

Tandem intercalation strategy for single-layer nanosheets as an effective alternative to conventional exfoliation processes

Sohee Jeong[1,*], Dongwon Yoo[1,*], Minji Ahn[1], Pere Miró[2], Thomas Heine[2] & Jinwoo Cheon[1]

Simple and effective generation of transition metal chalcogenides (TMCs) in a single-layer form has been a challenging task. Here we present a tandem molecular intercalation (TMI) as a new exfoliation concept for producing single-layer TMCs from multi-layer colloidal TMC nanostructures in solution phase. TMI requires tandem Lewis base intercalates, where short 'initiator' molecules first intercalate into TMCs to open up the interlayer gap, and the long 'primary' molecules then bring the gap to full width so that a random mixture of intercalates overcomes the interlayer force. Spontaneous exfoliation then yields single-layer TMCs. The TMI process is uniquely advantageous because it works in a simple one-step process under safe and mild conditions (that is, room temperature without sonication or H_2 generation). With the appropriate intercalates, we have successfully generated single-layer nanostructures of group IV (TiS_2, ZrS_2), group V (NbS_2) and VI (WSe_2, MoS_2) TMCs.

[1] Department of Chemistry, Yonsei University, Seoul 120-749, Korea. [2] Engineering and Science, Jacobs University Bremen, 28759 Bremen, Germany. * These authors contributed equally to this work. Correspondence and requests for materials should be addressed to J.C. (email: jcheon@yonsei.ac.kr).

Two-dimensional (2D) layered materials including transition metal chalcogenides (TMCs) with strong in-plane covalent bonds and weak interlayer van der Waals interactions exhibit a wide range of interesting anisotropic phenomena from charge transport to catalytic and optical properties[1–9]. Similar to graphene, such interesting effects are centred on single-layer TMCs and, as an example, when MoS_2, an indirect band gap semiconductor in bulk, is thinned to a single layer, a strong photoluminescence is observed due to the indirect-to-direct band gap change[6,7].

Single-layer or few-layer nanosheets can be obtained from bulk TMCs by several exfoliation techniques such as dry mechanical cleavage with Scotch tape[10,11] or solution-based exfoliation processes[12–26]. Intercalation of alkali metals by using elemental metals[12,13] (for example, potassium) or organo-alkali compounds (for example, butyllithium[14–16] and more recently sodium naphthalenide[17]) has been a widely used method for the exfoliation of layered materials. Such exfoliation and disintegration methods with K, air and EtOH-H_2O along with sonication of layered flakes have successfully generated monolayer nanoparticles of WS_2, BN and graphene[9,12,13]. Although quite effective, the use of reactive chemicals during exfoliation conditions and H_2 generation through multiple exfoliation processes has drawbacks[17]. While the electrochemical exfoliation method has been developed as a faster and controllable exfoliation protocol, complicated electrochemical set-ups present challenges for the mass production of single-layer nanosheets[18–21]. Alternatively, a sonication method in the presence of solvents[22,23] such as *N*-methyl-pyrrolidone has been demonstrated for the successful exfoliation of wide range of TMCs, but the products are often mixed with single- and multi-layer TMCs with potential degradation of samples due to the harsh sonication conditions required to overcome the interlayer interactions[24–26]. As stated here, current methodologies have their own advantages and disadvantages, but none of them are yet perfect[17,26], although it is true that these are effective for large-size layered nanostructures. The development of mild exfoliation strategies that can circumvent the key harsh process conditions (for example, sonication and H_2 gas formation) has been challenging and, therefore, new and better routes to obtain single-layer TMCs in a simple, efficient and reproducible manner are currently being pursued. In this study, we introduce the 'tandem molecular intercalation (TMI)' process as a mild and effective exfoliation strategy for single-layer 2D TMC nanosheets from multi-layer colloidal TMC nanostuctures. Our TMI process utilizes two different Lewis base intercalates for the initiative and primary roles. Relatively short length 'initiator' Lewis base molecules are introduced into the interlayer space of 2D TMCs to start the gap-widening process, and a simultaneous influx of 'primary' intercalate molecules with longer chains creates randomly mixed bilayers of intercalates to overcome the interlayer interactions and eventually generate single-layer nanosheets (Fig. 1). Although the host–guest chemistry for the intercalation of Lewis bases into bulk TMCs has been well documented[27,28], its successful extension to an exfoliation strategy has not been demonstrated yet.

Results

Intercalation of alkylamine in TiS_2 colloidal nanostructure. We first examined the group IV TiS_2 colloidal nanostructure (60 nm in diameter with a dozen layers) with an interlayer distance of 5.7 Å and with each single-layer TiS_2 comprising a S-Ti-S tri-atomic plate of 1T-type CdI_2 structure[29] (Fig. 2a). The intercalation and gap-widening capability of Lewis base molecular intercalates were investigated depending on the alkyl chain length of primary amines such as propyl-, butyl- and hexyl-amine (Fig. 2b)[29,30]. Here, an excess amount of intercalate is added to TiS_2 nanoparticle in dimethyl sulphoxide (DMSO) at room temperature, and the reaction mixture is stirred for 30 min to 5 h depending on the chain length of the intercalate molecules. Subsequently, the colloidal suspension is precipitated by centrifugation and washed with chloroform. The initial interlayer distance of pristine TiS_2 is 5.7 Å, but after intercalation with propyl- butyl- and hexyl-amine, it increases to 9.5, 10.5 and 14.3 Å, respectively, according to TEM analysis (Fig. 2c–f).

XRD analysis also confirms the widening interlayer distance (Fig. 2h). Interlayer gap dependent *c* axis peaks, such as (001), (002) and (003), are shifted to lower angles as the intercalate size increases. Pronounced effects are observed for the (001) peak position, where the 2θ angle changes from 15.5° to 9.0°, 8.7° and 6.3° for propylamine, butylamine and hexylamine, respectively. From the 2θ values, the interlayer distances are calculated to be 5.7, 9.8, 10.2 and 14.0 Å, and the expansion of the *c* axis is 4.1 Å for propylamine, 4.5 Å for butylamine and 8.3 Å for hexylamine (Fig. 2j), respectively. This (001) peak shift trend and TEM analysis confirm that the interlayer gap widening of TiS_2 is proportional to the alkylamine length. As the lengths of propyl-, butyl- and hexyl-amine are ~4.1, 5.1 and 7.7 Å (Fig. 2i), respectively, the observed expansion of the interlayer gap suggests a head-to-tail monolayer arrangement of alkylamines between the layers[31].

This alkylamine intercalation process proceeds through a monolayer arrangement at the initial stage, but after lengthened intercalation time, the intercalates start to adopt a bilayer arrangement between the layers. With the use of propylamine, the 2θ angle of the (001) peak position of TiS_2 shifts from 15.5° to 8.9° at 45 min and 6.4° at 7 h (Fig. 3a). At the intermediate time of 3 h, the two peaks at 8.9° and 6.4° are simultaneously observed. These XRD results suggest that the TiS_2 interlayer distance of 9.9 Å corresponds to a monolayer of propylamine, which subsequently changes to a bilayer formation (13.8 Å) through an intermediate state (3 h) (Fig. 3b). The intercalation modes of alkylamines follow monolayer to bilayer arrangements, and the interlayer gap widening can be controlled using different alkylamine intercalates. However, none of the layered TiS_2 show spontaneous exfoliation even though the London-dispersion-corrected DFT calculation estimates a critical interlayer distance of ~11 Å to overcome the interlayer van der Waals interaction between two layers of pristine TiS_2 (Fig. 3c). Our observation of non-exfoliation phenomenon is presumably due to the additional energy needed to overcome partially or fully interdigitated interactions between the alkyl chains of the intercalates (Fig. 3b)[30,32,33].

Exfoliation of TMCs via TMI method. Here, we devised the TMI concept, where two different lengths of intercalates are utilized for the exfoliation of layered nanostructures. This TMI process has distinctive features, namely a short 'initiator' alkylamine first expands the interlayer gap for efficient intercalation of the 'primary' long alkylamine; the difference in the chain length, then, generates a bilayer arrangement with empty space between intercalates to reduce van der Waals force between them; and finally spontaneous exfoliation occurs (Fig. 4a). We examined the TMI concept for the exfoliation of multi-layer TiS_2. By adding propylamine and hexylamine to colloidal TiS_2 nanostructure in DMSO at room temperature, the intercalation of propylamine is observed with an expanded interlayer distance of 9.5 Å in 30 min (Fig. 4b), and the presence of intercalated propylamine is

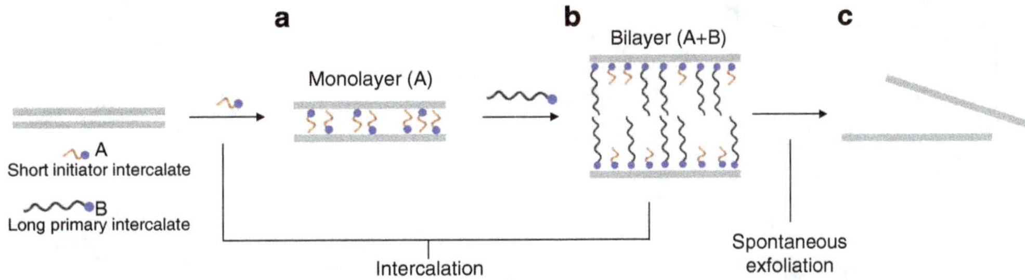

Figure 1 | Tandem molecular intercalation process for exfoliation of TMCs. The TMI process incorporates two different Lewis base intercalates. (**a**) Short 'initiator' intercalates (orange tail) are intercalated first into the interlayer gap, and (**b**) long 'primary' intercalates (black tail) then enter to form randomly mixed bilayers of intercalates and widen the interlayer gap. (**c**) Finally, TMCs with bilayer intercalates are spontaneously delaminated to single-layer nanosheets.

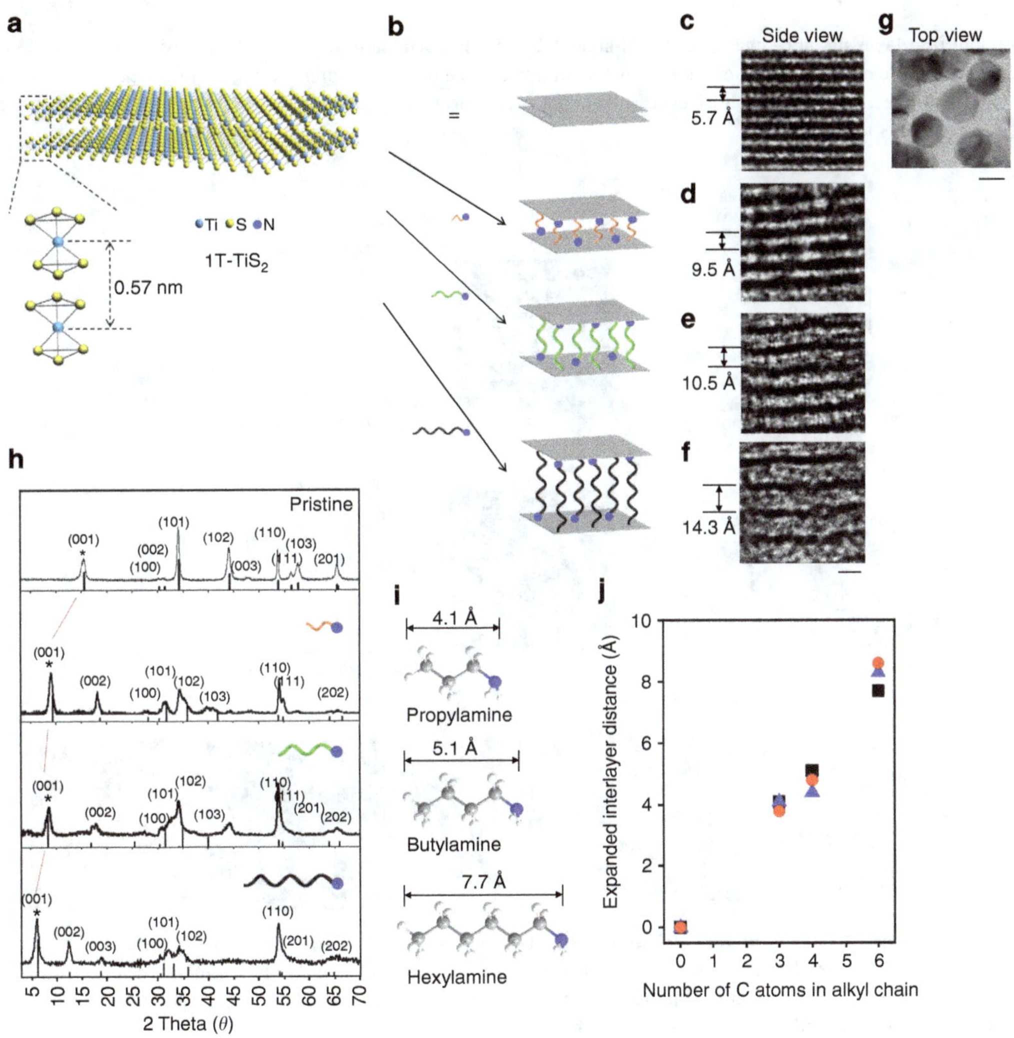

Figure 2 | Intercalate size-dependent interlayer expansion of multi-layer TiS_2. (**a**) Visualization of TiS_2 using ball and stick model to illustrate the 2D layered structure. (**b**) Schematic illustration of interlayer distance change of TiS_2 by intercalating alkylamine. Side view of layers in TEM images (**c**) before intercalation and after intercalation of (**d**) propylamine, (**e**) butylamine and (**f**) hexylamine. Scale bar, 1 nm. (**g**) Top-view TEM image of multi-layer TiS_2 before exfoliation. Scale bar, 50 nm. (**h**) XRD patterns of pristine TiS_2 and intercalated TiS_2 with propylamine, butylamine and hexylamine. Vertical lines indicate reference peaks of bulk and expanded TiS_2 (JCPDS #15-0853). (**i**) Calculated length between nitrogen and terminal hydrogen of propylamine (4.1 Å), butylamine (5.1 Å) and hexylamine (7.7 Å) by ChemBio 3D® Ultra program. (**j**) Expanded interlayer distance of TiS_2 plotted against the number of carbon (C) atoms in alkyl chain of intercalate (black square, calculated value; blue triangle, value observed by XRD; red circle, value observed by TEM).

confirmed by ^{1}H-NMR. Then, further expansion of the layer distance to 21 Å is observed in 6 h (Fig. 4c), and the coexistence of propylamine and hexylamine is confirmed by ^{1}H-NMR. At this stage, the layers spontaneously begin to separate, and single-layer TiS_2 nanosheets are obtained. Figure 4d,e show TEM images of single-layer TiS_2 nanosheets after exfoliation. A top-view high-resolution TEM (HRTEM) image of a single-layer TiS_2 nanosheet shows lattice fringes with interplanar spacings of 2.9 and 1.7 Å,

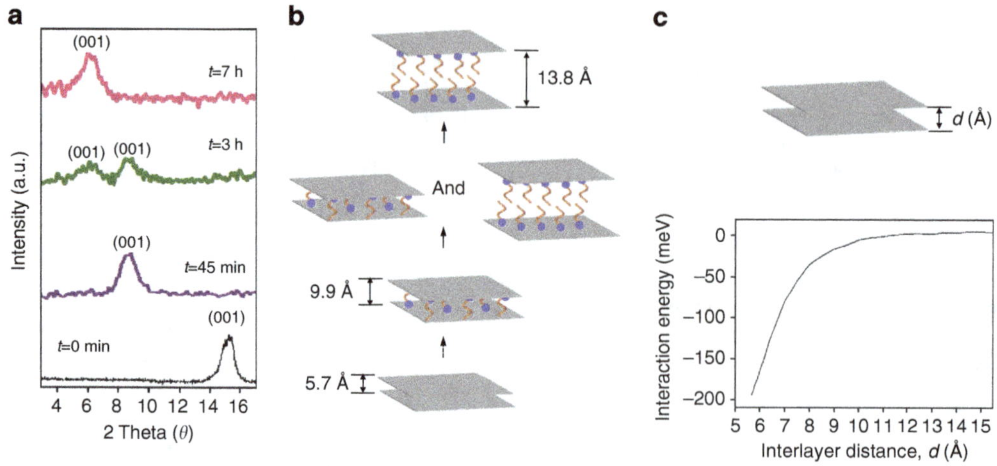

Figure 3 | Time-dependent interlayer distance change of multi-layer TiS_2 with propylamine. (**a**) XRD patterns of intercalated TiS_2 at 0 min, 45 min, 3 h and 7 h after treatment of propylamine. (**b**) Illustration of interlayer distance change of TiS_2 by intercalating propylamine. (**c**) Interlayer interaction energy calculated by density functional theory at the PBE-D3-BJ level. The critical interlayer distance for pristine TiS_2 exfoliation is estimated to be ~11.0 Å, where the interaction energy is close to $\sim k_BT$.

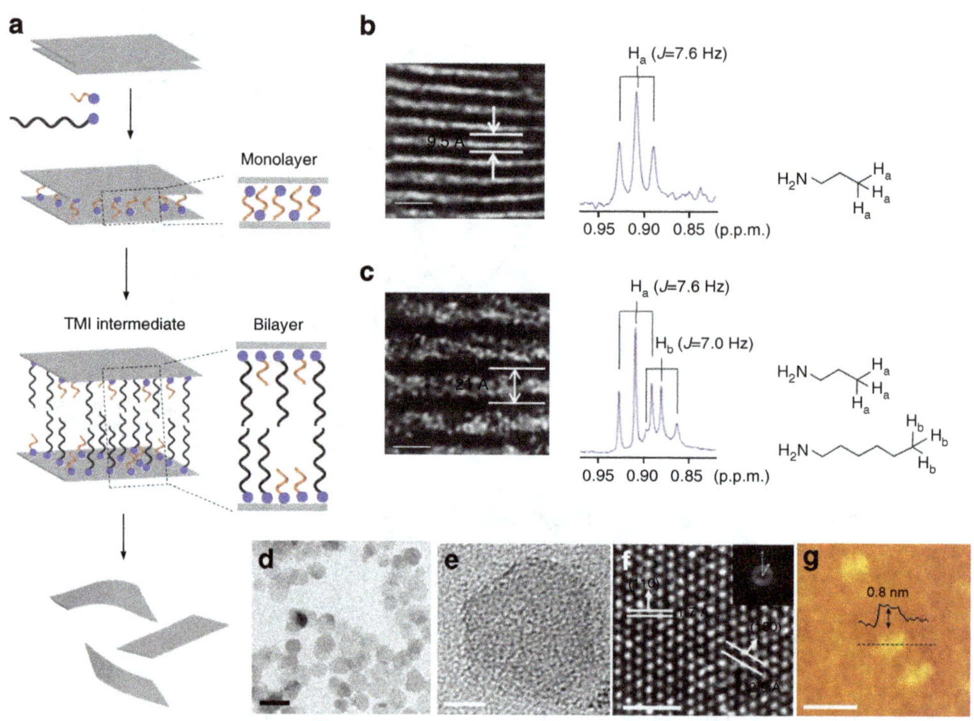

Figure 4 | Exfoliation of multi-layer TiS_2 nanostructure by TMI process. (**a**) Illustration of tandem molecular intercalates (TMI) process. Side-view TEM image and ^{1}H-NMR spectrum of TiS_2 intercalated with (**b**) propylamine and (**c**) both propylamine and hexylamine. (**d**) Top-view and (**e**) magnified TEM images of TiS_2 after exfoliation. (**f**) HRTEM image and FFT pattern (inset) of a single-layer TiS_2 nanosheet. (**g**) AFM image of single-layer TiS_2 nanosheets. Scale bars, 2 nm in **b** and **c**; 100 nm in **d** and **g**; 20 nm in **e**; 1 nm in **f**.

corresponding to the (100) and (110) planes of the TiS_2, respectively (Fig. 4f). The single-layer thickness of 8.0 Å is confirmed by AFM analysis (Fig. 4g). Additional XRD analysis shows single peak of (110) plane at 53.7° without any peaks in *c* axis direction, which confirms that final products are fully exfoliated single-layer TiS_2 (Supplementary Figs 1a and 2b). Such experimental observations are in good agreement with the calculations based on density-functional tight-binding theory, which confirms that mixed intercalates of different chain lengths decrease the total interaction energy below $2k_BT$ to have an effective exfoliation (Supplementary Fig. 3, Supplementary Tables 1–3).

Our tandem molecular exfoliation method can be applied to other colloidal nanoparticles of group IV, V and VI TMCs, such as ZrS_2, NbS_2 and WSe_2. First, the exfoliation of multi-layer ZrS_2 nanostructure (17 nm in lateral size and three layers) is examined (Fig. 5a,b). Although less explored than group VI counterparts, group IV TMCs also exhibit interesting single-layer properties such as enhanced thermoelectric property and electrical conductivity for TiS_2 and strain-driven indirect-to-direct band gap transition of ZrS_2 (refs 34–36). Similar to the case of TiS_2, propylamine and hexylamine are added to ZrS_2 nanostructure in DMSO with stirring at room temperature. After 4 h, the interlayer distance is expanded from 5.8 Å (Fig. 5b) to 20.5 Å (Fig. 5c).

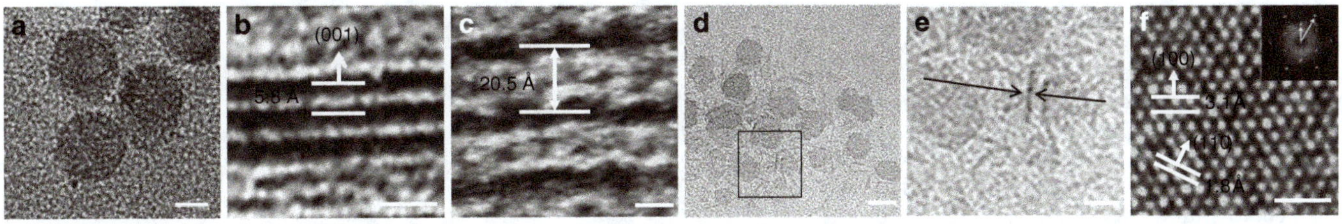

Figure 5 | Exfoliation of multi-layer ZrS_2 nanostructure. (**a**) Top-view and (**b**) side-view TEM image of three-layer ZrS_2 nanostructure before exfoliation. (**c**) Side-view TEM image of propylamine and hexylamine intercalated ZrS_2. (**d**) TEM image and (**e**) magnified TEM image of the single-layer ZrS_2 nanosheets in the box of **d**. A side-view of the single-layer ZrS_2 nanosheet between the arrows. (**f**) HRTEM image and FFT pattern (inset) of a single-layer ZrS_2 nanosheet. Scale bars, 10 nm in **a** and **e**; 1 nm in **b**, **c** and **f**; 20 nm in **d**.

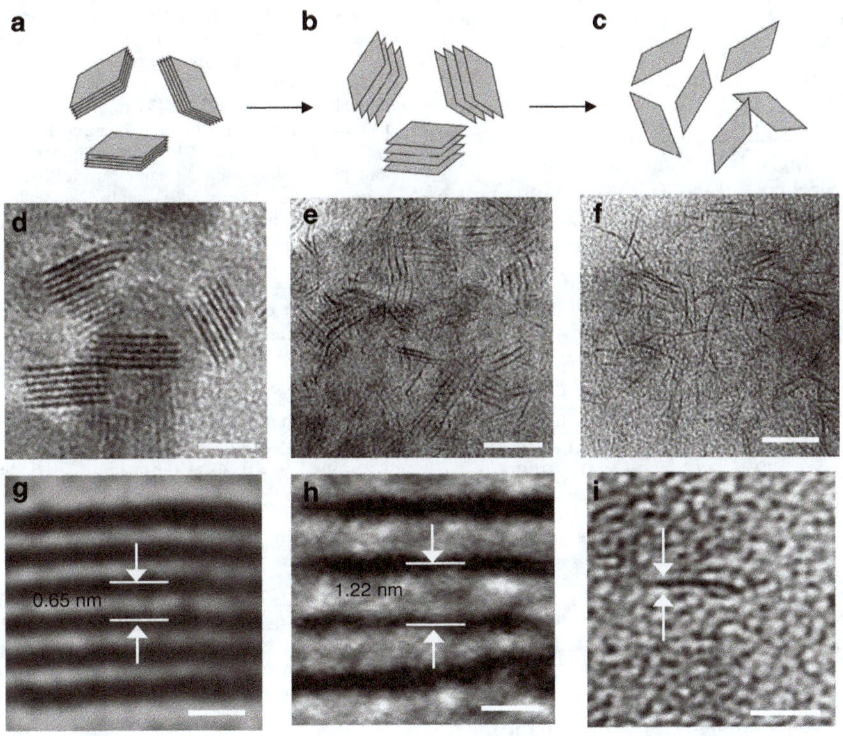

Figure 6 | Exfoliation of multi-layer WSe_2 nanostructure. (**a–c**) Schematic illustration of WSe_2 exfoliation processes. (**d–f**) Low-magnification and (**g–i**) magnified side-view TEM image of starting multi-layer WSe_2 nanostructures (**d**,**g**), WSe_2 intercalated with ethoxide in bilayer arrangement (**e**,**h**) and single-layer WSe_2 nanosheets (**f**,**i**). Scale bars, 5 nm in **d** and **i**; 10 nm in **e** and **f**; 1 nm in **g** and **h**.

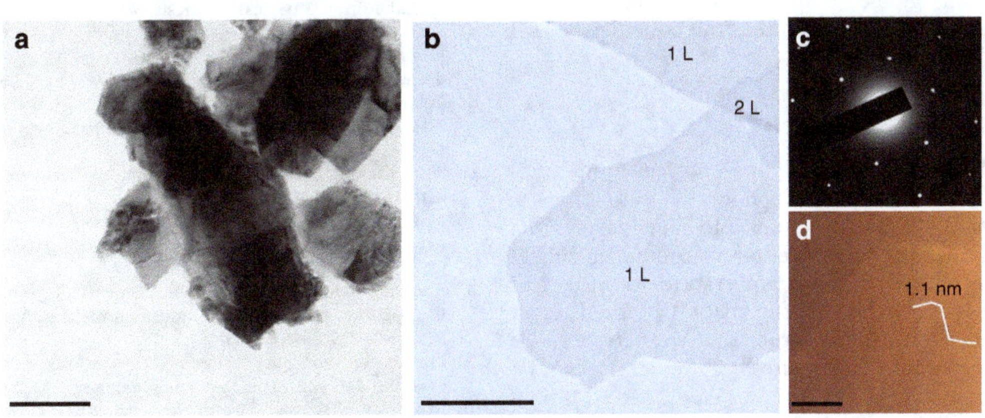

Figure 7 | Micron-size MoS_2 exfoliation by TMI method. (**a**) TEM image of micron size multi-layer MoS_2. (**b**) TEM image (pseudo-colour) of exfoliated single-layer MoS_2 (L, layer). (**c**) SAED pattern and (**d**) AFM image of exfoliated MoS_2. Scale bars, 1 µm in **a**; 0.5 µm in **b**; 0.1 µm in **d**.

After 9 h, single-layer ZrS_2 nanosheets are obtained after centrifugation (Fig. 5d). The thickness measured by TEM is ~6.0 Å, which is consistent with the estimated thickness of single-layer ZrS_2 (Fig. 5e). A top-view HRTEM image of a single-layer ZrS_2 nanosheet shows lattice fringes with interplanar spacings of 3.1 and 1.8 Å, corresponding to the (100) and (110) planes of the hexagonal 1T-ZrS_2, respectively (Fig. 5f and Supplementary Fig. 2c). This TMI method with propyl- and hexylamine intercalates is also effective to group V NbS_2 nanoparticles (Supplementary Fig. 4).

For the exfoliation of multi-layer WSe_2 nanostructure, tandem intercalates with high Lewis basicity are chosen. Two different-length intercalates of sodium ethoxide and sodium hexanolate are added to multi-layer WSe_2 nanostructure in DMSO. After stirring for 7 h at room temperature, single-layer WSe_2 nanosheets are isolated by centrifugation. During the course of intercalation, the expansion of the 6.5 Å interlayer spacing of WSe_2 (Fig. 6g) to 12.2 Å (Fig. 6h) suggests that ethoxide intercalation proceeds to form a bilayer arrangement between the layers. Figure 6f,i show low-magnification and magnified TEM images of single-layer WSe_2 nanosheets after exfoliation, where a side view of standing single-layer WSe_2 nanosheets is observed. This exfoliation process can be scaled up to sub-gram scale (Supplementary Fig. 5).

The choice of proper intercalates is critical to have their effective intercalation into TMCs. Intercalates act as Lewis base with highest occupied molecular orbital (HOMO) and TMCs act as Lewis acid with lowest unoccupied molecular orbital (LUMO). Therefore, as the HOMO–LUMO energy difference ($\Delta E_{HOMO-LUMO}$) is smaller, the more favoured their interaction becomes. Such HOMO–LUMO Lewis acid–base interaction can be generally applied to group IV and VI TMCs. LUMO of group VI TMCs (for example, MoS_2, $MoSe_2$, $MoTe_2$, WS_2, WSe_2 and WTe_2) is between -4.2 and -3.5 eV, which is in close energy level with HOMO (approximately -4.2 eV) of strong base alkoxides to have favoured interactions between them[37,38]. Similarly, LUMO of group IV TMCs (for example, TiS_2, ZrS_2, $ZrSe_2$, HfS_2 and $HfSe_2$) is ranging from -6.0 to -5.5 eV, which is close to HOMO (approximately -6.2 eV) of relatively weaker base alkylamines[37–39].

While our TMI method is most effective for the exfoliation of laterally small (<100 nm) TMC nanoparticles, it can be also extended to larger micron-size TMCs. For the exfoliation of micron-size TMCs (Fig. 7 and Supplementary Fig. 6), sodium ethoxide and hexanolate are added to a solution of MoS_2 in DMSO. After stirring 48 h, single-layer MoS_2 nanosheets over 1 μm are observed (Fig. 7). Similarly, micron-size multi-layer TMCs of WSe_2 and TiS_2 can also be exfoliated (Supplementary Figs 2a and 6). Although specific conditions (for example, time, type and concentration of intercalates, temperature and so on) should be further optimized for efficient exfoliation, this preliminary study indicates the potential of TMI method for large-size TMCs.

Discussion

In this study, we demonstrate that 'tandem molecular intercalation (TMI)' is an effective new concept for the exfoliation of multi-layer colloidal TMC nanostructures into single-layer TMC nanosheets. TMI is a relatively fast process for producing single-layer TMCs and is conducted at room temperature, with the clear merit of eliminating the occasional degradation of single-layer TMCs by avoiding harsh exfoliation processes. A variety of Lewis bases can be intercalated into TMC nanostructures, and group IV (TiS_2 and ZrS_2), group V (NbS_2) and VI (WSe_2, MoS_2) TMCs require different types of intercalates. A relatively weaker Lewis base, such as alklyamine, works nicely for group IV and V TMCs, whereas a stronger Lewis base, such as alkoxide, is required for group VI TMCs. In principle, TMI can serve as a general strategy for the exfoliation of wide range of colloidal TMC nanostructures under mild exfoliation conditions, namely, a room-temperature process without sonication or H_2 evolution.

Methods

Alkylamine chain length-dependent intercalation. Propylamine (0.07 ml, 0.91 mmol), pre-synthesized TiS_2 (ref. 40; 0.05 mmol) and DMSO (5 ml) are added to a 10 ml vial. The mixture is stirred for 30 min at room temperature. The mixture is then precipitated by centrifugation at 3,000 r.p.m. for 30 min, and the precipitates are washed three times with chloroform. The intercalation conditions for other alkylamines are identical except for the intercalation time (1 h for butylamine and 5 h for hexylamine).

Exfoliation of multi-layer TMC nanostructures. Multi-layer colloidal TiS_2 nanostructure (5.0 mg, 0.04 mmol)[40], propylamine (3.0 ml, 36.5 mmol), hexylamine (2.0 ml, 15.1 mmol) and DMSO (5 ml) are mixed to a 10 ml vial and stirred for 10 h at room temperature. Then, the mixture is centrifuged at 3,000 r.p.m. for 30 min to precipitate non-exfoliated materials. The supernatant is re-centrifuged at 15,000 r.p.m. for 50 min to precipitate exfoliated TMCs. Then, chloroform (10 ml) is added to dissolve the precipitate and the resulting purple solution is centrifuged at 17,500 r.p.m. for 10 min. This process repeats several times and then the final product is dried under vacuum at room temperature for 2 h to afford exfoliated TiS_2 (1.4 mg, 27%) as a purple powder.

The exfoliation of multi-layer ZrS_2 follows similar procedures. Multi-layer colloidal ZrS_2 nanostructure (5.0 mg, 0.03 mmol)[40], propylamine (2.5 ml, 30.4 mmol), hexylamine (1.0 ml, 7.6 mmol) and DMSO (5 ml) are mixed to a 10 ml vial and stirred for 9 h at room temperature. Single-layer ZrS_2 (1.6 mg, 32%) is obtained by following the same purification process of TiS_2.

For the exfoliation of WSe_2, multi-layer colloidal WSe_2 nanostructure (10.0 mg, 0.03 mmol), sodium ethoxide (1.50 g, 22.0 mmol), sodium hexanolate (0.50 g, 4.03 mmol) and DMSO (15 ml) are mixed. After 7 h stirring, the mixture is centrifuged at 6,000 r.p.m. for 30 min to precipitate the non-exfoliated materials. Then the supernatant is re-centrifuged at 15,000 r.p.m. for 60 min to precipitate exfoliated TMCs. The precipitate is washed several times with ethanol via centrifugation at 17,500 r.p.m. for 20 min to give exfoliated WSe_2 (3.3 mg, 33%).

Instruments. TEM and HRTEM analyses were performed using JEM 2100 at 200 kV and JEOL-ARM1300S at 1,250 kV, respectively. Atomic force microscopy (AFM) images were obtained in non-contact mode using a scanning probe microscope (Veeco). X-ray diffraction (XRD) studies were conducted using Rigaku-G2005304 equipped with a $Cu_{k\alpha}$ radiation source (30 kV, 15 mA). ^{1}H-NMR spectra were obtained using a 400 MHz FT-NMR Spectrometer (Bruker Biospin).

Density-functional theory calculations. The DFT calculations were performed using the Amsterdam Density-Functional (ADF2013-BAND) package[41,42]. We used the local VWN exchange-correlation potential with nonlocal Perdew-Burke–Ernzerhof exchange-correlation correction and empirical D3 treatment of London dispersion interactions (PBE-D3-BJ)[43].

References

1. Yoffe, A. D. Layer compounds. *Annu. Rev. Mater. Sci.* **3,** 147–170 (1973).
2. Frienda, R. H. & Yoffe, A. D. Electronic properties of intercalation complexes of the transition metal dichalcogenides. *Adv. Phys.* **36,** 1–94 (1987).
3. Jaramillo, T. *et al.* Identification of active edge sites for electrochemical H_2 evolution from MoS_2 nanocatalysts. *Science* **317,** 100–102 (2007).
4. Wang, Q. H., Kalantar-Zadeh, K., Kis, A., Coleman, J. N. & Strano, M. S. Electronics and optoelectronics of two-dimensional transition metal dichalcogenides. *Nat. Nanotechnol.* **7,** 699–712 (2012).
5. Lukowski, M. A. *et al.* Enhanced hydrogen evolution catalysis from chemically exfoliated metallic MoS_2 nanosheets. *J. Am. Chem. Soc.* **135,** 10274–10277 (2013).
6. Splendiani, A. *et al.* Emerging photoluminescence in monolayer MoS_2. *Nano Lett.* **10,** 1271–1275 (2010).
7. Mak, K. F., Lee, C., Hone, J., Shan, J. & Heinz, T. F. Atomically thin MoS_2: a new direct-gap semiconductor. *Phys. Rev. Lett.* **105,** 136805-1–136805-4 (2010).
8. Kuc, A., Zibouche, N. & Heine, T. Influence of quantum confinement on the electronic structure of the transition metal sulfide TS_2. *Phys. Rev. B* **83,** 245213-1–245213-4 (2011).
9. Lin, L. *et al.* Fabrication of luminescent monolayered tungsten dichalcogenides quantum dots with giant spin-valley coupling. *ACS Nano* **7,** 8214–8223 (2013).
10. Novoselov, K. S. *et al.* Two-dimensional atomic crystals. *Proc. Nat Acad. Sci. USA* **102,** 10451–10453 (2005).
11. Radisavljevic, B., Whitwick, M. B. & Kis, A. Integrated circuits and logic operations based on single-layer MoS_2. *ACS Nano* **5,** 9934–9938 (2011).
12. Lin, L. *et al.* Fabrication and luminescence of monolayered boron nitride quantum dots. *Small* **10,** 60–65 (2014).
13. Lin, L. & Zhang, S. Creating high yield water soluble luminescent graphene quantum dots via exfoliating and disintegrating carbon nanotubes and graphite flakes. *Chem. Commun.* **48,** 10177–10179 (2012).

14. Dines, M. B. Lithium interacalation via n-butyllithium of the layered transition metal dichalcogenides. *Mater. Res. Bull.* **10,** 287–292 (1975).
15. Joensen, P., Frindt, R. F. & Morrison, S. R. Single-layer MoS_2. *Mater. Res. Bull.* **21,** 457–461 (1986).
16. Eda, G. *et al.* Photoluminescence from chemically exfoliated MoS_2. *Nano Lett.* **11,** 5111–5116 (2011).
17. Zheng, J. *et al.* High yield exfoliation of two-dimensional chalcogenides using sodium naphthalenide. *Nat. Commun.* **5,** 2995 (2014).
18. Zeng, Z. *et al.* An effective method for the fabrication of few-layer-thick inorganic nanosheets. *Angew. Chem. Int. Ed.* **51,** 9052–9056 (2012).
19. Yuwen, L. *et al.* General synthesis of noble metal (Au, Ag, Pd, Pt) nanocrystal modified MoS_2 nanosheets and the enhanced catalytic activity of Pd–MoS_2 for methanol oxidation. *Nanoscale* **6,** 5762–5769 (2014).
20. Bissessur, R., Kanatzidis, M. G., Schindlerb, J. L. & Kannewurf, C. R. Encapsulation of polymers into MoS_2 and metal to insulator transition in metastable MoS_2. *J. Chem. Soc. Chem. Commun.* 1582–1585 (1993).
21. Gordon, R. A., Yang, D., Crozier, E. D., Jiang, D. T. & Frind, R. F. Structures of exfoliated single layers of WS_2, MoS_2, and $MoSe_2$ in aqueous suspension. *Phys. Rev. B* **65,** 125407-1–125407-9 (2002).
22. Coleman, J. N. *et al.* Two-dimensional nanosheets produced by liquid exfoliation of layered materials. *Science* **331,** 568–571 (2011).
23. Zhou, K.-G., Mao, N.-N., Wang, H.-X., Peng., Y. & Zhang, H.-L. A mixed-solvent strategy for efficient exfoliation of inorganic graphene analogues. *Angew. Chem. Int. Ed.* **50,** 10839–10842 (2011).
24. Smith, R. J. *et al.* Large-scale exfoliation of inorganic layered compounds in aqueous surfactant solutions. *Adv. Mater.* **23,** 3944–3948 (2011).
25. Mashtalir, O. *et al.* Intercalation and delamination of layered carbides and carbonitrides. *Nat. Commun.* **4,** 1716 (2013).
26. O'Neill, A., Umar, K. & Coleman, J. N. Preparation of high concentration dispersions of exfoliated MoS_2 with increased flake size. *Chem. Mater.* **24,** 2414–2421 (2012).
27. Gamble, F. R. *et al.* Intercalation complexes of Lewis bases and layered sulfides: a large of new superconductors. *Science* **174,** 493–497 (1971).
28. Levy, F. *Intercalated layered materials* (Springer Link, 1979).
29. Chianelli, R. R., Scanlon, J. C. & Thompson, A. H. Structure refinement of stoichiometric TiS_2. *Mater. Res. Bull.* **10,** 1379–1382 (1975).
30. Lagaly, G. Intercalation of alkylamines with different types of layered compounds. *Solid State Ion* **22,** 43–51 (1986).
31. Whittingham, M. S. & Chianelli, R. R. Layered compounds and intercalation chemistry: an example of chemistry and diffusion in solids. *J. Chem. Educ.* **57,** 569–574 (1980).
32. Ulman, A., Eilers, J. E. & Tillman, N. Packing and molecular orientation of alkanethiol monolayers on gold surfaces. *Langmuir* **5,** 1147–1152 (1989).
33. Sieval, A. B., Hout, B. V. D., Zuilhof, H. & Sudhölter, E. J. R. Molecular modeling of covalently attached alkyl monolayers on the hydrogen-terminated Si(111) surface. *Langmuir.* **17,** 2172–2181 (2007).
34. Zhang, R.-Z., Wan, C.-L., Wang, Y.-F. & Koumoto, K. Titanium sulphene: two-dimensional confinement of electrons and phonons giving rise to improved thermoelectric performance. *Phys. Chem. Chem. Phys.* **14,** 15641–15644 (2012).
35. Lin., C. *et al.* Hydrogen-incorporated TiS_2 ultrathin nanosheets with ultrahigh conductivity for stamp-transferrable electrodes. *J. Am. Chem. Soc.* **135,** 5144–5151 (2013).
36. Li, Y., Kang, J. & Li, J. Indirect-to-direct band gap transition of the ZrS_2 monolayer by strain: first-principles calculations. *RSC Adv.* **4,** 7396–7401 (2014).
37. Gong, C. *et al.* Band alignment of two-dimensional transition metal dichalcogenides: application in tunnel field effect transistors. *Appl. Phys. Lett.* **103,** 053513-1–053513-4 (2013).
38. NIST Standard Reference Database Number 101. http://cccbdb.nist.gov/ (accessed 25 August 2014).
39. Miro, P., Han, J. H., Cheon, J. & Heine, T. Hexagonal transition-metal chalcogenide nanoflakes with pronounced lateral quantum confinement. *Angew. Chem. Int. Ed.* **53,** 12624–12628 (2014).
40. Jeong, S., Yoo, D., Jang, J.-t., Kim, M. & Cheon, J. Well-defined colloidal 2-D layered transition-metal chalcogenide nanocrystals via generalized synthetic protocols. *J. Am. Chem. Soc.* **134,** 18233–18236 (2012).
41. Velde, G. T. & Baerends, E. J. Precise density-functional method for periodic structures. *Phys. Rev. B* **44,** 7888–7903 (1991).
42. Wiesenekker, G. & Baerends, E. J. Quadratic integration over the three-dimensional Brillouin zone. *J. Phys. Condens. Mater.* **3,** 6721–6742 (1991).
43. Perdew, J. P., Burke, K. & Ernzerhof, M. Generalized gradient approximation made simple. *Phys. Rev. Lett.* **77,** 3865–3868 (1996).

Acknowledgements

This work was financially supported by the National Creative Research Initiatives Program (2010-0018286).

Author contributions

S.J., D.Y. and J.C. designed the experiments and analysed the data. S.J. and M.A. performed the synthesis and characterization. S.J., D.Y. and J.C. wrote the manuscript. P.M. and T.H. performed the density-functional theory calculations. All the authors discussed the results and commented on the manuscript.

Additional information

Competing financial interests: The authors declare no competing financial interests.

17

Stable metal-organic frameworks containing single-molecule traps for enzyme encapsulation

Dawei Feng[1,*], Tian-Fu Liu[1,*], Jie Su[2], Mathieu Bosch[1], Zhangwen Wei[1], Wei Wan[2], Daqiang Yuan[3], Ying-Pin Chen[1,4], Xuan Wang[1], Kecheng Wang[1], Xizhen Lian[1], Zhi-Yuan Gu[1], Jihye Park[1], Xiaodong Zou[2] & Hong-Cai Zhou[1,4]

Enzymatic catalytic processes possess great potential in chemical manufacturing, including pharmaceuticals, fuel production and food processing. However, the engineering of enzymes is severely hampered due to their low operational stability and difficulty of reuse. Here, we develop a series of stable metal-organic frameworks with rationally designed ultra-large mesoporous cages as single-molecule traps (SMTs) for enzyme encapsulation. With a high concentration of mesoporous cages as SMTs, PCN-333(Al) encapsulates three enzymes with record-high loadings and recyclability. Immobilized enzymes that most likely undergo single-enzyme encapsulation (SEE) show smaller K_m than free enzymes while maintaining comparable catalytic efficiency. Under harsh conditions, the enzyme in SEE exhibits better performance than free enzyme, showing the effectiveness of SEE in preventing enzyme aggregation or denaturation. With extraordinarily large pore size and excellent chemical stability, PCN-333 may be of interest not only for enzyme encapsulation, but also for entrapment of other nanoscaled functional moieties.

[1] Department of Chemistry, Texas A&M University, College Station, Texas 77843-3255, USA. [2] Berzelii Centre EXSELENT on Porous Materials and Inorganic and Structural Chemistry, Department of Materials and Environmental Chemistry, Stockholm University, 106 91 Stockholm, Sweden. [3] State Key Laboratory of Structural Chemistry, Fujian Institute of Research on the Structure of Matter, Chinese Academy of Sciences, Fujian, Fuzhou 350002, P. R. China. [4] Department of Materials Science and Engineering, Texas A&M University, College Station, Texas 77842, USA. * These authors contributed equally to this work. Correspondence and requests for materials should be addressed to H.-C.Z. (email: zhou@chem.tamu.edu).

Metal-organic frameworks (MOFs) with large pores or channels are considered to be promising candidates for the immobilization of many catalysts, such as metal complexes, nanoparticles and enzymes[1-4]. Immobilization is particularly valuable for enzymes, since they are costly and can be difficult to reuse and recover[5]. A range of solid supports have been explored for enzymes, such as sol gels, hydrogels, organic microparticles, and porous or non-porous inorganic supports[5]. However, there have been disadvantages associated with these supports, including very low protein loading in the non-porous systems and denaturation of enzymes in sol gels due to their brittle and disordered structures[5-9]. Mesoporous silica nanoparticles, which have a large pore size, high porosity and a very ordered structure, have been demonstrated as promising enzyme supports[5]. However, due to their one-dimensional (1D) channel structure, insufficient interactions between the inner channel surface of mesoporous silica and enzymes result in significant leaching during multiple cycles of use. Moreover, aggregation still occurs inside the channel[5-9].

Recently, we studied MOFs containing predesigned single-molecule traps (SMTs) for carbon capture, wherein each SMT can trap one CO_2 molecule[10]. If the same concept is applied in enzyme immobilization, we can potentially diminish leaching and encapsulate each enzyme molecule in a SMT to eliminate aggregation, optimizing the catalytic environment. MOFs, which have ultra-high porosity, tunable pore shape and size and adjustable surface functionality, are an ideal platform for this novel single-enzyme encapsulation (SEE) or multiple-enzyme encapsulation (MEE) approach.

On the basis of structural rationalization, we herein report facile syntheses of a series of highly stable SMT-containing MOFs based on trivalent metal species, namely PCN-332(M) (M= Al(III), Fe(III), Sc(III), V(III), In(III)) and PCN-333(M) (M= Al(III), Fe(III), Sc(III)). PCN-333 exhibits the largest cage (5.5 nm) and one of the highest void volumes ($3.84\,cm^3\,g^{-1}$) among all reported MOFs. PCN-333(Al, Fe) shows high stability in aqueous solutions with pH values ranging from 3 to 9, making it an extraordinary candidate for enzyme encapsulation. Three enzymes with different sizes, horseradish peroxidase (HRP), cytochrome *c* (Cyt *c*) and microperoxidase-11 (MP-11), were selected. From a pure size-match point of view, HRP and Cyt *c* will most likely undergo SEE, whereas MP-11 will go through MEE due to size mismatch. Nevertheless, all three enzymes have been successfully encapsulated into PCN-333(Al) with extremely

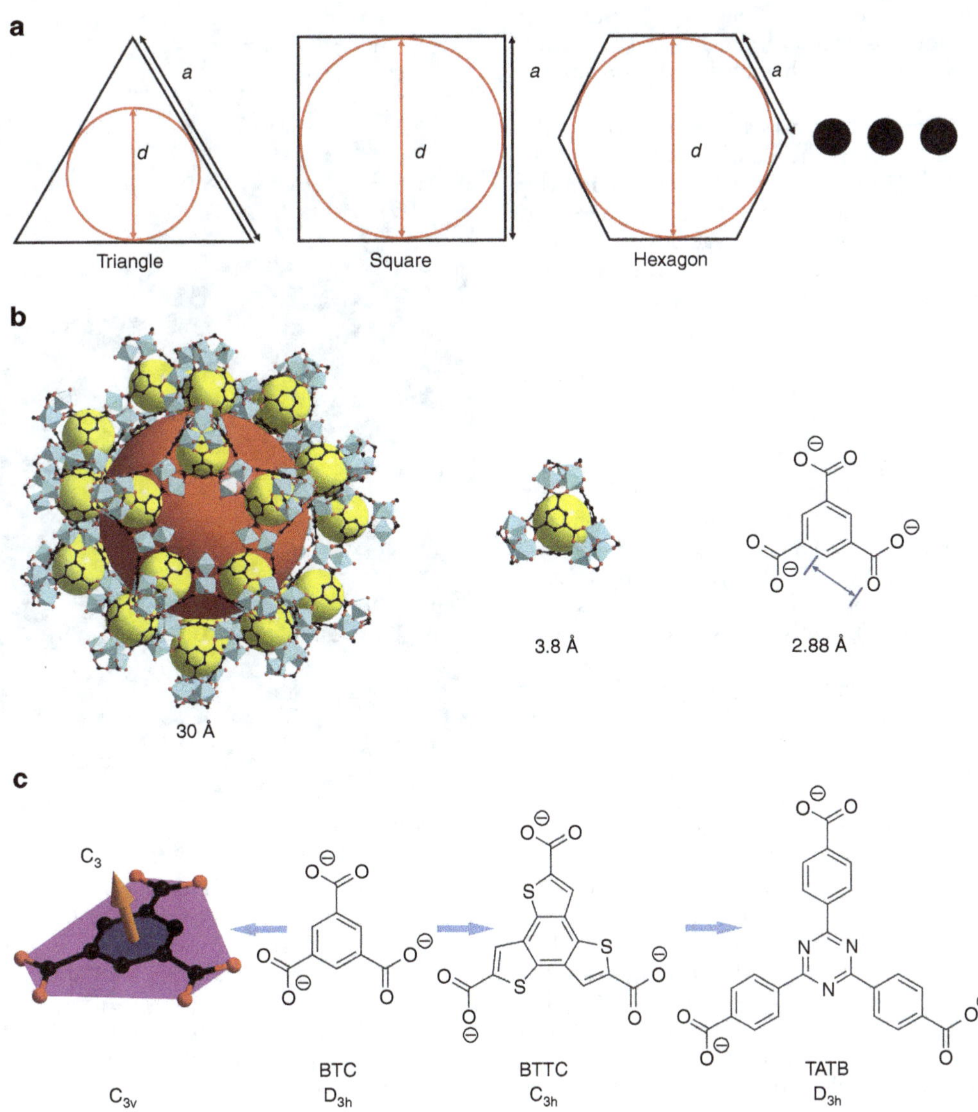

Figure 1 | Pore expansion with preserved structure (the isoreticular approach). (**a**) The relationship between the edge (*a*) and diameter (*d*) in regular polygons. (**b**) The largest cage, super tetrahedral unit and organic linker in MIL-100. (**c**) Idealized symmetries of benzenetricarboxylate (BTC) in MIL-100 and free BTC, BTTC and TATB.

high loading amounts among all reported solid supports. These immobilized enzymes either maintain or surpass their catalytic activities over the free enzymes, and exhibit smaller K_m and better catalytic performance in organic solvents. These immobilized enzymes show almost no leaching during catalysis and recycling, and maintain high catalytic activity.

Results

Custom-designed synthesis of PCN-332 and PCN-333. The largest pore reported in MOFs was based on a 1D hexagonal channel structure, which requires a complicated synthetic process to make an extremely long organic linker to reach enzymatic size[3]. Considering the structural feature of the 1D channel, it can only partially interact with immobilized enzymes even if the size is very compatible. As is in mesoporous silica, once the enzymes are incorporated, aggregation still exists. Moreover, after those channels are occupied, there is no other pathway allowing the fast diffusion of both reactants and products. In comparison, cages will act as SMTs affording strong interactions between the cages and the encapsulated enzymes. In addition, cages usually have small windows, which can efficiently encapsulate enzymes, diminishing aggregation, eliminating leaching and optimizing the catalytic environment. Unfortunately, there are very few reported MOFs with enzyme compatible cages, and even fewer can survive in aqueous media[3,11,12]. Thus, MOFs with larger cages and better stability are highly desired for enzyme encapsulation.

The extension of organic linkers, which mostly serve as the faces or edges of pores in MOFs, is the main approach to achieve large pores in MOFs[3,13–19]. However, the pore shape or symmetry plays a very important role in determining the overall pore size. Approximating the cross-section of a pore (cage or channel) as a regular polygon with n edges (Fig. 1a), the diameter d of the inscribed circle corresponding to the pore can be expressed as the following,

$$d = \frac{1}{\tan\left(\frac{\pi}{n}\right)} \times a \qquad (1)$$

where a is the edge of the regular polygon.

If a is considered as the edge of a pore cross-section, then $1/\tan(\pi/n)$ would be the magnification factor, which increases with n and determines how a certain change of a affects the overall pore diameter d. Therefore, by applying the isoreticular approach through ligand extension in MOFs, the pores with larger n values can be more efficiently enlarged than the pores with a smaller n value.

Hence, to construct MOFs with ultra-large cages and excellent chemical stability, we applied the isoreticular approach to the chemically highly stable mesoporous MOF, MIL-100 (ref. 20). The circular cross-section of the largest cage in MIL-100 is surrounded by 12 supertetrahedra, therefore n is very large ($n = 12$). In addition, MIL-100 is based on hard Lewis acidic trivalent metal species (Al^{3+}, Fe^{3+}, Cr^{3+}, V^{3+}, Sc^{3+}), which interact strongly with carboxylates and give rise to the high stability of the framework in aqueous environments[21–25].

Benzenetricarboxylate (Fig. 1b), the organic linker in MIL-100, has idealized D_{3h} symmetry as a free anion. In the framework, the ligand symmetry is reduced to C_{3v} by a bowl-shaped bending that increases the inherent energy of the ligand (the distance between the two planes is 0.288 Å; Fig. 1c, left). If the symmetry of the free linker were not D_{3h}, to form a similar structure, the energy of the linker would have been even higher in the framework, making the overall structure extremely energetically unfavored. To guarantee the formation of isoreticular structures of MIL-100, we selected two ligands that are most stable in a trigonal planar conformation, BTTC (benzo-tris-thiophene carboxylate) with a C_{3h} (pseudo-D_{3h} in crystallography) symmetry through the fused five-member ring and TATB (4,4',4''-s-triazine-2,4,6-triyl-tri-benzoate), with idealized D_{3h} symmetry arised from ligand planarity, which is enforced by conjugation (Fig. 1c).

Solvothermal reactions of H_3BTTC or H_3TATB with MCl_3 (M = Al, Fe, V, Sc, In) afforded octahedral crystals of PCN-332(M) (M = Al, Fe, Sc, V, In) or PCN-333(M) (M = Al, Fe, Sc), respectively (Fig. 2a, Supplementary Figs 1–16, Supplementary Data 1). High-resolution synchrotron powder X-ray diffraction (PXRD) collected at 17-BM, Argonne National Laboratory shows that PCN-333 is cubic with $a \approx 127$ Å. This was further confirmed by three-dimensional rotation electron diffraction (Supplementary Methods and Supplementary Figs 17–20). The space group of PCN-333(Al) was determined to be $Fd\text{-}3m$ by combining rotation electron diffraction and high-resolution transmission electron microscopy (HRTEM; Fig. 2c, Supplementary Methods, Supplementary Figs 20,21). Structural models of PCN-333 and PCN-332 were built based on the isoreticular structure of MIL-100. The structural models of PCN-333(Al) and PCN-332(Fe) were further refined against the

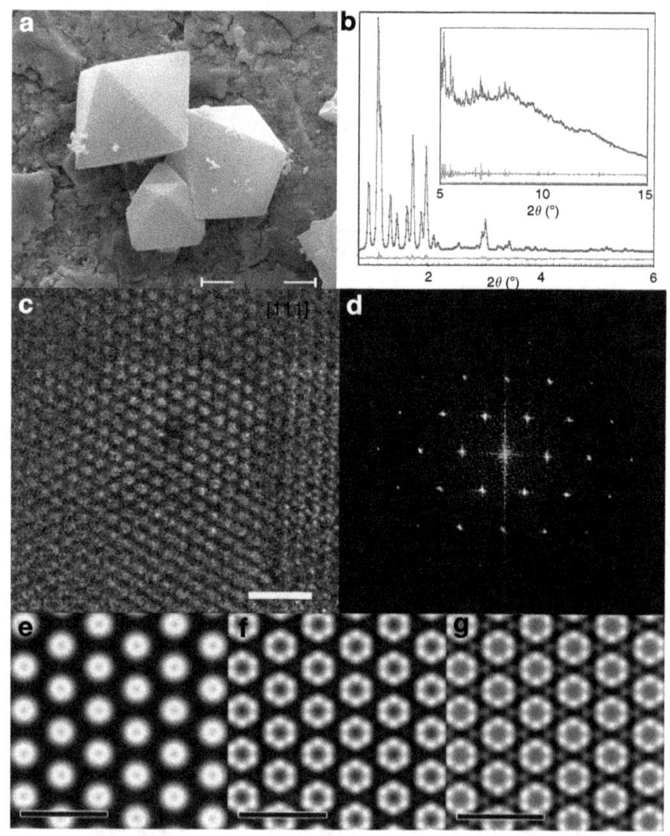

Figure 2 | Transmission- and scanning electron microscopy anlyses of PCN-333(Al). **(a)** SEM image of PCN-333(Al). Scale bar, 10 μm; **(b)** observed (blue), calculated (red) and difference (black) plot for the Rietveld refinement of PCN-333(Al); **(c)** HRTEM image of PCN-333(Al) taken along the [111] direction, showing the pore with the size of about 3.7 nm in projection. Scale bar, 20 nm; **(d)** Fourier transform of HRTEM image in **c**, from which the amplitudes and phases were extracted; **(e)** Symmetry-averaged map after imposing the *p6mm* symmetry on the amplitudes and phases; **(f)** Symmetry-averaged map (*p6mm*) calculated using the amplitudes from PXRD and phases from the HRTEM image; **(g)** Projected electrostatic potential map calculated using the structure factor amplitudes and the phases from the structural model. Scale bars in **e-g**, 10 nm.

synchrotron PXRD data by Rietveld refinement with soft restraints for the M–O bond distances and rigid body for the ligands (Fig. 2b and Supplementary Tables 1,2, Supplementary Fig. 5). A two-dimensional map is calculated by an inverse Fourier transformation from the amplitudes and phases of five symmetry-independent reflections with $d > 14$ Å extracted from the HRTEM image (Fig. 2c,e), in which channels along the [111] direction with a size around 3.7 nm can be clearly visualized. Owing to the beam damage of the sample and the damping effect of the objective lens (contrast transfer function), amplitudes obtained from the HRTEM image are attenuated. Replacing the amplitudes from the HRTEM image by those from the PXRD pattern gives an improved two-dimensional map (Fig. 2f), which matches better the projected potential map of PCN-333(Al) calculated from the structural model (Fig. 2g). This strongly indicates that the structural model is correct.

The structure of PCN-333 is built by sharing the vertices of the supertetrahedra, which consist of $M_3(\mu_3\text{-}O)(OH)(H_2O)_2$ units linked along the faces by the organic linkers (Fig. 3a). Consequently, two types of mesoporous cages, which can act as SMTs for encapsulation of an enzyme, are generated. A smaller dodecahedral cage is built of 20 supertetrahedra that are connected via vertex sharing to construct a cavity with an exclusive pentagonal window of 25.9 Å in diameter (Fig. 3b). A larger hexacaidecahedral (hexagonal-truncated trapezohedral) cage is surrounded by 24 supertetrahedra that form a sphere with not only the pentagonal windows, but also hexagonal windows with a diameter of 30 Å (Fig. 3b). The inner diameter is 11 Å for the super tetrahedral cage, 34 Å for the dodecahedral cage and 55 Å for the hexacaidecahedral cage (Fig. 3b). The largest cages in PCN-333 lie in a honeycomb arrangement in the [111] projection, which is consistent with the HRTEM result (Fig. 2). PCN-332 exhibits a similar structure as PCN-333, but with smaller cages due to the smaller linker (Fig. 3, Supplementary Methods).

Porosity and stability characterization. Owing to the small ligand size and strong coordination bond between the organic linker and metal struts, both the PCN-332 and the PCN-333 series can be activated directly on the removal of solvent, while many other MOFs with such high porosity must be carefully activated with supercritical CO_2 (refs 10,15–18). To assess the porosity of PCN-332 and PCN-333, we performed Ar sorption at 87 K and N_2 sorption at 77 K (Fig. 4a,b and Supplementary Figs 22–31 and Supplementary Methods). PCN-333(Al) shows a total Ar sorption of 2,950 $cm^3\,g^{-1}$ and N_2 uptake of 2,490 $cm^3\,g^{-1}$. The Brunauer–Emmett–Teller surface area is 4,000 $m^2\,g^{-1}$. Two steep increases at $p/p_0 = 0.3$ and 0.5 on the Ar adsorption isotherm correspond to two types of mesoporous cages in PCN-333(Al). The experimental void volume of PCN-333(Al) is 3.81 $cm^3\,g^{-1}$, which is in good agreement with that calculated from the structural model (3.85 $cm^3\,g^{-1}$). Such high porosity is comparable to those of the most porous MOFs[13,16–19,26,27]. The porosity of PCN-333(Fe) is similar to that of PCN-333(Al), while the porosity of PCN-333(Sc) is slightly lower due to its poorer crystallinity. With an isoreticular structure but a smaller organic linker, the porosity and Brunauer–Emmett–Teller surface area of PCN-332 are significantly lower than those of PCN-333.

Despite its high porosity, PCN-333 still shows excellent stability in aqueous solutions. PXRD shows that PCN-333(Al) and PCN-333(Fe) remain intact without loss of crystallinity after immersion in pure H_2O and in aqueous solutions with pH of 3 and 9 (Fig. 4d,f). Furthermore, the N_2 sorption isotherms are almost completely maintained through those treatments (Fig. 4c,e), which further demonstrates the stability of these frameworks. The high positive charge density (Z/r) on small M^{3+} (Al^{3+}, Fe^{3+}) ions causes a strong electrostatic interaction between metal nodes and carboxylate linkers, which endows the frameworks with not only excellent stability towards attack by reactive species, but also high thermal stability (Supplementary Figs 32–39).

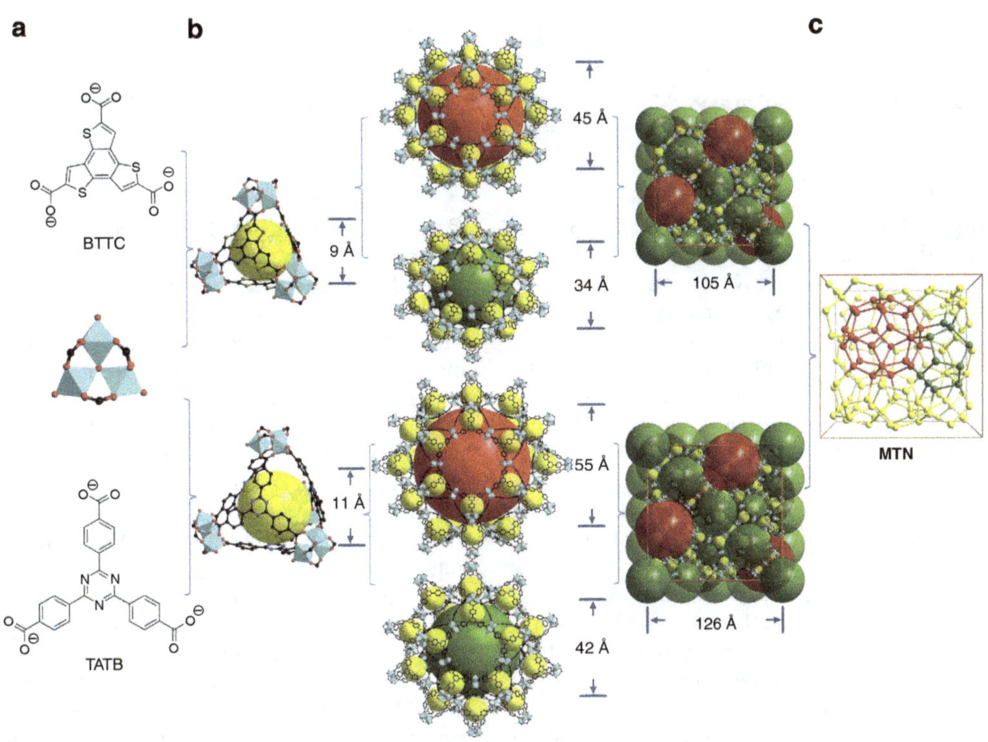

Figure 3 | Structure illustrations of PCN-332 and PCN-333. (**a**) Ligands used in PCN-332 and PCN-333. (**b**) Three different cages in PCN-332 and PCN-333. (**c**) Simplification of PCN-332 and PCN-333 into **MTN** topology.

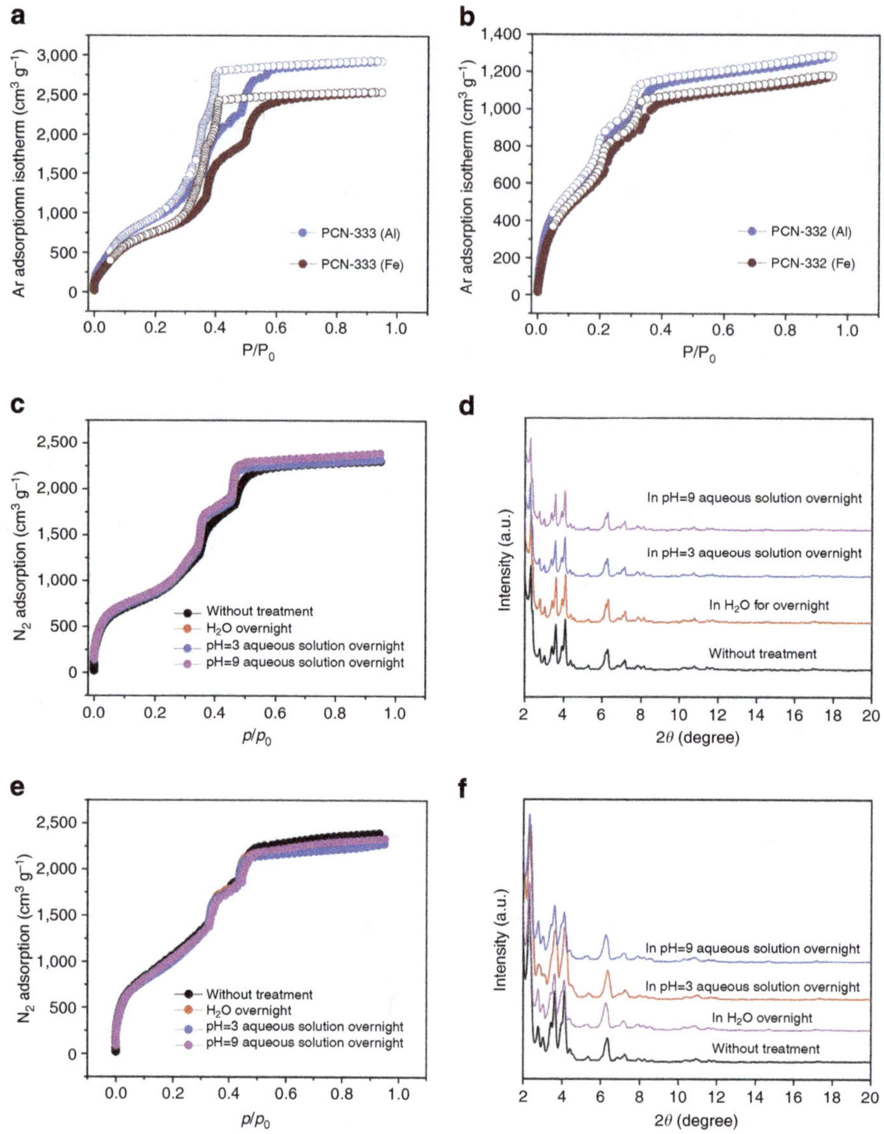

Figure 4 | Porosity and stability characterization of PCN-333 and PCN-332. (**a**) Ar sorption isotherms of PCN-333(Fe) and PCN-333(Al); (**b**) Ar sorption isotherms of PCN-332(Fe) and PCN-332(Al); (**c,d**) Stability tests of PCN-333(Al); (**e,f**) Stability tests of PCN-333(Fe).

PCN-333(Al) as support for different enzymes. PCN-333 is an extraordinary example that meets most prerequisites as a promising enzyme support: the ultra-large cavities can act as SMTs for enzyme encapsulation; the ultra-high porosity ensures high enzyme content after loading which guarantees the overall catalytic efficiency; and the SMTs can efficiently encapsulate enzymes, which can prevent both enzyme aggregation and leaching. In addition, it possesses excellent stability and is compatible with the working environment of an enzyme. A wide variety of enzymes, if smaller than the largest cage (5.5 nm) in PCN-333, can in principle be immobilized[5]. If the enzyme is smaller than the cage, but the cage cannot accommodate two enzymes due to size limitation, a SEE can be achieved.

We have selected three commercially available enzymes that are smaller than the largest cage in PCN-333(Al)—HRP, Cyt *c* and MP-11—for the encapsulation experiment and their catalytic performance and recyclability were also investigated.

To the best of our knowledge, all three enzymes have reached record-high loading among all enzyme immobilization cases[5]. The maximum loading is 1.0 $g\,g^{-1}$ (22.7 $\mu mol\,g^{-1}$) for HRP, 0.95 $g\,g^{-1}$ (77.0 $\mu mol\,g^{-1}$) for Cyt *c* and 0.89 $g\,g^{-1}$ (478 $\mu mol\,g^{-1}$) for MP-11 (Fig. 5a,b and Table 1). Taking into account the size of each enzyme, HRP (size: $4.0 \times 4.4 \times 6.8$ nm) can only be loaded into the large cage, while Cyt *c* (size: $2.6 \times 3.2 \times 3.3$) is suitable for both the large cage and the medium cage and the small cage with an inner diameter of 1.1 nm will be empty, allowing diffusion of solvent molecules, reactants and products. Assuming SEE, the maximum theoretical loading is 1.27 $g\,g^{-1}$ for HRP (assuming only the large cage acts as a SMT) and 1.04 $g\,g^{-1}$ for Cyt *c* (assuming both the large and medium cages act as SMTs), which are in good agreement with the experimental maximum loadings (Supplementary Methods). For MP-11, considering its much smaller size ($1.1 \times 1.7 \times 3.3$ nm), multiple enzymes may enter each meso-cage, leading to a much higher molar concentration than the other two enzymes.

Due to the favourable interaction between the cage and the enzyme, to reach a saturated loading of each enzyme in PCN-333(Al) is rapid. HRP takes about 40 min, Cyt *c* takes 30 min, while MP-11 takes about 10 min. However, for mesoporous silica, saturated enzyme loading usually takes at least several days to reach[28-30]. After the loading of enzymes, the crystallinity, porosity and thermal stability of PCN-333(Al) samples were characterized,

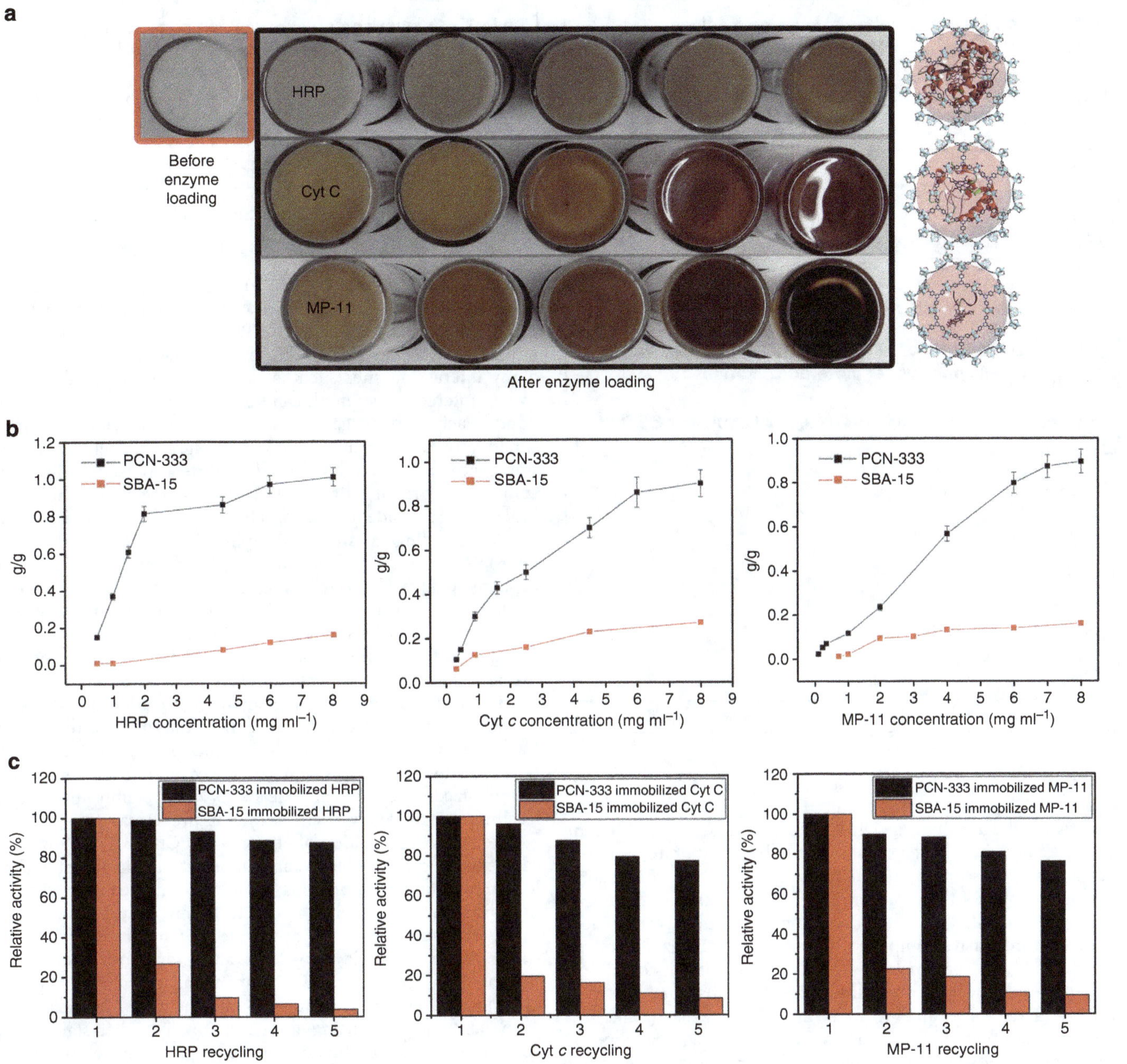

Figure 5 | Enzyme loading and catalytic recycles of different enzymes. (**a**) Colour variations of PCN-333(Al) when loaded with different enzymes at different concentrations. (**b**) Plots of the loading capacities of different enzymes in PCN-333(Al). The standard deviation (error bars) was obtained from three independent replicate experiment. (**c**) Catalytic activity of immobilized enzymes in each recycle test.

showing well-maintained framework (Supplementary Figs 40–54). Most importantly, the transmission electron microscopy–energy-dispersive X-ray spectroscopy elemental mapping of HRP@PCN-333(Al) sample indicates very well distributed HRP inside of the framework (Supplementary Fig. 67).

Enzyme leaching and loss of reactivity on reuse are the major problems for all other enzyme supports[5]. After the immobilization of the three enzymes in PCN-333(Al), their catalytic activity was then evaluated with corresponding reactions (Supplementary Figs 55–74).

Discussion

Oxidation of o-Phenylenediamine catalysed by HRP and oxidation of 2-2'-azino-bis(3-ethylbenzthiazoline-6-sulfonic acid) catalysed by Cyt *c* and MP-11 were performed, respectively, in acid-sodium citrate buffer with pH = 6 at room temperature. Compared with the free enzymes, enzymes immobilized in PCN-333(Al) show relatively smaller k_{cat} values, which may be caused by diffusion control. However, all three enzymes exhibit smaller K_m values, resulting in comparable k_{cat}/K_m values to the free enzymes (Table 2). The low K_m means that the enzyme requires a lower substrate concentration to achieve V_{max} and usually suggests that the enzyme may have a higher affinity with the substrate.

The free enzyme reacts more quickly than the encapsulated enzymes, though it requires significantly more concentration of substrate to reach its peak rate, under carefully controlled conditions designed to prevent it from being aggregated or denatured. However, in practical application, the reaction usually needs to be conducted in different media other than the buffer

Table 1 | Comparison of different enzyme loadings in PCN-333(Al) and SBA-15.

Enzyme	Max-loading in PCN-333(Al) ($g\,g^{-1}$)	Max-loading in PCN-333(Al) ($\mu mol\,g^{-1}$)	Max-loading in SBA-15 ($\mu mol\,g^{-1}$)
HRP	1.0	22.7	4.2
Cyt *c*	0.95	77.0	45
MP-11	0.89	478	20

Cyt c, cytochrome c; HRP, horseradish peroxidase; MP-11, microperoxidase-11.

Table 2 | Comparison of kinetic parameters for free and immobilized enzymes under buffered conditions.

Enzymes	K_m (mM)	k_{cat} (min^{-1})	k_{cat}/K_m ($min^{-1}\,mM^{-1}$)
HRP			
PCN-333 immobilized	0.84	1.54×10^4	1.833×10^4
Free	1.8	4.52×10^4	2.51×10^4
Cyt c			
PCN-333 immobilized	14.9	32.2	2.16
Free	48.3	65.8	1.36
MP-11			
PCN-333 immobilized	2.15	27.9	12.98
Free	3.28	80.2	24.45

Cyt c, cytochrome c; HRP, horseradish peroxidase; MP-11, microperoxidase-11.

solution. Therefore, we measured the specific activity to investigate the catalytic activity of SEE and MEE in different media. Cyt *c*, which is of a suitable size to be incorporated into the two kinds of mesocages in a 1:1 ratio, has most likely undergone SEE, and shows 14 times higher specific activity after immobilization in PCN-333(Al) than the free enzyme in water. To provide a more comprehensive comparison about the immobilized and free Cyt *c*, the kinetic parameters were investigated in this condition and the results indicate the immobilized Cyt *c* has a smaller K_m, and a higher K_{cat} and K_{cat}/K_m in water media (Table 3). Considering that Cyt *c* suffers severe self-aggregation or denaturation towards temperature or pH variation[31], such significant catalytic activity enhancement is evidence of SEE in PCN-333(Al). Immobilized HRP, which can also realize SEE, shows slightly higher specific activity (Supplementary Fig. 68) in water. In comparison, MP-11, the size of which is much smaller than the mesocages, suitable for MEE, shows much decreased activity after immobilization (Supplementary Fig. 68). Moreover, in several different organic solvent mixtures, the immobilized HRP performs better than the free HRP (Supplementary Fig. 71). These results show the effectiveness of SEE under harsh reaction conditions.

After the completion of each catalytic reaction, the immobilized enzymes were then recycled by a simple centrifuge. The catalytic reactivity remains almost unchanged for each PCN-333(Al) encapsulated enzyme after several cycles, while very significant decrease of the activity is observed for the same enzymes immobilized in SBA-15, a well-known type of mesoporous silica (Fig. 5c). The main reason for the excellent recyclability of PCN-333(Al) immobilized enzymes is the SEE or MEE of these enzymes: there is almost no leaching from these encapsulated enzymes during many cycles while high apparent

Table 3 | Comparison of kinetic parameters for free and immobilized Cyt c in aqueous media.

	K_m (mM)	k_{cat} (min^{-1})	k_{cat}/K_m ($min^{-1}\,mM^{-1}$)
PCN-333-immobilized Cyt *c*	6.8	197	28.97
Free Cyt *c*	109.4	109	1.00

Cyt c, cytochrome c.

leaching is observed in SBA-15. Leaching rate is usually thermodynamically related to the interaction strength between the enzyme and the support[5]. Physisorption is usually much weaker than covalent immobilization. However, when there are many interaction sites such as in the case of encapsulation, the overall interaction strength can be stronger than a single covalent bond. Each large cage in PCN-333 is surrounded by many supertetrahedra containing not only highly concentrated metal sites with polarized coordinating water molecules, but also conjugated organic linkers, which generate an amphiphilic inner sphere inside the cage. The cage can therefore act as a SMT, providing a strong trapping interaction to both the hydrophilic and hydrophobic parts of the enzyme, which explains the fast enzyme uptake and the negligible enzyme leaching after encapsulation. The strong interaction between the enzyme and the cages in PCN-333 indicates a large equilibrium constant if the loading process is considered as a reaction. The loading process is less reversible, which leads to fast completion of the forward reaction (enzyme loading). In addition, the small window on each large cage can dynamically reduce the enzyme leaching. In comparison, SBA-15, which has 1D channels with fully hydrophilic surfaces, can provide only limited interaction to immobilized enzymes and lead to very significant enzyme leaching during catalysis and recycling processes.

After the encapsulation of enzymes in PCN-333(Al), enzymatic catalysis in organic solvents, an important field in the biotechnological applications of proteins[32], can be conducted more easily by overcoming the solubility issues for both enzymes in organic media and substrates in aqueous media. The encapsulated HRP shows higher catalytic activity in tetrahydrofuran–water media than it does in pure water, and much higher activity than free HRP under the same condition (Supplementary Methods).

In summary, by using an exceptionally water-stable MOF containing mesocages that can act as single-molecule traps for the encapsulation of a single-enzyme molecule in one cage, essentially eliminating enzyme aggregation and leaching, immobilized enzymes with recyclability and high catalytic efficiency in harsh condition have been discovered. The single-enzyme encapsulation concept will revolutionize the application of enzymes in industry, enabling the application of enzymes in fundamental processes such as nitrogen fixation, hydrogen production and the conversion of natural gas into liquid fuels of high energy density with unprecedented efficiency.

Methods

Synthesis of PCN-332(Al). H_3BTTC (50 mg) and $AlCl_3\ 6H_2O$ (200 mg) were dissolved in 10 ml diethylformamide (DEF) or dimethylformamide (DMF), then 1.5 ml trifluoroacetic acid was added. The mixture was heated up at 150 °C in an oven for 12 h until white precipitate formed. The white precipitate was centrifuged and washed with fresh DMF for several times. Yield (based on ligand): ~70%.

Synthesis of PCN-332(Fe). H_3BTTC (50 mg) and anhydrous $FeCl_3$ (60 mg) were dissolved in 10 ml DEF or DMF, then 1.2 ml trifluoroacetic acid was added. The mixture was heated up at 150 °C in an oven for 12 h until brown precipitate formed. The brown precipitate was centrifuged and washed with fresh DMF for several times. Yield (based on ligand): ~85%.

Synthesis of PCN-332(V). H_3BTTC (50 mg) and anhydrous VCl_3 (60 mg) were dissolved in 10 ml DEF or DMF. The mixture was heated up at 150 °C in an oven for 4 h until green precipitate formed. The green precipitate was centrifuged and washed with fresh DMF for several times. Yield (based on ligand): ~75%.

Synthesis of PCN-332(Sc). H_3BTTC (50 mg) and $ScCl_3\ 6H_2O$ (200 mg) were dissolved in 10 ml DEF or DMF. The mixture was heated up at 150 °C in an oven for 4 h until white precipitate formed. The white precipitate was centrifuged and washed with fresh DMF for several times. Yield (based on ligand): ~80%.

Synthesis of PCN-332(In). H_3BTTC (10 mg) and $In(NO_3)_3\ xH_2O$ (20 mg) were dissolved in 2 ml DMF, then 0.2 ml HNO_3 (3 M in DMF) was added. The mixture was heated up at 150 °C in an oven for 12 h. The white precipitate was centrifuged and washed with fresh DMF for several times. Yield (based on ligand): ~85%.

Synthesis of PCN-333(Al). H_3TATB (50 mg) and $AlCl_3\ 6H_2O$ (200 mg) were dissolved in 10 ml DEF or DMF, then 1.0 ml trifluoroacetic acid was added. The mixture was heated up at 135 °C in an oven for 2 days until white precipitate formed. The white precipitate was centrifuged and washed with fresh DMF for several times. Yield (based on ligand): 80%.

Synthesis of PCN-333(Fe). H_3TATB (50 mg) and anhydrous $FeCl_3$ (60 mg) were dissolved in 10 ml DEF or DMF, then 0.5 ml trifluoroacetic acid was added. The mixture was heated up at 150 °C in an oven for 12 h until brown precipitate formed. The brown precipitate was centrifuged and washed with fresh DMF for several times. Yield (based on ligand): 85%.

Synthesis of PCN-333(Sc). H_3TATB (50 mg) and $ScCl_3\ 6H_2O$ (200 mg) were dissolved in 10 ml DEF or DMF. The mixture was heated up at 150 °C in an oven for 4 h until white precipitate formed. The white precipitate was centrifuged and washed with fresh DMF for several times. Yield (based on ligand): ~85%.

References

1. Zhou, H.-C., Yaghi, O. M. & Long, J. R. Introduction to metal–organic frameworks. *Chem. Rev.* **112,** 673–674 (2012).
2. Férey, G. *et al.* A chromium terephthalate-based solid with unusually large pore volumes and surface area. *Science* **309,** 2040–2042 (2005).
3. Deng, H. *et al.* Large-pore apertures in a series of metal-organic frameworks. *Science* **336,** 1018–1023 (2012).
4. Lykourinou, V. *et al.* Immobilization of MP-11 into a mesoporous metal-organic framework, MP-11@mesoMOF: a new platform for enzymatic catalysis. *J. Am. Chem. Soc.* **133,** 10382–10385 (2011).
5. Hudson, S., Cooney, J. & Magner, E. Proteins in mesoporous silicates. *Angew. Chem. Int. Ed.* **47,** 8582–8594 (2008).
6. Klibanov, A. M. Immobilized enzymes and cells as practical catalysts. *Science* **219,** 722–727 (1983).
7. Filice, M., Guisan, J. M., Terreni, M. & Palomo, J. M. Regioselective mono-deprotection of peracetylated carbohydrates. *Nat. Protoc.* **7,** 1783–1796 (2012).
8. Datta, S. *et al.* Enzyme immobilization: an overview on techniques and support materials. *Biotech.* **3,** 1–9 (2013).
9. Engstrom, K. *et al.* Co-immobilization of an enzyme and a metal into the compartments of mesoporous silica for cooperative tandem catalysis: an artificial metalloenzyme. *Angew. Chem. Int. Ed.* **52,** 14006–14010 (2013).
10. Li, J.-R. *et al.* Porous materials with pre-designed single-molecule traps for CO_2 selective adsorption. *Nat. Commun.* **4,** 1538 (2013).
11. Liu, W.-L. *et al.* Fast multipoint immobilized MOF bioreactor. *Chem. Eur. J* **20,** 8923–8928 (2014).
12. Liu, W.-L. *et al.* Novel trypsin–FITC@MOF bioreactor efficiently catalyzes protein digestion. *J. Mater. Chem. B* **1,** 928–932 (2013).
13. Furukawa, H. *et al.* Ultrahigh porosity in metal-organic frameworks. *Science* **329,** 424–428 (2010).
14. Farha, O. K. *et al.* Metal–organic framework materials with ultrahigh surface areas: Is the sky the limit? *J. Am. Chem. Soc.* **134,** 15016–15021 (2012).
15. Yuan, D. *et al.* An isoreticular series of metal–organic frameworks with dendritic hexacarboxylate ligands and exceptionally high gas-uptake capacity. *Angew. Chem. Int. Ed.* **49,** 5357–5361 (2010).
16. Koh, K., Wong-Foy, A. G. & Matzger, A. J. A crystalline mesoporous coordination copolymer with high microporosity. *Angew. Chem. Int. Ed.* **47,** 677–680 (2008).
17. Farha, O. K. *et al. De novo* synthesis of a metal–organic framework material featuring ultrahigh surface area and gas storage capacities. *Nat. Chem.* **2,** 944–948 (2010).
18. Chae, H. K. *et al.* A route to high surface area, porosity and inclusion of large molecules in crystals. *Nature* **427,** 523–527 (2004).
19. Furukawa, H., Cordova, K. E., O'Keeffe, M. & Yaghi, O. M. The chemistry and applications of metal-organic frameworks. *Science* **341,** 1230–1235 (2013).
20. Férey, G. *et al.* A hybrid solid with giant pores prepared by a combination of targeted chemistry, simulation, and powder diffraction. *Angew. Chem. Int. Ed.* **43,** 6296–6301 (2004).
21. Bloch, E. D. *et al.* Metal insertion in a microporous metal – organic framework lined with 2,2'-bipyridine. *J. Am. Chem. Soc.* **132,** 14382–14384 (2010).
22. Millange, F. *et al.* Synthesis, structure determination and properties of MIL-53as and MIL-53ht: the first Cr^{III} hybrid inorganic–organic microporous solids: $Cr^{III}(OH)\cdot\{O_2C\text{–}C_6H_4\text{–}CO_2\}\cdot\{HO_2C\text{–}C_6H_4\text{–}CO_2H\}_x$. *Chem. Commun.* 822–823 (2002).
23. Reinsch, H. *et al.* A new Al-MOF based on a unique column-shaped inorganic building unit exhibiting strongly hydrophilic sorption behaviour. *Chem. Commun.* **48,** 9486–9488 (2012).
24. Yang, S. *et al.* Selectivity and direct visualization of carbon dioxide and sulfur dioxide in a decorated porous host. *Nat. Chem.* **4,** 887–894 (2012).
25. Serre, C. *et al.* Role of solvent-host interactions that lead to very large swelling of hybrid frameworks. *Science* **315,** 1828–1831 (2007).
26. An, J. *et al.* Metal-adeninate vertices for the construction of an exceptionally porous metal-organic framework. *Nat. Commun.* **3,** 604 (2012).
27. Li, T. *et al.* Stepwise ligand exchange for the preparation of a family of mesoporous MOFs. *J. Am. Chem. Soc.* **135,** 11688–11691 (2013).
28. Qiao, S. Z. *et al.* Synthesis and bio-adsorptive properties of large-pore periodic mesoporous organosilica rods. *Chem. Mater.* **17,** 6172–6176 (2005).
29. Deere, J., Magner, E., Wall, J. G. & Hodnett, B. K. Mechanistic and structural features of protein adsorption onto mesoporous silicates. *J. Phys. Chem. B* **106,** 7340–7347 (2002).
30. Takahashi, H. *et al.* Catalytic activity in organic solvents and stability of immobilized enzymes depend on the pore size and surface characteristics of mesoporous silica. *Chem. Mater.* **12,** 3301–3305 (2000).
31. Vazquez-Duhalt, R. Cytochrome c as a biocatalyst. *J. Mol. Catal. B: Enzym.* **7,** 241–249 (1999).
32. Vazquez-Duhalt, R., Semple, K. M., Westlake, D. W. & Fedorak, P. M. Effect of water-miscible organic solvents on the catalytic activity of cytochrome c. *Enzyme Microb. Technol.* **15,** 936–943 (1993).

Acknowledgements

This work was supported as a part of the Methane Opportunities for Vehicular Energy (MOVE) Program under the Award Number DE-AR0000249, as part of the Center for Clean-Energy-Related Gas Separation, an Energy Frontier Research Center funded by the U. S. Department of Energy (DOE), Office of Science, Office of Basic Energy Sciences and as part of N000141310753 supported by the Office of Naval Research. Use of the Advanced Photon Source, an Office of Science User Facility operated for the U. S. Department of Energy (DOE) Office of Science by Argonne National Laboratory, was supported by the U. S. DOE under Contract No. DE-AC02-06CH11357. The structure characterization by PXRD and TEM was supported by the Swedish Research Council (VR), the Swedish Governmental Agency for Innovation Systems (VINNOVA) and the Knut & Alice Wallenberg Foundation through a grant for purchasing the TEM and the project grant 3DEM-NATUR. We also thank Zhenggang Xu and Dr Qian Peng from the Super-computing Facility at Texas A&M University and Laboratory for Molecular Simulation for offering access to computing time and valuable discussions; Dr Hung-Jue Sue and Dr Peng Liu from Mechanical Engineering at Texas A&M University for Thermo-gravimetric analysis; Dr Hansoo Kim in the Microscopy and Imaging Center at Texas A&M University for the TEM-DEX elemental mapping; Lizzie West for organizing other manuscript related works.

Author contributions

D.F., T.-F.L and H.-C.Z. planned and executed the synthesis and characterization. J.S., W.W. and X.Z. performed the structure characterization. M.B., Z.W., Y.-P.C., X.W., K.W., X.L., Z.-Y.G., J.P. and D.Y. participated in the preparation of the manuscript.

Additional information

Accession codes: The X-ray crystallographic coordinates for structures reported in this Article have been deposited at the Cambridge Crystallographic Data Centre (CCDC), under deposition number 999752 (PCN-332(Fe)) and 997911 (PCN-333(Al)). These data can be obtained free of charge from The Cambridge Crystallographic Data Centre via www.ccdc.cam.ac.uk/data_request/cif.

Competing financial interests: The authors declare no competing financial interests.

18

Self-catalysed aerobic oxidization of organic linker in porous crystal for on-demand regulation of sorption behaviours

Pei-Qin Liao[1], Ai-Xin Zhu[1,2], Wei-Xiong Zhang[1], Jie-Peng Zhang[1] & Xiao-Ming Chen[1]

Control over the structure and property of synthetic materials is crucial for practical applications. Here we report a facile, green and controllable solid–gas reaction strategy for on-demand modification of porous coordination polymer. Copper(I) and a methylene-bridged bis-triazolate ligand are combined to construct a porous crystal consisting of both enzyme-like O_2-activation site and oxidizable organic substrate. Thermogravimetry, single-crystal X-ray diffraction, electron paramagnetic resonance and infrared spectroscopy showed that the methylene groups can be oxidized by O_2/air even at room temperature via formation of the highly active Cu(II)-$O_2^{\cdot-}$ intermediate, to form carbonyl groups with enhance rigidity and polarity, without destroying the copper(I) triazolate framework. Since the oxidation degree or reaction progress can be easily monitored by the change of sample weight, gas sorption property of the crystal can be continuously and drastically (up to 4 orders of magnitude) tuned to give very high and even invertible selectivity for CO_2, CH_4 and C_2H_6.

[1] MOE Key Laboratory of Bioinorganic and Synthetic Chemistry, School of Chemistry and Chemical Engineering, Sun Yat-Sen University, Guangzhou 510275, P. R. China. [2] Faculty of Chemistry and Chemical Engineering, Yunnan Normal University, Kunming 650092, P. R. China. Correspondence and requests for materials should be addressed to J.-P.Z. (email: zhangjp7@mail.sysu.edu.cn).

Compared with conventional adsorbents, porous coordination polymers (PCPs) are unique for their diversified and tailorable coordination frameworks[1–11]. While tailoring structure/property generally refers to design/synthesis of new ligands and frameworks, some PCP prototypes can adopt several different metal ions and/or ligands, providing a rational strategy for adjusting the structure/property, albeit only in a limited degree[5,8,12–14]. A few of these prototypes can even form mixed-component (solid-solution) crystals with variable concentration/ratio of functional building blocks, which in principle allows the structure/property to be adjusted more continuously and precisely[15,16]. However, so far there is no rational strategy to directly monitor/control the composition of solid-solution frameworks during the synthesis process because complicated reaction environments involving solvents and/or liquid reactants are generally required for known direct-synthesis and post-synthetic modification (PSM)[3,7,17–21] methods. It should be noted that the reaction time/feeding ratio can be hardly used as a parameter for precise control over the reaction progress/framework composition, since the two variables have complicated relationships. Single-crystal X-ray diffraction is straightforward for direct visualization of the physical adsorption and chemical reaction events in PCPs[22], but retaining the sample single crystallinity after physical/chemical changes is always a great challenge[22–26], and crystallography is not a technique for quantitative analysis. If a PCP crystal could react with a gas (O_2 is a good yet difficult candidate) in the absence of assistant solvent/liquid, the sample weight or gas pressure can serve as an easily measurable parameter directly and linearly associated with its framework composition.

The selective and 'green' oxidation of organic molecules by air or dioxygen (O_2) is fundamental in the biosystem and has long been pursued in industry and chemical sciences[27–29]. Because the energy barrier for electron transfer from singlet organic substrates to the triplet O_2 is generally very high[30], harsh reaction conditions and/or efficient catalysts are usually necessary to activate the O_2 molecule for aerobic oxidization, so that the earth atmosphere can maintain a high O_2 concentration. Natural enzymes are well known as highly selective and efficient catalysts under mild conditions. For instance, the highly efficient O_2 activation centres in various copper proteins, such as galactose oxidase, tyrosinase and dopamine β-monooxygenase, have been extensively studied or used as structural models for synthetic catalysts.

Considering that PCPs generally lack sufficient robustness, reactivity and/or O_2-activation ability to allow aerobic oxidation of themselves, we designed and synthesized a porous metal azolate framework (MAF)[4] on the basis of Cu(I) and a methylene-bridged bis-triazolate ligand (Fig. 1). Although Cu(I)-based PCPs are very scarce because Cu(I) can be easily oxidized as Cu(II) by air to destroy the original metal–ligand connectivity[31], azolate derivatives have been demonstrated as suitable ligands for highly stable PCPs even with Cu(I) (ref. 4), and the coordination between Cu(I) and triazolate can be expected to give low-coordinated metal centres similar to those in the O_2-activating copper proteins. Furthermore, the methylene bridge in a diarylmethane-type ligand is obviously flexible and oxidizable (activated by two aromatic rings)[32], which could be oxidized in suitable catalytic conditions to form a more rigid and polar ketone group.

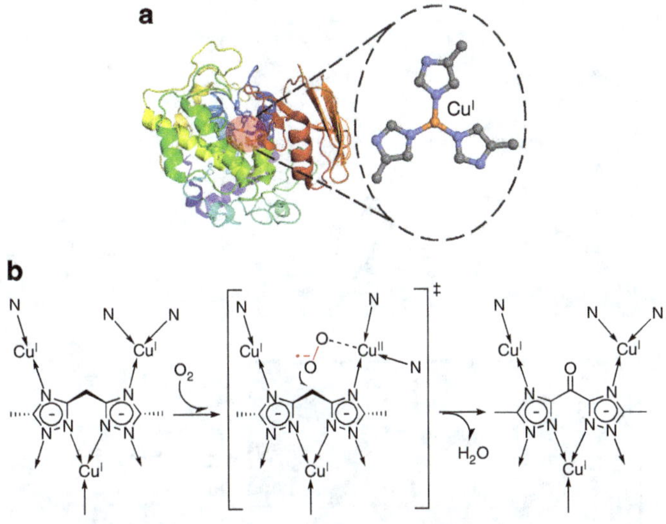

Figure 1 | Bioinspired aerobic oxidation strategy for multifunctional porous frameworks. (**a**) The O_2-activating site in a typical copper protein. (**b**) The possible coordination structure and reactivity of an extended network solid (porous crystal) consisting of Cu(I) and a methylene-bridged bistriazolate ligand.

Results

Preparation and characterization of the porous crystal. Colourless crystals of the titled compound, MAF-42, were synthesized in its large-pore (*lp*) form as $[Cu_4(btm)_2] \cdot C_6H_6$ (denoted as C_6H_6@MAF-42-*lp*, H_2btm = bis(5-methyl-1,2,4-triazolate-3-yl)methane) by solvothermal reaction of H_2btm and $[Cu(NH_3)_2]OH$ in a mixed aqueous ammonia/methanol/benzene solvent. Single-crystal X-ray diffraction analysis (Supplementary Table 1) showed that MAF-42-*lp* is a three-dimensional (3D) porous coordination framework composed of five independent (two of them locate at twofold axes with occupancies of 1/2) Cu(I) ions and two independent, fully deprotonated btm^{2-} ligands in a 2:1 molar ratio. As expected, all btm^{2-} ligands are six-coordinated and the average coordination number of Cu(I) ions is three[4]. Nevertheless, each Cu(I) ion either adopts the linear, distorted T-shaped or tetrahedral coordination geometry (Supplementary Fig. 1). The two independent ligands have different surrounding environments. The methylene group (C4) of one ligand is adjacent to a two- and two three-coordinated Cu(I) ions, while that (C11) of another ligand is only adjacent to two three- and a four-coordinated Cu(I) ions (Supplementary Table 2). The two triazolate rings of btm^{2-} are not coplanar because they are linked by a sp^3 hybridized methylene C atom. The 3D coordination framework can be regarded as a cross-packing structure of ribbon-like fragments (Supplementary Fig. 2), which retains large 1D rhombic channels (void 37.8%, cross-section size 4.8×7.1–6.0×10.8 Å^2) with disordered benzene molecules filled inside. It should be noted that the Cu(I) ions and methylene groups are fully exposed on the pore surface (Fig. 2a).

Thermogravimetry (TG) and powder X-ray diffraction (PXRD) measurements of C_6H_6@MAF-42-*lp* showed that the benzene molecules can be completely removed below 260 °C to form the small-pore (*sp*) phase $[Cu_4(btm)_2]$ (denoted as MAF-42-*sp* or simplified as MAF-42), which can be stable up to 410 °C (Supplementary Figs 3 and 4). Single-crystal structure of MAF-42-*sp* (Supplementary Table 1) showed significantly contracted unit cell (−22%) and channel size (void 14.1%, cross-section size 2.4×2.6–3.5×3.7 Å^2) because the cross angle of packed ribbons reduced from 72° to 53° (Fig. 2b)[5,8]. In MAF-42-*sp*, the Cu(I) ions and methylene groups are less exposed on the pore surface; however, their separations are similar to those in MAF-42-*lp* (Supplementary Table 2). While the coordination bond lengths were changed very little (Δ_{max} = 0.022 Å), the structural variation mainly occurred on the ligand conformations and coordination

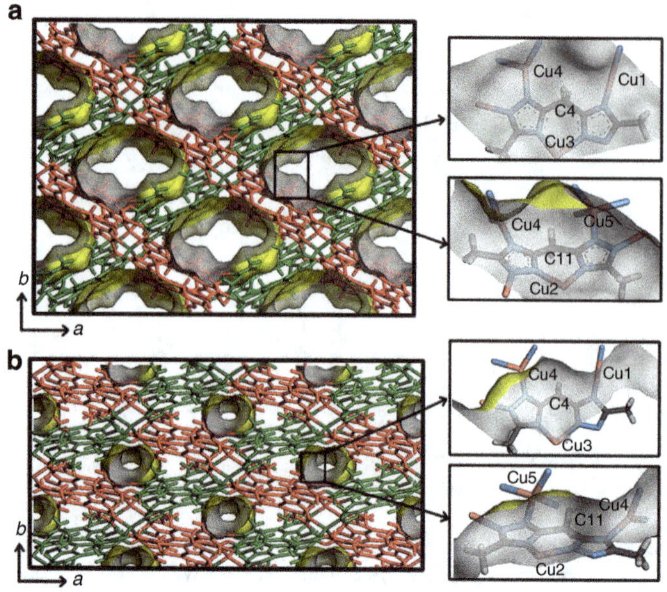

Figure 2 | Guest-induced framework breathing. (**a**) C_6H_6@MAF-42-*lp* (**b**) MAF-42-*sp*. The ribbon-like fragments of the coordination framework running along two directions are highlighted in green and red, respectively, and the pore surfaces of 1D channels are highlighted as yellow/grey curved planes. Desolvation leads to -28%, +10% and -22% changes in *a*, *b* and *V*, respectively.

geometries of the Cu(I) ions serving as the joints between the ribbon-like fragments ($\Delta_{max} = 7.3°$; Supplementary Tables 2 and 3). Actually, the ligand-bending directions are reversed, that is, the average torsion angle between two triazolate rings of btm^{2-} changes from 167.6° in MAF-42-*lp* to − 165.1° in MAF-42-*sp* (Supplementary Fig. 5). The structure transformation between C_6H_6@MAF-42-*lp* and MAF-42-*sp* can be reversibly triggered by adsorption/desorption of benzene vapour (Supplementary Fig. 4).

Self-catalysed aerobic oxidation. In air, colourless C_6H_6@MAF-42-*lp* turns brown and then black quickly (Fig. 3a), which should not be originated from Cu(I) to Cu(II) oxidation because the Cu(II) complex is usually blue or green. Electrospray ionization mass spectrometry of the demetalated samples showed that fresh C_6H_6@MAF-42-*lp* has only a signal at $m/z = 179$ (H_3btm^+), while a new peak at $m/z = 193$ corresponding to the expected oxidation product bis(5-methyl-1,2,4-triazol-3-yl)methanone (H_3btk^+) appeared after the sample was exposed in air (Supplementary Fig. 6). The infrared spectrum of the oxidized sample exhibits a strong band at 1641 cm^{-1} in the characteristic region of 1,750 ± 150 cm^{-1} for carbonyl groups (Supplementary Fig. 7). The single-crystal structure of a black crystal obtained by prolonged exposure (1 month) of C_6H_6@MAF-42-*lp* in air was measured (Supplementary Table 1). A residual electron peak appeared near one of the two crystallographically independent methylene groups that adjacent to the two-coordinated Cu(I) ion, which can be refined as an oxygen atom without any restriction, giving an occupancy of 0.28(5) and a C=O bond length of 1.20(5) Å (Supplementary Fig. 8)[33]. These observations demonstrated that the btm^{2-} ligands in C_6H_6@MAF-42-*lp* can be readily oxidized by air at room temperature, although the reaction rate is quite slow. The oxidation of MAF-42-*sp* (almost no colour change after exposed in air at room temperature for several days) is much slower, which may be ascribed to the even smaller pore size and less exposed active sites.

The TG analysis showed that heating microcrystalline MAF-42-*sp* in an O_2 flow at 418 K could be an optimized reaction condition, in which complete oxidation can be achieved after ~8,000 min (Fig. 3b). Meanwhile, mass spectrometry analysis of the effluent showed that water is produced as a byproduct during the oxidation process (Supplementary Fig. 9). Therefore, on the basis of the simple reaction equation (Fig. 1b), the oxidation degree of the crystals can be simply calculated and controlled by the heating time or by monitoring the sample weight. The PXRD pattern of the fully oxidized sample is similar to that of MAF-42-*sp* (Supplementary Fig. 10), indicating the retention of the original framework connectivity. X-ray photoelectron spectroscopy showed that, even on the particle surface of the oxidized sample, only very small amounts of Cu(I) ions have been oxidized to Cu(II) (Supplementary Fig. 11). While some oxidation reactions have been used to modify the organic linkers in PCP crystals, strong and environment-unfriendly oxidants (such as nitric acid and dimethyldioxirane), as well as organic solvents are necessary[17,18,34–36]. Compared with the conventional oxidation methods, the solvent-free aerobic oxidation is obviously much greener yet more difficult. At the same conditions, the free ligand H_2btm cannot be oxidized (Supplementary Fig. 12), highlighting the difficulty of covalent redox PSM and crucial role of enzyme-like Cu(I) centres. Compared with known direct-synthesis and PSM methods, the aerobic oxidation is also noteworthy for its controllable, solvent-free and solid–gas reaction mechanism, which can be conveniently used to tailor the adsorbent properties. Interestingly, the oxidized samples can be completely reduced back to C_6H_6@MAF-42-*lp* by hydrazine (Supplementary Fig. 13).

The structure of a highly oxidized single crystal $[Cu_4(btm)_{0.7}(btk)_{1.3}]$ (denoted as **O65** to reflect the oxidation degree, 65%) has been successfully measured (Supplementary Table 1). There are residual electron peaks near the two independent methylene groups, which were refined as oxygen atoms to give occupancies of 0.73(2) and 0.57(2) and C=O bond lengths of 1.236(10) and 1.234(11) Å, respectively (Fig. 3c,d). Again, the oxidation degree of methylene near the two-coordinated Cu(I) is higher. In **O65**, weak Cu(I)-carbonyl interactions can be observed (Supplementary Table 2), which are consistent with the relatively low wave number of carbonyl absorption in the infrared spectrum. Such unambiguous crystallography evidence is noteworthy because the harsh reaction condition for covalent PSM generally degrades crystallinity so that the modification degree is not high enough to be determined by crystal-structure analysis[23,37,38]. **O65** has a slightly smaller and distorted unit cell compared with MAF-42-*sp* ($\Delta V/V_1 = -0.6\%$, $\Delta\beta = -1.0°$), which can be ascribed to the generation of polar carbonyl groups on the pore surface and dipole–dipole attractive interactions within the host framework of **O65**. While the unit-cell volume follows C_6H_6@MAF-42-*lp* >> MAF-42-*sp* > **O65**, their organic ligands deviate from the planar conformation by 12.4, − 14.9 and − 12.9°, respectively (Supplementary Fig. 5). Obviously, oxidation enhances the conjugation degree and planarity of ligands, which are disadvantageous for framework contraction. To further shrink from MAF-42-*sp* to **O65**, the coordination bonding lengths ($\Delta_{max} = 0.044$ Å) and angles ($\Delta_{max} = 9.6°$) are forced to change more (Supplementary Tables 2 and 3). On the other hand, framework contraction also reduces the planarity of ligands in **O65**.

To gain more insight into the aerobic oxidation process, we tried to capture some reaction intermediates. In the single-crystal structure of MAF-42 loaded with O_2 (denoted as O_2@MAF-42) at low temperature of − 140 °C (Supplementary Table 1), although O_2 molecules are highly disordered, it can be seen that the primary adsorption sites are close to the two- and three-

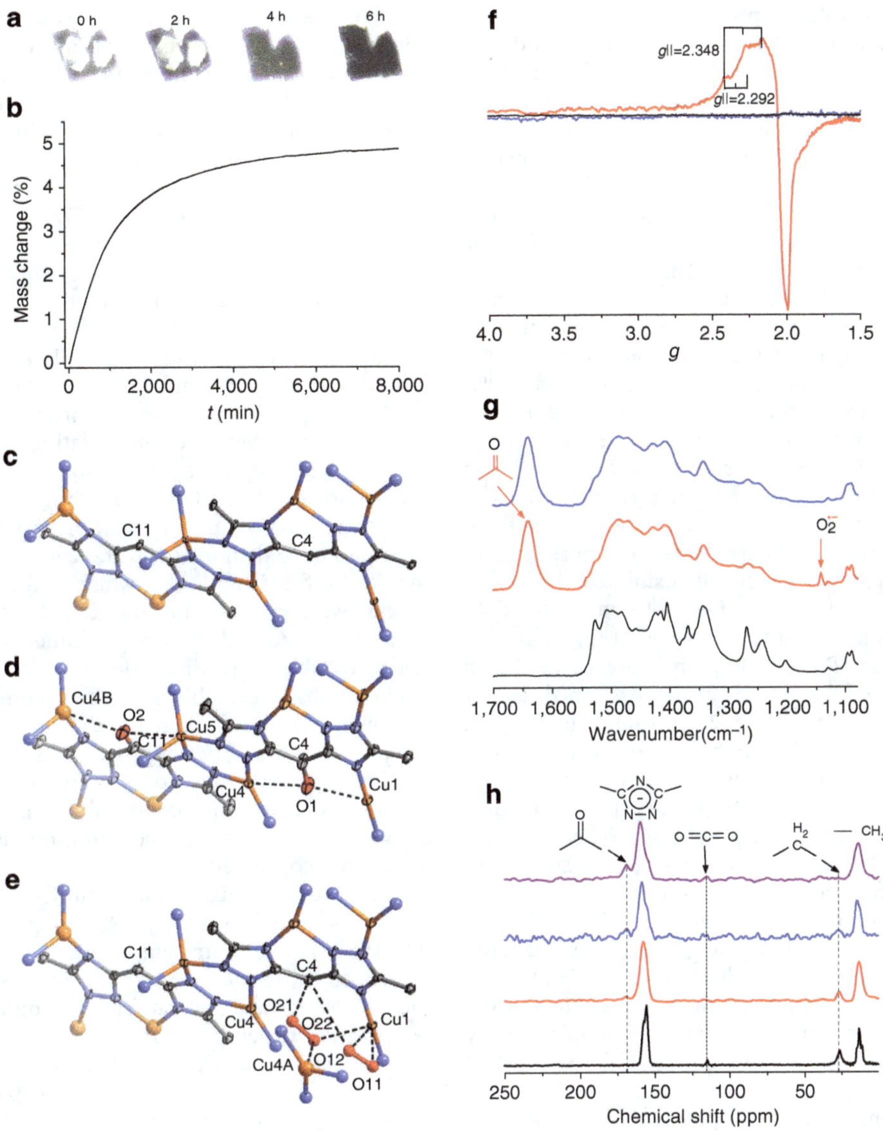

Figure 3 | Controlling the reaction degree and revealing the catalytic mechanism of aerobic oxidation. (**a**) Photographs of C_6H_6@MAF-42-*lp* exposed in air at room temperature for different time. (**b**) TG of MAF-42 in an O_2 flow at 418 K. (**c**–**e**) X-ray single-crystal structures of MAF-42, **O65** and O_2@MAF-42 (probability drawn at 30% for independent atoms; symmetric codes: A = 1 − *x*, − *y*, 1 − *z*; B = *x*, 1 − *y*, − 0.5 + *z*). The black dashed lines are drawn to highlight relatively short nonbonding interactions (Cu1···O11 3.04(4) Å, Cu1···O12 3.28(3) Å, Cu4···O22 3.73(6) Å), C4···O12 4.47(6) Å, C4···O21 3.46(6) Å, Cu1···O1 2.61(1) Å, Cu4···O2 2.79(1) Å). (**f**) EPR spectra of MAF-42 in O_2 measured at 298 K (black) and then at 473 K (red) and then back to 298 K (blue). (**g**) DR-FTIR spectra of MAF-42 measured at 473 K under a variable atmosphere from helium (black) to oxygen (red) and then back to helium (blue). (**h**) Solid-state ^{13}C NMR spectra of MAF-42 (black), **O53** (red), **O74** (blue) and **O100** (purple).

coordinated Cu(I) ions and the methylene group having a preference for oxidation (Fig. 3e and Supplementary Table 2). The presence of Cu(II) species during the aerobic oxidation reaction was confirmed by the axially symmetric signal with $g_{\parallel} = 2.292 \sim 2.348$ and the vertical signal with $g_{\perp} = 2.089$ in the *in situ* electron paramagnetic resonance (EPR) spectroscopy measured at 200 °C (Fig. 3f)[39]. Because the peak intensity at $g \sim 2.0$–2.3 originated from the Cu(II) ion is vastly greater than that of the radical $O_2^{\cdot -}$ at $g \sim 2$, the latter is difficult to assign. On the other hand, *in situ* diffuse reflectance-Fourier transform infrared spectroscopy showed absorption bands at 1,140 cm^{-1} characteristic for the $O_2^{\cdot -}$ species (Fig. 3g) and 458 cm^{-1} characteristic for the Cu–O coordination bond (Supplementary Fig. 14)[40]. The observation of physical adsorption at low temperature and chemical adsorption at high temperature is similar with some other heterogeneous reactions involving gas reactants and solid catalysts[41]. Therefore, some key stages of the aerobic oxidation mechanism have been observed, in which O_2 molecules first attack the low-coordinated Cu(I) ions, and then form the highly active intermediate Cu(II)-$O_2^{\cdot -}$ reactive enough to break the hydrocarbon C–H bond, and finally returned to Cu(I) and produced the carbonyl product and water. The reaction can be unambiguously assigned to the Cu(II)/Cu(I) redox couple and metal-centred four-electron oxidation mechanism, which are commonly expected for natural enzymes and small molecular complexes but have been scarcely confirmed[18,19]. The catalytic activity of the low-coordinated Cu(I) sites in MAF-42 for aerobic oxidation of guest reactants was further confirmed by its effectiveness and low activation energy for the reaction of $CO + O_2 = CO_2$ (Supplementary Methods, Supplementary Figs 4 and 15).

Regulation of gas adsorption properties. To reveal the structure–property relationship, MAF-42 was oxidized to different degrees by the above established aerobic oxidation method. By virtue of the solid–gas reaction mechanism, the products were obtained in quantitative yields and were used without further workup procedure. Solid-state ^{13}C nuclear magnetic resonance (NMR) spectra showed that the methylene peak at 27 p.p.m. and carbonyl peak at 169 p.p.m. gradually reduces and increases, respectively, as the oxidation degree increases (Fig. 3h). On the basis of the peak areas of the methylene (2.8 p.p.m.) and methyl (0.8 p.p.m.) groups observed in the solution ^{1}H NMR spectra of the DCl-digested samples (Supplementary Fig. 16), the oxidation degrees of the oxidized samples (hereafter denoted as **O53**, **O74** and **O100**, respectively) were calculated as 53%, 74% and 100%. PXRD and TG showed that oxidization enhances the hydrophilicity, decreases the thermal stability and reduces the unit-cell volumes of the samples (Supplementary Figs 17 and 18), being consistent with the generation of more polar carbonyl groups[4].

Single-component CO_2 isotherms were measured for MAF-42, **O53**, **O74** and **O100** at 195 K to characterize their porosity and framework flexibility/rigidity, which all exhibited hysteresis (Fig. 4a and Supplementary Fig. 19a). The isotherm of MAF-42 showed a direct transition from the nonporous (*np*) phase ($P/P_0<0.25$) to the *lp* phase ($P/P_0>0.40$). The absence of the *sp* phase in the isotherm of MAF-42 can be explained by its extremely small aperture size (2.2×2.6 Å^2) and inert pore surface. Differently, **O53** and **O74** exhibited the expected *sp*-to-*lp* transition, as their first-step saturation uptakes are coincident with the theoretical values calculated from the crystal structures of MAF-42 and **O65** (Supplementary Table 4). The increased CO_2-binding ability of the oxidized samples is consistent with their enhanced pore surface polarity. The second-step saturation uptake and the corresponding pore volume follow **O74** (128 and 0.23 $cm^3\,g^{-1}$) >**O53** (116 and 0.21 $cm^3\,g^{-1}$) >MAF-42 (105 and 0.19 $cm^3\,g^{-1}$), indicating that oxidization can expand the *lp* phase of the host frameworks, which was further quantified by the unit-cell volumes of guest-saturated samples (Supplementary Fig. 20). Framework expansion of the *lp* phase could be ascribed to the enhanced rigidity of the oxidized ligand, which usually increases the bridging length. In the *lp* phase, the dipole–dipole attraction effect (causing the shrinkage of the *sp* phase[42]) can be eliminated by insertion of guest in the channels[5,8].

More interestingly, when the oxidation degree increases, the gate-opening/closing pressure (defined as the intermediate pressure between two isotherm steps) increases during adsorption ($P/P_0=0.32$, 0.35 and 0.74 for MAF-42, **O53** and **O74**, respectively) but decreases during desorption ($P/P_0=0.25$, 0.21 and 0.14 for MAF-42, **O53** and **O74**, respectively) to widen the hysteresis loop, which demonstrates that oxidization increases the ligand rigidity and difficulty of phase transition and the oxidized samples are solid-solution frameworks rather than mechanical mixtures[15,43]. Further, **O100** showed one-step sorption isotherm with a small saturation uptake of *ca* 30 $cm^3\,g^{-1}$ (Fig. 4a and Supplementary Fig. 19a), confirming that it's *sp* phase has the most shrunk structure compared with the counterparts with lower oxidation ratios, and is too rigid to undergo the *sp*-to-*lp* transition. Therefore, the oxidized framework exhibited larger breathing amplitude, energy change and energy barrier during *sp*-to-*lp* transition. Only a couple of examples have demonstrated the control of framework flexibility by PSM of PCPs[42,44], in which different functional building blocks were added into the coordination framework, which always reduced the pore volume and the regulation is uncontrollable and discontinuous. In our case, the aerobic oxidation has negligible size effect (similar for methylene and carbonyl) but directly increases the ligand/framework rigidity and the pore volume of the *lp* phase, so that the oxidized framework exhibits enhanced adsorption capacity. More importantly, the solid–gas reaction mechanism allows modification ratio and gas sorption property to be continuously and conveniently monitored/controlled.

High-pressure single-component CO_2, C_2H_6 and CH_4 sorption isotherms were measured at 298 K (Fig. 4b–d and Supplementary Fig. 19b–d). The room-temperature CO_2 sorption isotherms are similar to those measured at 195 K, except that MAF-42 shows the *sp*-to-*lp* transition at room temperature. The abnormal temperature-dependent sorption behaviour of the *sp* phase of MAF-42 can be ascribed to thermal expansion (Supplementary Fig. 4) and/or decreased diffusion barrier[45]. The gradually increased gate-opening pressure (16, 18, 28 and >40 bar for

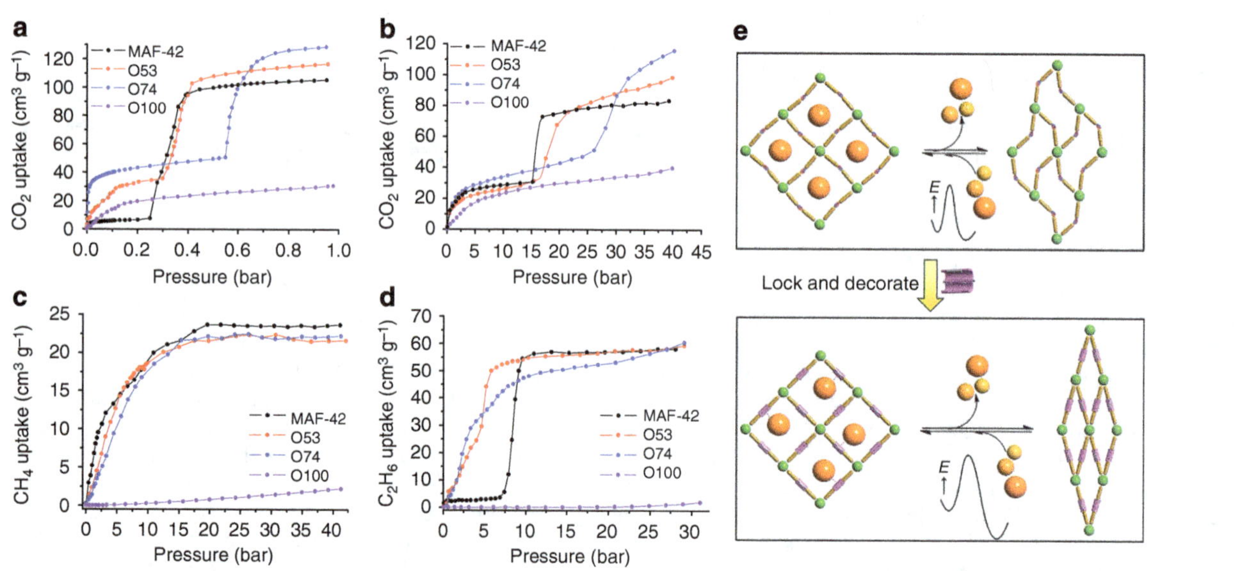

Figure 4 | Modulation of single-component gas adsorption properties. (**a**) CO_2 at 195 K. (**b**) CO_2 at 298 K. (**c**) CH_4 at 298 K. (**d**) C_2H_6 at 298 K. Lines are drawn to guide eyes. (**e**) Elucidation of the gas adsorption mechanism. The framework flexibility and pore surface polarity of a porous crystal are modulated by chemical modification of the organic linker. Specifically, for MAF-42, the hydrophobic and flexible methylene bridges are transformed to hydrophilic and rigid carbonyl ones via aerobic oxidation, and the transformation ratio is controlled by the oxidation degree.

MAF-42, **O53**, **O74** and **O100**, respectively) and decreased isotherm slope of the *sp*-to-*lp* transition are consistent with the increased framework rigidity of the oxidized samples (Fig. 4e).

For CH_4, MAF-42, **O53** and **O74** show type-I adsorption isotherms corresponding to the *sp* phase (Supplementary Table 4). Compared with CO_2, CH_4 has very low boiling point and polarity (Supplementary Table 5), meaning that it interacts with the host framework very weakly and can hardly open the narrow channel or induce the *sp*-to-*lp* transition. The variation of C_2H_6 sorption isotherms of MAF-42, **O53** and **O74** resembles that of CO_2 at 195 K. The gate-opening pressure is 8 bar for the *np*-to-*lp* transition of MAF-42, which is higher than those of 5 and 6 bar for the *sp*-to-*np* transitions of **O53** and **O74**, respectively. The nonporous *sp* phase of MAF-42 for C_2H_6 can be explained by the very large molecular size of the guest (Supplementary Table 5). Because of the over-shrinkage of the host framework, **O100** is virtually nonporous for the low-polarity gases CH_4 and C_2H_6.

The finely/drastically tunable gas sorption properties suggest their usefulness for on-demand gas separation. Mixed gas adsorption isotherms were measured to reveal the real gas adsorption selectivities of selected samples with promising isotherms. The CO_2/CH_4 selectivities of MAF-42 and **O100** at 1–10 bar (measured by mixed CO_2/CH_4 with 40:60 molar ratio) were observed as 28–14 and 700–600, respectively (Supplementary Methods, Fig. 5a–c), meaning that the performance of the crystal can be significantly improved by the aerobic oxidation treatment. The extremely high CO_2/CH_4 selectivity of **O100** should be suitable for purifying CH_4 from biogases (CH_4 45~65%, CO_2 30~50%)[46] and landfill gases (CH_4 35~55%, CO_2 40~45%)[47] by selective adsorption of CO_2.

More interestingly, the real CH_4/C_2H_6 selectivities (measured by mixed CH_4/C_2H_6 with 20:80 molar ratio) were observed to be 500–200 and 1/24–1/17 for MAF-42 and **O74**, respectively, indicating the catalytic oxidation can drastically invert the CH_4/C_2H_6 selectivity up to 4 orders of magnitude, which has not been realized by other adsorbents or methods (Supplementary Methods, Fig. 5d–f). The selective adsorption of CH_4 by MAF-42 at low pressure is due to its nonporous nature for C_2H_6 according to the molecular sieving effect. On the other hand, **O74** prefers adsorption of C_2H_6 because it is porous to both gases, and C_2H_6 has the larger molecular weight and quadrupole moment (Supplementary Table 5). This property may be useful for on-demand purification of different mixture gases (for example, natural gases usually contain CH_4 69~96% and C_2H_6 1~14%)[2,48].

Conventional adsorbents generally exhibit Langmuir-type isotherm and monotonically decreased (versus pressure) adsorption selectivity, which is not suitable for some gas separation applications operating at high pressures[49]. Some flexible PCPs showing gate-opening phenomena and stepped isotherms may be used to solve this problem; however, the real selectivities are usually much lower than those predicted by the single-component isotherms because the non-preferred guest can also enter the opened channel above the gate-opening pressure[50]. We observed that the real C_2H_6/CH_4 selectivity (measured by mixed C_2H_6/CH_4 with 75:25 molar ratio) of **O53** significantly increases from 92 to 140 at C_2H_6 partial pressure from 4.7 to 5.5 bar, respectively, and remains 113 at C_2H_6 partial pressure of 8.1 bar (Supplementary Methods and Supplementary Fig. 21). The abrupt increase in real adsorption selectivity occurs around the gate-opening pressure of single-component C_2H_6 isotherm at 4.7–5.2 bar. This observation indicates that the gate-opening is forced by filling the C_2H_6 molecules in the newly generated space. In other words, there is no space left for CH_4, preventing the commonly observed co-adsorption effect.

Discussion

In summary, we demonstrated that Cu(I) ions with enzyme-like O_2 activation ability can cooperate with a flexible and oxidizable bis-triazolate ligand to fabricate a porous crystal reactive towards molecular oxygen even at room temperature. The methylene bridge of the bis-triazolate ligand provides the key reactivity to molecular oxygen, as well as modifiable framework flexibility and pore surface polarity. Although the free ligand is inert to oxygen, the flexible methylene bridge near the low-coordinated Cu(I) centres in the coordination framework is reactive enough to be

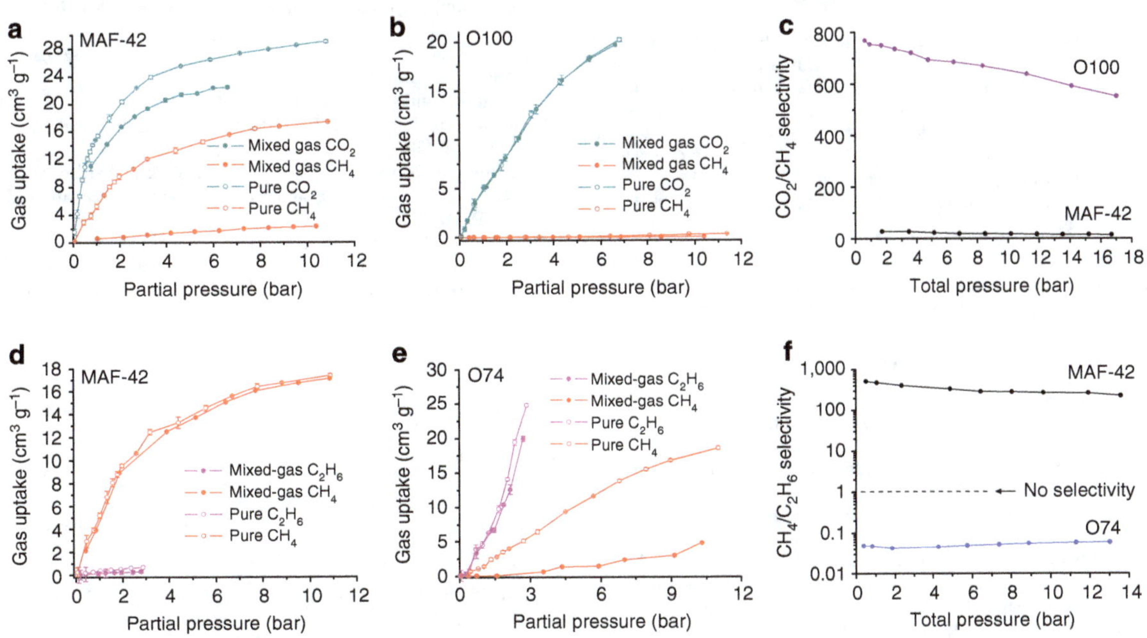

Figure 5 | Modulation of mixed-gas adsorption selectivity. (**a–c**) A ratio of 40:60 CO_2/CH_4 for MAF-42 and **O100**, as well as the corresponding CO_2/CH_4 selectivity. (**d–f**) A ratio of 80:20 CH_4/C_2H_6 for MAF-42 and **O74**, as well as the corresponding CH_4/C_2H_6 selectivity. All gas adsorption isotherms were measured at 298 K. Lines are drawn to guide eyes.

oxidized as a more rigid and polar carbonyl group. *In situ* structural and spectroscopic analyses confirmed that the low-coordinated Cu(I) centres behave like those in copperproteins during the aerobic oxidation. Compared with known material synthesis and processing methods, the solvent-free aerobic oxidation reaction has a series of advantages, such as green, quantitative yield, easy and precise monitor/control of modification degree, work-up procedure free, and so on. While conventional adsorbents separate adsorbates with fixed selectivities by either sorption affinity difference or molecular sieving effect, the porous crystal consisting of a reactive organic linker with modifiable flexibility and polarity offers a possibility to utilize both mechanisms, as demonstrated by the very high, drastically tunable and even switchable gas sorption selectivities. These results may enlighten future design and construction of multifunctional and controllable porous materials.

Methods

Materials. Commercially available reagents and solvents were used without further purification. The ligand H_2btm was synthesized according to the literature method[51].

Measurements. Infrared spectra were obtained from KBr pellets on a Bruker TENSOR 27 FT IR spectrometer in the 400- to 4,000-cm^{-1} region. Diffused reflection Fourier transform infrared spectroscopy (DR-FTIR) was performed on a Bruker VERTEX 70 spectrometer in the 400- to 4,000-cm^{-1} region. TG-mass spectra were performed on a hyphenated apparatus of NETZSCH STA 449 F3 Jupiter and NETZSCH QMS 403C Aedo. Elemental analyses (C, H and N) were performed with a Vario EL elemental analyser. TG analyses were performed by using a TA Q50 system. Solid-state ^{13}C NMR measurements were carried out at ambient temperature on a Bruker AVANCE 400 spectrometer. Solution ^{1}H NMR measurements were carried out at ambient temperature on a VARIAN Mercury-Plus 300 spectrometer, for which $\sim$20 mg of solid samples were digested with sonication in 550 µl of DCl (20 wt% in D_2O). PXRD patterns were collected (0.02° per step, 0.06 s per step except for otherwise stated) on a Bruker D8 Advance diffractometer (Cu Kα) at room temperature. Mass spectra were measured by a SHIMADZU LCMS-2010A equipment using an electrospray ionization source with MeOH as the mobile phase. Electron paramagnetic resonance measurements were performed at 9.7 GHz (X-band) using a Bruker BioSpin A300 spectrometer. The spin concentrations in the samples were determined from the second integral of the spectra using $CuSO_4 \cdot 5H_2O$ as a standard. The as-synthesized sample (weight of $\sim$100 – 200 mg) was placed in the self-made quartz tube and dried for 8 h at 260 °C under high vacuum to remove the remnant solvent molecules, sealed with back-filled O_2 before measurements.

Synthesis of materials. A mixture of aqueous ammonia (25%, 4 ml) solution of $[Cu(NH_3)_2]OH$ (0.025 mol l^{-1}), H_2btm (0.009 g, 0.05 mmol), methanol (3.0 ml) and benzene (1.0 ml) was sealed in a 15-ml Teflon-lined stainless reactor, which was heated in an oven at 160 °C for 72 h. The oven was cooled to room temperature at a rate of 5 °C h^{-1}. The resulting colourless block crystals were filtered, washed by ethanol and dried in air to give C_6H_6@MAF-42-*lp* (yield *ca* 83%). Guest-free MAF-42 was obtained by heating C_6H_6@MAF-42-*lp* at 260 °C under high vacuum for 24 h. The oxidized samples were obtained by heating MAF-42 at 418 K in the O_2 flow for different times.

X-ray crystallography. Single-crystal diffraction intensities were collected on a Bruker Apex CCD diffractometer with graphite-monochromated Mo Kα radiation or a Oxford Gemini S Ultra CCD diffractometer using mirror-monochromated Cu Kα radiation. Absorption corrections were applied by using the multiscan programme SADABS. The structures were solved by the direct method and refined with a full-matrix least-squares technique with the SHELXTL 6.10 programme package. The occupancies of carbonyl oxygen atoms and guest molecules were obtained by free refinement. Anisotropic thermal parameters were applied to all non-hydrogen atoms of host frameworks except for the oxygen atom in structure obtained by prolong exposure of C_6H_6@MAF-42-*lp* in air. Hydrogen atoms were generated geometrically. Crystal data for the complexes were summarized in Supplementary Table 1.

PXRD data for Pawley refinement were collected in 0.02° per step and 3 s per step. Indexing and Pawley refinement of the PXRD patterns was carried out by using the Reflex module of Material Studio 5.0. The patterns were indexed by the TREOR90 method with the aid of unit-cell parameters from single-crystal data. Pawley refinements were carried out with the cell parameters obtained from indexing in space group *C*2/*c*. Peak profiles, zero-shifts and unit-cell parameters were refined simultaneously. The peak profiles were refined by the Pseudo-Voigt function with Berar–Baldinozzi asymmetry correction parameters.

Gas sorption measurement and calculation. The single-component gas sorption isotherms were measured with automatic volumetric adsorption apparatuses (BELSORP-max and BELSORP-HP). The mixed gas adsorption isotherms were measured by volumetric adsorption – gas chromatograph instrument BELSORP-VC with a maximum measurement pressure of 15 bar. The as-synthesized sample (weight of $\sim$500 – 800 mg) was placed in the quartz tube and dried for 8 h at 260 °C to remove the remnant solvent molecules before measurements. CO_2 (99.999%), CH_4 (99.999%) and C_2H_6 (99.99%) were used for all measurements. The temperatures were controlled by an acetone – dry ice bath (195 K) or a water bath (298 K).

References

1. Herm, Z. R. *et al.* Separation of hexane isomers in a metal-organic framework with triangular channels. *Science* **340,** 960–964 (2013).
2. Li, J.-R., Kuppler, R. J. & Zhou, H.-C. Selective gas adsorption and separation in metal-organic frameworks. *Chem. Soc. Rev.* **38,** 1477–1504 (2009).
3. Cohen, S. M. Postsynthetic methods for the functionalization of metal–organic frameworks. *Chem. Rev.* **112,** 970–1000 (2012).
4. Zhang, J.-P., Zhang, Y.-B., Lin, J.-B. & Chen, X.-M. Metal azolate frameworks: from crystal engineering to functional materials. *Chem. Rev.* **112,** 1001–1033 (2012).
5. Férey, G. & Serre, C. Large breathing effects in three-dimensional porous hybrid matter: facts, analyses, rules and consequences. *Chem. Soc. Rev.* **38,** 1380–1399 (2009).
6. Bunck, D. N. & Dichtel, W. R. Mixed linker strategies for organic framework functionalization. *Chem. Eur. J.* **19,** 818–827 (2013).
7. Burrows, A. D. Mixed-component metal-organic frameworks (MC-MOFs): enhancing functionality through solid solution formation and surface modifications. *CrystEngComm* **13,** 3623–3642 (2011).
8. Horike, S., Shimomura, S. & Kitagawa, S. Soft porous crystals. *Nat. Chem.* **1,** 695–704 (2009).
9. Nugent, P. *et al.* Porous materials with optimal adsorption thermodynamics and kinetics for CO_2 separation. *Nature* **495,** 80–84 (2013).
10. Rabone, J. *et al.* An adaptable peptide-based porous material. *Science* **329,** 1053–1057 (2010).
11. Liao, P.-Q. *et al.* Strong and dynamic CO_2 sorption in a flexible porous framework possessing guest chelating claws. *J. Am. Chem. Soc.* **134,** 17380–17383 (2012).
12. Henke, S., Schneemann, A., Wütscher, A. & Fischer, R. A. Directing the breathing behavior of pillared-layered metal–organic frameworks via a systematic library of functionalized linkers bearing flexible substituents. *J. Am. Chem. Soc.* **134,** 9464–9474 (2012).
13. Schneemann, A. *et al.* Flexible metal-organic frameworks. *Chem. Soc. Rev.* **43,** 6062–6096 (2014).
14. Shekhah, O. *et al.* Made-to-order metal-organic frameworks for trace carbon dioxide removal and air capture. *Nat. Commun.* **5,** 4228 (2014).
15. Fukushima, T. *et al.* Solid solutions of soft porous coordination polymers: fine-tuning of gas adsorption properties. *Angew. Chem. Int. Ed.* **49,** 4820–4824 (2010).
16. Deng, H. *et al.* Multiple functional groups of varying ratios in metal-organic frameworks. *Science* **327,** 846–850 (2010).
17. Bernt, S., Guillerm, V., Serre, C. & Stock, N. Direct covalent post-synthetic chemical modification of Cr-MIL-101 using nitrating acid. *Chem. Commun.* **47,** 2838–2840 (2011).
18. Burrows, A. D., Frost, C. G., Mahon, M. F. & Richardson, C. Sulfur-tagged metal-organic frameworks and their post-synthetic oxidation. *Chem. Commun.* **0,** 4218–4220 (2009).
19. Jiang, H.-L., Feng, D., Liu, T.-F., Li, J.-R. & Zhou, H.-C. Pore surface engineering with controlled loadings of functional groups via click chemistry in highly stable metal–organic frameworks. *J. Am. Chem. Soc.* **134,** 14690–14693 (2012).
20. Karagiaridi, O. *et al.* Opening ZIF-8: a catalytically active zeolitic imidazolate framework of sodalite topology with unsubstituted linkers. *J. Am. Chem. Soc.* **134,** 18790–18796 (2012).
21. Li, T., Kozlowski, M. T., Doud, E. A., Blakely, M. N. & Rosi, N. L. Stepwise ligand exchange for the preparation of a family of mesoporous MOFs. *J. Am. Chem. Soc.* **135,** 11688–11691 (2013).
22. Zhang, J.-P., Liao, P.-Q., Zhou, H.-L., Lin, R.-B. & Chen, X.-M. Single-crystal X-ray diffraction studies on structural transformations of porous coordination polymers. *Chem. Soc. Rev.* **43,** 5789–5814 (2014).
23. Sato, H., Matsuda, R., Sugimoto, K., Takata, M. & Kitagawa, S. Photoactivation of a nanoporous crystal for on-demand guest trapping and conversion. *Nat. Mater.* **9,** 661–666 (2010).
24. Kawamichi, T., Kodama, T., Kawano, M. & Fujita, M. Single-crystalline molecular flasks: chemical transformation with bulky reagents in the pores of porous coordination networks. *Angew. Chem. Int. Ed.* **47,** 8030–8032 (2008).
25. Yang, S. *et al.* Selectivity and direct visualization of carbon dioxide and sulfur dioxide in a decorated porous host. *Nat. Chem.* **4,** 887–894 (2012).

26. Vaidhyanathan, R. *et al.* Direct observation and quantification of CO_2 binding within an amine-functionalized nanoporous solid. *Science* **330,** 650–653 (2010).
27. Punniyamurthy, T., Velusamy, S. & Iqbal, J. Recent advances in transition metal catalyzed oxidation of organic substrates with molecular oxygen. *Chem. Rev.* **105,** 2329–2364 (2005).
28. Que, L. & Tolman, W. B. Biologically inspired oxidation catalysis. *Nature* **455,** 333–340 (2008).
29. Li, L. *et al.* A crystalline porous coordination polymer decorated with nitroxyl radicals catalyzes aerobic oxidation of alcohols. *J. Am. Chem. Soc.* **136,** 7543–7546 (2014).
30. Piera, J. & Bäckvall, J.-E. Catalytic oxidation of organic substrates by molecular oxygen and hydrogen peroxide by multistep electron transfer—a biomimetic approach. *Angew. Chem. Int. Ed.* **47,** 3506–3523 (2008).
31. Grzywa, M. *et al.* CFA-2 and CFA-3 (Coordination Framework Augsburg University-2 and -3); novel MOFs assembled from trinuclear Cu(I)/Ag(I) secondary building units and 3,3',5,5'-tetraphenyl-bipyrazolate ligands. *Dalton Trans.* **42,** 6909–6921 (2013).
32. Barbara, E. *Ullmann's Encyclopedia of Industrial Chemistry* (Wiley-VCH Verlag GmbH & Co. KGaA, Weinheim, 2000).
33. Orpen, A. G. *et al.* Tables of bond lengths determined by X-ray and neutron diffraction. Part 2. Organometallic compounds and co-ordination complexes of the d- and f-block metals. *J. Chem. Soc. Dalton Trans.* S1–83 (1989).
34. Sippel, P. *et al.* Dielectric relaxation processes, electronic structure, and band gap engineering of MFU-4-type metal-organic frameworks: towards a rational design of semiconducting microporous materials. *Adv. Funct. Mater.* **24,** 3885–3896 (2014).
35. Yoon, J. W. *et al.* Controlled reducibility of a metal–organic framework with coordinatively unsaturated sites for preferential gas sorption. *Angew. Chem. Int. Ed.* **49,** 5949–5952 (2010).
36. Liu, T.-F. *et al.* Stepwise synthesis of robust metal–organic frameworks via postsynthetic metathesis and oxidation of metal nodes in a single-crystal to single-crystal transformation. *J. Am. Chem. Soc.* **136,** 7813–7816 (2014).
37. Deshpande, R. K., Minnaar, J. L. & Telfer, S. G. Thermolabile groups in metal–organic frameworks: suppression of network interpenetration, post-synthetic cavity expansion, and protection of reactive functional groups. *Angew. Chem. Int. Ed.* **49,** 4598–4602 (2010).
38. Wang, Z. & Cohen, S. M. Postsynthetic covalent modification of a neutral metal – organic framework. *J. Am. Chem. Soc.* **129,** 12368–12369 (2007).
39. Smeets, P. J., Woertink, J. S., Sels, B. F., Solomon, E. I. & Schoonheydt, R. A. Transition-metal ions in zeolites: coordination and activation of oxygen. *Inorg. Chem.* **49,** 3573–3583 (2010).
40. Vaska, L. Dioxygen-metal complexes: toward a unified view. *Acc. Chem. Res.* **9,** 175–183 (1976).
41. Otsuka, K. & Wang, Y. Direct conversion of methane into oxygenates. *Appl. Catal. A* **222,** 145–161 (2001).
42. Wang, Z. & Cohen, S. M. Modulating metal – organic frameworks to breathe: a postsynthetic covalent modification approach. *J. Am. Chem. Soc.* **131,** 16675–16677 (2009).
43. Coudert, F. X., Jeffroy, M., Fuchs, A. H., Boutin, A. & Mellot-Draznieks, C. Thermodynamics of guest-induced structural transitions in hybrid organic – inorganic frameworks. *J. Am. Chem. Soc.* **130,** 14294–14302 (2008).
44. Volkringer, C. & Cohen, S. M. Generating reactive MILs: isocyanate- and isothiocyanate-bearing mils through postsynthetic modification. *Angew. Chem. Int. Ed.* **49,** 4644–4648 (2010).
45. Wei, Y.-S. *et al.* Turning on the flexibility of isoreticular porous coordination frameworks for drastically tunable framework breathing and thermal expansion. *Chem. Sci.* **4,** 1539–1546 (2013).
46. Grande, C. A. & Rodrigues, A. E. Biogas to fuel by vacuum pressure swing adsorption i. behavior of equilibrium and kinetic-based adsorbents. *Ind. Eng. Chem. Res.* **46,** 4595–4605 (2007).
47. Pires, J., Bestilleiro, M., Pinto, M. & Gil, A. Selective adsorption of carbon dioxide, methane and ethane by porous clays heterostructures. *Sep. Purif. Technol.* **61,** 161–167 (2008).
48. He, Y., Zhou, W., Krishna, R. & Chen, B. Microporous metal-organic frameworks for storage and separation of small hydrocarbons. *Chem. Commun.* **48,** 11813–11831 (2012).
49. Horike, S., Inubushi, Y., Hori, T., Fukushima, T. & Kitagawa, S. A solid solution approach to 2D coordination polymers for CH_4/CO_2 and CH_4/C_2H_6 gas separation: equilibrium and kinetic studies. *Chem. Sci.* **3,** 116–120 (2012).
50. Hamon, L. *et al.* Co-adsorption and separation of $CO_2 - CH_4$ mixtures in the highly flexible MIL-53(Cr) MOF. *J. Am. Chem. Soc.* **131,** 17490–17499 (2009).
51. Bahceci, S., Yuksek, H. & Serdar, M. Reactions of amidines with some carboxylic acid hydrazides. *Indian J. Chem. B* **44,** 568–572 (2005).

Acknowledgements

This work was supported by the '973 Project' (2012CB821706 and 2014CB845602) and NSFC (21225105, 21290173 and 21473260). We thank Professor Hongbing Ji, Dr Zebao Rui and Mr Huayao Chen for the DR-FTIR measurement and Mr Guping Hu for the EPR measurement.

Author contributions

J.-P.Z. designed the research. P.-Q.L. and A.-X.Z. performed syntheses and measurements. P.-Q.L. and W.-X.Z. performed PXRD analyses. P.-Q.L., J.-P.Z. and X.-M.C. wrote the manuscript.

Additional information

Accession codes. The X-ray crystallographic coordinates for structures reported in this Article have been deposited at the Cambridge Crystallographic Data Centre (CCDC), under deposition number CCDC 1041776-1041780. These data can be obtained free of charge from The Cambridge Crystallographic Data Centre via www.ccdc.cam.ac.uk/data_request/cif.

Competing financial interests: The authors declare no competing financial interests.

19

Pseudocapacitance of MXene nanosheets for high-power sodium-ion hybrid capacitors

Xianfen Wang[1], Satoshi Kajiyama[1], Hiroki Iinuma[1], Eiji Hosono[2], Shinji Oro[3], Isamu Moriguchi[3], Masashi Okubo[1,4] & Atsuo Yamada[1,4]

High-power Na-ion batteries have tremendous potential in various large-scale applications. However, conventional charge storage through ion intercalation or double-layer formation cannot satisfy the requirements of such applications owing to the slow kinetics of ion intercalation and the small capacitance of the double layer. The present work demonstrates that the pseudocapacitance of the nanosheet compound MXene Ti_2C achieves a higher specific capacity relative to double-layer capacitor electrodes and a higher rate capability relative to ion intercalation electrodes. By utilizing the pseudocapacitance as a negative electrode, the prototype Na-ion full cell consisting of an alluaudite $Na_2Fe_2(SO_4)_3$ positive electrode and an MXene Ti_2C negative electrode operates at a relatively high voltage of 2.4 V and delivers 90 and 40 mAh g^{-1} at 1.0 and 5.0 A g^{-1} (based on the weight of the negative electrode), respectively, which are not attainable by conventional electrochemical energy storage systems.

[1] Department of Chemical System Engineering, School of Engineering, The University of Tokyo, Hongo 7-3-1, Bunkyo-ku, Tokyo 113-8656, Japan. [2] National Institute of Advanced Industrial Science and Technology, Umezono 1-1-1, Tsukuba 305-8568, Japan. [3] Graduate School of Engineering, Nagasaki University, Bunkyo-machi 1-14, Nagasaki 852-8521, Japan. [4] Elements Strategy Initiative for Catalysts and Batteries (ESICB), Kyoto University, Nishikyo-ku, Kyoto 615-8510, Japan. Correspondence and requests for materials should be addressed to A.Y. (email: yamada@chemsys.t.u-tokyo.ac.jp).

Electrochemical energy storage capable of more energy at high charge – discharge rates is receiving considerable attention because of the strong industrial demands for their widespread use in the smart grid. High-energy Li-ion batteries consisting of Li^+ (de)intercalation electrodes currently power most portable electronics, but the power density is not satisfactory for high-power applications[1,2]. High-power electrochemical capacitors make up many markets ranging from electronics to transportation and stationary applications, but the low energy density is a serious disadvantage[3,4]. This trade-off between the power and energy densities of electrochemical energy storage has been well recognized, which intrinsically originates from the charge storage mechanism of the electrodes: ion (de)intercalation in batteries and double-layer formation in electrochemical capacitors. In general, ion (de)intercalation stores charge more slowly than a double layer, whereas a double layer stores less charge than ion (de)intercalation[3]. Thus, the pseudocapacitance (or redox capacitance) has become an increasingly important mechanism for charge storage[5–10]. Pseudocapacitance occurs through (i) monolayer ion adsorption on the electrode surface with charge transfer, (ii) redox at the electrode surface ($RuO_2 \cdot xH_2O$, ref. 5) or (iii) ultrafast ion intercalation into the bulk (T-Nb_2O_5, ref. 6), whereby charge storage is partially liberated from the trade-off.

At present, urgent concerns over lithium resources have focused recent attention on Na-ion batteries owing to the abundance and low cost of sodium. The present set of negative-electrode materials for Na-ion batteries mainly consists of Na-ion intercalation, alloying or conversion materials such as hard carbon, expanded graphite, TiO_2, $Na_2Ti_3O_7$, phosphorus, Sb_2S_3, $Sn_{4+x}P_3$, SnS_2 and P2-$Na_{0.66}[Li_{0.22}Ti_{0.78}]O_2$ (refs 11–19). However, their energy and power densities severely suffer from the trade-off, and the large volume change accompanied by the reaction with the large Na ion generally results in poor cycle stability, both of which have spurred us to develop new negative electrodes based on pseudocapacitive charge storage for advanced Na-ion systems.

We have targeted MXenes as pseudocapacitor electrodes in this study. MXenes have been developed as a novel family of nanosheet compounds by Gogotsi, Barsoum and colleagues[20–25] MXenes are chemically derived from layered $M_{n+1}AX_n$ or MAX phases, where M is an early transition metal, A is an A-group element and X is C and/or N[26]. For example, a hydrofluoric-acid treatment of Ti_3AlC_2 selectively extracts the aluminum layer, resulting in an exfoliated nanosheet compound, the MXene Ti_3C_2 ($Ti_3C_2T_x$ where T_x denotes surface function groups), which has a surface terminated with OH, F and O. Importantly, the MXene has both a high electrical conductivity and large surface area; therefore, its application to pseudocapacitor electrodes holds great promise. Indeed, the pseudocapacitance of $Ti_3C_2T_x$, Ti_2CT_x, V_2CT_x and Nb_2CT_x in nonaqueous Li-ion electrolytes delivers a specific capacity $>100\,mAh\,g^{-1}$ below 2 V versus Li/Li^+ (refs 22,27–32).

Here, we report on Ti_2CT_x as a negative electrode material for Na-ion hybrid capacitors. The pseudocapacitance of Ti_2CT_x allows Na-ion hybrid capacitors to be liberated from the trade-off between high energy and high power.

Results

Electrochemical properties of Ti_2CT_x. Ti_2CT_x was synthesized by treating the MAX phase Ti_2AlC with a 10% HF aqueous solution. Powder X-ray diffraction (XRD) shows the expansion of the interlayer distance by the hydrofluoric acid (HF) treatment, suggesting successful removal of the Al layers by the HF treatment (Supplementary Fig. 1). The significant decrease in the amount of Al according to energy dispersive X-ray spectroscopy (EDX) supports the selective extraction of Al. A delamination process such as sonication was not conducted[22]; therefore, Ti_2CT_x in this work is a multilayered MXene, as imaged by scanning electron microscopy (Supplementary Fig. 2).

The Ti_2CT_x electrode in a nonaqueous Na^+ electrolyte behaves differently than in an aqueous Na^+ electrolyte or a nonaqueous Li^+ electrolyte. Figure 1a shows the cyclic voltammetry (CV) and corresponding *ex situ* XRD patterns of the MXene Ti_2C in a 1 M $NaPF_6$/ethylene carbonate (EC) – diethyl carbonate (DEC) electrolyte. Lukatskaya *et al.*[23] reported that $Ti_3C_2T_x$ placed in various aqueous salt (NaOH, KOH, LiOH and so on) solutions exhibits a downshift in the (002) diffraction peak in the XRD pattern, corresponding to spontaneous cation intercalation. In contrast, Ti_2CT_x immersed in a nonaqueous 1 M $NaPF_6$/EC – DEC electrolyte exhibits no downshift in the (002) diffraction peak by immersion into the electrolyte alone, which indicates that Na^+ intercalation does not occur spontaneously. The first cathodic scan exhibits an irreversible current below 1.0 V versus Na/Na^+ and an associated downshift in the (002) diffraction peak. As EDX confirms an increase in the Na content in Ti_2CT_x (Supplementary Table S1), Na^+ is definitely intercalated/adsorbed into/onto the MXene sheets electrochemically. On the basis of the shift in the (002) diffraction peak, the expansion of the interlayer distance is estimated as 2.5 Å. The expansion is also confirmed by transmission electron microscopy images (Fig. 1b,c). Because the expansion is consistent with the size of bare Na^+, Na^+ intercalation occurs after desolvation at the electrode/electrolyte interface during the initial reduction process (Fig. 2, activation).

After initial Na^+ intercalation, Ti_2CT_x exhibits pseudocapacitor behaviour. The subsequent CV cycles after the first cycle exhibit the stable rectangular-shaped CV below ~2.0 V,

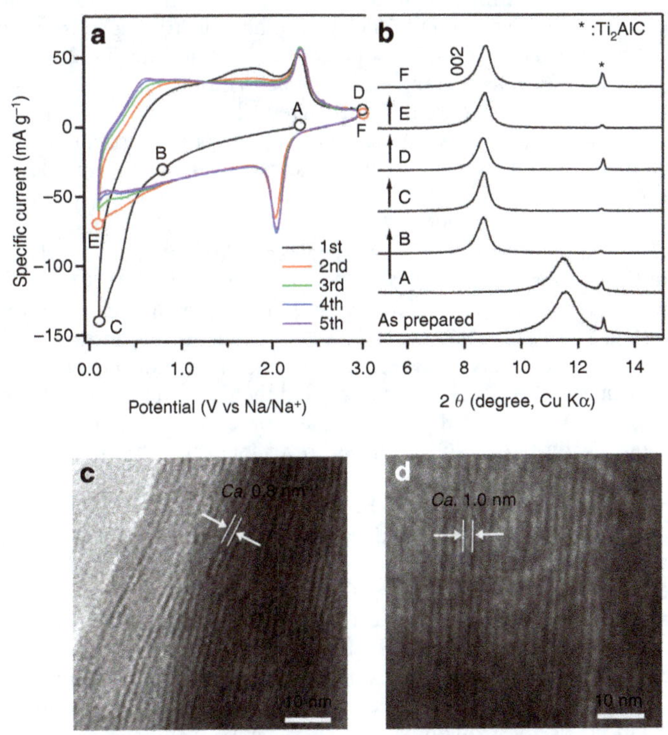

Figure 1 | Electrochemical properties and associated structural changes of the Ti_2CT_x. (**a**) CV for Ti_2CT_x in a 1 M $NaPF_6$/EC – DEC electrolyte at a scan rate of $0.2\,mV\,s^{-1}$. (**b**) *Ex situ* XRD patterns during the CV cycles. (**c**) Transmission electron microscopy (TEM) image of the pristine Ti_2CT_x. (**d**) TEM image of activated Ti_2CT_x after the first CV.

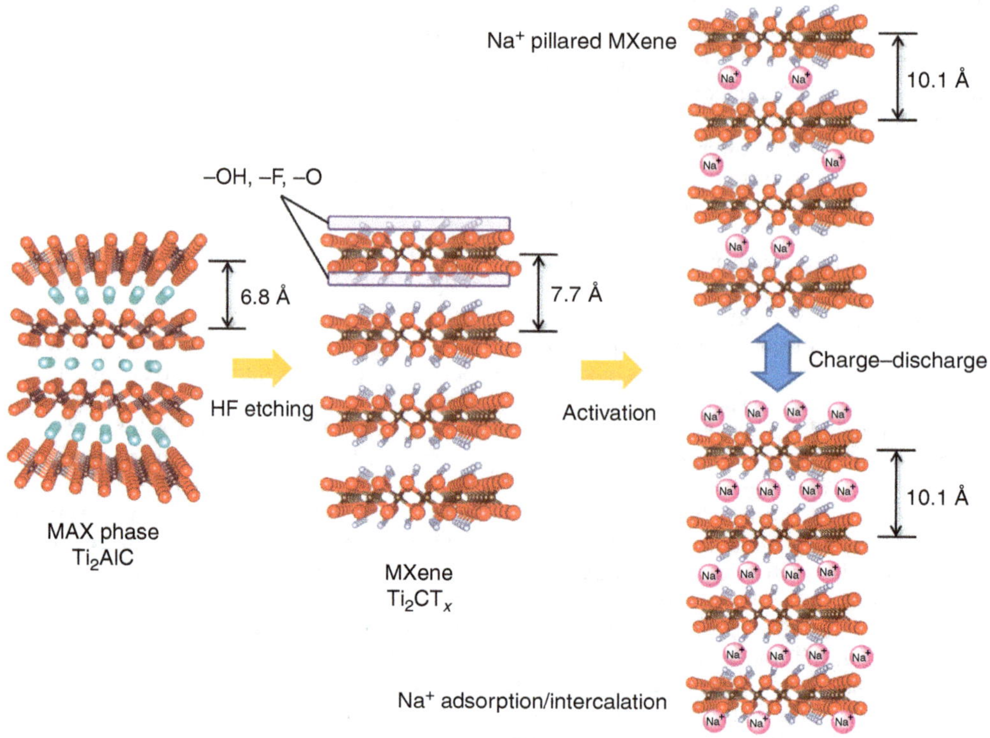

Figure 2 | Schematic illustration of the reaction mechanism of Ti_2CT_x by electrochemical activation. The MAX phase Ti_2AlC is transformed to the MXene Ti_2CT_x nanosheets by the hydrofluoric acid treatment. Ti_2CT_x exhibits expansion of the interlayer distance by the first Na^+ intercalation, then reversible Na^+ (de)intercalation occurs without significant interlayer distance change.

corresponding to typical capacitor behaviour. In addition, a reversible pair of relatively sharp cathodic/anodic peaks is observed at ~2.3 V in each cycle. Although the (002) diffraction peak does not shift during the cycles, EDX confirms a reversible change in the Na content, and X-ray photoelectron spectroscopy (XPS) detects a reversible spectral change in the Ti 2*p* XPS spectrum (Supplementary Fig. 3), which is indicative of a charge transfer reaction. Therefore, the current flow from Ti_2CT_x in the nonaqueous Na^+ electrolyte is attributable to the pseudocapacitance without exhibiting shrinkage/expansion of the interlayer distance.

As reported previously, $Ti_3C_2T_x$ and Ti_2CT_x in a nonaqueous Li^+ electrolyte exhibit shrinkage/expansion of the interlayer distance during the cycle[27,28]. For example, Ti_2CT_x in a nonaqueous Li^+ electrolyte exhibits a change in the interlayer distance between 18.7 and 18.0 Å (4% change). In contrast, bare Na^+ is intercalated between the MXene sheets for Ti_2CT_x in a nonaqueous Na^+ electrolyte to expand the interlayer distance from 7.7 to 10.1 Å during the first cathodic process (Fig. 2, activation). Then, activated Ti_2CT_x allows reversible Na^+ intercalation/deintercalation into the interlayer space as well as reversible Na^+ adsorption/desorption onto the surface of each layer/sheet, where the change in the interlayer distance is small within ~0.1 Å (Fig. 2 and Supplementary Table 2). Because EDX detects residual Na^+, even at 3.0 V (Supplementary Table 1), residual Na^+ may behave as a pillar between the MXene Ti_2C layers during the entire charge/discharge process, maintaining the interlayer distance at an almost constant value. Another possible explanation for the constant interlayer distance is that the expansion due to Na^+ intercalation is balanced by the Coulombic attraction between the MXene sheets and intercalated Na[33].

The CV curves of activated Ti_2CT_x measured at various scan rates (Supplementary Fig. 4) provide insights into the reaction kinetics. The *b*-value analysis ($i = av^b$; *i*: current; *v*: scan rate; *a*, *b*: constants) for the peak current at ~2.3 V gives a *b*-value of unity for $0.05 < v < 0.5\ \mathrm{mV\,s^{-1}}$, which is indicative of a surface-controlled reaction. Presumably, (de)solvation or charge transfer at the interface rather than slow Na^+ diffusion in the bulk determines the reaction rate.

Ti_2CT_x as negative-electrode material. Having confirmed that activated Ti_2CT_x is a promising candidate as a negative-electrode material for Na-ion hybrid capacitors, galvanostatic charge/discharge experiments were conducted. Here, charging is a cathodic process (Na^+ adsorption/intercalation), whereas discharging is an anodic process (Na^+ desorption/deintercalation). Figure 3a shows the charge – discharge curves measured between the cut-off voltages of 0.1 – 3.0 V versus Na/Na^+ at $20\ \mathrm{mA\,g^{-1}}$. The first charge exhibits a voltage plateau at ~0.6 V, delivering a relatively large capacity of $360\ \mathrm{mAh\,g^{-1}}$. The following first discharge curve has a capacitor-type slope in the range of 0.2 – 2.5 V but with a much smaller capacity, resulting in a relatively low Coulombic efficiency of 65% for the first cycle. However, after the initial few cycles for activation, Ti_2CT_x exhibits stable and efficient electrode performance. The charge/discharge profiles exhibit a capacitor-type slope in the range of 0.1 – 2.3 V (an average operating potential of 1.3 V), delivering a reversible capacity of $\sim 175\ \mathrm{mAh\,g^{-1}}$ with good cycle stability (Fig. 3b, 12 and 19% losses of the second capacity after 50 and 100 cycles, respectively). The average operating potential of 1.3 V is relatively high when used as the negative electrode but is beneficial for stable operation, retaining the capacity at a high rate and avoiding Na metal plating for safety.

The Na/Ti elemental ratio determined by EDX during the first two cycles allow us to estimate the capacity (Supplementary Fig. 5) for various possible nominal chemical formulae of Ti_2CT_x (Ti_2CO_2, $Ti_2C(OH)_2$ and Ti_2CF_2). The first experimental charge capacity ($359\ \mathrm{mAh\,g^{-1}}$) is larger than the estimated ones ($308 - 321\ \mathrm{mAh\,g^{-1}}$), whereas the first discharge capacity

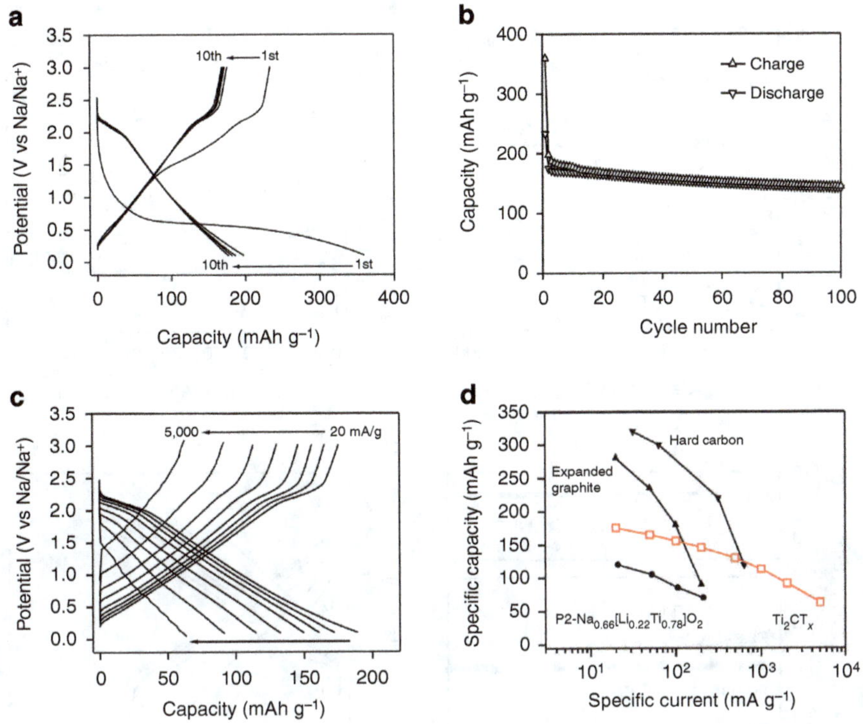

Figure 3 | Electrode performance of Ti_2CT_x in a Na^+ nonaqueous electrolyte. (**a**,**b**) Charge/discharge curves and cycle stability for Ti_2CT_x in a 1 M $NaPF_6$/EC – DEC electrolyte. The specific current is 20 mA g^{-1} with cut-off voltages in the range of 0.1 – 3.0 V versus (vs) Na/Na^+. (**c**) Charge/discharge curves at various rates. (**d**) Rate capability for Ti_2CT_x, hard carbon[11], expanded graphite[12] and P2-$Na_{0.66}[Li_{0.22}Ti_{0.78}]O_2$ (ref. 19).

(233 mAh g^{-1}) is close to the estimated ones (224 – 234 mAh g^{-1}). Thus, the initial irreversible capacity arises from the decomposition of the electrolyte as well as the residual Na^+ between the MXene layers (*vide supra*). The electrolyte decomposition during the initial cycles may form stable solid – electrolyte interface, which inhibits continuous electrolyte decomposition.

Figure 3c shows the charge/discharge profiles of Ti_2CT_x at various specific currents. The reversible capacity is retained, even at extremely high rates (156, 113 and 63 mAh g^{-1} at 200, 1,000 and 5,000 mA g^{-1}, respectively). Figure 3d compares the rate capability of Ti_2CT_x, hard carbon[11], expanded graphite[12] and P2-$Na_{0.66}[Li_{0.22}Ti_{0.78}]O_2$ (ref. 19). Although the carbon compounds (expanded graphite and hard carbon) deliver a high capacity of ~300 mAh g^{-1} at a low rate, the capacity retention at high rates is much lower than that of Ti_2CT_x. P2-$Na_{0.66}[Li_{0.22}Ti_{0.78}]O_2$ is an electrode material without a change in lattice volume during the cycle; therefore, the minimum structural change associated with (de)sodiation enables a high capacity retention at high rates. However, the theoretical capacity of P2-$Na_{0.66}[Li_{0.22}Ti_{0.78}]O_2$ is much smaller than that of Ti_2CT_x; Ti_2CT_x delivers much higher capacity at any charge/discharge rate. Therefore, Ti_2CT_x is a high-performance electrode material with a high capacity, stability, safety and a high power. It should be emphasized that the high rate capability was obtained with a thick electrode (thickness: 50 μm); therefore, its practical application to Na-ion hybrid capacitors is highly realistic.

Full cell with Ti_2CT_x and $Na_2Fe_2(SO_4)_3$. To further demonstrate the potential of Ti_2CT_x in Na-ion hybrid capacitors, a prototype full cell (Fig. 4a) was fabricated by utilizing Ti_2CT_x as a negative electrode and alluaudite $Na_2Fe_2(SO_4)_3$ as a positive electrode. Recently, our group has discovered alluaudite $Na_2Fe_2(SO_4)_3$ as a promising positive-electrode material that exhibits a specific capacity of 100 mAh g^{-1} at a high operating potential of 3.8 V versus Na/Na^+, the highest value among all Fe-based compounds, which enables the Na-ion full cell to achieve both high energy and high power densities[34,35]. The mass balance between the positive and negative electrodes was fixed to 4:1 to ensure full activation of Ti_2CT_x during the first charge.

Figure 4b shows the charge/discharge curves for the full cell measured at current densities of 150 and 600 mA g^{-1} (based on the weight of Ti_2CT_x). Note that the first cycle was conducted at 20 mA g^{-1} (based on the weight of Ti_2CT_x) to activate Ti_2CT_x. After the first cycle, the cell exhibits a voltage profile that mainly reflects the potential of Ti_2CT_x because the cell capacity is limited by the Ti_2CT_x negative electrode (Fig. 3a). The average operating voltage is 2.4 V, and the reversible capacities of the second cycle are 133 and 107 mAh g^{-1} at 150 and 600 mA g^{-1} (based on the weight of Ti_2CT_x), respectively, achieving theoretical energy densities of 320 and 260 Wh kg^{-1} at specific power densities of 360 and 1,440 W kg^{-1} (based on the weight of Ti_2CT_x), respectively.

Both the cycle stability and efficiency of the full cell are extremely high. As demonstrated in Fig. 4c, the loss of the discharge capacity after 100 cycles at 600 mA g^{-1} is only 4% of the second discharge capacity. Furthermore, the Coulombic efficiency for cycling at 600 mA g^{-1} reaches ~99.7% after a few initial cycles. The excellent cycle stability and efficiency may be attributed to the pseudocapacitive charge storage mechanism without a large structural change.

Figure 4d plots the charge/discharge profiles of the alluaudite $Na_2Fe_2(SO_4)_3$ – Ti_2CT_x full cell at various current rates, demonstrating the promising high-power performance of the cell (Supplementary Fig. 6). The full cell provides reversible capacities of 90 and 40 mAh g^{-1}, even at extremely high rates of 1,000 and 5,000 mA g^{-1} (based on the weight of Ti_2CT_x). Thus, by utilizing the pseudocapacitive negative electrode, the Na-ion hybrid capacitor overcomes the trade-off limit and potentially surpasses state-of-the-art high-power rechargeable batteries. Although the

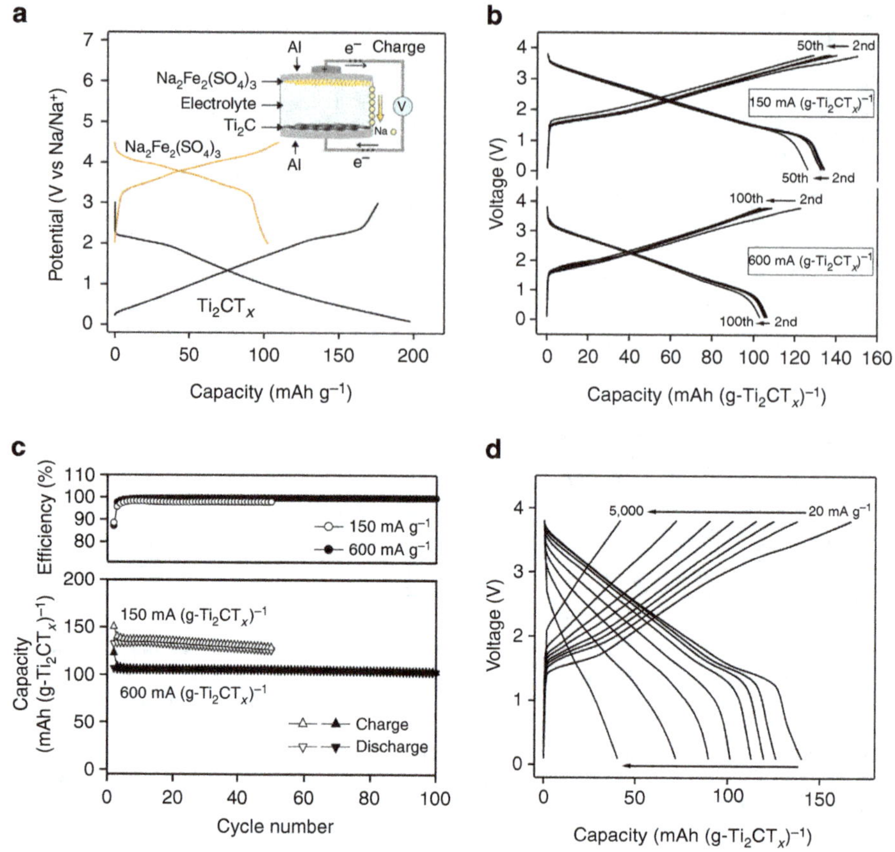

Figure 4 | Performance of the prototype Na-ion hybrid capacitor full cells. (**a**) Charge/discharge curves of Ti_2CT_x and alluaudite $Na_2Fe_2(SO_4)_3$ versus Na/Na^+; the specific currents are 30 and $6\,mA\,g^{-1}$, respectively. The inset is a schematic illustration of the Ti_2CT_x – alluaudite $Na_2Fe_2(SO_4)_3$ full cell. (**b**) Voltage profile of the Ti_2CT_x – alluaudite $Na_2Fe_2(SO_4)_3$ full cell at the specific currents of 150 and 600 mA $(g-Ti_2CT_x)^{-1}$. The specific capacity is normalized by the weight of Ti_2CT_x. (**c**) Cycle stability and Coulombic efficiency of the full cell with cutoff voltages in the range of 0.1 – 3.8 V at rates of 150 and 600 mA $(g-Ti_2CT_x)^{-1}$. (**d**) Charge/discharge profiles at various rates.

capacity for the total weight of the positive and negative electrode materials in the present prototype cell is limited ($26.6\,mAh\,g^{-1}$ at a low rate) owing to the excess amount of positive-electrode material, an improvement in the initial Coulombic efficiency could reduce the amount of positive-electrode material, and hence, greatly increase the capacity of the full cell.

Discussion

Finally, we note a positive perspective for further improvement in the Ti_2CT_x electrode. The gravimetric capacitance from CV of the Ti_2CT_x electrode in this work is higher than that of previously reported $Ti_3C_2T_x$ electrodes because of the light weight of Ti_2CT_x (Supplementary Fig. 7). However, very recently, Ghidui *et al.*[36] have reported a new method to synthesize $Ti_3C_2T_x$ ($Ti_3C_2T_x$ 'clay') using a solution of lithium fluoride and hydrochloric acid. The resulting $Ti_3C_2T_x$ 'clay' delivers a high volumetric capacitance of $900\,F\,cm^{-3}$ and a gravimetric capacitance of $245\,F\,g^{-1}$, which is much higher than those of the previously reported $Ti_3C_2T_x$ electrodes and the Ti_2CT_x electrode in this work. According to Ghidui *et al.*[36], the $Ti_3C_2T_x$ clay provides a much higher capacitance at higher rates owing to the improved accessibility of the interlayer spacing. Therefore, the synthetic procedure applied for the $Ti_3C_2T_x$ clay, that is, the LiF + HCl treatment, could further improve the performance of Ti_2CT_x.

In summary, the use of the pseudocapacitance of MXene nanosheets was demonstrated as an effective strategy for developing high-performance Na-ion hybrid capacitors. A pseudocapacitance with no significant structural changes can store more charge at a faster charge–discharge rate relative to the ion intercalation and double-layer mechanisms. In the present case, Ti_2CT_x operates as a pseudocapacitor electrode material at 1.3 V on average versus Na/Na^+ with a reversible capacity of $175\,mAh\,g^{-1}$. The alluaudite $Na_2Fe_2(SO_4)_3 - Ti_2CT_x$ prototype cell delivers a high specific energy of $260\,Wh\,kg^{-1}$ at a high specific power of $1.4\,kW\,kg^{-1}$ (based on the weight of Ti_2CT_x), which overcomes the trade-off limit associated with conventional electrochemical energy storage. By exploiting the compositional and structural versatility of MAX phases and the resulting optimized MXenes, we could further extend the possible sets of electrodes for advanced Na-ion hybrid capacitors.

Methods

Synthesis of Ti_2CT_x. Ti_2AlC was prepared by heating a precursor mixture of TiC (>99%, High Purity Chemicals, Japan), Ti (>99%, High Purity Chemicals, Japan) and Al (>99.9%, High Purity Chemicals, Japan) at 1,300 °C for 1 h in an Ar gas environment. Ti_2CT_x was synthesized by treating 1 g of Ti_2AlC powder in 10% HF aqueous solution (Wako) for 12 h at room temperature. The HF-treated powder was dried in vacuum at 60 °C for 24 h.

Materials characterization. Powder X-ray diffraction patterns were recorded on a Rigaku RINT-TTR III powder diffractometer with Cu Kα radiation in a step of 0.02° over a 2θ range of 10 – 80°. Scanning electron microscopy measurements were carried out on a JEOL 6510FA apparatus equipped with an energy dispersive X-ray (EDX) spectrometer for chemical analysis. Transmission electron microscopy was conducted on a JEOL JEM-2100 at 200 kV. X-ray photoelectron spectroscopy (XPS) data were collected using a ULVAC PHI 5000 VersaProbe spectrometer with monochromatized Al Kα radiation ($h\nu = 1{,}486.6\,eV$). The pressure in the analysis chamber during the measurements was maintained in the 10^{-7} Pa range. The powder sample was pressed onto conductive carbon, and peaks were recorded with a constant pass energy mode of 117 eV for survey investigation. High-resolution spectra were taken at a pass energy of 23 eV, with a step of 0.1 eV.

All binding energies were referenced to that of free carbon at 284.5 eV. The crystal structure was drawn using VESTA[37].

Electrochemical measurement. For electrochemical studies, the working electrode was fabricated by mixing 80 wt% Ti_2CT_x, 10 wt% acetylene black and 10 wt% polyvinylidene difluoride binder in a minimal amount of *N*-methylpyrrolidone (NMP) solvent. This slurry was pasted on an aluminum-foil current collector, and the as-obtained electrode sheet was dried overnight at 120 °C in vacuum. The electrode thickness and mass loading are $\sim 50\,\mu m$ and $\sim 1\,mg\,cm^{-2}$, respectively, for all electrochemical measurements (CV and charge–discharge). CR2032-type coin cells were assembled with Na metal as the counter electrode, a glass fibre filter (GB-100R, ADVANTEC) as the separator and 1 M $NaPF_6$ in ethylene carbonate-diethyl carbonate (EC – DEC, 1:1 v/v%) as the electrolyte. These cells were assembled inside an Ar-filled glove box (Miwa Inc., Japan) (dew point < – 100 K). They were galvanostatically (dis)charged in the potential range 0.1–3.0 V. The gravimetric capacitance in farads per gram from CV is given by $C = \frac{1}{\Delta V}\int \frac{j}{s}\, dV$, where C is the gravimetric capacitance, j is the specific current in amperes per gram, s is the scan rate in volts per second, V is the voltage in volts and ΔV is the voltage window.

Full cell fabrication. For full cell preparation, alluaudite $Na_2Fe_2(SO_4)_3$ was used as a positive electrode material. Alluaudite $Na_2Fe_2(SO_4)_3$ was synthesized according to the procedure described previously[34]. The full cell was prepared with the alluaudite $Na_2Fe_2(SO_4)_3$ electrode as a positive electrode, the Ti_2CT_x electrode as a negative electrode, a glass fibrr membrane as a separator, and 1 M $NaPF_6$/EC – DEC as an electrolyte. The cut-off voltages were 3.8 V for charging and 0.1 V for discharging.

References

1. Nazri, G. A. & Pistoia, G. *Lithium Batteries* (Springer, 2009).
2. Daniel, C. & Besenhard, J. O. *Handbook of Battery Materials* (Wiley, 2011).
3. Béguin, F. & Frackowiak, E. *Supercapacitors* (Wiley, 2013).
4. Conway, B. E. *Electrochemical Supercapacitors* (Kluwer, 1999).
5. Conway, B. E. Transition from "Supercapacitor" to "Battery" behavior in elctrochemical energy storage. *J. Electrochem. Soc.* **138,** 1539–1548 (1991).
6. Augustyn, V. *et al.* High-rate electrochemical energy storage through Li^+ intercalation pseudocapacitance. *Nat. Mater.* **12,** 518–522 (2013).
7. Zhu, Y. *et al.* Carbon-based supercapacitors produced by activation of graphene. *Science* **332,** 1537–1541 (2011).
8. Mefford, J. T., Hardin, W. G., Dai, S., Johnston, K. P. & Stevenson, K. J. Anion charge storage through oxygen intercalation in $LaMnO_3$ perovskite pseudocapacitor electrodes. *Nat. Mater.* **13,** 726–732 (2014).
9. Wu, C. *et al.* Two-dimensional vanadyl phosphate ultrathin nanosheets for high energy density and flexible pseudocapacitors. *Nat. Commun.* **4,** 2431 (2013).
10. Simon, P. & Gogotsi, Y. Materials for electrochemical capacitors. *Nat. Mater.* **7,** 845–854 (2008).
11. Ponrouch, A., Goñi, A. R. & Palacín, M. R. High capacity hard carbon anodes for sodium ion batteries in additive free electrolyte. *Electrochem. Commun.* **27,** 85–88 (2013).
12. Wen, Y. *et al.* Expanded graphite as superior anode for sodium-ion batteries. *Nat. Commun.* **5,** 4033 (2014).
13. Oh, S. M. *et al.* High electrochemical performances of microsphere C-TiO_2 anode for sodium-ion battery. *ACS Appl. Mater. Interfaces* **6,** 11295–11301 (2014).
14. Senguttuvan, P., Rousse, G., Seznec, V., Tarascon, J. M. & Palacín, M. R. $Na_2Ti_3O_7$: lowest voltage ever reported oxide insertion electrode for sodium ion batteries. *Chem. Mater.* **23,** 4109–4111 (2011).
15. Qian, J., Wu, X., Cao, Y., Ai, X. & Yang, H. High capacity and rate capability of amorphous phosphorus for sodium ion batteries. *Angew. Chem. Int. Ed.* **52,** 4633–4636 (2013).
16. Yu, D. Y. W. *et al.* High-capacity antimony sulphide nanoparticledecorated graphene composite as anode for sodium-ion batteries. *Nat. Commun.* **4,** 2922 (2013).
17. Li, W. *et al.* $Sn_{4+x}P_3$ @ amorphous Sn-P composites as anodes for sodium-ion batteries with low cost, high capacity, long life, and superior rate capability. *Adv. Mater.* **26,** 4037–4042 (2014).
18. Qu, B. *et al.* Layered SnS_2-reduced graphene oxide composite – a high-capacity, high-rate, and long-cycle life sodium-ion battery anode material. *Adv. Mater.* **26,** 3854–3859 (2014).
19. Wang, Y. *et al.* A zero-strain layered metal oxide as the negative electrode for long-life sodium-ion batteries. *Nat. Commun.* **3,** 2365 (2013).
20. Naguib, M. *et al.* Two-dimensional nanocrystals produced by exfoliation of Ti_3AlC_2. *Adv. Mater.* **23,** 4248–4253 (2011).
21. Naguib, M. *et al.* Two-dimensional transition metal carbides. *ACS Nano* **6,** 1322–1331 (2012).
22. Mashtalir, O. *et al.* Intercalation and delamination of layered carbides and carbonitrides. *Nat. Commun.* **4,** 1716 (2013).
23. Lukatskaya, M. R. *et al.* Cation intercalation and high volumetric capacitance of two-dimensional titanium carbide. *Science* **341,** 1502–1505 (2013).
24. Mashtalir, O., Naguib, M., Dyatkin, B., Gogotsi, Y. & Barsoum, M. W. Kinetics of aluminum extraction from Ti_3AlC_2 in hydrofluoric acid. *Mater. Chem. Phys.* **139,** 147–152 (2013).
25. Halim, J. *et al.* Transparent conductive two-dimensional titanium carbide epitaxial thin films. *Chem. Mater.* **26,** 2374–2381 (2014).
26. Barsoum, M. W. *MAX Phases* (Wiley, 2013).
27. Naguib, M. *et al.* MXene: a promising transition metal carbide anode for lithium-ion batteries. *Electrochem. Commun.* **16,** 61–64 (2012).
28. Come, J. *et al.* A non-aqueous asymmetric cell with a Ti_2C-based two-dimensional negative electrode. *J. Electrochem. Soc.* **159,** A1368–A1373 (2012).
29. Naguib, M. *et al.* New two-dimensional niobium and vanadium carbides as promising materials for Li-ion batteries. *J. Am. Chem. Soc.* **135,** 15966–15969 (2013).
30. Xie, Y. *et al.* Role of surface structure on Li-ion energy storage capacity of two-dimensional transition-metal carbides. *J. Am. Chem. Soc.* **136,** 6385–6394 (2014).
31. Xie, Y. *et al.* Prediction and characterization of MXene nanosheet anodes for non-lithium-ion batteries. *ACS Nano* **8,** 9606–9615 (2014).
32. Dall'Agnese, Y. *et al.* High capacitance of surface-modified 2D titanium carbide in acidic electrolyte. *Electrochem. Commun.* **48,** 118–122 (2014).
33. Levi, M. D. *et al.* Solving the capacitive paradox of 2D MXene using electrochemical quartz-crystal admittance and *in situ* electronic conductance measurements. *Adv. Energy Mater.* **5,** 1400815 (2015).
34. Barpanda, P., Oyama, G., Nishimura, S., Chung, S. C. & Yamada, A. A 3.8-V earth-abundant sodium battery electrode. *Nat. Commun.* **5,** 4358 (2014).
35. Ming, J., Barpanda, P., Nishimura, S., Okubo, M. & Yamada, A. An alluaudite $Na_{2+2x}Fe_{2-x}(SO_4)_3$ ($x = 0.2$) derivative phase as insertion host for lithium battery. *Electrochem. Commun.* **51,** 19–22 (2015).
36. Ghidiu, M., Lukatskaya, M. R., Zhao, M. Q., Gogotsi, Y. & Barsoum, M. W. Conductive two-dimensional titanium carbide 'clay' with high volumetric capacitance. *Nature* **516,** 78–81 (2014).
37. Momma, K. & Izumi, F. VESTA 3 for three-dimensional visualization of crystal, volumetric and morphology data. *J. Appl. Cryst.* **44,** 1272–1276 (2011).

Acknowledgements

This work was financially supported by the Ministry of Education, Culture, Sports, Science and Technology (MEXT), Japan under the 'Elemental Strategy Initiative for Catalysts and Batteries' (ESICB). M.O. was financially supported by the Murata Science Foundation.

Author contributions

M.O. and A.Y. conceived and directed the project. M.O., X.W., S.K., H.I., E.H., S.O. and I.M. synthesized Ti_2CT_x. M.O. and X.W. measured and analysed the electrochemical properties. S.K. measured the performance of the full cells. All authors contributed to the writing of the paper.

Additional information

Competing financial interests: The authors declare no competing financial interests.

20

An efficient molybdenum disulfide/cobalt diselenide hybrid catalyst for electrochemical hydrogen generation

Min-Rui Gao[1,*], Jin-Xia Liang[2,3,*], Ya-Rong Zheng[1], Yun-Fei Xu[1], Jun Jiang[1], Qiang Gao[1], Jun Li[2] & Shu-Hong Yu[1]

The electroreduction of water for sustainable hydrogen production is a critical component of several developing clean-energy technologies, such as water splitting and fuel cells. However, finding a cheap and efficient alternative catalyst to replace currently used platinum-based catalysts is still a prerequisite for the commercialization of these technologies. Here we report a robust and highly active catalyst for hydrogen evolution reaction that is constructed by *in situ* growth of molybdenum disulfide on the surface of cobalt diselenide. In acidic media, the molybdenum disulfide/cobalt diselenide catalyst exhibits fast hydrogen evolution kinetics with onset potential of −11 mV and Tafel slope of 36 mV per decade, which is the best among the non-noble metal hydrogen evolution catalysts and even approaches to the commercial platinum/carbon catalyst. The high hydrogen evolution activity of molybdenum disulfide/cobalt diselenide hybrid is likely due to the electrocatalytic synergistic effects between hydrogen evolution-active molybdenum disulfide and cobalt diselenide materials and the much increased catalytic sites.

[1] Division of Nanomaterials & Chemistry, Hefei National Laboratory for Physical Sciences at Microscale, Collaborative Innovation Center of Suzhou Nano Science and Technology, Department of Chemistry, University of Science and Technology of China, Hefei 230026, China. [2] Key Laboratory of Organic Optoelectronics and Molecular Engineering of Ministry of Education, Department of Chemistry, Tsinghua University, Beijing 100084, China. [3] Guizhou Provincial Key Laboratory of Computational Nano-Material Science, Guizhou Normal College, Guiyang 550018, China. * These authors contributed equally to this work. Correspondence and requests for materials should be addressed to S.-H.Y. (email: shyu@ustc.edu.cn).

Owing to their diffuse nature, electricity from renewable but intermittent energy (for example, solar and wind) must be stored durably for off-grid applications. Electrochemical water splitting to produce hydrogen (H_2) offers a promising and sustainable solution for this purpose by converting such electricity energy into stable chemical bonds[1,2]. Appropriate electrocatalysts, such as platinum (Pt) and its alloys, play a vital role in the H_2 evolution reaction (HER) because they can catalyse the conversion from a pair of protons and electrons to H_2 at high reaction rates and low overpotentials (η)[1–3]. However, the prohibitive cost and scarcity of Pt pose tremendous limitations to widespread use. Therefore, finding robust and efficient alternative catalysts that are geologically abundant is crucial to the future of 'hydrogen economy'.

Molybdenum disulfide (MoS_2), a widely used industrial catalyst for hydrodesulfurization[4], has recently demonstrated promise as effective HER catalyst based on both computational and experimental studies[5,6]. The HER activity was discovered to arise from the exposed (10–10) planes on edges of MoS_2, whereas the (0001) basal planes are catalytically inactive[5–7]. This understanding has led to great efforts to develop highly nanostructured MoS_2-based HER catalysts to maximize the number of edge sites, including crystalline[8–12] and amorphous materials[13–15], MoS_2-based hybrid materials[16–19] and molecular mimics[7]. Despite significant success, the design and fabrication of MoS_2-based HER electrocatalysts with satisfactory activity and stability remain a big challenge.

In recent years, we have been making efforts to explore efficient electrocatalysts by using Earth-abundant 3*d* metal (Co, Ni and so on) chalcogenides[20–27]. New NiSe nanofibres[24] and lamellar mesostructured $CoSe_2$/DETA (DETA, diethylenetriamine) nanobelts[25] were found to show decent HER activity in acidic electrolyte. Kong *et al.*[28] also observed good HER activities from various polycrystalline transition metal dichalcogenide (ME_2, M = Fe, Co, Ni; E = S, Se) films, especially from the $CoSe_2$. Recently, we found that the HER activity of $CoSe_2$ nanobelts can be improved greatly after anchoring Ni/NiO nanoparticles onto their surfaces[25]. The synergetic chemical coupling effects between $CoSe_2$ and grafted Ni/NiO were believed to contribute to the enhancement. Similar promoted performances have also been observed on MoS_2/graphene[16,17], MoO_3/MoS_2 (ref. 18) and MoS_2/Au (ref. 19) composite catalysts for H_2 production. These works point to the possibility to access new and efficient HER catalysts by combining the promising $CoSe_2$ and MoS_2.

We report here that a HER electrocatalyst based on quasi-amorphous MoS_2-coated $CoSe_2$ (denoted as $MoS_2/CoSe_2$) hybrid is highly active and stable in acidic electrolyte. Notably, without any noble metals, the $MoS_2/CoSe_2$ hybrid catalyst shows an onset potential close to commercial Pt catalyst (Johnson-Matthey, 20 wt% Pt/XC-72) and a small Tafel slope of ~36 mV per decade as well as no current loss after long-term chronoamperometry measurement, performing the best among the noble-metal-free HER electrocatalysts. These results suggest a strategy for designing non-noble metal catalysts with enhanced HER performance that is comparable to the state-of-the-art Pt-based catalysts.

Results

$MoS_2/CoSe_2$ hybrid catalyst. The $MoS_2/CoSe_2$ hybrid was prepared directly in a closed *N,N*-dimethylformamide (DMF)/hydrazine solvothermal system, where $(NH_4)_2MoS_4$ was used as a precursor for growing MoS_2 around the freshly made $CoSe_2$/DETA nanobelt substrates (Fig. 1a; Supplementary Fig. 1; see Methods for details of the synthesis). The MoS_2-coated $CoSe_2$ hybrid was shown by means of scanning electron microscopy and transmission electron microscopy (TEM; Fig. 2a–c), which revealed the compact graphene-like MoS_2 nanosheets grown on the surface of $CoSe_2$ with a partially free-standing branch-like feature (Supplementary Fig. 2). Substantial amino groups on the $CoSe_2$/DETA surface (Supplementary Fig. 3) serve as nucleation sites for coupling Mo precursor and subsequently reduced to MoS_2 on $CoSe_2$ (refs 21,23). A control experiment performed under identical synthesis conditions, but without $CoSe_2$ nanobelts, produced three-dimensional (3D) aggregates of MoS_2 sheets (Fig. 1b; Supplementary Fig. 4), suggesting that $CoSe_2$ could be a useful support for mediating the growth of loaded materials and constructing novel functional hybrids.

Figure 2d presents high-resolution TEM (HRTEM) images of the $MoS_2/CoSe_2$ hybrid. Layered MoS_2 (often less than five layers) nanosheets, with an interlayer separation of 0.63 nm, were grown intimately on the $CoSe_2$ substrate. High-crystalline $CoSe_2$ substrate with *d* spacing of 0.27 nm can be seen frequently through the interspace of grafted MoS_2 (see Supplementary Fig. 5 for additional images). This non-fully covered structure may take advantage of merits from both MoS_2 and $CoSe_2$ for catalysing H_2 evolution. Selected area electron diffraction (SAED) pattern (inset in Fig. 2c; Supplementary Fig. 6) revealed clear diffraction spots (marked by yellow arrows) from single-crystalline $CoSe_2$ support (JCPDS 9-0234) and also faint diffraction rings from grafted MoS_2 (JCPDS 77-1716). These barely recognizable diffraction rings indicate the quasi-amorphous structure of MoS_2, consistent with the X-ray diffraction (XRD) data of $MoS_2/CoSe_2$ with broadening diffraction peaks (Fig. 2e). Inasmuch as amorphous MoS_2 has recently been demonstrated to be effective HER catalysts[13–15] for its abundant defects and resultant more active edge sites[29], we thus infer that the quasi-amorphous MoS_2 modification may benefit the hybrid material to catalytically evolve H_2. Energy-dispersive X-ray spectrum (EDX; Fig. 2f) analysis further confirmed the formation of $MoS_2/CoSe_2$ hybrid with Co, Se, Mo and S as the principal elemental components (Cu and C peaks emanate from the carbon-coated TEM grid), agreeing with the X-ray photoelectron spectroscopy (XPS) results (Supplementary Fig. 7). Scanning TEM (STEM) and EDX elemental mappings revealed uniform spatial distribution of Co, Se, Mo and S over the marked detection range of the constructed hybrid material (Fig. 2g).

Catalytic hydrogen evolution. To assess the HER electrocatalytic activity, thin film of various catalysts was prepared on glassy carbon (GC) electrodes for cyclic voltammetry in H_2-saturated 0.5 M H_2SO_4 electrolyte (see Methods for experimental details). Potentials were measured vs saturated calomel electrode (SCE) and are reported vs reversible hydrogen electrode (RHE). The electrode was kept rotating (1,600 r.p.m.) during the measurements to remove *in situ*-emerged H_2 bubbles. Figure 3a shows that GC electrode guarantees a minimal background activity for H_2 evolution. A reductive sweep of $MoS_2/CoSe_2$ hybrid showed a low η of 11 mV for the HER, beyond which a sharp increase in cathodic current was observed, corresponding to catalytic H_2 evolution (Fig. 3a; Supplementary Fig. 9). By contrast, pure $CoSe_2$ nanobelts exhibited inferior HER activity with a larger onset potential of ~50 mV and lower catalytic current, while pure MoS_2 nanosheet aggregates only affected little HER activity. The HER kinetics of the above catalysts was probed by corresponding Tafel plots ($\log j \sim \eta$) (Fig. 3b). Tafel slope of ~36 mV per decade was measured for $MoS_2/CoSe_2$ hybrid, which is close to the value of 30 mV per decade for Pt/C catalyst and lower than that of 48 mV per decade for $CoSe_2$ and 101 mV per decade for 3D MoS_2 aggregates (Table 1). This Tafel slope is also comparable to or

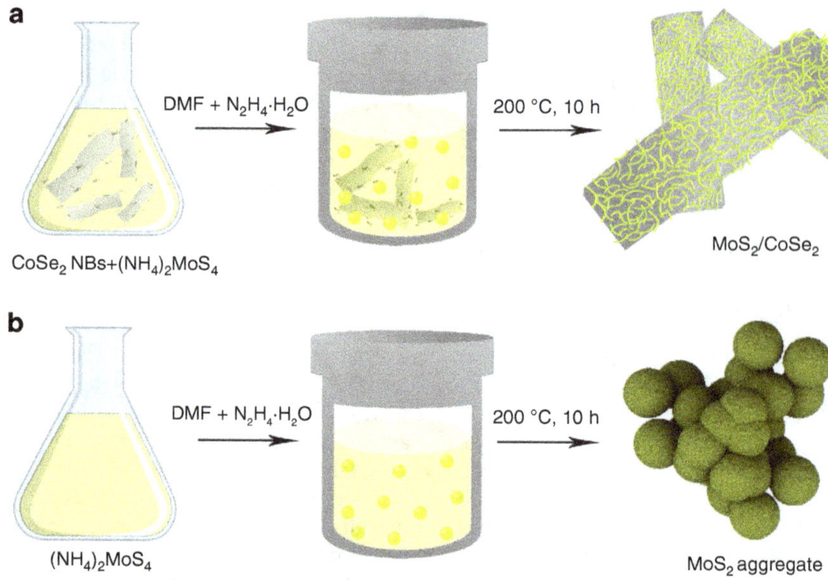

Figure 1 | Schematic illustration of the preparation of $MoS_2/CoSe_2$ hybrid. (**a**) Solvothermal synthesis with $CoSe_2$/DETA nanobelts as substrates for preparation of $MoS_2/CoSe_2$ hybrid. (**b**) Solvothermal synthesis without $CoSe_2$/DETA nanobelts leads to free MoS_2 nanosheet aggregates.

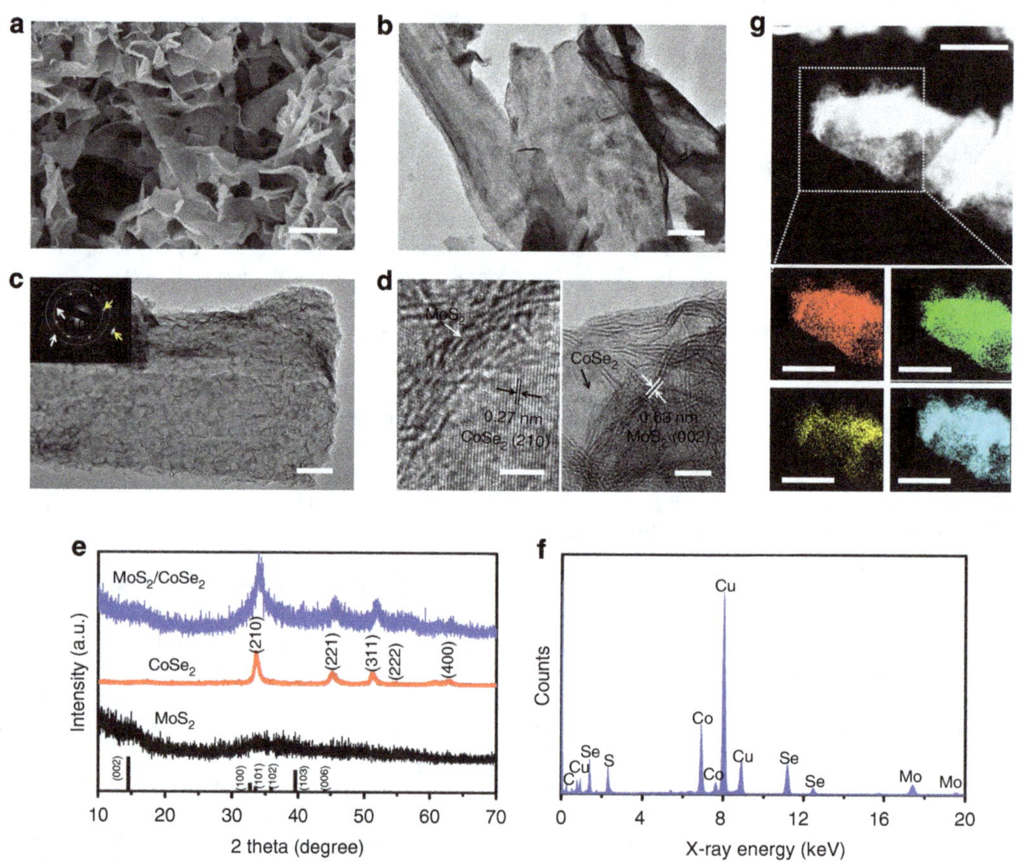

Figure 2 | Characterization of the $MoS_2/CoSe_2$ hybrid. (**a**) Scanning electron microscopy image of $MoS_2/CoSe_2$ hybrid. Scale bar, 800 nm. (**b**,**c**) TEM images with different magnifications of $MoS_2/CoSe_2$ hybrid. Scale bars, 200 and 50 nm, respectively. The inset in **c** shows corresponding SAED pattern. (**d**) HRTEM images of $MoS_2/CoSe_2$ hybrid showing distinguishable microstructures of MoS_2 and $CoSe_2$. Scale bars, 5 nm. (**e**,**f**) XRD patterns and EDX spectrum of the $MoS_2/CoSe_2$ hybrid, respectively. (**g**) STEM-EDX elemental mapping of $MoS_2/CoSe_2$ hybrid showing clearly the homogeneous distribution of Co (red), Se (green), Mo (yellow) and S (azure). Scale bars, 200 nm.

lower than that of all the reported noble-metal-free HER catalysts in the literature (Supplementary Table 1), demonstrating the superior HER kinetics of $MoS_2/CoSe_2$ hybrid. At high current densities (for example, $10\,mA\,cm^{-2}$ with η of 68 mV), $MoS_2/CoSe_2$ hybrid also represents a more efficient catalyst relative to other noble-metal-free HER catalysts (Supplementary Table 1). The HER inherent activity of these catalysts was evaluated by the exchange current density (j_0). The j_0 of $7.3 \times 10^{-2}\,mA\,cm^{-2}$ for $MoS_2/CoSe_2$ hybrid outperforms the values of $8.4 \times 10^{-3}\,mA\,cm^{-2}$ for pure $CoSe_2$ and $9.1 \times 10^{-4}\,mA\,cm^{-2}$ for pure

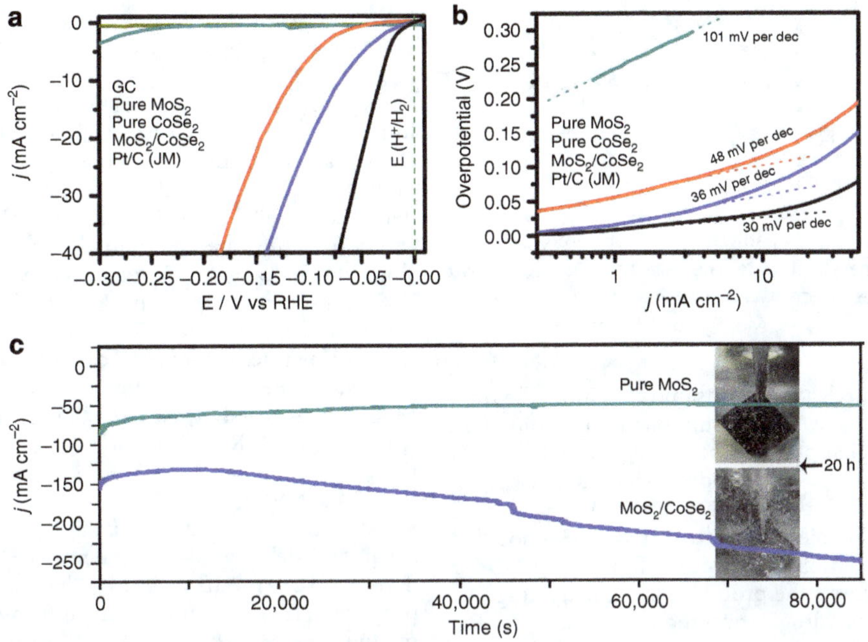

Figure 3 | Electrocatalytic hydrogen evolution of different catalysts. (**a**) Polarization curves for HER on bare GC electrode and modified GC electrodes comprising $MoS_2/CoSe_2$ hybrid, pure MoS_2, pure $CoSe_2$ and a high-quality commercial Pt/C catalyst. Catalyst loading is about 0.28 mg cm^{-2} for all samples. Sweep rate: 2 mV s^{-1}. (**b**) Tafel plot for the various catalysts derived from **a**. (**c**) Chronoamperometric responses ($j \sim t$) recorded on $MoS_2/CoSe_2$ hybrid and pure MoS_2 at a constant applied potential of −0.7 V vs SCE. The catalysts were deposited on CFP with the same loading of 1 mg cm^{-2}. Inset digital photos show the H_2 bubbles formed on MoS_2-modified CFP (up) and $MoS_2/CoSe_2$-modified CFP (down) at the time point of 20 h. All the measurements were performed in H_2-saturated 0.5 M H_2SO_4 electrolyte. dec, decade.

Table 1 | Comparison of catalytic parameters of different HER catalysts.

Catalyst	Onset potention (mV vs RHE)	Tafel slope (mV per decade)	Exchange current density (j_0, mA cm^{-2})*
MoS_2	−237	101	9.1×10^{-4}
$CoSe_2$	−50	48	8.4×10^{-3}
$MoS_2/CoSe_2$	−11	36	7.3×10^{-2}
Pt/C†	0	30	7.1×10^{-1}

HER, H_2 evolution reaction; RHE, reversible hydrogen electrode.
*j_0 were calculated from Tafel curves using extrapolation method.
†Johnson-Matthey, 20 wt% Pt/XC-72.

MoS_2 (Table 1) and also the j_0 value for most of the reported noble-metal-free HER catalysts (Supplementary Table 1). The high electrode kinetic metrics (including onset potential of −11 mV and the Tafel slope of 36 mV per decade) and large j_0 (only ~1 order of magnitude lower than the value of 7.1×10^{-1} mA cm^{-2} for Pt) highlight the exceptional H_2 evolving efficiency of the new $MoS_2/CoSe_2$ hybrid catalyst.

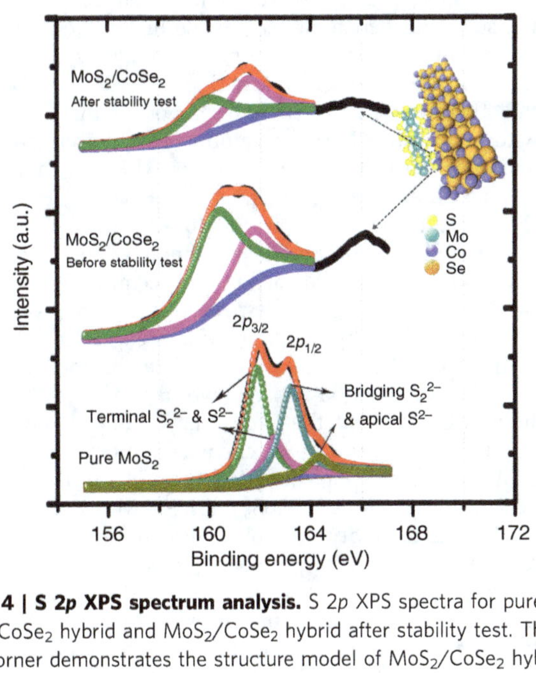

Figure 4 | S 2*p* XPS spectrum analysis. S 2*p* XPS spectra for pure MoS_2, $MoS_2/CoSe_2$ hybrid and $MoS_2/CoSe_2$ hybrid after stability test. The top right corner demonstrates the structure model of $MoS_2/CoSe_2$ hybrid.

Material stability. The long-term stability of $MoS_2/CoSe_2$ hybrid catalyst was assayed by means of chronoamperometry measurement ($j \sim t$) with a high catalyst loading of 1 mg cm^{-2} on carbon fibre paper (CFP). This quasi-electrolysis process was performed at a constant potential of −0.7 V vs SCE in 0.5 M H_2SO_4 for 24 h. As shown in Fig. 3c, the current density of $MoS_2/CoSe_2$-modified CFP electrode decreased gradually at the initial 3 h, which then increased quickly over 24 h of continuous operation. We hypothesize that the more efficient HER-active sites in our hybrid materials are the interfaces between MoS_2 and $CoSe_2$, where Co from the support materials could promote the HER kinetics by lowering the Gibbs free energy of adsorbed hydrogen (ΔG_H)[8,29,30]. The severe reducing condition at the initial 3 h caused degradation of external MoS_2 and allowed more electrolyte to access the MoS_2–$CoSe_2$ interfaces (Supplementary Fig. 11; Supplementary Note 1; Supplementary Table 2), yielding the increased HER current density. Remarkably, after 24 h of operation, XPS studies revealed no obvious chemical state change of HER-active S (Fig. 4) and the homogeneous elemental distribution was maintained (Supplementary Fig. 12; Supplementary Note 1), suggesting the robustness of the hybrid

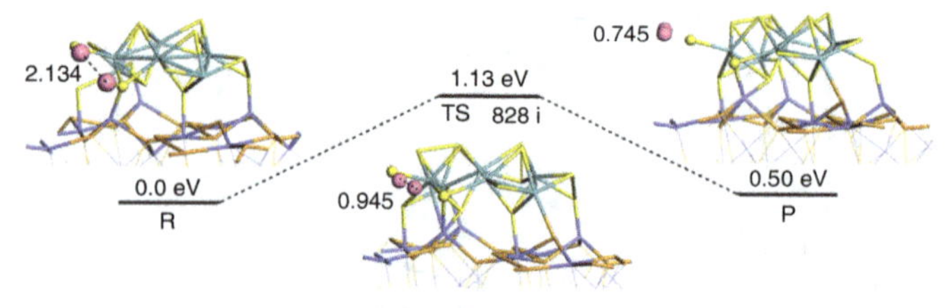

Figure 5 | HER mechanism. Reaction pathway of HER on $MoS_2/CoSe_2$ hybrid according to the Volmer–Tafel route. The calculated distance of two hydrogen atoms and energies are displayed in Å and eV. Blue, orange, azure, yellow and pink indicate Co, Se, Mo, S and H atoms, respectively.

catalyst. By comparison, under the exact same condition, the pure MoS_2 catalyst exhibited a slow but continuous decrease in HER activity. Furthermore, digital photo (inset in Fig. 3c) taken from $MoS_2/CoSe_2$-modified CFP electrode showed vigorous effervescence at 20 h (also Supplementary Movie 1), comparing favourably with the H_2 bubbles formed on free MoS_2-modified electrode. The exceptional long-term durability of our $MoS_2/CoSe_2$ hybrid catalyst suggests the promise for implementing this new catalyst into realistic hydrogen evolution electrode.

HER-enhanced mechanism. The experimentally observed high HER intrinsic activity ($j_0 = 7.3 \times 10^{-2}\,mA\,cm^{-2}$) of $MoS_2/CoSe_2$ hybrid catalyst prompted us to probe the enhanced mechanism. Generally, the value of ΔG_H is considered as a reasonable descriptor of HER activity for a wide variety of catalysts[3,30,31]. An optimum HER activity is suggested to be achieved at value of $\Delta G_H \approx 0$ (ref. 30). Lower ΔG_H will lead to very high surface coverage of H_{ads}, while higher ΔG_H will make the protons bonded too weakly on the catalyst surface, which both lead to the slow HER kinetics[3,30,31]. Previous density functional theory (DFT) calculations[5] showed that MoS_2 edge sites with unsaturated sulfur atoms can lower the ΔG_H approaching to 0 and thus are active for HER, which was later proved by Jaramillo *et al.*[6] experimentally. Recently, Hu and co-workers[13,14] found that amorphous MoS_2 films are particularly HER active and the increased coordinately and structurally unsaturated sulfur atoms led to the enhancement. Furthermore, first-row transition metal ions, especially Co, can promote the HER activity of MoS_2 by coupling with S-edges to lower their ΔG_H from 0.18 to 0.10 eV to afford a faster proton adsorption kinetics[8,29,30]. As to $MoS_2/CoSe_2$ hybrid catalyst, the quasi-amorphous MoS_2 around the $CoSe_2$ may bring in more active edge sites. Meanwhile, the $CoSe_2$ chemically interacts with MoS_2 by forming S–Co bond (Fig. 4 and inset), similar to the XPS peak observed in Co-promoted MoS_3 film[32], which can further improve the HER activity of MoS_2. By contrast, no such XPS peak was found in Fig. 4 for pure MoS_2 catalyst. XPS analyses of the S 2*p* region also exhibit a dramatically decreased electron-binding energy (by ~1.3 eV) after growing MoS_2 on the $CoSe_2$, suggesting the formation of more terminal S_2^{2-} and S^{2-} ions, which are HER active (Fig. 4)[14,19]. Therefore, the $CoSe_2$ nanobelts, a material with decent HER activity by itself, can not only chemically couple with MoS_2 to promote the HER activity, but also serve as an effective support for mediating the growth of MoS_2 to form more terminal S_2^{2-} and S^{2-}. Meanwhile, anchored MoS_2 may also boost the HER-active sites of $CoSe_2$ and these electrocatalytic synergistic effects[33] together lead to the high HER performance of $MoS_2/CoSe_2$ hybrid catalyst.

For the HER in acidic media, two separate pathways (the Volmer–Tafel or the Volmer–Heyrovsky mechanism) have been proposed for reducing H^+ to H_2 (refs 30,34). Specifically, the two distinct mechanisms involve three principal steps, referring to the Volmer (electrochemical hydrogen adsorption: $H_3O^+ + e^- \rightarrow H_{ads} + H_2O$), the Heyrovsky (electrochemical desorption: $H_{ads} + H_3O^+ + e^- \rightarrow H_2\uparrow + H_2O$) and the Tafel (chemical desorption: $H_{ads} + H_{ads} \rightarrow H_2\uparrow$) reactions[30,34]. Tafel slope, an intrinsic property of electrocatalysts, could be used to probe the elementary steps involved in the H_2 evolution. For example, HER kinetic models suggest that Tafel slope of about 120, 40 or 30 mV per decade will be obtained if the Volmer, Heyrovsky or Tafel reaction is the rate-determining step, respectively[34]. The Tafel slope down to ~36 mV per decade for $MoS_2/CoSe_2$ in 0.5 M H_2SO_4 is the lowest value measured till now for MoS_2-based HER catalysts (Supplementary Table 1), even approaching to that of ~30 mV per decade for Pt/C catalysts. This Tafel slope suggests a Tafel-step-determined Volmer–Tafel mechanism works in the $MoS_2/CoSe_2$ catalyst, where the synergistic $CoSe_2$ substrate with decent HER activity presumably contributes to this HER mechanism.

Considering that at least one decade of linearity in Tafel extrapolation at relatively large η is desirable to ensure an accurate Tafel analysis, the calculated Tafel slopes (Fig. 3b) and derived HER mechanism may be inconclusive. To prove the proposed HER mechanism, we performed a computational study on the new $MoS_2/CoSe_2$ catalyst to gain detailed insights into the adsorption, activation and reaction processes (Fig. 5; see Supplementary Methods for computational details). DFT calculations suggested that reactant (**R**) contains two H atoms adsorbed on different sites of the optimized $MoS_2/CoSe_2$ model with H–H distance of 2.134 Å. Two approached H atoms on the side S atoms of MoS_2 cluster then formed a transition state (**TS**) with H–H distance of 0.945 Å and imaginary vibration frequency of $828i\,cm^{-1}$. After crossing the **TS**, product (**P**) with a weakly absorbed H_2 molecule (0.745 Å) was formed. The calculated activation barrier of 1.13 eV (30.7 kcal mol^{-1}) for the Tafel-step reaction on $MoS_2/CoSe_2$ hybrid, which can be overcome at a slightly higher η, approaches that of Pt (111) electrode[35], agreeing well with our experimentally observed fast HER kinetics.

Discussion

In conclusion, we demonstrate that an effective and robust hydrogen evolution catalyst can be made by marrying inexpensive transition metal chalcogenides. The new $MoS_2/CoSe_2$ catalyst shows exceptional HER catalytic properties in acidic electrolyte with onset potential of mere −11 mV, a small Tafel slope of 36 mV per decade and a high exchange current density, representing the first non-noble metal catalyst that approaches the performance of state-of-the-art Pt/C catalyst. Inspired by the Nature using transition metals as catalytically active site to construct effective HER catalysts (for example, hydrogenase enzymes)[5,36], we believe that our study here will facilitate the development of newly efficient HER catalysts based on transitional metal chalcogenides.

Methods

Synthesis of $CoSe_2$/DETA nanobelts and $MoS_2/CoSe_2$ hybrid. All chemicals are of analytical grade and were used as received without further purification. First, ultrathin lamellar mesostructured $CoSe_2$/DETA nanobelts were prepared by our recently developed method[20]. Briefly, 0.249 g $Co(OAc)_2 \cdot H_2O$ and 0.173 g Na_2SeO_3 were added into a mixed solution (40 ml) with a volume ratio of $V_{DETA}/V_{DIW} = 2:1$ (DIW, deionized water). The obtained wine solution was then transferred into a 50 ml Taflon-lined autoclave, which was sealed and maintained at 180 °C for 16 h. The resulting black floccules were collected by centrifugation (4,000 r.p.m. for 5 min) and washed by absolute ethanol for three times, and the resulting $CoSe_2$/DETA nanobelts were dried for next use. To prepare the $MoS_2/CoSe_2$ hybrid, 10 mg freshly made $CoSe_2$/DETA nanobelts and 10 mg $(NH_4)_2MoS_4$ were dispersed in 10 ml DMF and sonicated for 15 min under ambient conditions. Then, 0.05 ml $N_2H_4 \cdot H_2O$ was added into the suspension. After sonicating for another 15 min to dissolve completely, the mixed solution was transferred into a 50 ml Teflon-lined autoclave, which was sealed and heated in an oven at 200 °C for 10 h and then cooled to room temperature naturally. The resulting black product was collected by centrifugation (7,000 r.p.m. for 8 min), then washed at least four times by distilled water and absolute ethanol to remove ions and possible remnants, and dried under vacuum at 80 °C for 6 h.

Synthesis of free 3D MoS_2 sheet aggregates. The synthetic procedure of free 3D MoS_2 sheet aggregates is the same with that for preparing $MoS_2/CoSe_2$ hybrid, the only difference is that no $CoSe_2$/DETA nanobelt substrates were added during the synthesis.

Characterization. The samples were characterized by different analytic techniques. X-ray powder diffraction (XRD) was carried out on a Rigaku D/max-rA X-ray diffractometer with Cu Ka radiation ($\lambda = 1.54178$ Å); TEM images, HRTEM images, SAED and an energy-disperse X-ray spectrum (EDS) were taken with a JEOJ-2010 transmission electron microscope with an acceleration voltage of 200 kV. STEM and EDX elemental mapping were performed on JEOL ARM-200F. The X-ray photoelectron spectra (XPS) were recorded on an ESCALab MKII XPS using Mg Ka radiation exciting source. The Fourier transform infrared spectra were measured on a Bruker Vector-22 FT-IR spectrometer at room temperature. Nitrogen sorption was determined by Brunauer, Emmett and Teller measurements with an ASAP-2020 surface area analyzer (see Supplementary Fig. 8 for results). The degradation of external MoS_2 around $CoSe_2$ was determined by the inductively coupled plasma–atomic emission spectrometry analysis using an Atomscan Advantage (Thermo Ash Jarrell Corporation, USA) spectrometer.

Electrocatalytic study. Electrochemical measurements were performed at room temperature using a rotating disk working electrode made of GC (PINE, 5 mm diameter, 0.196 cm^2) connected to a Multipotentiostat (IM6ex, ZAHNER elektrik, Germany). The GC electrode was polished to a mirror finish (No. 40-6365-006, Gamma Micropolish Alumina, Buehler; No.40-7212, Microcloth, Buehler) and thoroughly cleaned before use. Pt wire and SCE were used as counter and reference electrodes, respectively. The potentials reported in our work were vs the RHE through RHE calibration described below.

The preparation method of the working electrodes containing investigated catalysts can be found as follows. In short, 5 mg of catalyst powder was dispersed in 1 ml of 3:1 v/v DIW/isopropanol mixed solvent with 40 µl of Nafion solution (5 wt%, Sigma-Aldrich), then the mixture was ultrasonicated for about 30 min to generate a homogeneous ink. Next, 10 µl of the dispersion was transferred onto the GC disk, leading to the catalyst loading ~0.28 mg cm^{-2}. Finally, the as-prepared catalyst film was dried at room temperature. For comparison, bare GC electrode that has been polished and cleaned was also dried for electrochemical measurement.

Before the electrochemical measurement, the electrolyte (0.5 M H_2SO_4) was degassed by bubbling pure hydrogen for at least 30 min to ensure the H_2O/H_2 equilibrium at 0 V vs RHE at a rotation rate of 1,600 r.p.m. The polarization curves were obtained by sweeping the potential from -0.7 to -0.2 V vs SCE at room temperature and 1,600 r.p.m. (to remove the *in situ*-formed H_2 bubbles on the RDE), with a sweep rate of 2 mV s^{-1}. The electrochemical impedance spectroscopy measurement was performed in the same configuration at open circuit potential over a frequency range from 100 kHz to 5 mHz at the amplitude of the sinusoidal voltage of 5 mV and room temperature (see Supplementary Fig. 10 for results). $MoS_2/CoSe_2$ and pure MoS_2-coated CFPs (Toray, 1 cm^2, catalyst loading 1 mg) were used as working electrodes to collect chronoamperometry data at the applied potential of -0.7 V vs SCE. The polarization curves were replotted as overpotential (η) vs log current (log j) to get Tafel plots for assessing the HER kinetics of investigated catalysts. By fitting the linear portion of the Tafel plots to the Tafel equation ($\eta = b \log(j) + a$), the Tafel slope (b) can be obtained. All data were reported without iR compensation.

RHE calibration. In all measurements, we used SCE as the reference electrode. It was calibrated with respect to RHE. The calibration was performed in the high-purity hydrogen-saturated electrolyte with a Pt foil as the working electrode. Cyclic voltammetry was run at a scan rate of 1 mV s^{-1}, and the average of the two potentials at which the current crossed 0 was taken to be the thermodynamic potential for the hydrogen electrode reaction. In 0.5 M H_2SO_4 solution, $E_{RHE} = E_{SCE} + 0.28$ V.

DFT calculations. The computational modelling of the adsorption, activation and reaction processes involved in HER on $MoS_2/CoSe_2$ was performed by periodic DFT with the Vienna *Ab-initio* Simulation Package (VASP). $MoS_2/CoSe_2$ model-building details (Supplementary Figs 13–15), HER mechanism and relevant references are provided in the Supplementary Methods.

References

1. Walter, M. G. *et al.* Solar water splitting cells. *Chem. Rev.* **110,** 6446–6473 (2010).
2. Cook, T. R. *et al.* Solar energy supply and storage for the legacy and nonlegacy worlds. *Chem. Rev.* **110,** 6474–6502 (2010).
3. Greeley, J., Jaramillo, T. F., Bonde, J., Chorkendorff, I. B. & Norskov, J. K. Computational high-throughput screening of electrocatalytic materials for hydrogen evolution. *Nat. Mater.* **5,** 909–913 (2006).
4. Prins, R., Debeer, V. H. J. & Somorjai, G. A. Structure and function of the catalyst and the promoter in Co-Mo hydrodesulfurization catalysts. *Catal. Rev. Sci. Eng.* **31,** 1–41 (1989).
5. Hinnemann, B. *et al.* Biornimetic hydrogen evolution: MoS_2 nanoparticles as catalyst for hydrogen evolution. *J. Am. Chem. Soc.* **127,** 5308–5309 (2005).
6. Jaramillo, T. F. *et al.* Identification of active edge sites for electrochemical H_2 evolution from MoS_2 nanocatalysts. *Science* **317,** 100–102 (2007).
7. Karunadasa, H. I. *et al.* A molecular MoS_2 edge site mimic for catalytic hydrogen generation. *Science* **335,** 698–702 (2012).
8. Bonde, J., Moses, P. G., Jaramillo, T. F., Norskov, J. K. & Chorkendorff, I. Hydrogen evolution on nano-particulate transition metal sulfides. *Faraday Discuss.* **140,** 219–231 (2008).
9. Kibsgaard, J., Chen, Z. B., Reinecke, B. N. & Jaramillo, T. F. Engineering the surface structure of MoS_2 to preferentially expose active edge sites for electrocatalysis. *Nat. Mater.* **11,** 963–969 (2012).
10. Lukowski, M. A. *et al.* Enhanced hydrogen evolution catalysis from chemically exfoliated metallic MoS_2 nanosheets. *J. Am. Chem. Soc.* **135,** 10274–10277 (2013).
11. Kong, D. S. *et al.* Synthesis of MoS_2 and $MoSe_2$ films with vertically aligned layers. *Nano Lett.* **13,** 1341–1347 (2013).
12. Xie, J. F. *et al.* Defect-rich MoS_2 ultrathin nanosheets with additional active edge sites for enhanced electrocatalytic hydrogen evolution. *Adv. Mater.* **25,** 5807–5813 (2013).
13. Merki, D., Fierro, S., Vrubel, H. & Hu, X. L. Amorphous molybdenum sulfide films as catalysts for electrochemical hydrogen production in water. *Chem. Sci.* **2,** 1262–1267 (2011).
14. Vrubel, H., Merki, D. & Hu, X. L. Hydrogen evolution catalyzed by MoS_3 and MoS_2 particles. *Energy Environ. Sci.* **5,** 6136–6144 (2012).
15. Benck, J. D., Chen, Z. B., Kuritzky, L. Y., Forman, A. J. & Jaramillo, T. F. Amorphous molybdenum sulfide catalysts for electrochemical hydrogen production: insights into the origin of their catalytic activity. *ACS Catal.* **2,** 1916–1923 (2012).
16. Liao, L. *et al.* MoS_2 formed on mesoporous graphene as a highly active catalyst for hydrogen evolution. *Adv. Funct. Mater.* **23,** 5326–5333 (2013).
17. Li, Y. G. *et al.* MoS_2 nanoparticles grown on graphene: an advanced catalyst for the hydrogen evolution reaction. *J. Am. Chem. Soc.* **133,** 7296–7299 (2011).
18. Chen, Z. B. *et al.* Core-shell MoO_3-MoS_2 nanowires for hydrogen evolution: a functional design for electrocatalytic materials. *Nano Lett.* **11,** 4168–4175 (2011).
19. Wang, T. Y. *et al.* Enhanced electrocatalytic activity for hydrogen evolution reaction from self-assembled monodispersed molybdenum sulfide nanoparticles on an Au electrode. *Energy Environ. Sci.* **6,** 625–633 (2013).
20. Gao, M. R., Yao, W. T., Yao, H. B. & Yu, S. H. Synthesis of unique ultrathin lamellar mesostructured $CoSe_2$-amine (protonated) nanobelts in a binary solution. *J. Am. Chem. Soc.* **131,** 7486–7487 (2009).
21. Gao, M. R., Xu, Y. F., Jiang, J., Zheng, Y. R. & Yu, S. H. Water oxidation electrocatalyzed by an efficient $Mn_3O_4/CoSe_2$ nanocomposite. *J. Am. Chem. Soc.* **134,** 2930–2933 (2012).
22. Gao, M. R. *et al.* A methanol-tolerant Pt/$CoSe_2$ nanobelt cathode catalyst for direct methanol fuel cells. *Angew. Chem. Int. Ed.* **50,** 4905–4908 (2011).
23. Gao, M. R. *et al.* In situ controllable synthesis of magnetite nanocrystals/$CoSe_2$ hybrid nanobelts and their enhanced catalytic performance. *J. Mater. Chem.* **20,** 9355–9361 (2010).
24. Gao, M. R. *et al.* Mixed-solution synthesis of sea urchin-like NiSe nanofiber assemblies as economical Pt-free catalysts for electrochemical H_2 production. *J. Mater. Chem.* **22,** 13662–13668 (2012).
25. Xu, Y. F., Gao, M. R., Zheng, Y. R., Jiang, J. & Yu, S. H. Nickel/nickel(II) oxide nanoparticles anchored onto cobalt(IV) diselenide nanobelts for the electrochemical production of hydrogen. *Angew. Chem. Int. Ed.* **52,** 8546–8550 (2013).
26. Gao, M. R., Jiang, J. & Yu, S. H. Solution-based synthesis and design of late transition metal chalcogenide materials for oxygen reduction reaction (ORR). *Small* **8,** 13–27 (2012).
27. Gao, M. R., Xu, Y. F., Jiang, J. & Yu, S. H. Nanostructured metal chalcogenides: synthesis, modification, and applications in energy conversion and storage devices. *Chem. Soc. Rev.* **42,** 2986–3017 (2013).
28. Kong, D. S., Cha, J. J., Wang, H. T., Lee, H. R. & Cui, Y. First-row transition metal dichalcogenide catalysts for hydrogen evolution reaction. *Energy Environ. Sci.* **6,** 3553–3558 (2013).
29. Merki, D. & Hu, X. L. Recent developments of molybdenum and tungsten sulfides as hydrogen evolution catalysts. *Energy Environ. Sci.* **4,** 3878–3888 (2011).
30. Parasons, R. The rate of electrolytic hydrogen evolution and the heat of adsorption of hydrogen. *Trans. Faraday Soc.* **54,** 1053–1063 (1958).

31. Norskov, J. K. *et al.* Trends in the exchange current for hydrogen evolution. *J. Electrochem. Soc.* **152,** J23–J26 (2005).
32. Merki, D., Vrubel, H., Rovelli, L., Fierro, S. & Hu, X. L. Fe, Co, and Ni ions promote the catalytic activity of amorphous molybdenum sulfide films for hydrogen evolution. *Chem. Sci.* **3,** 2515–2525 (2012).
33. Chen, W. F. *et al.* Hydrogen-evolution catalysts based on non-noble metal nickel-molybdenum nitride nanosheets. *Angew. Chem. Int. Ed.* **51,** 6131–6135 (2012).
34. Conway, B. E. & Tilak, B. V. Interfacial processes involving electrocatalytic evolution and oxidation of H_2, and the role of chemisorbed H. *Electrochim. Acta* **47,** 3571–3594 (2002).
35. Skulason, E. *et al.* Modeling the electrochemical hydrogen oxidation and evolution reactions on the basis of density functional theory calculations. *J. Phys. Chem. C* **114,** 18182–18197 (2010).
36. DuBois, M. R. & DuBois, D. L. The roles of the first and second coordination spheres in the design of molecular catalysts for H_2 production and oxidation. *Chem. Soc. Rev.* **38,** 62–72 (2009).

Acknowledgements

We acknowledge the funding support from the National Basic Research Program of China (Grants 2010CB934700, 2013CB933900, 2014CB931800 and 2013CB834603), the National Natural Science Foundation of China (Grants 21431006, 91022032, 91227103, 21061160492 and J1030412) and the Chinese Academy of Sciences (Grant KJZD-EW-M01-1).

Author contributions

S.-H.Y. and M.-R.G. conceived the idea. M.-R.G., Y.-R.Z. and Y.-F.X. planned and performed the experiments, collected and analysed the data. J.-X.L. and J.L. performed the DFT calculations. J.J. and Q.G. assisted with the experiments and characterizations. M.-R.G. and S.-H.Y. co-wrote the manuscript. All authors discussed the results and commented on the manuscript.

Additional information

Competing financial interests: The authors declare no competing financial interests.

21

Graphene oxide/metal nanocrystal multilaminates as the atomic limit for safe and selective hydrogen storage

Eun Seon Cho[1], Anne M. Ruminski[1], Shaul Aloni[1], Yi-Sheng Liu[2], Jinghua Guo[2] & Jeffrey J. Urban[1]

Interest in hydrogen fuel is growing for automotive applications; however, safe, dense, solid-state hydrogen storage remains a formidable scientific challenge. Metal hydrides offer ample storage capacity and do not require cryogens or exceedingly high pressures for operation. However, hydrides have largely been abandoned because of oxidative instability and sluggish kinetics. We report a new, environmentally stable hydrogen storage material constructed of Mg nanocrystals encapsulated by atomically thin and gas-selective reduced graphene oxide (rGO) sheets. This material, protected from oxygen and moisture by the rGO layers, exhibits exceptionally dense hydrogen storage (6.5 wt% and 0.105 kg H_2 per litre in the total composite). As rGO is atomically thin, this approach minimizes inactive mass in the composite, while also providing a kinetic enhancement to hydrogen sorption performance. These multilaminates of rGO-Mg are able to deliver exceptionally dense hydrogen storage and provide a material platform for harnessing the attributes of sensitive nanomaterials in demanding environments.

[1]The Molecular Foundry, Materials Sciences Division, Lawrence Berkeley National Laboratory, 1 Cyclotron Road, Berkeley, California 94720, USA. [2]The Advanced Light Source, Lawrence Berkeley National Laboratory, Berkeley, California 94720, USA. Correspondence and requests for materials should be addressed to J.J.U. (email: jjurban@lbl.gov).

The established environmental impacts resulting from fossil fuels have stimulated urgent efforts to decarbonize our fuel sources. Hydrogen is the ultimate carbon-free energy carrier—it possesses the highest energy density among chemical fuels and water is the sole combustion product. Although major car manufacturers have made commitments to hydrogen as a 'fuel of the future'[1], hydrogen storage for FCEVs (fuel cell electric vehicles) currently relies on compressed gas tanks[2]. These are unable to meet long-term storage targets and severely compromise on-board occupancy. Metal hydrides for solid-state hydrogen storage are one of the few materials capable of providing sufficient storage density required to meet these long-term targets. However, simultaneously meeting gravimetric, volumetric, thermodynamic and kinetic requirements has proven challenging, owing to the strong binding enthalpies for the metal hydride bonds, long diffusion path lengths and oxidative instability of zero-valent metals. Although nano-structuring has been shown to optimize binding enthalpies[3], synthesis and oxidative stabilization of metallic nanocrystals remains a challenge[4]. Protection strategies against oxidization and sintering of nanocrystals often involve embedding these crystals in dense matrices, which add considerable 'dead' mass to the composite, in turn decreasing gravimetric and volumetric density. Thus, although metal hydrides show the most promise for non-cryogenic applications, it remains true that no single material has met all of these essential criteria[5,6].

Here we demonstrate mixed dimensional reduced graphene oxide (GO)/Mg nanocrystal hybrids as a novel high-performance materials platform for solid-state hydrogen storage. After the first report of the preparation of individual graphene sheets in 2004 (refs 7,8), its unique optoelectronic properties attracted great attention. GO, formerly considered just a precursor for the synthesis of graphene, has begun to find independent applications in water purification and gas separations due to its hydrophilicity, chemical structure and atomistic pore size diameters[9,10]. For example, GO membranes have recently been explored as materials for crucial gas separation challenges. Interestingly, these studies have shown extreme permeability for H_2 relative to other atmospheric gases such as O_2 and N_2, thus providing a potential avenue for use as an atomically thin, selective barrier layer for sensitive hydrogen storage materials. Furthermore, related studies have shown that reduction of GO to form reduced GO (rGO) further results in a dramatic decrease in water permeance, while maintaining desirable gas permeability characteristics[11]. In this study, we have prepared mixed dimensional laminates of two-dimensional (2D) rGO filled with Mg nanocrystals for hydrogen storage applications (Fig. 1a). In this composite, rGO serves as the atomic limit for barrier layer materials in functional composites, providing maximum environmental protection for the least possible amount of inactive mass—theoretically, for single sheet rGO protection, up to 98 wt% of the composite mass can be active Mg—thus, this rGO composite approach yields the greatest performance in selective hydrogen permeability and kinetic enhancement in hydrogen storage. As illustrated in Fig. 1a, rGO sheets function as a protective layer against Mg nanocrystal oxidation by preventing the permeation of O_2 and H_2O, while still allowing hydrogen to easily penetrate, diffuse along the layers and be released. Moreover, beyond the crucial gas barrier behaviour, we demonstrate that the rGO layers add functionality to the laminates by reducing the activation energies associated with hydrogen absorption and desorption, key kinetically limiting steps for traditional metal hydride systems. Several studies have shown that carbon-based materials such as carbon fibres, nanotubes and graphite exhibit a beneficial catalytic effect on the kinetics and cyclability of hydrogen absorption and desorption of metal hydrides[12–14]. Although there are other reports using graphitic materials in composites for Li-ion battery applications, to our knowledge there have been no reports that take advantage of both the unique catalytic properties and high variability in gas permeability of rGO to synergistically yield new functionality. For the nanolaminate system presented, rGO layers are ideal encapsulating materials: they provide atomically thin structure to minimize added mass, catalytically enhanced rate-limiting hydrogen absorption/desorption events and protective barriers to prevent degradation of Mg nanocrystals.

Results

Synthesis and characterization of rGO-Mg multilaminates. The majority of reported composites consisting of metals and carbon materials are prepared via ball milling or solidification with either polymers or carbon frameworks. However, ball-milled materials are notoriously polydisperse, which introduces corresponding inhomogeneity in properties. Moreover, such energy-intensive processes can intrinsically introduce unwanted morphological disruptions and chemical inhomogeneities, all of which detract from performance. By contrast, we have developed a direct, one-pot co-reduction, thus simultaneously forming both pristine, monodisperse nanocrystals and the desired rGO without energy-intensive processing or ligand chemistries. Observing that current approaches for reduction of GO and reduction of metal precursors to form Mg nanocrystals both rely on similar methodologies[4,15–17], we synthesized rGO-Mg nanocomposites via a facile solution-based co-reduction method. In this process, the Mg^{2+} precursor is stabilized by GO and both of them are reduced by lithium naphthalenide. The obtained rGO-Mg was characterized via transmission electron microscopy (TEM) and X-ray diffraction (XRD), as shown in Fig. 1b,c. Mg nanocrystals were 3.26 nm diameter ($\pm$0.87 nm) based on TEM images (Supplementary Figs 1 and 2), presenting fine monodisperse nanocrystals, compared with other metal hydrides prepared by conventional method such as ball milling. Although Scherrer analysis of the XRD peak width indicates larger crystallite size of ~15 nm (Supplementary Table 1), hundreds of TEM images over dozens of samples consistently show 3.26-nm-sized Mg nanocrystals. The observation of individual nanocrystals in TEM was challenging due to the low electron density of Mg, although in our materials Mg nanocrystals of diameter 3–4 nm are present and confirmed by multiple characterization methods. However, it is probable that there exist some clusters and networks of Mg particles in the sample as the XRD results imply the possible existence of a minor fraction comprising agglomerations of nanocrystals; as we discuss later, the H_2 storage properties are most consistent with ~few nanometre-sized crystallites. The area of the peaks in electron energy loss spectroscopy (EELS) measurements (Mg $\ggg$ C > O) indicated a high density of Mg within the composite (Fig. 1d). In addition, despite containing a highly dense packing of Mg nanocrystals, the nanolaminates were observed to be remarkably environmentally stable. To investigate the limits of stability, rGO-Mg samples were exposed to air and characterized over time by XRD and TEM (Fig. 1c and Supplementary Fig. 1); remarkably, even after 3 months of air exposure, the nanocrystals remained almost entirely zero-valent crystalline Mg, while showing invasion of only a low-intensity $Mg(OH)_2$ peak (Supplementary Fig. 3 and an additional corrosion test result in Supplementary Fig. 4 and Supplementary Discussion). Moreover, to show the reliability offered by this approach, we completely exposed the sample to air and then demonstrated hydrogen cycling. This is not possible with any other hydride technology with comparable storage density. Although none of the techniques used for

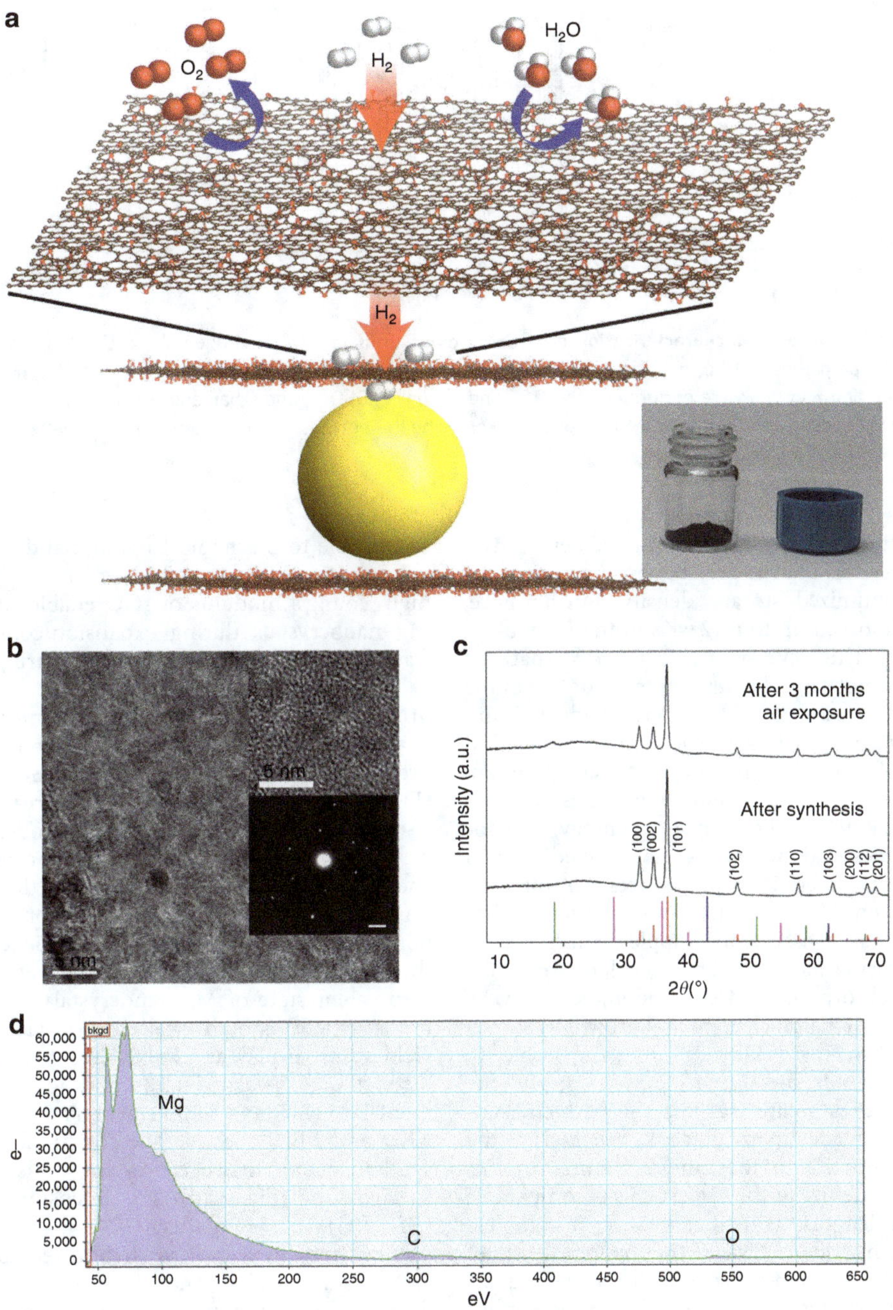

Figure 1 | Mixed dimensional rGO-Mg laminates for stable and energetically-dense hydrogen storage. (**a**) Illustrations depicting the structure of the rGO-Mg nanolaminates where rGO layers prevent O_2 and H_2O from penetrating, while allowing the diffusion of H_2 (inset: photograph of rGO-Mg in air). (**b**) TEM images of rGO-Mg showing the high density of Mg nanocrystals with no visible aggregates. The upper inset is a high-resolution image and the lower inset is diffraction pattern where the hexagonal dots are matched to Mg (100), corresponding to a *d*-spacing of 2.778 Å (JCPDS 04-0770). (More TEM images and analysis are provided in Supplementary Figs 1 and 2.) (**c**) XRD spectra of rGO-Mg after synthesis and after 3 months in air with indices of peaks (JCPDS 04-0770; the bottom bars represent an XRD pattern of hexagonal Mg (hcp) (red), tetragonal MgH_2 (pink), hexagonal $Mg(OH)_2$ (green) and cubic MgO (blue)). (**d**) EELS spectrum of a representative rGO-Mg composite flake suspended over a hole in the TEM support grid. The spectrum shows a dominant Mg *L*-edge peak and a carbon *K*-edge peak, indicating a large quantity of Mg crystals within the rGO support.

characterization show evidence for extensive formation of oxides of Mg (MgO or $Mg(OH)_2$), it is probable that a thin, self-terminating layer of oxide or sub-oxide may form on the nanocrystals; however, this oxide does not measurably have an impact on performance (Fig. 1d and Supplementary Fig. 5).

Hydrogen capacity and kinetic analysis. Hydrogen absorption and desorption characteristics of the rGO-Mg composite were tested using a Sieverts PCT-Pro instrument at 15 bar H_2 and 0 bar, respectively, as shown in Fig. 2a. Hydrogen uptake was immediate and the formation of MgH_2 was confirmed by XRD (Fig. 2c) and electron diffraction (Supplementary Fig. 1). The hydrogen absorption capacity of the composite was 6.5 wt% and 0.105 kg H_2 per litre in the total composite, which is the highest capacity reported (calculated on a full composite mass basis) using metal hydrides under the comparable conditions. This corresponds to 7.56 wt% H_2 in Mg nanocrystals, which is 99.5% of the theoretical value (7.6 wt%)[5]. Given the atomically thin

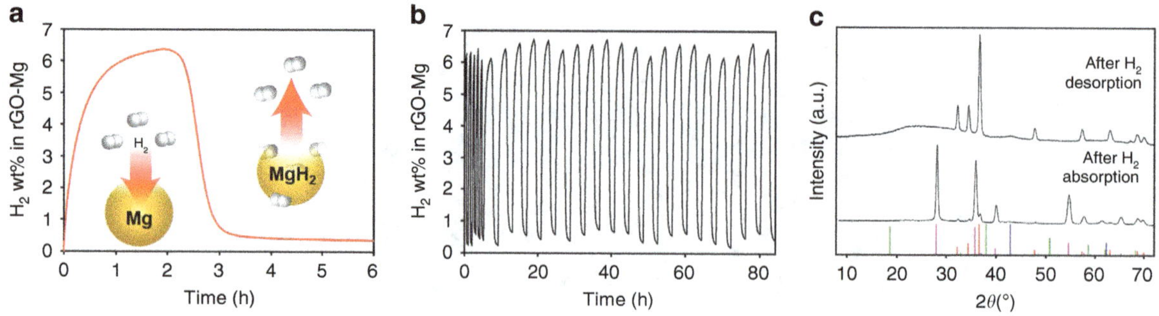

Figure 2 | Hydrogen absorption/desorption characterization of rGO-Mg multilaminates. (**a**) Hydrogen absorption/desorption (at 200 °C and 15 bar H_2/300 °C and 0 bar) for the prepared rGO-Mg multilaminates. (**b**) Hydrogen absorption/desorption cycling of rGO-Mg multilaminates that were first exposed to air overnight. The first 5 cycles were performed at 250 °C and 15 bar H_2/350 °C and 0 bar, and the additional 20 cycles at 200 °C and 15 bar H_2/300 °C and 0 bar. (**c**) XRD spectra of rGO-Mg after absorption/desorption (the bottom bars represent the XRD patterns of Mg (red), MgH_2 (pink), $Mg(OH)_2$ (green) and MgO (blue)).

nature of the encapsulation, these nanocomposites achieve denser packing of metal nanocrystals than is possible by any competing approach, leading to optimized storage density. Furthermore, hydrogen also readily desorbed up to 6.12 wt% in the composite, thus demonstrating excellent reversibility. To verify that the hydrogen absorption was not caused by the presence of GO in the composite, control studies using only GO were conducted and exhibited minimal (<0.2 wt% in GO) absorption at 200 °C and 250 °C (Supplementary Fig. 6). This is a negligible contribution, given that the amount of GO in the multilaminates is <2 wt% overall. To analyse the kinetics, the activation energy (E_a) for hydrogen absorption/desorption was determined from measurements at three different temperatures, fitting the result with the Johnson–Mehl–Avrami model (details are described in Supplementary Discussion, Supplementary Figs 7 and 8, and Supplementary Table 2)[18]. The E_a values were 60.8 and 92.9 kJ mol^{-1} for absorption and desorption, respectively, consistent with one-dimensional nucleation and growth as shown previously[4,19]. Remarkably, these kinetics are comparable to transition metal-catalysed bulk metal-hydride systems, and the overall capacity and kinetics greatly surpass the best environmentally robust samples made up to date[4,17]. We ascribe the performance of our composite to the unique features of this multilaminate: the nanoscale size of the Mg crystals is comparable to molecular diffusion lengths, which enables near-complete conversion to the metal hydride (99.5% of the theoretical value) and the interaction of the Mg nanocrystals with the rGO layers protects against invasion of oxygen, while enabling rapid surface diffusion of hydrogen, which enhances kinetics[20]. Indeed, rGO-Mg hydrogen absorption/desorption is faster than Mg-polymer composites containing nanocrystals of similar size (Supplementary Fig. 9 and Supplementary Discussion). Consistent with previous studies, the diffusion of hydrogen atoms was facilitated by the interaction between Mg and carbon layers, enhancing both the hydrogen capacity and kinetics of Mg (additional data in Supplementary Fig. 10 and Supplementary Discussion)[21,22]. Cycle tests were performed at 250 °C/350 °C for 5 cycles and at 200 °C/300 °C for additional 20 cycles (Fig. 2b). Importantly, each cycle test was conducted successively to better mimic real-world fueling/re-fueling conditions without manual evacuation in between each cycle (additional experimental results in Supplementary Fig. 11). The capacity and kinetics were well preserved during further cycles, although a slight decrease in the capacity was observed. Noticeably, the Mg nanocrystal size and size distributions were well preserved after several absorption/desorption cycles without sintering or grain growth (Supplementary Fig. 1). Although bulk metal hydrides are susceptible to mechanical fracture and cracking due to the large volume expansion on hydriding (*ca.* 33% from Mg to MgH_2), the high Young's modulus of rGO enables it to robustly encase the Mg nanocrystals during expansion/contraction events without fracture and prevent macroscale sintering.

Structural analysis of rGO-Mg composite. X-ray absorption near-edge structure (XANES) measurements were performed to probe the interactions between rGO and Mg nanocrystals (Fig. 3a). Compared with GO, increased intensity of carbon *K*-edge at 288.4 and 290.3 eV were observed, corresponding to carbon atoms attached to oxygen or other oxygen-containing chemical species. From this, we infer that the multilaminates are uniquely stabilized by the formation of interfacial Mg–O–C bonds forged during synthesis[23]. We believe this to be the basis of the exceptional stability of these multilaminates. In addition, the zero-valent state of Mg nanocrystals was verified at Mg *L*-edge measurement as both total electron yield and total fluorescence yield scans display an explicit Mg metal peak (49.8 eV) without distinct MgO peaks (Fig. 3b and additional analysis in Supplementary Fig. 5 and Supplementary Discussion), consistent with XRD and EELS spectra. The structural evolution of GO during synthesis and hydrogen cycling was studied using Raman spectroscopy (Fig. 3c,d). The intensity ratio of *D* and *G* peaks ($I(D)/I(G)$) increased after rGO-Mg synthesis, indicating that the average domain size of sp^2 hybridized regions was decreased as GO was reduced[24]. The 2D peak, whose position and shape depends on the number of graphene layers, shifted to lower frequency (2,701 cm^{-1} to 2,685 cm^{-1}) and its full width at half maximum also decreased on the formation of rGO-Mg (Fig. 3d). This suggests that few, if any, isolated multilayers of rGO exist in the composite, and that most rGO layers are actively wrapping Mg nanocrystals[25,26] (additional analysis in Supplementary Fig. 12 and Supplementary Discussion). No change was observed in the Raman spectra of freshly synthesized rGO-Mg in comparison with samples studied after hydrogen cycling. Importantly, $I(D)/I(G)$ ratios remained consistent as well (1.370 after synthesis and 1.337 after cycling), indicating that the defect density, a key attribute of rGO responsible for selective hydrogen transport, was well maintained even after several hydrogen absorption and desorption cycles[25]. In addition, the chemical environment of GO and rGO-Mg were investigated via X-ray photoelectron spectroscopy (Fig. 4). Peaks associated with oxygen-containing functional groups in the GO diminished after the formation of rGO-Mg, confirming reduction of GO[27]. The rGO-Mg composite contained an additional peak at 282.5 eV, which is attributed to the interaction between carbon species and

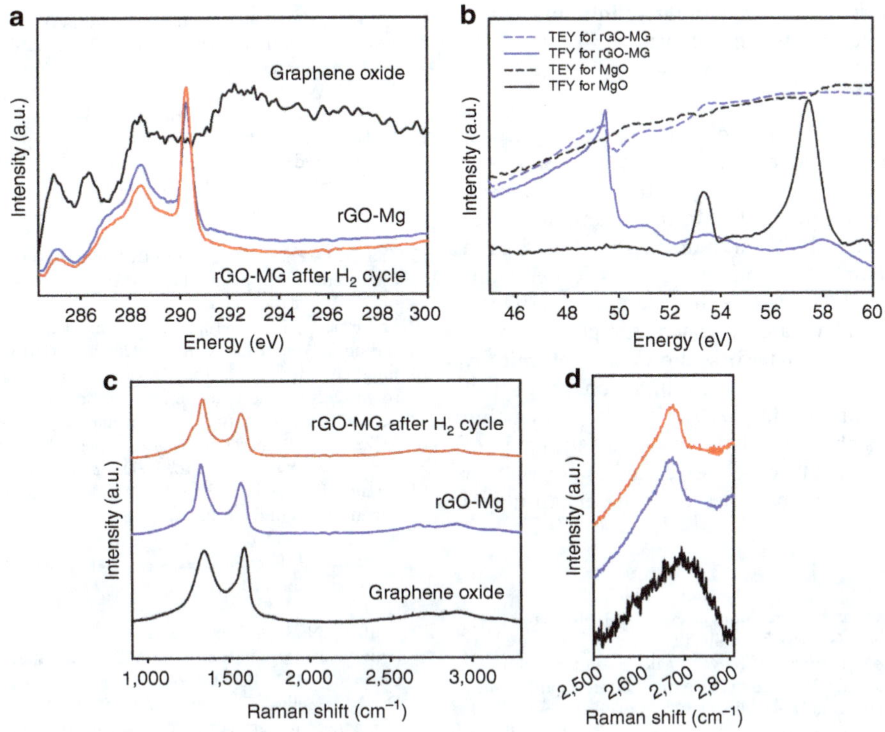

Figure 3 | XANES and Raman spectral analysis of GO and rGO-Mg multilaminates before and after hydrogen cycling. XANES spectra of (**a**) GO, rGO-Mg after synthesis and after cycling at carbon *K*-edge, and (**b**) rGO-Mg after synthesis and MgO at Mg *L*-edge. Raman spectra of (**c**) GO, rGO-Mg after synthesis and after H_2 cycling, and (**d**) the 2D peak region.

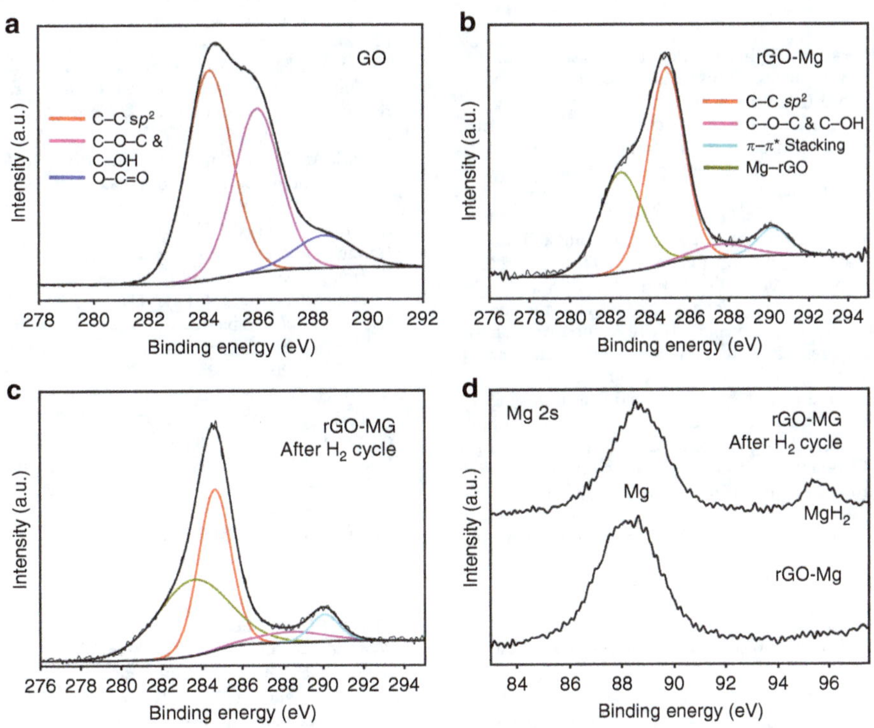

Figure 4 | X-ray photoelectron spectra (XPS) of rGO-Mg after synthesis and after hydrogen cycling. XPS spectra (C, 1 s) of (**a**) GO, (**b**) rGO-Mg after synthesis and (**c**) rGO-Mg after H_2 cycling. (**d**) XPS pattern (Mg, 2 s) for rGO-Mg after synthesis and after H_2 cycling.

metal particles[28,29], corresponding to the interaction of rGO and Mg nanocrystals. Furthermore, a prominent π–π^* stacking peak was observed at 290.1 eV, resulting from Mg nanocrystal wrapping, which was also observed by TEM (Supplementary Fig. 1). In the Mg 2s spectrum, one additional peak appears in the higher energy region after hydrogen cycling, implying a new chemical state, consistent with MgH_2 (ref. 30). The Mg and MgH_2 contents were 91.45% and 8.55%, respectively, after completing cycle test based on Mg 2s spectrum, explaining a slight amount of the remaining hydride phase in the end of cycling.

Discussion

We have developed a facile method of preparing the densest possible loading of reactive nanocrystals safely into a composite material, a crucial step forward for enhancing the energy density of nanomaterials. As a result, our rGO-Mg multilaminates offer exceptional environmental stability and unsurpassed hydrogen storage capability, exceeding that offered by any other non-cryogenic reversible material. We believe that these results suggest the possibility of practical solid-state hydrogen storage and use in the near future. Furthermore, this work shows that atomically thin 2D materials can be used to simultaneously protect nanocrystals from ambient conditions, while also imparting new functionality. Such stable mixed dimensional laminates of zero-valent nanocrystalline metals can be extended to a variety of additional applications, including batteries, catalysis, encapsulants and energetic materials.

Methods

Synthesis of rGO-Mg nanocomposite. The composites of rGO-Mg were synthesized in an argon glove box. GO was ball milled for 10 min before use, to break it down to GO platelets so that it can effectively make a complex with bis(cyclopentadienyl) magnesium (Cp_2Mg). To prepare the lithium naphthalenide solution, naphthalene (2.40 g, 0.0187 mol) was dissolved in 120 ml of tetrahydrofuran (THF), followed by the immediate addition of Li metal (0.36 g, 0.0253 mol), leading to a dark green solution. GO (6.25 mg) was dispersed in 12.5 ml of THF under Ar, sealed in a container and sonicated for 1.5 h. A Cp_2Mg solution (2.31 g, 0.015 mol, in 22.5 ml of THF) was added to GO solution afterwards, stirring for 30 min. The resulting GO/Cp_2Mg solution was added to the lithium naphthalenide solution and magnetically stirred for 2 h. The product was centrifuged (10,000 r.p.m., 20 min) and washed with THF (10,000 r.p.m., 20 min) twice, followed by vacuum drying overnight.

Characterization and instrumentation. High-resolution TEM was performed using JEOL 2100-F Field-Emission Analytical Transmission Electron operated at 120 kV and equipped with Oxford INCA energy-dispersive electron X-ray spectrometer and Tridiem Gatan imaging Filter and spectrometer. The powder samples were dispersed on lacey carbon grids from THF solutions. Elemental analysis of the EELS and energy-dispersive spectra was performed using Digital Micrograph software (Gatan Inc.). XRD patterns were acquired with a Bruker AXS D8 Discover GADDS X-Ray Diffractometer, using Cu Kα radiation ($\lambda = 0.154$ nm). Hydrogen absorption/desorption measurement was performed, using a HyEnergy PCT Pro-2000 at 15/0 bar of H_2 at different temperatures. XANES spectroscopy was performed on Beamline 8.0.1.3 and 4.0.3 at the Advanced Light Source. The energy resolution at carbon *K*-edge and Mg *L*-edge was set to 0.1 eV and the experimental chamber had a base pressure of at most 1×10^{-8} torr. An highly ordered pyrolytic graphite (HOPG) reference sample was measured before and after all XANES experiments for energy calibration. The XANES spectra were recorded using total electron yield and total fluorescence yield detection modes. The Raman spectra of GO and rGO-Mg samples were collected using Horiba Jobin Yvon LabRAM ARAMIS automated scanning confocal Raman microscope with a 532-nm excitation source and X-ray photoelectron spectra were obtained via PHI 5400 X-ray Photoelectron Spectroscopy System with Al Kα. The Mg content in the composite was determined by inductively coupled plasma-optical emission spectroscopy at the Advanced Light Source Life Sciences Division and Environmental.

References

1. Tollefson, J. Fuel of the future? *Nature* **464,** 1262–1264 (2010).
2. Schlapbach, L. & Zuttel, A. Hydrogen-storage materials for mobile applications. *Nature* **414,** 353–358 (2001).
3. Shao, H. Y., Xin, G. B., Zheng, J., Li, X. G. & Akiba, E. Nanotechnology in Mg-based materials for hydrogen storage. *Nano Energy* **1,** 590–601 (2012).
4. Jeon, K. J. *et al.* Air-stable magnesium nanocomposites provide rapid and high-capacity hydrogen storage without using heavy-metal catalysts. *Nat. Mater.* **10,** 286–290 (2011).
5. Aguey-Zinsou, K. F. & Ares-Fernandez, J. R. Hydrogen in magnesium: new perspectives toward functional stores. *Energy Environ. Sci.* **3,** 526–543 (2010).
6. Yang, J., Sudik, A., Wolverton, C. & Siegel, D. J. High capacity hydrogen storage materials: attributes for automotive applications and techniques for materials discovery. *Chem. Soc. Rev.* **39,** 656–675 (2010).
7. Novoselov, K. S. *et al.* Electric field effect in atomically thin carbon films. *Science* **306,** 666–669 (2004).
8. Geim, A. K. Graphene status and prospects. *Science* **324,** 1530–1534 (2009).
9. Joshi, R. K. *et al.* Precise and ultrafast molecular sieving through graphene oxide membranes. *Science* **343,** 752–754 (2014).
10. Kim, H. W. *et al.* Selective gas transport through few-layered graphene and graphene oxide membranes. *Science* **342,** 91–95 (2013).
11. Li, H. *et al.* Ultrathin, molecular-sieving graphene oxide membranes for selective hydrogen separation. *Science* **342,** 95–98 (2013).
12. Adelhelm, P. & de Jongh, P. E. The impact of carbon materials on the hydrogen storage properties of light metal hydrides. *J. Mater. Chem.* **21,** 2417–2427 (2011).
13. Yao, X. D. *et al.* Metallic and carbon nanotube-catalyzed coupling of hydrogenation in magnesium. *J. Am. Chem. Soc.* **129,** 15650–15654 (2007).
14. Liu, X. F., Peaslee, D. & Majzoub, E. H. Tailoring the hydrogen storage properties of $Li_4BN_3H_{10}$ by confinement into highly ordered nanoporous carbon. *J. Mater. Chem. A* **1,** 3926–3931 (2013).
15. Jung, H., Yang, S. J., Kim, T., Kang, J. H. & Park, C. R. Ultrafast room-temperature reduction of graphene oxide to graphene with excellent dispersibility by lithium naphthalenide. *Carbon* **63,** 165–174 (2013).
16. Rieke, R. D. Preparation of organometallic compounds from highly reactive metal powders. *Science* **246,** 1260–1264 (1989).
17. Ruminski, A. M., Bardhan, R., Brand, A., Aloni, S. & Urban, J. J. Synergistic enhancement of hydrogen storage and air stability via Mg nanocrystal-polymer interfacial interactions. *Energy Environ. Sci.* **6,** 3267–3271 (2013).
18. Fernandez, J. F. & Sanchez, C. R. Rate determining step in the absorption and desorption of hydrogen by magnesium. *J. Alloy Compd.* **340,** 189–198 (2002).
19. Norberg, N. S., Arthur, T. S., Fredrick, S. J. & Prieto, A. L. Size-dependent hydrogen storage properties of Mg nanocrystals prepared from solution. *J. Am. Chem. Soc.* **133,** 10679–10681 (2011).
20. Denis, A., Sellier, E., Aymonier, C. & Bobet, J. L. Hydrogen sorption properties of magnesium particles decorated with metallic nanoparticles as catalyst. *J. Alloy Compd.* **476,** 152–159 (2009).
21. Du, A. J., Smith, S. C., Yao, X. D. & Lu, G. Q. Catalytic effects of subsurface carbon in the chemisorption of hydrogen on a Mg(0001) surface: an ab-initio study. *J. Phys. Chem. B* **110,** 1814–1819 (2006).
22. Yao, X. D. *et al.* Mg-based nanocomposites with high capacity and fast kinetics for hydrogen storage. *J. Phys. Chem. B* **110,** 11697–11703 (2006).
23. Wang, H. L. *et al.* An ultrafast nickel-iron battery from strongly coupled inorganic nanoparticle/nanocarbon hybrid materials. *Nat. Commun.* **3,** 917 (2012).
24. Ha, H. W., Kim, I. Y., Hwang, S. J. & Ruoff, R. S. One-pot synthesis of platinum nanoparticles embedded on reduced graphene oxide for oxygen reduction in methanol fuel cells. *Electrochem. Solid State Lett.* **14,** B70–B73 (2011).
25. Ferrari, A. C. & Basko, D. M. Raman spectroscopy as a versatile tool for studying the properties of graphene. *Nat. Nanotechnol.* **8,** 235–246 (2013).
26. Ferrari, A. C. *et al.* Raman spectrum of graphene and graphene layers. *Phys. Rev. Lett.* **97,** 187401 (2006).
27. Ganguly, A., Sharma, S., Papakonstantinou, P. & Hamilton, J. Probing the thermal deoxygenation of graphene oxide using high-resolution *in situ* X-ray-based spectroscopies. *J. Phys. Chem. C* **115,** 17009–17019 (2011).
28. Jia, L. S. *et al.* Visible-light-induced photocatalyst based on C-doped LaCoO3 synthesized by novel microorganism chelate method. *Catal. Commun.* **10,** 1230–1234 (2009).
29. Wang, H. Q., Wu, Z. B. & Liu, Y. A simple two-step template approach for preparing carbon-doped mesoporous TiO_2 hollow microspheres. *J. Phys. Chem. C* **113,** 13317–13324 (2009).
30. Friedrichs, O., Kolodziejczyk, L., Sanchez-Lopez, J. C., Lopez-Cartes, C. & Fernandez, A. Synthesis of nanocrystalline MgH_2 powder by gas-phase condensation and in situ hydridation: TEM, XPS and XRD study. *J. Alloy Compd.* **434,** 721–724 (2007).

Acknowledgements

Work at the Molecular Foundry and the Advanced Light Source was supported by the Office of Science, Office of Basic Energy Sciences, of the U.S. Department of Energy under Contract Number DE-AC02-05CH11231. We thank Yi-De Chuang for XANES experimental support. This material is based on work supported by the Department of Energy (DOE) through the Bay Area Photovoltaic Consortium (BAPVC) under Award Number DE-EE0004946 and also in part under the US-India Partnership to Advance Clean Energy-Research (PACE-R) for the Solar Energy Research Institute for India and the United States (SERIIUS), funded jointly by the U.S. Department of Energy (Office of Science, Office of Basic Energy Sciences, and Energy Efficiency and Renewable Energy, Solar Energy Technology Program, under Subcontract DE-AC36-08GO28308 to the National Renewable Energy Laboratory, Golden, Colorado) and the Government of India, through the Department of Science and Technology under Subcontract IUSSTF/JCERDC-SERIIUS/2012 dated 22 November 2012. We sincerely appreciate Jeong Yun Kim and Jayoung Kim for assisting graphic work.

Author contributions

J.J.U. and E.S.C. conceived and designed the experiments. E.S.C. synthesized and characterized the composite and performed the experiments. S.A. contributed towards TEM and EELS acquisition and analysis. Y.-S.L. and J.-H.G. performed XANES experiments. E.S.C., A.M.R. and J.J.U. analysed the data and wrote the manuscript. All authors discussed the results and commented on the manuscript.

Additional information

Competing financial interests: The authors declare no competing financial interests.

22

Hierarchy of bond stiffnesses within icosahedral-based gold clusters protected by thiolates

Seiji Yamazoe[1,2], Shinjiro Takano[1], Wataru Kurashige[3], Toshihiko Yokoyama[4], Kiyofumi Nitta[5], Yuichi Negishi[3] & Tatsuya Tsukuda[1,2]

Unique thermal properties of metal clusters are believed to originate from the hierarchy of the bonding. However, an atomic-level understanding of how the bond stiffnesses are affected by the atomic packing of a metal cluster and the interfacial structure with the surrounding environment has not been attained to date. Here we elucidate the hierarchy in the bond stiffness in thiolate-protected, icosahedral-based gold clusters $Au_{25}(SC_2H_4Ph)_{18}$, $Au_{38}(SC_2H_4Ph)_{24}$ and $Au_{144}(SC_2H_4Ph)_{60}$ by analysing Au L_3-edge extended X-ray absorption fine structure data. The Au–Au bonds have different stiffnesses depending on their lengths. The long Au–Au bonds, which are more flexible than those in the bulk metal, are located at the icosahedral-based gold core surface. The short Au–Au bonds, which are stiffer than those in the bulk metal, are mainly distributed along the radial direction and form a cyclic structural backbone with the rigid Au–SR oligomers.

[1] Department of Chemistry, School of Science, The University of Tokyo, 7-3-1 Hongo, Bunkyo-ku, Tokyo 113-0033, Japan. [2] Elements Strategy Initiative for Catalysts and Batteries (ESICB), Kyoto University, Katsura, Kyoto 615-8520, Japan. [3] Department of Applied Chemistry, Faculty of Science, Tokyo University of Science, 1-3 Kagurazaka, Shinjuku-ku, Tokyo 162-8601, Japan. [4] Department of Materials Molecular Science, Institute for Molecular Science, Myodaiji, Okazaki, Aichi 444-8585, Japan. [5] Japan Synchrotron Radiation Research Institute, SPring-8, 1-1-1 Koto, Sayo, Hyogo 679-5198, Japan. Correspondence and requests for materials should be addressed to T.T. (email: tsukuda@chem.s.u-tokyo.ac.jp).

Metal nanoparticles (NPs) exhibit specific thermal properties and phase transition behaviour that are quite different from the corresponding bulk metal[1–4]. For example, the melting point of a metal NP is significantly depressed with respect to that of the bulk metal[5]. The melting of a metal NP starts with premelting of the surface layer, which then expands towards the inner core and leads to a complete transition to the liquid[6]. These size-specific thermal behaviours are explained in terms of the large surface energy of a metal NP. In other words, a crucial factor that governs thermal behaviours is the hierarchy of the bonding within NPs; metal–metal bonds on the surface are generally softer than those within the core[3,7,8]. The thermal behaviours of metal NPs are not only affected by their size, but also by the atomic packing of the core[3,7–16] and the interaction with the environment, such as an organic ligand[17] or solid support[3,18,19]. For example, the vibrational spectrum of a metal NP with an icosahedral (Ih) structure has a component with a vibrational frequency that is higher than those of other structures such as cuboctahedra and decahedra, which suggests the formation of stiff bonds within the NPs[7]. The influence of surface adsorbates on the thermal properties of metal NPs has been ascribed to a change in the bond stiffness; Pt–Pt bonds of small Pt NPs are softened by hydrogen adsorption, but stiffened by oxidation[19] or capping with a polymer[17]. However, an atomic-level understanding of how the bond stiffnesses are affected by a variety of structural parameters has not been attained to date because of the experimental difficulties in defining the atomic packing of a metal NP and the interfacial structure with the surrounding environment.

Recently, a series of thiolate (RS)-protected gold clusters $Au_n(SR)_m$ with well-defined compositions have gained attention as ideal platforms to study the structure–property correlation and the size-dependent evolution of properties[20]. Among others, $Au_n(SR)_m$ with $(n, m) = (25, 18)$, $(38, 24)$ and $(144, 60)$ have been studied most extensively as prototypical systems. Single-crystal X-ray diffraction (XRD) analysis revealed that $Au_{25}(SR)_{18}$ and $Au_{38}(SR)_{24}$ have icosahedral Au_{13} and bi-icosahedral Au_{23} cores, respectively, which are protected by combinations of $-SR-(Au-SR)_2-$ and $-SR-Au-SR-$ oligomers (Fig. 1)[21–23]. The atomic structure of $Au_{144}(SR)_{60}$ has not yet been determined by X-ray crystallography, but has been theoretically predicted as composed of a hollow icosahedral Au_{114} core protected by 30 $-SR-Au-SR-$ oligomers (Fig. 1)[24,25]. In addition, it has been identified that there are distinct Au–Au bonds with different lengths in the range of 2.7–3.3 Å in these clusters (Supplementary Fig. 1). The $Au_n(SR)_m$ clusters $(n = 25, 38, 144)$ provide an ideal opportunity to study the hierarchy of the bond stiffness within Au clusters with well-defined atomic structures and surface modification.

In this study, the stiffnesses of the Au–Au and Au–S bonds in $Au_n(PET)_m$ (phenylethanethiolate (PET) = PhC_2H_4S) are examined using Au L_3-edge extended X-ray absorption fine structure (EXAFS) analysis, which is a powerful tool to study the local structure of the ligand-protected gold clusters[26,27]. The temperature dependence of the Debye–Waller (DW) factors of individual bonds are analysed in the framework of the Einstein model, in which a metal cluster is treated as an ensemble of quantum harmonic oscillators with individual Einstein temperatures that depend on the bonding. The hierarchy in the bond stiffness is elucidated. The Au–S bonds are much stiffer than Au–Au bonds in all the clusters and there are two types of Au–Au bonds; one is stiffer and the other is softer than those in the bulk Au. A major portion of the shorter and stiffer Au–Au bonds is distributed within the Ih-based Au core as radial bonds. The stiff Au–Au bonds distributed on the surface of the core are connected by the rigid Au–S oligomers to form a ring structure in all the clusters. These ring structures may act as a rigid framework to enhance the thermal stability of the thiolate-protected Au clusters.

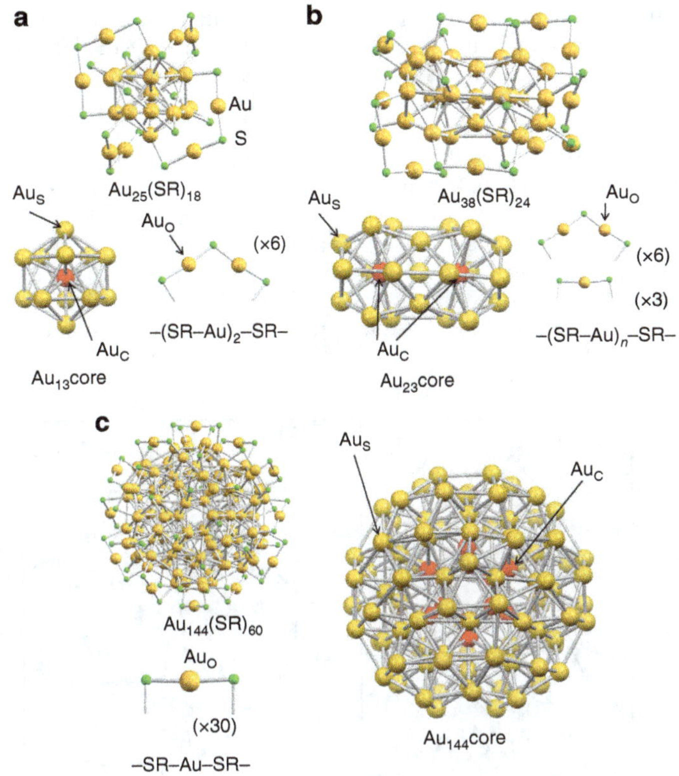

Figure 1 | Geometric structures of $Au_n(SR)_m$ samples. (**a**) Au-S framework of $Au_{25}(SR)_{18}$ determined by single crystallography[22], (**b**) Au-S framework of $Au_{38}(SR)_{24}$ determined by single crystallography[23] and (**c**) Au-S framework of $Au_{144}(SR)_{60}$ predicted by density functional theory calculation (ref. 24). Au_C, Au_S and Au_O represent Au atoms bonded only to Au atoms, those bonded both to Au and S atoms, and those bridged by two S atoms, respectively for $Au_{25}(SR)_{18}$ and $Au_{38}(SR)_{24}$. The definition of AuC, Au_S, and Au_O for $Au_{144}(SR)_{60}$ is given in the text. *R* is omitted for simplicity.

Results

Structural analysis. Figure 2a shows EXAFS oscillations of $Au_{25}(PET)_{18}$ measured at 300 and 8 K, respectively. EXAFS oscillation is clearly observed up to the k range of 21 Å^{-1} at 8 K, whereas it is damped in the k range of > 14 Å^{-1} at 300 K. Given that the EXAFS oscillation in the high k region mainly originates from heavy atoms, the large amplitude in Fig. 2a suggests that thermal fluctuation of the Au–Au bonds in $Au_{25}(PET)_{18}$ is significantly suppressed at 8 K. Figure 2b represents Fourier transformed (FT)-EXAFS spectra of $Au_{25}(PET)_{18}$ at 300 and 8 K, respectively. Most notably, the peak due to Au–Au bonds is clearly observed in the bond length (r) range of 2.1–3.0 Å at 8 K, in addition to that for Au–S bonds in the r range of 1.5–2.0 Å, whereas the Au–Au peak is hardly discernible at 300 K (Fig. 2b). Similar temperature dependence has been reported in the literature[26] and is attributed to suppression of the thermal fluctuation of Au–Au bonds at low temperature. Previous single crystallographic study revealed that the Au–Au bonds of $Au_{25}(PET)_{18}$ are classified into three groups according to their lengths: the Au–Au(x) bonds ($x = 1$–3) assigned mainly to the Au_C–Au_S, Au_S–Au_S and Au_S–Au_O bonds, respectively (see Fig. 1 and Supplementary Note 1; Supplementary Fig. 1a). The average coordination number (CN) and r values calculated for the

Au–Au(x) bonds (x = 1–3) are summarized in Table 1. Based on this information, curve-fitting analysis of the FT-EXAFS data for $Au_{25}(PET)_{18}$ at 8 K was conducted by assuming a single Au–S and three types of Au–Au bonds with different r values. However, we found that contribution from the longest Au–Au bonds is significantly smaller than the other two Au bonds due to larger DW factor (see Supplementary Discussion, Supplementary Fig. 2, and Supplementary Table 1). Thus, the FT-EXAFS data was analysed by assuming a single Au–S and two types of Au–Au bonds, Au–Au(S) and Au–Au(L). Their CN and r values thus obtained are summarized in Table 1. The CN and r values for the Au–S, Au–Au(S) and Au–Au(L) bonds obtained by EXAFS are in good agreement with those of the Au–S, Au–Au(1) and Au–Au(2) bonds determined by single crystal XRD data, respectively. This agreement indicates that low-temperature EXAFS measurements allow quantitative probing of Au–Au bonds within the Au_{13} core: Au_C–Au_S and Au_S–Au_S bonds. The FT-EXAFS spectrum simulated assuming DW factors of 0.0027, 0.0040 and 0.0037 Å^2 (Supplementary Fig. 3) for Au_C, Au_S and Au_O, respectively, corresponded well with the experimental spectrum (Supplementary Fig. 4).

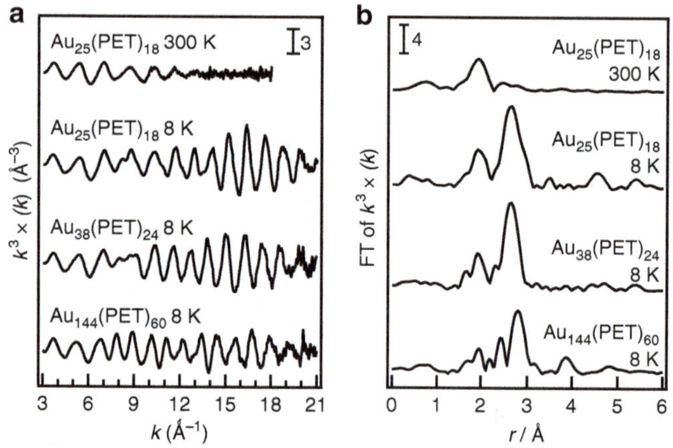

Figure 2 | EXAFS oscillations and FT-EXAFS spectra. (**a**) EXAFS oscillations for $Au_{25}(PET)_{18}$ measured at 300 and 8 K, $Au_{38}(PET)_{24}$ at 8 K, and $Au_{144}(PET)_{60}$ at 8 K and (**b**) the corresponding FT-EXAFS spectra.

Similar temperature dependence was observed in the Au L_3-edge EXAFS spectrum for $Au_{38}(PET)_{24}$ measured at 8 K (Fig. 2a), where a clear oscillation was observed in the k range of 3.0–21.0 Å^{-1}. The FT-EXAFS spectrum (Fig. 2b) at 8 K has peaks for the Au–S (1.5–2.0 Å) and Au–Au (2.1–3.0 Å) bonds. Single-crystal XRD measurements of $Au_{38}(PET)_{24}$ show that 38 Au atoms are categorized into three different sites, Au_C, Au_S and Au_O, as in the case of $Au_{25}(PET)_{18}$ (Fig. 1b). The length distribution of Au–Au bonds obtained from the single crystal structure[23] is shown in Supplementary Fig. 1b. The Au–Au bonds were classified into three groups with border distances of 2.90 and 3.05 Å: Au–Au(x) bonds (x = 1–3) although there is larger ambiguity in the determination of the border distances than in the case of $Au_{25}(PET)_{18}$ (Supplementary Fig. 1a). The Au–Au(x) bonds (x = 1–3) are composed of mainly Au_C–Au_S, Au_S–Au_S and Au_S–Au_O bonds, respectively. The average CN and r values calculated for the Au–Au(x) bonds (x = 1–3) are summarized in Table 1. The FT-EXAFS spectrum was analysed by assuming a single Au–S and two types of Au–Au bonds with different lengths, Au–Au(S) and Au–Au(L), as in the case of $Au_{25}(PET)_{18}$. The CN and r values for the Au–S, Au–Au(S) and Au–Au(L) bonds obtained by curve-fitting analysis (Table 1) quantitatively agree with those for the Au–S, Au–Au(1) and

Table 1 | Structural parameters of $Au_n(PET)_m$ obtained by curve-fitting analysis of Au L_3-edge FT-EXAFS and single-crystal XRD data.

n, m	Method	Bonds	CN*	r (Å)†	σ^2 (Å^2)‡	R (%)§	σ_S^2 (Å^2)‖	θ_E (K)¶
25, 18	EXAFS at 8 K	Au-S	1.6 (2)	2.319 (4)	0.0037 (18)	10.6	0.0017 (3)	429 (38)
		Au-Au (S)	1.5 (4)	2.770 (3)	0.0027 (11)		0.0008 (6)	137 (10)
		Au-Au (L)	1.5 (6)	2.936 (6)	0.0040 (24)		0.0020 (6)	101 (4)
	Single crystal XRD#	Au-S	1.4	2.33				
		Au-Au (1)	1.4	2.78				
		Au-Au (2)	1.9	2.95				
		Au-Au (3)	2.9	3.16				
38, 24	EXAFS at 8 K	Au-S	1.2 (2)	2.315 (4)	0.0030 (10)	10.7	0.0012 (5)	416 (57)
		Au-Au (S)	2.8 (4)	2.788 (1)	0.0041 (4)		0.0028 (12)	153 (11)
		Au-Au (L)	1.6 (9)	2.982 (15)	0.0090 (49)		0.0066 (5)	106 (15)
	Single crystal XRD**	Au-S	1.3	2.33				
		Au-Au (1)	2.7	2.82				
		Au-Au (2)	1.8	2.98				
		Au-Au (3)	2.2	3.21				
144, 60	EXAFS at 8 K	Au-S	0.9 (2)	2.326 (8)	0.0048 (29)	12.7	0.0024 (9)	381 (45)
		Au-Au (S)	1.2 (5)	2.733 (9)	0.0035 (20)		0.0021 (5)	148 (11)
		Au-Au (L)	6.0 (7)	2.870 (4)	0.0059 (13)		0.0041 (6)	128 (9)
	DFT††	Au-S	0.8	2.34				
		Au-Au (1)	1.5	2.77				
		Au-Au (2)	5.9	2.92				
		Au-Au (3)	1.5	3.24				

*Coordination number.
†Bond length.
‡Debye-Waller factor.
§$R = (\Sigma(k^3\chi^{data}(k) - k^3\chi^{fit}(k))^2)^{1/2} / (\Sigma(k^3\chi^{data}(k))^2)^{1/2}$.
‖Static component of Debye-Waller factor at 8 K.
¶Einstein temperature.
#ref. 22.
**ref. 23.
††ref. 24.

Au–Au(2) bonds, respectively. The FT-EXAFS spectrum simulated using DW factors for each of the Au sites shown in Supplementary Fig. 3 reproduced the experimental spectrum (Supplementary Fig. 5).

These analyses demonstrate that low-temperature EXAFS data provides quantitative structural information on the Au bonds within the Au cores of $Au_n(SR)_m$. Following this successful characterization, we next conducted Au L_3-edge EXAFS measurement of $Au_{144}(PET)_{60}$ at 8 K, of which the structure has not yet been determined by single-crystal XRD. Au L_3-edge EXAFS oscillation was clearly observed in the k range as large as 21.0 Å^{-1}, as shown in Fig. 2a. The FT-EXAFS spectrum of $Au_{144}(PET)_{60}$ (Fig. 2b) shows a small peak for the Au–S bond in the range of 1.5–2.0 Å and a large peak for the Au–Au bond in the range of 2.1–3.0 Å. The curve-fitting analysis indicates that the data can be reproduced by assuming Au–S (CN = 0.9, r = 2.326 Å), Au–Au(S) (CN = 1.2, r = 2.733 Å) and Au–Au(L) (CN = 6.0, r = 2.870 Å) bonds, as shown in Table 1. The Au–Au bonds are slightly shorter than those of $Au_{25}(PET)_{18}$ and $Au_{38}(PET)_{24}$, whereas the Au–S bond length is comparable to those of $Au_{25}(PET)_{18}$ and $Au_{38}(PET)_{24}$. According to density functional theory (DFT) calculations[24], $Au_{144}(SCH_3)_{60}$ has a hollow icosahedral Au_{12} core. The absence of the central Au atom may be a cause of the reduction in the Au–Au bond lengths compared with those of $Au_{25}(PET)_{18}$ and $Au_{38}(PET)_{24}$. Supplementary Figure 1c plots the bond lengths between the nearest neighbour Au atoms in $Au_{144}(SCH_3)_{60}$ obtained by DFT calculations[24]. When Au–Au bonds are divided into three groups Au–Au(x) (x = 1–3) with border distances of 2.83 and 3.10 Å, the experimentally obtained CN and r values of the Au–S, Au–Au(S) and Au–Au(L) bonds are in quantitative agreement with those of the Au–S, Au–Au(1) and Au–Au(2) bonds, respectively, as shown in Table 1. The EXAFS analysis strongly supports the $Au_{144}(SCH_3)_{60}$ model structure[24]. We then examined which Au–Au bonds contribute mainly to the Au–Au(1) and Au–Au(2) bonds. When the Au atoms of $Au_{144}(SCH_3)_{60}$ are divided into three groups: 12 Au atoms at the hollow icosahedral core (Au_C), 102 Au atoms located at the middle and the outermost layers (Au_S) and 30 Au atoms at the oligomer (Au_O) (Fig. 1), the Au–Au(1) and Au–Au(2) bonds are assigned mainly to the Au_C–Au_C and Au_S–Au_S bonds, respectively. The EXAFS oscillation is again dominated by the Au–Au bonds in the core and the shell, whereas the contribution from the bonds with Au_O is negligibly small. The FT-EXAFS spectrum simulated using DW factors for each of the Au sites shown in Supplementary Fig. 3 corresponded well with the experimental spectrum (Supplementary Fig. 6).

Einstein temperatures for Au–S and Au–Au bonds. The DW factors for the Au–S, Au–Au(L) and Au–Au(S) bonds of $Au_n(PET)_m$ (n = 25, 38, 144) were determined by least-squares fit analysis while keeping the CN values the same as those in Table 1, and the results are plotted as a function of temperature in Fig. 3. Details of the analysis are given in Methods section. Figure 3 shows that the DW factors for both the Au–Au(S) and Au–Au(L) bonds increase monotonically with the temperature, whereas those of Au–S bonds remain almost constant in the temperature range of 8–300 K in all clusters. The DW factors for the Au–Au(S) bonds are smaller than those for the Au–Au(L) bonds and are comparable to that for the Au–Au bond in the bulk Au (Supplementary Fig. 7).

The stiffnesses of the Au–S bonds and two types of Au–Au bonds of $Au_n(PET)_m$ (n = 25, 38, 144) were evaluated within the framework of the Einstein model. In the framework of the Einstein model, we assume three independent quantum oscillators with different Einstein frequencies (ω_E) for the Au–S, Au–Au(L) and Au–Au(S) bonds. The Einstein temperatures (θ_E) were determined by fitting the temperature dependence of the DW factors for the Au–S, Au–Au(L) and Au–Au(S) bonds. Details of the fitting procedure are given in Methods section. The best-fit results are shown in Fig. 3, and the optimized θ_E and σ_S^2 values are listed in Table 1.

Discussion

In this study, the hierarchy of the bond stiffnesses within $Au_n(PET)_m$ (n = 25, 38, 144) was examined using Au L_3-edge EXAFS. EXAFS results provided quantitative structural information (r and CN values) of two types of Au–Au bonds (Au–Au(S) and Au–Au(L)) within the Ih-based Au cores in addition to the Au–S bonds. The temperature dependence of the DW factors of the individual bonds was analysed within the framework of the Einstein model. Einstein temperatures for the two types of Au–Au bonds were determined to be much lower than that of the Au–S bonds in all clusters, which suggests that thiolate-protected Au clusters can be viewed as soft Au cores capped by rigid Au–SR staples. This is in agreement with the results of molecular dynamic (MD) simulation on a model system $Au_{25}(SH)_{18}^-$ that the –SR–(Au–SR)$_2$– units are rigid and confine the elastic Au core internally[28].

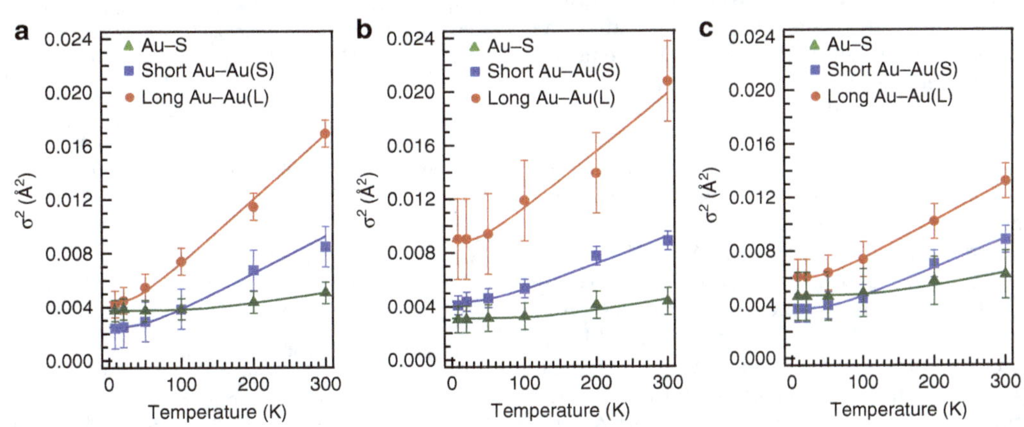

Figure 3 | Temperature dependence of the DW factors. Temperature dependence of the DW factors (σ^2) for Au-S, Au-Au(S) and Au-Au(L) bonds. (**a**) $Au_{25}(PET)_{18}$, (**b**) $Au_{38}(PET)_{24}$ and (**c**) $Au_{144}(PET)_{60}$. Solid lines represent the fitting results. Error bars represent the s.d. in the curve-fitting procedure of FT-EXAFS.

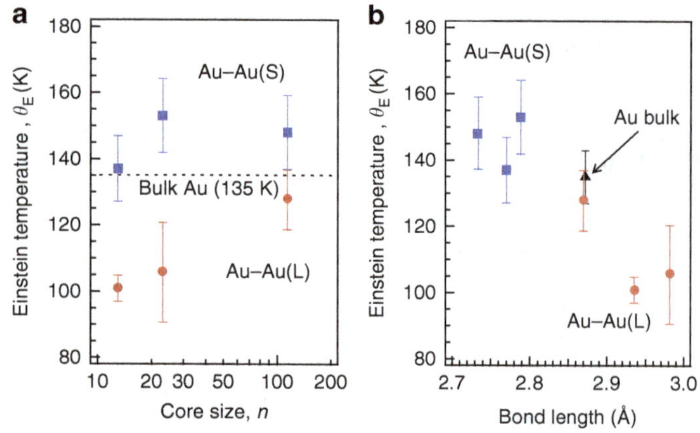

Figure 4 | Einstein temperature. (**a**) Einstein temperature (θ_E) as a function of the core size and (**b**) bond length for Au-Au(S) (square) and Au-Au(L) (circle) bonds for core size of $n = 13$, 23 and 114. θ_E and Au-Au bond distance for bulk Au are 135 K and 2.872 Å, respectively. Error bars represent the s.d. in the fitting procedure of θ_E.

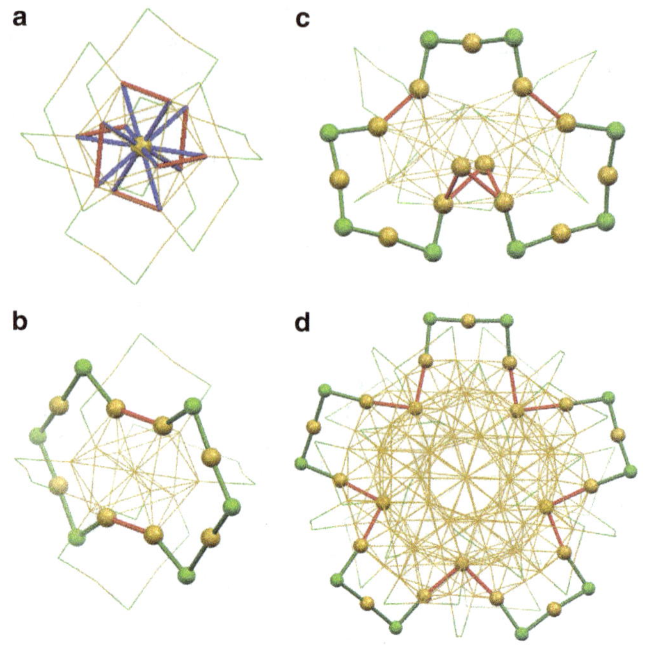

Figure 5 | Network structures of stiff bonds. (**a**) Distribution of stiff Au-Au bonds in $Au_{25}(PET)_{18}$. Red and blue bonds represent Au_S-Au_S and Au_C-Au_S bonds, respectively. Ring structures in (**b**) $Au_{25}(PET)_{18}$, (**c**) $Au_{38}(PET)_{24}$ and (**d**) $Au_{144}(PET)_{60}$.

More interestingly, two classes of Au-Au bonds with different stiffnesses were identified. Figure 4a shows the Einstein temperatures of the Au-Au(S) and Au-Au(L) bonds as a function of the Au core sizes (13, 23, 114) of $Au_n(PET)_m$ and that of Au-Au bonds in the bulk Au (135 ± 8 K)[29]. This plot indicates that the Au-Au(S) and Au-Au(L) bonds are respectively stiffer and more flexible than the Au-Au bonds in the face-centred cubic bulk Au, regardless of the cluster size. To the best of our knowledge, this result is the first experimental verification of hierarchy of Au-Au bond stiffnesses in thiolate-protected Au clusters. The correlation suggests that the bond stiffness is related to the bond length. To confirm this hypothesis, Einstein temperatures of the Au-Au bonds were plotted as a function of their lengths in Fig. 4b. The Au-Au bond becomes stiffer with a reduction of the bond length. The Au-Au(S) bonds that are shorter and stiffer than Au-Au bonds in the bulk Au are specific to the Ih core because Au-Au(S) bonds are formed along the radial direction in the icosahedral Au core. The formation of Au-Au bonds in $Au_n(PET)_m$ that are stiffer than those in the bulk Au is consistent with the theoretical prediction that the vibrational density of states distribution of the Ih cluster has a tail towards a higher frequency than that of bulk Au[4,7,8]. The Au-Au(L) bonds between the surface Au atoms are more flexible on average than those in the bulk Au, even though they are bonded with the rigid Au-SR oligomers. The above consideration suggests that the Au core of the thiolate-protected Au clusters tends to deform more easily along the lateral direction than in the radial direction. This is consistent with theoretical prediction based on the MD simulation of $Au_{25}(SH)_{18}^-$ that Au core atoms prefer vibrating in the tangential directions as opposed to the radial direction[28].

Close inspection of the distribution of Au-Au(S) bonds reveals that the stiff Au-Au bonds are distributed not only in the centre of the core, but also on the surface of the core. For example, 6 Au_S-Au_S bonds and 12 Au_C-Au_S bonds in $Au_{25}(PET)_{18}$ are categorized as the Au-Au(S) bonds, as shown in Fig. 5a. By bridging these stiff Au-Au surface bonds with the $-SR-(Au-SR)_2-$ oligomers, a ring structure is formed in $Au_{25}(PET)_{18}$ (Fig. 5b). Similar ring structures can be found in $Au_{38}(PET)_{24}$ and $Au_{144}(PET)_{60}$ (Fig. 5c,d). Tlahuice-Flores *et al.*[30] also suggested the formation of circular networks ($Au_{20}Cl_{10}$) with short Au-Au and Au-Cl bonds as subunits in $Au_{144}Cl_{60}^{2+}$, which is modelled by replacement of the SR ligands of $Au_{144}(SR)_{60}$ with Cl. These results suggest that thiolate-protected Au clusters contain rigid ring subunits that consist of short Au-Au bonds at the surface and Au-SR staples. The formation of these rigid ring structures may act as frameworks to enhance the thermal stability of the thiolate-protected Au clusters.

Methods

Chemicals. Hydrogen tetrachloroaurate tetrahydrate ($HAuCl_4 \cdot 4H_2O$) was purchased from Tanaka Kikinzoku. phenylethanethiol, tetraoctylammonium bromide (TOABr), sodium tetrahydroborate ($NaBH_4$), methanol, acetone, dichloromethane and toluene were obtained from Wako Pure Chemical Industries. All chemicals were used without further purification. Deionized water with a resistivity above 18.2 MΩ cm was used.

Synthesis and characterization. $Au_{25}(PET)_{18}$ was synthesized by the methods similar to that in the literature[31]. First, $HAuCl_4 \cdot 4H_2O$ (0.75 mmol) was dissolved in 25 ml tetrahydrofuran solution containing TOABr (0.76 mmol) at room temperature. After stirring for 15 min, phenylethanethiol (4.7 mmol) was added to this solution and the solution was stirred for 15 min. A cold aqueous solution (5.8 ml) containing $NaBH_4$ (8.7 mmol) was then rapidly added to the solution and then the solution was stirred at room temperature. After 12 h, tetrahydrofuran solvent was evaporated and the remaining red brown powder was washed with methanol to remove excess thiol and other byproducts. The $Au_{25}(PET)_{18}$ cluster was extracted from the dried sample using acetonitrile.

$Au_{38}(PET)_{24}$ was synthesized by the methods similar to that in the literature[32]. First, $HAuCl_4 \cdot 4H_2O$ (0.50 mmol) and glutathione (0.25 mmol) were dissolved in acetone (20 ml). This solution was stirred for 15 min at room temperature and for another 15 min at 0 °C. A cold aqueous solution (6.0 ml) containing $NaBH_4$ (5.0 mmol) was then rapidly added to this solution and the solution was stirred in ice bath. After 15 min, the acetone was discarded and the residue was dissolved in water (6.0 ml). This solution was then added to the mixture of toluene (2.0 ml) and ethanol (0.30 ml) containing phenylethanethiol (15 mmol). The resulting solution was stirred at 80 °C. After 12 h, the organic solution was evaporated and the resulting black powder was washed by methanol to remove byproducts. The $Au_{38}(PET)_{24}$ thus obtained was further purified by the gel permeation chromatography.

$Au_{144}(PET)_{60}$ was synthesized by the methods similar to that in the literature[33]. First, $HAuCl_4 \cdot 4H_2O$ (0.60 mmol) and TOABr (0.69 mmol) were dissolved in methanol (30 ml) at room temperature. After stirring for 5 min, phenylethanethiol (3.7 mmol) was added to this solution and the solution was stirred for 15 min. A cold aqueous solution (10 ml) containing $NaBH_4$ (6.0 mmol) was then rapidly added to this solution and the solution was stirred at room temperature. After 4 h, the black precipitate was separated by centrifugation and washed repeatedly with methanol or acetonitrile to remove excess thiol, byproducts and the other-sized clusters. The $Au_{144}(PET)_{60}$ thus obtained was further purified by gel permeation chromatography.

Chemical compositions and purities of the $Au_n(PET)_m$ samples were confirmed by a high-resolution Fourier-transform mass spectrometer with an electrospray ionization source (Bruker, Solarix). A 1-mg ml^{-1} dispersion of $Au_n(SR)_m$ in toluene/acetonitrile (1:1, v-v) was electrosprayed at a flow rate of 800 µl per hour. Ultraviolet-visible-near infrared (UV-Vis-NIR) spectra of $Au_n(PET)_m$ in toluene were recorded with a spectrometer (Jasco, V-670).

EXAFS measurements. Au L_3-edge EXAFS measurements were conducted using the BL01B1 beamline at the SPring-8 facility of the Japan Synchrotron Radiation Research Institute. The incident X-ray beam was monochromatized by a Si(311) double crystal monochromator. A solid sample of each $Au_n(PET)_m$ cluster was diluted with boron nitride powder, pressed into a pellet, and then mounted on a copper holder attached to a cryostat. All EXAFS spectra were measured in the transmission mode using ionization chambers at 8–300 K. The X-ray energy was calibrated using Cu foil. The EXAFS analysis was conducted using REX2000 Ver. 2.5.9 program (Rigaku Co.) as follows. The χ spectra were extracted by subtracting the atomic absorption background using cubic spline interpolation and were normalized to the edge height. The k^3-weighted χ spectra within the k ranges of 3.0–21.0 Å^{-1} for structural analysis and of 3.0–16.0 Å^{-1} for analysis of temperature dependence of the DW factor were Fourier-transformed into r space. The curve-fitting analysis was performed for Au–S and Au–Au bonds over the r range of 1.5–3.1 Å. In the curve-fitting analysis, the phase shifts and back-scattering amplitude functions for Au–S and Au–Au were extracted from Au_2S (ICSD#78718) and Au metal (ICSD#44362), respectively, using the FEFF8 program[34] by setting σ^2 to 0.0036. This σ^2 value did not significantly affect the phase shift and back-scattering amplitude functions. EXAFS spectra for $Au_n(PET)_m$ were also simulated using the FEFF8 (ref. 34) programs. Reported crystal structures for $Au_{25}(PET)_{18}$ (ref. 22) and $Au_{38}(PET)_{24}$ (ref. 23) and the calculated structure for $Au_{144}(SCH_3)_{60}$ (ref. 24) were used for the EXAFS simulations. The theoretical bond lengths for the $Au_{144}(SCH_3)_{60}$ structure were scaled by 0.974 because the Perdew, Burke and Ernzerhof approximation used to the exchange-correlation functional overestimates the Au–Au bond lengths[24]. Different DW factors for Au atoms at different sites, as shown in Supplementary Fig. 3, were used for EXAFS simulations because the DW factors for Au atoms are dependent on the sites shown in Table 1.

Evaluation of DW factors. The EXAFS equation is generally expressed as:

$$\chi(k) = S_0^2 \sum \frac{\mathrm{CN}}{kr^2} f(k;\pi)\exp(-2\sigma^2 k^2)\sin(2kr + \delta(k) - \frac{4}{3}C_3k^3), \quad (1)$$

where S_0^2, CN, k, r, $f(k;\pi)$, σ^2, $\delta(k)$ and C_3 are passive electron reduction factor, coordination number, photoelectron energy, bond distance, back-scattering amplitude function, DW factor, phase shift, and the third cumulant, respectively[35]. Supplementary Figure 9 shows FT-EXAFS spectra for $Au_n(PET)_m$ at various temperatures obtained from the temperature-dependent EXAFS data ($3.0 \le k \le 16.0$ Å^{-1}) (Supplementary Fig. 8). It was confirmed by the analysis of bulk Au that the temperature dependence of C_3 must be taken into account in addition to the σ^2 and r values (Supplementary Fig. 7)[14] to reproduce the FT-EXAFS spectra. C_3, r and σ^2 values for individual Au–S and Au–Au bonds were determined by least-squares fit analysis while keeping the CN values the same as those in Table 1, and the results are plotted as a function of temperature in Supplementary Fig. 10. In all clusters, the DW factors for both the Au–Au(S) and Au–Au(L) bonds increase monotonically with the temperature, whereas those of Au–S bonds remain almost constant in the temperature range of 8–300 K (Supplementary Fig. 10). The DW factors for the Au–Au(S) bonds are smaller than those for the Au–Au(L) bonds and are comparable to that for the Au–Au bond in the bulk Au (Supplementary Figs 7 and 10).

Evaluation of Einstein temperatures. The DW factor σ^2 is composed of a dynamic component σ_D^2 and a static component σ_S^2, which arises from the thermal atomic oscillation and the temperature-independent structural disorder, respectively:

$$\sigma^2 = \sigma_S^2 + \sigma_D^2, \quad (2)$$

We assume within the framework of the Einstein model three independent quantum oscillators with different Einstein frequencies (ω_E) for the Au–S and Au–Au bonds. The Einstein temperature (θ_E) is defined as:

$$\theta_E = \frac{h\omega_E}{2\pi k_B}, \quad (3)$$

where h and k_B are the Planck and Boltzmann constants, respectively. According to the Einstein model, the dynamic DW factor σ_D^2 can be fitted using the equation:

$$\sigma_D^2 = \frac{h^2}{8\pi^2 \mu k_B \theta_E} \coth \frac{\theta_E}{2T}, \quad (4)$$

where μ and T represent the reduced mass of adjacent atoms and the temperature, respectively. The θ_E values were determined by fitting the temperature dependence of the DW factors for the Au–S, Au–Au(L) and Au–Au(S) bonds using equations (2)–(4). The best-fit results are shown in Fig. 3, and the optimized θ_E and σ_S^2 values are listed in Table 1.

References

1. Bai, H. Y., Luo, J. L. & Jin, D. Particle size and interfacial effect on the specific heat of nanocrystalline Fe. *J. Appl. Phys.* **79,** 361–364 (1996).
2. Schmidt, M., Kusche, R., Issendorff, B. V. & Haberland, H. Irregular variations in the melting point of size-selected atomic clusters. *Nature* **393,** 238–240 (1998).
3. Narvaez, G. A., Kim, J. & Wilkins, W. Effects of morphology on phonons of nanoscopic silver grains. *Phys. Rev. B* **72,** 155411 (2005).
4. Araujo, L. L., Kluth, P., Azevedo, G. M. & Ridgway, M. C. Vibrational properties of Ge nanocrystals determined by EXAFS. *Phys. Rev. B* **74,** 184102 (2006).
5. Buffat, P. & Borel, J.-P. Size effect on the melting temperature of gold particles. *Phys. Rev. A* **13,** 2287–2298 (1976).
6. Cleveland, C. L., Luedtke, W. D. & Landman, U. Melting of gold clusters: icosahedral precursors. *Phys. Rev. Lett.* **81,** 2036–2039 (1998).
7. Sauceda, H. E., Salazar, F., Pérez, L. A. & Garzón, I. L. Size and shape dependence of the vibrational spectrum and low-temperature specific heat of Au nanoparticles. *J. Phys. Chem. C* **117,** 25160–25168 (2013).
8. Ortigoza, M. A., Heid, R., Bohnen, K.-P. & Rahman, T. S. Anomalously soft and stiff modes of transition-metal nanoparticles. *J. Phys. Chem. C* **118,** 10335–10347 (2014).
9. Yang, C. C., Xiao, M. X., Li, W. & Jiang, Q. Size effects on Debye temperature, Einstein temperature, and volume thermal expansion coefficient of nanocrystals. *Solid State Commun.* **139,** 148–152 (2006).
10. Cuenya, B. R. *et al.* Size-dependent evolution of the atomic vibrational density of states and thermodynamic properties of isolated Fe nanoparticles. *Phys. Rev. B* **86,** 165406 (2012).
11. Yokoyama, T., Kimoto, S. & Ohta, T. Temperature-dependent EXAFS study on supported silver and palladium clusters. *Jpn J. Appl. Phys.* **28,** L851–L853 (1989).
12. Cuenya, B. R. *et al.* Thermodynamic properties of Pt nanoparticles: size, shape, support, and adsorbate effects. *Phys. Rev. B* **84,** 245438 (2011).
13. Cuenya, B. R. *et al.* Anomalous lattice dynamics and thermal properties of supported size- and shape-selected Pt nanoparticles. *Phys. Rev. B* **82,** 155450 (2010).
14. Comaschi, T., Balerna, A. & Mobilio, S. Temperature dependence of the structural parameters of gold nanoparticles investigated with EXAFS. *Phys. Rev. B* **77,** 075432 (2008).
15. Dubiel, M. *et al.* Temperature dependence of EXAFS cumulants of Ag nanoparticles in glass. *J. Phys. Conf. Ser.* **190,** 012123 (2009).
16. Sauceda, H. E., Pelayo, J. J., Salazar, F., Pérez, L. A. & Garzón, I. L. Vibrational spectrum, caloric curve, low-temperature het capacity, and Debye temperature of sodium clusters: the Na_{139}^+ case. *J. Phys. Chem. C* **117,** 11393–11398 (2013).
17. Giovanetti, L. J. *et al.* Anomalous vibrational properties induced by surface effects in capped Pt nanoparticles. *J. Phys. Chem. C* **111,** 7599–7604 (2009).
18. Kang, J. H., Menard, L. D., Nuzzo, R. G. & Frenkel, A. I. Unusual non-bulk properties in nanoscale materials: thermal metal-metal bond contraction of γ-alumina-supported Pt catalysts. *J. Am. Chem. Soc.* **128,** 12068–12069 (2006).
19. Sanchez, S. I. *et al.* The emergence of nonbulk properties in supported metal clusters: negative thermal expansion and atomic disorder in Pt nanoclusters supported on γ-Al_2O_3. *J. Am. Chem. Soc.* **131,** 7040–7054 (2009).
20. Tsukuda, T. & Häkkinen, H. *Protected Metal Clusters: from Fundamentals to Applications* (Elsevier, 2015).
21. Heaven, M. W., Dass, A., White, P. S., Holt, K. M. & Murray, R. W. Crystal structure of the gold nanoparticle $[N(C_8H_{17})_4][Au_{25}(SCH_2CH_2Ph)_{18}]$. *J. Am. Chem. Soc.* **130,** 3754–3755 (2008).
22. Zhu, M., Aikens, C. M., Hollander, F. J., Schatz, G. C. & Jin, R. Correlating the crystal structure of a thiol-protected Au_{25} cluster and optical properties. *J. Am. Chem. Soc.* **130,** 5883–5885 (2008).
23. Qian, H., Echenhoff, W. T., Zhu, Y., Pintauer, T. & Jin, R. Total structure determination of thiolate-protected Au_{38} nanoparticles. *J. Am. Chem. Soc.* **132,** 8280–8281 (2010).
24. Lopez-Acevedo, O., Akola, J., Whetten, R. L., Grönbeck, H. & Häkkinen, H. Structure and bonding in the ubiquitous icosahedral metallic gold cluster $Au_{144}(SR)_{60}$. *J. Phys. Chem. C* **113,** 5035–5038 (2009).
25. Weissker, H.-Ch. *et al.* Information on quantum states pervades the visible spectrum of the ubiquitous $Au_{144}(SR)_{60}$ gold nanocluster. *Nat. Commun.* **5,** 3785 (2014).
26. Zhang, P. X-ray spectroscopy of gold-thiolate nanoclusters. *J. Phys. Chem. C* **118,** 25291–25299 (2014).
27. Chevrier, D. M., Yang, R., Chatt, A. & Zhang, P. Bonding properties of thiolate-protected gold nanoclusters and structural analogs from X-ray absorption spectroscopy. *Nanotechnol. Rev.* **4,** 193–206 (2015).

28. Mäkinen, V. & Häkkinen, H. Density functional theory molecular dynamics study of the $Au_{25}(SH)_{18}^-$ cluster. *Eur. Phys. J. D* **66,** 310 (2011).
29. Kluth, P. *et al.* Vibrational properties of Au and Cu nanocrystals formed by ion implantation. *AIP Conf. Proc.* **882,** 731–733 (2007).
30. Tlahuice-Flores, A., Black, D. M., Bach, S. B. H., Yose-Yacamán, M. & Whetten, R. L. Structure & bonding of the gold-subhalide cluster *I*-$Au_{144}Cl_{60}^{[z]}$. *Phys. Chem. Chem. Phys.* **15,** 19191–19195 (2013).
31. Dharmaratne, A. C., Krick, T. & Dass, A. Nanocluster size evolution studied by mass spectrometry in room temperature $Au_{25}(SR)_{18}$ synthesis. *J. Am. Chem. Soc.* **131,** 13604–13605 (2009).
32. Qian, H., Zhu, Y. & Jin, R. Size-focusing synthesis, optical and electrochemical properties of monodisperse $Au_{38}(SC_2H_4Ph)_{24}$ nanoclusters. *ACS Nano* **3,** 3795–3803 (2009).
33. Qian, H. & Jin, R. Ambient synthesis of $Au_{144}(SR)_{60}$ nanoclusters in methanol. *Chem. Mater.* **23,** 2209–2217 (2011).
34. Ankudinov, A. L., Ravel, B., Rehr, J. J. & Conradson, S. D. Real-space multiple-scattering calculation and interpretation of X-ray-absorption near-edge structure. *Phys. Rev. B* **58,** 7565–7576 (1998).
35. Tranquada, J. M. & Ingalls, R. Extended X-ray-absorption fine-structure study of anharmonicity in CuBr. *Phys. Rev. B* **28,** 3520–3528 (1983).

Acknowledgements

The EXAFS measurements were conducted under an approval of JASRI (Proposal Nos 2015A1590, 2014B1430, 2013B1659, 2013B1421, 2012B1421 and 2012B1986). We thank Professor Hannu Häkkinen (University of Jyväskylä) for providing atomic coordinates of $Au_{144}(SCH_3)_{60}$ obtained by DFT calculations. This research was financially supported by the Funding Program for Next Generation World-leading Researchers (NEXT Program, GR-003), 'Elements Strategy Initiative to Form Core Research Center', a Grant-in-Aid for Scientific Research (No. 26248003) from the MEXT (Ministry of Education, Culture, Sports, Science and Technology) and a Grant for Basic Science Research Projects from The Sumitomo Foundation.

Author contributions

S.Y. and T.T. designed this study. Y.N. and W.K. synthesized and characterized the gold cluster samples. K.N. developed a sample holder for low-temperature XAFS measurements. S.Y. and S.T. conducted the XAFS measurements. S.Y. analysed the experimental and theoretical spectra with the help of T.Y. All authors took part in the writing of this manuscript.

Additional information

Competing financial interests: The authors declare no competing financial interests.

23

A porous proton-relaying metal-organic framework material that accelerates electrochemical hydrogen evolution

Idan Hod[1,2], Pravas Deria[1], Wojciech Bury[1,3], Joseph E. Mondloch[1], Chung-Wei Kung[1,2,4], Monica So[1], Matthew D. Sampson[5], Aaron W. Peters[1,2], Cliff P. Kubiak[5], Omar K. Farha[1,2,6] & Joseph T. Hupp[1,2,7]

The availability of efficient hydrogen evolution reaction (HER) catalysts is of high importance for solar fuel technologies aimed at reducing future carbon emissions. Even though Pt electrodes are excellent HER electrocatalysts, commercialization of large-scale hydrogen production technology requires finding an equally efficient, low-cost, earth-abundant alternative. Here, high porosity, metal-organic framework (MOF) films have been used as scaffolds for the deposition of a Ni-S electrocatalyst. Compared with an MOF-free Ni-S, the resulting hybrid materials exhibit significantly enhanced performance for HER from aqueous acid, decreasing the kinetic overpotential by more than 200 mV at a benchmark current density of 10 mA cm^{-2}. Although the initial aim was to improve electrocatalytic activity by greatly boosting the active area of the Ni-S catalyst, the performance enhancements instead were found to arise primarily from the ability of the proton-conductive MOF to favourably modify the immediate chemical environment of the sulfide-based catalyst.

[1] Department of Chemistry, Northwestern University, 2145 Sheridan Road, Evanston, Illinois 60208, USA. [2] Argonne-Northwestern Solar Energy Research (ANSER) Center, Northwestern University, Evanston, Illinois 60208, USA. [3] Department of Chemistry, Warsaw University of Technology, Noakowskiego 3, 00-664 Warsaw, Poland. [4] Department of Chemical Engineering, National Taiwan University, Number 1, Sector 4, Roosevelt Road, Taipei 10617, Taiwan. [5] Department of Chemistry and Biochemistry, University of California San Diego, 9500 Gilman Drive, La Jolla, California 92093-0358, USA. [6] Faculty of Science, Department of Chemistry, King Abdulaziz University, Jeddah 21589, Saudi Arabia. [7] Argonne National Laboratory, 9700 South Cass Avenue, Argonne, Illinois 60439, USA. Correspondence and requests for materials should be addressed to O.K.F. (email: o-farha@northwestern.edu) or to J.T.H. (email: j-hupp@northwestern.edu).

Substantial effort is being invested in the design and development of efficient electrocatalytic hydrogen evolution technologies, as these hold promise for a future carbon-free energy economy[1–4]. In acid solutions, the hydrogen evolution reaction (HER) entails the electrochemical reduction of protons: $2H^+ + 2e^- \rightarrow H_2$. Catalysts are needed to decrease the otherwise enormous kinetic overpotential required to drive the reaction at suitably high current densities, *J*. For solar energy applications, where the corresponding oxidation reaction is the four-electron conversion of water to O_2, a commonly cited target *J* value is $10\,mA\,cm^{-2}$ (ref. 5). Provided that catalyst poisons are absent, selected noble metal electrodes (especially platinum electrodes) can support hydrogen evolution at high current densities under low kinetic overpotentials[6]. However, realization of large-scale hydrogen production requires the development of alternative low-cost electrocatalysts containing only highly abundant elements. A variety of candidate catalyst materials, including metal alloys[7], carbides[8], phosphides[9,10], borides[8] and nitrides[11], have shown promising HER activity and stability in acidic solutions. In addition, several metal sulfides and selenides (MS/MS_2/MSe/MSe_2 where M = Co, W, Mo, Fe and Ni)[12–16] have likewise recently been observed to display promising catalytic behaviour for hydrogen evolution. Nevertheless, substantial improvements in catalytic performance are needed to match the kinetic superiority of platinum electrodes.

Complementary to changing the composition of the electrocatalyst, a useful strategy for obtaining a significant leap in performance is to employ a porous supporting scaffold and create a composite electrode[17,18]. (A third approach, illustrated by various enzymes and their artificial mimics, is to modify the environment of the catalytically active site so as to provide proximal H-bond-donating or -accepting sites and/or proton-relay and -delivery channels)[19–21]. Porous scaffolds can boost an electrocatalyst's performance, that is, decrease overpotentials, by: (a) boosting the areal density of catalytically active surface sites, (b) improving the electronic conductivity of the electrode and/or (c) facilitating proton delivery. In addition, it has been shown that for some materials, certain crystal faces are much more catalytically active than others; the most striking examples perhaps are edge versus plane sites of layered, two-dimensional molybdenum or tungsten dichalcogenides[12,16,22]. By decreasing the lateral dimensions of crystallites constituting these catalysts, both a greater fraction and a greater absolute number of high-activity sites can be exposed[14]. The limit of this approach (not explored here) would be the synthesis of materials in small-cluster form, where the majority of exterior atoms, rather than only a tiny fraction, possesses the appropriate chemical coordination and geometric arrangement for high catalytic activity.

We reasoned that porous, electrode-surface-immobilized, metal-organic framework (MOF) crystallites might be ideal templates for synthesis of high-areal-density versions of inorganic electrocatalysts. In particular, we hypothesized that electrodeposition of nickel sulfide within the channels of the supported MOF might yield the desired composite or hybrid, nanostructured catalysts. Although underexplored as an HER electrocatalyst, especially in acidic environments, the electrodeposition of nickel sulfide is reasonably well understood[23,24]. Furthermore, a variety of sulfides of other first-row transition metals have been shown to be electrocatalytically competent, at modest kinetic overpotentials, for hydrogen evolution from water at low pHs[12–16]. As a candidate template or scaffold, we chose NU-1000 (ref. 25), a mesoporous MOF that can be grown solvothermally on conductive glass as a film comprising rod-shaped crystallites of a few microns length and a fraction of a micron width[26]. NU-1000 is one member of a growing family of porous MOFs featuring hexa-zirconium oxo/hydroxo/aquo clusters as nodes[27–31] and displaying exceptional resistance to hydrolysis[32–34], even at elevated temperature and/or in the presence of hydronium or hydroxide ions. Thus, this material was anticipated to be sufficiently chemically stable to function as a component of a hybrid HER electrocatalyst. It should be noted that various MOFs that anchor or encapsulate noble-metal nanoparticles or biomimetic molecular catalysts have recently been used in combination with molecular chromophores and sacrificial reductants (hole scavengers) to demonstrate photochemical hydrogen evolution[35–40].

Here we report on the electrochemical assembly of a surface immobilized hybrid of Ni-S (we intentionally use Ni-S (rather than NiS) to denote a material comprising nickel and sulfur, but of initially unspecified metal:non-metal stoichiometry; we subsequently specify more precisely the composition of the form of Ni-S electrodeposited here) and NU-1000, and on the performance of the hybrid as an HER electrocatalyst in water at pH 1. The hybrid assembly yields a benchmark cathodic current of $10\,mA\,cm^{-2}$ (together with H_2 bubbles) at iR-corrected overpotentials (η-values) as low as 238 mV. Under the same conditions, and at the same benchmark current density, the observed overpotential with flat, MOF-free Ni-S films as electrocatalysts is 443 mV.

Results

Electrocatalyst synthesis. Illustrations of the crystal structure of NU-1000, its $Zr_6(\mu_3\text{-}O)_4(\mu_3\text{-}OH)_4(OH)_4(OH_2)_4$ nodes[41] and its organic linkers are presented in Fig. 1. The crystal structure shows the existence of both triangular and hexagonal one-dimensional channels; together with smaller apertures (*ca.* 8 Å diameter) along the channel walls, they are responsible for the high porosity of the material (the diameters of the one-dimensional channels are *ca.* 10 and 31 Å. In addition, around 20% of the hexagonal pores contain an extra node at the hexagon centre. The extra node is connected to the rest of the MOF via additional $TBAPy^{4-}$ linkers. For simplicity, these are omitted from Fig. 1. N_2 adsorption measurements (bulk samples) show that the MOF void volume is *ca.* 71% of the total volume). Figure 2a shows a scanning electron microscopy (SEM) image of a NU-1000_Ni-S electrode (2 min of Ni-S deposition, see also Supplementary Fig. 1 for experimental conditions), featuring the hexagonal rod-shaped crystals that are typical for NU-1000, whereas Fig. 2b shows a cross-sectional SEM view of a NU-1000_Ni-S electrode. X-ray Photoelectron Spectroscopy (XPS) measurements of the NU-1000_Ni-S film (Supplementary Fig. 2) confirm the presence of both nickel and sulfur. The observed S 2*p* binding energy is *ca.* 162 eV, implying sulfur incorporation as S^{2-}. The presence of a peak at 168 eV is indicative of sulfur in a higher oxidation state, presumably oxy-sulfur species that might well arise from surface oxidation (air oxidation) of Ni-S[15]. The Ni 2*p* region shows a broad peak at a binding energy of centred around 860 eV, which can be attributed to Ni^{2+} as well as to other nickel oxidation states[42].

Elemental analysis of the hybrid films using both energy dispersive X-ray spectroscopy (EDS) and inductively coupled plasma-optical emission spectroscopy (ICP-OES), corroborate and quantify the conclusion that the hybrid films contain Ni and S. The EDS-obtained ratio of Ni:S was 1.7:1.0. ICP-OES yields a similar Ni:S ratio of 1.6:1.0, suggesting Ni_3S_2 as a primary deposition product (see Supplementary Table 1). EDS cross-sectional mapping of an MOF rod revealed, however, that only trace amount of nickel and sulfur are present in the extended portion of the MOF rods (Supplementary Fig. 3). Instead, these elements are confined and concentrated as a planar film, as illustrated in the bottom section of Fig. 3. As the MOFs

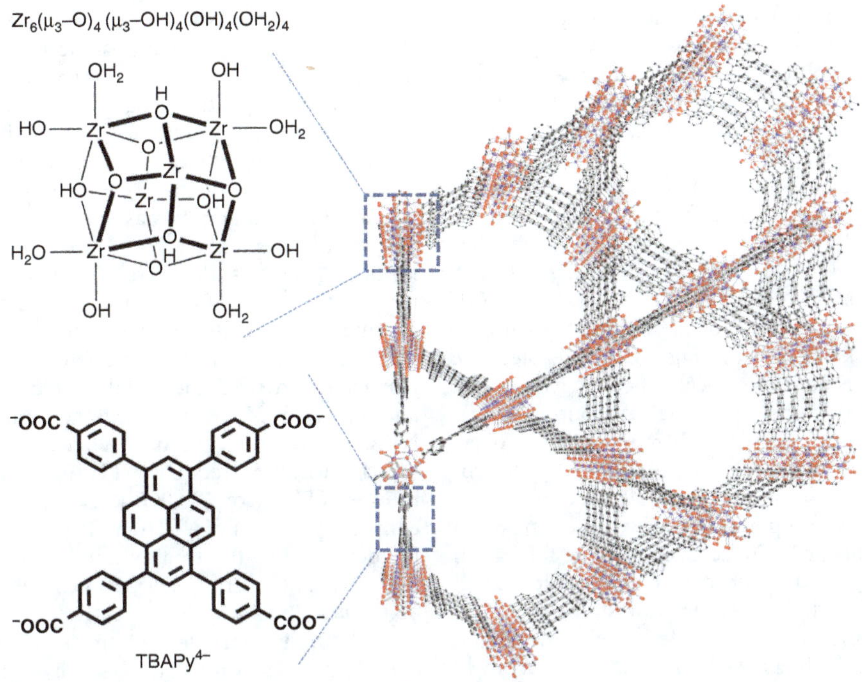

Figure 1 | Schematic representation. Schematic representation of NU-1000's crystal structure.

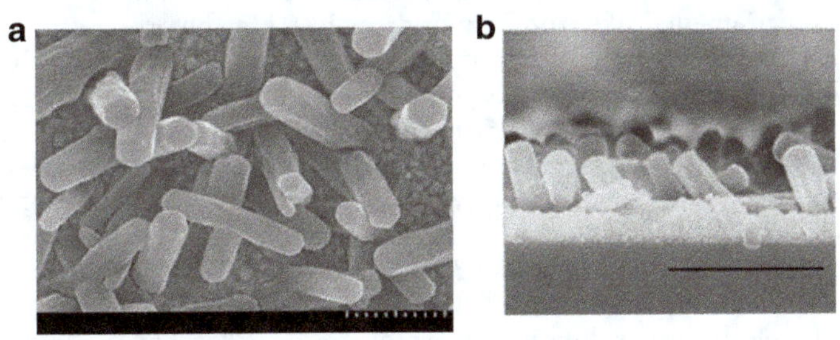

Figure 2 | SEM images of NU-1000_Ni-S films. (**a**) SEM image of an NU-1000_Ni-S film showing the typical hexagonal rod-shaped crystals of NU-1000 on top of the FTO substrate (scale bar, 1 μm). (**b**) Cross-sectional SEM image of NU-1000_Ni-S film (scale bar, 2 μm).

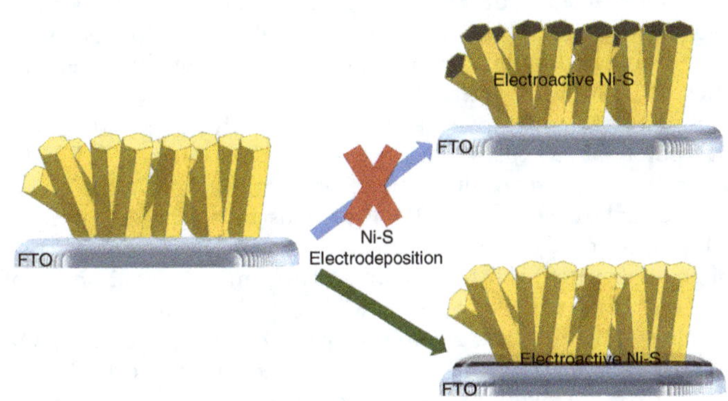

Figure 3 | Illustration of Ni-S electrodeposition to create the NU-1000_Ni-S hybrid system. The initial aim was to use the MOF as a template for depositing high surface area Ni-S rods (top section). However, Ni-S was deposited as a flat layer (bottom section).

themeslves are insulating in the region of potential where catalytic evolution of H_2 occurs, the MOF-distributed traces of nickel and sulfur are electrochemically inaccessible and, therefore, unable to participate in electrocatalysis (HER electrocatalysis by metal sulfides generally requires proximal pairs of sulfur atoms (ions). The loading of sulfur in the extended region of the MOF, however, is <0.5 sulfur ions per hexa-zirconium node. Thus, even if the MOF could be rendered conductive, it is unlikely that these sulfur atoms would be catalytic).

Powder X-ray diffraction (XRD) measurements of putative FTO_NU-1000 films confirm that the observed crystallites consist of NU-1000 (Supplementary Fig. 4). Notably, the MOF's crystal structure remains intact. However, no peaks attributable to nickel sulfide are observed, suggesting that the material deposited is amorphous, in agreement with previous reports[23].

More detailed information about the composition of the electrodeposited material was obtained via Raman spectroscopy. As can be seen in Supplementary Fig. 5, the Raman spectrum of a Ni-S MOF hybrid film contains the signature vibrational peaks of Ni_3S_2 (176, 190, 217, 300, 319 and 341 cm^{-1})[43]. In addition, it is important to note that small peaks are present, corresponding to NiS_2 (ref. 44) (276, 447 and 465 cm^{-1}) and NiS (379 cm^{-1})[45]. The Raman spectroscopy results combined with the EDS and ICP-OES elemental analysis establish that the main electrodeposition product is Ni_3S_2.

HER catalysis. As suggested by the top portion of Fig. 3, our initial aim was simply to boost the metal-sulfide electrode surface area by using the MOF as a porous template for the electrodeposition of mesoscale nickel sulfide rods. In contrast to our expectations, however, the MOF did not function as a physical template and, consequently, did not increase the electrode surface area. Instead, Ni-S deposited as a flat film that in-filled the bottoms of the MOF rods, while coating both the bare conductive glass platform (fluorine-doped tin oxide, FTO) and the portions of FTO residing at the bottom of each MOF channel (see Fig. 3).

Notably, these channels are wide enough to readily allow permeation by the electrolyte solution in a hydrogen-evolution cell. As such, the MOF channels help to define the local chemical environment of the portion of the deposited Ni-S that is in contact with the electrolyte solution. As will be discussed further below, it is this local environment that alters the catalytic activity of the metal-sulfide electrode and yields the observed substantial decrease in HER overpotential.

To assess the electrocatalytic properties of the hybrid assembly towards hydrogen evolution, we studied four types of electrodes: bare FTO, a MOF scaffold grown on FTO substrate (FTO_NU-1000), Ni-S deposited on FTO substrate (FTO_Ni-S) and Ni-S deposited on FTO-supported NU-1000 (NU-1000_Ni-S). *J–V* curves and Tafel plots (*V* versus log *J*) of the different systems in aqueous 0.1 M HCl (pH 1) are shown in Fig. 4a. As expected, the bare FTO electrode exhibits poor electrocatalytic activity towards the HER, with a high reaction onset potential of around 600 mV versus Reversible Hydrogen Electrode (RHE) and an overpotential of *ca.* 1 V at $J = 10\ mA\ cm^{-2}$. For Ni-S directly electrodeposited on FTO (only), the overpotential at $10\ mA\ cm^{-2}$ is *ca.* 560 mV lower, clearly illustrating its ability to function as an electrocatalyst. Addition of the MOF scaffold, that is, formation of the hybrid assembly, further decreases the overpotential $10\ mA\ cm^{-2}$ to 238 mV (this was the lowest iR-corrected overpotential observed based on several electrodes. The values occasionally ranged as a high as 280 mV). Finally, a control experiment with FTO_NU-1000, but no metal sulfide, yields a kinetic overpotential at $10\ mA\ cm^{-2}$ of *ca.* 640 mV. Although the value in isolation is unremarkable, the *ca.* 360 mV decrease in η relative to the MOF-free bare FTO electrode is both substantial and unexpected. The isolated MOF contains no components capable of redox mediation and there are no obvious sites for adsorption and stabilization of potential reaction intermediates such as neutral hydrogen atoms.

Returning to the hybrid Ni-S catalyst, the observed kinetic overpotential at $10\ mA\ cm^{-2}$ compares reasonably well with recently reported results for several other non-noble metal electrocatalysts for the HER in aqueous acid. For example, an NiS_2 electrocatalyst showed only $4\ mA\ cm^{-2}$ at around 240 mV of overpotential, Ni_3S_2 exhibit $10\ mA\ cm^{-2}$ at 213 mV of overpotential, whereas CoS_2 and FeS_2 yielded $10\ mA\ cm^{-2}$ at η-values of *ca.* 230 and 260 mV, respectively (all in 0.5 M H_2SO_4, pH 0; it is noteworthy that pH 1 was used in this work)[13,24]. In neutral (pH 7) solutions, η-values of <160 mV, at $10\ mA\ cm^{-2}$, have been reported for electrodeposited and subsequently annealed CoS[46].

The Ni-S mass loading on NU-1000_Ni-S films during electrodeposition was measured *in situ* using electrochemical quartz crystal microbalance (EQCM) techniques (see Methods section). As can be seen in Supplementary Fig. 6a, after 2 min of electrodeposition, Ni-S loading in NU-1000_Ni-S is $28\ \mu g\ cm^{-2}$ ($0.116\ \mu mol\ cm^{-2}$). Potentiostatic electrolysis of NU-1000_Ni-S for a period of over 2 h at overpotential of 210 mV was done, while recording the charge flowing through the system (Supplementary Fig. 6b). From this charge, we have been able to estimate the amount of H_2 formed during the electrolysis (taking into account the H_2 Faradic yield of 93%, as determined by gas chromatography based on electrolysis at $2\ mA\ cm^{-2}$ for 3 h; see Supplementary Fig. 7). We note that following previous reports, we suggest that Faradaic efficiency falls short of 100% due to H_2 bubble formation on the surface of the electrode[15,47]. If we assume that only the portion of the metal sulfide that is in contact with the solution is catalytic (that is, the electrolyte does not permeate and swell the metal sulfide), and if we assume that an adjacent pair of nickel-coordinated sulfide ion constitutes one HER catalytic site, we obtain an active site normalized (upper limit) turnover frequency (TOF) of $34\ s^{-1}$ and a turnover number (TON) of 273,000 (based on 8,000 s of electrolysis at a constant potential of − 210 mV versus RHE; see Methods section for details). For the total amount of Ni-S in the NU-1000_Ni-S film ($0.116\ \mu mol\ cm^{-2}$), TOF and TON values were calculated to be $0.208\ s^{-1}$ and 1668, respectively.

As a preliminary test of the stability of the NU-1000_Ni-S combination, galvanostatic electrolysis measurements were conducted at $10\ mA\ cm^{-2}$, in aqueous HCl at pH 1 (Fig. 5a). Notably, the overpotential needed to produce the imposed current remains constant (± 20 mV) for 2 h of measurement, indicating that the NU-1000_Ni-S system can sustain its catalytic activity for relatively long periods of time. Power XRD measurements taken on a hybrid-catalyst film before and after 2 h of galvanostatic electrolysis at $10\ mA\ cm^{-2}$ show that the MOF scaffold retains its crystallinity under electrocatalysis working conditions (Fig. 5b). To further assess the stability of the hybrid film during catalysis, we recorded a ultraviolet–visible spectrum of the pH 1 electrolyte solution after 2 h of galvanostatic electrolysis at $10\ mA\ cm^{-2}$ (Supplementary Fig. 8), monitoring the wavelengths at which 1,3,6,8-tetrakis(p-benzoic acid)pyrene (H_4TBAPy) linkers exhibit maximum absorbance (~390 nm). The electrolysis solution displays no traces of leached NU-1000 linker, indicating that the MOF remains intact over the course of 2 h of catalysis. (For comparison, a spectrum of a digested NU-1000_Ni-S film is presented, showing a peak at 390 nm attributable to the linker.)

Origin(s) of enhanced catalytic performance by hybrid assemblies. It is well known that in certain cases Tafel slopes can serve as indicators of the rate-determining step in the HER, as well as

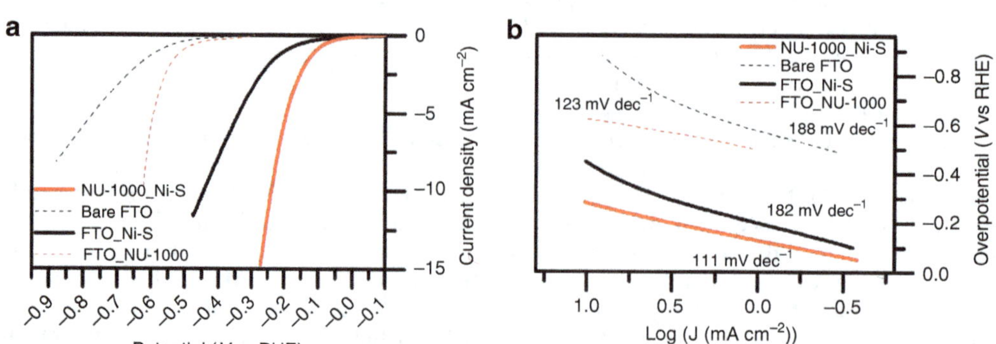

Figure 4 | Comparison between the electrocatalytic HER performance. Four types of electrodes: bare FTO (dotted black), FTO_NU-1000 (dotted red), FTO_Ni-S (black) and NU-1000_Ni-S (red). (**a**) *J-V* curves. (**b**) Tafel plots. The listed values of 182 and 188 mV dec^{-1} for the HER Tafel slopes of the bare FTO and FTO_Ni-S samples refer to the low current density region only. The slopes clearly are larger at higher current densities.

indicate the existence of other kinetically relevant effects[48]. Figure 4b shows Tafel plots (V versus log J plots) for the four types of electrodes examined here. For the MOF-free electrodes, with or without Ni-S, the slopes of the plots are near 180 mV per decade of current density, whereas for the NU-1000-functionalized electrodes the slopes are near 120 mV dec^{-1}. Values substantially >120 mV typically are indicative of the presence of an uncompensated resistive element—for example, the resistance of the catalyst itself[49].

If we assume that in the presence of the MOF scaffold, complicating factors are absent and the Tafel slope reflects only interfacial kinetics, the observed values of *ca.* 120 mV dec^{-1} imply that the HER is governed by either the Volmer–Heyrovsky or Volmer–Tafel mechanism (see Supplementary Note 1 for HER mechanistic details). Both mechanisms involve rate-limiting reduction of a proton to yield a catalyst-adsorbed H atom. For the Volmer–Heyrovsky mechanism, this step is followed by fast electrochemical reduction of a second proton and concomitant formation of an H–H bond. For the Volmer–Tafel pathway, the initial step is followed by fast formation of an H–H bond between a pair of surface-bound hydrogen atoms to form H_2. It is noteworthy that the latter mechanism requires the catalyst to offer an immediately proximal pair of H-atom adsorption sites—for example, a pair of surface-exposed sulfide ions.

Still to be answered is why MOF-Ni-S hybrid catalyst formation decreases the overpotential for Ni-S-catalysed H_2 evolution at 10 mA cm^{-2} by more than 200 mV. Our initial hypothesis, dispelled by SEM and EDS results mentioned above, was that catalyst electrodeposition within the MOF scaffold would greatly boost its surface area. Nevertheless, the presence of the MOF might still serve to roughen the electrodeposited film and therby increase its effective surface area. Cyclic voltammetry (CV) measurements of electrochemical currents in the voltage range positive of the hydrogen evolution region, that is, capacitive currents associated with electrical double-layer charging, are expected to scale as the solution-accessible surface area of the electrocatalyst[50]. Comparison of these currents over a range of voltammetric sweep rates (Fig. 6c) indicates remarkably similar surface areas. Indeed, the electroactive surface area for the hybrid assembly is only about 1.5 times greater than that for the simple Ni-S film, that is, far too little to account for the change in overpotential.

If the number of electrocatalytically active sites is little changed by introduction of the MOF scaffold, we are left with alterations in the local reaction environment as the most probable source of the observed reactivity enhancement. The putative local environment effects could conceivably take the form of electronic modulation of the properties of Ni-S, perhaps involving the creation of highly catalytic sites at MOF-engendered grain boundaries. Such effects could be especially important for two-dimensional layered compounds such as MoS_2, but seem less likely to enormously influence the activity of an amorphous, three-dimensional material such as Ni-S.

To gain insight into the possible importance of electronic effects on the hybrid system HER performance, we employed electrochemical impedance spectroscopy with FTO_Ni-S and NU-1000_Ni-S electrodes, taken under HER working conditions. Nyquist plots show a resistance element which is attributed to solution voltage drops, in series with one semicircle

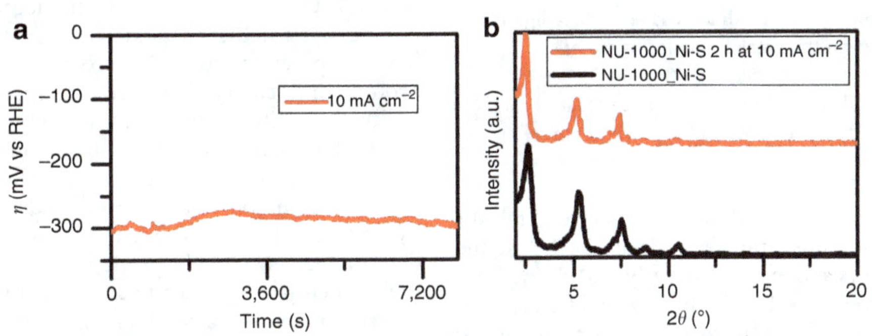

Figure 5 | Stability analysis of NU-1000_Ni-S under HER working conditions. (**a**) Galvanostatic electrolysis measurement of NU-1000_Ni-S system under a constant current of 10 mA cm^{-2}, showing the voltage stability of the system for more than 2 h. (**b**) Powder XRD plots for an NU-1000_Ni-S film before and after galvanostatic electrolysis measurement under a constant current of 10 mA cm^{-2} for 2 h. The plots show that framework crystallinity is retained.

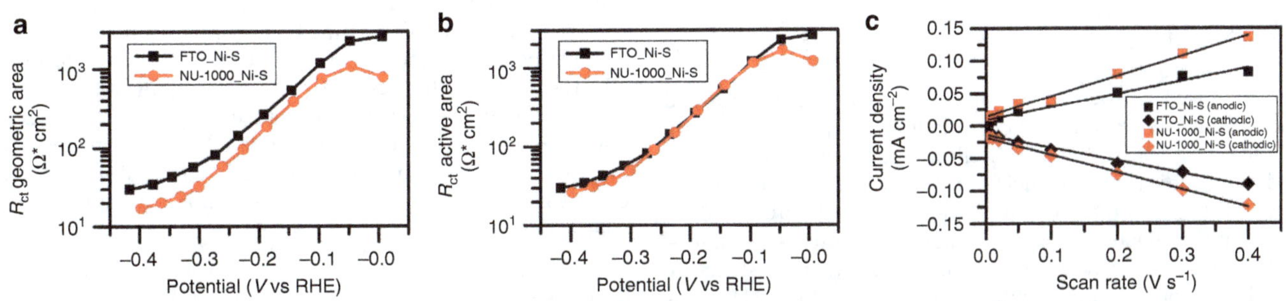

Figure 6 | EIS analysis of the MOF's scaffold impact on the electronic properties of Ni-S. (**a**) A plot of geometric area normalized charge transfer resistance (R_{ct}) versus applied potential, comparing between FTO_NiS and NU-1000_Ni-S electrodes. (**b**) A plot of electroactive surface area normalized R_{ct} versus applied potential, exhibiting similar R_{ct} values for both FTO_Ni-S and NU-1000_Ni-S electrodes. (**c**) A plot of double-layer capacitive current versus CV scan rate. The slope of each curve is relative to the amount of electroactive surface area of the system.

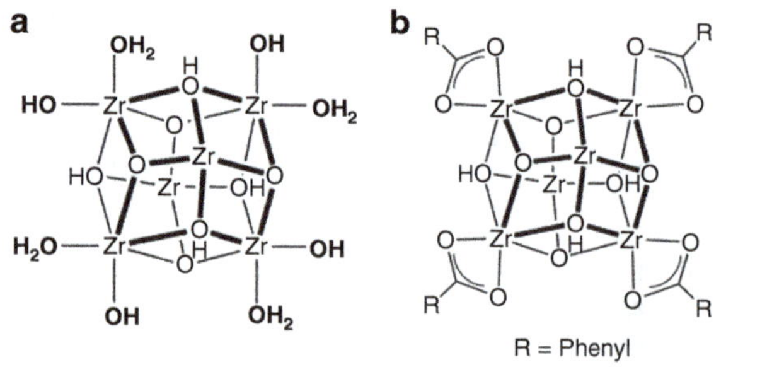

Figure 7 | Effect of Zr_6 based node's terminal -OH and -OH_2 ligands on HER activity. (**a**) NU-1000's Zr_6 based node, containing the terminal -OH and -OH_2 ligands (marked in red). (**b**) Schematic representation of a benzoate-modified NU-1000 Zr_6 based node. The benzoate replaces the node's terminal -OH and -OH_2 ligands. Omitted, for simplicity, are atoms associated with coordinated linkers.

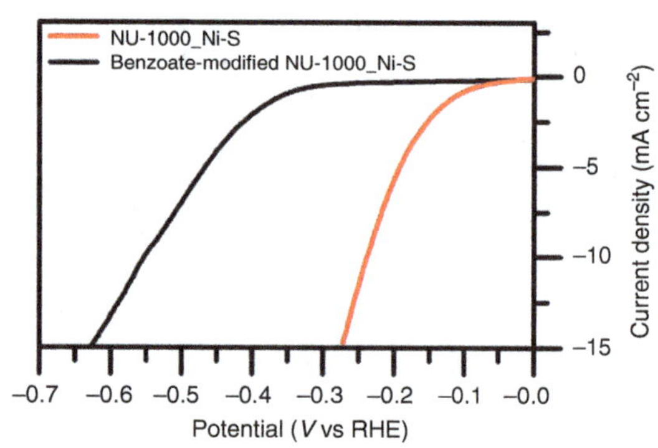

Figure 8 | Effect of Zr_6 based node's terminal -OH and -OH_2 ligands, *J-V* curves. Comparison between the catalytic performance of NU-1000_Ni-S (red) and benzoate-modified NU-1000_Ni-S (black). Substitution of the terminal -OH and -OH_2 ligands by benzoate ligands significantly reduces the HER catalytic activity of the hybrid film.

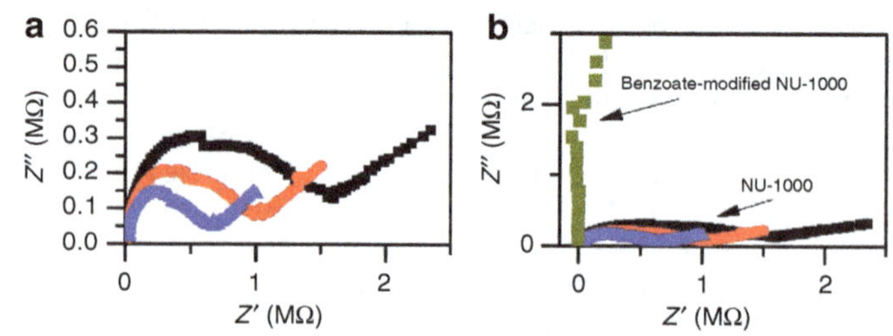

Figure 9 | Proton conductivity measuremets. Nyquist plots of AC impedance measurements of proton conductivity of (**a**) NU-1000 in contact with humid air. As exposure time increases, the arc size decreases. (**b**) Comparison between Nyquist plots of NU-1000 and benzoate-modified NU-1000. The observed large resistance for the benzoate-modified sample is equivalent to low conductivity.

corresponding to the parallel contribution of both catalyst's chemical capacitance and charge transfer resistance (R_{ct}) at the catalyst/solution interface (see Supplementary Fig. 9)[15]. Figure 6 presents plots of log R_{ct} versus V for the two types of electrodes without (Fig. 6a) and with (Fig. 6b) correction for the minor difference in catalyst surface area. The plots show in striking manner that the interfacial electron-transfer step, in isolation, is not influenced by the presence of the MOF scaffold.

We next considered the possibility that catalytically important modification of the active-site environment could be associated with the solution side of the interface. The aquo- and hydroxo-rich nodes of the MOF (see Fig. 7a) could, for example, alter the local proton activity, as distinct from the bulk solution activity, or the MOF's nodes might facilitate local proton delivery and/or long-range proton transport.

Although we lack specific insight into about how the aquo- and hydroxo-functionalized MOF node might assist Ni-S in catalysing the HER, we reasoned that largely eliminating these ligands could provide an indication of their importance. We assembled a variant of NU-1000_Ni-S using 'as-synthesized' NU-1000 in which benzoate ligands replace the node's terminal -OH and -OH_2 ligands[25,41,51,52] (see Fig. 7b for Zr_6-based node structure illustration and Supplementary Fig. 10 for ^{1}H-NMR characterization of benzoate-modified NU-1000, showing the incorporation of four benzoates per Zr_6 node). As shown by the *J–V* comparisons in Fig. 8, benzoate substitution eliminates the co-catalytic behaviour of the MOF scaffold, now requiring an overpotential of 553 mV at 10 mA cm^{-2}. SEM images (top view and cross-section) of the benzoate-modified NU-1000 show that the MOF's crystal morphology has not changed and the film exhibits similar inter-rod spacing for the electrodeposition of Ni-S (Supplementary Fig. 11a). In addition, EQCM data shown on Supplementary Fig. 11b reveal similar mass loadings of Ni-S in the benzoate-modified NU-1000 film (26.7 μg cm^{-2}) compared with aquo- and hydroxo-functionalized NU-1000 (28 μg cm^{-2}). The results in Fig. 8 clearly demonstrate the importance of the node's terminal -OH and -OH_2 ligands to the HER performance of the hybrid system. As one would anticipate, re-installing benzoate ligands on the MOF nodes and thereby displacing node-coordinated aqua and hydroxo ligands[51–53] reverse the catalytic enhancement and yield HER *J–V* behaviour similar to that of the benzoate-containing 'as-synthesized' version of the hybrid metal-sulfide/MOF system; see Supplementary Fig. 12.

To gauge whether the aquo-ligated version of the MOF scaffold has the potential to assist proton transport, we examined its proton conductivity via impedance spectroscopy during and after infiltration with water from a high humidity atmosphere. A substantial literature on proton conductivity in MOFs now exists[54–57]. One means of engendering conductivity in MOFs is by the presentation of Bronsted acids or bases to an infiltrating hydroxylic solvent. Although pK_a and pK_b values have yet to be determined for the poly-protic nodes of NU-1000, clearly there is the potential for acid/base reactivity. Figure 9a shows representative Nyquist plots for a measurement of a pellet of NU-1000. The plots are characterized by a single arc, attributable to proton conductivity. The diameter of the arc decreases with time, reflecting progressive uptake of water and indicating increasing conductivity. The conductivity obtained at equilibrium is 2×10^{-7} S cm^{-1}. Although of only modest magnitude, the observed conductivity implies the ability of the scaffold to assist in proton delivery to proximal catalytic sites. The extent to which such assistance is of value in aqueous 0.1 M HCl remains to be determined. In comparison with NU-1000, Nyquist plots for a measurement of a pellet of benzoate-modified NU-1000 (Fig. 9b) revealed such a significant increase in the arc's diameter (larger resistance or smaller conductivity) that no meaninful data fitting could be obtained. These measurements confirm that the node's terminal -OH and/or -OH_2 ligands are essential for obtaining measurable proton conductivity.

Discussion

In conclusion, hybrid electrocatalysts consisting of electrodeposited Ni-S and solvothermally grown, electrode-supported NU-1000, a mesoporous, acid-stable MOF, display very good activity for hydrogen evolution. In aqueous HCl at pH 1, the hybrid material can deliver a catalytic current of 10 mA cm^{-2} at an iR-corrected overpotential of just 238 mV—a sizable (*ca.* 200 mV) decrease in overpotential relative to MOF-free Ni-S films as electrocatalysts. The hybrid probably catalyses the reaction via the Volmer–Tafel mechanism, as evidenced in part by Tafel slopes near 120 mV dec^{-1} for plots of log *J* versus overpotential. Despite its substantial porosity and high surface area, the MOF scaffold does not exert its co-catalytic effect by boosting the electroactive surface area of the subsequently deposited nickel sulfide. Instead, it serves to modify the immediate environment of the electrocatalyst, rendering it more favourable for local proton delivery and/or transport. Thus, the aquo- and hydroxo-rich nodes of the MOF appear to contribute in a manner that is reminiscent of organic acids or bases in proximity to metal-ion-based cofactors in enzymes that catalyse proton-coupled electron-transfer reactions, a concept that has likewise been exploited in the design of abiotic molecular catalysts for both hydrogen[58–61] and oxygen[62] evolution, as well as CO_2 reduction[63].

Methods

Chemicals. All chemicals, benzoic acid (Aldrich, 99.5%), zirconyl chloride octahydrate (Aldrich, 98%), *N,N*-dimethylformamide (DMF) (Macron, 99.8%), hydrochloric acid (HCl, 36.5%–38.0%, Macron), acetone (Macron), $NiCl_2$ (Aldrich, 98%) and thiourea (Sigma, 99%) were used as received without further purification. Deionized water was used throughout the work. The chemicals used for the synthesis of the H_4TBAPy linkers were the same as those reported in our previous work[25].

Instrumentation. Thin-film XRD patterns were measured on a Rigaku ATX-G thin-film diffraction workstation.

SEM images and EDS mapping were measured on a Hitachi SU8030 SEM.

For ICP-OES experiments, two samples of the NU-1000_Ni-S thin film were scraped from their substrates and collected into a microwave vial (4 ml). Then, 0.25 ml concentrated H_2O_2 and 0.75 ml concentrated H_2SO_4 were added. The vial was capped and irradiated in a microwave oven at 150 °C for 5 min. The resultant clear solution was diluted to 25 ml with nanopure water and analysed via ICP-OES (Varian Vista MPX instrument). Ni, S and Zr concentration were calculated from external stock solutions.

Raman spectroscopy measurements were made using an Acton TriVista Confocal Raman Spectroscopy System. Sample irradiation was done with a 514.5-nm laser. The acquisition time was 60 s and the reported spectrum was obtained by averaging ten runs.

All CV and impedance spectroscopy experiments were performed on a Solarton Analytical Modulab Potentiostat equipped with an FRA Impedance module. A three-electrode electrochemical setup was used, with a platinum mesh counter electrode and Ag/AgCl/KCl (sat'd) electrode as reference electrode. Electrochemical data were measured in aqueous 0.1 M HCl (pH 1) solutions and were adjusted to RHE scale by adding (0.197 + 0.059 × pH) V to the measured potential. iR corrections were made to the obtained *i–V* curves according to the series resistances measured on the same electrochemical setup using impedance spectroscopy.

EQCM experiments were conducted with a MAXTEK RQCM system. The mass loading of Ni-S during electrodeposition was measured on a NU-1000-covered (by the recently described Electrophoretic deposition), tin-oxide-coated gold QCM crystal (6 MHz).

The Faradiac efficiency for H_2 was determined by galvanostatic electrolysis at 2 mA cm^{-2} for 3 h. The experiment was carried out in a 60-ml Gamry 5-neck cell equipped with 3 Ace-Thred ports to hold each electrode and two joints capable of being sealed with septa for gas sparging. This setup included the NU-1000_Ni-S as the working electrode (*ca.* 0.5 cm^2 surface area), a Pt wire counter electrode (flame annealed with a butane torch before use and separated from the bulk solution by fine glass frit) and a Ag/AgCl reference electrode (leakless assembly, eDAQ). Outside of the electrolyte solution, a bare portion of the FTO working electrode was attached to a Cu wire by a minimal amount of non-conductive, chemically resistant epoxy, to attach the FTO glass to the potentiostat leads. A BASi Epsilon potentiostat was used to apply constant current and record potential. The electrolysis experiment was carried out in 30 ml of total electrolyte solution (0.1 M HCl in water). The solution was constantly stirred throughout the experiment. Gas analysis was performed using 1 ml sample injections on a Hewlett-Packard 7890A Series gas chromatograph with two molsieve columns (30 m × 0.53 mm ID × 25 μm film). Gas chromatography calibration curve were made by sampling known volumes of H_2 gas.

Upper limit TON and TOF were estimated in the following manner: we assume that only the portion of the metal sulfide that is in contact with the solution is catalytic (that is, the electrolyte does not permeate and swell the metal sulfide). In addition, we assume that an adjacent pair of nickel-coordinated sulfide ions in the Ni_3S_2 catalyst constitutes one HER catalytic site. Using the reported sulfur to sulfur distance in Ni_3S_2 (3.5 Å) as well as the ionic radius of S (1.7 Å), a rough estimate of catalytic site area in cm^2 was made (assuming a rectangular shaped site):

Catalytic site length $(3.5 \times 10^{-8} + (1.7 \times 10^{-8} \times 2)) \times$ catalytic site width $(1.7 \times 10^{-8} \times 2) =$ catalytic site area (2.34×10^{-15} cm^2). The number of catalytic sites per 1 cm^2 is $4.26 \times 10^{+14}$; hence, there are 7.07×10^{-10} moles of catalytic sites per 1 cm^2. The number of moles of produced H_2 during 8,000 s of electrolysis (taking into account 93% H_2 Faradaic efficiency) was calculated to be 1.94×10^{-4}.

As a result, upper limit TON = moles of H_2/moles of catalytic sites (273,000). Upper limit TOF = upper limit TON/time of electrolysis (34 s^{-1}).

TON and TOF values based on the total amount of Ni-S in the film (28 μg cm^{-2}, 1.16×10^{-7} moles) are 1,668 and 0.208, respectively.

Impedance spectroscopy measurements under HER working conditions were made using an AC voltage of 20 mV, with a frequency range of 500 kHz–50 mHz.

Growth of NU-1000 thin films. The growth of NU-1000 films on glass-supported, transparent and electronically conductive, FTO electrodes ('FTO_NU-1000') was done using a slightly modified version of a previously reported solvothermal route[26]. The FTO glass substrate (15 Ω^{-2}, Hartford Glass), with a size of 2.5 × 1.25 cm, was washed in soapy water, ethanol and acetone by sonication for 15 min sequentially. Thereafter, the substrate was dried and soaked in a solution of 0.5 mM H_4TBAPy in DMF at room temperature for 12 h. The detailed synthesis of the H_4TBAPy has been reported in our previous work[1]. The substrate was then cleaned with DMF and dried. Benzoic acid (2.7 g) and 105 mg of zirconyl chloride octahydrate were added into 8 ml of DMF and ultrasonically dissolved in a 20-ml screw-thread sample vial (Cole-Parmer, 28 mm × 57 mm), equipped with a urea cap and polytetrafluoroethylene (PTFE) foam-backed liner. Thereafter, the solution was placed into an oven at 80 °C for 2 h. After cooling down the solution to room temperature, 40 mg of H_4TBAPy was added into this solution and the mixture was sonicated for 20 min. The as-prepared FTO substrate was then placed into the solution, with the conducting side facing down to the bottom. Subsequently, the vial was placed on the bottom of a gravity convection oven (VWR symphony) with the temperature set at 90 °C; the bottom of the oven provided a temperature gradient inside the vial, which is required for the growth of the MOF thin film. The vial was taken out from the oven after 13 h of reaction and the FTO substrate was taken out from the vial. After removing all the precipitations on the back side of the substrate, the substrate was washed with DMF; a uniform pale yellow MOF thin film could be observed on the front side of the FTO substrate. Benzoates (modulator) coordinated in the obtained MOF thin film was then removed by the following activation process: 0.5 ml of 8 M hydrochloric acid aqueous solution was mixed with 13 ml of DMF; 0.05 ml of the obtained solution was then mixed with 49.95 ml of DMF to form the diluted acidic solution. The MOF thin film was then soaked in the diluted acidic solution in a 100 °C oven for 4 days. Then, the film was washed with acetone for several times and soaked in acetone for 1 day. After drying the film in air, the FTO_NU-1000 thin film was then obtained. In a few cases, the HCl treatment was omitted and the benzoate-coordinating version of the material was used instead.

Electrodeposition of Ni-S. Potentiostatic electrodeposition of Ni-S on FTO_NU-1000 ('NU-1000_Ni-S') and bare FTO ('FTO_Ni-S') films was conducted according to a slightly modified version of a previously published procedure[23]. Briefly, an aqueous solution containing 10 mM $NiCl_2$ and 0.5 M thiourea as nickel and sulfur sources, respectively, was used as the deposition bath. The electrodeposition was carried out in a standard three-electrode setup containing either FTO_NU-1000 or bare FTO as a working electrode, Ag/AgCl as a reference electrode and FTO as a counter electrode. To determine the potential at which the deposition should be done, CV measurements of the deposition solution with and without a sulfur source were recorded (see Supplementary Fig. 1). It is clear that on addition of the sulfur source, the reduction peak for Ni^{2+}/Ni^0 (Ni deposition) is shifted to more anodic potentials, from − 1.35 to − 1.1 V versus Ag/AgCl, where the shift is a result of the reaction between the deposited Ni and thiourea (sulfur source) to generate Ni-S[23]. As a consequence, all subsequent Ni-S electrodepositions were carried out at a fixed potential of − 1.1 V versus Ag/AgCl.

Proton conductivity measurements. To measure the MOF's proton conductivity, disk pellets of NU-1000 (only) or benzoate modified NU-1000 were prepared, with dimensions of 7 mm diameter and 3 mm thickness. Each side of the pellet was coated with conductive silver epoxy, which was used to anchor a pair of tin-coated copper wires as electrical contacts. Then, the pellet was placed in an oven set to 60 °C for half an hour, to cure the epoxy. The pellet was exposed to H_2O vapour at ambient temperature and its impedance spectroscopy response based on an AC signal of 20 mV was recorded over the frequency range of 500 kHz to 0.5 Hz.

References

1. Meyer, T. J. Chemical approaches to artificial photosynthesis. *Accounts. Chem. Res.* **22,** 163–170 (1989).
2. Lewis, N. S. & Nocera, D. G. Powering the planet: chemical challenges in solar energy utilization. *Proc. Natl Acad. Sci. USA* **103,** 15729–15735 (2006).

3. Turner, J. A. Sustainable hydrogen production. *Science* **305,** 972–974 (2004).
4. Gray, H. B. Powering the planet with solar fuel. *Nat. Chem.* **1,** 7–7 (2009).
5. Walter, M. G. *et al.* Solar water splitting cells. *Chem. Rev.* **110,** 6446–6473 (2010).
6. Merki, D. & Hu, X. L. Recent developments of molybdenum and tungsten sulfides as hydrogen evolution catalysts. *Energ. Environ. Sci.* **4,** 3878–3888 (2011).
7. Brown, D. E., Mahmood, M. N., Turner, A. K., Hall, S. M. & Fogarty, P. O. Low overvoltage electrocatalysts for hydrogen evolving electrodes. *Int. J. Hydrogen Energ.* **7,** 405–410 (1982).
8. Vrubel, H. & Hu, X. L. Molybdenum boride and carbide catalyze hydrogen evolution in both acidic and basic solutions. *Angew. Chem. Int. Ed.* **51,** 12703–12706 (2012).
9. Popczun, E. J. *et al.* Nanostructured nickel phosphide as an electrocatalyst for the hydrogen evolution reaction. *J. Am. Chem. Soc.* **135,** 9267–9270 (2013).
10. Feng, L. G., Vrubel, H., Bensimon, M. & Hu, X. L. Easily-prepared dinickel phosphide (Ni_2P) nanoparticles as an efficient and robust electrocatalyst for hydrogen evolution. *Phys. Chem. Chem. Phys.* **16,** 5917–5921 (2014).
11. Chen, W. F. *et al.* Hydrogen-evolution catalysts based on non-noble metal nickel-molybdenum nitride nanosheets. *Angew. Chem. Int. Ed.* **51,** 6131–6135 (2012).
12. Jaramillo, T. F. *et al.* Identification of active edge sites for electrochemical H-2 evolution from MoS_2 nanocatalysts. *Science* **317,** 100–102 (2007).
13. Kong, D. S., Cha, J. J., Wang, H. T., Lee, H. R. & Cui, Y. First-row transition metal dichalcogenide catalysts for hydrogen evolution reaction. *Energ. Environ. Sci.* **6,** 3553–3558 (2013).
14. Kibsgaard, J., Chen, Z. B., Reinecke, B. N. & Jaramillo, T. F. Engineering the surface structure of MoS_2 to preferentially expose active edge sites for electrocatalysis. *Nat. Mater.* **11,** 963–969 (2012).
15. Tran, P. D. *et al.* Novel cobalt/nickel-tungsten-sulfide catalysts for electrocatalytic hydrogen generation from water. *Energ. Environ. Sci.* **6,** 2452–2459 (2013).
16. Voiry, D. *et al.* Enhanced catalytic activity in strained chemically exfoliated WS_2 nanosheets for hydrogen evolution. *Nat. Mater.* **12,** 850–855 (2013).
17. Li, Y. G. *et al.* MoS_2 nanoparticles grown on graphene: an advanced catalyst for the hydrogen evolution reaction. *J. Am. Chem. Soc.* **133,** 7296–7299 (2011).
18. Hinnemann, B. *et al.* Biornimetic hydrogen evolution: MoS_2 nanoparticles as catalyst for hydrogen evolution. *J. Am. Chem. Soc.* **127,** 5308–5309 (2005).
19. Layfield, J. P. & Hammes-Schiffer, S. Hydrogen tunneling in enzymes and biomimetic models. *Chem. Rev.* **114,** 3466–3494 (2014).
20. Cornish, A. J., Gartner, K., Yang, H., Peters, J. W. & Hegg, E. L. Mechanism of proton transfer in [FeFe]-hydrogenase from *Clostridium pasteurianum*. *J. Biol. Chem.* **286,** 38341–38347 (2011).
21. Thorp, H. H., Sarneski, J. E., Brudvig, G. W. & Crabtree, R. H. Proton-coupled electron-transfer in $[(Bpy)2mn(O)2mn(Bpy)2]^{3+}$. *J. Am. Chem. Soc.* **111,** 9249–9250 (1989).
22. Bollinger, M. V. *et al.* One-dimensional metallic edge states in MoS_2. *Phys. Rev. Lett.* **87,** 196803–196804 (2001).
23. Sun, H. C. *et al.* Dye-sensitized solar cells with NiS counter electrodes electrodeposited by a potential reversal technique. *Energ. Environ. Sci.* **4,** 2630–2637 (2011).
24. Jiang, N. *et al.* Electrodeposited nickel-sulfide films as competent hydrogen evolution catalysts in neutral water. *J. Mater. Chem. A* **2,** 19407–19414 (2014).
25. Mondloch, J. E. *et al.* Vapor-phase metalation by atomic layer deposition in a metal-organic framework. *J. Am. Chem. Soc.* **135,** 10294–10297 (2013).
26. Kung, C. W. *et al.* Metal-organic framework thin films composed of free-standing acicular nanorods exhibiting reversible electrochromism. *Chem. Mater.* **25,** 5012–5017 (2013).
27. Katz, M. J. *et al.* A facile synthesis of UiO-66, UiO-67 and their derivatives. *Chem. Commun.* **49,** 9449–9451 (2013).
28. Katz, M. J. *et al.* Simple and compelling biomimetic metal-organic framework catalyst for the degradation of nerve agent simulants. *Angew. Chem. Int. Ed.* **53,** 497–501 (2014).
29. DeCoste, J. B. *et al.* Stability and degradation mechanisms of metal-organic frameworks containing the Zr6O4(OH)(4) secondary building unit. *J. Mater. Chem. A* **1,** 5642–5650 (2013).
30. Furukawa, H. *et al.* Water adsorption in porous metal-organic frameworks and related materials. *J. Am. Chem. Soc.* **136,** 4369–4381 (2014).
31. Morris, W. *et al.* Synthesis, structure, and metalation of two new highly porous zirconium metal-organic frameworks. *Inorg. Chem.* **51,** 6443–6445 (2012).
32. Wu, H., Yildirim, T. & Zhou, W. Exceptional mechanical stability of highly porous zirconium metal-organic framework UiO-66 and its important implications. *J. Phys. Chem. Lett.* **4,** 925–930 (2013).
33. Vermoortele, F., Ameloot, R., Vimont, A., Serre, C. & De Vos, D. An amino-modified Zr-terephthalate metal-organic framework as an acid-base catalyst for cross-aldol condensation. *Chem. Commun.* **47,** 1521–1523 (2011).
34. Mondloch, J. E. *et al.* Are Zr_6-based MOFs water stable? Linker hydrolysis vs. capillary-force-driven channel collapse. *Chem. Commun.* **50,** 8944–8946 (2014).
35. Silva, C. G., Luz, I., Xamena, F. X. L. I., Corma, A. & Garcia, H. Water stable Zr-benzenedicarboxylate metal-organic frameworks as photocatalysts for hydrogen generation. *Chem. Eur. J.* **16,** 11133–11138 (2010).
36. He, J. *et al.* Significantly enhanced photocatalytic hydrogen evolution under visible light over CdS embedded on metal-organic frameworks. *Chem. Commun.* **49,** 6761–6763 (2013).
37. Wang, J. L., Wang, C. & Lin, W. B. Metal-organic frameworks for light harvesting and photocatalysis. *ACS Catal.* **2,** 2630–2640 (2012).
38. Kataoka, Y. *et al.* Photocatalytic hydrogen production from water using porous material [Ru-2(p-BDC)(2)](n). *Energ. Environ. Sci.* **2,** 397–400 (2009).
39. Wang, C., deKrafft, K. E. & Lin, W. B. Pt nanoparticles@photoactive metal-organic frameworks: efficient hydrogen evolution via synergistic photoexcitation and electron injection. *J. Am. Chem. Soc.* **134,** 7211–7214 (2012).
40. deKrafft, K. E., Wang, C. & Lin, W. B. Metal-organic framework templated synthesis of Fe_2O_3/TiO_2 nanocomposite for hydrogen production. *Adv. Mater.* **24,** 2014–2018 (2012).
41. Planas, N. *et al.* Defining the proton topology of the Zr-6-based metal-organic framework NU-1000. *J. Phys. Chem. Lett.* **5,** 3716–3723 (2014).
42. Nesbitt, H. W., Legrand, D. & Bancroft, G. M. Interpretation of Ni2p XPS spectra of Ni conductors and Ni insulators. *Phys. Chem. Miner.* **27,** 357–366 (2000).
43. Cheng, Z., Abernathy, H. & Liu, M. L. Raman spectroscopy of nickel sulfide Ni_3S_2. *J. Phys. Chem. C.* **111,** 17997–18000 (2007).
44. de las Heras, C. & Agullo-Rueda, F. Raman spectroscopy of $NiSe_2$ and $NiS_{2-x}Se_x$ ($0<x<2$) thin films. *J. Phys. Condens. Mater.* **12,** 5317–5324 (2000).
45. Guillaume, F., Huang, S., Harris, K. D. M., Couzi, M. & Talaga, D. Optical phonons in millerite (NiS) from single-crystal polarized Raman spectroscopy. *J. Raman Spectrosc.* **39,** 1419–1422 (2008).
46. Sun, Y. J. *et al.* Electrodeposited cobalt-sulfide catalyst for electrochemical and photoelectrochemical hydrogen generation from water. *J. Am. Chem. Soc.* **135,** 17699–17702 (2013).
47. Zeng, K. & Zhang, D. K. Recent progress in alkaline water electrolysis for hydrogen production and applications. *Prog. Energ. Combust.* **36,** 307–326 (2010).
48. Bockris, J. O. M. & Potter, E. C. The mechanism of the cathodic hydrogen evolution reaction. *J. Electrochem. Soc.* **99,** 169–186 (1952).
49. Vrubel, H., Moehl, T., Gratzel, M. & Hu, X. L. Revealing and accelerating slow electron transport in amorphous molybdenum sulphide particles for hydrogen evolution reaction. *Chem. Commun.* **49,** 8985–8987 (2013).
50. McCrory, C. C. L., Jung, S. H., Peters, J. C. & Jaramillo, T. F. Benchmarking heterogeneous electrocatalysts for the oxygen evolution reaction. *J. Am. Chem. Soc.* **135,** 16977–16987 (2013).
51. Deria, P., Bury, W., Hupp, J. T. & Farha, O. K. Versatile functionalization of the NU-1000 platform by solvent-assisted ligand incorporation. *Chem. Commun.* **50,** 1965–1968 (2014).
52. Deria, P. *et al.* Perfluoroalkane functionalization of NU-1000 via solvent-assisted ligand incorporation: synthesis and CO_2 adsorption studies. *J. Am. Chem. Soc.* **135,** 16801–16804 (2013).
53. Hod, I. *et al.* Bias-switchable permselectivity and redox catalytic activity of a ferrocene-functionalized, thin-film metal-organic framework compound. *J. Phys. Chem. Lett.* **6,** 586–591 (2015).
54. Shigematsu, A., Yamada, T. & Kitagawa, H. Wide control of proton conductivity in porous coordination polymers. *J. Am. Chem. Soc.* **133,** 2034–2036 (2011).
55. Kundu, T., Sahoo, S. C. & Banerjee, R. Alkali earth metal (Ca, Sr, Ba) based thermostable metal-organic frameworks (MOFs) for proton conduction. *Chem. Commun.* **48,** 4998–5000 (2012).
56. Horike, S., Umeyama, D. & Kitagawa, S. Ion conductivity and transport by porous coordination polymers and metal-organic frameworks. *Accounts. Chem. Res* **46,** 2376–2384 (2013).
57. Taylor, J. M., Dawson, K. W. & Shimizu, G. K. H. A water-stable metal-organic framework with highly acidic pores for proton-conducting applications. *J. Am. Chem. Soc.* **135,** 1193–1196 (2013).
58. Wiedner, E. S., Appel, A. M., DuBois, D. L. & Bullock, R. M. Thermochemical and mechanistic studies of electrocatalytic hydrogen production by cobalt complexes containing pendant amines. *Inorg. Chem.* **52,** 14391–14403 (2013).
59. Wiese, S. *et al.* Hydrogen production using nickel electrocatalysts with pendant amines: ligand effects on rates and overpotentials. *ACS Catal.* **3,** 2527–2535 (2013).
60. Liu, T. B., DuBois, D. L. & Bullock, R. M. An iron complex with pendent amines as a molecular electrocatalyst for oxidation of hydrogen. *Nat. Chem.* **5,** 228–233 (2013).

61. Dutta, A. *et al.* Minimal proton channel enables H-2 oxidation and production with a water-soluble nickel-based catalyst. *J. Am. Chem. Soc.* **135,** 18490–18496 (2013).
62. Chng, L. L., Chang, C. J. & Nocera, D. G. Catalytic O-O activation chemistry mediated by iron hangman porphyrins with a wide range of proton-donating abilities. *Org. Lett.* **5,** 2421–2424 (2003).
63. Costentin, C., Drouet, S., Robert, M. & Saveant, J. M. A local proton source enhances CO_2 electroreduction to CO by a molecular Fe catalyst. *Science* **338,** 90–94 (2012).

Acknowledgements

We thank the following for postdoctoral or graduate fellowship support: the Fulbright Commission (I.H.), the Graduate Students Study Abroad Program sponsored by National Science Council, Taiwan (C.W.K.), the National Defense Science and Engineering Graduate (NDSEG) Fellowship programme (M.S.) and a grant from the Air Force Office of Scientific Research, MURI programme (Award Number FA9550-10-1-0572) (M.D.S). J.T.H and O.K.F acknowledge that this work was supported as part of the ANSER Center, an Energy Frontier Research Center funded by the U.S. Department of Energy, Office of Science, Office of Basic Energy Sciences under Award Number DE-SC0001059.

Author contributions

J.T.H., O.K.F and I.H. conceived and designed the experiments. P.D. and W.B. participated in the discussions and analysis of data. J.E.M. and A.W.P. performed the ICP measurements. C.W.K. assisted in SEM and EDS analysis. M.S. performed the XPS measurements, while M.D.S. and C.P.K. conducted and analysed the H_2 Faradaic efficiency measurements. I.H., J.T.H and O.K.F wrote the manuscript.

Additional information

Competing financial interests: The authors declare no competing financial interest.

24

Supramolecular-jack-like guest in ultramicroporous crystal for exceptional thermal expansion behaviour

Hao-Long Zhou[1], Yue-Biao Zhang[1,2], Jie-Peng Zhang[1] & Xiao-Ming Chen[1]

The dynamic behaviours of host frameworks and guest molecules have received much attention for their great relevance with smart materials, but little has been developed to control or understand the host–guest interplay. Here we show that the confined guest can utilize not only molecular static effects but also bulk dynamic properties to control the host dynamics. By virtue of the three-dimensional hinge-like framework and quasi-discrete ultramicropores, a flexible porous coordination polymer exhibits not only drastic guest-modulation effect of the thermal expansion magnitude (up to $422 \times 10^{-6}\,K^{-1}$) and even the anisotropy but also records positive/negative thermal expansion coefficients of $+482/-218 \times 10^{-6}\,K^{-1}$. Moreover, single-crystal X-ray diffraction analyses demonstrate that the jack-like motion of the guest supramolecular dimers, being analogous to the anisotropic thermal expansion of bulk van der Waals solids, is crucial for changing the flexibility mode and thermal expansion behaviour of the crystal.

[1] MOE Key Laboratory of Bioinorganic and Synthetic Chemistry, School of Chemistry and Chemical Engineering, Sun Yat-Sen University, Guangzhou 510275, China. [2] School of Physical Science and Technology, ShanghaiTech University, 100 Haike Road, Shanghai 201210, China. Correspondence and requests for materials should be addressed to J.-P.Z. (email: zhangjp7@mail.sysu.edu.cn).

The remarkable framework flexibility is one of the most important advantages of porous coordination polymers (PCPs) or metal–organic frameworks, which allows the host structures to be altered by external stimuli[1]. While chemical stimuli (mainly change of guest) have been extensively used as the driving forces[2–5], physical stimuli such as light irradiation and temperature change have been scarcely explored[6–10]. Recently, flexible PCPs are emerging as an ideal platform for realizing abnormal temperature-induced structural dynamic behaviours, such as negative thermal expansion (NTE)[11–14] and very large thermal expansion (especially for $|\alpha| > 100 \times 10^{-6}\,K^{-1}$; refs 15–20), which have great potential for compensation of the small positive thermal expansion (PTE, $0 < \alpha < 20 \times 10^{-6}\,K^{-1}$) of common solids, designing sensitive thermomechanical actuators, and so on[21–29]. Besides framework flexibility and porosity, the designable framework connectivity of PCP[30–32] offers an additional approach for engineering the magnitude and anisotropy of thermal expansion[14–16,33,34]. It should be noted that, for its response in different dimensions, anisotropic thermal expansion is attractive for microdevices and information materials[29,35]. More importantly, just like other physical properties[36–39], thermal expansion of PCP can be readily guest-dependent, albeit it is still very challenging[16–19,40,41]. As the most remarkable examples, Kepert and coworkers achieved isotropic PTE ($\alpha = +10.0 \times 10^{-6}\,K^{-1}$) to NTE ($\alpha = -33.5 \times 10^{-6}\,K^{-1}$) transition via guest removal in $[Cd(CN)_2]\cdot xCCl_4$[41], and Barbour and coworkers showed that the PTE of [Zn(OH)(niba)]·alcohol (Hniba = 4-(1H-naphtho[2,3-d]imidazol-1-yl)benzoic acid) can be increased by $\sim 90 \times 10^{-6}\,K^{-1}$ via changing the alcohol guest molecules[18], in which the microporous structures have been demonstrated to be crucial for realizing the guest-modulation effect, since they maximize host–guest interaction around the guest molecules[17,18,40,41].

On the other hand, the confinement effect of the porous host allows the formation of low-dimensional guest aggregations with abnormal/appealing physical properties[17,39,42]. The overwhelmingly large thermal expansion of the fluidic guest (usually beyond 100 times that of common solids) might be utilized to further drive the thermal expansion of the porous host. When the anisotropic structures of the host framework, pore environment and/or the low-dimensional guest aggregation are appropriately coupled, more exciting microscopic/macroscopic phenomenon could be expected. A key to these goals is to visualize, understand and control the delicate host–guest interplay under external stimulus, in which *in situ* single-crystal X-ray diffraction (SCXRD) is the most convenient, straightforward and reliable method[43–45]. Nevertheless, retaining the sample single-crystallinity after guest removal/absorption/exchange and temperature cycling treatments is always a great challenge, and the extremely large mobility of guests also impedes determination of their positions[16,17]. Actually, even when the cross-section size of a one-dimensional channel is small enough to fit a single guest molecule well, the guest chain may still show liquid-like mobility[17]. To enable direct observation of thermal expansion of guest aggregations, discrete, isolated or 0D pores with suitable sizes for accommodation of two or a little bit more guest molecules should be optimal[46,47]. However, materials with such pores are generally difficult to change guests, preventing rational modulation of the physical property[20,48,49].

In this work, we report a PCP possessing not only a three-dimensional (3D) hinge-like structure with multimode flexibility but also quasi-0D ultramicropores suitable for adsorption and confinement of pairs of different amide-type guests. SCXRD studies of this material under temperature- and guest-stimuli reveal the largest NTE and PTE among crystalline solids to the best of our knowledge. We also report a drastic guest-modulation effect, and even a change of the main thermal expansion direction since the steric hindrance and thermal expansion effects of the guest dimers are large enough to alter the flexibility mode of the host framework (Fig. 1).

Results

Synthesis and structure of the porous crystal. Solvothermal reaction of $Cd(NO_3)_2$, 3-(pyridin-4-yl)benzoic acid (H34pba) and 4-(pyridin-4-yl)benzoic acid (H44pba) in mixed solvent ethanol/water at 90 °C afforded colourless needle-shaped crystals [Cd(34pba)(44pba)] (**1**, MCF-82). SCXRD analysis showed that **1** crystallizes in the monoclinic space group $P2_1/c$ (Fig. 2, Supplementary Fig. 1, Supplementary Table 1 and Supplementary Data 1), containing one Cd^{2+} cation, one $34pba^-$ ligand and one $44pba^-$ ligand in its asymmetric unit. A pair of Cd^{2+} ions are bridged by two exobidentate carboxylate groups from two $34pba^-$ ligands and further coordinated by two chelating carboxylate groups from two $44pba^-$ ligands and four monodentate pyridyl groups from two $34pba^-$ and two $44pba^-$ ligands to form a dinuclear $Cd_2(RCOO)_4(L_{py})_4$ (L_{py} = pyridyl group) unit, which interconnects with eight neighbours through four $34pba^-$ and four $44pba^-$ ligands to form a 3D coordination network with a uninodal 8-connected **bcu** topology (Fig. 2 and Supplementary Fig. 2)[50–52]. Due to the presence of long bridging ligands, the crystal contains a quasi-0D pore system, in which olive-like cavities with cross-sectional size of $6.0 \times 7.0 \times 9.7\,Å^3$ are interconnected through narrow necks (smaller than the diameter of a hydrogen atom) with diameter of 1.7 Å along the *a*-axis.

Thermal expansion of the porous crystal. Variable-temperature SCXRD revealed giant temperature-induced crystal deformation property for **1** (Fig. 3, Supplementary Fig. 3, Supplementary Tables 1 and 2 and Supplementary Data 1). While the *a*- and *c*-axes decrease by 2.2 and 1.6% from 112 to 300 K, respectively, its *b*-axis has a significant increase of 9.0%, giving rise to totally 5.8% increase of the unit-cell volume. Concomitantly, its void ratio increases from 20.4 to 24.8%. Such large temperature-induced crystal deformation has not been observed in coordination networks[15–19,25]. Comparison of the single-crystal structures at 112 and 300 K showed that, similar to other flexible

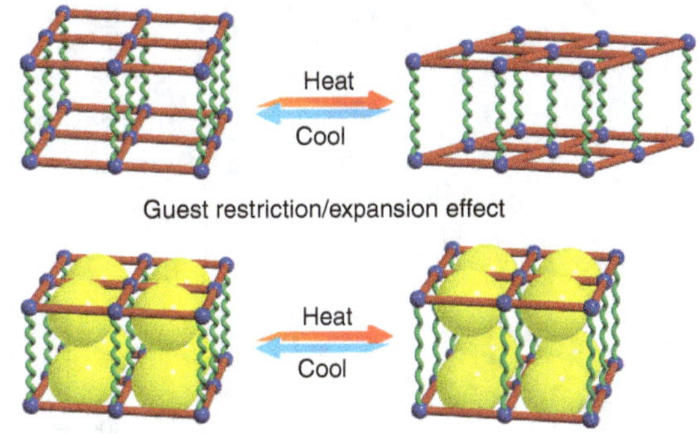

Figure 1 | Guest-controlled multimode framework flexibility. The red squares and green spires represent two possible responsive modes in a porous crystal encoded with multimode framework flexibility. Without guest, thermal expansion of the host framework is mainly determined by the hinge action across the red fences, which can be restricted by the steric hindrance effect of guest (yellow spheres). Further, the anisotropic thermal expansion effect of the guest dimers (significant change of separation between two yellow spheres), behaving like a jack, can force the host framework to deform in another direction along the green spires.

PCPs, the framework distortions mainly arise from variation of the coordination angles rather than the coordination bonds (Supplementary Fig. 2 and Supplementary Tables 3 and 4). As defined by its **bcu** topology, the flexible 8-connected building unit can be regarded as a framework hinge and the whole framework can be considered as a 3D hinged rhombic fence consisting of many two-dimensional (2D) rhombic fences running across four crystallographic planes[15,53]. Although this framework topology is highly symmetric (cubic in its highest symmetric form), which allows the crystal to deform in any direction, the real deformation manner should depend on the anisotropy of the coordination network and pore system, as well as the distribution of the guest molecules. As the pores are running along the *a*-axis of **1**, it can be understood that the 3D coordination framework can distort easily on the *bc*-plane. Actually, there are 2D hinged fence-like motifs (consisting of dinuclear clusters and bent 34pba⁻ ligands) running across the *bc*-plane, whose distortion causes the acute expansion of the *b*-axis and simultaneous contraction of the *c*-axis, when the temperature increases. The relative small contraction of the *c*-axis can be attributed to the nonplanar shape of the *bc*-plane hinged fence, which transfers some contraction effect to the *a*-axis.

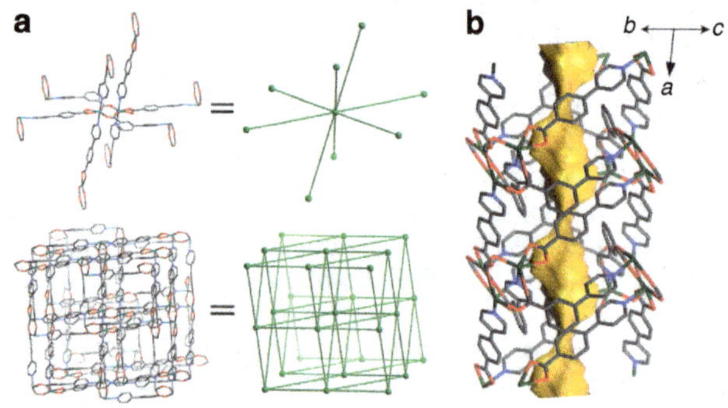

Figure 2 | X-ray single-crystal structure of the ultramicroporous framework. (**a**) Perspective view of 3D coordination framework of **1** along the *a*-axis (the aromatic rings of the 34pba⁻ and 44pba⁻ ligands are highlighted in blue and yellow, respectively). (**b**) Solvent-accessible pore surface structure of **1** viewed along the [011] direction, in which large cavities ($6.0 \times 7.0 \times 9.7$ Å^3) are interconnected by very narrow necks ($d \approx 1.7$ Å).

Host–guest behaviour of the porous crystal. Gas adsorption measurements at low temperatures (Supplementary Fig. 4) showed that while **1** cannot adsorb N_2 and Ar with kinetic diameters of 3.64 and 3.54 Å, respectively, at 77 K, it can readily adsorb the smaller O_2 (3.47 Å) at the same condition. Nevertheless, the O_2 isotherm exhibits obvious hysteresis and the saturated uptake (32 cm^3 g^{-1}) is significantly lower than the value empirically calculated from the crystal structure (146 cm^3 g^{-1}), indicating that the adsorption/desorption is very slow due to the high-energy barrier for gas diffusion in the quasi-0D ultramicropores. On the other hand, being similar with some flexible porous coordination polymers[3,16], **1** can reversibly adsorb/release *N,N*-dimethylformamide (DMF) and *N,N*-dimethylacetamide (DMA) vapour at room temperature to give [Cd(34pba)(44pba)]·DMF (**1**·DMF) and [Cd(34pba)(44pba)]·DMA (**1**·DMA) without destroying the single-crystallinity (Supplementary Figs 5–7), indicating that the host framework possesses notable framework flexibility, especially for solvent molecules. SCXRD analyses of **1**·DMF and **1**·DMA at room temperature showed that they are isomorphic with **1**. Their unit-cell volumes are only 1.4% and 2.5% larger than that of **1** (Supplementary Fig. 1, Supplementary Table 1 and Supplementary Data 1), which are relatively small among flexible PCPs[2–4]. The most important structural feature of **1**·DMF and **1**·DMA is that each cavity is occupied by a pair of DMF/DMA guest molecules arranged in a centrosymmetric fashion with their molecular planes parallel with each other. Due to the size

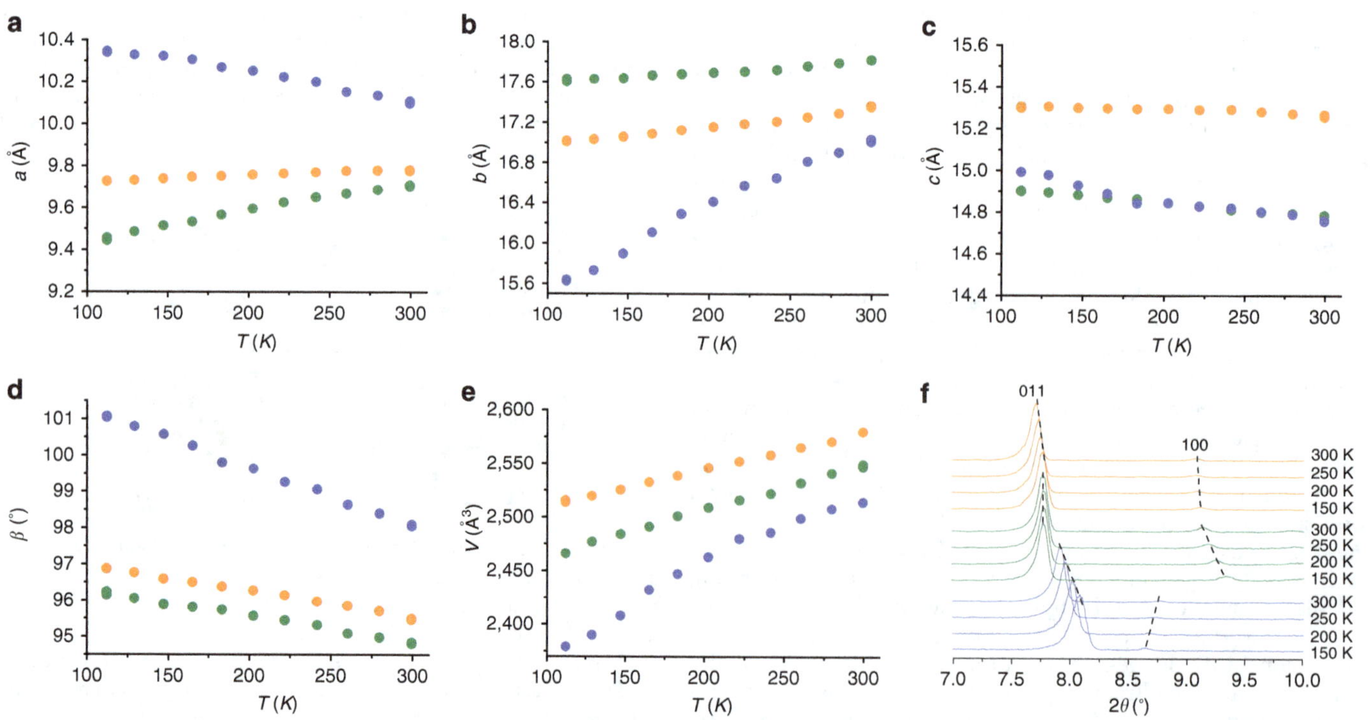

Figure 3 | Guest-dependent and guest-induced crystal deformation. (**a**–**e**) Temperature-dependent unit-cell parameters of **1** (blue), **1**·DMF (green) and **1**·DMA (orange) determined by SCXRD. (**f**) PXRD patterns of **1** (blue), **1**·DMF (green) and **1**·DMA (orange).

difference of the guests, the DMF and DMA supramolecular dimers are distinct in their configurations[54,55]. Specifically, the molecular plane of DMA has an inclined angle of 40.0° with the *bc*-plane, while that of DMF is only 17.3°, because the small pore forces the larger guest molecule to rotate to avoid significant steric hindrance. As a result of these molecular orientations, the DMF dimer exhibits much larger overlap as compared with those in the DMA dimer (Supplementary Fig. 8 and Supplementary Table 5). Consequently, adjacent DMA dimers form weak interaction ($CH_3 \cdots CH_3$ 4.15 Å), while the DMF dimers are well isolated from each other (shortest separation O···O 5.71 Å).

Supramolecular-jack-like motion of the guest aggregations. Variable-temperature powder X-ray diffraction (PXRD) showed that (Fig. 3), while the (011) peaks of the three compounds gradually move to lower angles in different extents during temperature increase, the (100) peaks of **1** and **1**·DMF/**1**·DMA move in different directions, indicating that the thermal responses of their *a*-axes contrast with each other. SCXRD revealed that (Fig. 3, Supplementary Fig. 3, Supplementary Tables 1 and 2 and Supplementary Data 1), while the *a*-axis of **1** decreases by 2.2% from 112 to 300 K, those of **1**·DMF and **1**·DMA increase by 2.8 and 0.7%, respectively. Moreover, the unit-cell volume and the *b*-/*c*-axis of **1** changed more largely than those of **1**·DMF and **1**·DMA.

The restricted hinge motion of the *bc*-plane can be explained by the intrinsic steric hindrance effects of the guest molecules. When the quasi-0D cavities, that is, the apertures of the 2D hinged fences on the *bc*-plane, are filled by guest molecules, the flexibility of the hinged fence is reduced, giving smaller temperature-induced deformation on the *b*-/*c*-axis. Though DMF is smaller than DMA, the temperature-induced hinge motion on the *bc*-plane of **1**·DMF is the smallest one among the three materials, because the DMF molecules are almost parallel with the *bc*-plane, which exhibit strong steric hindrance effect with the host framework and restrain the hinge action across this plane (Supplementary Fig. 9).

More importantly, the reversed directions of changes of the *a*-axis can be assigned to the thermal expansion of the guest supramolecular aggregation (Fig. 4, Supplementary Figs 8 and 10 and Supplementary Table 6). In **1**·DMF, the interplanar separation within a dimer significantly increases by 0.45 Å from 112 to 300 K, while this value is only 0.15 Å in **1**·DMA, which highlights that the thermal expansion of the guest aggregation mainly occurs at the directions of the supramolecular interactions, and the tighter supramolecular contacts between the guest molecules can more effectively response to their local thermal vibrations. Comparison of their temperature-dependent electron density maps can clearly illustrate dynamic behaviours of the guest molecules. On the other hand, the closest interdimer separation decreases by 0.09 Å in **1**·DMF but increases by 0.17 Å in **1**·DMA. Since the interdimer interaction is ignorable in the crystal **1**·DMF, the separation does not expand with increasing temperature. Instead, the large thermal expansion within the DMF dimer reduces the interdimer separation. Nevertheless, the thermal expansion of the dimer mainly transfers to or expands the flexible host framework via host–guest interaction, so that the interdimer separation decreases by much <0.45 Å. In the case of **1**·DMA, the interdimer supramolecular contact is close enough to show thermal expansion, and the intradimer thermal expansion is not significant enough to compress the interdimer separation. As defined by the directions of the guest–guest interactions, the thermal expansions of the guests are highly anisotropic in the crystals, which is relatively parallel with the *a*-axis of **1**·DMF or has a considerable inclined angle with the *a*-axis of **1**·DMA. Therefore, the *a*-axis of **1**·DMA expands much less than

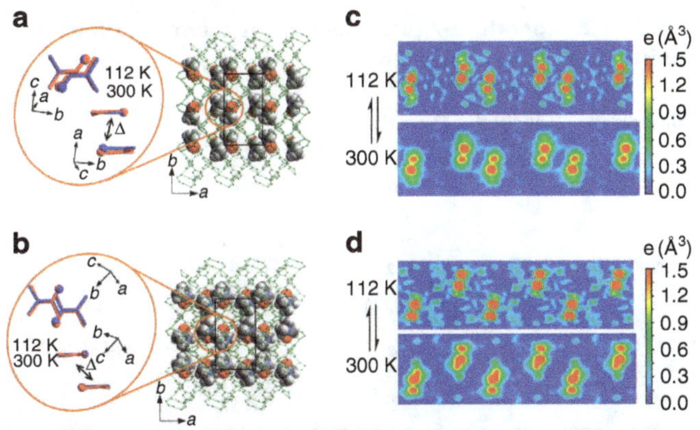

Figure 4 | Temperature-induced structural transformations of the guest supramolecular aggregations. (**a**,**b**) Crystal structures of (**a**) **1**·DMF and (**b**) **1**·DMA. The host frameworks and the guest molecules are shown in green stick and multicolour space-filling modes (carbon, grey; nitrogen, blue; oxygen, red; hydrogen atoms are omitted for clarity), respectively. The enlarged insets compare the relative positions of the two guest molecules (stick mode with oxygen atoms highlighted as spheres) within the supramolecular dimer at 112 (blue) and 300 (red) K, along two representative directions (perpendicular and parallel to the molecular planes). The double-headed arrows represent the vectors of the guest-guest movements from 112 to 300 K within each dimer. (**c**,**d**) Temperature-dependent electron density maps of (**c**) **1**·DMF and (**d**) **1**·DMA.

1·DMF, since the thermal expansion of DMA supramolecular aggregations contribute to both the *a*- and *b*-axes. What's more, the void volume of **1**·DMF (from 23.4 to 27.0%) has a larger increase than that of **1**·DMA (from 27.5 to 28.8%), further indicating that the DMF dimer with much larger overlap can have stronger thermal expansion effect. While the static effects (that is, steric hindrance and supramolecular attraction to the host) of individual guest molecules have been observed to affect the thermal expansion behaviours of a few host frameworks[18,40,41], this is the first structural evidence of thermal expansion of guest aggregations, which performs like supramolecular jacks to force the deformation of the host framework. It should be also mentioned that the relatively low crystallographic symmetry and suitable pore size (for a pair of amide molecules), as well as the multimode framework flexibility of **1** are decisive for the unambiguous observation of these interesting phenomena.

Principal axial thermal expansion of the crystal. Since **1**, **1**·DMF and **1**·DMA crystallize in the monoclinic crystal system, the axial thermal expansion coefficients of the principal axes were calculated using the programme *PASCal* (Table 1 and Supplementary Table 7)[56]. The principal X_2-axis is the same as the original crystallographic *b*-axis, while the principal X_1- and X_3-axes approximate the original crystallographic *a*- and *c*-axes or the [101] and [−101] directions. All the unit-cell parameters of **1**, **1**·DMF and **1**·DMA change linearly or approximately linearly against temperature, meaning that their axial thermal expansion coefficients are virtually constant in the measured temperature range. Such behaviour is critical for precise thermomechanical actuators and sensors. **1**·DMF shows huge PTE along the X_1-axis ($\alpha_1 = +171 \times 10^{-6}\,K^{-1}$) and moderate PTE and NTE along the X_2- and X_3-axes ($\alpha_2 = +60 \times 10^{-6}\,K^{-1}$, $\alpha_3 = -56 \times 10^{-6}\,K^{-1}$), respectively. When DMF is replaced by DMA, the thermal expansion coefficients of the X_1- and X_2-axes are approximately halved and doubled ($\alpha_1 = +85 \times 10^{-6}\,K^{-1}$, $\alpha_2 = +103 \times 10^{-6}\,K^{-1}$), respectively, while that of the X_3-axis is almost unchanged ($\alpha_3 = -51 \times 10^{-6}\,K^{-1}$). More remarkably,

Table 1 | Temperature-/guest-dependent thermal expansion coefficients.

Compound	Principal axis	Direction	α ($\times 10^{-6}$ K^{-1})	β_V ($\times 10^{-6}$ K^{-1})
1	X_1	~a	+61(1)	+319(13)
	X_2	b	+482(12)	
	X_3	~c	−218(3)	
1·DMF	X_1	~a	+171(4)	+175(2)
	X_2	b	+60(3)	
	X_3	~c	−56(2)	
1·DMA	X_1	~a	+85(1)	+138(2)
	X_2	b	+103(4)	
	X_3	~c	−51(3)	

α and β_V represent axial and volumetric thermal expansion coefficients, respectively.

although the PTE of the X_1-axis of **1** ($\alpha_1 = +61 \times 10^{-6}$ K^{-1}) is smaller than those of **1**·DMF and **1**·DMA, its X_2- and X_3-axes exhibit giant PTE of $\alpha_2 = +482 \times 10^{-6}$ K^{-1} and NTE of $\alpha_3 = -218 \times 10^{-6}$ K^{-1}, respectively, which are much larger than for other known crystalline materials (Supplementary Table 8). It is noteworthy that the guest-modulation effect among **1**, **1**·DMF and **1**·DMA are exceptionally enormous, which change the PTE and NTE magnitudes by up to 422×10^{-6} K^{-1} and 167×10^{-6} K^{-1}, respectively. Also interestingly, the X_2-axial PTE coefficient follows the order **1** > **1**·DMA > **1**·DMF, while the X_1-axial PTE coefficient follows the opposite order **1**·DMF > **1**·DMA > **1**. Consequently, the thermal expansion coefficients of the X_2- and X_3-axes of **1** are much larger than that of its X_1-axis, whereas in **1**·DMF, the X_1-axis shows the largest thermal expansion coefficient among three principal axes, meaning that the thermal expansion of the guest dimer can change the main thermal expansion axis or thermal expansion anisotropy of the crystal. Anisotropic physical property, as one of the most important features of crystalline materials, is predominantly determined by the crystallographic symmetry. The guest-dependent anisotropy of thermal expansion of isomorphic **1**, **1**·DMF and **1**·DMA (same crystallographic symmetry) demonstrates the possibility of controlling and utilizing the host–guest interplay for developing multi-responsive smart materials.

Discussion

In summary, by using mixed linear and bent ligands, we successfully constructed an ultramicroporous flexible framework with a highly connected 3D hinge-like structure and multimode flexibility, which exhibits the largest PTE and NTE coefficients among framework solids. Remarkably, the quasi-0D ultramicropores in this material can accommodate dimers of different small amide molecules, giving unprecedentedly significant guest-modulation effects on both the magnitude and anisotropy of thermal expansion. By virtue of the robustness, low symmetry and suitable pore size of the crystal, the guest- and temperature-induced structural transformations were directly visualized by PXRD and SCXRD, which revealed that the confined guest dimers can show anisotropic thermal expansion behaviours similar with their bulk forms and force the host framework to drastically change not only the coefficients but also the main axis of thermal expansion. Generally, since the structural transformations of the host framework and the guest aggregations are always observed simultaneously, there is still no rational approach to analyse their relationship or identify which one is the original force. In this context, the multimode framework flexibility of the host framework, which has potential to deform in different ways under different stimuli, is crucial for unambiguous elucidation of the mechanism. Of course, these supramolecular-jack-like dimers and similar guest aggregations should be also effective and applicable for other flexible porous solids, even with only single-mode flexibility.

Methods

Materials. The ligands 3-(pyridin-4-yl)benzoic acid (H34pba) and 4-(pyridin-4-yl)benzoic acid (H44pba) were synthesized according to literature[17,32]. Other reagents were commercially available and used without further purification.

Measurements. Elemental analyses were performed using a Vario EL elemental analyser. The FT-IR (KBr pellet) spectra were recorded in the range of 400–4,000 cm^{-1} on a Bruker TENSOR 27 FT-IR spectrometer. Thermogravimetric analyses were carried out using a TA Instruments Q50 thermogravimetric analyser under N_2 at a rate of 10 °C/min. PXRD patterns were recorded on a Bruker D8-Advance diffractometer using Cu Kα radiation and a LynxEye detector at room temperature except otherwise stated. Gas sorption isotherms were measured on a volumetric adsorption apparatus (Micromeritics ASAP 2020 M Physisorption Analyzer). Ultrahigh-purity-grade (purity > 99.999%) N_2, O_2 and Ar gases were used in all adsorption measurements. As-synthesized samples were degassed under high vacuum at 493 K for 3 h before measurement. Temperatures were maintained by liquid nitrogen.

Synthesis. H34pba (0.40 g, 2.0 mmol) and H44pba (0.40 g, 2.0 mmol) were dissolved in EtOH (80 ml) using a 250-ml scintillation vial, added with a water solution of $Cd(NO_3)_2$ (0.05 mol l^{-1}, 40 ml). The mixture was then sealed with a screw cap and heated to 90 °C for 24 h. Colourless needle-like crystals of **1**·EtOH·H_2O were obtained (yielded 0.81 g, ~80%). Then the samples were filtered and briefly dried in air to give crystals of **1**. Elemental analysis calcd (%) for [Cd(34pba)(44pba)]·3H_2O ($C_{24}H_{21}N_2O_7Cd$): C 51.31, H 3.77, N 4.99; found: C 51.20, H 4.35, N 4.68. FT-IR (KBr, cm^{-1}): 3,426(m), 3,062(w), 2,920(w), 1,610(s), 1,556(s), 1,396(s), 1,276(w), 1,224(w), 1,072(w), 1,014(w), 858(m), 837(m), 765(m), 736(m), 678(m), 624(w), 559(w) and 503(m). When single crystals of **1** were sealed in a test tube saturated with DMF/DMA vapour for 3 days, some single crystals of **1**·DMF/**1**·DMA were obtained.

X-ray single-crystal structure analyses. Diffraction data were collected on an Agilent SuperNova CCD diffractometer with graphite-monochromated Cu Kα radiation. Single crystals of **1**·DMF and **1**·DMA were mounted directly on the top of a glass fibre using minimum amount of glue. A single crystal of **1** was sealed in high vacuum in a glass capillary. The test temperature was controlled by dry nitrogen flow using a Cryostream Plus cooler system, and corrected by a thermal couple at the crystal position. Though thermogravimetry curves of **1**·DMF and **1**·DMA showed no weight loss below 350 K, the unit-cell parameters of **1**·DMF and **1**·DMA did change above 340 K under the temperature-controlling nitrogen flow after ~2 h, which can be attributed to the guest escaping. Nevertheless, under the same measurement conditions, no change of unit-cell parameter or chemical composition occurred for single crystals of **1**·DMF and **1**·DMA below 300 K even for one day. The unit-cell parameters at each measurement temperature used for calculation of the thermal expansion coefficients were determined by 30 diffraction images, while those for crystal-structure refinements were determined by 600–900 diffraction images. The reversibility of temperature-induced crystal deformations of the samples were confirmed by the coincidence of data points over a heating-cooling cycle. Absorption corrections were applied by using the multi-scan programme *CrysAlisPro*[57]. The crystal structures were solved through direct methods and developed by difference Fourier techniques using the *SHELXTL* software packages[58]. All hydrogen atoms were added geometrically, and anisotropic displacement parameters were used to refine all non-hydrogen atoms.

References

1. Horike, S., Shimomura, S. & Kitagawa, S. Soft porous crystals. *Nat. Chem.* **1,** 695–704 (2009).
2. Serre, C. *et al.* Role of solvent-host interactions that lead to very large swelling of hybrid frameworks. *Science* **315,** 1828–1831 (2007).
3. Zhang, J. P. & Chen, X. M. Exceptional framework flexibility and sorption behavior of a multifunctional porous cuprous triazolate framework. *J. Am. Chem. Soc.* **130,** 6010–6017 (2008).
4. Warren, J. E. *et al.* Shape selectivity by guest- driven restructuring of a porous material. *Angew. Chem. Int. Ed.* **53,** 4592–4596 (2014).
5. Keene, T. D. *et al.* Solvent-modified dynamic porosity in chiral 3D kagome frameworks. *Dalton Trans.* **42,** 7871–7879 (2013).
6. Nagarkar, S. S., Desai, A. V. & Ghosh, S. K. Stimulus-responsive metal-organic frameworks. *Chem. Asian J.* **9,** 2358–2376 (2014).
7. Sakata, Y. *et al.* Shape-memory nanopores induced in coordination frameworks by crystal downsizing. *Science* **339,** 193–196 (2013).
8. Yanai, N. *et al.* Guest-to-host transmission of structural changes for stimuli-responsive adsorption property. *J. Am. Chem. Soc.* **134,** 4501–4504 (2012).

9. Park, J. *et al.* Reversible alteration of CO_2 adsorption upon photochemical or thermal treatment in a metal-organic framework. *J. Am. Chem. Soc.* **134,** 99–102 (2012).
10. Henke, S., Schneemann, A. & Fischer, R. A. Massive anisotropic thermal expansion and thermo-responsive breathing in metal-organic frameworks modulated by linker functionalization. *Adv. Funct. Mater.* **23,** 5990–5996 (2013).
11. Rowsell, J. L. C., Spencer, E. C., Eckert, J., Howard, J. A. K. & Yaghi, O. M. Gas adsorption sites in a large-pore metal-organic framework. *Science* **309,** 1350–1354 (2005).
12. Dubbeldam, D., Walton, K. S., Ellis, D. E. & Snurr, R. Q. Exceptional negative thermal expansion in isoreticular metal-organic frameworks. *Angew. Chem. Int. Ed.* **46,** 4496–4499 (2007).
13. Wu, Y. *et al.* Negative thermal expansion in the metal-organic framework material Cu_3(1,3,5-benzenetricarboxylate)$_2$. *Angew. Chem. Int. Ed.* **47,** 8929–8932 (2008).
14. Wu, Y., Peterson, V. K., Luks, E., Darwish, T. A. & Kepert, C. J. Interpenetration as a mechanism for negative thermal expansion in the metal-organic framework $Cu_3(btb)_2$ (MOF-14). *Angew. Chem. Int. Ed.* **53,** 5175–5178 (2014).
15. DeVries, L. D., Barron, P. M., Hurley, E. P., Hu, C. H. & Choe, W. 'Nanoscale lattice fence' in a metal-organic framework: interplay between hinged topology and highly an isotropic thermal response. *J. Am. Chem. Soc.* **133,** 14848–14851 (2011).
16. Wei, Y.-S. *et al.* Turning on the flexibility of isoreticular porous coordination frameworks for drastically tunable framework breathing and thermal expansion. *Chem. Sci.* **9,** 1539–1546 (2013).
17. Zhou, H. L. *et al.* Direct visualization of a guest-triggered crystal deformation based on a flexible ultramicroporous framework. *Nat. Commun.* **4,** 2534 (2013).
18. Grobler, I., Smith, V. J., Bhatt, P. M., Herbert, S. A. & Barbour, L. J. Tunable anisotropic thermal expansion of a porous zinc(II) metal-organic framework. *J. Am. Chem. Soc.* **135,** 6411–6414 (2013).
19. Yang, C., Wang, X. P. & Omary, M. A. Crystallographic observation of dynamic gas adsorption sites and thermal expansion in a breathable fluorous metal-organic framework. *Angew. Chem. Int. Ed.* **48,** 2500–2505 (2009).
20. Cai, W. & Katrusiak, A. Giant negative linear compression positively coupled to massive thermal expansion in a metal-organic framework. *Nat. Commun.* **5,** 4337 (2014).
21. Mary, T. A., Evans, J. S. O., Vogt, T. & Sleight, A. W. Negative thermal expansion from 0.3 to 1050 Kelvin in ZrW_2O_8. *Science* **272,** 90–92 (1996).
22. Evans, J. S. O. Negative thermal expansion materials. *J. Chem. Soc., Dalton Trans.* 3317–3326 (1999).
23. Barrera, G. D., Bruno, J. A. O., Barron, T. H. K. & Allan, N. L. Negative thermal expansion. *J. Phys. Condens. Mat* **17,** R217–R252 (2005).
24. Han, S. S. & Goddard, W. A. Metal-organic frameworks provide large negative thermal expansion behavior. *J. Phys. Chem. C* **111,** 15185–15191 (2007).
25. Goodwin, A. L. *et al.* Colossal positive and negative thermal expansion in the framework material $Ag_3[Co(CN)_6]$. *Science* **319,** 794–797 (2008).
26. Das, D., Jacobs, T. & Barbour, L. J. Exceptionally large positive and negative anisotropic thermal expansion of an organic crystalline material. *Nat. Mater.* **9,** 36–39 (2010).
27. Panda, M. K. *et al.* Colossal positive and negative thermal expansion and thermosalient effect in a pentamorphic organometallic martensite. *Nat. Commun.* **5,** 4811 (2014).
28. Deng, H. X., Olson, M. A., Stoddart, J. F. & Yaghi, O. M. Robust dynamics. *Nat. Chem.* **2,** 439–443 (2010).
29. Browne, W. R. & Feringa, B. L. Making molecular machines work. *Nat. Nanotechnol.* **1,** 25–35 (2006).
30. Wang, C., Zhang, T. & Lin, W. Rational synthesis of noncentrosymmetric metal-organic frameworks for second-order nonlinear optics. *Chem. Rev.* **112,** 1084–1104 (2012).
31. Zhang, Z. J., Wojtas, L. & Zaworotko, M. J. Organic-inorganic hybrid polyhedra that can serve as supermolecular building blocks. *Chem. Sci.* **5,** 927–931 (2014).
32. Zhang, Y. B. *et al.* Geometry analysis and systematic synthesis of highly porous isoreticular frameworks with a unique topology. *Nat. Commun.* **3,** 642 (2012).
33. Collings, I. E. *et al.* Homologous critical behavior in the molecular frameworks $Zn(CN)_2$ and Cd(imidazolate)$_2$. *J. Am. Chem. Soc.* **135,** 7610–7620 (2013).
34. Collings, I. E., Tucker, M. G., Keen, D. A. & Goodwin, A. L. Geometric switching of linear to area negative thermal expansion in uniaxial metal-organic frameworks. *CrystEngComm* **16,** 3498–3506 (2014).
35. van Oosten, C. L., Bastiaansen, C. W. M. & Broer, D. J. Printed artificial cilia from liquid-crystal network actuators modularly driven by light. *Nat. Mater.* **8,** 677–682 (2009).
36. Halder, G. J., Kepert, C. J., Moubaraki, B., Murray, K. S. & Cashion, J. D. Guest-dependent spin crossover in a nanoporous molecular framework material. *Science* **298,** 1762–1765 (2002).
37. Takashima, Y. *et al.* Molecular decoding using luminescence from an entangled porous framework. *Nat. Commun.* **2,** 168 (2011).
38. Talin, A. A. *et al.* Tunable electrical conductivity in metal-organic framework thin-film devices. *Science* **343,** 66–69 (2014).
39. Zhao, H. X. *et al.* Transition from one-dimensional water to ferroelectric ice within a supramolecular architecture. *Proc. Natl Acad. Sci. USA* **108,** 3481–3486 (2011).
40. Goodwin, A. L., Chapman, K. W. & Kepert, C. J. Guest-dependent negative thermal expansion in nanoporous Prussian Blue analogues $M^{II}Pt^{IV}(CN)_6{\cdot}x\{H_2O\}$ ($0 \leq x \leq 2$; M = Zn, Cd). *J. Am. Chem. Soc.* **127,** 17980–17981 (2005).
41. Phillips, A. E., Goodwin, A. L., Halder, G. J., Southon, P. D. & Kepert, C. J. Nanoporosity and exceptional negative thermal expansion in single-network cadmium cyanide. *Angew. Chem. Int. Ed.* **47,** 1396–1399 (2008).
42. Kitaura, R. *et al.* Formation of a one-dimensional array of oxygen in a microporous metal-organic solid. *Science* **298,** 2358–2361 (2002).
43. Zhang, J. P., Liao, P. Q., Zhou, H. L., Lin, R. B. & Chen, X. M. Single-crystal X-ray diffraction studies on structural transformations of porous coordination polymers. *Chem. Soc. Rev.* **43,** 5789–5814 (2014).
44. Vaidhyanathan, R. *et al.* Direct observation and quantification of CO_2 binding within an amine-functionalized nanoporous solid. *Science* **330,** 650–653 (2010).
45. Shinohara, K. *et al.* A chiral wedge molecule inhibits telomerase activity. *J. Am. Chem. Soc.* **132,** 3778–3782 (2010).
46. Kawamichi, T., Haneda, T., Kawano, M. & Fujita, M. X-ray observation of a transient hemiaminal trapped in a porous network. *Nature* **461,** 633–635 (2009).
47. Inokuma, Y. *et al.* X-ray analysis on the nanogram to microgram scale using porous complexes. *Nature* **495,** 461–466 (2013).
48. Chen, B. L., Xiang, S. C. & Qian, G. D. Metal-organic frameworks with functional pores for recognition of small molecules. *Acc. Chem. Res.* **43,** 1115–1124 (2010).
49. Lin, J. B. *et al.* Chemical/physical pressure tunable spin-transition temperature and hysteresis in a two-step spin crossover porous coordination framework. *Inorg. Chem.* **51,** 9423–9430 (2012).
50. Long, D. L. *et al.* Non-natural eight-connected solid-state materials: a new coordination chemistry. *Angew. Chem. Int. Ed.* **43,** 1851–1854 (2004).
51. Long, D. L., Blake, A. J., Champness, N. R., Wilson, C. & Schroder, M. Unprecedented seven- and eight-connected lanthanide coordination networks. *Angew. Chem. Int. Ed.* **40,** 2444–2513 (2001).
52. Mehlana, G., Bourne, S. A., Ramon, G. & Ohrstrom, L. Concomitant metal organic frameworks of cobalt(II) and 3-(4-Pyridyl)benzoate: optimized synthetic conditions of solvatochromic and thermochromic systems. *Cryst. Growth Des.* **13,** 633–644 (2013).
53. Ogborn, J. M., Collings, I. E., Moggach, S. A., Thompson, A. L. & Goodwin, A. L. Supramolecular mechanics in a metal-organic framework. *Chem. Sci.* **3,** 3011–3017 (2012).
54. Wernsdorfer, W., Aliaga-Alcalde, N., Hendrickson, D. N. & Christou, G. Exchange-biased quantum tunnelling in a supramolecular dimer of single-molecule magnets. *Nature* **416,** 406–409 (2002).
55. Yamauchi, K., Takashima, Y., Hashidzume, A., Yamaguchi, H. & Harada, A. Switching between supramolecular dimer and nonthreaded supramolecular self-assembly of stilbene amide-α-cyclodextrin by photoirradiation. *J. Am. Chem. Soc.* **130,** 5024–5025 (2008).
56. Cliffe, M. J. & Goodwin, A. L. *PASCal*: a principal axis strain calculator for thermal expansion and compressibility determination. *J. Appl. Cryst.* **45,** 1321–1329 (2012).
57. *CrysAlisPRO*, Oxford Diffraction. (Agilent, Technologies, UK Ltd, West Lothian, UK, 2011).
58. Sheldrick, G. M. *SHELX-97: Programs for Crystal Structure Analysis* (University of Göttingen, 1997).

Acknowledgements

This work was supported by the '973 Project' (2012CB821706 and 2014CB845602) and NSFC (21225105, 21290173 and 21473260).

Author contributions

J.P.Z. designed the research; H.L.Z. performed synthesis and measurements; J.P.Z. and H.L.Z. analysed data; all authors discussed the results, contributed to the writing of manuscript and commented on it.

Additional information

Accession codes. The X-ray crystallographic coordinates for structures reported in this Article have been deposited at the Cambridge Crystallographic Data Centre (CCDC), under deposition number CCDC 1049147–1049152. These data can be obtained free of charge from The Cambridge Crystallographic Data Centre via www.ccdc.cam.ac.uk/data_request/cif.

Competing financial interests: The authors declare no competing financial interests.

25

Organic-inorganic supramolecular solid catalyst boosts organic reactions in water

Pilar García-García[1], José María Moreno[1], Urbano Díaz[1], Marta Bruix[2] & Avelino Corma[1,3]

Coordination polymers and metal-organic frameworks are appealing as synthetic hosts for mediating chemical reactions. Here we report the preparation of a mesoscopic metal-organic structure based on single-layer assembly of aluminium chains and organic alkylaryl spacers. The material markedly accelerates condensation reactions in water in the absence of acid or base catalyst, as well as organocatalytic Michael-type reactions that also show superior enantioselectivity when comparing with the host-free transformation. The mesoscopic phase of the solid allows for easy diffusion of products and the catalytic solid is recycled and reused. Saturation transfer difference and two-dimensional ^1H nuclear Overhauser effect NOESY NMR spectroscopy show that non-covalent interactions are operative in these host-guest systems that account for substrate activation. The mesoscopic character of the host, its hydrophobicity and chemical stability in water, launch this material as a highly attractive supramolecular catalyst to facilitate (asymmetric) transformations under more environmentally friendly conditions.

[1] Instituto de Tecnología Química, UPV-CSIC, Universidad Politécnica de Valencia, Avenida de los Naranjos s/n, E-46022 Valencia, Spain. [2] Instituto de Química Física Rocasolano, CSIC, Serrano 119, 28006 Madrid, Spain. [3] King Fahd University of Petroleum and Minerals, PO Box 989, 31261 Dhahran, Saudi Arabia. Correspondence and requests for materials should be addressed to A.C. (email: acorma@itq.upv.es).

Metal-organic frameworks (MOFs) emerged as a new type of porous crystalline solids with tailorable structures and functions. The high versatility and possibilities that these materials offer are making them very attractive for specific applications such as gas storage[1–3], separations[4,5], luminescent[6], conducting[7], and magnetic materials[8], drug delivering systems[9] and catalysis[10].

MOF-type metal-organic structures are frequently obtained as 3D frameworks based on the coordination binding between metallic nodes and rigid bi- or multi-podal linkers[11,12]. Preparation of modifiable layered metal-organic materials formed by individual sheets, spatially organized through electrostatic interactions is, however, not described in the literature, as far as we know. Only incipient studies related with the formation of MOF nanosheets by modulation of growth kinetics have been reported[13]. Nevertheless, the existence of 3D conventional MOFs which exhibited lamellar structural tendency in the spatial distribution of their structural building units could be indicative of the possibility to obtain metal-organic structures by assembly of lamellar units. In particular, from a topological point of view, this type of MOFs are based on infinite inorganic M-O-M chains, formed by only discrete metal clusters, separated by aryl dicarboxylate linkers perpendicularly located to individual metallic nodes alignment[14,15]. This topology can be observed in different well-known MOFs such as MIL-53(Al) (ref. 16), DUT-4, DUT-5 (ref. 17), DUT-8 (refs 18,19), MOF-Zn-DABCO (ref. 20), Cu-(tpa)[21] or MIL-68(Al) (ref. 22), in which it is possible to distinguish the 2D lamellar inorganic sub-networks.

Notably, many enzymes can promote chemical reactions via hydrophobic cavities wherein the cumulative influence of not only the active site in the enzyme but also the non-covalent bonds interactions account for substrate activation. Several synthetic host–guest systems have been used that copy these remarkable properties such as cyclodextrins[23–25], cucurbiturils[26,27], (metalla)crown ethers[28], calixarenes[29], carcerands[30] and zeolites[31]. Solvation effects[32,33] have also been studied and proved to account for rate acceleration in chemical reactions in such a way that approach enzyme-like catalysis. Lately, self-assembled hosts[34–37] have demonstrated to function as 'molecular flasks' to entail unusual reactions, or unique chemical phenomena. However, the development and engineering of larger and more sophisticated host structures that can incorporate several reaction components is still a challenge. An appealing strategy for supramolecular catalysis involves the design of a host system able to accommodate a given (ideally any) homogeneous catalyst and also the two reacting components, facilitating the reaction of the two partners even further and/or allowing for more environmentally friendly reaction conditions, such as lower reaction temperature, lower homogeneous catalyst loading or the use of water as a green solvent. Here we report the initial steps towards this design plan. Taking into consideration the lamellar tendency of MIL-53 structures, we prepare a novel metal-organic-type material, Al-ITQ-HB, which is based on ordered individual aluminium clusters-type sheets, similar to the sub-networks detectable in MIL-53-type materials. Key in our design is the use of a specific organic spacer, 4-heptylbenzoic acid (HB), with only one reactive carboxylate group that interacts with inorganic metallic nodes through stable coordination bonds. This particular organic spacer contains hydrocarbonated tails, which control the separation between metallic nanosheets, inhibit the 3D growth observed in conventional MOFs, favouring the formation of mesoscopic non-ordered phases with well-defined metal-organic monolayers[38]. We use this material as a supramolecular hybrid structure in cases where the hydrophobic part acts to concentrate reactants and acts as hydrophobic pockets wherein molecules are activated when performing reactions even in aqueous media. The material markedly accelerates condensation reactions in water in the absence of acid or base catalyst, as well as organocatalytic Michael-type reactions that also show superior enantioselectivity when comparing with the host-free transformation.

Results

Synthesis and characterization of the metal-organic material. We started with a MIL-53 (Al)-type material and decided to synthesize 2D nanosheets (see Fig. 1). This was achieved by inhibiting the 3D growth in the MIL-53 (Al) crystals and favouring the formation of mesoscopic phases that fitted our

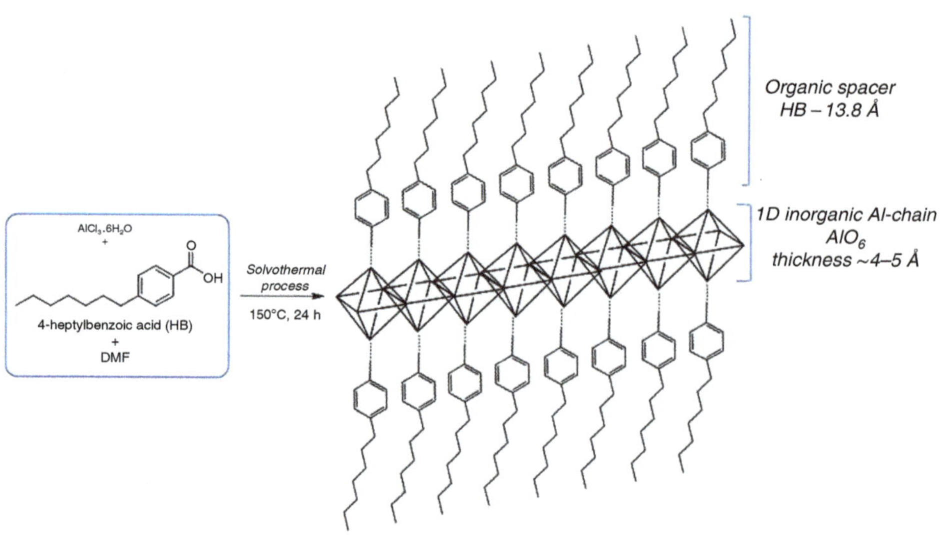

Figure 1 | Synthetic route followed to obtain mesoscopic-type hybrid material. The synthetic procedure was replicated >10 times. Representation of Al-ITQ-HB as individual organic–inorganic nanosheet is shown. The individual nanosheets would be formed by consecutive corner-sharing octahedral (AlO_6) units, conforming 1D inorganic chains, which are separated by alkyl benzene monocarboxylate ligands, located on both sides of metallic nodes. So, the coordinative association between inorganic chains and organic spacers in the opposite sides of metallic nodes forms each individual organic–inorganic layer which would be the basis of this type of metal-organic mesoscopic structures. DMF, dimethylformamide.

catalytic purposes. In our case, *para*-heptyl benzene monocarboxylate compound (HB) was used as molecular spacers (Fig. 1). This organic spacer was utilized instead of conventional rigid aryl dicarboxylate linkers, such as terephthalic acid, normally used in the preparation of the major part of MOF-type structures. Solvothermal process in the presence of aluminium chloride, dimethylformamide and HB as inorganic source, solvent and organic spacers, respectively, facilitated the preparation of the mesoscopic hybrid material whose walls were based on metal-organic layers with octahedral aluminium units separated by the hydrocarbonated tails acting as spacers, perpendicularly located to inorganic nodes. Representation of individual organic–inorganic layers is also drawn in Fig. 1, highlighting the role of organic spacers (HB) as effective growing inhibitors of the standard 3D metal-organic structure.

The X-ray diffraction pattern of the as-synthesized hybrid material Al-ITQ-HB (Supplementary Fig. 1), showed that mesoscopic organization could be achieved being clearly observable a *(100)* low-angle diffraction peak ($d_{100} = \sim 35$ Å) which is characteristic of mesoporous solids with short-range hexagonal ordering. However, *(100)* diffraction band could also be distinctive of 2D lamellar frameworks formed by individual sheets perpendicularly disposed to axis-*a*, indicating that a certain regularity would exist in the separation (basal space of ~ 35 Å) between successive piled organic–inorganic nanosheets. In any case, the analysed hybrid solid did not exhibit a very high structural order, as deduced from the low intensity and broadness of the low-angle *(100)* diffraction band together with the absence of repetitive *(h00)* diffraction bands.

The mesoscopic long-order organization was observed from TEM microscopy, being clearly detected the presence of mesoporous which were distributed along plate-like crystals (Fig. 2a,b, Supplementary Fig. 2). The micrographs also showed the existence of areas formed by short sheets which would evidence the dual nature of the Al-ITQ-HB materials, that is, lamellar and mesoporous due to the presence of metal-organic nanolayers, probably assembled around non-ordered mesocavities. However, the low structural periodicity exhibited by this type of hybrid solids along the channel direction would hinder the easy detection of the lamellar sub-domains from electronic microscopy. Despite that, the lamellar nature of the hybrid sample was confirmed through the exfoliation phenomenon achieved in presence of non-polar solvents (Fig. 2c,d, Supplementary Fig. 3).

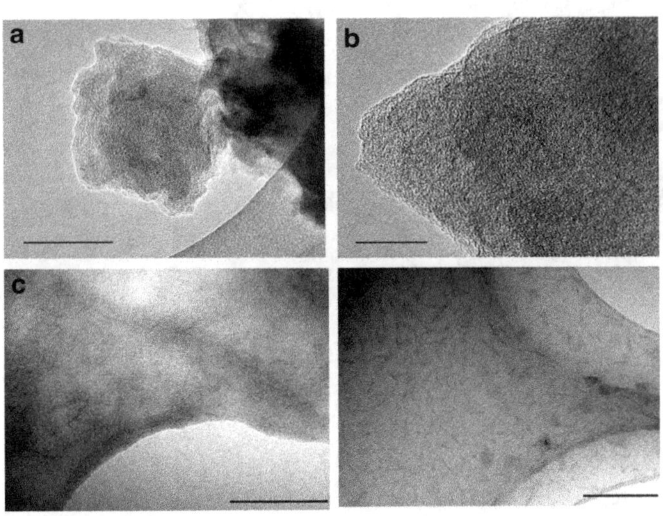

Figure 2 | HRTEM images of Al-ITQ-HB sample. (**a**) As-synthesized material. Scale bar, 100 nm. (**b**) As-synthesized material. Scale bar, 20 nm. Mesoporous cavities are clearly detected. A pore-size distribution around 20 Å in diameter is observed. (**c,d**) material after post-synthesis treatment with dichloromethane generating a stable solution. Scale bars, 200 nm. The high hydrophobicity exhibited by Al-ITQ-HB solid, due to the elevated organic content, facilitated the interlayer penetration of solvent molecules with the consequent delamination effect, being observable individual nanolayers dispersed in the stable solution.

From the elemental CHNS analysis (Supplementary Table 1), it was possible to estimate the amount of organic content present in the hybrid material, Al-ITQ-HB. The results indicated that the organic counterpart contribution was around 40 wt%, corresponding to alkylaryl fragments. Furthermore, the practical absence of nitrogen content in the solid showed that most of the dimethylformamide, used as solvent during the solvothermal synthesis, was removed during the successive washing steps.

The weight loss with temperature and the corresponding derivative (TGA and DTA curves; Supplementary Fig. 4) for the mesoscopic metal-organic hybrid material allowed establishing, not only the amount of organic spacers incorporated in the Al-ITQ-HB solid, but also their hydrothermal stability. After elimination of the major part of hydration water and residual dimethylformamide (both detected at around 80–150 °C), it was possible to see a main weight loss, between 450 and 600 °C, assigned to the presence of *para*-alkyl benzene monocarboxylate molecules (HB) used as spacers. It is important to remark the weight loss observed between 250 and 400 °C assigned to the decomposition (dehydration phenomenon) of $AlO_4(OH)_2$ units present in the 1D inorganic chains[39], which would confirm again the presence of lamellar organic–inorganic sub-domains integrated into the mesoscopic framework. In this temperature range, contribution of alkyl chains from incorporated organic linkers HB, would also be included into this weight loss.

The ^{13}C CP/MAS NMR spectrum of Al-ITQ-HB (Supplementary Fig. 5B) confirms the total integrity of the organic spacer (HB) after the solvothermal synthesis processes since all carbon atoms were unequivocally assigned. In addition, the presence and integrity of alkyl benzene monocarboxylate units located in the walls of Al-ITQ-HB solid was confirmed by infrared spectroscopy (Supplementary Fig. 6). On the other hand, ^{27}Al MAS NMR spectrum (Supplementary Fig. 7) of the Al-ITQ-HB sample, presented a defined peak at 6.7 p.p.m. chemical shift that is characteristic of aluminium (oxo)hydroxide species, showing a regular octahedral environment of metallic atoms. This result would confirm the formation of 1D inorganic chains of (AlO_6) octahedral units in the assembled organic–inorganic nanolayers[40].

Textural properties for Al-ITQ-HB solid were analysed from Argon adsorption isotherm (Supplementary Fig. 8), being estimated a reduced surface area and porous volume (S_{BET} $\sim 35\ m^2\,g^{-1}$, V_{TOTAL} $\sim 0.11\ cm^3\,g^{-1}$) due to, probably, the big amount of organic content ($\sim$40 wt%) present in the hybrid material that hinders the correct gas adsorption along the mesocavities. The marked hydrophobic character of the material would reinforce this phenomenon. However, the Hörvath–Kawazoe pore-size distribution (Supplementary Fig. 9) showed that the majority of pores were centred at $\sim$20–25 Å, this value being coincident with that estimated from electronic microscopy, remarking the mesoscopic nature of the Al-ITQ-HB material.

CO adsorption at low temperature (100 K) was monitored by FTIR spectroscopy to gather information relating Lewis acidic sites in both samples, MIL-53 (Al) and Al-ITQ-HB (Supplementary Figs 10 and 11). An absorption band is observed at 2,158 cm^{-1} that already indicates that the Lewis acidity of these materials is very weak. This being confirmed by the fact that a complete disappearance of this band is observed at 100 K by subsequent treatment of the sample in vacuum.

Knoevenagel condensation reaction. The extremely flexible hybrid material obtained could act as a supramolecular catalyst in cases where the hydrocarbonated alkylaryl spacers would constitute hydrophobic pockets when working in aqueous media. Indeed, we found that Al-ITQ-HB efficiently promotes the Knoevenagel condensation of 2-naphthaldehyde (**1a**) with Meldrum's acid (**2**) in H_2O:$CHCl_3$ in the absence of base or acid catalysts. Under the optimized conditions (Supplementary Tables 2 and 3), the condensation product (**3a**) was formed in 91% yield after 10 h (Table 1, entry 1). The reaction conditions are very mild (neutral, room temperature) and the reaction proceeds catalytically (30 mol% of Al-ITQ-HB) as the product easily diffuse through the mesoporous channels liberating the host to embrace another pair of reacting components. Without Al-ITQ-HB, **1a** gave only small amount of the **3a** (4%) under otherwise identical conditions. Carrying out the reaction at higher molarity of $CHCl_3$ as the sole solvent in the absence of Al-ITQ-HB, afforded the product in low yield of 10%, (Supplementary Table 4), allowing us to conclude that rate enhancement is not a mere consequence of the concentration rise. Al-ITQ-HB is not simply increasing molarity of reagents but also promotes the condensation in water. Control experiment with conventional MIL-53(Al) showed barely formation of product (17%). The hydrophobic cavity of Al-ITQ-HB is, therefore, crucial for the reaction because the individual MOF components Al(OH)$(C_2H_3O_2)_2$ or HB acid did not promote the transformation substantially (Supplementary Table 4). Micellar catalysis was also tested by using the equivalent amount of sodium 4-heptylbenzoate that resulted in low product formation (19% yield under otherwise identical conditions, Supplementary Table 4). Rate acceleration appeared more obvious when following the Knoevenagel condensation reaction over time (Supplementary Figs 12 and 13). We observe that initial reaction rate in the case of Al-ITQ-HB in the presence of water (Supplementary Fig. 12) is 540 times higher than when no additive is used. Initial reaction rate is also superior for Al-ITQ-HB versus utilization of HB acid (54 times higher), sodium 4-heptylbenzoate (54 times higher) or MIL-53 (Al) (216 times higher). Rise in reaction rate for Al-ITQ-HB was also observed in organic solvents (such as $CHCl_3$, Supplementary Fig. 13) wherein initial rate is 105 times superior when using Al-ITQ-HB versus no additive utilization.

Substrate scope of the Knoevenagel reaction is shown in Table 1 (see also Supplementary Methods). Under the very mild conditions, a variety of aromatic aldehydes underwent efficient condensation when using Al-ITQ-HB. Control experiments without Al-ITQ-HB showed that reaction hardly occurred. Once again, 3D MIL-53 (Al) showed no relevant conversion to the **3a** highlighting the prominent role of the hydrophobic pocket and demonstrating that substrate activation is not a mere consequence of Lewis acidic activation by the framework, or the influence of π-stacking interaction of the starting aromatic aldehyde with the framework. The method proceeded better with aromatic aldehydes having electron donating groups (Table 1, entry 4), although rate acceleration is observed in all cases.

After completion of the reaction, the supramolecular catalyst Al-ITQ-HB could be recovered by simple centrifugation and can be reused for the next run with a slight deterioration of the catalytic activity (72% conversion). PXRD showed that the recycled Al-ITQ-HB maintain its mesoscopic nature after use, being observable the characteristic low-angle diffraction band (Supplementary Fig. 14).

Multicomponent reaction. The activity of Al-ITQ-HB was also tested in the multicomponent reaction of isatin, dimedone and malononitrile for the synthesis of the spirooxindole product (Fig. 3). This transformation proceeds under micellar catalysis in water by means of sodium stearate (10 mol%) (ref. 41). We found out that Al-ITQ-HB efficiently promoted the three-component reaction in an organic solvent such as chloroform, while the product is barely formed when using conventional MIL-53 (Al)

Table 1 | Al-ITQ-HB catalysed Knoevenagel condensation of aldehydes 1 with Meldrum's acid 2*.

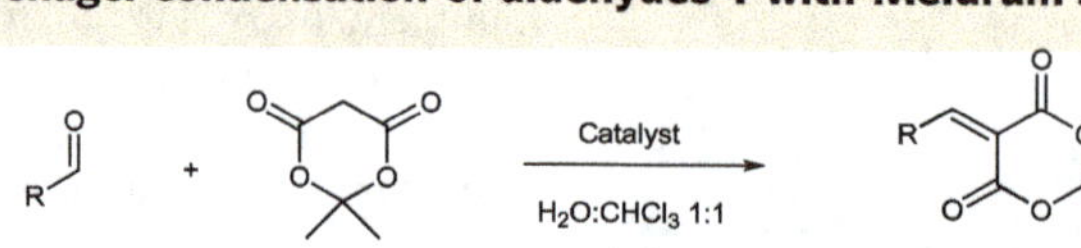

1 2 3

Entry	1		Time (h)	Yield†		
				With Al-ITQ-HB	Without Cat	With MIL-53(Al)
1	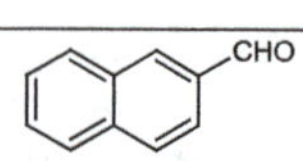	1a	10	91	4	17
2	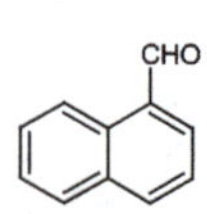	1b	24	91	15	15
3	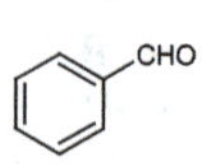	1c	8	86	11	15
4	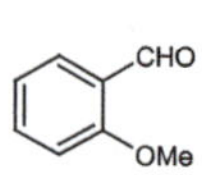	1d	8	93	15	21
5	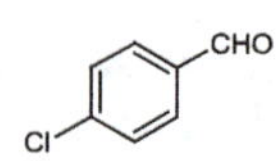	1e	24	62	9	30

*Reaction conditions: aldehyde (**1**) (0.1 mmol), Meldrum's acid (**2**) (0.1 mmol) and catalyst (30 mol%) in H_2O:$CHCl_3$ 1:1 (0.5 M) at room temperature.
†^{1}H NMR yields. Yields are the average of at least two runs.

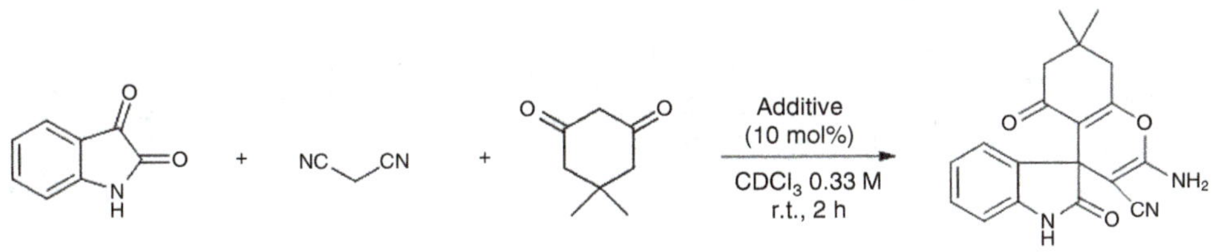

Additive	Initial rate	Yield (%)
Al-ITQ-HB	2.3	91
MIL-53 (Al)	--	10
HB (2.5 mol%)	--	--[a]
HB (10 mol%)	--	--[a]
---	--	--[a]
Na stearate (10 mol% in H_2O, 3 h)[b]		61[b]

[a]Desired product not detected by ^{1}H NMR
[b]Ref. 41

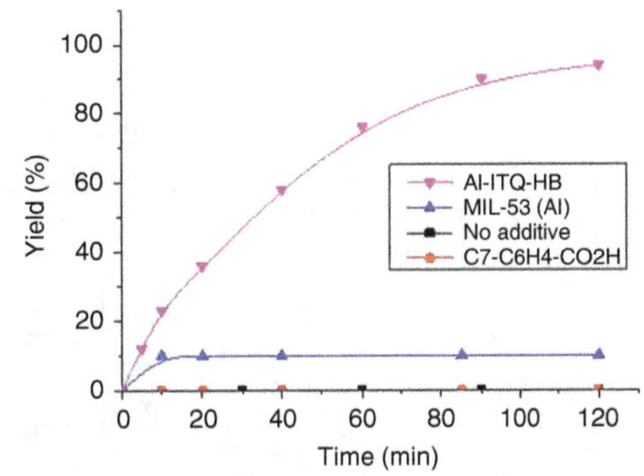

Figure 3 | Al-ITQ-HB catalysed multicomponent reaction for the synthesis of spirooxindole product. Reaction conditions: isatin (0.2 mmol), dimedone (0.2 mmol), malononitrile (0.2 mmol) and additive (10 mol% except noted) in $CDCl_3$ (0.33 M) at room temperature. ^{1}H NMR yields (d6-DMSO) are reported (internal standard: Ph_3CH).

and not detected by ^{1}H NMR after 2 h reaction time when using HB acid or no additive at all (Fig. 3).

While micellar catalysis by means of sodium stearate gave comparable results as our Al-ITQ-HB on the bases of initial reaction rates (Supplementary Fig. 16), the attained yield after 2 h reaction time is higher for Al-ITQ-HB. Once again, in the presence of water, Al-ITQ-HB probed rate enhancement as compared with the MOF-free reaction under otherwise similar conditions (Supplementary Fig. 17). Superiority of Al-ITQ-HB is also shown versus utilization of the 3D MIL-53 (Al) or HB acid (Supplementary Fig. 17).

Organocatalytic Michael-type reactions. Given the excellent results attained in rate acceleration in the above condensation reactions and the mesoscopic nature of Al-ITQ-HB, we envisioned that this framework could also be applied to facilitate asymmetric organocatalytic reactions. Organocatalysis have become a very productive field or research that well complements metal catalysis and biocatalysis in offering new opportunities for (enantio)selective transformations[42]. Probably, the main limitations in organocatalysis are the fact of low turnover numbers and slow reaction rates often affecting procedures. As an example, Michael addition of isobutyraldehyde to nitrostyrene catalysed by several primary amine bifunctional organocatalysts (similar to **4** in Fig. 4) have shown to require loadings of 20–30 mol%, to achieve good to high product formation after relatively long reaction time of 2 days (refs 43,44). In particular, 10 mol% of bifunctional amino-urea catalyst **4** provided the addition product in 14% yield and high enantioselectivity after 24 h reaction time (see equation 1 in Fig. 4). Remarkably, the combination of organocatalyst **4** with our supramolecular host catalyst Al-ITQ-HB provided substantial rate acceleration while maintaining the same degree of asymmetric induction. By this means, high yield (of 96%) can be achieved with relatively low organocatalyst loading, at room temperature, without the aid of noxious additives and after reasonable reaction time. Further analysis of the Al-ITQ-HB mediated reaction proved the role of the hydrophobic pocket in the host catalyst. Similar trends are observed here as in the case of the above depicted Knoevenagel condensation reaction. The presence of additives such as 1-phenylheptane, HB acid or MIL-53(Al) had no influence on the reaction rate (Supplementary Table 5). While host catalyst, Al-ITQ-HB, also shows rate enhancement in bare organic solvents such as toluene and CH_2Cl_2, higher product yields are attained in the presence of water evidencing the interplay of the hydrophobic effect that accounts partly for the observed efficiency. It was found that micellar catalysis exerted similar effect as our hydrophobic Al-ITQ-HB (Supplementary Table 5). However, while micellar medium required the so called critical micellar concentration to be operative, with our system the rate enhancement is observed even at low host catalyst load (Supplementary Table 5). The use of brine as reaction media instead of plain water showed advantage in the host-free reaction, although it did not provide further reaction acceleration in the presence of Al-ITQ-HB.

Following the organocatalytic Michael-type reaction over time shows the magnitude of the rate acceleration (Supplementary Fig. 18). We observe an important increase in the reaction rate in the case of Al-ITQ-HB versus utilization of HB acid, sodium 4-heptylbenzoate or no catalyst at all.

The observed efficiency achieved by combining host Al-ITQ-HB with the chiral bifunctional organocatalyst prompted us to evaluate the generality of this effect in other organocatalytic asymmetric transformations. Indeed, analogous rate enhancement is observed in the Michael addition of dimethylmalonate to 4-phenyl-3-buten-2-one (see equation 2 in Fig. 4). What it is more, higher enantioselectivity is observed when performing the reaction in the presence of host Al-ITQ-HB, evidencing that the supramolecular catalyst is indeed hosting the reacting 3-component system with non-covalent bond interactions playing a decisive role in rate acceleration and asymmetric induction. Furthermore, supramolecular host catalyst Al-ITQ-HB showed rate as well as enantioselectivity enhancement in the Michael addition of nitromethane to 4-phenyl-3-buten-2-one (see equation 3 in Fig. 4), establishing a considerable generality for the hydrophobic host system to facilitate organic reactions in aqueous media.

NMR spectroscopy study of host–guest interactions. Intrigued by the way of action of solid Al-ITQ-HB in exerting the observed

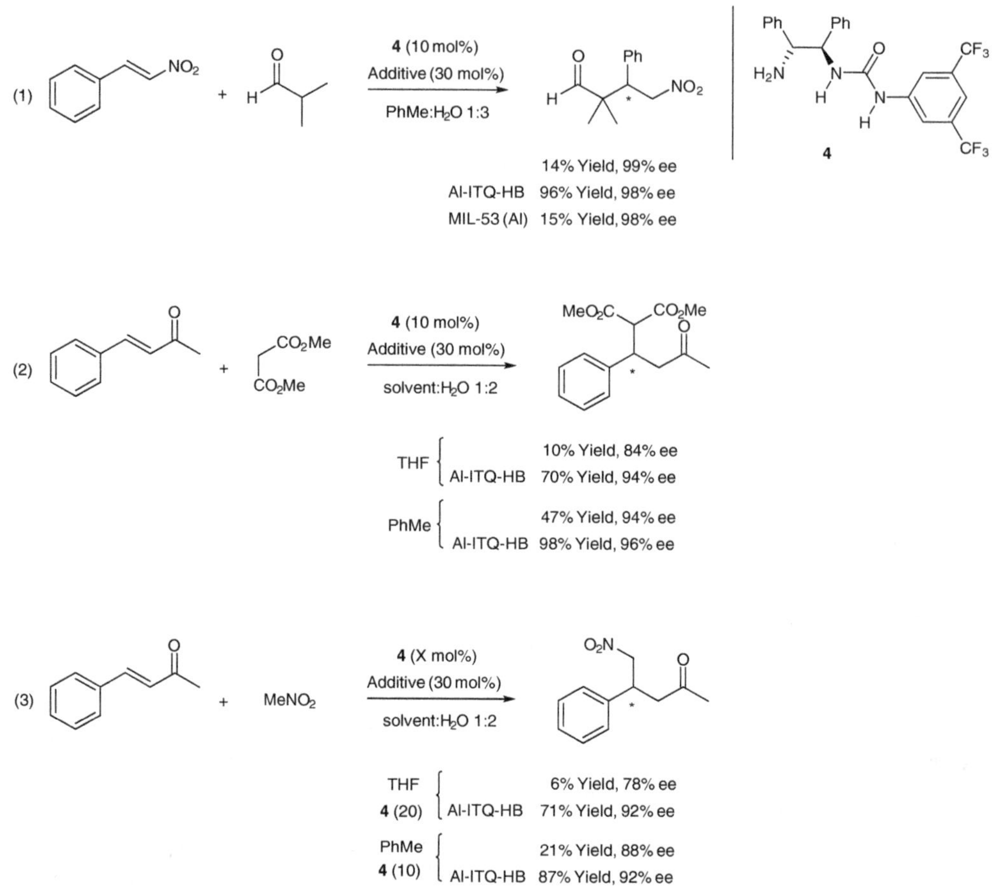

Figure 4 | Enantioselective organocatalytic Michael reactions promoted by Al-ITQ-HB. Organocatalyst **4** was used in the amount of 10 mol% in all cases except noted. Additives used were Al-ITQ-HB (results highlighted in blue) or MIL-53 (Al) (results highlighted in red) in the amount of 30 mol%. All the experiments were performed at room temperature for 24 h. Yields are the average of at least two runs.

rate acceleration in the varied reactions shown above, we completed this study with information from techniques that provide a direct and detailed view on the underlying interactions at the molecular level. Solution NMR spectroscopy could provide such a view since intermolecular interactions can be studied through their effect on the chemical shift, relaxation times and translational diffusion coefficients of the various species. In the case of protein receptors and ligand binding, specific NMR approaches are routinely applied to provide information about the binding process and conformations. Due to the high flexibility of our solid Al-ITQ-HB and its high organic content, we wondered whether such solution NMR-based approaches could also be utilized in our heterogeneous system for the *in situ* characterization of the molecular interactions. In fact, 2D NOESY NMR experiments were carried out on suspension mixtures of naphthaldehyde **1a** + Al-ITQ-HB, organocatalyst **4** + Al-ITQ-HB and nitrostyrene + Al-ITQ-HB (Supplementary Figs 19, 20 and 21, respectively). In all three cases, it is shown through space correlation between the aromatic protons of the substrates **1a**, **4** and nitrostyrene with the hydrocarbon chain of Al-ITQ-HB. What it is more, changes in the chemical shifts and an increase in signal linewidths are observed in the ^{1}H NMR spectrum of organocatalyst **4** (in $CDCl_3$) when forming the suspension mixture with Al-ITQ-HB (Supplementary Fig. 22) evidencing the interactions between both systems and probably since Al-ITQ-HB may be disrupting the self-aggregation of the organocatalyst.

Saturation transfer difference (STD) NMR spectroscopy is a powerful technique to investigate ligand binding to proteins[45]. Working with an excess of ligand, it consist in saturating some resonances of the protein target in the on-resonance experiment taking care not to affect the ligand resonances. This selective saturation subsequently spreads through the entire network of dipolar-coupled protons in the protein via spin diffusion. When a ligand binds to the protein, part of the saturation is transferred onto its protons. Since each protein undergoes multiple binding events during the saturation time, a sizable fraction of the ligands is affected, leading to a reduction of the ligand resonance intensity. This is easily characterized by subtracting a reference, the off-resonance experiment wherein saturation is applied outside the frequency range where ligand and protein resonances occur. The difference spectrum, therefore, yields non-zero intensities only for binding ligands. Furthermore, as ligand protons in close contact with the protein receive more saturation than more distant ones, the relative intensity of the ligand resonances can be interpreted in terms of a binding epitope. Despite the broad information that this technique could provide in supramolecular systems, it has been almost exclusively utilized in protein–ligand studies and only a few studies have been performed on synthetic hosts[46,47]. Here we demonstrate the potential of STD NMR for the investigation of interactions between hybrid metal-organic Al-ITQ-HB and several substrates. A suspension mixture of **1a** and Al-ITQ-HB was analysed (Fig. 5a). Saturation was performed at the resonance of the methyl group (0.9 p.p.m.) in the Al-ITQ-HB far away from resonances of aromatic protons of naphthaldehyde, whereas off-resonance frequency was − 150 p.p.m. (Supplementary Methods). Pulse sequences followed were those reported in literature for protein–ligand systems[48]. Figure 5b shows the result

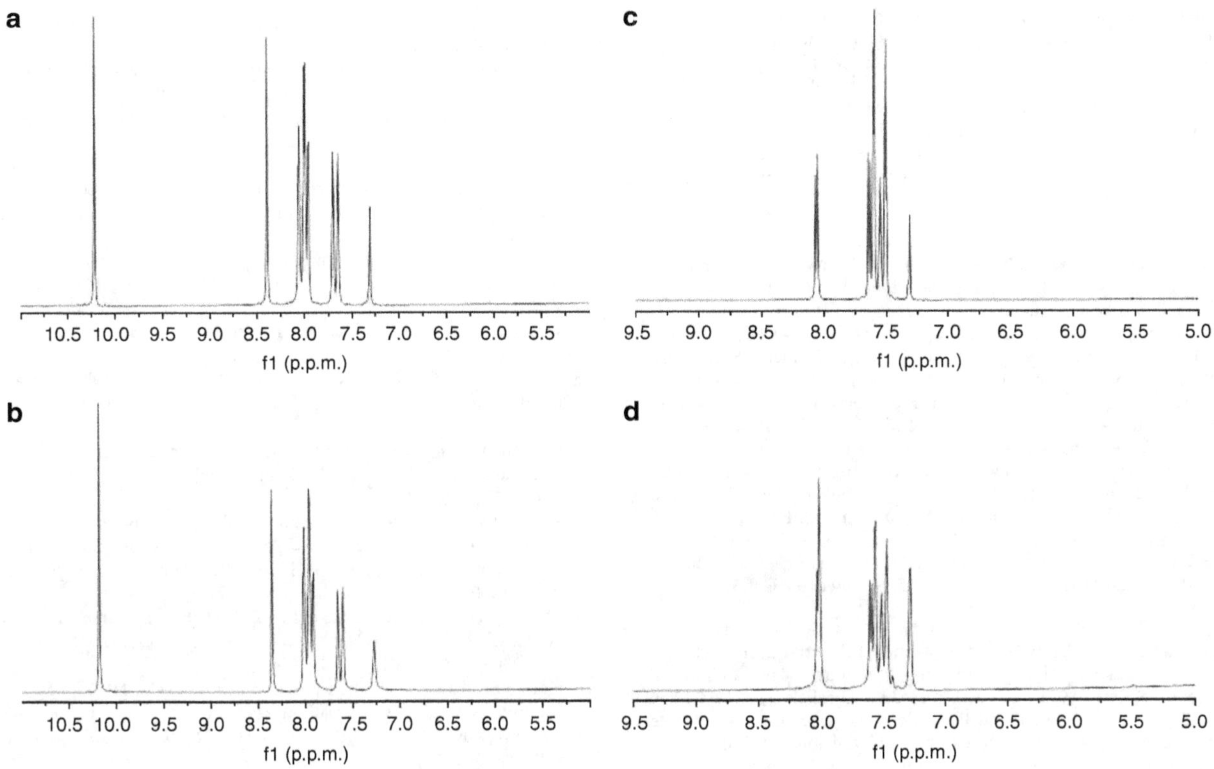

Figure 5 | STD spectra (STD_{off}—STD_{on}) of substrate and Al-ITQ-HB suspensions. (**a**) ^{1}H NMR reference spectrum of naphthaldehyde **1a** and Al-ITQ-HB suspension. (**b**) STD NMR of naphthaldehyde **1a** and Al-ITQ-HB suspension with saturation of methyl signal (0.9 p.p.m.) of Al-ITQ-HB. All of the naphthaldehyde resonances appear in the STD experiment with equal intensity as in the reference ^{1}H NMR demonstrating that naphthaldehyde is indeed binding to Al-ITQ-HB. (**c**) ^{1}H NMR reference spectrum of nitrostyrene and Al-ITQ-HB suspension. (**d**) STD NMR of nitrostyrene and Al-ITQ-HB suspension with saturation of methyl signal (0.9 p.p.m.) of Al-ITQ-HB. All of the nitrostyrene resonances appear in the STD experiment with equal intensity as in the reference ^{1}H NMR demonstrating that nitrostyrene is indeed binding to Al-ITQ-HB.

of the STD experiment to the mixture shown in Fig. 5a. All of the naphthaldehyde resonances appear in the STD experiment with equal intensity as in the reference ^{1}H NMR (Fig. 5a) demonstrating that naphthaldehyde is indeed binding to the supramolecular host catalyst. The major difference here with respect to the protein–ligand systems relays on the fact that we do not have a one-molecule site that binds to one protein-binding site, but rather a hydrophobic surface area that is large, with respect to the individual molecules, providing many locations and modes for binding. Similar results were attained when the binding substrate chosen was nitrostyrene (Fig. 5c = reference ^{1}H NMR and Fig. 5d = STD NMR experiment). Further details on this study and blank experiments are shown in the Supplementary Figs 23–27. These results show that non-covalent interactions are operative in these host–guest systems that together with the above demonstrated hydrophobic effect and the higher effective molarity account for the observed substrate activation.

Discussion

We have constructed a mesoscopic and lamellar metal-organic material that was successfully used as a host catalyst for the condensation reactions in water under neutral conditions. Activity displayed appears to emulate enzyme's way of action in natural systems, wherein the hydrophobic pocket and non-covalent interactions account for the extended efficiency. The mesoscopic host material can also shelter a chiral organocatalyst and two reacting components, facilitating asymmetric transformations under more environmentally friendly conditions, meaning lower catalyst loading, room-temperature conditions and water as a green solvent. The mesoscopic phase of the solid allows for easy diffusion of starting materials and products resulting in procedures promoted by a catalytic amount of the solid. Although the formed lipophilic products prefer the hydrophobic environment in the catalyst system over the aqueous solvent, the obtained products are easily extracted and the solid could be recycled and reused. Furthermore, we have shown that STD NMR is a valuable technique for the *in situ* solution characterization of intermolecular interactions between substrates and dispersed hybrid metal-organic materials, allowing for detection of binding ligands.

Methods

Synthesis of the mesoscopic hybrid material Al-ITQ-HB. General experimental information can be found in the Supplementary Methods. Al-ITQ-HB material was synthesized from equimolar quantities of $AlCl_3.6H_2O$ (3.1 mmol) and HB acid (3.1 mmol) which were dissolved in two different solutions with 15 ml dimethylformamide in each. The two solutions were mixed and the resulting slurry (pH = 2.5) was introduced into a stainless steel autoclave, being heated at 150 °C for 24 h under autogeneous pressure and static conditions. Once cooled to room temperature, the solution (pH = 5.5) was filtered and the collected powder was washed with distilled water. Then, the sample was stirred in methanol for 24 h to efficiently remove the remaining unreacted linker and dimethylformamide solvent molecules. Finally, the material was isolated and dried under vacuum at room temperature.

Knoevenagel condensation reaction. General experimental information can be found in the Supplementary Methods. The mesoscopic hybrid material, Al-ITQ-HB (8.0 mg) was placed in a 1-ml glass vessel. Aldehyde **1a** (0.1 mmol, 15.6 mg) and compound **2** (0.1 mmol, 14.4 mg) were then added. Chloroform (100 µl) and water (100 µl) were subsequently added and the reaction mixture was left to stir vigorously at room temperature for 10 h. The product was extracted with EtOAc (3 × 1 ml), and the solvent evaporated *in vacuo* to give **3a** (Supplementary Methods) in 91% yield as determined by analysis by ^{1}H NMR.

Multicomponent reaction. The mesoscopic hybrid material, Al-ITQ-HB (5.3 mg, 10 mol%) or the corresponding additive (as indicated in the Supplementary Figs 15–17) was placed in a 1-ml glass vessel. Isatin (29.4 mg, 0.2 mmol), dimedone (28 mg, 0.2 mmol) and Ph_3CH (12.2 mg, 0.05 mmol) were then added. Subsequently, a stock solution of malononitrile in $CDCl_3$ was then added (0.6 ml, 0.2 mmol of malononitrile) and the reaction mixture was left to stir vigorously at room temperature. Evolution of the reaction was followed by ^1H NMR by taking aliquots of the reaction mixture and dissolving the product in d6-DMSO.

Michael addition of isobutyraldehyde to *trans*-β-nitrostyrene. General experimental information can be found in the Supplementary Methods. The mesoscopic hybrid material, Al-ITQ-HB (24 mg) and organocatalyst **4** (14.1 mg, 0.03 mmol) were placed in a 2-ml glass vessel. *trans*-β-Nitrostyrene (45.0 mg, 0.3 mmol) was then added. Toluene (150 μl) and water (450 μl) were subsequently added followed by isobutyraldehyde (81 μl, 0.9 mmol) and the reaction mixture was left to stir vigorously at room temperature for 24 h. The product was extracted with EtOAc (5 × 1 ml), and host catalyst separated by centrifugation. Volatile components were then removed under reduced pressure and the crude product was purified by column chromatography using hexane/ethyl acetate (90/10) as eluent to yield addition product (Supplementary Fig. 28 and Supplementary Methods) in 96% yield (63.7 mg, 0.29 mmol). Enantiomeric excess was determined to be 98% using HPLC on a chiral stationary phase. (Kromasil 5-Cellucoat, *n*-hexane/isopropanol = 70:30, 1 ml min^{-1}, 220 nm, t (major) = 10.1 min and t (minor) = 13.6 min (Supplementary Fig. 31).

Michael addition of dimethylmalonate to 4-phenyl-3-buten-2-one. The mesoscopic hybrid material, Al-ITQ-HB (24 mg, 0.09 mmol, 30 mol%) and organocatalyst **4** (14.1 mg, 0.03 mmol) were placed in a 2-ml glass vessel. 4-Phenyl-3-buten-2-one (43.9 mg, 0.3 mmol) was then added. Toluene (150 μl) and water (300 μl) were subsequently added followed by dimethylmalonate (69 μl, 0.6 mmol) and the reaction mixture was left to stir vigorously at room temperature for 24 h. The product was extracted with EtOAc (5 × 1 ml), and host catalyst separated by centrifugation. Volatile components were then removed under reduced pressure and the crude product was purified by column chromatography using hexane/ethyl acetate as eluent to yield addition product (Supplementary Fig. 30 and Supplementary Methods) in 98% yield (81.8 mg, 0.29 mmol). Enantiomeric excess was determined to be 96% using HPLC on a chiral stationary phase. (Kromasil 5-amycoat, *n*-hexane/isopropanol = 90:10, 1 ml min^{-1}, 220 nm, t (minor) = 9.79 min and t (mayor) = 11.3 min (Supplementary Fig. 33).

Michael addition of nitromethane to 4-phenyl-3-buten-2-one. The mesoscopic hybrid material, Al-ITQ-HB (24 mg, 0.09 mmol, 30 mol%) and organocatalyst **4** (14.1 mg, 0.03 mmol) were placed in a 2-ml glass vessel. 4-Phenyl-3-buten-2-one (43.9 mg, 0.3 mmol) was then added. Toluene (150 μl) and water (300 μl) were subsequently added followed by nitromethane (150 μl) and the reaction mixture was left to stir vigorously at room temperature for 24 h. The product was extracted with EtOAc (5 × 1 ml), and host catalyst separated by centrifugation. Volatile components were then removed under reduced pressure and the crude product was purified by column chromatography using hexane/ethyl acetate as eluent to yield addition product (Supplementary Fig. 29 and Supplementary Methods) in 87% yield (54.1 mg, 0.26 mmol). Enantiomeric excess was determined to be 92% using HPLC on a chiral stationary phase. (Daicel Chiralcel OJ, *n*-hexane/isopropanol = 40:60, 1.5 ml min^{-1}, 220 nm, t (minor) = 9.3 min and t (mayor) = 13.3 min (Supplementary Fig. 32).

References

1. Li, B. *et al.* A porous metal-organic framework with dynamic pyrimidine groups exhibiting record high methane storage working capacity. *J. Am. Chem. Soc.* **136,** 6207–6210 (2014).
2. Getman, R. B., Bae, Y.-S., Wilmer, C. E. & Snurr, R. Q. Review and analysis of molecular simulations of methane, hydrogen, and acetylene storage in metal–organic frameworks. *Chem. Rev.* **112,** 703–723 (2012).
3. Suh, M. P., Park, H. J., Prasad, T. K. & Lim, D.-W. Hydrogen storage in metal–organic frameworks. *Chem. Rev.* **112,** 782–835 (2012).
4. Li, B., Wen, H.-M., Zhou, W. & Chen, B. Porous metal-organic frameworks for gas storage and separation: what, how, and why? *J. Phys. Chem. Lett.* **5,** 3468–3479 (2014).
5. Li, J.-R., Sculley, J. & Zhou, H.-C. Metal–organic frameworks for separations. *Chem. Rev.* **112,** 869–932 (2012).
6. Cui, Y., Yue, Y., Qian, G. & Chen, B. Luminescent functional metal–organic frameworks. *Chem. Rev.* **112,** 1126–1162 (2012).
7. Yoon, M., Suh, K., Natarajan, S. & Kim, K. Proton conduction in metal–organic frameworks and related modularly built porous solids. *Angew. Chem. Int. Ed.* **52,** 2688–2700 (2013).
8. Kurmoo, M. Magnetic metal-organic frameworks. *Chem. Soc. Rev.* **38,** 1353–1379 (2009).
9. Horcajada, P. *et al.* Metal–organic frameworks in biomedicine. *Chem. Rev.* **112,** 1232–1268 (2012).
10. Liu, J. *et al.* Applications of metal-organic frameworks in heterogeneous supramolecular catalysis. *Chem. Soc. Rev.* **43,** 6011–6061 (2014).
11. Rowsell, J. L. C. & Yaghi, O. M. Metal–organic frameworks: a new class of porous materials. *Micropor. Mesopor. Mat.* **73,** 3–14 (2004).
12. Eubank, J. F. *et al.* The next chapter in MOF pillaring strategies: trigonal heterofunctional ligands to access targeted high-connected three dimensional nets, isoreticular platforms. *J. Am. Chem. Soc.* **133,** 17532–17535 (2011).
13. Rodenas, T. *et al.* Metal-organic framework nanosheets in polymer composite materials for gas separation. *Nat. Mater.* **14,** 48–55 (2015).
14. Chang, Z. *et al.* Rational construction of 3D pillared metal–organic frameworks: synthesis, structures, and hydrogen adsorption properties. *Inorg. Chem.* **50,** 7555–7562 (2011).
15. Cheetham, A. K., Rao, C. N. R. & Feller, R. K. Structural diversity and chemical trends in hybrid inorganic-organic framework materials. *Chem. Commun.* 4780–4795 (2006).
16. Loiseau, T. *et al.* A rationale for the large breathing of the porous aluminum terephthalate (MIL-53) upon hydration. *Chem. Eur. J.* **10,** 1373–1382 (2004).
17. Senkovska, I. *et al.* New highly porous aluminium based metal-organic frameworks: Al(OH)(ndc) (ndc = 2,6-naphthalene dicarboxylate) and Al (OH)(bpdc) (bpdc = 4,4′-biphenyl dicarboxylate). *Micropor. Mesopor. Mat.* **122,** 93–98 (2009).
18. Klein, N. *et al.* Structural flexibility and intrinsic dynamics in the M2(2,6-ndc)2(dabco) (M = Ni, Cu, Co, Zn) metal-organic frameworks. *J. Mater. Chem.* **22,** 10303–10312 (2012).
19. Hoffmann, H. C. *et al.* High-pressure *in situ* 129Xe NMR spectroscopy and computer simulations of breathing transitions in the metal–organic framework Ni2(2,6-ndc)2(dabco) (DUT-8(Ni). *J. Am. Chem. Soc.* **133,** 8681–8690 (2011).
20. Gu, J.-M., Kim, W.-S. & Huh, S. Size-dependent catalysis by DABCO-functionalized Zn-MOF with one-dimensional channels. *Dalton Trans.* **40,** 10826–10829 (2011).
21. Carson, C. G. *et al.* Synthesis and structure characterization of copper terephthalate metal–organic frameworks. *Eur. J. Inorg. Chem.* **2009,** 2338–2343 (2009).
22. Yang, Q. *et al.* Probing the adsorption performance of the hybrid porous MIL-68(Al): a synergic combination of experimental and modelling tools. *J. Mater. Chem.* **22,** 10210–10220 (2012).
23. Li, H. *et al.* Visible light-driven water oxidation promoted by host-guest interaction between photosensitizer and catalyst with a high quantum efficiency. *J. Am. Chem. Soc.* **137,** 4332–4335 (2015).
24. Hapiot, F., Bricout, H., Menuel, S., Tilloy, S. & Monflier, E. Recent breakthroughs in aqueous cyclodextrin-assisted supramolecular catalysis. *Catal. Sci. Technol.* **4,** 1899–1908 (2014).
25. Harada, A., Takashima, Y. & Nakahata, M. Supramolecular polymeric materials via cyclodextrin-guest interactions. *Acc. Chem. Res.* **47,** 2128–2140 (2014).
26. Cong, H. *et al.* Substituent effect of substrates on cucurbit[8]uril-catalytic oxidation of aryl alcohols. *J. Mol. Catal. A Chem.* **374-375,** 32–38 (2013).
27. Masson, E., Ling, X., Joseph, R., Kyeremeh-Mensah, L. & Lu, X. Cucurbituril chemistry: a tale of supramolecular success. *RSC Adv.* **2,** 1213–1247 (2012).
28. Song, F.-T., Ouyang, G.-H., Li, Y., He, Y.-M. & Fan, Q.-H. Metallacrown ether catalysts containing phosphine-phosphite polyether ligands for Rh-catalyzed asymmetric hydrogenation—enhancements in activity and enantioselectivity. *Eur. J. Org. Chem.* **2014,** 6713–6719 (2014).
29. Rebilly, J.-N. & Reinaud, O. Calixarenes and resorcinarenes as scaffolds for supramolecular metallo-enzyme mimicry. *Supramol. Chem.* **26,** 454–479 (2014).
30. Ajami, D., Liu, L. & Rebek, Jr J. Soft templates in encapsulation complexes. *Chem. Soc. Rev.* **44,** 490–499 (2015).
31. Corma, A. & Garcia, H. Supramolecular host-guest systems in zeolites prepared by ship-in-a-bottle synthesis. *Eur. J. Inorg. Chem.* **2004,** 1143–1164 (2004).
32. Kemp, D. S., Cox, D. D. & Paul, K. G. Physical organic chemistry of benzisoxazoles. IV. Origins and catalytic nature of the solvent rate acceleration for the decarboxylation of 3-carboxybenzisoxazoles. *J. Am. Chem. Soc.* **97,** 7312–7318 (1975).
33. Thorn, S. N., Daniels, R. G., Auditor, M. T. & Hilvert, D. Large rate accelerations in antibody catalysis by strategic use of haptenic charge. *Nature* **373,** 228–230 (1995).
34. Yoshizawa, M., Klosterman, J. K. & Fujita, M. Functional molecular flasks: new properties and reactions within discrete, self-assembled hosts. *Angew. Chem. Int. Ed.* **48,** 3418–3438 (2009).

35. Yoshizawa, M., Tamura, M. & Fujita, M. Diels-Alder in aqueous molecular hosts: unusual regioselectivity and efficient catalysis. *Science* **312,** 251–254 (2006).
36. Murase, T., Nishijima, Y. & Fujita, M. Cage-catalyzed knoevenagel condensation under neutral conditions in water. *J. Am. Chem. Soc.* **134,** 162–164 (2012).
37. Zhao, C., Toste, F. D., Raymond, K. N. & Bergman, R. G. Nucleophilic substitution catalyzed by a supramolecular cavity proceeds with retention of absolute stereochemistry. *J. Am. Chem. Soc.* **136,** 14409–14412 (2014).
38. Choi, M. *et al.* Stable single-unit-cell nanosheets of zeolite MFI as active and long-lived catalysts. *Nature* **461,** 246–249 (2009).
39. Loiseau, T. *et al.* MIL-96, a porous aluminum trimesate 3D structure constructed from a hexagonal network of 18-membered rings and μ3-Oxo-centered trinuclear units. *J. Am. Chem. Soc.* **128,** 10223–10230 (2006).
40. Bezverkhyy, I. *et al.* MIL-53(Al) under reflux in water: formation of γ-AlO(OH) shell and H2BDC molecules intercalated into the pores. *Micropor. Mesopor. Mat.* **183,** 156–161 (2014).
41. Wang, L.-M. *et al.* Sodium stearate-catalyzed multicomponent reactions for efficient synthesis of spirooxindoles in aqueous micellar media. *Tetrahedron* **66,** 339–343 (2010).
42. List, B. *Science of Synthesis: Asymmetric Organocatalysis 1, Lewis Base and Acid Catalysts* (Georg Thieme Verlag, 2012).
43. He, T., Gu, Q. & Wu, X.-Y. Highly enantioselective Michael addition of isobutyraldehyde to nitroalkenes. *Tetrahedron* **66,** 3195–3198 (2010).
44. Avila, A., Chinchilla, R., Fiser, B., Gómez-Bengoa, E. & Nájera, C. Enantioselective Michael addition of isobutyraldehyde to nitroalkenes organocatalyzed by chiral primary amine-guanidines. *Tetrahedron Asymmetry* **25,** 462–467 (2014).
45. Meyer, B. & Peters, T. NMR spectroscopy techniques for screening and identifying ligand binding to protein receptors. *Angew. Chem. Int. Ed.* **42,** 864–890 (2003).
46. Szczygiel, A., Timmermans, L., Fritzinger, B. & Martins, J. C. Widening the view on dispersant – pigment interactions in colloidal dispersions with saturation transfer difference NMR spectroscopy. *J. Am. Chem. Soc.* **131,** 17756–17758 (2009).
47. Basilio, N., Martín-Pastor, M. & García-Río, L. Insights into the structure of the supramolecular amphiphile formed by a sulfonated calix[6]arene and alkyltrimethylammonium surfactants. *Langmuir* **28,** 6561–6568 (2012).
48. Mayer, M. & Meyer, B. Characterization of ligand binding by saturation transfer difference NMR spectroscopy. *Angew. Chem. Int. Ed.* **38,** 1784–1788 (1999).

Acknowledgements

This work was funded by ERC-AdG-2014-671093-SynCatMatch and the Generalitat Valenciana (Prometeo). M.B. acknowledges the funding: CTQ2014-52633-P. The Severo Ochoa program (SEV-2012-0267) is thankfully acknowledged.

Author contributions

A.C. conceived and directed the project. J.M.M. and U.D. synthesized and collected the characterization data for material Al-ITQ-HB. A.C., U.D. and J.M.M. analysed the synthesis and characterization data for material Al-ITQ-HB. A.C. and P.G.-G. planned, conducted and analysed the data for the catalytic experiments. P.G.-G. and M.B. planned, conducted and analysed the data for the STD and NOESY NMR experiments. P.G.-G., U.D. and A.C. wrote the manuscript. All authors discussed the results and commented on the manuscript.

Additional information

Competing financial interests: The authors declare no competing financial interests.

26

Understanding of multimetallic cluster growth

Stefan Mitzinger[1], Lies Broeckaert[1,2], Werner Massa[1], Florian Weigend[2] & Stefanie Dehnen[1]

The elucidation of formation mechanisms is mandatory for understanding and planning of synthetic routes. For (bio-)organic and organometallic compounds, this has long been realized even for very complicated molecules, whereas the formation of ligand-free inorganic molecules has widely remained a black box to date. This is due to poor structural relationships between reactants and products and the lack of structurally related intermediates—due to the comparably high coordination flexibility of involved atoms. Here we report on investigations of the stepwise formation of multimetallic clusters, based on a series of crystal structures and complementary quantum-chemical studies of $(Ge_2As_2)^{2-}$, $(Ge_7As_2)^{2-}$, $[Ta@Ge_6As_4]^{3-}$, $[Ta@Ge_8As_4]^{3-}$ and $[Ta@Ge_8As_6]^{3-}$. The study makes use of efficient quantum-chemical tools, enabling the first detailed screening of the energy hypersurface along the formation of ligand-free inorganic species for a semi-quantitative picture. The results can be generalized for an entire family of multimetallic clusters.

[1] Fachbereich Chemie and Wissenschaftliches Zentrum für Materialwissenschaften, Universität Marburg, Hans-Meerwein-Straße, D-35043 Marburg, Germany. [2] Institut für Nanotechnologie, Karlsruher Institut für Technologie, Hermann-von-Helmholtz Platz 1, D-76344 Eggenstein-Leopoldshafen, Germany. Correspondence and requests for materials should be addressed to F.W. (email: florian.weigend@kit.edu) or to S.D. (email: dehnen@chemie.uni-marburg.de).

The evaluation of reaction mechanisms is not only useful but also essential for understanding, planning and optimizing chemical reactions in a reasonable and also efficient and economical way. In organic chemistry, this is a highly common procedure that allowed for the development of the retro-synthetic approach for systematic access of complex target molecules from simpler precursor fragments in the 1980s (ref. 1). However, neither this procedure nor any kind of systematic mechanistic study has so far been applied to the formation of inorganic molecules. Especially for polynuclear complexes or clusters, the formation mechanisms have widely remained unexplored to date. This is usually due to a poor structural relationship between reactants and product molecules. Furthermore, the flexibility of the metal atoms within a cluster regarding coordination numbers and geometry allows for a relatively quick re-organization, which usually prohibits monitoring of the processes and the detection of structurally related intermediates—at least in homoatomic cases, which lack any kind of 'tracer' atom.

Metal clusters in general have been subject of countless studies over the past decades, the more as monodisperse species came into sight as veritable and controllable quantum dots[2,3], which have also been used to generate novel nanostructured solids[4–6]. In particular, multimetallic clusters have attracted much attention by chemists and physicists in the recent past as they represent monodisperse intermetallic particles with superatom characteristics[7] and fine-tunable opto-eletronic as well as magnetic properties[8–10]. For this, these clusters can be viewed as molecular models to doped metals or bimetallic, catalytically active nanostructures[11,12].

However, the design of such clusters has appeared to be very challenging, since for reactions in condensed matter, the very first steps in the formation starting out from atomic, molecular or solid state precursors have essentially remained unexplored so far. It would thus be of great benefit to elucidate these processes to make such particles more generally available. A profound knowledge of all evolutionary steps would allow for overcoming the challenges of a reproducible synthesis, controlling the shape and size as well as the fine-tuning of chemical and physical properties.

A recently active investigated class of corresponding multi-metallic compounds are intermetalloid clusters, that is, main-group (semi-)metal cages embedding transition metal atoms[13–16]. A large variety of different cluster structures has been presented over the last two decades, with ever larger and more complex architectures, that are usually obtained in solution by reaction of homoatomic or heteroatomic Zintl anions with transition metal complexes. Here, in some cases, intermediate complexes have been isolated that allowed for some understanding of the stepwise release of organic ligands from the used transition metal complexes[17–21], but it was not possible to trace back the complicated re-arrangement processes of the smaller Zintl anions in the presence of transition metal atoms.

The lack of knowledge regarding the relationship between reactants and products becomes particularly obvious for non-deltahedral cluster architectures including group 14 metal atoms[11,22–27], since all known precursors have been deltahedral main-group element polyanions so far.

Herein, we report on the synthesis and isolation of a series of compounds containing heterometallic or intermetalloid poly-anions of different sizes, $(Ge_2As_2)^{2-}$, $(Ge_7As_2)^{2-}$, $[Ta@Ge_8As_4]^{3-}$ and $[Ta@Ge_8As_6]^{3-}$. During these studies, we additionally crystallized an unprecedented intermediate, $[Ta@Ge_6As_4]^{3-}$, of the corresponding intermetalloid cluster anions. This result prompted us to investigate the reaction by a systematic exploration of the energy hypersurface with recently developed tools for the search of low-lying minima in mixed-metallic systems[28] and for the optimization of transition pathways[29]. This way, we shed light on the complex formation processes behind non-deltahedral multimetallic clusters in a semi-quantitative manner.

Results

Experimental findings. The study started out with the synthesis of a precursor compound with a binary anion, $[K([2.2.2]crypt)]_2(Ge_2As_2) \cdot en$ (**1**), which was carried out according to the syntheses of homologous or isoelectronic compounds[30–34], by fusion of K, Ge and As (1:1:1) in Ta tubes at high temperatures, slow cooling to room temperature and subsequent extraction with *en*/[2.2.2]crypt at room temperature. The only difference from previous procedures was the application of a somewhat higher maximum temperature, 950 °C instead of 600 °C at the fusion step, which we initially chose in regard of the higher melting temperature of germanium as compared with those of tin or lead, and which turned out to be necessary to gain Ta atoms from the bulk (see below). As in many other cases before, we also crystallized a compound with a 9-atom binary polyanion from the extraction solution, $[K([2.2.2]crypt)]_2(Ge_7As_2)$ (**2**), and we detected a known homoatomic 10-atom cage $(Ge_{10})^{2-}$ (**A**)[35], along with **1** (compound **2** was also gained starting from a K/Ge/As phase generated in silica ampoules, indicating that Ta is not needed for its synthesis). However, the procedure also afforded two further compounds, which comprise ternary intermetalloid clusters with Ta atoms inside, $[K([2.2.2]crypt)]_3[Ta@Ge_6As_4] \cdot 2tol$ (**3**) and $[K([2.2.2]crypt)]_5[K([2.2.2]crypt)(en)][Ta@Ge_8As_4]_{1.21}[Ta@Ge_8As_6]_{0.79} \cdot en$ (**4**).

Obviously, the higher temperature at the beginning of the synthesis and the particular elemental combination enabled the reaction with the tube material and subsequent incorporation of Ta^{5+} (as confirmed by X-ray absorption spectroscopy (EDX) analyses of the resulting solid). The highly polarizing nature of the hard Ta^{5+} ion, which increases the covalent character of the bonds between transition metal and main-group (semi-) metal atoms, is made responsible in turn for the possibility to isolate fragmentary/intermediate complexes such as the anion in **3**. According structures are not likely to be isolated in the presence of less-polarizing cations, such as the Ln^{3+} series used before.

We thereupon developed a systematic access to these phases by addition of Ta powder. All four compounds were isolated from the extraction solution in single-crystalline form, and experimentally characterized by energy dispersive EDX (Supplementary Fig. 1, Supplementary Tables 1 and 2), X-ray diffraction (Fig. 1, Supplementary Figs 2–13, Supplementary Tables 3–7) and electrospray ionization mass spectrometry (ESI-MS, Supplementary Figs 14–26). The elucidation of single-crystal structures was not trivial here as it faced the following complications: (I) indistinguishability of Ge and As atoms by MoKα radiation, inhibiting the experimental assignment of Ge/As atomic positions, (II) positional disorder of the anions (**2** and **4**) and (III) co-crystallization of diverse anions of different structures (**3**) or composition (**4**), in the latter case arising along with complication (II).

From crystal structures to a sketched formation mechanism. Besides $[K([2.2.2]crypt]^+$ cations, the compounds comprise a tetrahedral $(Ge_2As_2)^{2-}$ anion (**1**), a nine-atom anion $(Ge_7As_2)^{2-}$ (**2**) with the topology of the well-known $(Ge_9)^{4-}$ cage[13], two different isomers of $[Ta@Ge_6As_4]^{3-}$ with so far unprecedented 10-atom architectures (**3**), as well as the clusters $[Ta@Ge_8As_4]^{3-}$ and $[Ta@Ge_8As_6]^{3-}$ (**4**). One of the co-crystallizing anions in **4** is based on a 12-atom cage with a rare non-deltahedral topology,

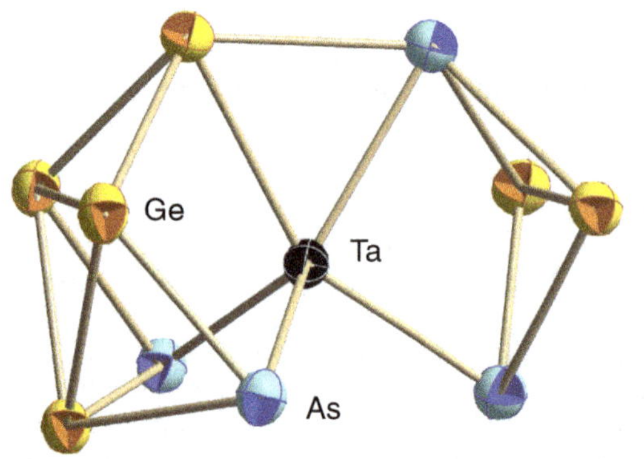

Figure 1 | Molecular structure of the predominant anionic component in 3. Thermal ellipsoids are shown at 50% probability. Selected distances [Å]: Ge/As-Ge/As 2.4770(7)-2.761(1); Ta-Ge/As 2.496(1)-2.719(1). Owing to non-distinguishability of Ge and As atoms for X-rays, the refinement was done based on the assignment of Ga or As atoms resulting from DFT calculations. The minority component isomer is shown in Supplementary Fig. 7b.

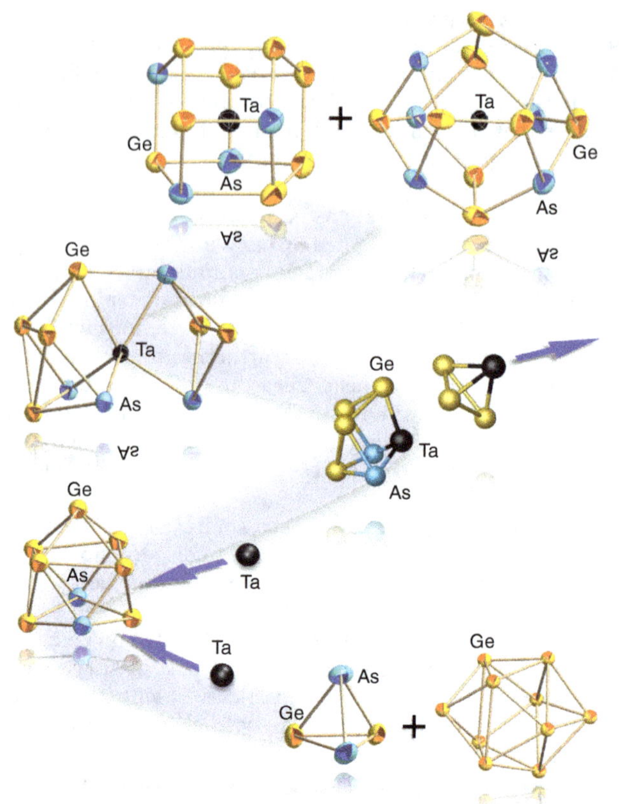

Figure 2 | Outline of the stepwise formation of non-deltahedral intermetalloid Ta/Ge/As clusters. The shown pathway starts out from the $(Ge_2As_2)^{2-}$ anions in **1** and the $(Ge_{10})^{2-}$ anion[40] (bottom) under consideration of all isolable species, hence the anions in **2** (2nd from bottom), in **3** (2nd from top) and in **4** (top), plus the calculated, hypothetic anions $[Ta@Ge_4As_2]^-$ and $(TaGe_3)^-$ (centre). The minimum structures of the latter are shown in a different style of representation for easier tracking.

reported recently by Goicoechea and co-workers for the binary cluster $[Ru@Ge_{12}]^{3-}$ (ref. 36), and for $[V@Ge_8As_4]^{3-}$ produced in our lab[37]. The Ge/As shell of the second anion in **4** adopts the stable 14-atom enneahedron of main-group metals, which was observed recently for several lanthanide-centered examples, $[Ln@Sn_xBi_{14-x}]^{4-}$ (Ln/x = Eu/6; La/7, Ce/7)[25,26]. While the topology of the latter is known, the corresponding cluster in **4** comprises the smallest 14-atom cage known to date (Ø 5.7…5.8 Å), and it is the second intermetalloid cluster of this type that does not embed a lanthanide cation but a group 5 metal ion besides $[Nb@Ge_8As_6]^{3-}$ (ref. 37). Hence, whereas the topologies of the anions in **1**, **2** and **4** are known for other elemental compositions, neither the composition nor the structures of the anions in **3** have been known so far. Figure 1 shows the molecular structure of the predominant component (90%) of the two isomeric anions in **3** (for the minority component isomer, see Supplementary Fig. 7b). The crystal structures of all other anions are presented in Fig. 2 and in Supplementary Figs 3,5,10 and 11. Full crystallographic details can be gained from Supplementary Table 3 and Supplementary Data 1–4.

According to these results, the uncommon cluster structure can be described as being composed of two parts, a 6-atom (Ge_4As_2) unit and a 4-atom (Ge_2As_2) unit, attached to a Ta atom. As will be discussed in the following, we claim this anion to be the key species for the formation of non-deltahedral, multimetallic clusters. This leads us to the following general formation protocol (equation (1)) including all observed anionic compositions, as well as Ta metal (represented by atoms here):

$$
\begin{aligned}
&(\mathbf{Ge_2As_2})^{2-} + (\mathbf{Ge_{10}})^{2-} + \mathbf{Ta} \\
&\rightarrow (\mathbf{Ge_7As_2})^{2-} + \mathbf{Ta} \\
&\rightarrow [\mathbf{Ta@Ge_6As_4}]^{3-} \\
&\rightarrow [\mathbf{Ta@Ge_8As_6}]^{3-} + [\mathbf{Ta@Ge_8As_4}]^{3-} \qquad (1)
\end{aligned}
$$

to use the information gained so far for understanding of the formation mechanism along the steps indicated in the above equation, we systematically investigated the energy hypersurface of all of the given anions. This was realized with the programme system Turbomole[38] by a genetic algorithm (GA)[39–43] based on density functional theory (DFT) with the functional by Becke and Perdew, BP86 (refs 44, 45), and polarized split-valence basis sets, def-SVP (refs 46-48), and further employing the conductor-like screening model (COSMO)[49] with standard settings for charge compensation. Within the GA, one is faced with the problem of assigning Ge and As to the atomic positions in the energetically best way. This (re)assignment of atoms to places was done by first-order perturbation theory (RP), with the difference in nuclear charges of the two elements as perturbation parameter[50,51]. The resulting procedure, called GA-RP[28] was applied to all systems with the settings given in ref. 28. The resulting local minimum structures for each isomer were subsequently re-optimized employing the functional by Tao, Perdew, Staroverov and Scuseria, TPSS (ref 52), with polarized triple zeta valence basis sets[53] and Dirac-Hartree-Fock effective core potentials[54], dhf-TZVP, which for main-group element clusters yields results close to that of coupled-cluster calculations[55]. We note in passing that the influence of different types of functionals is small, barrier heights with BP86 are lower by typically 3.8 kJ mol^{-1}, maximal by 9.1 kJ mol^{-1}; differences to the functional by Perdew, Burke and Ernzerhof (PBE)[56] or to the hybrid functional by Tao, Perdew, Starovereov and Scuseria (TPSSh)[57] functionals are even smaller, see Supplementary Tables 8 and 9. The global minima obtained from the GA-RP procedures possess topologies that are identical with those of the experimentally determined structures throughout, and all structural parameters agree well. For the species $(Ge_2As_2)^{2-}$/$(Ge_7As_2)^{2-}$/$[Ta@Ge_6As_4]^{3-}$/$[Ta@Ge_8As_6]^{3-}$/$[Ta@Ge_8As_4]^{3-}$ the mean deviations amount to 0.027/0.019/0.012/0.017/0.033 Å, the maximum deviations to 0.062/0.097/0.037/0.074/0.104 Å. This indicates both the reliability of the procedure for their identification and the reliability of TPSS/dhf-TZVP for the description of the energy hypersurface at 0 K (disregarding zero-point energy). For all cases of interest, connecting pathways between isomers were calculated with an iterative method based on a local quadratic approximation of the energy hypersurface[29]. Subsequently, all minima and maxima of two of these pathways were re-optimized, which leads to a small increase of energies for barriers (by ca. $2\,kJ\,mol^{-1}$). We convinced ourselves that all transitions in fact represent reaction coordinates. This was done by distorting each maximum along its imaginary mode in positive and negative direction and subsequently optimizing the distorted structures (minimum search). In all cases we verified that the two resulting structures are identical to the minimum structures right and left to the maximum. Moreover, thermal corrections from energies at 0 K to free enthalpies were calculated from partition sums within the standard harmonic oscillator approximation[58] for each molecule in the gas phase; vibrational frequencies were used non-scaled. The resulting data are given in Supplementary Figs 27 and 28, Supplementary Tables 10 and 11 for $T=298$, 500, 700 and 900 K. For these temperatures, mean changes in barrier heights amount to $+0.6/+3.8/+6.4/+10.1\,kJ\,mol^{-1}$, maximum changes to $+3.2/+7.7/+12.7/+18.3\,kJ\,mol^{-1}$, respectively. The overall changes of the barrier heights resulting from the bare inspection of the pathway to those obtained when considering the above effects (individual optimization of minima/maxima, functional dependence and temperature dependence) thus are estimated to be smaller than $10\,kJ\,mol^{-1}$ at room temperature and below $20\,kJ\,mol^{-1}$ at $T=900$ K; changes for relative energies of minima are even smaller. These—rather moderate—effects thus were not considered in the following and reaction barriers were taken directly from the optimized reaction pathways.

Furthermore, the GA-RP procedure was used to determine the global minimum structures of further fragments that resulted from hypothetic withdrawal of atoms from the experimentally observed larger clusters. Two of such hypothetical fragments, $[Ta@Ge_4As_2]^-$ and $(TaGe_3)^-$, turned out to be reasonable candidates for missing links in the reaction cascade, considered to form on interaction of $(Ge_7As_2)^{2-}$ with Ta atoms. They were, however, not detected in the experimental studies, presumably owing to their (high) reactivity. Figure 2 provides an illustration of the stepwise cluster formation as suggested based on the preliminary results regarding all of the named species and their topologies.

Deeper insight into the formation pathway. Although the relative amounts of isolated products are dependent to some extent on the K:Ge:As ratio of the reactants (besides an approximately constant amount of released Ta tube material of ca. 4%), the fusing temperature, the amount of solvent, the extraction time and the layering technique, it was not possible so far to isolate the separate species stepwise in real time. Mass spectrometry experiments indicate that the final clusters that co-crystallize in compound **4** can be already detected in the first spectrum recorded after injection of fresh extracts of the solid precursor phase. This indicates that either all of the detected species are formed side by side during cooling of the melt (the Ta-containing ones on contact of the K/Ge/As alloy with the Ta tube material or Ta powder), or that those reactions that occur in solution are rather quick. As explained and quantified below, the first steps leading from the anion in **1** through that in **2** to the first Ta-containing anion, $[Ta@Ge_4As_2]^-$, require higher temperatures, as they finally come along with the release of Ta atoms from Ta metal. The formation of the anion in **3** can follow in the solid or in solution at room temperature, given that *en* would sufficiently stabilize the $[Ta@Ge_4As_2]^-$ species. The last step, leading from the anion in **3** to the anions in **4**, is very likely a reaction in solution. For the processes being considerably rapid, we were definitely lucky to find a system that enabled the crystallization of reactants and diverse synthons along with the final products. Still, as more experimental information cannot be gained here due to the complicated and fast rearrangement processes within complex reaction mixtures, and owing to the lack of energetic information, a complementary quantum-chemical analysis of the formation process was necessary for a refined and at least semi-quantitative picture.

In this comprehensive study, we did not only consider all species mentioned above plus $(Ge_3As)^{3-}$ and $(As_2)^{2-}$ as additional leaving groups, but also the energetically higher isomers of the involved anions, which resulted from the GA-RP procedure; although the latter were not observed in stable crystals, they may play a role as reactive intermediates. Reaction energies were calculated for those steps of the reaction with atom- and charge-balance for educts and products, since only for these cases, the dependency of the calculated energy differences on the dielectric constant chosen in the COSMO approach is reasonably small (Supplementary Table 12). Reaction pathways were calculated with an iterative method based on a local quadratic approximation of the energy hypersurface[29].

Here, we will focus on three general aspects: First, the initial step that leads to the formation of the ubiquitous nine-atomic anion $(Ge_7As_2)^{2-}$ from the four-atomic and ten-atomic precursor anions $(Ge_2As_2)^{2-}$ and $(Ge_{10})^{2-}$. Second, the formation of the $[Ta@Ge_6As_4]^{3-}$ anion in **3** from $(Ge_7As_2)^{2-}$ through $[Ta@Ge_4As_2]^-$. These steps are apparently intuitive regarding the development of the structures (Fig. 2), thus we will use this part of the mechanism mainly to indicate the role of the transition metal atom. The final step to form $[Ta@Ge_8As_4]^{3-}$ and $[Ta@Ge_8As_6]^{3-}$ from $[Ta@Ge_6As_4]^{3-}$ lacks an according structural relationship. Hence, we explored the availability and the role of possible isomers of $[Ta@Ge_6As_4]^{3-}$ in the third part of the following discussion.

We note in advance that the formation pathway that we derived from our combined experimental and extensive quantum-chemical study represents only one of several possible routes, but obviously a very plausible one, as it is based on 'smooth' inter- or intramolecular re-arrangements; these are feasible with energies that are significantly smaller than those of uncompensated bond-breaking, like breaking an As–As bond in As_4, which amounts to $179\,kJ\,mol^{-1}$ (calculated as $(E(As_4)-4\,E(As))/6$) at level TPSS/dhf-TZVP), or breaking a Ge–Ge bond ($188\,kJ\,mol^{-1}$)[59]. The presented route leads to a topological relation between all structures therein, and as it also bridges to all previously known non-deltahedral clusters (see below).

Ta-independent first step. The co-existence of $(E^{14}{}_2E^{15}{}_2)^{2-}$, $(E^{14}{}_{10})^{2-}$ and $(E^{14}{}_7E^{15}{}_2)^{2-}$ anions in solutions of ternary $K/E^{14}/E^{15}$ phases[25,26,30,33,34] is very obvious, but has heretofore never been explained. Thus, we inspected this step in detail, which at the same time represents the initial step in the intermetalloid cluster formation pathway shown in Fig. 2. The energy difference for the formation of two $(Ge_7As_2)^{2-}$ anions from $(Ge_{10})^{2-}$ and two $(Ge_2As_2)^{2-}$ anions cannot be calculated reliably owing to missing charge balance (see above). For that, we performed the calculation with a total charge of 4– throughout, thus formally considering a partially oxidized precursor anion '$(Ge_2As_2)^-$'. The latter, which actually represents the electronic configuration of the species found under ESI-MS conditions, was calculated to be a global minimum with an elongated Ge…Ge

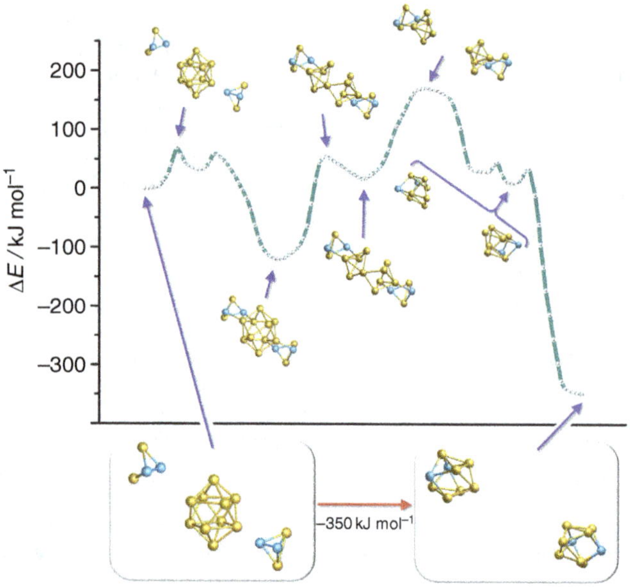

Figure 3 | Formation of two $(Ge_7As_2)^{2-}$ anions from $\{(Ge_2As_2)\ldots(Ge_{10})\ldots(Ge_2As_2)\}^{4-}$. The pathway was modelled based on the initial arrangement of three separate anions (bottom left), the final separate $(Ge_7As_2)^{2-}$ anions (bottom right) and an intermediate, local minimum structure, in which the two $(Ge_2As_2)^{-}$ species are attached to the $(Ge_{10})^{2-}$ cage in topologically reasonable way regarding the products (marked with an asterisk). Selected structures along the pathway are shown.

edge, according with the fact that the HOMO of $(Ge_2As_2)^{2-}$ represents the Ge–Ge bond. Although we cannot exclude that the oxidation takes place later on during this step, an early oxidation right at the beginning seems to be reasonable in terms of easier approach of the anionic species.

The reaction pathway from an initial arrangement of three separate species $\{(Ge_2As_2)\ldots(Ge_{10})\ldots(Ge_2As_2)\}^{4-}$ to the final arrangement of two separate anions $\{(Ge_7As_2)\ldots(Ge_7As_2)\}^{4-}$ (Fig. 3) was calculated[29] using 53 intermediate structures at level TPSS/dhf-TZVP (refs 52–54), including one intermediate optimized local minimum structure, in which the two (Ge_2As_2) species are attached to the (Ge_{10}) cage. An asterisk in Fig. 3 marks this intermediate. The chosen type of attachment is the one that requires minimum re-arrangement for the $(Ge_7As_2)^{2-}$ products (that is, further types of attachment cannot be excluded, but would cause higher barriers). The reaction pathway is summarized as follows. The intermediate is more stable than the initial arrangement by ca. 120 kJ mol^{-1}, and achievable by a small barrier of ca. 70 kJ mol^{-1} (and a second, smaller one). The final arrangement in turn is more stable than the intermediate by ca. 240 kJ mol^{-1}, but separated from the latter by two barriers, amounting to 180 kJ mol^{-1} and 160 kJ mol^{-1}, respectively (besides some further, much smaller ones). The first of the two barriers comes along with a significant break-up of the central Ge_{10} cage. The reason for a still rather small barrier in spite of the loss of a total of five Ge–Ge contacts can be explained by changes in nature and quality of the Ge–Ge bonds. The marked isomer exhibits electron-deficient multi-centre bonding, as reflected by an average coordination number of 4.8 for the 10 Ge atoms in the central part, whereas for the higher-energy state these atoms are in a nearly electron-precise situation with an average coordination number of 3.8 (including the exo-Ge–As bonds for both cases). Thus, the decreasing number of bonds is accompanied by an increasing bond order for the remaining bonds. The second barrier mainly reflects the cleavage of the remaining Ge–Ge bond between the two newly formed fragments. The dissociation energy is in the range of a Ge–Ge single bond (188 kJ mol^{-1}) (ref. 59). These newly formed fragments show a rather low-average coordination number (3.3), which increases to 3.6 as two pairs of Ge atoms approach each other (second image from right in Fig. 3) and finally to 4.0 in the nine-atom cages; the As atoms are three-bonded throughout the path before they turn into four-bonded ones during the final process. The entire reaction is exothermal by ca. 350 kJ mol^{-1}, and the barriers are conquerable in the high-temperature flux.

Inclusion of Ta and formation of the key anion in 3. Due to its shape, the $[Ta@Ge_6As_4]^{3-}$ anion in **3** is most reasonably described as $[Ta@Ge_4As_2]^{-}\cdot(Ge_2As_2)^{2-}$. This may be viewed as the result of an insertion of the $[Ta@Ge_4As_2]^{-}$ part of the anion in **3** into the As–As bond (that is, the LUMO) of the original $(Ge_2As_2)^{2-}$ tetrahedron. During this step, Ta is thus formally oxidized from a +3 to a +5 state. The fusion of these two parts to the anion in **3** is exothermic by 331 kJ mol^{-1} (Supplementary Table 12). The formation of the preceding species, $[Ta@Ge_4As_2]^{-}$, is intuitively considered as the reaction of Ta atoms with $(Ge_7As_2)^{2-}$ during the high-temperature reaction, as the former can be derived from the latter by replacement of a $(Ge_3)^{2-}$ ring with Ta; the $(Ge_3)^{2-}$ unit is thereby trapped by a second Ta atom to form $(TaGe_3)^{-}$ under 1-e$^-$ oxidation; this species was not experimentally isolated; it should be reactive enough to immediately react with half an equivalent of $(As_2)^{2-}$, released in the formation of $[Ta@Ge_8As_4]^{3-}$ (see final step below) to form the experimentally proven $(Ge_3As)^{3-}$ anion (Supplementary Fig. 19). The formation of $[Ta@Ge_4As_2]^{-}$ and $(TaGe_3)^{-}$ from $(Ge_7As_2)^{2-}$ and two Ta atoms is exothermic by 903 kJ mol^{-1} (Supplementary Table 12). Of course, the Ta atoms need to be liberated previously from the bulk phase, that is, the wall of the reaction tube. The enthalpy of atomization of Ta metal amounts to 782 kJ mol^{-1} per atom[59], which may be the upper limit. However, for a surface atom, this energy is much smaller, in particular when considering defects like edges or corners. It should be significantly smaller than the dissociation energy of (formally) quadruple-bonded Ta_2, 504 kJ mol^{-1} (energy difference to separated atoms at TPSS/dhf-TZVP level), so that the overall reaction will be exothermic.

If the subsequent formation of **3** takes place in solution, that is, during or shortly on the extraction step, a chelating solvent—such as *en*—is absolutely essential to intermediately stabilize the quoted interim species like $[Ta@Ge_4As_2]^{-}$ or $(TaGe_3)^{-}$ by coordinating to the valence-unsaturated Ta atom; according *en* complexes are presumably not stable enough for detection (at least under the conditions that we applied in our measurements), but indeed, other solvents than *en* did not lead to any identifiable products in this study. Alternatively, all initial steps might already occur in the flux, such that the anions in **1–3** co-exist in the solid prior to extraction.

Independent from the very reaction conditions, the described process involves species with stepwise increasing formal oxidation state at the Ta atoms (0, +3 and +5), which is in line with the idea of successive redox cascades involving the transition metal atom.

Isomerization processes and final step. The formation of the two clusters in **4**, $[Ta@Ge_8As_4]^{3-}$ and $[Ta@Ge_8As_6]^{3-}$, is not possible starting from $[Ta@Ge_6As_4]^{3-}$ in its experimentally observed shape (see Fig. 1) as neither the topology nor the Ge/As distribution match. Instead, higher-energy isomers of $[Ta@Ge_6As_4]^{3-}$ with more suitable topologies and Ge/As distributions with regard to the larger clusters should be the reactive ones in subsequent reactions with $(Ge_2As_2)^{2-}$ (to obtain

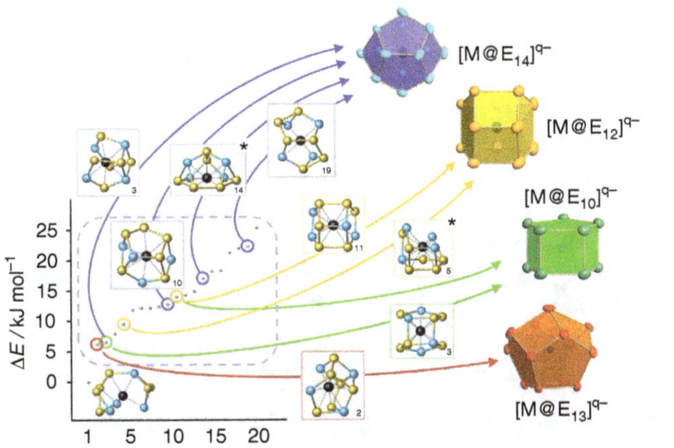

Figure 4 | Illustration of the low-energy re-arrangement of the anion in 3 to form isomers. In addition to the global minimum structure, the energies (dashed blue box) and selected structures of the 19 isomers following in energy are given. The drawn isomers represent exact fragments of the known non-deltahedral intermetalloid cluster topologies (for references see text). Two isomers are marked with an asterisk, symbolizing those that are discussed as direct precursors to the cluster with 12 atoms (isomer 5, yellow) or 14 atoms (isomer 14, blue), respectively.

the minimum structure of $[Ta@Ge_8As_4]^{3-}$ and $(As_2)^{2-}$ or the minimum structure of $[Ta@Ge_8As_6]^{3-}$ and $2e^-$). In principle, these isomers might form as kinetic products directly from the precursors $(Ge_2As_2)^{2-}$ and $(Ge_{10})^{2-}$, or the $(Ge_7As_2)^{2-}$ nine-atom cage in an alternative way than shown here; however, as this hypothesis requires much more assumptions and is not based on further experimental proof, we rather suggest a re-arrangement starting out from the experimentally observed global minimum structure of $[Ta@Ge_6As_4]^{3-}$.

For this purpose, we inspected the lowest 20 isomers resulting from the GA-RP procedure. These isomers are higher than the minimum by ca. 5–25 kJ mol^{-1}, thus their temporary existence is plausible. Many of them are fragments of regular $[M@E_n]$ anionic polyhedra ($n = 10$–14), that is, they have topologies fitting to all experimentally known non-deltahedral intermetalloid clusters containing tetrel atoms, as illustrated in Fig. 4. This finding extends the meaningfulness of this work to a more general understanding of multimetallic cluster formation.

As mentioned above, for a subsequent reaction to $[Ta@Ge_8As_4]^{3-}$ or to $[Ta@Ge_8As_6]^{3-}$, additionally a matching Ge/As distribution is required. This is the case, for the fifth-stable isomer of $[Ta@Ge_6As_4]^{3-}$ (isomer 5; +9 kJ mol^{-1}; marked with a yellow asterisk in Fig. 4), which is the lowest energy isomer representing an exact fragment of the experimentally found topology and global minimum Ge/As distribution of $[Ta@Ge_8As_4]^{3-}$. The latter (and additionally the reactive intermediate $(As_2)^{2-}$, see above) is formed from this fragment with $(Ge_2As_2)^{2-}$ in a near isoenergetic step (−1 kJ mol^{-1}). Similarly, the fourteenth-stable isomer (isomer 14; + 17 kJ mol^{-1}; marked with a blue asterisk in Fig. 4) is identical to the upper part of the most favourable $[Ta@Ge_8As_6]^{3-}$ structure. Hence, for completion of this cluster, only a (Ge_2As_2) unit is lacking; since the latter is originally provided in its dianionic form, an oxidation has to take place. As this reaction thus yields different charges for educts and products, the corresponding energy could not be reliably calculated (see above). However, similar energetics can be expected as for the reaction towards $[Ta@Ge_8As_4]^{3-}$.

For the described steps, it remained to find possible pathways between the global minimum of $[Ta@Ge_6As_4]^{3-}$, our 'start structure', and the two most promising reactive species, the fifth and the fourteenth-stable isomer, called 'end structures'. Here, we started from an initial pathway consisting of 14 intermediate structures obtained from interpolation between start and end structure, followed by an iterative optimization; in each iteration the gradients are calculated for each of the 14 structures. The number of possible pathways between any two isomers of $[Ta@Ge_6As_4]^{3-}$ amounts to $6! \cdot 4! = 17{,}280$, as there are 6! ways to connect corresponding the 6 Ge atoms of the start and end structure, and similarly 4! ways for the 4 As atoms. For the two isomerizations, initial pathways were calculated for all possibilities; however, the costly optimization procedure was only carried out for those with interatomic distances not shorter than 1.6 A for all atoms in all interpolated structures of the initial pathways. The discarded pathways will most probably not be favourable in the end, and furthermore, quantum-chemical calculations for atom distances far away from equilibrium are often problematic. For the remaining ca. 700 pathways (for each of the two isomerizations), 35 iterations were carried out with economic DFT settings BP86 (refs 44,45)/def-SVP (refs 46–48). Thereupon, the pathways with lowest barriers were refined with a larger number (53) of intermediate structures at the more accurate level TPSS/dhf-TZVP (refs 52–54) until convergence. For each of the two isomerizations, the finally resulting pathway with lowest barriers is shown in Fig. 5. Isomer 5 (fragment of $[Ta@Ge_8As_4]^{3-}$) can be reached from the global minimum via three barriers with heights of 76/63/46 kJ mol^{-1}. For isomer 14 (fragment of $[Ta@Ge_8As_6]^{3-}$), the first barrier amounts to 92 kJ mol^{-1}, followed by five barriers of 14/50/48/72/40 kJ mol^{-1}. All barriers are obviously high enough to allow for (low yield) crystallization of the $[Ta@Ge_6As_4]^{3-}$ anion in **3**, but also low enough to be finally overcome at the isomerization into suitable fragments as precursors of the two clusters in **4**. Interestingly, no direct pathway was found between isomer 5 and isomer 14, thus indicating that the subsequent reactions into the global minimum structures of $[Ta@Ge_8As_4]^{3-}$ or $[Ta@Ge_8As_6]^{3-}$, respectively (Supplementary Figs 29 and 30), are straight forward then.

One of the most important lessons we learned from these investigations is surprisingly obvious in its retrospective. Different from the preliminary assumption, and probably different from further suggestion on cluster formation pathways to be found in the literature, we deduce the following, general statement from our findings: clusters that are observed in crystalline 'intermediates' are not necessarily directly involved in further cluster growth, but they can play a key role in the reaction cascade upon isomerization, such as found in the present case. Here, several intermediate compositions co-exist in nearly isoenergetic isomeric forms, with higher-energy isomers being naturally more reactive—thus not being detectable or at least not isolable, and most probably being the species that are actually involved in the reaction cascade, whereas the global minimum species rather represent thermodynamic sinks. We note in passing that the exact tracing of the re-arrangement between the respective isomers was facilitated by the presence of two different elements in the present case.

With respect to the known charges of the involved species, a cascade of redox steps needs to be taken into consideration. A final overview of atom and electron balances along the pathway is provided in Fig. 6. Following our assumptions, we can explain the whereabouts of all electrons during the complex process—except for two electrons to be released at the initial step, which does not seem to be problematic regarding the variety of small polyanionic by-products formed alongside the reaction, detectable by means of ESI-mass spectrometry (Supplementary Figs 14,18 and 24).

In summary, our findings strongly suggest that the transition metal atoms come into play early in the cluster formation process,

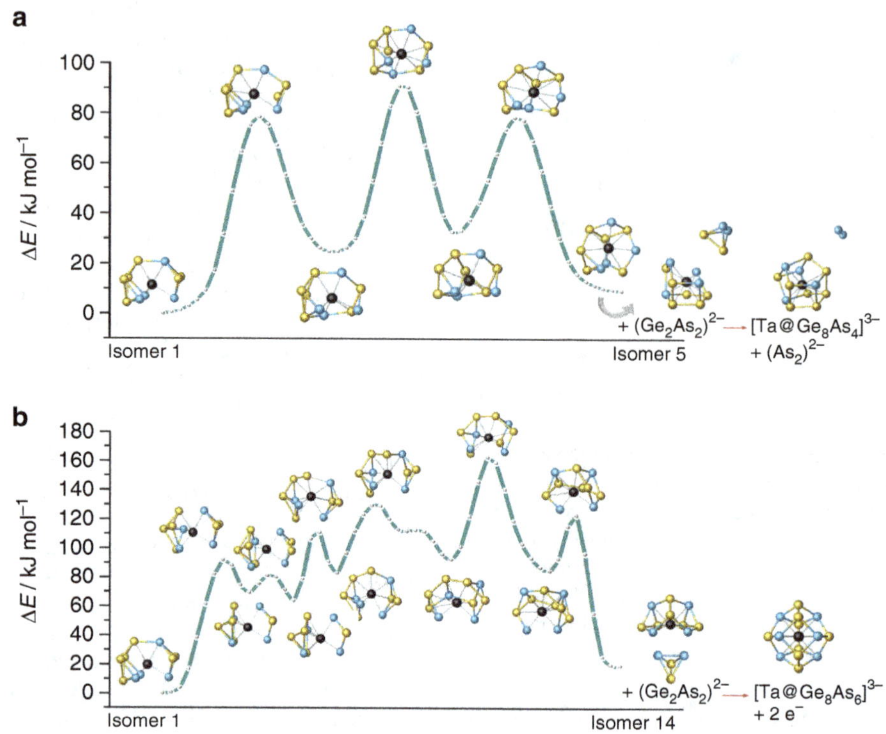

Figure 5 | Stepwise re-arrangements of the most stable anion isomer in 3. (**a**) Re-arrangement into isomer 5. (**b**) Rearrangement into isomer 14. The final topologies are suitable for uptake of another $(Ge_2As_2)^{2-}$ unit from **1** to form the clusters $[Ta@Ge_8As_4]^{3-}$ (top) or $[Ta@Ge_8As_6]^{3-}$ (bottom) in **4**. Detailed explanations are given in the text.

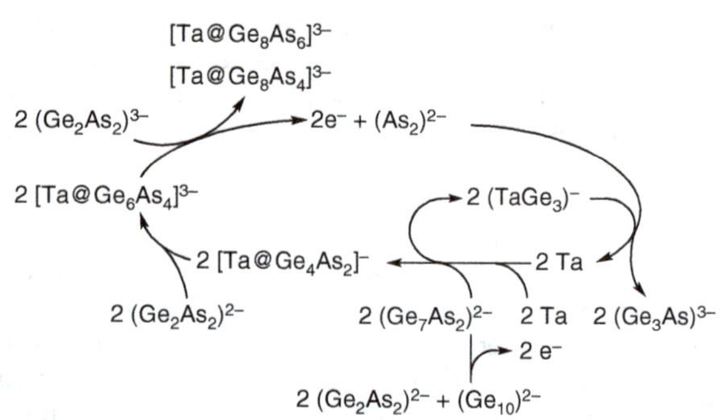

Figure 6 | Formation cycle for the generation of Ta/Ge/As clusters. The scheme gives an overview of atom and electron balances along the pathway for the formation of multimetallic clusters $[Ta@Ge_8As_6]^{3-}$ and $[Ta@Ge_8As_4]^{3-}$ starting out from $(Ge_2As_2)^{2-}$, $(Ge_{10})^{2-}$ and Ta.

that they are redox-active if delivered as neutral atoms, and that they are needed to weaken the bonds between the main-group atoms for easier fragmentation in their oxidized form. This clearly serves to explain the role of the transition metal atoms in such reactions in general, such that they can be viewed as catalysts triggering bond-breaking and bond-formation steps during the observed re-arrangements. Different from the behaviour of a common catalyst, at least a fraction of the transition metal atoms rather act as templates, such that they are finally surrounded by a rather stable, closed cluster shell, while another fraction seems to behave like a veritable catalyst, being involved in the formation of small fragments, which are anticipated to leave the formation pathway for being re-used again. Hence, we show that at least for the considered cluster family, the main-group atom shell is rather built stepwise around the transition metal atom instead of a transition metal atom entering a pre-formed, entire cluster shell.

Discussion

With this study, we contributed to a more detailed and at least semi-quantitative understanding of ligand-free, multimetallic cluster formation by providing deeper insight into possible processes. This was possible by a combination of new experimental findings on binary and ternary anionic clusters with comprehensive quantum-chemical investigations applying a new methodology (GA-RP)[28]. We identified several isomers of a previously unknown key synthon, which are topologically identical with exact fragments of all known non-deltahedral intermetalloid clusters containing tetrel atoms. Additionally, the study rationalized why all of the reactions of $(E^{14}{}_2E^{15}{}_2)^{2-}$ anions pass the $(E^{14}{}_7E^{15}{}_2)^{2-}$ step, and why and how these, as well as reactions that start out from $(E^{14}{}_9)^{4-}$, can proceed toward anions of the quoted type and composition by thorough screening of the energy hypersurface. This allowed shedding light on the formation of multimetallic clusters in general, not only for the compounds presented herein.

Our study still leaves room for further extension. In this regard, alternative pathways have to be explored that lead into other cluster families, such as the deltahedral ones. The uncountable variety of isomers with larger atom numbers and their re-arrangements into each other via energy barriers are likewise part of an exhaustive ongoing study, which has now become realistic by the development of the efficient methodology. The technique is applicable to all other systems dealing with binary or multinary cluster compounds, hence allowing for a detailed insight into cluster formation pathways. This way, we hope to finally contribute to a comprehensive understanding of inorganic reaction mechanisms in general, and to illuminate the former 'black box'.

Methods

General synthetic methods. All manipulations and reactions were performed under dry Ar atmosphere by using standard Schlenk or glovebox techniques. All solvents were dried and freshly distilled prior to use. [2.2.2]crypt (4,7,13,16,21,24-Hexaoxa-1,10-diazabicyclo[8.8.8]hexacosane, purchased as Kryptofix 222 from Merck) was dried *in vacuo* for at least 18 h. The synthesis of ternary phases $K_1Ge_1As_x$ ($x = 0.5$, 1) were approached by fusing K, Ge and As in the respective stoichiometric ratios in a silica glass ampoule with an oxygen torch or in a tantalum ampoule within an oven at 950 °C, respectively.

Fusion reactions. The generation of a solid mixture with the nominal composition of 'KGeAs' was approached by combining K, Ge and As in equimolar amounts in a tantalum ampoule, which was sealed by arc-welding within the glove box. The ampoule was then placed in an oven for 48 h with initial heating to 950 °C and subsequent slow cooling to room temperature (heating and cooling rates of 50 K h^{-1}). The resulting solid was thoroughly ground in a mortar prior to further use As confirmed by EDX spectroscopy results (Supplementary Fig. 1a and Supplementary Table 1), the solid product contains ~4 atom-% of Ta. For this, the precursor phase will be denoted as 'KGeAs:Ta' (precursor phase 1) in the following. The synthesis of the 'KGeAs' solid was also performed in a silica glass ampoule with an oxygen torch (precursor phase 2). Results of EDX spectroscopy are provided in Supplementary Fig. 1b and Supplementary Table 1.

Conjoint synthesis of 1–4. A total of 150 mg (0.81 mmol) of 'KGeAs:Ta' (precursor phase 1) and 460 mg (1.22 mmol) of [2.2.2]crypt were weighed out into a Schlenk tube. Then 1,2-diaminoethane (*en*, 4 ml) was added to result in a dark red suspension. The reaction mixture was allowed to stir for 2 days. The liquid was filtered through a standard glass frit, yielding a red solution that was carefully layered by toluene (*tol*, 7 ml). Crystals of **1** form within 3 days. After 10 days, four distinct kinds of crystals (**1–4**) can be identified in the Schlenk tube (Supplementary Fig. 2). Due to the variety of crystals obtained in this reaction specific yields for **1–4** could not be determined with certainty. The overall yield of crystalline material is ~19% (based on [2.2.2]crypt). $[K([2.2.2]crypt)]_5[K([2.2.2]crypt)(en)][Ta@Ge_8As_4]_{1.21}[Ta@Ge_8As_6]_{0.79} \cdot en$ (**4**) is the major product of this reaction according to visual inspection of the Schlenk tube (~60% of crystalline material).

Conjoint synthesis of 1 and 2. For proving that the presence of Ta is not mandatory for the formation of salts of the $(Ge_2As_2)^{2-}$ and $(Ge_7As_2)^{2-}$ anion, 150 mg (0.81 mmol) of 'KGeAs' prepared in a silica glass ampoule (precursor phase 2) and 460 mg (1.22 mmol) of [2.2.2]crypt were weighed out into a Schlenk tube and then suspended in *en* (4 ml). The reaction mixture was allowed to stir for 1 day. The orange mixture was filtered through a standard glass frit, and the resulting solution was carefully layered by *tol* (7 ml). After 7 days, black-looking, block-like crystals of **2** as well as orange plates of **1** were obtained. On cutting the dark blocks into a size suitable for single-crystal X-ray diffraction, they split off into orange plates (Supplementary Fig. 2, centre), which had only been agglomerated. Due to the crystal mixture, specific yields for **1** and **2** could not be determined with certainty. The overall yield of crystalline material is ~24% (based on [2.2.2]crypt).

Energy dispersive X-ray spectroscopy analyses. EDX analyses were performed to support the elemental composition that was suggested based on the single-crystal X-ray diffraction experiments. These were carried out using an EDX-device Voyager 4.0 of Noran Instruments coupled with an electron microscope CamScan CS 4DV. Data acquisition was performed with an acceleration voltage of 20 kV and 100 s accumulation time. The radiation emitted by the atoms was analysed: K-K, Ge-K, As-K and Ta-M/L. To minimize surface effects in the measurement, the K-lines were preferably used to calculate the elemental composition.

The spectra measured on the precursor phases 1 and 2 are shown in Supplementary Fig. 1, the results are summarized in Supplementary Table 1. Supplementary Table 2 summarizes the results for the precursor phases and compounds **1–4**, respectively. The results of the EDX investigations confirm the element ratios of the investigated substances within the expected accuracy.

Crystallographic study of 1–4. Single-crystal X-ray diffraction data were collected on STOE imaging plate systems IPDS2 or IPDS2T, using graphite-monochromized Mo-Kα radiation ($\lambda_{Mo-K\alpha} = 0.71073$ Å) at 100 K. The structures were solved by direct methods, using SHELXS-97 (ref. 60) or SIR2011 (ref. 61), and refined by full-matrix-least-squares methods against F^2 with SHELXL-2013 software[60]. Crystal data: (**1**) $C_{38}H_{80}As_2Ge_2K_2N_6O_{12}$, $M_r = 1186.3$, triclinic, space group $P1$, $a = 10.9738(4)$ Å, $b = 11.9313(4)$ Å, $c = 12.6146(6)$ Å, $\alpha = 118.021(3)°$, $\beta = 108.361(4)°$, $\gamma = 96.476(3)°$, $V = 1{,}315.07(10)$ Å^3, $Z = 1$, $\rho_{calc} = 1.498$ g cm^{-3}, $\mu(Mo_{K\alpha}) = 2.608$ mm^{-1}, 23,683 reflections were measured, 10,275 of which were unique, $R(int) = 0.065$, R_1 ($I > 2\sigma(I)$) $= 0.0454$, wR_2 (all data) $= 0.1189$, S (all data) $= 1.075$. (**2**) $C_{36}H_{72}As_2Ge_7K_2N_4O_{12}$, $M_r = 1{,}489.14$, trigonal, space group $P\bar{3}c1$, $a = 11.8653(3)$ Å, $c = 22.3848(9)$ Å, $V = 2729.2(2)$ Å^3, $Z = 2$, $\rho_{calc} = 1.812$ g cm^{-3}, $\mu(Mo_{K\alpha}) = 5.212$ mm^{-1}, 3,581 reflections were measured, 1,942 of which were unique, $R(int) = 0.035$, R_1 ($I > 2\sigma(I)$) $= 0.0521$, wR_2 (all data) $= 0.1048$, S (all data) $= 0.958$. (**3**) $C_{68}H_{124}As_4Ge_6K_3N_6O_{18}Ta$, $M_r = 2347.19$, monoclinic, space group $P2_1/n$, $a = 13.7543(2)$ Å, $b = 28.6077(5)$ Å, $c = 22.9845(4)$ Å, $\beta = 92.489(1)°$, $V = 9{,}035.4(3)$ Å^3, $Z = 4$, $\rho_{calc} = 1.725$ g cm^{-3}, $\mu(Mo_{K\alpha}) = 4.829$ mm^{-1}, 148,425 reflections were measured, 19,185 of which were unique, $R(int) = 0.070$, R_1 ($I > 2\sigma(I)$) $= 0.0346$, wR_2 (all data) $= 0.0838$, S (all data) $= 0.884$. (**4**) $C_{56}H_{116}As_{4.79}Ge_8K_3N_8O_{18}Ta$, $M_r = 2{,}427.40$, triclinic, space group $P\bar{1}$, $a = 16.6082(5)$ Å, $b = 22.6960(8)$ Å, $c = 23.9068(8)$ Å, $\alpha = 94.960(3)°$, $\beta = 94.036(3)°$, $\gamma = 91.402(3)°$, $V = 8951.2(5)$ Å^3, $Z = 4$, $\rho_{calc} = 1.801$ g cm^{-3}, $\mu(Mo_{K\alpha}) = 5.819$ mm^{-1}, 64,236 reflections were measured, 31,253 of which were unique, $R(int) = 0.057$, R_1 ($I > 2\sigma(I)$) $= 0.0735$, wR_2 (all data) $= 0.1973$, S (all data) $= 0.990$. Further details are given in Supplementary Table 3. Cambridge Crystallographic Data Centre deposition codes are provided in the Accession codes section.

Details of the structure determination of 1. The structure of $[K([2.2.2]crypt)]_2[Ge_2As_2] \cdot en$ (**1**) comprises two independent $[K[2.2.2]crypt)]^+$ cations, a tetrahedral $(Ge_2As_2)^{2-}$ anion (Supplementary Fig. 3) and a disordered *en* molecule. The space group is indeed $P1$ with pseudo-symmetry $P\bar{1}$. The tetrahedral $(Ge_2As_2)^{2-}$ anion breaks the centro-symmetry. At $Z = 1$ it would appear disordered on the atomic positions of a cube in case of true centro-symmetry, which is not the case here. Since Ge and As cannot be distinguished by X-ray diffraction using Mo-Kα radiation, all four sites are half-occupied by Ge and As. Supplementary Table 4 summarizes interatomic distances and angles. The cations form a honeycombe-like packing with channels along a in which the $(Ge_2As_2)^{2-}$ anions are aligned together with the disordered *en* molecules (Supplementary Fig. 4).

Details of the structure determination of 2. Compound $[K([2.2.2]crypt)]_2[Ge_7As_2]$ (**2**) crystallizes in space group $P\bar{3}c1$ with the $[K[2.2.2]crypt)]^+$ cation on a threefold axis (site 3d) and the $(Ge_7As_2)^{2-}$ anion on the 2a site with symmetry 32 (D_3). The Ge/As occupation 7:2 as confirmed by mass spectrometry (MS) and density functional theory (DFT) investigations was fixed for each cluster atom. Around the 2a site, 24 metal positions could be located. This can be explained by orientational disorder of a 9-atom anion cluster (Supplementary Fig. 5a) over three positions generated by the threefold axis along [001] with Ge/As1 common for two orientations (Supplementary Fig. 5b). The large displacement ellipsoids let us assume that the C_2 axis (or C_4 axis with indistinguishable Ge/As) of the $(Ge_7As_2)^{2-}$ anion does not coincide exactly with the C_2 axes of the crystallographic 2a site. Thus the disorder model is really an overlay of six instead of three orientations and the refined geometry may appear adulterated, therefore. Supplementary Table 5 summarizes interatomic distances and angles. Similar to compound **1**, but with trigonal symmetry, the cations form a honeycombe-like packing with the disordered anions in channels along the c axis (Supplementary Fig. 6).

Details of the structure determination of 3. The structure of $[K([2.2.2]crypt)]_3[Ta@Ge_6As_4] \cdot 2tol$ (**3**) in space group $P2_1/n$ is built by a $[Ta@Ge_6As_4]^{3-}$ anion showing statistical overlay of two isomers with occupations of 87.7(1) and 12.3(1)% (Supplementary Fig. 7, Supplementary Table 6), three independent [K([2.2.2]crypt)] cations (Supplementary Fig. 8) and two toluene solvent molecules. The interpretation of the disorder model of the anion isomers and attribution of Ge and As to the atom sites has been done based on theoretical calculations. The large anisotropic displacement parameters of Ge7A, Ge8A and Ge9A suggest additional orientational disorder of the tetrahedral $(Ge_3As)^{3-}$ group of isomer 2. The bond lengths in this region may therefore be adulterated. The left hand side of the isomer might derive from an intermediately formed $(Ge_6As_3)^-$ nine-atom cage, homologues of which had been observed in previous studies with Sn/Bi or Pb/Bi anions[25,26,34]. The formation of $(Ge_6As_3)^-$, in turn, is attributed on another relative orientation of $(Ge_2As_2)^{2-}$ and $(Ge_{10})^{2-}$ during the respective attack in the first step of the reaction cascade. The complicated packing of cations, anions and toluene molecules in compound **3** is shown in Supplementary Fig. 9.

Details of the structure determination of 4. The triclinic centrosymmetric structure of $[K([2.2.2]crypt)]_5[K([2.2.2]crypt)(en)][Ta@Ge_8As_4]_{1.21}[Ta@Ge_8As_6]_{0.79} \cdot en$ (**4**) shows severe disorder effects on the two independent anion positions as well as at some of the six independent cations. The anions are on two independent sites. On site one, superposition of a $[Ta@Ge_8As_4]^{3-}$ cluster (89.4%) and a $[Ta@Ge_8As_6]^{3-}$ cluster (10.6%) has been found. The occupations by Ge and/or As were taken according to information from EDX spectroscopy, mass spectrometry, the most stable configurations of DFT calculations and their probable disordered orientations (Supplementary Fig. 10). Bond lengths are given in Supplementary Table 7. On site 2, a $[Ta@Ge_8As_6]^{3-}$ cluster is dominating (68.5%) superimposed by a $[Ta@Ge_8As_4]^{3-}$ cluster in two different orientations (12.9 and 18.6%). By this complicated disorder with many approximately common positions, the individual geometrical data appear adulterated and are not listed, therefore. The structures are given in Supplementary Fig. 11. Five of the six cations are [K[2.2.2](crypt)] cations like in the structures of compounds **1–3**. The refinement of these [2.2.2]crypt molecules in the presence of many heavy atoms was performed using geometrical restraints on the bond lengths and 1,3- distances. The anisotropic displacement ellipsoids show sometimes irregular shape as they include disorder effects (Supplementary Fig. 12a–d). They were refined with restraints to avoid too anisotropic displacement parameters. For one [2.2.2]crypt ligand with strong disorder, no sensible disorder model could be established. Its contribution was subtracted, therefore, by the back Fourier transform method from the data set. A sixth cation has an *en* molecule coordinated to K^+ in addition to the [2.2.2]crypt ligand (Supplementary Fig. 12e). In addition, a non-coordinated *en* molecule was located. The packing of molecules in the structure of compound **4** is shown in Supplementary Fig. 13.

Electrospray ionization mass spectrometry investigations. ESI-MS measurements have been performed on a Finnigan LTQ-FT spectrometer by Thermo Fischer Scientific in the negative ion mode: Spray voltage 3.90 kV, capillary temperature 300 °C, capillary voltage −11 V, tube lens voltage −140 V, sheath gas flow rate 25 arb, sweep gas flow rate 0 arb. As it is common for Zintl anions and intermetallic cluster anions, the observed fragments have been detected as oxidized, singly charged species.

ESI-MS of the *DMF/en*/[2.2.2]crypt extract of 'KGeAs'. The ESI(−) mass spectrum of the extract of 'KGeAs' (precursor phase 2, prepared in a silica glass ampoule) in DMF/*en* in presence of [2.2.2]crypt was measured after 18 h of extraction time (Supplementary Fig. 14). The study confirms the concurrent presence of the singly charged cluster species (Ge_2As_2H)—(m/z = 296.69, Supplementary Fig. 15), obviously formed under ESI-MS conditions by protonation of the anion in **1**, as well as $(Ge_7As_2)^-$ (m/z = 658.29, Supplementary Fig. 16) side by side with $(Ge_{10})^-$ (m/z = 726.22, Supplementary Fig. 17) in the same solution. The concurrent presence of these polyanions strongly supports the assumption that the 9-atom cluster $(Ge_7As_2)^{2-}$ is the product of the reaction of two $(Ge_2As_2)^{2-}$ clusters with one $(Ge_{10})^{2-}$. Also adducts of K^+ and $[K([2.2.2]crypt)]^+$ were detected: $(Ge_7As_2K)^-$ (m/z = 697.26), ($Ge_{10}C_{18}H_{36}N_2O_6K$; m/z = 1141.44).

ESI-MS of the *en*/[2.2.2]crypt extract of 'KGeAs:Ta' . The ESI(−) mass spectra of the extract of 'KGeAs:Ta' (precursor phase 1, prepared in a Ta ampoule) in *en* in the presence of [2.2.2]crypt was measured after 3 h of extraction time. The overview spectrum (Supplementary Fig. 18) confirms the presence of a variety of clusters in solution. Remarkable is the presence anions of **1** and intermediate **3** in the solution after just 3 h of extraction. Due to fragmentation of the high-mass isotopic patterns and increasing degradation of the solution during injection the observed intensity of the isotopic pattern decreased during the measurement. In the reaction mixture, the presence of $(Ge_2As_2H)^-$ (m/z = 296.69) beside $(Ge_3As)^-$ (m/z = 292.68) was confirmed (Supplementary Fig. 19). The isotopic patterns of $(Ge_6As_4Ta)^-$ (m/z = 916.17), $(Ge_6As_4TaK)^-$ (m/z = 955.13) and $(Ge_6As_4TaK_2)^-$ (m/z = 994.09, Supplementary Figs 20–22) were identified as well as these of $[K([2.2.2]crypt)]^+$ adducts. In addition, the isotopic pattern of $(Ge_5As_3C_{18}H_{36}N_2O_6K_3)^-$ (m/z = 1082.5) was identified (Supplementary Fig. 23).

ESI-MS of a *DMF/en* solution of 4. In the spectrum, various anions with and without $[K([2.2.2]crypt)]^+$ were identified (Supplementary Fig. 24). The following species, representing oxidized clusters of $[Ta@Ge_8As_4]^{3-}$ and $[Ta@Ge_8As_6]^{3-}$ were found: $(Ge_8As_4TaC_{36}H_{72}N_4O_{12}K_2)^-$ (m/z = 1892.46, Supplementary Fig. 25) and $(Ge_8As_6TaC_{18}H_{36}N_2O_6K)^-$ (m/z = 1627.07, Supplementary Fig. 26).

Quantum-chemical methods. All calculations were done with the programme system TURBOMOLE (ref. 38). Global minimum searches were carried out with a DFT-based GA (refs 42,43) extended[28] by an atom-to-place re-assignment step[50,51] with the following settings: Population size $P = 20$ structures; cross-over after optimization of 10 structures, leading to the formation of $P = 10$ new (child) structures. The mutation probability was set to 1%. The procedure was stopped after 30 generations. Optimizations of reaction pathways were done with an iterative method based on a local quadratic approximation of the energy hypersurface[29]. For these steps bases of polarized split-valence quality, def-SV(P)[46,47] (with an effective core potential of Wood-Boring type for Ta (ref. 48) and the generalized gradient approximation DFT functional (BP86) by Becke[44] and Perdew[45] were chosen for reasons of economy. The resulting best structures and best pathways were re-optimized using more flexible polarized triple zeta valence basis sets dhf-TZVP (ref. 53; with an effective core potential of the Dirac-Hartree-Fock type for Ta)[54] and the meta-generalized gradient approximation functional by Tao, Perdew, Staroverov and Scuseria (TPSS)[52]. For comparison, also PBE[56] and TPSSh[57] functionals were applied, with the results given in Supplementary Tables 8, 9 and 12. For all cases, the negative charges of the clusters were compensated by using the COSMO[49], with the dielectric constant ε set to infinity (default). Thermochemical data were calculated from partition sums within the standard harmonic oscillator approximation for molecules in the gas phase[58]. The vibrational frequencies were used non-scaled. The RI approximation was used throughout[62]. Molecules were visualized with the programme CYLView[63].

References

1. Corey, E. J. The logic of chemical synthesis: multistep synthesis of complex carbogenic molecules (nobel lecture). *Angew. Chem. Int. Ed.* **30,** 455–465 (1991).
2. Okrut, A. *et al.* Selective molecular recognition by nanoscale environments in a supported iridium cluster catalyst. *Nat. Nanotechnol.* **9,** 459–465 (2014).
3. Bals, S. *et al.* Atomic scale dynamics of ultrasmall germanium clusters. *Nat. Commun.* **3,** 897 (2012).
4. Qian, M. *et al.* Cluster-assembled materials: toward nanomaterials with precise control over properties. *ACS Nano* **4,** 235–240 (2009).
5. Armatas, G. S. & Kanatzidis, M. G. Mesostructured Germanium with cubic pore Symmetry. *Nature* **441,** 1122–1125 (2006).
6. Guloy, A. *et al.* A guest-free germanium clathrate. *Nature* **443,** 320–323 (2006).
7. Luo, Z. & Castleman, A. W. Special and general superatoms. *Acc. Chem. Res.* **47,** 2931–2940 (2014).
8. Grubisic, A., Ko, Y. J., Wang, H. & Bowen, K. H. Photoelectron spectroscopy of lanthanide-silicon cluster anions $LnSi_n^-$ ($3 \leq n \leq 13$; Ln = Ho, Gd, Pr, Sm, Eu, Yb): prospect for magnetic silicon-based clusters. *J. Am. Chem. Soc.* **131,** 10783–10790 (2009).
9. Wagner, F. R., Noor, A. & Kempe, R. Ultrashort metal–metal distances and extreme bond orders. *Nat. Chem.* **1,** 529–536 (2009).
10. Kong, X.-J., Long, L.-S., Zheng, Z., Huang, R.-B. & Zheng, L.-S. Keeping the ball rolling: fullerene-like molecular clusters. *Acc. Chem. Res.* **43,** 201–209 (2010).
11. Korber, N. The Shape of germanium clusters to come. *Angew. Chem. Int. Ed.* **48,** 3216–3217 (2009).
12. Zhou, S., McIlwrath, K., Jackson, G. & Eichhorn, B. W. Enhanced CO tolerance for hydrogen activation in au–pt dendritic heteroaggregate nanostructures. *J. Am. Chem. Soc.* **128,** 1780–1781 (2006).
13. Fässler, T., Scharfe, S., Kraus, F., Stegmaier, S. & Schier, A. Zintl Ions, cage compounds, and intermetalloid clusters of group 14 and group 15 elements. *Angew. Chem. Int. Ed.* **50,** 3630–3670 (2011).
14. Butovskii, M. V. *et al.* Molecules containing rare-earth atoms solely bonded by transition metals. *Nat. Chem.* **2,** 741–744 (2010).
15. Lips, F. & Dehnen, S. $[Zn_6Sn_3Bi_8]^{4-}$: expanding the intermetalloid zintl anion concept to ternary systems. *Angew. Chem. Int. Ed.* **48,** 6435–6438 (2009).
16. Lips, F., Clérac, R. & Dehnen, S. $[Pd_3Sn_8Bi_6]^{4-}$: A 14-vertex Sn/Bi cluster embedding a Pd_3 triangle. *J. Am. Chem. Soc.* **133,** 14168–14171 (2011).
17. Wang, J.-Q., Stegmaier, S., Wahl, B. & Fässler, T. Step-by-step synthesis of the endohedral stannaspherene $[Ir@Sn_{12}]^{3-}$ via the capped cluster anion $[Sn_9Ir(cod)]^{3-}$. *Chem. Eur. J.* **16,** 1793–1798 (2010).
18. Moses, M. J., Fettinger, J. & Eichhorn, B. W. Charged molecular alloys: synthesis and characterization of the binary anions $Pd_7As_{16}^{4-}$ and $Pd_2As_{14}^{4-}$. *J. Am. Chem. Soc.* **124,** 5944–5945 (2002).
19. Kesanli, B. *et al.* Cluster growth and fragmentation in the highly fluxional platinum derivatives of Sn_9^{4-}: synthesis, characterization, and solution dynamics of $Pt_2@Sn_{17}^{4-}$ and $Pt@Sn_9H^{3-}$. *J. Am. Chem. Soc.* **129,** 4567–4574 (2007).
20. Schnöckel, H. Metalloid Al- and Ga-clusters: a novel dimension in organometallic chemistry linking the molecular and the solid-state areas? *Dalton Trans.* 3131–3136 (2005).
21. Benda, C. B., Waibel, M. & Fässler, T. F. On the formation of intermetalloid clusters: titanocene(III)diammin as a versatile reactant toward nonastannide zintl clusters. *Angew. Chem. Int. Ed.* **54,** 522–526 (2015).
22. Wang, J.-Q., Stegmaier, S. & Fässler, T. $[Co@Ge_{10}]^{3-}$: an intermetalloid cluster with archimedean pentagonal prismatic structure. *Angew. Chem. Int. Ed.* **48,** 1998–2002 (2009).
23. Zhou, B., Denning, M. S., Kays, D. L. & Goicoechea, J. M. Synthesis and isolation of $[Fe@Ge_{10}^{3-}]$: a pentagonal prismatic zintl ion cage encapsulating an interstitial iron atom. *J. Am. Chem. Soc.* **131,** 2802–2803 (2009).
24. Krämer, T. *et al.* Structural trends in ten-vertex endohedral clusters, $M@E_{10}$ and the synthesis of a new member of the family, $[Fe@Sn_{10}]^{3-}$. *Dalton Trans.* **42,** 12120–12129 (2013).
25. Lips, F., Clérac, R. & Dehnen, S. $[Eu@Sn_6Bi_8]^{4-}$: a mini-fullerane-type zintl anion containing a lanthanide ion. *Angew. Chem. Int. Ed.* **50,** 960–964 (2010).
26. Lips, F. *et al.* Doped semimetal clusters: ternary, intermetalloid anions $[Ln@Sn_7Bi_7]^{4-}$ and $[Ln@Sn_4Bi_9]^{4-}$ (Ln = La, Ce) with adjustable magnetic properties. *J. Am. Chem. Soc.* **134,** 1181–1191 (2012).
27. Weinert, B., Weigend, F. & Dehnen, S. Subtle impact of atomic ratio, charge and lewis basicity on structure selection and stability: the Zintl anion $[(La@In_2Bi_{11})(\mu\text{-}Bi)_2(La@In_2Bi_{11})]^{6-}$. *Chem. Eur. J.* **18,** 13589–13595 (2012).
28. Weigend, F. Extending DFT-based genetic algorithms by atom-to-place reassignment via perturbation theory: a systematic and unbiased approach to structures of mixed-metallic clusters. *J. Chem. Phys.* **141,** 134103 (2014).
29. Plessow, P. Reaction path optimization without NEB springs or interpolation algorithms. *J. Chem. Theory Comput.* **9,** 1305–1310 (2013).
30. Critchlow, S. C. & Corbett, J. D. Heteropolyatomic anions of the post transition metals. Synthesis and structure of the ditindibismuthide(2−) anion, $Sn_2Bi_2^{2}$. *Inorg. Chem.* **21,** 3286–3290 (1982).
31. Corbett, J. D. Polyatomic Zintl anions of the post-transition elements. *Chem. Rev.* **85,** 383–397 (1985).
32. Xu, L. i. & Sevov, S. C. Heteroatomic deltahedral clusters of main-group elements: synthesis and structure of the Zintl ions $[In_4Bi_5]^{3-}$, $[InBi_3]^{2-}$, and $[GaBi_3]^{2-}$. *Inorg. Chem.* **39,** 5383–5389 (2000).

33. Lips, F., Schellenberg, I., Pöttgen, R. & Dehnen, S. The subtle influence of binary versus homoatomic Zintl ions: the phenyl-ligated trimetallic cage $[Sn_2Sb_5(ZnPh)_2]^{3-}$. *Chem. Eur. J.* **15,** 12968–12973 (2009).
34. Ababei, R. *et al.* Making practical use of the pseudo-element concept: an efficient way to ternary intermetalloid clusters by an isoelectronic Pb^-–Bi combination. *Chem. Commun.* **48,** 11295–11297 (2012).
35. Belin, C., Mercier, H. & Angilella, V. Homoatomic polyanions of post transition elements. synthesis and structure of salts containing the decagermanide Ge_{10}^{2-} and the paramagnetic Ge_9^{3-} anions. *New J. Chem.* **15,** 931–938 (1991).
36. Espinoza-Quintero, G., Duckworth, J. C. A., Myers, W. K., McGrady, J. E. & Goicoechea, J. M. Synthesis and characterization of $[Ru@Ge_{12}]^{3-}$: an endohedral 3-connected cluster. *J. Am. Chem. Soc.* **136,** 1210–1213 (2014).
37. Mitzinger, S., Broeckaert, L., Massa, W., Weigend, F. & Dehnen, S. $[V@Ge_8As_4]^{3-}$ and $[Nb@Ge_8As_6]^{3-}$: unprecedented encapsulation of electron-poor transition metal atoms. *Chem. Commun.* **51,** 3866–3869 (2015).
38. TURBOMOLE 6.6, TURBOMOLE GmbH 2014. University of Karlsruhe and Forschungszentrum Karlsruhe, 1989-2007, TURBOMOLE GmbH since 2007.
39. Wales, D. J. & Doye, J. P. K. Global optimization by basin-hopping and the lowest energy structures of Lennard-Jones clusters containing up to 110 atoms. *J. Phys. Chem. A* **101,** 5111–5116 (1997).
40. Froemming, N. S. & Henkelman, G. Optimizing core-shell nanoparticle catalysts with a genetic algorithm. *J. Chem. Phys.* **131,** 234103 (2009).
41. Hartke, B. Geometry optimization of clusters using genetic algorithms. *J. Phys. Chem.* **97,** 9973–9976 (1993).
42. Deaven, D. M. & Ho, K. M. Molecular geometry optimization with a genetic algorithm. *Phys. Rev. Lett.* **75,** 288–291 (1995).
43. Sierka, M. *et al.* Unexpected structures of aluminum oxide clusters in the gas phase. *Angew. Chem. Int. Ed.* **46,** 3372–3375 (2007).
44. Becke, A. D. Density-functional exchange-energy approximation with correct asymptotic behavior. *Phys. Rev. A* **38,** 3098–3100 (1988).
45. Perdew, J. P. Density-functional approximation for the correlation energy of the inhomogeneous electron gas. *Phys. Rev. B* **33,** 8822–8824 (1986).
46. Schäfer, A., Horn, H. & Ahlrichs, R. Fully optimized contracted Gaussian basis sets for atoms Li to Kr. *J. Chem. Phys.* **97,** 2571 (1992).
47. Eichkorn, K., Weigend, F., Treutler, O. & Ahlrichs, R. Auxiliary basis sets for main row atoms and transition metals and their use to approximate coulomb potentials. *Theor. Chem. Acc.* **97,** 119–124 (1997).
48. Andrae, D., Haeussermann, U., Dolg, M., Stoll, H. & Preuss, H. Energy-adjusted ab initio pseudopotentials for second and third row transition elements. *Theor. Chim. Acta* **77,** 123–141 (1990).
49. Klamt, A. & Schüürmann, G. COSMO: a new approach to dielectric screening in solvents with explicit expressions for the screening energy and its gradient. *J. Chem. Soc. Perkin Trans.* **2,** 799–805 (1993).
50. Weigend, F., Schrodt, C. & Ahlrichs, R. Atom distributions in binary atom clusters: a perturbational approach and its validation in a case study. *J. Chem. Phys.* **121,** 10380–10384 (2004).
51. Weigend, F. & Schrodt, C. Atom-type assignment in molecules and clusters by perturbation theory—a complement to X-ray structure analysis. *Chem. Eur. J.* **11,** 3559–3564 (2005).
52. Tao, J., Perdew, J. P., Staroverov, V. N. & Scuseria, G. E. Climbing the density functional ladder: nonempirical meta-generalized gradient approximation designed for molecules and solids. *Phys. Rev. Lett.* **91,** 146401 (2003).
53. Weigend, F. & Baldes, A. Segmented contracted basis sets for one- and two-component Dirac-Fock effective core potentials. *J. Chem. Phys.* **133,** 174102 (2010).
54. Figgen, D., Peterson, K. A., Dolg, M. & Stoll, H. Correlation consistent basis sets: the transition metals Hf—Pt. *J. Chem. Phys.* **130,** 164108 (2009).
55. Drebov, N., Weigend, F. & Ahlrichs, R. Structures and properties of neutral gallium clusters: a theoretical investigation. *J. Chem. Phys.* **135,** 044314 (2011).
56. Perdew, J. P., Burke, K. & Ernzerhof, M. Generalized gradient approximation made simple. *Phys. Rev. Lett.* **77,** 3865–3868 (1996).
57. Staroverov, V. N., Scuseria, G. E., Tao, J. & Perdew, J. P. Comparative assessment of a new nonempirical density functional: molecules and hydrogen-bonded complexes. *J. Chem. Phys.* **119,** 12129–12137 (2003).
58. McQuarrie, D. A. & Simon, J. D. *Molecular Thermodynamics* ISBN 978-1891389054 (University Science Books, 1999).
59. James, A. M. & Lord, M. P. in *Macmillan's Chemical and Physical Data* (Macmillan, 1992).
60. Sheldrick, G. M. A short history of SHELX. *Acta Crystallogr.* **A64,** 112–122 (2008).
61. Burla, M. C. *et al.* SIR2011: a new package for crystal structure determination and refinement. *J. Appl. Crystallogr.* **45,** 357–361 (2012).
62. Weigend, F. Accurate coulomb-fitting basis sets for H to Rn. *Phys. Chem. Chem. Phys.* **8,** 1057–1065 (2006).
63. Legault, C. Y. CYLView. 1.0b (Université de Sherbrooke, 2009).

Acknowledgements

This work was supported by the Alexander von Humboldt Foundation, the Friedrich-Ebert-Stiftung and the Deutsche Forschungsgemeinschaft (DFG). We thank R. Riedel for X-ray data collection.

Author contributions

S.M. conceived and performed the synthetic experiments and further analyses (XRD, EDX spectroscopy and ESI-mass spectrometry). W.M. solved and refined the structures. L.B. and F.W. planned and performed all quantum-chemical investigations. S.D. co-planned and supervised the work. All authors co-wrote the paper.

Additional information

Accession codes: The X-ray crystallographic coordinates for structures reported in this Article have been deposited at the Cambridge Crystallographic Data Centre, under deposition numbers 1016136-1016139. These data can be obtained free of charge from The Cambridge Crystallographic Data Centre via www.ccdc.cam.ac.uk/data_request/cif.

Competing financial interests: The authors declare no competing financial interests.

27

Porous molybdenum carbide nano-octahedrons synthesized via confined carburization in metal-organic frameworks for efficient hydrogen production

Hao Bin Wu[1,*], Bao Yu Xia[1,*], Le Yu[1], Xin-Yao Yu[1] & Xiong Wen (David) Lou[1]

Electrochemical water splitting has been considered as a promising approach to produce clean and sustainable hydrogen fuel. However, the lack of high-performance and low-cost electrocatalysts for hydrogen evolution reaction hinders the large-scale application. As a new class of porous materials with tunable structure and composition, metal-organic frameworks have been considered as promising candidates to synthesize various functional materials. Here we demonstrate a metal-organic frameworks-assisted strategy for synthesizing nano-structured transition metal carbides based on the confined carburization in metal-organic frameworks matrix. Starting from a compound consisting of copper-based metal-organic frameworks host and molybdenum-based polyoxometalates guest, mesoporous molybdenum carbide nano-octahedrons composed of ultrafine nanocrystallites are successfully prepared as a proof of concept, which exhibit remarkable electrocatalytic performance for hydrogen production from both acidic and basic solutions. The present study provides some guidelines for the design and synthesis of nanostructured electrocatalysts.

[1]School of Chemical and Biomedical Engineering, Nanyang Technological University, 62 Nanyang Drive, Singapore 637459, Singapore. * These authors contributed equally to this work. Correspondence and requests for materials should be addressed to X.W.L. (email: xwlou@ntu.edu.sg).

The rapid growth of global energy consumption and the associated environmental issues have triggered the urgent demand for renewable and clean energy sources. Electrochemical water splitting driven by solar energy has been considered as an attractive approach to produce hydrogen (H_2) fuel, a sustainable, secure and environmentally benign energy vector[1–3]. Efficient water splitting requires high-performance electrocatalysts to promote the hydrogen evolution reaction (HER). Platinum (Pt) has been identified as the most active HER catalyst, whereas its high cost and low abundance hinder the large-scale application[4]. Therefore, numerous efforts have been devoted to search for noble metal-free HER catalysts[5–18]. Transition metal carbides, such as molybdenum (Mo) and tungsten (W) carbides have been under investigation for decades in the fields of catalysis in view of their high similarity to Pt-group metals[19,20], and have been recently suggested as promising electrocatalysts for HER[21,22]. Particularly, β-phase molybdenum carbide (β-Mo_2C) has been demonstrated as a highly active HER catalyst even as bulky particles[23], and the performance could be further improved by constructing proper nanostructures[24–27]. However, controllable synthesis of nanostructured metal carbides with small nanocrystallites and desirable porosity towards high electrocatalytic activity still remains as a great challenge, due to the difficulty to achieve uniform carburization and the inevitable coalescence of nanoparticles at high reaction temperature.

In recent years, metal-organic frameworks (MOFs) have emerged as a new class of porous materials with widespread applications in gas storage/separation, catalysis, sensing and drug delivery[28–30]. Moreover, syntheses of functional materials from MOFs have drawn fast-growing interests as well. The periodically porous and hybrid structure of MOFs offer unique benefits for the fabrication of carbon and/or metal-based nanostructured materials. Specifically, MOFs-derived porous carbon, metal oxides, metal/carbon and metal oxide/carbon nanocomposites have been reported by using MOFs as both the precursor and template[31–36]. For example, MOFs-derived nanoporous carbon materials exhibit exceptionally high surface area and uniform porosity, which are largely originated from the ordered and porous structure of MOFs[32,33,37]. We have previously used Prussian blue cubic microcrystals to prepare various iron oxide-based hollow microboxes with complex shell structures and compositions[38,39]. This category of MOFs-derived materials has been recently extended to iron carbide[40]. However, in most of these studies, MOFs are exclusively used as the sole precursor. Although the huge family of MOFs has covered a wide range of metal species, these conventional MOFs-derived strategies are mainly based on a limited number of MOFs that are easily obtainable and/or with controllable morphology. Consequently, the reported MOFs-derived metal-based materials are typically limited to a handful of elements (for example, Zn, Cu, Co, Fe and so on).

In this work, we develop a MOFs-assisted strategy for synthesizing porous molybdenum carbide octahedral nanoparticles (denoted as MoC_x nano-octahedrons) that consist of very small nanocrystallites as electrocatalysts for efficient hydrogen production. Distinct from previous studies, the present synthesis strategy relies on the *in situ* and confined carburization reaction between the organic ligands (or their derived carbon-based species) of MOFs and guest polyoxometalates (POMs) that reside in the pores of the MOFs host. The introduction of guest metal species into the MOFs host as co-precursor enables easy synthesis of early transition metal (for example, Mo, W and V) carbides, which are difficult to obtain from a single MOFs source. Meanwhile, these non-coordinating POMs are also uniformly distributed and surrounded by organic ligands in atomic scale, thus guaranteeing *in situ* and homogeneous carburization reaction that produces small carbide nanocrystallites. In addition, the carburization process would be confined within the carbonaceous matrix derived from organic ligands of MOFs, which effectively prevents the agglomeration and coalescence of *in situ*-generated carbide nanocrystallites. As a proof of concept, we demonstrate the synthesis of molybdenum carbide using this MOFs-assisted approach in view of its promising application in catalysis (for example, for HER) and the easy encapsulation of Mo-based POMs in a particular MOFs host as discussed shortly. Interestingly, the as-prepared molybdenum carbide is in a η-MoC phase, which is unexpected at a relatively low carburization temperature of 800 °C and has not been well investigated for electrocatalytic hydrogen production[41]. Benefiting from the porous and robust structure, as well as the ultrafine primary nanocrystallites, the as-prepared molybdenum carbide nano-octahedrons exhibit remarkable electrocatalytic activity for HER in both acidic and basic conditions. The present strategy is also applicable to synthesize W and Mo-W carbides, and could be extended to other early transition metals as well.

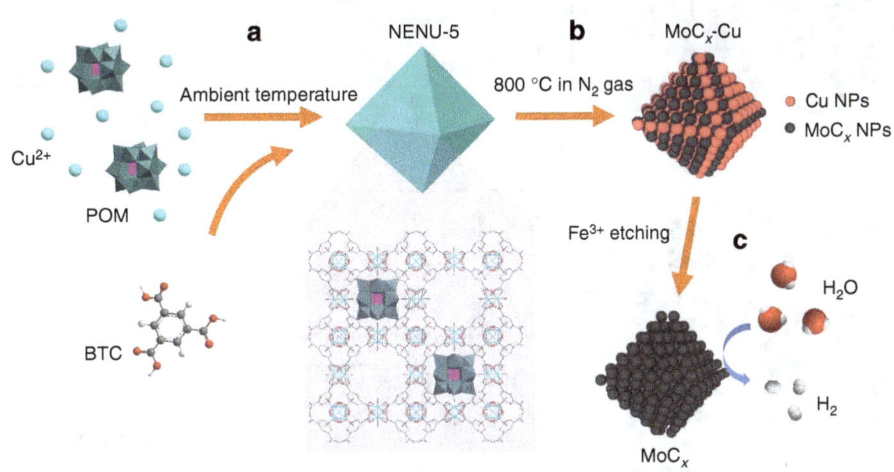

Figure 1 | Schematic illustration of the synthesis procedure for porous MoC_x nano-octahedrons. (**a**) Synthesis of NENU-5 nano-octahedrons with Mo-based POMs residing in the pores of HKUST-1 host. (**b**) Formation of MoC_x-Cu nano-octahedrons after annealing at 800 °C. (**c**) Removal of metallic Cu nanoparticles by Fe^{3+} etching to produce porous MoC_x nano-octahedrons for electrocatalytic hydrogen production.

Results

MOFs-assisted synthesis strategy for molybdenum carbide. The overall synthesis route to prepare porous MoC_x octahedral sub-micrometre-sized particles (denoted as MoC_x nano-octahedrons) as efficient HER catalysts is illustrated in Fig. 1. We choose a unique MOFs-based compound as the precursor with a formula of $[Cu_2(BTC)_{4/3}(H_2O)_2]_6[H_3PMo_{12}O_{40}]$ (NENU-5; BTC= benzene-1,3,5-tricarboxylate), which is based on a well-studied Cu-based MOF [HKUST-1; $Cu_3(BTC)_2(H_2O)_3$] with Mo-based Keggin-type POMs ($H_3PMo_{12}O_{40}$) periodically occupying the largest pores[42]. In this work, NENU-5 nano-octahedrons are synthesized by a facile and scalable co-precipitation method at ambient temperature. The as-prepared NENU-5 nano-octahedrons containing substantial amount of Mo are directly heated at 800 °C in N_2 gas flow to produce MoC_x-Cu. During the annealing process, the Mo-based POMs react with carbonaceous species derived from BTC ligands to form MoC_x nanocrystallites, meanwhile Cu^{2+} clusters are reduced to metallic Cu. Finally, MoC_x nano-octahedrons composed of small nanocrytallites are obtained by etching the metallic Cu nanoparticles with aqueous solution of $FeCl_3$ ($2\ Fe^{3+} + Cu \rightarrow Cu^{2+} + 2\ Fe^{2+}$), and used as electrocatalysts for HER.

Synthesis of porous MoC_x nano-octahedrons. Figure 2a shows typical field-emission scanning electron microscopy (FESEM) images of the as-prepared NENU-5 particles in octahedral or slightly truncated octahedral shape with a sub-micrometre size of ~800 nm. The size of the as-prepared NENU-5 particles can be easily tuned by varying the addition amount of L-glutamic acid that slows down the nucleation rate (Supplementary Fig. 1). Powder X-ray diffraction (XRD) pattern shown in Fig. 2b (upper curve) confirms the phase purity as well as the excellent crystallinity of the NENU-5 nano-octahedrons. The XRD pattern of HKUST-1 is also provided for comparison (lower curve in Fig. 2b), which exhibits different diffraction peaks especially below 10°. Moreover, the successful incorporation of Mo-based POMs in the HKUST-1 framework can be visually verified by the green colour of the product, distinct from the blue colour of pristine HKUST-1 (insets of Fig. 2b)[43]. The chemical composition is further examined by energy-dispersive X-ray spectroscopy (EDX). The spectrum (Fig. 2c) evidently shows the presence of substantial amount of Mo in the as-prepared NENU-5 nano-octahedrons. Moreover, a Mo/Cu atomic ratio of ~0.7 suggested by EDX quantitative analysis is lower than the theoretical value of 1 in NENU-5. Therefore, some vacancies of POMs exist in the as-prepared NENU-5 nano-octahedrons, which are probably due to the low synthesis temperature and/or the blocking of pores by other species from the solution. Such vacancies are also responsible for the low intensity of characteristic XRD peaks in small angle region[43]. Moreover, the vacancies of POMs would reduce the yield of MoC_x and possibly result in less uniform distribution of MoC_x nanocrystallites, which should be minimized in future studies.

After the annealing and etching processes, the MoC_x sample is obtained as a black powder (inset of Fig. 2d). FESEM image (Fig. 2d) reveals that the octahedral shape of the particles is well retained, while the surface becomes slightly rougher. The complete removal of Cu particles by Fe^{3+} etching is confirmed by both XRD and EDX analyses. As shown in Fig. 2e, the three strong and sharp peaks from metallic Cu in the XRD pattern of MoC_x-Cu sample (upper curve) are no longer observed in the pattern of MoC_x sample (lower curve). Moreover, EDX spectrum of the MoC_x sample confirms the main composition of Mo and C, and also excludes the presence of Cu (Fig. 2f). Surprisingly, the XRD pattern of MoC_x sample shows distinct results from previous reports that typically produce β-Mo_2C by annealing mixtures of molybdenum salts and organic compounds at similar temperatures[25–27,44]. The pattern can be satisfactorily assigned to hexagonal η-MoC phase (Supplementary Fig. 2) that is usually produced at much higher temperatures[45,46] or in the presence of NiI_2 with short reaction duration[41]. In our system, the η-MoC phase can be produced between 750 and 850 °C with little alteration in the XRD patterns (Supplementary Fig. 3). This unusual result implies the distinct characteristics of chemical reactions confined in the MOFs matrix, which is also exemplified

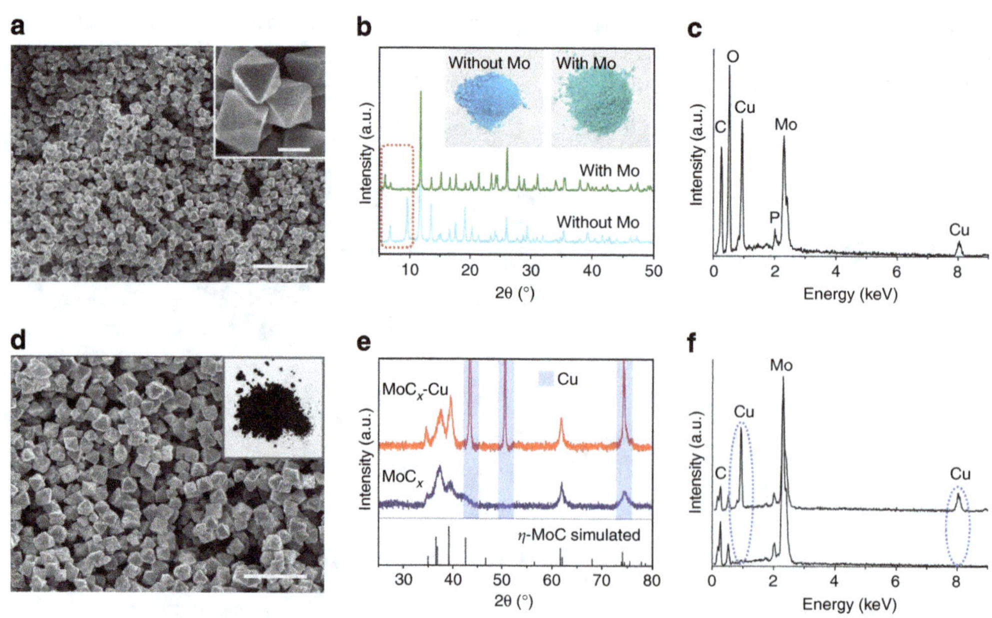

Figure 2 | Characterizations of precursors and MoC_x nano-octahedrons. (**a**) FESEM image (inset: magnified image; scale bar, 500 nm) of the as-prepared NENU-5 nano-octahedrons; scale bar, 5 μm. (**b**) XRD patterns (inset: digital photos) of NENU-5 (with Mo) and HKUST-1 (without Mo). (**c**) EDX spectrum of the as-prepared NENU-5 nano-octahedrons. (**d**) FESEM image (inset: digital photo) of porous MoC_x nano-octahedrons; scale bar, 2 μm. (**e**) XRD patterns and (**f**) EDX spectrums of MoC_x-Cu and MoC_x nano-octahedrons.

by the uncommon synthesis of pure brookite-phase TiO_2 through replication of MOFs[47]. Considering the likelihood to form substoichiometric η-MoC_{1-x} phases[46], and the difficulty to determine the exact chemical composition due to the presence of extra amorphous carbon (as discussed shortly), the as-prepared molybdenum carbide is denoted as MoC_x in this work. Moreover, the diffraction peaks are significantly broadened, suggesting the very small size of nanocrystallites due to the effective inhibition of coalescence and crystal growth during the confined carburization process. Meanwhile, excessive growth of Cu particles in the MoC_x-Cu sample as suggested by the XRD pattern (see upper curve in Fig. 2e) still occurs during the annealing process, which is probably related to the relatively low melting point of Cu.

Structural characterizations of MoC_x nano-octahedrons. The structure of the as-prepared MoC_x nano-octahedrons is further examined by transmission electron microscopy (TEM) as depicted in Fig. 3. The sample appears as rhombic or cubic particles under TEM observation as the projections of octahedral particles from different directions (Fig. 3a–c). A closer examination on the MoC_x nano-octahedrons reveals the highly porous texture throughout the whole particle. Each MoC_x nano-octahedron is composed of numerous small nanocrystallites, and the polycrystalline nature is confirmed by selected-area electron diffraction (SAED) pattern as shown in the inset of Fig. 3c. An interesting observation is that some large particles appear at the corners of the octahedral particles. This is probably due to the higher surface activity and stress in these regions that cause easy collapse of the MOFs matrix and subsequent growth of MoC_x nanocrystallites during the high-temperature reaction. Indeed, our initial attempt to prepare MoC_x using much smaller NENU-5 nanoparticles (shown in Supplementary Fig. 1a) with high surface activity results in strongly aggregated particles with a poorly crystallized β-Mo_2C phase that cannot be well dispersed into suspension for subsequent electrochemical measurements (Supplementary Fig. 4). Therefore, a robust secondary structure with a moderate size (for example, sub-micrometre-sized NENU-5 particles) is essential to successfully carry out the confined carburization reaction while providing large exposed surface for catalytic purpose.

A closer TEM examination on the edge of a MoC_x nano-octahedron gives more details of the nanostructure. Figure 3d clearly shows that numerous MoC_x clusters (darker area with visible lattice fringes indicated by green circles) are embedded in amorphous carbon matrix. Judging from their crystal lattices, the size of primary MoC_x nanocrystallites is typically within 5 nm, although some seem to slightly aggregate and appear as larger clusters. The presence of amorphous carbon is also verified by thermogravimetric analysis and Raman spectrum (Supplementary Fig. 5), which might play important roles in prohibiting the growth of MoC_x nanocrystallites and stabilizing the octahedral particles. A representative high-resolution (HR) TEM image (Fig. 3e) clearly shows lattice fringes with an interplanar distance of 0.24 nm, corresponding to the (006) planes of η-MoC. The uniform distribution of Mo and C elements is illustrated by EDX elemental mappings shown in Fig. 3f. Moreover, the MoC_x nano-octahedrons exhibit a highly mesoporous structure, as evidenced by a high specific Brunauer–Emmett–Teller surface area of $147\,m^2\,g^{-1}$ and abundant mesopores mainly distributed in the range of 4–10 nm (Supplementary Fig. 6). The mesopores in the MoC_x nano-octahedrons are obviously larger than the pores in the pristine HKUST-1, which would be related to the substantial mass loss (that is, loss of C, H, O, Cu elements) during the carburization process. Such a porous structure with high uniformity is largely inherited from the ordered porous architecture of the MOFs precursor, which would benefit the application in electrocatalysis.

Electrocatalytic performance for HER. The as-prepared porous MoC_x nano-octahedrons are evaluated as electrocatalysts for HER in both acidic and basic aqueous solutions. The porous MoC_x nano-octahedrons exhibit optimal performance with a mass loading of $0.8\,mg\,cm^{-2}$ on a glassy carbon (GC) disk electrode (Supplementary Fig. 7), while the catalysts prepared with shorter carburization time or at higher temperature show slightly inferior performance (Supplementary Fig. 8). The representative

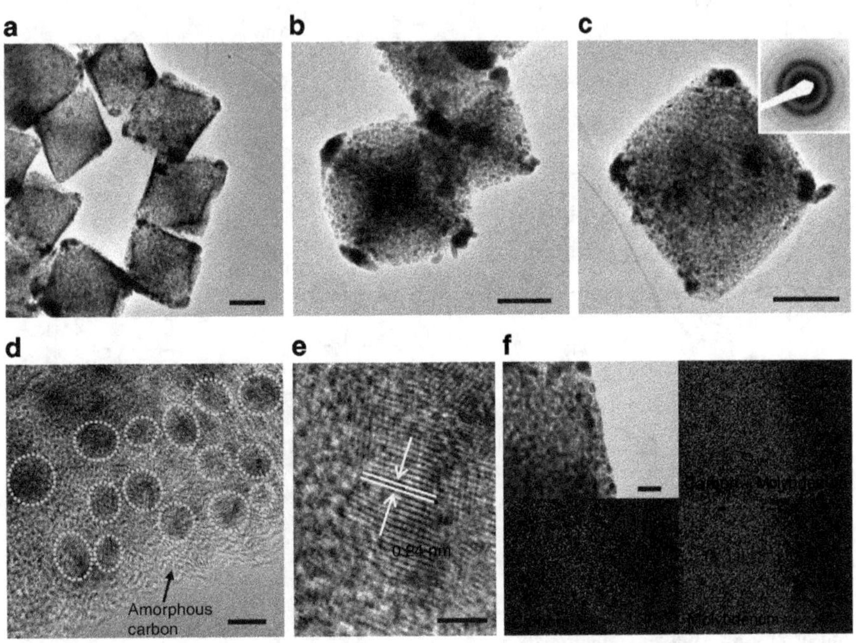

Figure 3 | Characterizations of MoC_x nano-octahedrons. (**a–c**) TEM images (scale bar, 200 nm; inset of c: SAED pattern), (**d**) magnified TEM image (scale bar, 5 nm), (**e**) HRTEM image (scale bar, 2 nm) and (**f**) elemental mapping (red: carbon; blue: molybdenum; scale bar, 50 nm) of porous MoC_x nano-octahedrons.

polarization curve (current density is based on geometric area of the electrode) and Tafel plot of the MoC_x electrocatalyst in 0.5 M H_2SO_4 are shown in Fig. 4a,b, respectively, along with the performance of the benchmark Pt/C catalyst (40 wt% Pt on carbon black from Johnson Matthey, mass loading of 0.8 mg cm^{-2}) for reference. As expected, the Pt/C catalyst exhibits excellent catalytic activity with an onset potential of ~0 V in acidic electrolyte. Meanwhile, the as-prepared MoC_x electrocatalyst also shows a small onset potential of ~25 mV, estimated from the low current density region of the Tafel plot (Supplementary Fig. 9), beyond which the cathodic current increases rapidly[48]. To achieve current densities (j) of 1 and 10 mA cm^{-2}, the MoC_x electrocatalyst requires overpotentials (η) of ~87 and 142 mV, respectively. Tafel plots depicted in Fig. 4b show a small Tafel slope of 53 mV per decade for MoC_x nano-octahedrons, higher than 29 mV per decade for the Pt/C catalyst. By extrapolating the Tafel plot, the exchange current density of MoC_x nano-octahedrons can be calculated as 0.023 mA cm^{-2}. Figure 4c,d shows the electrocatalytic properties of the MoC_x nano-octahedrons and Pt/C in basic condition. Although the Pt/C catalyst exhibits a small onset potential close to 0 V compared with ~80 mV for the MoC_x nano-octahedrons (Supplementary Fig. 10), the MoC_x electrocatalyst outperforms the Pt/C catalyst for $\eta \geq 220$ mV with rapidly rising cathodic current. Small overpotentials of 92 and 151 mV are required for the MoC_x nano-octahedrons to drive $j = 1$ and 10 mA cm^{-2}, respectively. In addition, the MoC_x nano-octahedrons exhibit a smaller Tafel slope (59 mV per decade) than the Pt/C catalyst (113 mV per decade) as shown in Fig. 4d, along with an exchange current density of ~0.029 mA cm^{-2}. The above comparison is based on the same loading mass of catalysts, which better reflects their performance in practical application and can be directly translated into their relative mass activity. Alternatively, we further compare the current density based on the mass of active materials (MoC for MoC_x nano-octahedrons and Pt for Pt/C) and the turnover frequency assuming all metal atoms are involved in the HER process (Supplementary Fig. 11). Similar trends are observed. Specifically, the Pt/C catalyst possesses overwhelming advantage in acidic media, while in basic media the activity of MoC_x nano-octahedrons approaches that of Pt/C catalyst at high

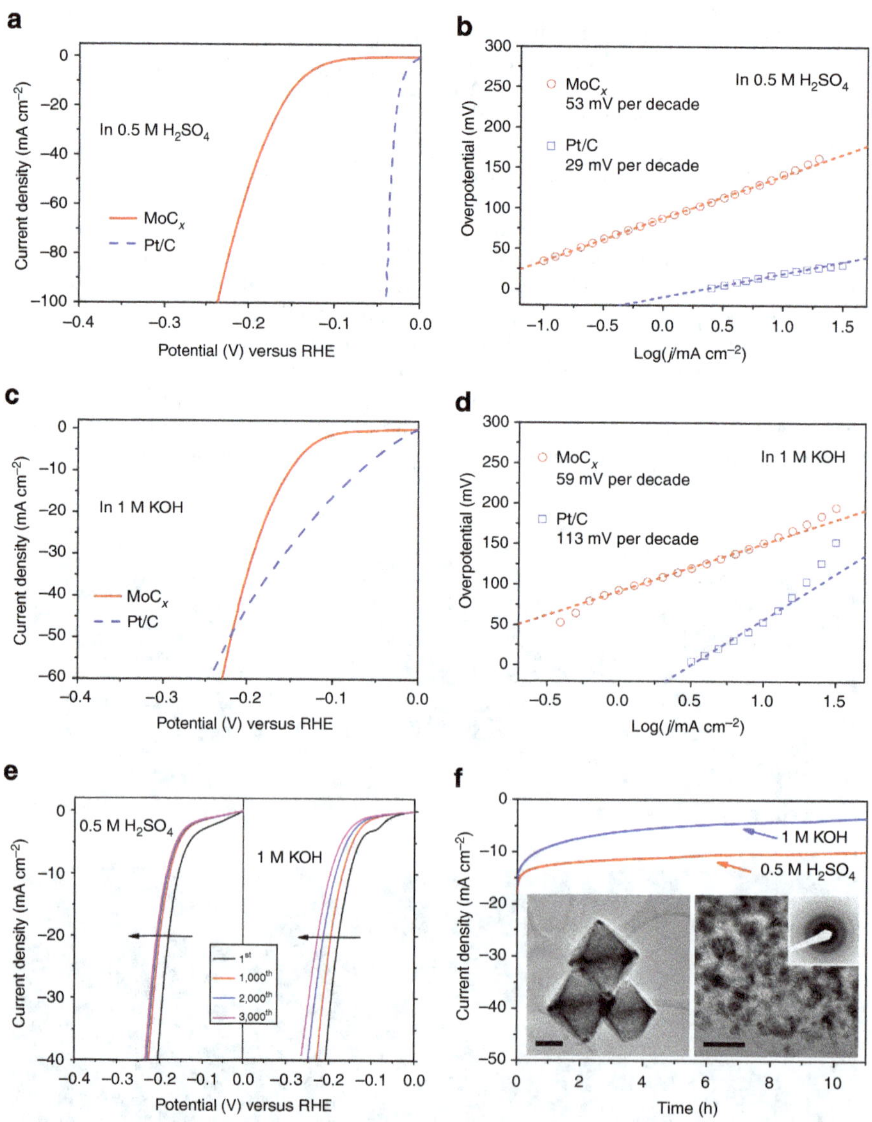

Figure 4 | HER performance of porous MoC_x nano-octahedrons. (**a**) Polarization curve at 2 mV s^{-1} and (**b**) Tafel plots in 0.5 M H_2SO_4. (**c**) Polarization curve at 2 mV s^{-1} and (**d**) Tafel plots in 1 M KOH. (**e**) Polarization curves after continuous potential sweeps at 50 mV s^{-1} in 0.5 M H_2SO_4 (left) and 1 M KOH (right). (**f**) Time-dependent current density curves under $\eta = 170$ mV in 0.5 M H_2SO_4 and $\eta = 180$ mV in 1 M KOH (insets: TEM images and SAED pattern after 5,000 potential sweeps in 0.5 M H_2SO_4. Scale bars, left inset (200 nm) and right inset (20 nm)).

Table 1 | Summary of HER activities of porous MoC_x nano-octahedrons.

Electrolyte	Onset η (mV versus RHE)	η at $j=1\,mA\,cm^{-2}$ (mV versus RHE)	η at $j=10\,mA\,cm^{-2}$ (mV versus RHE)	Tafel slope (mV per decade)	Exchange current density ($mA\,cm^{-2}$)
0.5 M H_2SO_4	~25	87	142	53	0.023
1 M KOH	~80	92	151	59	0.029

HER, hydrogen evolution reaction.

overpotential. The electrochemical properties of the MoC_x nano-octahedrons are summarized in Table 1, demonstrating the remarkable electrocatalytic HER activity in both acidic and basic solutions.

To better understand the origin of such high electrocatalytic performance, we further compare our MoC_x nano-octahedrons with the irregular MoC_x nanoparticles (denoted as MoC_x NPs, as shown in Supplementary Fig. 4 after grinding) as a reference sample, which exhibits similar composition but without well-defined nanostructure. Polarization curves in Supplementary Fig. 12 clearly reveal the much inferior performance of MoC_x NPs with $\eta \approx 230$ mV to drive $j=1\,mA\,cm^{-2}$ in both acidic and basic solutions. To reveal whether the high activity comes from increased surface area, we compare the apparent electrochemical surface area (ECSA) of these two electrocatalysts by measuring the double-layer capacitance (C_{dl}), which is typically used to represent the ECSA (Supplementary Fig. 13)[12,18]. Surprisingly, the MoC_x NPs actually possess quite similar C_{dl}, equivalent to ECSA, compared with MoC_x nano-octahedrons. Although the high surface area of MoC_x nano-octahedrons would obviously result in certain advantages when compared with bulky or low surface area materials, this is not the sole reason accounting for the high electrocatalytic activity.

Electrochemical impedance spectroscopy (EIS) analysis is performed on MoC_x nano-octahedrons and MoC_x NPs. Consistent with the previous studies, the EIS Nyquist plots of MoC_x nano-octahedrons in both acidic and basic solutions exhibit two time constants (Supplementary Fig. 14). The first one at high frequency is related to the surface porosity of the electrode; the second one at low frequency, which depends on the overpotential, reflects the charge transfer process during HER[25,27]. Generally speaking, the charge-transfer resistance (R_{ct}) shows strong correlation with the electrochemical performance. Thus, the Nyquist plots of MoC_x nano-octahedrons and MoC_x NPs at given overpotentials (that is, $\eta=90$ and 190 mV) in 0.5 M H_2SO_4 are compared and fitted to an equivalent electrical circuit with two time constants (Supplementary Fig. 15). It can be seen that for both samples the R_{ct} substantially reduces at high overpotential. However, the value for MoC_x nano-octahedrons is much smaller (about one order of magnitude lower) than that for MoC_x NPs at the same overpotential, in line with their different HER activity. The small charge transfer resistance would be mainly related to the synergistic effect and strong interaction between the MoC_x nanocrytallites and the continuous and *in situ*-incorporated carbon matrix, which ensures the facile electron transfer in the porous MoC_x nano-octahedrons. Together with the above analysis of ECSA, we speculate that the high electrochemical activity of MoC_x nano-octahedrons is due to their improved electronic/chemical properties and/or the exposure of more active sites, which are related to their unique mesoporous structure and small primary nanocrystallites. Nevertheless, more in-depth investigations would be necessary to reveal the detailed mechanism involved.

To assess the durability of the MoC_x electrocatalyst, accelerated linear potential sweeps are conducted repeatedly on the electrodes at a scan rate of $50\,mV\,s^{-1}$ as shown in Fig. 4e. In acidic condition, the polarization curves show a small shift of ~25 mV at $j=10\,mA\,cm^{-2}$ in the initial 1,000 sweeps, and then appear rather stable afterwards. On the other hand, the MoC_x nano-octahedrons exhibit a continuous but small loss of activity in basic condition on repeated potential sweeps, implying some minor corrosion of electrocatalyst in the basic electrolyte. It should be noted that loss of electrocatalyst from the electrode on rapid rotation might also account for some degradation of the performance. The stability of the electrocatalyst is also evaluated by prolonged electrolysis at constant potentials, as shown in Fig. 4f. In line with the above studies, the current density of MoC_x nano-octahedrons generally remains stable in 0.5 M H_2SO_4 for more than 10 h, whereas small degradation is observed in 1 M KOH on long-term operation. We have also examined the MoC_x nano-octahedrons under TEM observation after continuous linear potential sweeps in acidic (insets of Fig. 4f) and basic (Supplementary Fig. 16) media. The nanostructure and crystallinity are well retained after the degradation measurement, again corroborating the good stability in acidic environment. However, some corrosion of the MoC_x nano-octahedrons (especially the amorphous carbon) occurs during the potential sweeps in 1 M KOH. Such corrosion would cause the disintegration of the MoC_x nano-octahedrons and loss of active materials, which is expected to account for the small but continuous degradation in basic condition.

Discussion

The MOFs-assisted strategy presented in this work is facile and easily reproducible to synthesize porous MoC_x particles composed of a few nanometer-sized nanocrystallites. Compared with other synthesis methods, such as solid–gas phase reaction[49] and pyrolysis of composites containing metal and carbon sources[24,25], this MOFs-assisted approach guarantees homogeneous and efficient reaction, as well as uniform mesoporosity of the carbide product, originating from the unique crystalline structure with atomically hybridized MOFs matrix and Mo-based POMs. In addition, the highly localized and confined carburization process produces small nanocrystallites, which are embedded in an amorphous carbon matrix and prohibited from excessive growth. More importantly, the present strategy can be easily extended to synthesize tungsten carbide and molybdenum–tungsten mixed carbide (Supplementary Fig. 17), and potentially applicable to other early transition metals as well (Supplementary Note 1).

In virtue of the unique nanostructure, the porous MoC_x nano-octahedrons exhibit excellent electrocatalytic activity for HER. In acidic aqueous electrolyte, the performance is among the best reported when compared with many representative noble metal-free electrocatalysts, such as various molybdenum-based compounds, transition metal dichalcogenides and phosphides (Supplementary Table 1). The HER performance in basic condition is also compared favourably with many HER catalysts (Supplementary Table 2). In particular, the electrocatalytic activity of MoC_x nano-octahedrons in acidic media is comparable to the state-of-the-art β-Mo_2C-based electrocatalysts with extra graphitic carbon supports (for example, graphene and/or carbon nanotubes)[25,26,50]. However, such high HER activity has not yet been achieved on other phases of molybdenum carbide[41]. Considering the sub-micrometre size of the MoC_x nano-octahedrons, such catalytic activity is truly impressive. The high HER performance of our MoC_x nano-octahedrons might be attributed to the following aspects. First, molybdenum carbides possess exceptional intrinsic HER activity, which is probably related to their Pt-like electronic and chemical properties[22]. Second, the small size of primary MoC_x nanocrystallites and the high porosity render large electrochemical active surface and more importantly, perhaps more active sites to the MoC_x nano-octahedrons. Third, the uniform morphology and mesoporous structure are expected to facilitate the charge and mass transport within these relatively large octahedral particles, thus promoting the hydrogen production process. Moreover, the amorphous carbon matrix might grant high robustness of the octahedral particles and provide extra protection for the ultrafine MoC_x nanocrystallites with high surface energy. Nevertheless, the active materials in the centre part of MoC_x nano-octahedrons might not be fully utilized. Thus, further improvement is highly expected by optimizing the size/porosity of the particles and/or incorporating supports (for example, carbon nanotubes, graphene sheets and integrated current collectors).

In summary, we report a novel MOFs-assisted strategy for synthesizing nanostructured MoC_x nano-octahedrons as a highly active electrocatalyst for HER. This strategy relies on the confined and *in situ* carburization reaction occurring in a unique MOFs-based compound (NENU-5) consisting of a Cu-based MOFs (HKUST-1) host and guest Mo-based Keggin POMs resided in pores, which enables the uniform formation of metal carbide nanocrystallites without coalescence and excess growth. The as-prepared MoC_x nano-octahedrons consist of ultrafine nanocrytallites with an unusual η-MoC phase embedded in an amorphous carbon matrix, and possess a uniform and highly mesoporous structure. Benefiting from the desirable nanostructure, these porous MoC_x nano-octahedrons exhibit remarkable electrocatalytic activity for HER in both acidic and basic solutions with good stability. Moreover, such a strategy could be applicable for synthesizing other nanostructured transition metal carbides, thus opening new opportunities to develop high-performance functional materials for various applications.

Methods

Synthesis of NENU-5 nano-octahedrons. All chemicals were purchased and used without further purification. In a typical synthesis, 1 mmol of copper (II) acetate monohydrate (Sigma-Aldrich), 0.5 mmol of L-glutamic acid (Sigma-Aldrich) and 0.3 g of phosphomolybdic acid hydrate (Sigma-Aldrich) were dissolved in 40 ml of deionized water and stirred at ambient condition for 20 min. After that, 0.67 mmol of 1,3,5-benzenetricarboxylic acid (Merck) completely dissolved in 40 ml of ethanol was poured into the above solution under continuous stirring. The solution immediately turns turbid due to the rapid formation of NENU-5 nanocrystals. After stirring for 14 h at ambient condition, the green precipitate was collected by centrifugation and washed twice with ethanol. The product was dried at 70 °C overnight for further experiment and characterizations. The size of the NENU-5 particles can be easily tuned by varying the added amount of L-glutamic acid.

Synthesis of porous MoC_x nano-octahedrons. The NENU-5 nano-octahedrons were heated in a tube furnace under N_2 flow with a ramp rate of 2 °C min^{-1}, maintained at 800 °C for 6 h and then cooled down naturally. The as-prepared sample was denoted as MoC_x-Cu. The copper particles were removed by dispersing the sample in 0.1 M $FeCl_3$ aqueous solution at ambient condition for 2 h. The resulting porous MoC_x nano-octahedrons were collected by centrifugation, repeatedly washed with deionized water and then dried at 70 °C overnight.

Materials characterizations. The morphologies and structures of the products were characterized with FESEM (JEOL, JSM-6700 F, 5 kV) and TEM (JEOL, JEM-2010 and JEM-2100F, 200 kV). The chemical compositions and elemental mapping of the samples were analysed by EDX attached to JSM-7600F (FESEM, JEOL, 15 kV) and JEM-2100 F. The crystallographic information was collected by powder XRD (Bruker D8 Advance diffractometer with Cu Kα radiation $\lambda = 1.5406$ Å). The N_2 adsorption–desorption isotherms were collected using a Quantachrome Instruments Autosorb AS-6B at liquid-nitrogen temperature. Raman spectrum was collected using a Renishaw system 1000 micro-Raman spectroscope (Renishaw, UK). Thermogravimetric analysis was performed under air flow with a temperature ramp of 10 °C min^{-1}.

Electrochemical measurements. The electrocatalytic activity was evaluated in a three-electrode configuration using a rotating disk electrode (PINE Research Instrumentation, at a rotation speed of 1500 r.p.m.) with an Autolab potentiostat/galvanostat (Model PGSTAT-72637) workstation at ambient temperature. A GC disk electrode (5 mm in diameter) served as the support for the working electrode. The catalyst suspension was prepared by dispersing 10 mg of catalyst in 2 ml of solution containing 1.9 ml of ethanol and 100 µl of 0.5 wt% Nafion solution followed by ultrasonication for 20 min. Then the catalyst suspension was pipetted using a micropipettor on the GC surface. The working electrode was dried at ambient temperature. A saturated calomel electrode (SCE) was used as the reference electrode and a graphite rod was used as the counter electrode. Potentials were referenced to a reversible hydrogen electrode (RHE): $E(RHE) = E(SCE) + (0.242 + 0.059\,pH)V$. Linear sweep voltammetry was recorded in 0.5 M H_2SO_4 (pH = 0.3) and 1 M KOH (pH = 14) at a scan rate of 2 mV s^{-1} to obtain the polarization curves. The long-term stability tests were performed by continuous linear sweep voltammetry scans from −0.2 to −0.6 V (versus SCE, in 0.5 M H_2SO_4) and −1.0 to −1.4 V (versus SCE, in 1 M KOH) at a sweep rate of 50 mV s^{-1}. All the data presented were corrected for iR losses and background current. To evaluate the ECSA, cyclic voltammograms were obtained from −0.2 to 0.2 V (versus SCE, in 0.5 M H_2SO_4) with sweep rates of 5, 10, 20, 50, 100 mV s^{-1}. EIS was performed at various overpotentials with frequency from 0.1 to 100,000 Hz and an amplitude of 10 mV.

References

1. Lewis, N. S. & Nocera, D. G. Powering the planet: chemical challenges in solar energy utilization. *Proc. Natl Acad. Sci. USA* **103,** 15729–15735 (2006).
2. Walter, M. G. *et al.* Solar water splitting cells. *Chem. Rev.* **110,** 6446–6473 (2010).
3. Cook, T. R. *et al.* Solar energy supply and storage for the legacy and nonlegacy worlds. *Chem. Rev.* **110,** 6474–6502 (2010).
4. Gray, H. B. Powering the planet with solar fuel. *Nat. Chem.* **1,** 7–7 (2009).
5. Liu, Q. *et al.* Carbon nanotubes decorated with CoP nanocrystals: a highly active non-noble-metal nanohybrid electrocatalyst for hydrogen evolution. *Angew. Chem. Int. Ed.* **53,** 5427–5430 (2014).
6. Xie, J. *et al.* Controllable disorder engineering in oxygen-incorporated MoS_2 ultrathin nanosheets for efficient hydrogen evolution. *J. Am. Chem. Soc.* **135,** 17881–17888 (2013).
7. Wang, H. *et al.* Electrochemical tuning of vertically aligned MoS_2 nanofilms and its application in improving hydrogen evolution reaction. *Proc. Natl Acad. Sci. USA* **110,** 19701–19706 (2013).
8. Sun, Y. *et al.* Electrodeposited cobalt-sulfide catalyst for electrochemical and photoelectrochemical hydrogen generation from water. *J. Am. Chem. Soc.* **135,** 17699–17702 (2013).
9. Kibsgaard, J., Chen, Z., Reinecke, B. N. & Jaramillo, T. F. Engineering the surface structure of MoS_2 to preferentially expose active edge sites for electrocatalysis. *Nat. Mater.* **11,** 963–969 (2012).
10. Voiry, D. *et al.* Enhanced catalytic activity in strained chemically exfoliated WS_2 nanosheets for hydrogen evolution. *Nat. Mater.* **12,** 850–855 (2013).
11. Popczun, E. J., Read, C. G., Roske, C. W., Lewis, N. S. & Schaak, R. E. Highly active electrocatalysis of the hydrogen evolution reaction by cobalt phosphide nanoparticles. *Angew. Chem. Int. Ed.* **53,** 5427–5430 (2014).
12. Faber, M. S. *et al.* High-performance electrocatalysis using metallic cobalt pyrite (CoS_2) micro- and nanostructures. *J. Am. Chem. Soc.* **136,** 10053–10061 (2014).
13. Vrubel, H., Merki, D. & Hu, X. Hydrogen evolution catalyzed by MoS_3 and MoS_2 particles. *Energy Environ. Sci.* **5,** 6136–6144 (2012).
14. Cao, B., Veith, G. M., Neuefeind, J. C., Adzic, R. R. & Khalifah, P. G. Mixed close-packed cobalt molybdenum nitrides as non-noble metal electrocatalysts for the hydrogen evolution reaction. *J. Am. Chem. Soc.* **135,** 19186–19192 (2013).
15. Li, Y. *et al.* MoS_2 nanoparticles grown on graphene: an advanced catalyst for the hydrogen evolution reaction. *J. Am. Chem. Soc.* **133,** 7296–7299 (2011).
16. Gong, M. *et al.* Nanoscale nickel oxide/nickel heterostructures for active hydrogen evolution electrocatalysis. *Nat. Commun.* **5,** 4695 (2014).
17. Popczun, E. J. *et al.* Nanostructured nickel phosphide as an electrocatalyst for the hydrogen evolution reaction. *J. Am. Chem. Soc.* **135,** 9267–9270 (2013).
18. Tian, J., Liu, Q., Cheng, N., Asiri, A. M. & Sun, X. Self-supported Cu_3P nanowire arrays as an integrated high-performance three-dimensional cathode for generating hydrogen from water. *Angew. Chem. Int. Ed.* **53,** 9577–9581 (2014).

19. Hwu, H. H. & Chen, J. G. Surface chemistry of transition metal carbides. *Chem. Rev.* **105,** 185–212 (2004).
20. Oyama, S. T. Preparation and catalytic properties of transition metal carbides and nitrides. *Catal. Today* **15,** 179–200 (1992).
21. Michalsky, R., Zhang, Y.-J. & Peterson, A. A. Trends in the hydrogen evolution activity of metal carbide catalysts. *ACS Catal.* **4,** 1274–1278 (2014).
22. Chen, W.-F., Muckerman, J. T. & Fujita, E. Recent developments in transition metal carbides and nitrides as hydrogen evolution electrocatalysts. *Chem. Commun.* **49,** 8896–8909 (2013).
23. Vrubel, H. & Hu, X. Molybdenum boride and carbide catalyze hydrogen evolution in both acidic and basic solutions. *Angew. Chem. Int. Ed.* **51,** 12703–12706 (2012).
24. Chen, W.-F. *et al.* Biomass-derived electrocatalytic composites for hydrogen evolution. *Energy Environ. Sci.* **6,** 1818–1826 (2013).
25. Chen, W. F. *et al.* Highly active and durable nanostructured molybdenum carbide electrocatalysts for hydrogen production. *Energy Environ. Sci.* **6,** 943–951 (2013).
26. Pan, L. F. *et al.* Molybdenum carbide stabilized on graphene with high electrocatalytic activity for hydrogen evolution reaction. *Chem. Commun.* **50,** 13135–13137 (2014).
27. Liao, L. *et al.* A nanoporous molybdenum carbide nanowire as an electrocatalyst for hydrogen evolution reaction. *Energy Environ. Sci.* **7,** 387–392 (2014).
28. Furukawa, H., Cordova, K. E., O'Keeffe, M. & Yaghi, O. M. The chemistry and applications of metal-organic frameworks. *Science* **341,** 1230444 (2013).
29. Kuppler, R. J. *et al.* Potential applications of metal-organic frameworks. *Coord. Chem. Rev.* **253,** 3042–3066 (2009).
30. James, S. L. Metal-organic frameworks. *Chem. Soc. Rev.* **32,** 276–288 (2003).
31. Torad, N. L. *et al.* Direct synthesis of MOF-derived nanoporous carbon with magnetic Co nanoparticles toward efficient water treatment. *Small* **10,** 2096–2107 (2014).
32. Torad, N. L. *et al.* Facile synthesis of nanoporous carbons with controlled particle sizes by direct carbonization of monodispersed ZIF-8 crystals. *Chem. Commun.* **49,** 2521–2523 (2013).
33. Jiang, H.-L. *et al.* From metal–organic framework to nanoporous carbon: toward a very high surface area and hydrogen uptake. *J. Am. Chem. Soc.* **133,** 11854–11857 (2011).
34. Yang, S. J. *et al.* Preparation and exceptional lithium anodic performance of porous carbon-coated ZnO quantum dots derived from a metal–organic framework. *J. Am. Chem. Soc.* **135,** 7394–7397 (2013).
35. Xu, X., Cao, R., Jeong, S. & Cho, J. Spindle-like mesoporous α-Fe_2O_3 anode material prepared from MOF template for high-rate lithium batteries. *Nano Lett.* **12,** 4988–4991 (2012).
36. Chaikittisilp, W. *et al.* Synthesis of nanoporous carbon–cobalt-oxide hybrid electrocatalysts by thermal conversion of metal–organic frameworks. *Chem. Eur. J.* **20,** 4217–4221 (2014).
37. Wu, H. B. *et al.* Embedding sulfur in MOF-derived microporous carbon polyhedrons for lithium–sulfur batteries. *Chem. Eur. J.* **19,** 10804–10808 (2013).
38. Zhang, L., Wu, H. B., Madhavi, S., Hng, H. H. & Lou, X. W. Formation of Fe_2O_3 microboxes with hierarchical shell structures from metal–organic frameworks and their lithium storage properties. *J. Am. Chem. Soc.* **134,** 17388–17391 (2012).
39. Zhang, L., Wu, H. B. & Lou, X. W. Metal–organic-frameworks-derived general formation of hollow structures with high complexity. *J. Am. Chem. Soc.* **135,** 10664–10672 (2013).
40. Hou, Y. *et al.* Metal – organic framework-derived nitrogen-doped core-shell-structured porous Fe/Fe_3C@C nanoboxes supported on graphene sheets for efficient oxygen reduction reactions. *Adv. Energy Mater.* **4,** 1400337 (2014).
41. Wan, C., Regmi, Y. N. & Leonard, B. M. Multiple phases of molybdenum carbide as electrocatalysts for the hydrogen evolution reaction. *Angew. Chem. Int. Ed.* **53,** 6407–6410 (2014).
42. Sun, C.-Y. *et al.* Highly stable crystalline catalysts based on a microporous metal – organic framework and polyoxometalates. *J. Am. Chem. Soc.* **131,** 1883–1888 (2009).
43. Liu, Y. *et al.* Facile synthesis of a nanocrystalline metal–organic framework impregnated with a phosphovanadomolybdate and its remarkable catalytic performance in ultradeep oxidative desulfurization. *ChemCatChem* **5,** 3086–3091 (2013).
44. Yan, Z. *et al.* MoC-graphite composite as a Pt electrocatalyst support for highly active methanol oxidation and oxygen reduction reaction. *J. Mater. Chem. A* **2,** 4014–4022 (2014).
45. Shohoji, N. *et al.* Formation of hexagonal η-MoC_{1-x} phase at a temperature lower than 1660 °C by solar radiation heating under presence of excess free carbon. *Int. J. Refract. Met. H. Mater.* **25,** 220–225 (2007).
46. Hugosson, H. W. *et al.* Theory of phase stabilities and bonding mechanisms in stoichiometric and substoichiometric molybdenum carbide. *J. Appl. Phys.* **86,** 3758–3767 (1999).
47. Hall, A. S., Kondo, A., Maeda, K. & Mallouk, T. E. Microporous brookite-phase titania made by replication of a metal–organic framework. *J. Am. Chem. Soc.* **135,** 16276–16279 (2013).
48. Chen, W.-F. *et al.* Hydrogen-evolution catalysts based on non-noble metal nickel–molybdenum nitride nanosheets. *Angew. Chem. Int. Ed.* **51,** 6131–6135 (2012).
49. Lee, J. S., Oyama, S. T. & Boudart, M. Molybdenum carbide catalysts: I. Synthesis of unsupported powders. *J. Catal.* **106,** 125–133 (1987).
50. Youn, D. H. *et al.* Highly active and stable hydrogen evolution electrocatalysts based on molybdenum compounds on carbon nanotube–graphene hybrid support. *ACS Nano* **8,** 5164–5173 (2014).

Acknowledgements

We are grateful to the Ministry of Education (Singapore) for financial support through the AcRF Tier-1 funding (RG 12/13, M401120000).

Author contributions

H.B.W. and X.W.L. conceived the idea and co-wrote the manuscript. H.B.W. carried out the synthesis. H.B.W. and B.Y.X. carried out the electrochemical evaluation. L.Y. & X.-Y.Y. helped in materials characterizations.

Additional information

Competing financial interests: The authors declare no competing financial interests.

28

Synthesis of tetra- and octa-aurated heteroaryl complexes towards probing aromatic indoliums

Jun Yuan[1,*], Tingting Sun[2,*], Xin He[1], Ke An[2], Jun Zhu[2] & Liang Zhao[1]

Polymetalated aromatic compounds are particularly challenging synthetic goals because of the limited thermodynamic stability of polyanionic species arising from strong electrostatic repulsion between adjacent carbanionic sites. Here we describe a facile synthesis of two polyaurated complexes including a tetra-aurated indole and an octa-aurated benzodipyrrole. The imido trinuclear gold(I) moiety exhibits nucleophilicity and undergoes an intramolecular attack on a gold(I)-activated ethynyl to generate polyanionic heteroaryl species. Their computed magnetic properties reveal the aromatic character in the five-membered ring. The incorporation of the aurated substituents at the nitrogen atom can convert non-aromaticity in the parent indolium into aromaticity in the aurated one because of hyperconjugation. Thus, the concept of hyperconjugative aromaticity is extended to heterocycles with transition metal substituents. More importantly, further analysis indicates that the aurated substituents can perform better than traditional main-group substituents. This work highlights the difference in aromaticity between polymetalated aryls and their organic prototypes.

[1] Key Laboratory of Bioorganic Phosphorus Chemistry and Chemical Biology (Ministry of Education), Department of Chemistry, Tsinghua University, Beijing 100084, China. [2] State Key Laboratory of Physical Chemistry of Solid Surfaces and Collaborative Innovation Center of Chemistry for Energy Materials (iChEM), Fujian Provincial Key Laboratory of Theoretical and Computational Chemistry and Department of Chemistry, College of Chemistry and Chemical Engineering, Xiamen University, Xiamen 361005, China. * These authors contributed equally to this work. Correspondence and requests for materials should be addressed to J.Z. (email: jun.zhu@xmu.edu.cn) or to L.Z. (email: zhaolchem@mail.tsinghua.edu.cn).

Aryl- and heteroaryl-Au(I) compounds have recently attracted substantial attention because of their crucial intermediate role in versatile gold-catalysed transformations for the formation of new carbon–carbon and carbon–heteroatom bonds[1–5]. Structural characterization and reactivity studies of the organogold intermediates facilitated comprehension of the reaction mechanisms and provided inspiration for developing new synthetic methodologies[6–9]. Moreover, great interest has also arisen in the exploration of rich reactivity of aryl-Au(I) compounds. They can serve as an organometallic partner for metal-catalysed (for example, Pd and Ni) cross-coupling reactions[10], as a nucleophile to react with diverse appropriate electrophiles[11], and as a key intermediate to undergo a Au(I/III)-catalysed C-H activation and cross-coupling reaction in the presence of oxidants[12,13]. In addition, due to the attractive aurophilic interaction[14,15] the d^{10} gold(I) atom holds a great potential to aggregate into a polynuclear cluster. This structural characteristic perplexes the mechanism studies of gold-catalysed reactions, especially when bi- or polynuclear gold catalysts were used[16,17]. Therefore, the structure and reactivity studies of polyaurated organometallics may not only lead to new reagents for use in synthesis but also help understand the reaction mechanisms of Au-catalysed reactions. Furthermore, detailed structural characterization of polymetalated aryls may also provide a new insight into our current ideas about bonding between metal ions and organic species. However, relative to numerous reported mono-metalated aryls[18–20] derived from direct electrophilic metalation or halogen-lithium-metal exchange reactions[21], to date polymetalated aryl compounds still represent a very rare class of molecules[22,23] that are synthetically hard to reach by conventional methods. The formidable challenge of this synthesis lies in the difficult generation of multitopic anions and the limited thermodynamic stability of polyanionic species due to strong electrostatic repulsion between adjacent carbanionic sites.

On the other hand, the concept of hyperconjugative aromaticity was first proposed in 1939 by Mulliken[24], who considered that the saturated CH_2 group in cyclopentadiene could contribute 'pseudo' 2π electrons by hyperconjugation to the four olefinic π electrons, thus leading to aromatic cyclopentadiene with 6π electrons. Later, Schleyer, Nyulászi and O'Ferrall extended this concept and found that electropositive substituents (for example, SiH_3, GeH_3 and SnH_3) led to enhanced hyperconjugative aromaticity whereas electronegative substituents (for example, F, Cl) resulted in anti-aromatic five-membered rings[25–27]. Very recently, Houk revealed that the transition state distortion energies for the Diels–Alder reactions of 5-substituted cyclopentadienes with ethylene or maleic anhydride were directly related to the hyperconjugative aromaticity[28]. We found that hyperconjugation could play an important role in triplet state aromaticity[29]. However, all the previous studies on hyperconjugative aromaticity are mainly limited to carbocycles and all the substituents considered are main-group elements only.

In this contribution, we disclose a facile synthesis of two cationic polymetalated heteroaryl compounds including a tetra-aurated indolium and an octa-aurated benzodipyrrole complex. We find that the μ_3-imido aurated species exhibit good nucleophilicity and undergo an intramolecular attack on a Au(I)-activated ethynyl to generate polyanionic heteroaryls. The polyaurated heteroaryls exhibit reasonable stability upon exposure to air. The theoretical investigation reveals that the incorporation of the aurated substituents at the nitrogen atom can not only convert non-aromaticity in the parent indolium into aromaticity in aurated one due to the hyperconjugation but also lead to the most aromatic indolium. Thus, our findings not only extend the concept of hyperconjugative aromaticity to transition metal-involved heterocycles and accounts for the excellent stability of two polyaurated heteroaryls but also reveal better performance of transition metal-involved substituents over the traditional main-group ones on hyperconjugative aromaticity. In contrast to metallabenzene complexes where a metal atom is incorporated directly into a ring[30–32], the polymetalated heteroaryl compounds reported in this work represent a class of metalla-aromatics where the aromaticity is introduced by the hyperconjugation of transition metal substituents.

Results

Synthesis of tetra- and octa-aurated heteroaryl complexes. It is well-known that the gold(I) centre can act as an excellent soft Lewis acid to activate carbon–carbon multiple bonds. Many mononuclear organogold(I) compounds have been synthesized by the reactions of allene/alkene/alkyne substrates and nucleophiles with stoichiometric gold(I) compounds[18–20,33]. For example, σ- and π-activation of the ethynyl group of *o*-alkynylanilines by a gold(I) species led to the formation of 2- and 3-aurated indole complexes, respectively[34,35]. In view of the acknowledged isolobality of the proton and the LAu^+ fragment (L = phosphine or carbene ligands)[36], we therefore assume that an aurated imido species in an *o*-alkynylaniline derivative may act like an amine group to undergo a similar annulation reaction to react with the π- and/or σ-activated ethynyl group.

Sharp and co-workers have previously reported that a primary amine group readily reacts with the μ_3-oxo trinuclear gold(I) compound $[(PPh_3Au)_3(\mu_3\text{-}O)](BF_4)$ (**[Au₃O]**) to form a μ_3-imido trinuclear gold(I) compound[37,38]. According to this synthesis, we then carried out the reaction of **[Au₃O]** (1 equiv.) with *o*-(trimethylsilylethynyl)aniline **1** (trimethylsilyl = TMS) aiming at the synthesis of the gold(I) imido complex **3** (Fig. 1). The reaction mixture experienced a colour change from light brown to yellow. Diffusion of diethyl ether to the resulting yellow solution yielded colourless crystals of complex **2**. Structural analysis by X-ray crystallography (*vide infra*) revealed that **2** has an unexpected tetra-aurated indolyl structure instead of the NAu_3 cluster aggregate. This unusual transformation can be rationalized in the light of the following experimental facts, and divided as two steps: the auration of the amino group in **1** and the subsequent nucleophilic attack of the aurated imido on an Au(I)-activated carbon–carbon triple bond.

The electrospray ionization mass spectroscopy (ESI-MS) analysis of the reaction mixture of **1** and **[Au₃O]** revealed the strongest signal at $m/z = 1564.2502$ that can be ascribed to the 1+ species of **3** (~1564.2548), substantiating that the μ_3-imido trinuclear gold(I) species **3** indeed forms in the **1**-to-**2** transformation process (Supplementary Fig. 1). In order to clarify what kind of activation (σ-, π- or σ,π-mode)[39] involved in the **1**-to-**2** transformation, we employed the inner alkyne **5** to react with **[Au₃O]** based on the synthetic procedure for **2**. The μ_3-imido trinuclear gold(I) complex **6** was obtained as a crystalline solid in almost quantitative yield (crystal structure is shown in Supplementary Fig. 2). No cyclization products were found in this transformation. The obtaining of **6** not only provides further evidence for the formation of the NAu_3 species but also suggests the necessity of the σ-activation in the **1**-to-**2** transformation. Since the reaction of **1** and **[Au₃O]** is too rapid to isolate the σ-activated complex **4**, we next conducted the reaction between a simplified analogue PhC≡CTMS and **[Au₃O]**. The ^{1}H-NMR monitoring explored the formation of the known complex $PhC{\equiv}CAuPPh_3$ in the reaction mixture, suggesting that **[Au₃O]** can break the C–Si bond in the TMS-C≡C moiety and engender the formation of a σ-bonded gold acetylide. This result affords a rationale for the σ-activation of **1** in the **1**-to-**2** transformation

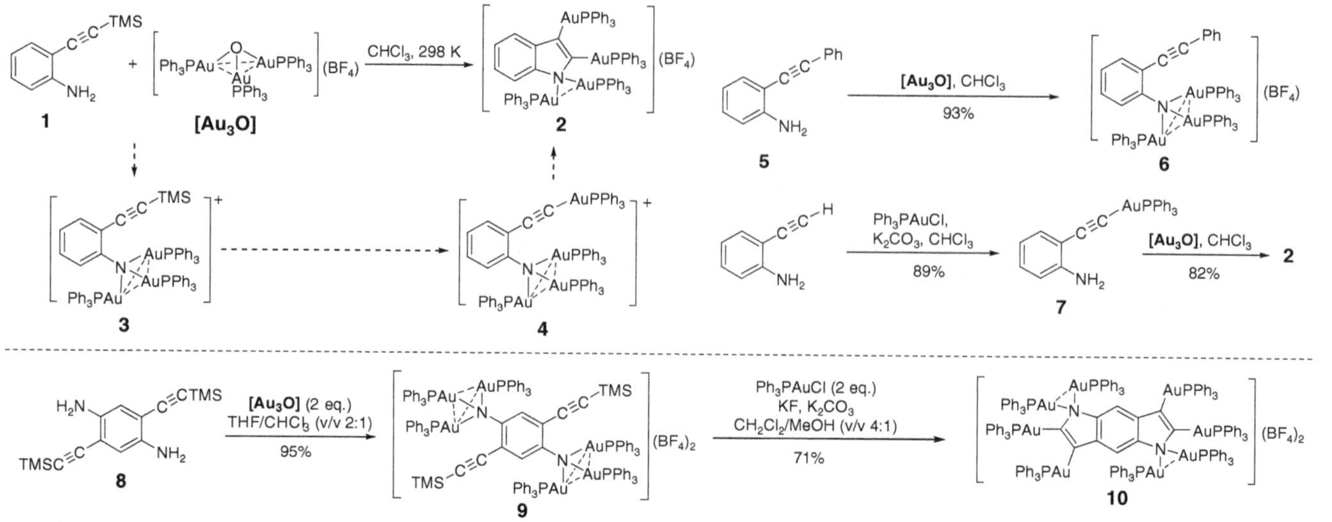

Figure 1 | Synthesis of tetra-aurated indole 2 and octa-aurated benzodipyyrole 10. The structures of complexes **2**, **6**, **9** and **10** have been determined by X-ray crystallography.

that should include the **[Au$_3$O]**-induced C–Si bond cleavage as well. In addition, we found the σ-aurated complex **7**, which derived from the reaction of 2-ethynylaniline with PPh$_3$AuCl, can react with **[Au$_3$O]** to produce **2** quantitatively within 5 min as evidenced in the ^{1}H-NMR monitoring (Supplementary Fig. 3). The **7**-to-**2** transformation supports the necessity of the σ-activation and meanwhile embodies the remarkable nucleophilicity of the aurated imido moiety.

We then carried out the reaction of the ditopic starting material **8** with **[Au$_3$O]** to purposefully synthesize the poly-aurated benzodipyrrole complex **10** based on the synthetic procedure for complex **2** (Fig. 1). However, this reaction gave a mixture of several gold-containing species which were observed in NMR spectroscopy and ESI-MS analysis. This outcome may result from several incomplete transformations including the auration of NH$_2$, the σ-metalation of TMS-C≡C and the cyclization between the imido and the ethynyl group as similar as the **1**-to-**2** transformation. Fortunately, in a mixed solvent of tetrahydrofuran (THF) and chloroform (v/v = 2:1) the reaction of **8** with **[Au$_3$O]** (2 equiv.) produced a yellow precipitate of complex **9**. This transformation is quantitative, so further separation is not necessary. X-ray crystallographic analysis disclosed two μ_3-imido trinuclear gold(I) moieties in the crystal structure of **9** (Supplementary Fig. 4). In view of the prerequisite σ-activation for the **7**-to-**2** transformation, we then conducted the reaction of **9** with Ph$_3$PAuCl (2 equiv.) in the presence of potassium carbonate and fluoride in order to replace the two TMS groups by two gold(I)-containing units. Amazingly, this transformation directly resulted in the formation of the octa-aurated benzodipyrrole complex **10**, which was characterized by X-ray crystallography (*vide infra*).

Molecular structures of the tetra-aurated indole complex **2** and the octa-aurated benzodipyrrole complex **10** were unambiguously determined by X-ray crystallographic analysis and confirmed by elemental analysis, multinuclear NMR spectroscopy and ESI-MS. As shown in Fig. 2a, crystal structure of **2** is composed of a tetra-aurated indolyl trianion, four coordinative AuPPh$_3$ units plus a BF$_4^-$ counter anion. The nitrogen and carbon atoms of the indolyl moiety in **2** are coplanar. Therein, carbon atoms C1 and C2 each is coordinated by a AuPPh$_3$ unit with the Au–C bond lengths in the range of 2.01(3)–2.05(3) Å, comparable with the values in reported mononuclear gold(I)-indole complexes[34,35]. Both gold atoms (Au3 and Au4) maintain the typical linear geometry with the P–Au–C bond angles of 175.4(9)° and 175.6(7)°, respectively. The indolyl nitrogen atom N1 is bonded by two geminal gold atoms (N1-Au1 = 2.17 (2) Å and N1-Au2 = 2.08 (2) Å), which together with C1 and C8 constitute a distorted tetrahedral bonding geometry for N1. The two gold atoms are held together by a strong aurophilic interaction[14,15] with the Au···Au distance of 2.862(2) Å. Although the whole structure of **2** was determined by X-ray crystallography and its composition was further substantiated by elemental analysis and ESI-MS; however, due to crystal twinning it is difficult to acquire the precise dimension of the indolyl skeleton in **2**. The mean deviation of all C–C and C–N bond lengths is as large as 0.04 Å.

Crystal structure of **10** can be described as a higher congener of **2** (Fig. 2b). To the best of our knowledge, complex **10** represents the first aurated benzodipyrrole compound. The negatively charged carbon and nitrogen atoms in the benzodipyrrole skeleton are mono- or di-aurated as similar as in **2**. An inversion centre is located in the centroid of the benzodipyrrole skeleton. Thus, two identical tetra-aurated pyrrolyl units are included in the crystal structure of **10**. The Au–C bond lengths in **10** lie in the range of 2.019(11)–2.036(8) Å, a bit shorter than the Au–N distances in the range of 2.082(11)–2.094(10) Å. The Au1···Au2 distance of 2.879(1) Å is comparable with the value in complex **2**. Although there are a number of examples containing the μ_3-imido trinuclear gold(I) cluster[37,38,40,41], the geminally diaurated fashion of a heteroaryl nitrogen atom as shown in **2** and **10** has not been reported before. In view of the isolobality of the proton and the LAu$^+$ fragment[36], such diaurated indolyl structure can be considered as an analogue of indolium. However, scrutiny of the dimension of the benzodipyrrole moiety in **10** reveals that the cyclic five-membered ring exhibits very good aromaticity with the C–N and C–C bond lengths being equalized and comparable with the parent indole.

Complexes **2** and **10** have very good stability upon exposure to air and moisture. In solution, these two complexes can also keep their structures intact as evidenced in ESI-MS and NMR. In the ESI mass spectrum of **2**, an isotopically well-resolved peak at m/z = 1950.2641 corresponding to the $[\mathbf{2} - \mathrm{BF_4}]^+$ species was observed (Supplementary Fig. 5). The ^{1}H-NMR spectrum of **2** clearly showed two multiplets at 7.12 and 7.82 p.p.m., which can be assigned to the four protons on the benzene ring of the indolyl skeleton (Supplementary Fig. 6). ESI mass spectrum of **10**

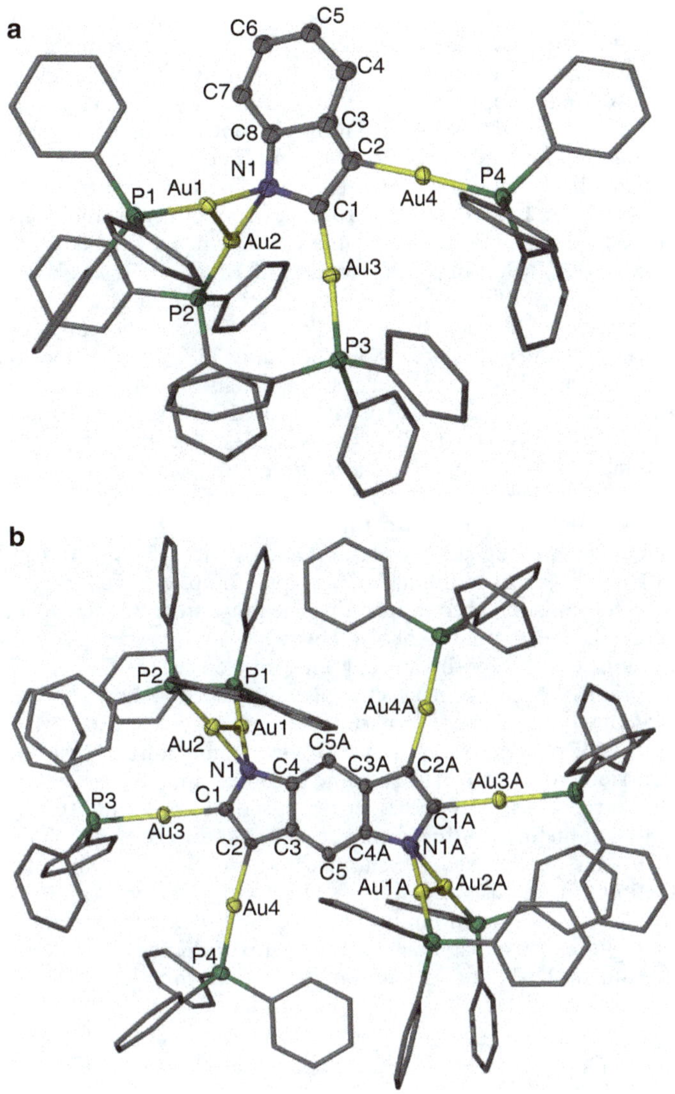

Figure 2 | Crystal structures of heteroaryl-gold(I) complexes. (**a**) tetra-aurated complex **2** and (**b**) octa-aurated complex **10.** Hydrogen atoms and tetrafluoroborate counter anions are omitted for clarity. Dimension of the benzodipyrrolium moiety in **10** (Å): N1-C1 1.414(14); N1-C4 1.470(14); C1-C2 1.344(13); C2-C3 1.417(12); C3-C4 1.387(14); C3-C5 1.393(13); and C4-C5A 1.381(13).

revealed a strong and isotopically well-resolved peak at $m/z = 1911.7428$ corresponding to the $[\mathbf{10} - 2BF_4]^{2+}$ species (Supplementary Fig. 7). Its ^{1}H-NMR spectrum gave a sharp singlet at 8.29 p.p.m. due to two isolated proton atoms on the central benzene ring of the benzodipyrrole moiety (Supplementary Fig. 8). ^{31}P-NMR spectra of **2** and **10** revealed two similar resonance peaks at around 33.6 and 39.4 p.p.m. (Supplementary Fig. 9), which can be assigned to the $AuPPh_3$ units attached on the nitrogen and carbon atoms.

DFT computations on aurated heteroaryl complexes. The equalized bond lengths and good stability of the polyaurated heteroaryl complexes **2** and **10** prompt us to examine whether the auration has a special influence on the aromaticity of the heteroaryl moieties. Density functional theory (DFT) computations[42] have then been carried out to help understand the aromaticity of the tetra- and octa-aurated heteroaryl complexes. For simplified model complexes **2'** and **10'**, the PH_3 ligands was

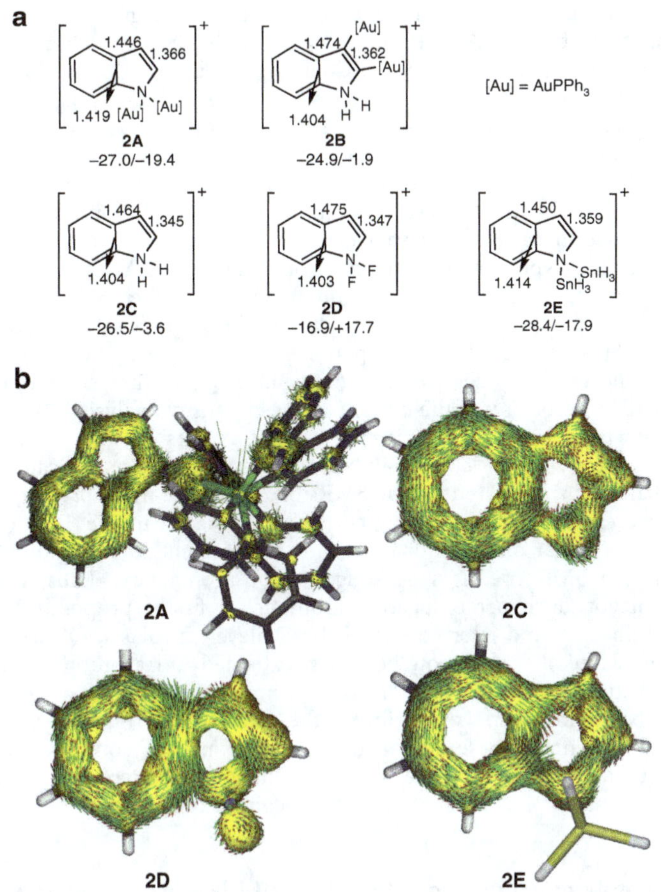

Figure 3 | Hyperconjugative (anti)aromaticity in indolium derivatives. (**a**) The selected C-C bond lengths (Å), $NICS(1)_{zz}$ values (p.p.m.) on the rings of **2A-2E**. The $NICS(1)_{zz}$ values given before and after the '/' are those computed at 1 Å above the geometrical centres of six- and five-membered rings, respectively. (**b**) ACID isosurfaces of **2A**, **2C**, **2D** and **2E** by the π contribution. Current density vectors are plotted onto the ACID isosurface of 0.03 to indicate dia- and para-tropic ring currents. The magnetic field vector is orthogonal with respect to the ring plane and points upward.

used to replace the PPh_3 ligands of the cationic moiety of **2** and **10**. The nucleus-independent chemical shift (NICS) computations[43,44] show that the $NICS(1)_{zz}$ values[29,45] for the five-membered ring (5MR) in **2'** and **10'** are −16.6 and −14.2 p.p.m., respectively (Supplementary Fig. 10). In general, negative values indicate aromaticity and positive values account for anti-aromaticity. These $NICS(1)_{zz}$ values in the 5MR are in sharp contrast to that of the parent indolium **2C** (− 3.6 p.p.m., Fig. 3), indicating an aromaticity switch from non-aromatic to aromatic. Note that the Au · · · Au distance in **2'** is 3.303 Å, which is much longer than the experimental value. Thus, the PPh_3 ligands including dispersion corrections have been considered for the optimization of complexes **2A** and **2B**. Indeed, the Au · · · Au distance in **2A** becomes 2.889 Å, indicating that the large ligands and dispersion corrections play an important role in reproducing the aurophilic interaction. Further NICS calculations on complexes **2A** and **2B** suggest that it is the aurated substituents on the nitrogen atom rather than those on the sp^2-hybridized carbon atom that are responsible for the aromaticity change. Specifically, the $NICS(1)_{zz}$ value in the 5MR of **2A** is close to that of **2'**, whereas complexes **2B** has the similar $NICS(1)_{zz}$ value with the parent indolium **2C**.

Previous study by Schleyer and Nyulászi[25] indicated that distannylcyclopentadiene is nearly as aromatic as furan according to various aromaticity indices. To evaluate the magnitude of the aromaticity in **2A**, we performed the NICS calculation on distannylindolium **2E**. The NICS(1)$_{zz}$ value in the 5MR of **2E** is −17.9 p.p.m., which is less negative than that of **2A**, suggesting higher aromaticity in **2A** over **2E**. Thus complex **2A** could be considered as the most aromatic indolium reported so far. The highest aromaticity of **2A** is also supported by highest delocalized C–C bonds in the 5MR. Specifically, the C–C bond length alternation (0.080 Å) is smallest among **2A**-**2E**. The important role of hyperconjugation in aromaticity of **2A** is further supported by the anti-aromaticity in the 5MR of difluoroindolium **2D**, indicated by the significant positive NICS(1)$_{zz}$ value (+ 17.7 – p.p.m.).

The hyperconjugatvie aromaticity in **2A** is further supported by the anisotropy of the current-induced density analysis[46,47]. As shown in Fig. 3b, the current density vectors plotted on the anisotropy of the current-induced density isosurface indicate a diatropic ring current in the 5MRs of **2A** and **2E** in the π system whereas the paratropic ring current is displayed in the 5MR of **2D**. No clear diatropic or paratropic ring current could be found in the 5MR of **2C**, suggesting its non-aromaticity. Thus, the concept of hyperconjugative aromaticity, firstly proposed by Mulliken[24] and later confirmed by Schleyer, Nyulászi, O'Ferrall and Houk[25–28], has now been first extended to transition metal substituted heterocycles[45]. More importantly, the aurated substituents can outperform the traditional main group substituents. These calculation results nicely confirm the aromatic essence of the geminally diaurated heteroaryls and account for the excellent stability of complexes **2** and **10**.

Photoluminescent properties of aurated heteroaryl complexes. Previous studies have explored that the heavy-atom effect of gold significantly influences the spin–orbit coupling between states of different spin multiplicity, thus enhancing intersystem crossing from singlet to triplet excited states and facilitating radiative decay of the triplet excited state[48,49]. We therefore embarked on photophysical studies on four structurally well-defined polyaurated complexes **2**, **6**, **9** and **10**. **2** and **6** have similar absorption bands below 375 nm. In contrast, the absorption spectra of **9** and **10** extend to ~475 nm (Supplementary Fig. 11). This bathochromic shift could be rationalized by the extended π system from **2** and **6** to **9** and **10**. In addition, the red shift is also reproduced qualitatively by time-dependent DFT calculations of **2'** and **10'** (the simplified complex of **10** with PH_3 in place of PPh_3). Specifically, the calculated intense absorption bands of complexes **2'** and **10'** at $\lambda = 321$ and 382 nm are consistent with the bands observed at $\lambda = 307$ and 355 nm of **2** and **10**, respectively. These computed absorption bands can be assigned to the electronic transitions HOMO→LUMO + 1 and HOMO→LUMO + 2 (HOMO is denoted as highest occupied molecular orbital whereas LUMO refers to lowest unoccupied molecular orbital, Supplementary Fig. 12), respectively, which are clearly a ligand-to-metal excitation nature.

Upon excitation at 325 nm (for **2** and **6**) and 345 nm (for **9** and **10**) at 298 K in dichloromethane, all four complexes are luminous with the emission spectra spanning the range from 350 to 650 nm (Fig. 4). The luminescence intensity of all four complexes is diminished by exposure to oxygen (Supplementary Fig. 13). In addition, the emission spectra of **2**, **9** and **10** show vibronic structures with a mean peak-to-peak separation of 1,010, 1,190 and 1,247 cm^{-1}, respectively. At low temperature (77 K), the emission from 450 to 550 nm becomes more intense and structured emission appears in the spectra of **6**, **9** and **10**. The emission lifetimes of the strongest emission for every complex at 77 K were collected and summarized in Table 1. The long lifetimes in microsecond-scale together with the oxygen-quenching of the emission and the large Stokes shifts indicate a phosphorescence parentage from a triplet-excited state. time-dependent DFT calculations on the model complex **2'** in the

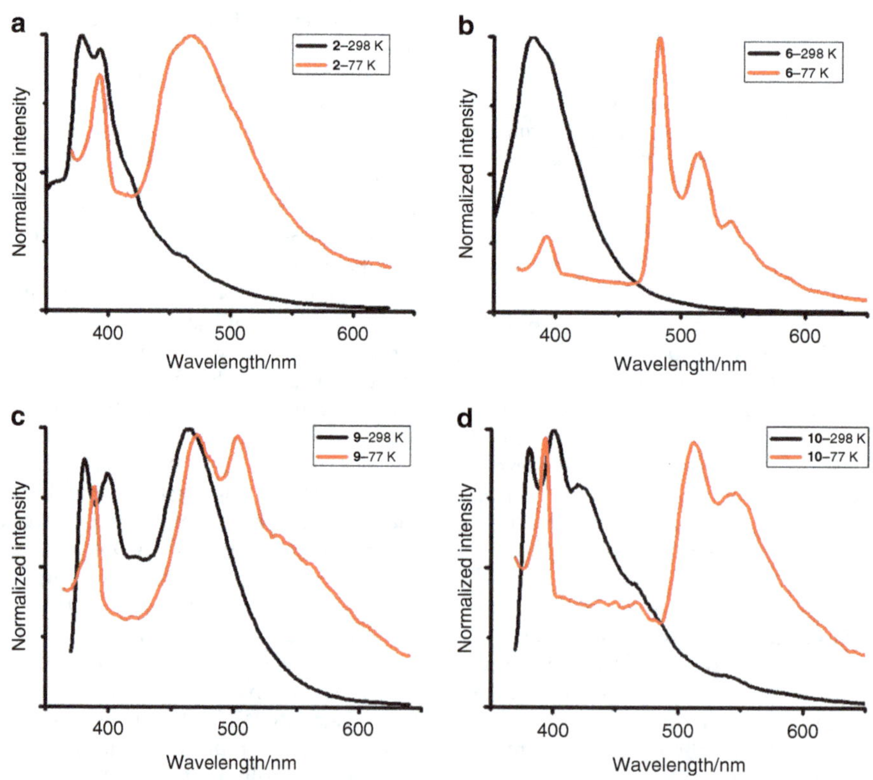

Figure 4 | Emission spectra of polynuclear gold complexes. Emission spectra of (**a**) **2**, (**b**) **6**, (**c**) **9** and (**d**) **10** collected at 298 and 77 K in dichloromethane ($c = 5 \times 10^{-6}$ M). Excitation: 325 nm (for **2** and **6**) and 345 nm (for **9** and **10**).

Table 1 | Photophysical parameters.

Complex	Emission		
	T/K	λ_{max}/nm (τ_o/μs)	Φ_{em}*
2	298	378, 393, 464	0.04
	77	393, 468(219)†	
6	298	381	0.37
	77	393, 482(609)†, 515, 540	
9	298	380, 398, 421, 462	0.14
	77	389, 470(62)‡, 505, 567	
10	298	381, 400, 421, 468, 541	0.02
	77	394, 512(53)‡, 546	

*Measured at 298 K using quinine sulfate dihydrate as a standard.
†Measured in a 2-methyltetrahydrofuran glass at 77 K.
‡Measured in a EtOH-MeOH-CH_2Cl_2 (40:20:1 v/v) glass at 77 K.

lowest triplet state (T_1) indicate that the calculated emission of **2'** at λ_{max} = 383, 412 and 474 nm are consistent with the bands observed at λ = 378, 393 and 464 nm of complex **2**, respectively. In addition, calculations reveal that the three emission bands are ligand-to-metal, metal-centred and ligand-centred, respectively (Supplementary Fig. 14). Gray and co-workers have once proposed a plausible excited-state hypothesis accounting for the temperature-dependent emission of aurated aryls[49]. It includes the decay of a singlet excited state to an excimer, which catalyses or pumps the formation of a triplet. High temperatures suppress excimer formation, resulting in emission difference. Furthermore, quantum yield measurements at 298 K revealed that the imido trinuclear gold(I) complexes **6** (37%) and **9** (14%) have higher yields than **2** (4%) and **10** (2%). This difference is possibly ascribed to the crowded arrangement of the $AuPPh_3$ units in **6** and **9**, which impedes non-radiative decay processes by blocking molecular rotations and vibrations of PPh_3 ligands.

Discussion

The synthesis of the polyaurated heteroaryls **2** and **10** is unprecedented and unusual. Stepwise transformations from 2-ethynylaniline through **7**-to-**2** and from **8** through **9**-to-**10** highlight two alternative pathways towards the synthesis of polyaurated aryls no matter which step (the auration of the amine group and the σ-metalation of the TMS-substituted ethynyl group) takes place first. A common prerequisite for the synthesis of **2** and **10** is the σ-activation of the carbon–carbon triple bond, which accounts for the unsuccessful annulation of internal alkyne substrates such as **5**. Considering the isolobality of the proton and the LAu^+ fragment, the μ_3-imido aurated species is analogous to an ammonium species that is lack of nucleophilicity. We hypothesize that the NAu_3 species in **3** and **9** exhibit nucleophilicity on condition that the trinuclear gold cluster aggregate dissociates into low nuclearity species. This speculation is partially evidenced by the transformation of **9** to **10**. In this transformation process, the stoichiometric ratio of **9** to Ph_3PAuCl is 1:2. The number of [$AuPPh_3$] units in **9** plus two equivalents Ph_3PAuCl is equal to that of **10**, while the number of the nitrogen-bonded gold unit decreases from three in **9** to two in **10**. This result suggests the release of a $AuPPh_3$ unit from the imido NAu_3 aggregate. This $AuPPh_3$ unit may translocate to finally attach on a carbon centre of the resulting heteroaryls. If replacing Ph_3PAuCl (2 equiv.) by the same amount of Cy_3PAuCl (Cy = cyclohexyl) in the **9**-to-**10** transformation, the ESI-MS of the reaction mixture revealed several Au_8-benzodipyrrole species containing different ligand combinations such as [$6PPh_3 + 2PCy_3$], [$5PPh_3 + 3PCy_3$] and [$4PPh_3 + 4PCy_3$], suggesting the dissociation and intermolecular exchange of phosphine ligands (Supplementary Fig. 15).

On the other hand, hyperconjugation, a weak interaction in chemistry, has a strong effect on the aromaticity. Simply tuning the substituents can lead to an aromaticity switch from non-aromaticity to (anti)aromaticity. Previous studies on hyperconjugative aromaticity mainly focus on the carbocycles. Examples on heterocycles are particularly rare and all the substituents reported so far contain main group elements only. Our calculations first demonstrate that the substituent containing the transition metal gold is also able to work for hyperconjugative aromaticity. More importantly, the introduction of *d*-orbitals from the transition metal can have a stronger effect on the hyperconjugative aromaticity in comparison with that of *p*-orbitals from traditional main group elements, thus opening an avenue to the design of various novel metalla-aromatics.

In summary, we have demonstrated a facile synthetic method of producing polyaurated heteroaryls. Two alternative pathways by activating the ethynyl group at first and then metalating the amine group or *vice versa* both accomplished the synthesis of unprecedented polyaurated indole and benzodipyrrole complexes. Theoretical analysis revealed an unexpected aromaticity change due to the hyperconjugation caused by the introduction of aurated substituents, which can perform better than the traditional main group substituents, extending the concept of hyperconjugative aromaticity to the transition metal-involved heterocycles. This study showcases a promising approach to acquire more complicated and intricate organometallics by forming saturated polymetalated units. Structure and reactivity comprehension of these polymetalated organometallics foresees deep research in bonding nature and mechanistic studies and stimulates the advancement of more efficient and versatile synthetic methods.

Methods

General. All commercially available reagents were used as received. Flash column chromatography was performed on silica gel (100–200). The solvents used in this study were dried by 4 Å molecular sieves. ^{1}H-NMR, ^{13}C-NMR and ^{31}P-NMR spectra were recorded by using 400 MHz spectrometers. Chemical shifts are reported in ppm versus tetramethylsilane with either tetramethylsilane or the residual solvent resonance used as an internal standard. Synthetic procedures for complexes **6**, **7** and **9** are summarized in Supplementary Methods. ESI-MS spectra of complexes **6** and **9** are shown in Supplementary Figs 16 and 17, respectively. ^{1}H-, ^{31}P- and ^{13}C-NMR spectra of complexes **7** (Supplementary Figs 18–20), **6** (Supplementary Figs 21–23), **2** (Supplementary Figs 24–26), **9** (Supplementary Figs 27–29) and **10** (Supplementary Figs 30–32) are shown in Supplementary Information.

Synthesis of 2. Complex **7** (10 mg, 0.017 mmol) and oxotris((triphenylphosphine)gold) tetrafluoroborate (25.7 mg, 0.017 mmol) were dissolved in $CHCl_3$ (2 ml) with stirring at room temperature over 20 min. Addition of hexane into the reaction mixture afforded a white solid of **2**. Yield: 82% (30 mg, 0.014 mmol). Single crystals of **2** were obtained by vapour diffusion of diethyl ether into a $CHCl_3$ solution of **2**. ^{1}H-NMR (400 MHz, $CDCl_3$): δ 7.87–7.80 (m, 2H); 7.48 (dd, J = 12.3, 7.8 Hz, 10H); 7.38 (dd, J = 12.4, 7.5 Hz, 20H); 7.30 (d, J = 7.7 Hz, 12H); 7.15–7.11 (m, 2H); and 7.05 (td, J = 7.8, 1.8 Hz, 18H). ^{13}C-NMR (100 MHz, $CDCl_3$): δ 134.2, 134.1, 132.1, 131.6, 130.5, 129.9, 129.5, 129.4, 129.3, 129.2, 128.6, 122.7, 121.9, 120.1 and 114.7. ^{31}P-NMR (162 MHz, $CDCl_3$): δ 39.43, 33.61. IR (KBr, cm^{-1}): 3052, 1964, 1890, 1812, 1667, 1615, 1587, 745, 710 and 692; HR-MS (ESI): calcd. for $[M-BF_4]^+$ ($C_{80}H_{64}Au_4NP_4$) 1950.2646, found 1950.2641. Elemental Analysis: Calcd. for ($C_{80}H_{64}Au_4BF_4NP_4 + 1/2(CHCl_3)$): C, 46.85; H, 3.37; and N, 0.65. Found: C, 46.88; H, 3.31; and N, 0.78.

Synthesis of 10. To a mixture of **9** (32.3 mg, 0.010 mmol) and 2 equiv. of PPh_3AuCl (9.9 mg, 0.020 mmol) in dichloromethane (2 ml) were added a methanolic solution (0.5 ml) of potassium fluoride (2.3 mg, 0.030 mmol) dropwise and 20 equiv. of K_2CO_3 (27.6 mg, 0.20 mmol). The mixture was stirred overnight in the dark. The solvent was removed by vacuum and the residue was redissolved in CH_2Cl_2. The resulting mixture was filtered through celite. Addition of Et_2O to the filtrate afforded a yellow solid of **10**. Yield: 71% (28.4 mg, 0.0071 mmol). Single crystals of **10** were obtained by vapour diffusion of diethyl ether into a $CHCl_3$ solution of **10** after 2 days. ^{1}H-NMR (400 MHz, $CDCl_3$): δ 8.29 (s, 2H); 7.48-7.36 (m, 64H); and 7.09-6.97 (m, 56H). ^{13}C-NMR (100 MHz, $CDCl_3$): δ 134.1, 131.9, 131.4, 130.6, 130.0, 129.2 and 105.8. ^{31}P-NMR (162 MHz, $CDCl_3$): δ 39.48, 33.56.

IR (KBr, cm^{-1}): 3049, 1964, 1890, 1812, 1585, 746, 709 and 692; HR-MS (ESI): calcd. for $[M-2BF_4]^{2+}$ ($C_{154}H_{122}Au_8N_2P_8$) 1911.7428; found 1911.7428. Elemental analysis: Calcd. for ($C_{154}H_{122}Au_8B_2F_8N_2P_8 + CH_2Cl_2$): C, 45.98; H, 3.22; and N, 0.68. Found: C, 45.67; H, 3.12; and N, 0.85.

X-ray crystallographic analysis. Data for complexes **2**, **6**, **9** and **10** (CCDC-1435045) were collected at 173 K with Mo-Kα radiation ($\lambda = 0.71073$ Å) on a Rigaku Saturn 724+ CCD diffractometer with frames of oscillation range 0.5°. All structures were solved by direct methods, and non-hydrogen atoms were located from difference Fourier maps. All non-hydrogen atoms were subjected to anisotropic refinement by full-matrix least-squares on F^2 by using the SHELXTL program unless otherwise noticed[50]. All crystal structure figures were drawn by using X-seed program[51]. Structural refinement details are summarized in Supplementary Methods. See also Supplementary Data 1.

Photophysical measurements. Photophysical measurement of low-temperature glass samples was carried out with the solid sample loaded in a quartz tube inside a quartz-walled Dewar flask. Liquid nitrogen was placed into the Dewar flask for low temperature (77 K). Luminescence quantum yields for solution samples were measured by the optical dilute method. All values are relative to quinine sulfate in 1.0 M H_2SO_4 as being 0.55 as the reference. Luminescent decay experiments were measured on an Edinburgh FLS920 spectrometer with μF920 microsecond flash lamp (pulse width < 2 μs) as light source.

References

1. Hashmi, A. S. K. Gold-catalyzed organic reactions. *Chem. Rev.* **107,** 3180–3211 (2007).
2. Gorin, D. J., Sherry, B. D. & Toste, F. D. Ligand effects in homogeneous Au catalysis. *Chem. Rev.* **108,** 3351–3378 (2008).
3. Li, Z., Brouwer, C. & He, C. Gold-catalyzed organic transformations. *Chem. Rev.* **108,** 3239–3265 (2008).
4. Corma, A., Leyva-Pérez, A. & Sabater, M. J. Gold-catalyzed carbon-heteroatom bond-forming reactions. *Chem. Rev.* **111,** 1657–1712 (2011).
5. Dorel, R. & Echavarren, A. M. Gold(I)-catalyzed activation of alkynes for the construction of molecular complexity. *Chem. Rev.* **115,** 9028–9072 (2015).
6. Cheong, P. H.-Y., Morganelli, P., Luzung, M. R., Houk, K. N. & Toste, F. D. Gold-catalyzed cycloisomerization of 1,5-allenynes via dual activation of an ene reaction. *J. Am. Chem. Soc.* **130,** 4517–4526 (2008).
7. Hashmi, A. S. K. *et al.* Simple gold-catalyzed synthesis of benzofulvenes-gem-diaurated species as 'instant dual-activation' precatalysts. *Angew. Chem. Int. Ed.* **51,** 4456–4460 (2012).
8. Gómez-Suárez, A. & Nolan, S. P. Dinuclear gold catalysis: are two gold centers better than one? *Angew. Chem. Int. Ed.* **51,** 8156–8159 (2012).
9. Ye, L., Wang, Y., Aue, D. H. & Zhang, L. Experimental and computational evidence for gold vinylidenes: generation from terminal alkynes via a bifurcation pathway and facile C-H insertions. *J. Am. Chem. Soc.* **134,** 31–34 (2012).
10. Hirner, J. J., Shi, Y.-L. & Blum, S. A. Organogold reactivity with palladium, nickel, and rhodium: transmetalation, cross-coupling, and dual catalysis. *Acc. Chem. Res.* **44,** 603–613 (2011).
11. Boogaerts, I. I. F. & Nolan, S. P. Carboxylation of C-H bonds using N-heterocyclic carbene gold(I) complexes. *J. Am. Chem. Soc.* **132,** 8858–8859 (2010).
12. Cambeiro, X. C., Boorman, T. C., Lu, P. & Larrosa, I. Redox-controlled selectivity of C-H activation in the oxidative cross-coupling of arenes. *Angew. Chem. Int. Ed.* **52,** 1781–1784 (2013).
13. Ball, L. T., Lloyd-Jones, G. C. & Russell, C. A. Gold-catalyzed oxidative coupling of arylsilanes and arenes: origin of selectivity and improved precatalyst. *J. Am. Chem. Soc.* **136,** 254–264 (2014).
14. Pyykkö, P. Theoretical chemistry of gold. *Angew. Chem. Int. Ed.* **43,** 4412–4456 (2004).
15. Schmidbaur, H. & Schier, A. Aurophilic interactions as a subject of current research: an up-date. *Chem. Soc. Rev.* **41,** 370–412 (2012).
16. Sherry, B. D., Maus, L., Laforteza, B. N. & Toste, F. D. Gold(I)-catalyzed synthesis of dihydropyrans. *J. Am. Chem. Soc.* **128,** 8132–8133 (2006).
17. Jin, L. *et al.* Trinuclear gold clusters supported by cyclic(alkyl)(amino)carbene ligands: mimics for gold heterogeneous catalysts. *Angew. Chem. Int. Ed.* **53,** 9059–9063 (2014).
18. Johnson, A., Laguna, A. & Gimeno, M. C. Axially chiral allenyl gold complexes. *J. Am. Chem. Soc.* **136,** 12812–12815 (2014).
19. Heng, W. Y., Hu, J. & Yip, J. H. K. Attaching gold and platinum to the rim of pyrene: a synthetic and spectroscopic study. *Organometallics* **26,** 6760–6768 (2007).
20. Hashmi, A. S. K., Braun, I., Rudolph, M. & Rominger, F. The role of gold acetylides as a selectivity trigger and the importance of gem-diaurated species in the gold-catalyzed hydroarylating-aromatization of arene-diynes. *Organometallics* **31,** 644–661 (2012).
21. Winter, C. H., Seneviratne, K. N. & Bretschneider-hurley, A. Permetalated aromatic compounds. *Comment Inorg. Chem.* **19,** 1–23 (1996).
22. Wei, Q.-H., Zhang, L.-Y., Yin, G.-Q., Shi, L.-X. & Chen, Z.-N. Luminescent heteronuclear $Au^{I}_{5}Ag^{I}_{8}$ complexes of $\{1,2,3\text{-}C_6(C_6H_4R\text{-}4)_3\}^{3-}$ (R = H, CH_3, Bu^t) by cyclotrimerization of arylacetylides. *J. Am. Chem. Soc.* **126,** 9940–9941 (2004).
23. Heckler, J. E., Zeller, M., Hunter, A. D. & Gray, T. G. Geminally diaurated gold(I) aryls from boronic acids. *Angew. Chem. Int. Ed.* **51,** 5924–5928 (2012).
24. Mulliken, R. S. Intensities of electronic transitions in molecular spectra IV. Cyclic dienes and hyperconjugation. *J. Chem. Phys.* **7,** 339–352 (1939).
25. Nyulászi, L. & Schleyer, P. v. R. Hyperconjugative π-aromaticity: how to make cyclopentadiene aromatic. *J. Am. Chem. Soc.* **121,** 6872–6875 (1999).
26. Lawlor, D. A. *et al.* Hyperaromatic stabilization of arenium ions: cyclohexa- and cycloheptadienyl cations—experimental and calculated stabilities and ring currents. *J. Am. Chem. Soc.* **133,** 19729–19742 (2011).
27. Fernández, I., Wu, J. I. & Schleyer, P. v. R. Substituent effects on 'hyperconjugative' aromaticity and antiaromaticity in planar cyclopolyenes. *Org. Lett.* **15,** 2990–2993 (2013).
28. Levandowski, B. J., Zou, L. & Houk, K. N. Schleyer hyperconjugative aromaticity and Diels–Alder reactivity of 5-substituted cyclopentadienes. *J. Comput. Chem.* **37,** 117–123 (2016).
29. Sun, H., An, K. & Zhu, J. Triplet state aromaticity: NICS criterion, hyperconjugation, and charge Effects. *Chem. Asian J.* **11,** 234–240 (2016).
30. Landorf, C. W. & Haley, M. M. Recent advances in metallabenzene chemistry. *Angew. Chem. Int. Ed.* **45,** 3914–3936 (2006).
31. Dalebrook, A. F. & Wright, L. J. Metallabenzenes and metallabenzenoids. *Adv. Organometal. Chem.* **60,** 93–177 (2012).
32. Fernández, I., Frenking, G. & Merino, G. Aromaticity of metallabenzenes and related compounds. *Chem. Soc. Rev.* **44,** 6452–6463 (2015).
33. Alcaide, B. & Almendros, P. Progress in allene chemistry. *Chem. Soc. Rev.* **43,** 2886–2887 (2014).
34. Gimeno, A. *et al.* Competitive gold-activation modes in terminal alkynes: an experimental and mechanistic study. *Chem. Eur. J.* **20,** 683–688 (2014).
35. Zeng, X., Kinjo, R., Donnadieu, B. & Bertrand, G. Serendipitous discovery of the catalytic hydroammoniumation and methylamination of alkynes. *Angew. Chem. Int. Ed.* **49,** 942–945 (2010).
36. Raubenheimer, H. & Schmidbaur, H. Gold chemistry guided by the isolobality concept. *Organometallics* **31,** 2507–2522 (2012).
37. Ramamoorthyt, V. & Sharp, P. R. Late-transition-metal μ-oxo and μ-imido complexes. 6. Gold(I) imido complexes. *Inorg. Chem.* **29,** 3336–3339 (1990).
38. Ramamoorthy, V., Wu, Z., Yi, Y. & Sharp, P. R. Preparation and decomposition of gold(I) hydrazido complexes: gold cluster formation. *J. Am. Chem. Soc.* **114,** 1526–1527 (1992).
39. Hashmi, A. S. K. Isolable vinylgold intermediates–first access to phantoms of homogeneous gold catalysis. *Gold Bull.* **42,** 275–279 (2009).
40. Shan, H., Yang, Y., James, A. J. & Sharp, P. R. Dinitrogen bridged gold clusters. *Science* **275,** 1460–1462 (1997).
41. Sharp, P. R. Oxo and imido ligands in late transition metal chemistry. *J. Chem. Soc. Dalton Trans.* 2647–2657 (2000).
42. Frisch, M. J. *et al. Gaussian 09, Revision D.01* (Gaussian, Inc., 2013).
43. Schleyer, P. v. R., Maerker, C., Dransfeld, A., Jiao, H. & Hommes, N. J. R. v. E. Nucleus-independent chemical shifts: a simple and efficient aromaticity probe. *J. Am. Chem. Soc.* **118,** 6317–6318 (1996).
44. Chen, Z., Wannere, C. S., Corminboeuf, C., Puchta, R. & Schleyer, P. v. R. Nucleus-independent chemical shifts (NICS) as an aromaticity criterion. *Chem. Rev.* **105,** 3842–3888 (2005).
45. Fallah-Bagher-Shaidaei, H., Wannere, C. S., Corminboeuf, C., Puchta, R. & Schleyer, P. v. R. Which NICS aromaticity index for planar π rings is best? *Org. Lett.* **8,** 863–866 (2006).
46. Herges, R. & Geuenich, D. Delocalization of electrons in molecules. *J. Phys. Chem. A.* **105,** 3214–3220 (2001).
47. Geuenich, D., Hess, K., Kçhler, F. & Herges, R. Anisotropy of the induced current density (ACID), a general method To quantify and visualize electronic delocalization. *Chem. Rev.* **105,** 3758–3772 (2005).
48. Partyka, D. V., Esswein, A. J., Zeller, M., Hunter, A. D. & Gray, T. G. Gold(I) pyrenyls: excited-state consequences of carbon-gold bond formation. *Organometallics* **26,** 3279–3282 (2007).
49. Gao, L. *et al.* Mono- and di-gold(I) naphthalenes and pyrenes: syntheses, crystal structures, and photophysics. *Organometallics* **28,** 5669–5681 (2009).
50. Sheldrick, G. M. *SHELXL-97* (Univ. Göttingen (1997).
51. Atwood, J. L. & Barbour, L. J. Molecular graphics: from science to art. *Cryst. Growth Des.* **3,** 3–8 (2003).

Acknowledgements
Financial support by NNSFC (21132005, 21421064, 21522206, and 21573179), MOST (2013CB834501) and Top-Notch Young Talents Program of China is gratefully acknowledged. We are grateful to Profs. Mei-Xiang Wang, Qing-Zheng Yang, Lian Duan and Ming-Tian Zhang for data measurements and helpful discussions.

Author contributions
L.Z. and J.Z. conceived and supervised the project. The synthetic experiments and photophysical measurements were carried out by J.Y. and X.H. T.S. and K.A. performed the theoretical calculation. L.Z. and J.Z. co-wrote the manuscript. All authors discussed the results and commented on the manuscript.

Additional information
Accession codes: The X-ray crystallographic coordinates for structures reported in this article have been deposited at the Cambridge Crystallographic Data Centre (CCDC), under deposition number CCDC-1435042 (**2**), 1435041 (**6**), 1435040 (**9**) and 1435045 (**10**). These data can be obtained free of charge from the Cambridge Crystallographic Data Centre via www.ccdc.cam.ac.uk/data_request/cif.

Competing financial interests: The authors declare no competing financial interests.

Permissions

All chapters in this book were first published in NC, by Nature Publishing Group; hereby published with permission under the Creative Commons Attribution License or equivalent. Every chapter published in this book has been scrutinized by our experts. Their significance has been extensively debated. The topics covered herein carry significant findings which will fuel the growth of the discipline. They may even be implemented as practical applications or may be referred to as a beginning point for another development.

The contributors of this book come from diverse backgrounds, making this book a truly international effort. This book will bring forth new frontiers with its revolutionizing research information and detailed analysis of the nascent developments around the world.

We would like to thank all the contributing authors for lending their expertise to make the book truly unique. They have played a crucial role in the development of this book. Without their invaluable contributions this book wouldn't have been possible. They have made vital efforts to compile up to date information on the varied aspects of this subject to make this book a valuable addition to the collection of many professionals and students.

This book was conceptualized with the vision of imparting up-to-date information and advanced data in this field. To ensure the same, a matchless editorial board was set up. Every individual on the board went through rigorous rounds of assessment to prove their worth. After which they invested a large part of their time researching and compiling the most relevant data for our readers.

The editorial board has been involved in producing this book since its inception. They have spent rigorous hours researching and exploring the diverse topics which have resulted in the successful publishing of this book. They have passed on their knowledge of decades through this book. To expedite this challenging task, the publisher supported the team at every step. A small team of assistant editors was also appointed to further simplify the editing procedure and attain best results for the readers.

Apart from the editorial board, the designing team has also invested a significant amount of their time in understanding the subject and creating the most relevant covers. They scrutinized every image to scout for the most suitable representation of the subject and create an appropriate cover for the book.

The publishing team has been an ardent support to the editorial, designing and production team. Their endless efforts to recruit the best for this project, has resulted in the accomplishment of this book. They are a veteran in the field of academics and their pool of knowledge is as vast as their experience in printing. Their expertise and guidance has proved useful at every step. Their uncompromising quality standards have made this book an exceptional effort. Their encouragement from time to time has been an inspiration for everyone.

The publisher and the editorial board hope that this book will prove to be a valuable piece of knowledge for researchers, students, practitioners and scholars across the globe.

List of Contributors

Xiao Liang, Connor Hart, Quan Pang and Linda F. Nazar
Department of Chemistry, University of Waterloo, 200 University Avenue West, Waterloo, Ontario, Canada N2L 3G1

Arnd Garsuch and Thomas Weiss
BASF SE, 67056 Ludwigshafen, Germany

Zoel Codolà, Miquel Costas and Julio Lloret-Fillol
Institut de Química Computacional i Catàlisi (IQCC) and Departament de Química, Universitat de Girona, Campus de Montilivi, 17071 Girona, Spain

Laura Gómez
Department of Chemistry and Center for Metals in Biocatalysis, University of Minnesota, 207 Pleasant St S.E., Minneapolis, Minnesota 55455, USA

Scott T. Kleespies and Lawrence Que Jr
Serveis Tècnics de Recerca (STR), Parc Cientific i Tecnològic, Universitat de Girona, 17003 Girona, Spain

Mirian E. Casco, Joaquín Silvestre-Albero, Manuel Martínez-Escandell and Francisco Rodríguez-Reinoso
Laboratorio de Materiales Avanzados, Departamento de Química Inorgánica-Instituto Universitario de Materiales, Universidad de Alicante, Ctra. San Vicente-Alicante s/n, E-03690 San Vicente del Raspeig, Spain

Anibal J. Ramírez-Cuesta
Oak Ridge National Laboratory, Chemical and Engineering Materials Division, 1 Bethel Valley Road, Oak Ridge, Tennessee 37831, USA

Fernando Rey and Jose L. Jordá
Instituto de Tecnología Química, Universidad Politécnica de Valencia-CSIC, Avda. de los Naranjos s/n, 46022 Valencia, Spain

Atul Bansode and Atsushi Urakawa
Institute of Chemical Research of Catalonia (ICIQ), 43007 Tarragona, Spain

Inma Peral
ALBA Light Source, 08290 Cerdanyola del Vallés, Barcelona, Spain

Shin-ichiro Noro
Research Institute for Electronic Science, Hokkaido University, Sapporo 001-0020, Japan
Graduate School of Environmental Earth Science, Hokkaido University, Sapporo 060-0810, Japan
PRESTO, Japan Science and Technology Agency (JST), 4-1-8 Honcho, Kawaguchi 332-0012, Japan
Creative Research Institute (CRIS), Hokkaido University, Sapporo 001-0021, Japan

Junya Mizutani
Graduate School of Environmental Earth Science, Hokkaido University, Sapporo 060-0810, Japan

Yuh Hijikata
Institute of Transformative Bio-Molecules (WPI-ITbM), Nagoya University, Chikusa-ku, Nagoya 464-8602, Japan

Ryotaro Matsuda
Institute for Integrated Cell-Material Science (WPI-iCeMS), Kyoto University, Kyoto 615-8510, Japan

Hiroshi Sato and Susumu Kitagawa
Institute for Integrated Cell-Material Science (WPI-iCeMS), Kyoto University, Kyoto 615-8510, Japan

Yasutaka Inubushi
Synthesis Research Laboratory, Kurashiki Research Center, Kuraray Co. Ltd., Sakazu, Kurashiki, Okayama 710-0801, Japan.
EVAL Research and Development Department, EVAL Division, Vinyl Acetate Company, Kuraray Co. Ltd., 7471, Tamashimaotoshima, Kurashiki, Okayama 713-8550, Japan

Kunihisa Sugimoto
Research and Utilization Division, Japan Synchrotron Radiation Research Institute, Hyogo 679-5198, Japan

Kazuya Kubo and Takayoshi Nakamura
Research Institute for Electronic Science, Hokkaido University, Sapporo 001-0020, Japan
Graduate School of Environmental Earth Science, Hokkaido University, Sapporo 060-0810, Japan

Katsumi Kaneko
Institute for Integrated Cell-Material Science (WPI-iCeMS), Kyoto University, Kyoto 615-8510, Japan

Daniel Franz, Elisabeth Irran and Shigeyoshi Inoue
Institut für Chemie, Anorganische und Analytische Chemie, Technische Universität Berlin, Straße des 17 Juni 135, Sekr. C2, 10623 Berlin, Germany

Tibor Szilvási
Department of Inorganic and Analytical Chemistry, Budapest University of Technology and Economics, Szent Gellért tér 4, 1111 Budapest, Hungary

Yong-Sheng Wei, Mei Zhang, Pei-Qin Liao, Rui-Biao Lin, Tai-Yang Li, Guang Shao, Jie-Peng Zhang and Xiao-Ming Chen
MOE Key Laboratory of Bioinorganic and Synthetic Chemistry, School of Chemistry & Chemical Engineering, Sun Yat-Sen University, Guangzhou 510275, China

Yang-Gang Wang
Institute for Interfacial Catalysis, Pacific Northwest National Laboratory, Richland, Washington 99352, USA
Department of Chemistry, Tsinghua University, Beijing 100084, China

Donghai Mei, Vassiliki-Alexandra Glezakou and Roger Rousseau
Institute for Interfacial Catalysis, Pacific Northwest National Laboratory, Richland Washington 99352, USA

Jun Li
Department of Chemistry, Tsinghua University, Beijing 100084, China
Environmental Molecular Sciences Laboratory, Pacific Northwest National Laboratory, Richland, Washington 99352, USA

Panchao Yin, Fadi Haso, Lang Hu, Baofang Zhang and Tianbo Liu
Department of Polymer Science, The University of Akron, Akron, Ohio 44325-3909, USA
Department of Chemistry, Lehigh University, Bethlehem, Pennsylvania 18015, USA

Zhi-Ming Zhang, Yanqing Gao, Yang-Guang Li and En-Bo Wang
Key Laboratory of Polyoxometalate Science of Ministry of Education, Northeast Normal University, Changchun, Jilin 130024, China

Hongjin Lv, John Bacsa, Yu Hou and Craig L. Hill
Department of Chemistry, Emory University, Atlanta, Georgia, 30322, USA

Tao Li
X-ray Science Division, Advanced Photon Source, Argonne National Laboratory, Argonne, Illinois 60439, USA

Yongge Wei
Department of Chemistry, Tsinghua University, Beijing 100084, China

Kirsi Salorinne and Xi Chen
Department of Chemistry, Nanoscience Center, University of Jyväskylä, FI-40014 Jyväskylä, Finland

Sami Malola
Department of Physics, Nanoscience Center, University of Jyväskylä, FI-40014 Jyväskylä, Finland

O. Andrea Wong, Christopher D. Rithner and Christopher J. Ackerson
Department of Chemistry, Colorado State University, Fort Collins, Colorado 80523, USA

Hannu Häkkinen
Department of Chemistry, Nanoscience Center, University of Jyväskylä, FI-40014 Jyväskylä, Finland
Department of Physics, Nanoscience Center, University of Jyväskylä, FI-40014 Jyväskylä, Finland

Pei-Qin Liao, Wei-Xiong Zhang, Jie-Peng Zhang and Xiao-Ming Chen
MOE Key Laboratory of Bioinorganic and Synthetic Chemistry, School of Chemistry and Chemical Engineering, Sun Yat-Sen University, Guangzhou 510275, P.R. China

Yucong Jiao and Xianfeng Zhang
State Key Laboratory of Molecular Engineering of Polymers and Department of Macromolecular Science, Fudan University, Shanghai 200433, China
Department of Chemistry and Collaborative Innovation Center of Chemistry for Energy Materials, Fudan University, Shanghai 200433, China

Dandan Han and Angang Dong
Department of Chemistry and Collaborative Innovation Center of Chemistry for Energy Materials, Fudan University, Shanghai 200433, China

Yi Ding, Guannan Guo, Jianhua Hu and Dong Yang
State Key Laboratory of Molecular Engineering of Polymers and Department of Macromolecular Science, Fudan University, Shanghai 200433, China

Jesús Ferrando-Soria, Eufemio Moreno Pineda, Antonio Fernandez, Samantha A. Magee, Iñigo J. Vitorica-Yrezabal, Floriana Tuna, Grigore A. Timco, Eric J.L. McInnes and Richard E.P. Winpenny
School of Chemistry and Photon Science Institute, The University of Manchester, Oxford Road, Manchester M13 9PL, UK

Alessandro Chiesa, Stefano Carretta and Paolo Santini
Dipartimento di Fisica e Scienze della Terra, Università di Parma, viale delle Scienze 7/a, Parma 43123, Italy

Qineng Xia, Zongjia Chen, Yi Shao, Xueqing Gong, Haifeng Wang, Xiaohui Liu and Yanqin Wang
Key Laboratory for Advanced Materials, Research Institute of Industrial Catalysis, East China University of Science and Technology, Shanghai 200237, China

Stewart F. Parker
ISIS Facility, STFC Rutherford Appleton Laboratory, Chilton, Oxfordshire OX11 0QX, UK

Xue Han
School of Chemistry, University of Nottingham, Nottingham, NG7 2RD, UK
School of Chemistry, University of Manchester, Manchester M13 9PL, UK

Sihai Yang
School of Chemistry, University of Manchester, Manchester M13 9PL, UK

Mary S. Collins, Matthew E. Carnes, Bryan P. Nell and Darren W. Johnson
Department of Chemistry and Biochemistry and the Materials Science Institute, University of Oregon, Eugene, Oregon 97403-1253, USA

Lev N. Zakharov
Center for Advanced Materials Characterization in Oregon (CAMCOR), 1241 University of Oregon, Eugene, Oregon 97403-1241, USA

Jeff Joseph A. Celaje, Zhiyao Lu, Elyse A. Kedzie, Nicholas J. Terrile, Jonathan N. Lo and Travis J. Williams
Donald P. and Katherine B. Loker Hydrocarbon Institute and Department of Chemistry, University of Southern California, Los Angeles, California 90089-1661, USA

Sohee Jeong, Dongwon Yoo, Minji Ahn and Jinwoo Cheon
Department of Chemistry, Yonsei University, Seoul 120-749, Korea

Pere Miró and Thomas Heine
Engineering and Science, Jacobs University Bremen, 28759 Bremen, Germany

Dawei Feng, Tian-Fu Liu, Mathieu Bosch, Zhangwen Wei, Xuan Wang, Kecheng Wang, Xizhen Lian, Zhi-Yuan Gu and Jihye Park
Department of Chemistry, Texas A&M University, College Station, Texas 77843-3255, USA

Jie Su, Wei Wan and Xiaodong Zou
Berzelii Centre EXSELENT on Porous Materials and Inorganic and Structural Chemistry, Department of Materials and Environmental Chemistry, Stockholm University, 106 91 Stockholm, Sweden

Daqiang Yuan
State Key Laboratory of Structural Chemistry, Fujian Institute of Research on the Structure of Matter, Chinese Academy of Sciences, Fujian, Fuzhou 350002, P. R. China

Ying-Pin Chen and Hong-Cai Zhou
Department of Chemistry, Texas A&M University, College Station, Texas 77843-3255, USA
Department of Materials Science and Engineering, Texas A&M University, College Station, Texas 77842, USA

Pei-Qin Liao, Wei-Xiong Zhang, Jie-Peng Zhang and Xiao-Ming Chen
MOE Key Laboratory of Bioinorganic and Synthetic Chemistry, School of Chemistry and Chemical Engineering, Sun Yat-Sen University, Guangzhou 510275, P. R. China

Ai-Xin Zhu
MOE Key Laboratory of Bioinorganic and Synthetic Chemistry, School of Chemistry and Chemical Engineering, Sun Yat-Sen University, Guangzhou 510275, P. R. China
Faculty of Chemistry and Chemical Engineering, Yunnan Normal University, Kunming 650092, P. R. China

XianfenWang, Satoshi Kajiyama and Hiroki Iinuma
Department of Chemical System Engineering, School of Engineering, The University of Tokyo, Hongo 7-3-1, Bunkyo-ku, Tokyo 113-8656, Japan

Eiji Hosono
National Institute of Advanced Industrial Science and Technology, Umezono 1-1-1, Tsukuba 305-8568, Japan

Shinji Oro and Isamu Moriguchi
Graduate School of Engineering, Nagasaki University, Bunkyo-machi 1-14, Nagasaki 852-8521, Japan

Masashi Okubo and Atsuo Yamada
Department of Chemical System Engineering, School of Engineering, The University of Tokyo, Hongo 7-3-1, Bunkyo-ku, Tokyo 113-8656, Japan
Elements Strategy Initiative for Catalysts and Batteries (ESICB), Kyoto University, Nishikyo-ku, Kyoto 615-8510, Japan

Min-Rui Gao, Ya-Rong Zheng, Yun-Fei Xu, Jun Jiang, Qiang Gao and Shu-Hong Yu
Division of Nanomaterials & Chemistry, Hefei National Laboratory for Physical Sciences at Microscale, Collaborative Innovation Center of Suzhou Nano Science and Technology, Department of Chemistry, University of Science and Technology of China, Hefei 230026, China

Jin-Xia Liang
Key Laboratory of Organic Optoelectronics and Molecular Engineering of Ministry of Education, Department of Chemistry, Tsinghua University, Beijing 100084, China

Guizhou Provincial Key Laboratory of Computational Nano-Material Science, Guizhou Normal College, Guiyang 550018, China

Jun Li
Key Laboratory of Organic Optoelectronics and Molecular Engineering of Ministry of Education, Department of Chemistry, Tsinghua University, Beijing 100084, China

Eun Seon Cho, Anne M. Ruminski, Shaul Aloni and Jeffrey J. Urban
The Molecular Foundry, Materials Sciences Division, Lawrence Berkeley National Laboratory, 1 Cyclotron Road, Berkeley, California 94720, USA

Yi-Sheng Liu and Jinghua Guo
The Advanced Light Source, Lawrence Berkeley National Laboratory, Berkeley, California 94720, USA

Seiji Yamazoe and Tatsuya Tsukuda
Department of Chemistry, School of Science, The University of Tokyo, 7-3-1 Hongo, Bunkyo-ku, Tokyo 113-0033, Japan
Elements Strategy Initiative for Catalysts and Batteries (ESICB), Kyoto University, Katsura, Kyoto 615-8520, Japan

Shinjiro Takano
Department of Chemistry, School of Science, The University of Tokyo, 7-3-1 Hongo, Bunkyo-ku, Tokyo 113-0033, Japan

Wataru Kurashige and Yuichi Negishi
Department of Applied Chemistry, Faculty of Science, Tokyo University of Science, 1-3 Kagurazaka, Shinjuku-ku, Tokyo 162-8601, Japan

Toshihiko Yokoyama
Department of Materials Molecular Science, Institute for Molecular Science, Myodaiji, Okazaki, Aichi 444-8585, Japan

Kiyofumi Nitta
Japan Synchrotron Radiation Research Institute, SPring-8, 1-1-1 Koto, Sayo, Hyogo 679-5198, Japan

Idan Hod and Aaron W. Peters
Department of Chemistry, Northwestern University, 2145 Sheridan Road, Evanston, Illinois 60208, USA
Argonne-Northwestern Solar Energy Research (ANSER) Center, Northwestern University, Evanston, Illinois 60208, USA

Pravas Deria, Joseph E. Mondloch and Monica So
Department of Chemistry, Northwestern University, 2145 Sheridan Road, Evanston, Illinois 60208, USA

Wojciech Bury
Department of Chemistry, Northwestern University, 2145 Sheridan Road, Evanston, Illinois 60208, USA
Department of Chemistry, Warsaw University of Technology, Noakowskiego 3, 00-664 Warsaw, Poland

Chung-Wei Kung
Department of Chemistry, Northwestern University, 2145 Sheridan Road, Evanston, Illinois 60208, USA
Argonne-Northwestern Solar Energy Research (ANSER) Center, Northwestern University, Evanston, Illinois 60208, USA
Department of Chemical Engineering, National Taiwan University, Number 1, Sector 4, Roosevelt Road, Taipei 10617, Taiwan

Matthew D. Sampson and Cliff P. Kubiak
Department of Chemistry and Biochemistry, University of California San Diego, 9500 Gilman Drive, La Jolla, California 92093-0358, USA

Omar K. Farha
Department of Chemistry, Northwestern University, 2145 Sheridan Road, Evanston, Illinois 60208, USA
Argonne-Northwestern Solar Energy Research (ANSER) Center, Northwestern University, Evanston, Illinois 60208, USA
Faculty of Science, Department of Chemistry, King Abdulaziz University, Jeddah 21589, Saudi Arabia

Joseph T. Hupp
Department of Chemistry, Northwestern University, 2145 Sheridan Road, Evanston, Illinois 60208, USA
Argonne-Northwestern Solar Energy Research (ANSER) Center, Northwestern University, Evanston, Illinois 60208, USA
Argonne National Laboratory, 9700 South Cass Avenue, Argonne, Illinois 60439, USA

Hao-Long Zhou, Jie-Peng Zhang and Xiao-Ming Chen
MOE Key Laboratory of Bioinorganic and Synthetic Chemistry, School of Chemistry and Chemical Engineering, Sun Yat-Sen University, Guangzhou 510275, China

Yue-Biao Zhang
MOE Key Laboratory of Bioinorganic and Synthetic Chemistry, School of Chemistry and Chemical Engineering, Sun Yat-Sen University, Guangzhou 510275, China
School of Physical Science and Technology, ShanghaiTech University, 100 Haike Road, Shanghai 201210, China

Pilar García-García, José María Moreno and Urbano Díaz
Instituto de Tecnología Química, UPV-CSIC, Universidad Politécnica de Valencia, Avenida de los Naranjos s/n, E-46022 Valencia, Spain

Marta Bruix
Instituto de Química Física Rocasolano, CSIC, Serrano 119, 28006 Madrid, Spain

Avelino Corma
Instituto de Tecnología Química, UPV-CSIC, Universidad Politécnica de Valencia, Avenida de los Naranjos s/n, E-46022 Valencia, Spain
King Fahd University of Petroleum and Minerals, PO Box 989, 31261 Dhahran, Saudi Arabia

Stefan Mitzinger, Werner Massa and Stefanie Dehnen
Fachbereich Chemie and Wissenschaftliches Zentrum für Materialwissenschaften, Universität Marburg, Hans-Meerwein-Straße, D-35043 Marburg, Germany

Lies Broeckaert
Fachbereich Chemie and Wissenschaftliches Zentrum für Materialwissenschaften, Universität Marburg, Hans-Meerwein-Strae, D-35043 Marburg, Germany
Institut für Nanotechnologie, Karlsruher Institut für Technologie, Hermann-von-Helmholtz Platz 1, D-76344 Eggenstein-Leopoldshafen, Germany

Florian Weigend
Institut für Nanotechnologie, Karlsruher Institut für Technologie, Hermann-von-Helmholtz Platz 1, D-76344 Eggenstein-Leopoldshafen, Germany

Hao Bin Wu, Bao Yu Xia, Le Yu, Xin-Yao Yu and Xiong Wen (David) Lou
School of Chemical and Biomedical Engineering, Nanyang Technological University, 62 Nanyang Drive, Singapore 637459, Singapore

Jun Yuan, Xin He and Liang Zhao
Key Laboratory of Bioorganic Phosphorus Chemistry and Chemical Biology (Ministry of Education), Department of Chemistry, Tsinghua University, Beijing 100084, China

Tingting Sun, Ke An and Jun Zhu
State Key Laboratory of Physical Chemistry of Solid Surfaces and Collaborative Innovation Center of Chemistry for Energy Materials (iChEM), Fujian Provincial Key Laboratory of Theoretical and Computational Chemistry and Department of Chemistry, College of Chemistry and Chemical Engineering, Xiamen University, Xiamen 361005, China

Index

R

S

T

U

V

W

X

Z

www.ingramcontent.com/pod-product-compliance
Lightning Source LLC
LaVergne TN
LVHW060058220926
848484LV00008BA/2236